PRINCIPLES OF PHYSICAL COSMOLOGY

PRINCETON SERIES IN PHYSICS

Edited by Philip W. Anderson, Arthur S. Wightman, and Sam B. Treiman
(published since 1976)

PRINCIPLES OF PHYSICAL COSMOLOGY

P.J.E. Peebles

PRINCETON UNIVERSITY PRESS
PRINCETON, NEW JERSEY

Published by Princeton University Press, 41 William Street, Princeton, New Jersey 08540
In the United Kingdom: Princeton University Press, Chichester, West Sussex

Library of Congress Cataloging-In-Publication Data

Peebles, P.J.E. (Phillip James Edwin)
   Principles of physical cosmology / P.J.E. Peebles
      p.   cm. — (Princeton series in physics)
   Includes bibliographical references and index.
   ISBN 0-691-07428-3. — ISBN 0-691-01933-9 (pbk.)
   1. Cosmology. 2. Astrophysics. I. Title. II. Series.
QB981.P424   1993
523.1--dc20   92-33370

Printed in the United States of America

10 9 8 7 6 5 4 3 2

hi Al, Les, Ellen, Marion

CONTENTS

III. Topics in Modern Cosmology 361

PREFACE

This book is meant to be a survey of the state of physical cosmology, including the network of observational and theoretical elements that establish the subject as a mature physical science, and the more notable attempts to improve and extend the picture.

Much of the excitement in cosmology centers on recent observational advances and on the remarkable variety of ideas for how we might interpret the observations, new and old. The ideas that prove to be of lasting interest are likely to build on the framework of the now standard world picture, the hot big bang model of the expanding universe. The full extent and richness of this picture is not as well understood as I think it ought to be, even among those making some of the most stimulating contributions to the flow of ideas. In part this is because the framework has grown so slowly, over the course of some seven decades, and sometimes in quite erratic ways. Compare the rate of development of cosmology to that of particle physics—six decades ago we already had the modern concept of the expanding universe but only the beginning elements of a quantum field theory. In a subject that develops in such a slow way, things tend to be forgotten, sometimes to be rediscovered. Now that cosmology has become popular there is the additional hazard that it can be difficult to pick out the pattern of well-established results against the exciting distraction of all the new but possibly transient ideas. Thus I think there is a need for a survey of the results that seem to be of lasting use for cosmology, including a conservative assessment of notable recent developments as well as a compilation of the equally important older elements we may tend to overlook. This book contains my choices for the list.

My guideline has been to concentrate on results that seem likely still to be of general interest a decade from now. In many cases the choices are easy. Hubble's law for the cosmological redshift as a function of distance has been a central element in cosmology since the late 1920s. It continues to agree with the improving observational tests, and I see no reason to doubt that this will continue to be the case in the next ten years. The situation is very different for the inflation scenario of what happened in the very early universe. Here we have no useful observational evidence for or against the picture, but there is a strong case for including it. The idea has been with us for ten years, and its popularity has made the problems it is meant to address quite visible, yet no credible alternative has emerged. The problems are very real, so unless or until a reasonable alternative is found, or the concept somehow is shown to be wrong, we must expect that inflation will continue to guide theoretical explorations of the physics of the very early universe. On the other hand, the varieties of models for how inflation might have been implemented seem much less worth recording, for we know that at best one of the variants now under discussion can be on the right track, and the record suggests there will be more to consider.

My previous attempt at a summary of the memorable results in this field appeared some two decades ago, in *Physical Cosmology*. Since then I have become persuaded that a survey of cosmology really ought to include general relativity theory. Many aspects of the standard cosmological model can be derived by symmetry arguments along with a few basic results from general relativity. However, the prospects for meaningful tests of the evolving spacetime geometry, including the classical cosmological tests and an increasingly interesting range of variants, look considerably better now that galaxies actually are observed at redshifts well above unity, where the observational effects of the cosmological model are pronounced. Of course, the effects of evolution are pronounced too, and untangling the astronomy and cosmology is likely to be a major occupation in the coming decade. Assuming the needed astronomy does start to make sense, attention will turn toward the meaning of the cosmological tests in the real universe with its clumpy mass distribution, and this will involve increasingly detailed applications of the theory of our spacetime geometry. Accordingly, this book departs from *Physical Cosmology* in giving considerably greater emphasis to general relativity theory.

The first part of the book presents the basic elements in the style of *Physical Cosmology*, finessing most of general relativity theory. I still like this approach, because I see no reason to burden the physics and astronomy with an unneeded theoretical superstructure. I feel reasonably confident that I am satisfying the ten-year guideline throughout this part, because most of the theoretical concepts are a good deal older than that and show little evidence of evolution, and the newer observations used in this part seem secure enough.

The subject of the second part is relativistic cosmology. I hope the presentation is complete in the sense that one can understand the results of general relativity theory relevant to cosmology from what is presented here, without recourse to the standard treatises (except of course for second opinions). However, this is not meant to be a textbook on relativity, for it includes only the basic elements relevant to physical cosmology, and it omits advanced topics, such as the relativistic singularity theorems and the physics of black holes, which certainly are relevant but beyond the intended level of this book.

The third part deals with the nature and evolution of the intergalactic medium, the measures of the large-scale patterns in the distribution and motion of galaxies and mass, and the theories for the origin of this structure. There have been great advances in the structure problem in the last two decades. We have a rich and rapidly growing fund of evidence from observations, at high redshifts and low, and from the measurements of the radiation backgrounds, and we have a remarkable variety of ideas from inflation and cosmic fields on what it all might mean. I have been struck by the low level of interaction between the observational and theoretical branches of the effort. This does follow an old and honorable tradition in cosmology, but I am betting that the approach is now inefficient and will not last. My thoughts on how a closer interaction of theory and observation might be developing are based on arguments that in some cases are not much better than guesses, and in the best of cases far from unambiguous, so I have to expect that many will prove to have violated the ten-year rule.

I expect the violations will be most pronounced in the last two sections, where I present my impression of where the subject seems to be headed. The reader should be aware that my arguments are based on readings of observational evidence that in many cases is enigmatic and may not even be right. I have been struck by the differences in the weights and even signs attached to these clues by people whose judgments I respect, but this has to be so in an active and still very open subject. I apologize to colleagues whose models and ideas have been slighted. Perhaps the best compensation is the sure knowledge that many of my choices will fall by the wayside, and some of theirs will be seen to be on the path to a truer understanding.

The reader should also take note of other features of this book. In an attempt to make it useful for a broad class of readers I have arranged the presentation in sections, most of which begin with basic concepts and move toward subsections on more technical details. I imagine people will tend to read through each section or subsection until the going gets tough, then move on to the next, and return if it becomes apparent that some later parts really might be relevant for the purpose at hand, because that is the way I read books such as this one. The text does sprawl, but then so does the subject.

Part of the sprawl is a tendency for the discussion of a topic to spread over several sections, according to what seems to me to be the long-term logical organization. For example, the geometry of spacetime around a stationary cosmic string is discussed in section 11, the physics of a cosmic string is dealt with in section 16, and sections 25 and 26 detail some of the lessons one learns from the attempts to use cosmic strings in a theory of galaxy formation. I have tried to indicate where related discussions are to be found when the connection seems particularly useful. I hope the index and table of contents will be an adequate guide to where a wanted topic is to be found among all the others.

I have placed heavy emphasis on order-of-magnitude estimates based on convenient simplifications, rather than the details of computations. Computers are important but can be a trap; they have made it easy to put a gloss on a numerical result, as it always has been for analytic work. The art of estimates is not dying and need not be hidden. I learned as a youth, and still believe, that you shouldn't trust a numerical result you can't understand from sensible estimates. There are counterexamples, but I think they are not common.

It should be no surprise, given the prejudices I have expressed, that I have chosen not to devote much space to details of specific models where there is little immediate prospect for observational applications. Such models can be very useful as illustrations of concepts, and may even be pointing in the right direction. I think I can identify some that surely are headed the wrong way, but have less strong feelings about the many cases where I can't find anything wrong. In general, however, the interesting thing really is the underlying physics, for when one understands that it usually is easy to work out the models as they roll in. My goal has been to trace derivations back to undergraduate physics, avoiding wherever I can the dread words, "it can be shown that." Where a derivation parallels something already worked through, or otherwise seems reasonably easy and educational, I have left the completion as the traditional "exercise for the student." I have been informed that more formal problem sets would be useful. I agree, but feel they would have lowered the tone. We are joined in a search for understanding, not grades.

My treatment of references follows a similar philosophy. I refer to the original or first exploratory papers, where I think I know them, and to reviews that seem likely to be of lasting interest, but not to the most recent paper in a rapidly developing topic where a better, more authoritative one surely will appear shortly after this book. For popular topics, the place to start reading about recent developments is in the published proceedings of conferences. We all complain about the preparation of contributions to proceedings when not much has changed since the last meeting, but the results are a valuable guide to what people are thinking.

Finally, my friends have warned me that I have not done justice to the rich details of the astronomy or the particle physics relevant to modern cosmology. Maybe this is evidence that I have achieved a rough balance, but the truer explanation is that I am not competent to offer reasoned assessments of the state of either subject. I have tried to survey the fundamental elements of the physics of cosmology at about the level of the physics we encounter in a good undergraduate education (and mostly forget, but we can recover it). For the most part this limit does not much matter; we are dealing with deep mysteries of Nature, not methods of physics. Perhaps the largest gap is in the treatment of the particle-physics candidates for dark-matter. If it should happen that a specific dark matter particle is experimentally identified, you would not be able to understand its provenance from this book. However, the search for these particles is a young subject that is to be found in many recent conference proceedings, review articles, and books. In particular, if a detection were announced in the media I would turn first for references to *The Early Universe* by Kolb and Turner. The details of the quantum field theories that would be the foundation for a fully successful and complete inflationary theory of the early universe are not to be found here, but again I do not consider this a serious omission because there are several books on inflation, all recent because the subject is new. An entertaining introduction is Linde's *Particle Physics and Inflationary Cosmology*. The physics of cosmological phase transitions and the resulting remnants is described by Vilenkin and Shellard in *Cosmic Strings and Other Topological Defects*.

Several conventions should be explained here. Astronomers like base ten for logarithms, so I write

$$\log_{10}(x) = \log(x),$$
$$\log_e(x) = \ln(x).$$

The curious base for the logarithmic system of measures of optical energy fluxes, $10^{2/5}$, is described in section 3.

As discussed in section 5, the linear relation between the distance to a galaxy and the redshift of its spectrum is much better established than the constant of proportionality, Hubble's constant, so it is written as

$$H_o = 100h \text{ km s}^{-1} \text{ Mpc}^{-1}.$$

This useful practice allows us to express how a result depends on the uncertain dimensionless number h. Human nature being what it is, the convention often is "improved" by using other values for the factor with dimensions that multiplies h. Sometimes this is indicated; thus it is a reasonable presumption that h_{50} is the value of H_o in units of $50 \text{ km s}^{-1} \text{ Mpc}^{-1}$. The confusion is reduced by quoting distances in terms of the cosmological recession velocity, but still we need h to express timescales or predicted mass densities.

The parameter H_o really has units of reciprocal time, but in many topics that is not the familiar way to express it. My choices of units and symbols follow what I observe to be standard practices, in the hope that the advantages of familiarity outweigh the confusion of usages. Thus Stefan's constant for blackbody radiation is written as a, or as a_B (for Boltzmann) where it might be confused with the cosmological expansion parameter, $a(t)$. I rely on context to distinguish Boltzmann's constant k from the wave number, and the Hubble parameter h from Planck's constant. In discussions of relativity I follow standard practice in setting the velocity of light c to unity but keeping Newton's gravitational constant G as a quantity with dimensions, while in the physics of the very early universe I often follow the practice of setting $\hbar$ and c to unity and replacing G with the Planck mass $m_{\text{pl}} = G^{-1/2}$.

When getting numerical results I am inclined to put all the units back in, but again I use standard conventions, which leads to a remarkable variety of units. Thus in this book you can find lengths measured in units of the Planck length $(l_{\text{pl}} = (G\hbar)^{1/2}c^{-3/2} = 1.6 \times 10^{-33}$ cm), Fermis, Ångstroms, microns, millimeters, centimeters, kilometers (but only in the velocity unit, km s^{-1}), parsecs (1 pc = 3.086×10^{18} cm, the distance at which one second of arc subtends the mean separation between the Earth and Sun), kiloparsecs, megaparsecs, and for really large distances gigaparsecs and the Hubble length $= 3h^{-1}$ Gpc $= 3000h^{-1}$ Mpc.

The optimistic reading is that the abundance of conventions for symbols and units reminds us of our rich heritage. Undoubtedly it also leads to confusion and inefficiency. (What is the integrated amount of time people have spent programming computers to deal with a negative angle such as $\delta = -0° \ 30'$?) Practices do evolve. Thus it is becoming unusual to encounter light years as a unit of distance (though of course in many parts of relativity and particle physics distances are expressed in terms of light travel time), femtometers conveniently replace Fermis (1 fm $= 10^{-13}$ cm), micrometers replace microns $(1 \mu m = 1\mu = 10^{-4}$ cm), and it is increasingly common to encounter wavelengths measured in nanometers instead of Ångstroms (1 nm = 10 Å). Perhaps science will evolve more coherent practices, but those who would speed the process have to deal with the fact we are captives of habit; we choose the units we think our audience is expecting, and these soon are the units that seem right and proper to us. My guess is that notation in physical science never will be fully rationalized, nor will the world evolve via advances in communications to a global village, for I don't see that people tend to operate that way.

There is ambiguity about what is meant by the "standard model" in cosmology: in everyday use it can refer to the well-established elements of the theory, or to plausible but speculative attempts to fill in details. In this book, the standard model is the nearly homogeneous and isotropic expansion of the universe, according to general relativity theory, that traces back to a state hot and dense enough to have produced the light elements (by the process discussed in section 6). Weinberg brought the phrase, "the standard model," to cosmology from particle physics, where it signifies a theory that is considered well established because it has survived nontrivial tests. The use of the word "model" is appropriate, because the picture in known to be an incomplete approximation to what is really happening, and there certainly is the chance that there is something very wrong with the picture. The word "standard" is meant to signify that it has survived an impressive variety of nontrivial tests that have left no known credible alternatives, as will be argued at length in the first part of this book. It is a long tradition in cosmology to use the word "model" to refer to a specific theory and choice of parameters. Thus the Einstein-de Sitter model is a solution to Einstein's field equation, which assumes that the significant dynamical actors are the mass density and the expansion rate (with negligibly small cosmological constant and space curvature). This world model gives definite testable relations among the present rate of expansion of the universe, the maximum ages of the galaxies, and the mean mass density. As will be discussed beginning in section 15, there are nontrivial theoretical reasons to suspect that this model may be a useful approximation to our universe. However, there is not yet a strong empirical case for it, and in section 26 I will list the observational problems that may or may not prove to be only apparent. Unless or until this is clarified I will not count the Einstein-de Sitter solution as part of the standard model.

I shall mean by a model a theoretical scheme that offers testable predictions, as does the Einstein-de Sitter model. That leaves the many less well specified ideas for how we might improve the standard model by tightening the options or enlarging the boundaries. In regular usage a scenario is an outline for a proposed sequence of events. I follow my book *Physical Cosmology* by classifying as a scenario a promising or otherwise sensible set of ideas, perhaps even with some observational basis, but one that is not yet definite enough to yield testable predictions by which the scheme might be falsified. An example is the inflation scenario reviewed in section 17. It offers an elegant resolution to a deep puzzle, the origin of the remarkable large-scale isotropy of the observable universe, and the physics has a distinguished pedigree from particle theory. But it is difficult to see how to test the idea. The one definite prediction of inflation as now understood is that space sections at constant world time have negligibly small curvature, but since that condition was one of the elements that went into the invention of inflation it is a little dangerous to count it as a prediction. Thus, inflation has to be considered an elegant and influential scenario, but not part of the standard model.

The simplest model of inflation, and the cold dark matter model that naturally follows from it, are often called parts of the new paradigm in cosmology. That is appropriate in the sense that these are the patterns for research which many people followed in the last decade. But others rightly point to the fact that there is not much empirical evidence for these concepts, and there are other ideas to consider, as discussed at length in sections 25 and 26. My impression is that it is too soon to declare any paradigms in cosmology beyond the standard model.

Another somewhat confusing usage is the name "the big bang" for the standard model. It is not appropriate, because it connotes a spatially isolated event, an explosion, that marked the start of everything. As we will discuss in the following sections, none of this is part of the standard cosmological model. But the name has a very evident appeal, and I expect people will continue to use it.

I have used a few special abbreviations. The isotropic 3K thermal microwave-submillimeter cosmic background radiation plays such a central role and is mentioned so often that we need a shorthand notation. There is no preferred choice; I use CBR. My book *Large-Scale Structure of the Universe* contains a lot of details about the measurement and theory of evolution of density fluctuations, much of which need not be repeated here. The reference is to LSS. For a few oldies but goodies I refer to *Physical Cosmology* as PC. I use the expression "rms" for the root mean square value.

Finally, there has to be a convention for approximate equalities. There are cases in cosmology where it is meaningful to quote numerical values to several significant figures, and cases where we would be proud to be within a few orders of magnitude of the right value. With a few special exceptions I use just two symbols. The expression $a = b$ means the statement is true to better than a factor of two, by definition or construction or a reasonably convincing measurement. The expression $a \sim b$ means the relation is uncertain at least to a factor of two, either because I have not bothered to work the computation in greater detail or because it is not known how to do it better.

The big bang cosmology is six decades old, and I am startled to realize I have been studying this world model for nearly half that time. It never was my plan; in fact, my first reaction to cosmology was one of surprise that grown people could seriously care about such a schematic physical theory. I think I stuck with it because I enjoyed working in such uncrowded and fertile ground. Now the subject is crowded, at least by the standards of just a decade ago, but too exciting to leave. I have presented histories of pioneering contributions to cosmology, as best I understand them, but my memory is not always reliable and my knowledge is limited. I am sure also that I have on occasion lapsed into the habit of describing discoveries as they ought to have happened. I know best the parts I was involved in, and on occasion indulge in personal recollections. I think such stories play an important role in illustrating the way the physical sciences operate, but bear in mind that this certainly is not a formal history of cosmology, and that the subsections are meant to facilitate selective reading.

First approximations to this book evolved out of the cosmology course I have on occasion taught at Princeton University. I know of nothing better than the prospect of facing students to focus the mind on the essential elements of a subject. I am grateful to the Aspen Center for Physics, where the June conferences on astrophysics added to my education, and the special atmosphere stimulated reflection and the frank and thorough exchange of ideas. I am particularly indebted to John Bahcall and the Princeton Institute for Advanced Study for their hospitality during the 1990–91 academic year that gave me the opportunity and the stimulation to write the first draft.

The opinions expressed at such length in this volume have grown out of several decades in the community of cosmology, in exchanges ranging from loud debates to the slightest of hints. This is not meant to shift the blame for the wrong and uninteresting opinions away from its rightful place, but rather to emphasize that science moves in a complex flow through enormous varieties of perturbations large and small. First on the list of those whose influence on me is most evident are my professor of continuing education, Bob Dicke, my other most influential advisers to things practical and physical, Ed Groth and Dave Wilkinson, and my counselors on things astronomical, John Bahcall and Jerry Ostriker. A large number of people helped in the preparation of this book; I must select for special mention the heroic contributions of Bharat Ratra and Michael Strauss in reading and improving vast sections of the text. For generous help in creating and straightening the figures and text and concepts, often at the expense of considerable time and effort, I am grateful to Richard Ellis and Tom Shanks at the University of Durham; Jim Condon and Juan Uson at NRAO; Art Wolfe at UCSD; Jeremy Mould at CIT; George Efstathiou, Steve Maddox, and Will Saunders at Oxford; Renzo Sancisi at Groningen; Bill Oegerle at STScI; John Hoessel at the University of Wisconsin; Masataka Fukugita at Kyoto University; John Scalo at the University of Texas; Harry van der Laan at ESO; Adrian Melott at the University of Kansas; Tod Lauer at KPNO; Craig Hogan at the University of Washington; Marc Davis at UC Berkeley; Ray Soneira at Sonera Technologies; Josh Frieman at Fermilab; Dave Schramm at the University of Chicago; Margaret Geller and John Huchra at Harvard; Keith Jahoda, John Mather, and Rick Shafer at NASA Goddard SFC; Steve Boughn at Haverford College; Avery Meiksin at CITA; and in the Princeton community Neta Bahcall, Ruth Daly, Russell Kulsrud, Avi Loeb, Alison Peebles, David Spergel, Martin Terman, Ed Turner, Neil Turok, and David Weinberg. To the many other colleagues who influenced this work, even when we agreed to disagree, I extend my thanks and best wishes in the ongoing research that will make this version of cosmology obsolete.

PRINCIPLES OF PHYSICAL COSMOLOGY

I. The Development of Physical Cosmology

1. The Standard Cosmological Model

Physical cosmology is the attempt to make sense of the large-scale nature of the material world around us, by the methods of the natural sciences. It is to be hoped that those who love physical science will take pleasure in cosmology as an example of the art. It operates under the special restrictions of astronomy that allow one to look but not touch, but according to the rules and procedures that have proved to be so wonderfully successful in sister fields from stellar astronomy to particle physics. I will argue throughout this first part of the book that cosmology as an enterprise in physical science really has made substantial progress, though the advances certainly have moved around considerable gaps in our understanding, as will be discussed in part 3.

Behind physics is the more ancient and honorable tradition of attempts to understand where the world came from, where it is going, and why. Cosmology inherits this tradition, in part by design, in larger part because that is where the astronomy and physics have led us. We have believable evidence that the universe is expanding, the space between the galaxies opening up, and that this expansion traces back to a hot dense phase, the big bang. The expansion may reverse in the future, and the world as we know it end in a collapse to a hot dense big crunch. An alternative is a universe that continues to expand indefinitely, to arbitrarily low mean density, but with most of the matter trapped in galaxies and clusters of galaxies that eventually contract (through loss of energy by evaporation of stars and gravitational radiation) and end up as black holes, in a series of little crunches. This is exciting stuff, and it has served a useful purpose in physical cosmology in keeping us all occupied with speculations on how the world ought to begin and end, as we sort through the evidence of what really is happening.

It is remarkable that physical science can be used for this purpose, that elements of Nature can be analyzed and seen to operate in a predictable way within the rules of a rigid mathematical physical theory, but we have abundant success stories, as in quantum physics and general relativity theory. There is a specific reason why one might have been particularly doubtful about the enterprise in cosmology. One usually deals with the physical world as a hierarchy of structures, including quarks in atomic nuclei of atoms and molecules on the small-scale end, while on the large-scale end we see that the Solar System is in the Milky Way galaxy of stars, which is part of the Local Group of galaxies, which in turn is on the outer edge of the Local Supercluster. (These large-scale structures are discussed in section 3.) A discipline in physical science generally isolates a thin slice of this hierarchy. The analysis must take account of the interactions from other levels, but one attempts to find approximations that reflect the essential features of the interactions while keeping the problem simple enough that a quantitative analysis is feasible. Thus in the study of turbulent flow of water in a stream one takes account of the fact that water consists of molecules of atoms by assigning coefficients of viscosity and thermal conductivity that are so small the values are not even very important, and the fact that the stream is flowing on a planet in a galaxy is well modeled by assigning a uniform local gravitational acceleration. Newtonian fluid dynamics is the challenging part of this problem.

Returning to cosmology, one might ask how the scheme of isolating essential and simple parts of a problem that has proved so successful in science might be applied to the physics of the universe itself rather than a specific one of its components. The answer lies in the idea that the universe is remarkably simple in the large-scale average, close to homogeneous and isotropic. There certainly are prominent fluctuations from the mean, but the evidence to be reviewed in sections 3 and 7 is that they are confined to scales less than a few hundred million light years (where 1 megaparsec ~ 3 million light years), and that the average over these fluctuations reveals a universe that looks much the same in any direction, and would appear much the same when viewed from any other position, as if there were no preferred center and no edges (within the part causality allows us to check). Modern cosmology is based on this simple characterization of the universe. That is, as in any physical science, we are dealing with a slice of the hierarchy of the material world, in this case the behavior of the nearly homogeneous large-scale distribution of matter and the evolution of the departures from homogeneity that we recognize as galaxies and clusters and superclusters of galaxies.

Since a homogeneous mass distribution is easy to characterize, and it is not so difficult to deal with departures from a mean distribution, it may not be surprising that some progress is feasible within this picture. It is reasonable to ask whether the progress might be circular, whether cosmologists have only invented a problem that is easy to solve. The evidence that there is more to it than that has to be indirect, for there is no way we can consult intelligent life in distant galaxies, to discover whether their observations of the large-scale structure of the universe agree with ours and whether they would agree with our interpretations of the observations. But the tactic of validation by indirect inference is familiar and demonstrated to be wonderfully successful in other fields of science, and we should not hesitate to try it here. No one has seen a quark, yet the weight of evidence from high-energy physics compels belief in these peculiar objects as a useful working approximation to reality. The weight of evidence in cosmology is not nearly as great, but, as will be argued in the following sections, far from negligible. Precision measurements of the angular distribution of the X-ray background from distant sources, of the centimeter through submillimeter electromagnetic radiation backgrounds, and deep galaxy counts, all directly show that in the large-scale average the universe is quite close to isotropic around our position. These observations, which are discussed in section 3, tell us either that we are at a very special place at the center of a spherically symmetric universe, or that the observable universe is close to homogeneous. The former seems unlikely; the billions of galaxies outside the Milky Way look like equally good homes for observers, and it would be surprising if we were in one of the privileged few from which the universe appears isotropic. The latter interpretation, that the universe is homogeneous, leads to the prediction of Hubble's law—that the apparent recession velocity of a galaxy is proportional to its distance—for that is the only expansion law allowed by homogeneity (section 5). The expanding world model in turn predicts the thermal spectrum of the centimeter-submillimeter wavelength background radiation (section 6). If we add general relativity theory, we arrive at a reasonable picture for the mass and age of the universe, and for the abundances of the light elements as remnants of the universe when it was young. That is, we are seeing in cosmology a developing network of interconnected results. This network is what suggests that we really are on the path to a believable approximation to reality.

The main elements of the standard world picture are summarized as follows.

1. The mass distribution is close to homogeneous in the large-scale average. The constraint from the Sachs-Wolfe effect to be discussed in section 21 is that fluctuations in the mass distribution averaged over volumes comparable in size to the Hubble length $\sim 4000\,\mathrm{Mpc}$ in equation (1.2) below are bounded by $\delta M/M \lesssim 10^{-4}$. The mass fluctuations become large, $\delta M/M \sim 1$, when the smoothing radius is reduced to about one percent of the Hubble length (sections 3, 19, and 21).

2. The universe is expanding, in the sense that the mean distance l between conserved particles is increasing with time at the rate

$$\frac{dl}{dt} = H_o l .\tag{1.1}$$

This effect is discussed in section 5. The constant of proportionality is time-dependent; the present value is Hubble's constant H_o. At the Hubble length,

$$L_H = c/H_o \sim 4000\,\mathrm{Mpc} ,\tag{1.2}$$

this expression for the recession velocity extrapolates to the velocity of light, and we need a more detailed treatment. The relativistic theory of cosmology and its possible tests are discussed in part 2.

3. The dynamics of the expanding universe are described by Einstein's general relativity theory. One might well ask what the universe is expanding into, or where the space opening up between the particles came from. As will be discussed in sections 5 and 12, general relativity theory gives us a mathematically consistent, and, as far as is known, experimentally and observationally successful description of the expanding universe without admitting such questions. With general relativity theory, we are assuming local physics is the same everywhere and at all times. As will be discussed here and in later sections, that has to be wrong at early enough epochs, because the standard world picture extrapolates back to a singular state in which conventional physics becomes undefined. Deciding how and when the physics departed from the standard expanding world model is one of the puzzles to be discussed in part 3.

4. The universe expanded from a hot dense state where its mass was dominated by thermal blackbody radiation. The set of clues that led to this concept are chronicled in section 6.

The familiar name for this picture, the "big bang" cosmological model, is unfortunate because it suggests we are identifying an event that triggered the expansion of the universe, and it may also suggest the event was an explosion localized in space. Both are wrong. The universe we observe is inferred to be close to homogeneous, with no evidence for a preferred center that might have been the site of an explosion. The standard cosmological picture deals with the universe as it is now and as we can trace its evolution back in time through an interlocking network of observation and theory. We have evidence from the theory of the origin of the light elements that the standard model successfully describes the evolution back to a time when the mean distance between conserved particles was some ten orders of magnitude smaller than it is now. If it is found that still earlier epochs left evidence that can be analysed and used to test our ideas, then that may be incorporated in the standard model or some extension of it. If there were an instant, at a "big bang," when our universe started expanding, it is not in the cosmology as now accepted, because no one has thought of a way to adduce objective physical evidence that such an event really happened.

A less vivid but maybe less misleading name for the expanding world picture is the "standard model." Weinberg (1972) brought the phrase to cosmology from particle physics. The use of the term "model" has a long history in cosmology, as in the books by Tolman (1934) and Milne (1935). It is meant to express the fact that the theory cannot be the final answer, if there is one, because it has unresolved puzzles. The open questions are more numerous in cosmology than in a typical mature physical science, and it is only prudent to bear in mind that one of these problems, or maybe something yet to be discovered, may point to something fundamentally wrong with the world picture. As the observational checks have accumulated, this has come to seem less likely; if there is a serious flaw in the standard world picture it has been subtly hidden. Thus the word "standard" is meant to express the fact that there is a very significant body of evidence indicating that the hot big bang model is a useful approximation to the real world.

Under active discussion are many ideas on how the parameters of the standard expanding world picture might be made more definite, or the boundaries of the picture expanded. An example is the Einstein-de Sitter cosmological model, in which there are negligibly small values for the parameters in equation (5.18) for the curvature of space sections at constant time and for a cosmological constant. This model is the simplest and most natural homogeneous and isotropic solution to Einstein's field equations. It predicts definite testable relations among the present values of the mean mass density, the age of the expanding universe, and the present rate of expansion. It is not yet part of the standard model, because there is not much evidence that the predictions are consistent with the still quite uncertain observations. A second example is the inflation scenario discussed in section 17. This is an elegant and influential idea for how we might resolve the puzzle of what happened before the universe started expanding, whatever that means. But inflation is not tested, and it is not even easy to see how it could be falsified, so it is not part of the standard model.

Central to the standard model is the underlying physics. As we have indicated, the standard model assumes general relativity theory, with the condition that local physics is everywhere the same, based on laboratory results and what can be inferred from what is observed elsewhere under more extreme conditions. That is, we are going to extrapolate the physics that is known to be successful until it is seen to fail. We noted that the extrapolation does fail if applied far enough back in time, for the standard model traces back to a singular state, and the failure certainly could show up well before that. The point was given particular emphasis in a remarkable survey of physics by Dicke (1970), who was motivated in part by the Mach's principle discussed in the next section, and by Bondi (1960), in connection with the steady-state cosmology presented in section 7. The failure of conventional physics must be kept in mind, but it should be noted also that the extrapolation out in space and back in time is by no means without empirical support. For example, we will see in section 20 that the gravitational lensing of background galaxies by the mass concentrations in clusters of galaxies, as analyzed in general relativity theory, is consistent with the masses derived from the motions of the galaxies and from the plasma pressure within the clusters. The relevant length scale here—the impact parameter at the cluster—is ten orders of magnitude larger than that of the precision tests of general relativity in the Solar System and in binary pulsar systems, a remarkable extrapolation. There is a problem: the net mass seen in the stars in the galaxies in a cluster is an order of magnitude less than the masses derived from gravitational lensing or from the dynamics of the cluster plasma and galaxy motions. Perhaps this is a sign that the gravity physics is starting to fail on the scale of clusters of galaxies. But since the extrapolation seems to give consistent results for dynamics and lensing, the first possibility to consider, and the one emphasized in section 18, is that the gravity physics is reliable and the bulk of the mass in a cluster is in a form not readily detected, apart from its gravitational effect. The edge of the observable universe is another four orders of magnitude beyond the size of the central part of a cluster, and the analysis of what is happening there still is a considerable extrapolation of the physics. The way to decide whether the analysis is successful is through the cosmological tests discussed in sections 13 and 14.

The spectra of distant objects indicate local physics is much the same everywhere: the radio, optical, and X-ray lines, and the continuum between them, have the right arrangement and sensible-looking shapes, consistent with the standard physics of gas and stars. Some elements of these observations are reviewed in sections 5, 23, and 24. Again, we know physics must have been different far enough back in time because the standard variety leads back to a singularity, and it is a sensible bet that the physics of the very early universe left observable effects yet to be discovered, perhaps exotic forms of matter that might account for the cluster masses (section 18), or cosmic field topological defects that might have caused the clusters to form (section 16), or the magnetic fields seen in young galaxies, or even the process of formation of the young galaxies. These, however, are topics for research outside the standard model at the time this is written.

Since the evidence is that atoms and gas clouds and stars are much the same on the other side of the observable universe, it seems reasonable to expect that there are planets. If there were a civilization on one of these planets and this organization took an interest in such matters, would it be led to decide with us that the universe is expanding and cooling from a denser and hotter state? It does not seem overly presumptuous to imagine that a civilization operating in a fluid atmosphere, and maybe even using the atmosphere for heat engines or flight, would have our concepts of thermodynamic temperature and physical time, for they are fundamental to the rules by which a gas or fluid is known to behave in our laboratories and in stars near and far. And it seems to be a reasonable bet that if the optical and radio bands were clear, and the civilization cared to use them, they would notice with us that there are galaxies of stars, and between them there is a sea of thermal radiation. The evidence is that they also could deduce with us that the material content of the universe is expanding and cooling on a timescale of about 10^{10} years. The reader is invited to review this conclusion after reading the first part of this book. And of course we would all welcome the chance to ask the question.

Two aspects of the methods of physical science applied in cosmology have tended to lead to confusion. First, how can we hope to understand the character of the universe in all its variety from the exceedingly limited observations we can bring to the problem? Second, how can we conclude that we really have succeeded in characterizing the universe when there are in cosmology so many elementary questions yet to be answered?

Ellis (1980, 1985) has given particular emphasis to the ideal of an empirical cosmology constructed as far as possible from the observations. In the limit one might even imagine deriving boundary conditions for the structure of the $3+1$ dimension spacetime of our world from the data that in principle are available on the $2+1$ dimension null light cone of our observations in the optical, radio, and other radiation bands, along with whatever fossils are left in the thin timelike tube occupied by the material near our world line in the galaxy. The studies of how one might approach this program certainly are of great interest and importance, but as a practical matter we must live with the fact that observations at great distance always will be schematic; there is no way we could hope to get initial value data on our past light cone precise enough for an empirical construction of a world model whose evolution we could trust even a modest way back into the past, or whose structure we could believe at the depths of the most distant observed objects. The consolation, if there is one, is that the ideal of a world picture pieced together from data alone is not a realistic model for any branch of physical science. The time is long past when people claimed with Newton that they framed no hypotheses, that they worked from the empirical to the theoretical. That never was the whole truth, and it is very far from the way science works now. Quantum physics could never have been derived as an empirical picture, because the basic elements, operators and their state vectors, are in principle not observable. Fortunately we are not required to decide here whether the operators and state vectors have some sort of physical reality or are only aids to computation. We do learn from this extraordinarily successful physical science the tactic of validation by indirect tests. That is the way science is done and the way cosmology operates. My purpose in this part of the book is to argue that the results so far are encouraging, and there is the promise of considerably more to come.

The conventional attitude to the open puzzles of cosmology also follows the examples from sister physical sciences. Perhaps classical thermodynamics ranks as a complete closed physical science, but the list of problem-free theories certainly is short. By definition, in active sciences there are open questions whose significance is at best dimly grasped. They may be pointing to flaws in the framework, or perhaps only to a lack of suitable perception in the application of established principles. Principles can be overthrown, as in the spectacular replacements of the classical world picture with the quantum and relativity principles, but in a mature science this is, by definition, rare; it would be silly if we were allowed to invent a new law of physics for each new phenomenon. In cosmology the density of puzzles is high, and with it the chance that we have overlooked something fundamental. For example, we have no convincing theory for the origin of galaxies. It would be exciting news if it could be shown that the existence of these objects is incompatible with the standard cosmological picture as outlined above. So far that has not happened, and the candidate theories discussed in part 3, while far from convincing, do serve as "existence proofs" that seem to support the proposition that galaxies could exist in our world picture. A related puzzle, and a capital discovery waiting to be made, is the identification of the nature of the dark mass in the outskirts of galaxies, or a demonstration of how we have misapplied the dynamics from which its presence is inferred (section 18). An example on a grander scale is the expansion of the universe. In the standard picture the expansion traces back at a finite time t_o in the past to a singular state of arbitrarily high density, where the theory fails in the sense that there is no way it can predict what happened before that, what the universe was like one day prior to the time t_o. Does the fact that our world picture is so manifestly incomplete mean that it is wrong? As the evidence has accumulated for the success of the picture at times not so close to t_o, it has come to seem likely that we do have a useful approximation. But we certainly have to remember that a better theory, perhaps inflation, is needed to understand what really happened "before t_o," whatever that means.

In sum, the hopeful interpretation of the abundance of puzzles in cosmology is that this is an active observationally driven field with many well-specified problems to consider, and a great many chances to learn something new.

2.   Mach's Principle and the Cosmological Principle

Physical sciences develop in seemingly chaotic ways, by paths that are at best dimly seen at the time, and leave traditions that may seem mysterious and even irrational. That is why the history of ideas is an important part of any science, and particularly worth examining in cosmology, where the subject has evolved over several generations. The choice of where to begin has to be somewhat arbitrary, since any step in science builds on earlier ones. We start with the puzzle of inertia because it is fairly easy to trace the connection to Einstein's bold idea that the universe is homogeneous in the large-scale average, what Milne (1935) called Einstein's "cosmological principle."

In Newtonian mechanics one defines a set of preferred motions in space, the inertial reference frames, by the condition that a freely moving body has a constant velocity. In this theory, accelerations and rotations have an absolute character. For example, if a pail of water is rotating about its vertical axis on the Earth, we observe that the water is forced toward the outside of the pail; the more nearly level the surface of the water, the smaller the rate of rotation. The question naturally arises: What determines the motion of the reference frame that is not rotating? Is inertial motion an absolute property of the world, or could it be that inertia is determined by some field such as the aether that at the turn of this century was invoked for a mechanical interpretation of electromagnetism?

Ernst Mach argued that it is absurd to think that the inertial reference frames reflect an absolute property of nature, and perhaps equally bad to invoke an otherwise unobservable aether. Instead, he found it preferable to think that inertial frames are determined relative to the motion of the rest of the matter in the universe. Thus Mach (1893) wrote: "Newton's experiment with the rotating vessel of water simply informs us, that the relative motion of the water with respect to the sides of the vessel produces *no* noticeable centrifugal forces, but that such forces *are* produced by its relative rotation with respect to the mass of the Earth and the other celestial bodies. No one is competent to say how the experiment would turn out if the sides of the vessel increased in thickness and mass until they were ultimately several leagues thick." Einstein adopted, as Mach's principle, the idea that inertial frames of reference are determined by the distribution and motion of the matter in the universe.

The following summarizes some of the ideas needed to understand Einstein's implementation of Mach's principle. They are discussed in more detail in part 2.

An event in spacetime is an idealized instant of time at a definite position in space. The event is labeled by time and position coordinates t, x, y, and z, written collectively as x^i with $i = 0$, 1, 2, and 3. Consistent with Mach's principle, in general relativity theory the coordinates x^i for an event have no absolute significance; they are just arbitrary (but continuous and single-valued) labels. The difference dx^i between the coordinate labels of neighboring events in spacetime is given an invariant meaning by the expression for the line element connecting the events,

$$ds^2 = \sum_{i,j} g_{ij}\, dx^i dx^j$$

$$= g_{ij}\, dx^i dx^j \, .$$

(2.1)

In the second line, and throughout this book, we follow Einstein's summation convention — that repeated upper and lower indices are summed over their range, here 0 to 3. The metric tensor field $g_{ij}(x)$ in this expression is a four-by-four set of functions of position $x = x^j$ in spacetime. The metric tensor is symmetric,

$$g_{ij}(x) = g_{ji}(x) \, ,$$

(2.2)

so there are ten independent functions.

The interval ds^2 is assumed to have a definite coordinate-independent value for a given pair of events. Since the coordinate differences dx^i depend on the arbitrary assignment of coordinate labels, the metric tensor field $g_{ij}(x)$ also must depend on the coordinates. In locally Minkowski coordinates the coordinate labels are arranged so that at a chosen event in spacetime the metric tensor is equal to the Minkowski form,

$$g_{ij} = \eta_{ij} \equiv \begin{bmatrix} 1 & 0 & 0 & 0 \\ 0 & -1 & 0 & 0 \\ 0 & 0 & -1 & 0 \\ 0 & 0 & 0 & -1 \end{bmatrix}, \qquad \frac{\partial g_{ij}}{\partial x^k} = 0 \, .$$

(2.3)

As the second equation indicates, in locally Minkowski coordinates all first derivatives of g_{ij} vanish at the chosen event. In this coordinate system the line element (2.1) in the neighborhood of the event has the form,

$$ds^2 = dt^2 - dx^2 - dy^2 - dz^2, \tag{2.4}$$

familiar from special relativity. (Here, and in all the discussions of relativity, units are chosen so the velocity of light is unity. When we turn to measurements we usually put back the velocity of light as an explicit factor.)

Recall that in special relativity ds^2 in the line element (2.4) is unchanged by a rotation in space or by a Lorentz velocity transformation. The invariance of ds^2 in equation (2.1) generalizes this to a general coordinate labeling of a spacetime that need not be flat. The interpretation of ds^2 also follows special relativity. If the Minkowski line element ds^2 in equation (2.4) is positive, then ds is the proper time interval between the two events as measured by an observer who moves from one event to the other. If ds^2 is negative, then $|ds|$ is the proper spatial distance between the events as measured by an observer who is moving so the events appear to happen simultaneously. In either case, it is assumed that the result of the measurement is independent of the choice of measuring device. The same interpretation is applied to the interval ds in equation (2.1) in a general coordinate labeling of a curved spacetime: it is the proper time or distance between the events as measured by any idealized physical clock or measuring rod.

Because the locally Minkowski coordinate system defined by equation (2.3) is chosen to agree with the proper lengths and times a freely moving observer would set up in the neighborhood of an event in spacetime, it may not be surprising (and will be shown in section 9) that this coordinate system is equivalent to the local inertial coordinates of Newtonian mechanics for the neighborhood. Einstein's field equation in general relativity theory is a set of differential equations relating the metric tensor field $g_{ij}(x)$ and the matter distribution. That is, the matter distribution acts as a source for the metric tensor, the metric tensor describes the geometry of spacetime through the prescription for proper length and time intervals in equation (2.1), and the metric tensor determines the locally Minkowski coordinate systems that are locally inertial reference frames. There are cases where the results are consistent with Einstein's interpretation of Mach's principle. For example, in Mach's massive water bucket experiment the thick walls of the vessel would contribute to g_{ij} a term that makes the inertial frame for the water precess relative to the distant stars, perhaps just as Mach would have expected. However, it is easy to find other solutions to Einstein's field equations that violate Mach's ideas. Let us consider an "anti-Machian" solution and then what to do about it.

The metric tensor for a flat spacetime, with g_{ij} everywhere equal to the Minkowski form η_{ij} in equation (2.3), is the unique nonsingular spherically symmetric solution to Einstein's field equations in the absence of any matter. It is reasonable that $g_{ij} = \eta_{ij}$ should be a solution to Einstein's field equations, because we know that in our neighborhood of galaxies the mass density is small and the flat spacetime approximation with $g_{ij}(x)$ closely equal to η_{ij} gives a good description of the geometry of our nearby spacetime. However, this solution also allows us to postulate that a particle can move arbitrarily far from all the rest of the matter in the universe, to regions where the matter density makes a quite negligible contribution to Einstein's field equations. In this limiting case, the static solution to the field equations is arbitrarily close to $g_{ij}(x) = \eta_{ij}$. But this metric tensor defines definite inertial frames for the particle. That is, we have a solution to the field equation that defines inertial frames arbitrarily far from the rest of the matter in the universe. That certainly is contrary to Mach's ideas.

The way around the dilemma within general relativity theory is to add boundary conditions to eliminate the unwanted anti-Machian solution. A fascinating view of the evolution of Einstein's thinking on this point is to be found in the series of three papers by the Dutch astronomer Willem de Sitter, who was among the first to recognize the importance of Einstein's general relativity theory. In the second paper in the series, de Sitter (1916) mentions Einstein's idea that at great distances from material bodies the components of the metric tensor might degenerate to the "natural" values,

$$
g_{ij} = \begin{bmatrix} 0 & 0 & 0 & \infty \\ 0 & 0 & 0 & \infty \\ 0 & 0 & 0 & \infty \\ \infty & \infty & \infty & \infty \end{bmatrix}. \tag{2.5}
$$

Any transformation to new spatial coordinates that are accelerated relative to the original ones (with the time coordinate x^0 left unchanged) leaves this form unchanged. That is, this singular metric tensor does not define inertial motion. In Einstein's hypothesis, the presence of the matter in our neighborhood would cause the metric tensor to assume a nonsingular form that with a suitable choice of coordinates can be brought to the locally Minkowski form η_{ij} (eq. [2.3]), defining inertial motion.

De Sitter argued that the matter responsible for the transition of the metric tensor from the degenerate form (2.5), to the Minkowski form could not be any of the observed fixed stars or spiral nebulae, for if these bodies were near the regions where the metric tensor makes the transition from $g_{ij} \sim \eta_{ij}$ to the form (2.5), one would expect the strong space curvature to cause a pronounced gravitational frequency shift in the spectra of these objects, contrary to the observations. Thus the transition in form of the metric tensor would have to be due to the space curvature produced by "hypothetical" unobserved masses. This is not dissimilar to the hypothetical "dark matter" invoked in present-day cosmology to close the universe, but de Sitter objected, writing that, "to my mind, at least, the hypothetical masses are quite as objectionable as absolute space, if not more so. If we believe in these supernatural masses, which control the whole physical universe without ever having been observed," then, de Sitter argued, the theory appears to be no more appealing than the aether picture for the origin of inertia. Fascinating examples of the exchanges between de Sitter and Einstein on these issues are given by Kerszberg (1990), and Kahn and Kahn (1975).

Einstein soon abandoned the hypothetical masses in favor of another scheme that has proved to be of considerably more lasting interest. He proposed that the mass distribution in the universe is homogeneous in the large-scale average, and that the mass causes space to curve so as to close in on itself, in a three-dimensional analog of the closed two-dimensional surface of the balloon illustrated in figure 5.2. This would make the spatial volume of the universe finite, but there would be no hypothetical matter at the boundaries of the universe, and indeed no boundaries. And it would be impossible for a particle to move arbitrarily far from the rest of the universe, for wherever it moved it would encounter matter and a nonsingular metric tensor that defines local inertial reference frames that move with the mean motion of the matter in the universe.

Having discussed his ideas with de Sitter, Einstein surely was aware that he was making a bold assumption, because to astronomers at that time the universe usually meant the system of stars in our Milky Way galaxy. An example is Eddington's (1914) book, *Stellar Motions and the Structure of the Universe*. As Eddington describes, counts of stars as a function of brightness and position in the sky had shown that the Milky Way is bounded, shaped like a flattened disk. The radius of the disk was considerably underestimated, and we were put much closer to the center than we really are, because it was not fully appreciated that interstellar dust obscures the distant parts of our galaxy. However, the basic picture remains that we are in an island universe of stars. Because Einstein wrote of the distribution of stars, it is not clear whether he considered the idea that the diffuse objects called spiral nebulae might be other island universes, what are now called galaxies, comparable in size to the Milky Way. In the now standard cosmological model it is the mean distribution of the galaxies, averaged over the fluctuations in their clustering, that traces the nearly homogeneous large-scale distribution of mass.

Einstein's idea won rapid acceptance. This was in part because Hubble's counts of galaxies to fainter limits did continue to reveal increasingly large numbers of galaxies, as one would expect for an unbounded distribution, and probably also in part because of Einstein's prestige, and perhaps also because homogeneity greatly simplifies the mathematical problem of solving the field equations of general relativity. There were cautionary remarks, as from de Sitter (1931), who wrote: "It should not be forgotten that all this talk about the universe involves a tremendous extrapolation, which is a very dangerous operation." But more typical was Einstein's (1933) comment: "Hubble's research has, furthermore, shown that these objects [the galaxies] are distributed in space in a statistically uniform fashion, by which the schematic assumption of the theory of a uniform mean density receives experimental confirmation." From an empirical viewpoint de Sitter's would have been the more sensible position. It was known that galaxies tend to appear in groups and clusters, and, as Charlier (1908, 1922) noted, it would not be unreasonable to imagine that the hierarchy of clustering continues to indefinitely large scales. As will be discussed further in the next section and in section 7, this model predicts that Hubble would continue to find galaxies as his surveys probed to increasing distances, but the mean number density would decrease as the limits of the surveys expanded through the levels of the hierarchy. Einstein (1922) granted that this picture is consistent with general relativity theory, but he rejected it as inconsistent with Mach's principle, for it would mean that in most of space the mass density is arbitrarily small yet inertia is defined by a nearly Minkowski line element.

The idea of a homogeneous world model offers the new possibility that the universe is spatially homogeneous but anisotropic, in the sense that the matter velocity field defined by the mean streaming motion of the galaxies has shear, and perhaps also rotation relative to locally inertial reference frames, as in Gödel's (1949) remarkable rotating world model. Rotation of inertial frames relative to the mean motion of the matter surely goes against Einstein's interpretation of Mach's principle, and the same is true of shear, for there is nothing in the mass distribution to tell the shear in which direction to point. In any case, Einstein's concept as it came to be embodied in the cosmological principle is that in the large-scale average the visible parts of our universe are isotropic and homogeneous.

Milne's (1935) term, "Einstein's cosmological principle," is appropriate in the sense that the conditions of homogeneity and isotropy do greatly restrict the range of possible cosmologies, as Milne was among the first to appreciate. However, the cosmological principle is not to be compared to the uncertainty principle in quantum mechanics, which is demanded by the basic elements of the theory, because general relativity theory as it usually is understood does not demand that the universe be homogeneous and isotropic in the large-scale average, and it is hard to see how our existence would be threatened by a universe considerably more lumpy than what we observe. An example of an apparently viable world picture that violates the cosmological principle in the sense Einstein seems to have intended it is the inflation scenario (section 17). In this picture we happen to live not too close to the edge of a bounded, nearly uniform part of a universe that is quite chaotic on very large scales.

In the now standard cosmological model, the cosmological principle reduces to a consistency condition, that the large-scale departures from a homogeneous and isotropic mass distribution and motion have to be consistent with observable consequences: the anisotropy of radiation backgrounds, the counts of objects as a function of direction and distance, and the peculiar motions derived from Doppler shifts. For example, galaxy counts averaged within spherical regions of radius $30h^{-1}$ Mpc fluctuate around the large-scale mean by about 30%. If the rms deviations from homogeneity in the mass distribution were similar, $\delta\rho/\rho \sim 0.3$, and if the mass fluctuations were uncorrelated at larger separations, the gravitational acceleration caused by the irregular mass distribution would be predicted to cause deviations from a homogeneous velocity field that are comparable to what is observed (section 21). If the fluctuations in the mass distribution were much larger than this, it would be a crisis for the standard model. (If the mass fluctuations were appreciably smaller, it could be less serious, because one might hope to find a nongravitational source for the observed motions, as in explosions.) On larger scales, the Sachs-Wolfe effect on the anisotropy of the thermal background radiation (sections 6 and 21) requires that the fluctuations in the mass distribution averaged within regions of radius comparable to the Hubble length in equation (1.2) are no greater than about $\delta\rho/\rho \sim 1 \times 10^{-4}$. There are no near-term prospects for comparing this prediction to the actual fluctuations in the galaxy distribution on such large scales; we can only conclude that if the standard model is valid, the observable universe has to have been constructed to be remarkably close to homogeneous and isotropic on scales comparable to the Hubble length, and that no believable contradiction to this condition has been found.

Might the cosmological principle be elevated to a physical principle that has to be true? We should bear in mind that although some may have been glad to accept the cosmological principle because it simplifies the mathematics, and others because it agrees with the observations, Einstein was motivated by something quite different: the idea that a universe that is not homogeneous and isotropic in the large-scale average is absurd. Since the argument has proved successful, perhaps it is telling us something deep about the nature of the universe.

3. The Realm of the Nebulae

With the discovery that the spiral nebulae are galaxies of stars, coequals of the Milky Way, people were led to the exploration of a new level in the hierarchy of structure in the physical universe, what Sanford (1917) and Hubble (1936) called the realm of the spiral nebulae. Since then cosmology has grown to include the study of other systems, such as the radiation backgrounds and the gas clouds seen as absorption features in the lines of sight to distant quasars, but common luminous galaxies comparable to the Milky Way remain the centerpiece. An appreciable fraction of the mass of the universe at the present epoch is in and around these galaxies. They are abundant enough to give a useful indication of the structure of the mass distribution, bright enough to be visible at great distances, where one might hope to see the general relativistic departures from the static flat spacetime of special relativity, and possibly stable enough to act as permanent markers of how mass concentrations were placed in the remote past.

The purpose of this section is to present some highlights of the ways people came to see that the spiral nebulae are island universes of stars, and to discuss some elements of the systematics of the structures and spatial distribution of the galaxies.

Discovery

Figure 3.1 shows the distribution across the sky of the brighter of the diffuse nebulae as they were known at about the time Einstein introduced the cosmological principle (Charlier 1922). Some of these objects are gas clouds in our galaxy, but most are spiral nebulae, that is, spiral galaxies similar to the Milky Way. A few are elliptical galaxies, whose images tend to be dominated by a smooth concentration of older stars rather than by the prominent disk of gas and dust and younger stars seen in most spiral galaxies.

We are close to the central plane of the disk of the Milky Way galaxy, so the plane of the Milky Way appears in the sky as a great circle, and it appears in this map as the central horizontal line. The sphere of the sky is represented in this flat map with equal areas in the sky appearing as equal areas in the map. The polar axis of the galactic coordinate system is normal to the disk of the galaxy, so the plane of the disk is at polar angle (colatitude) $\theta = 90°$. The galactic latitude b is measured from the equator, with $b = 90° - \theta$. The horizontal lines in the figure

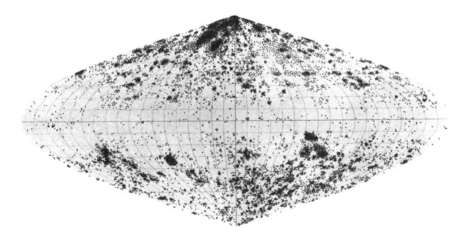

Figure 3.1. Map of the nebulae, from Charlier (1922). Most of the objects plotted in this map are extragalactic nebulae, that is, galaxies of stars. The absence of galaxies in the zone of avoidance along the horizontal centerline of the map is caused by the obscuration by dust near the plane of our Milky Way galaxy. The strong concentration near the north pole, at the top of the map, is the Virgo cluster, the nearest of the rich clusters of galaxies.

are lines of constant latitude at $10°$ spacing. Some further details of this galactic coordinate system are given at the end of this section.

Quite evident in the figure is the zone of avoidance along the equator, where very few nebulae are seen (and many of these are gas clouds in our galaxy, not extragalactic nebulae). If the spiral nebulae were independent galaxies, they would not be expected to prefer the poles of the Milky Way. However, even when this map was made it was known that there are regions in the plane of the Milky Way where few stars are seen, and it was suggested, we now know correctly, that these dark patches in the Milky Way are caused by interstellar dust clouds. This same dust of course would obscure the extragalactic nebulae in the zone of avoidance (e.g., Sanford 1917). It was noted that there are many more nebulae of small angular size than large, as one would expect if the nebulae were more or less uniformly distributed around us. For example, Curtis (1918) wrote: "It is my belief that all the many thousands of nebulae not definitely to be classified as diffuse or planetary are true spirals, and that the very minute spiral nebulae appear as featureless discs or ovals because of their small size. Were the Great Nebula in Andromeda situated five hundred times as far away as at present, it would appear as a structureless oval about $0'.2$ long, with a very bright center, and not to be distinguished from the thousands of very small, round or oval nebulae found wherever the spirals are found."

The Andromeda Nebula (also known as M31, for its place in the Messier list of bright nebulae) is the nearest large spiral galaxy outside the Milky Way. Öpik (1922) obtained the first useful estimate of the distance to this galaxy by the following argument.

By 1920 the luminosity of our Milky Way galaxy had been estimated from star counts and statistical estimates of star distances, and the mass had been estimated from star velocities. Öpik (1922) put the ratio of the mass to the luminosity at

$$\frac{\mathcal{M}}{L} \sim 3\frac{\mathcal{M}_\odot}{L_\odot}. \tag{3.1}$$

The units are solar masses and solar luminosities; these and some other units in common use in cosmology are summarized at the end of this section. The mass-to-light ratio is close to unity when expressed in solar units because the bright parts of the Milky Way are dominated by stars with masses and luminosities roughly similar to that of the Sun. (The contributions to the mass and light densities in our stellar neighborhood, as a function of star mass, are shown in table 18.1 in part 3.)

Öpik noted that if the spiral nebulae were other galaxies of stars they might be expected to have similar mass-to-light ratios, and that this could be used to infer the distances to the nebulae. The velocity of rotation of the disk in the Andromeda Nebula is measured from the Doppler shifts in the wavelengths of emission lines from the disk gas. Because the Doppler shifts across the disk are close to symmetric relative to the center of the galaxy, it is reasonable to assume that the disk material is moving in circular orbits, at linear speed v_c at the edge of the disk, at radius r. Then the gravitational acceleration g at the edge of the disk is

$$\frac{GM}{r^2} = \frac{GM}{(\theta D)^2} = g = \frac{v_c^2}{r} = \frac{v_c^2}{\theta D}. \tag{3.2}$$

The first equation assumes the mass M of the galaxy is spherically distributed, but the gravitational acceleration is not much different for a flattened mass distribution. The radius has been written as $r = \theta D$, where θ is the observed angular radius of the disk and D is the unknown distance to the galaxy. If the luminosity of the galaxy is L, the observed energy flux f of the light from the galaxy is

$$f = \frac{L}{4\pi D^2}. \tag{3.3}$$

The result of combining these two equations is

$$D \sim \frac{v_c^2 \theta}{4\pi Gf} \frac{L}{M}. \tag{3.4}$$

Using the observed values of v_c, θ and f, and the assumed value of the mass-to-light ratio in equation (3.1), Öpik found[1]

$$D_{\text{Öpik}} = 450\,\text{kpc}. \tag{3.5}$$

The modern value, based on observations of variable stars with known intrinsic luminosities, is $D = 770 \pm 30\,\text{kpc}$. Öpik's number is close because, as he had expected, the mass-to-light ratio in the brighter parts of a giant galaxy is close to a universal value, not far from equation (3.1). Some typical numbers are given in section 18.

Öpik pointed out that his result puts M31 well outside our galaxy. (We are near the edge of the Milky Way, at about 8 kpc from the center.) With the distance in equation (3.5), the mass of M31 from equation (3.2) is comparable to that of our galaxy. Thus Öpik concluded that M31 is another island universe of stars, as others had speculated before him.

Öpik's result was confirmed by Hubble's (1925, 1926a) identification of Cepheid variable stars in the Andromeda Nebula and its companion galaxy, the

---

[1] One parsec is the distance at which one second of arc subtends one astronomical unit, which is the mean distance between the Earth and Sun. This is convenient because as the Earth moves around the Sun, a star at one parsec distance from us is seen to move relative to the distant stars in an ellipse with semimajor axis equal to one second of arc. As usual, one kiloparsec = 1 kpc = 10^3 pc and one megaparsec = 1 Mpc = 10^6 pc. A handy working number is 1 pc $\sim$ 3 light years.

Triangulum Nebula M33. (These galaxies are named after the constellations in which they appear in the sky; Cepheid variables are named after their prototype, δ Cephei, in the constellation Cepheus.) Cepheid variable stars had already been identified in the Magellanic Clouds. These are dwarf satellites of the Milky Way, near enough that they were seen to be systems of stars, and with reasonably secure distance estimates from the comparison of stars in the Magellanic Clouds with those in our galaxy. It was known through the work of Leavitt (1912 and earlier references therein) that in this class of variable stars the mean intrinsic luminosity is correlated with the period of the variation in luminosity, so Hubble could use the inverse square law to find the ratio of distances to M31 or M33 and the Magellanic Clouds by comparing the observed brightnesses of Cepheids with the same period and hence the same luminosity. Using Shapley's (1925) estimate of the distance to the Magellanic Clouds, Hubble found

$$D_{\text{Hubble}} = 300 \, \text{kpc} \tag{3.6}$$

for M31 and M33, comparable to Öpik's result in equation (3.5). This is smaller than the modern value, because Shapley underestimated the distance to the Magellanic Clouds by a factor close to two, and Hubble's value for the ratio of the distances of M31 and M33 to the distance to the Magellanic Clouds was small by another 30% . But that is a detail. The main point is that the astronomers had hit on believable evidence that the spiral nebulae are island universes comparable to the Milky Way.

Hubble's Test for the Space Distribution

It was natural that people should ask whether the galaxies are uniformly distributed, because that is what Einstein had postulated for the world matter. Perhaps more convincing to astronomers is that, as we have noted, it was known that there are many more spiral nebulae of small angular size than large, as one would expect if the nebulae were homogeneously spread through space. Hubble (1926b) translated this observation into a quantitative test by adapting a method previously used to study the space distribution of stars.

Suppose for simplicity that space is static and the geometry Euclidean, and that all galaxies have the same intrinsic luminosity, L. The last assumption will be removed in a moment; the relativistic corrections for the expansion and space curvature of the standard cosmological model are dealt with in part 2.

In static Euclidean space a galaxy of luminosity L at distance r appears at apparent brightness (observed energy flux)

$$f = \frac{L}{4\pi r^2} . \tag{3.7}$$

If all galaxies had the same luminosity, all galaxies brighter than f would be closer than r. The volume of space in a steradian of the sky out to distance r is

$V = r^3/3$, so if galaxies were distributed uniformly on average, with mean number density n, the mean number of galaxies per steradian brighter than f would be

$$N(>f) = nV = \frac{nr^3}{3} = \frac{n}{3}\left(\frac{L}{4\pi f}\right)^{3/2}. \tag{3.8}$$

The last step follows from the inverse square law in equation (3.7). In reality, galaxies have a broad spread of luminosities. We can take that into account by letting n_i be the mean number density of galaxies in luminosity class L_i. Then equation (3.8) applies to each class of galaxies, and the sum over luminosity classes gives the total count of galaxies as a function of observed energy flux f,

$$N(>f) = \frac{1}{3}\sum_i n_i \left(\frac{L_i}{4\pi f}\right)^{3/2} \propto f^{-3/2}. \tag{3.9}$$

It is a standard convention in astronomy to express L and f in logarithmic measures of absolute and apparent magnitudes. The apparent magnitude m of an object with received energy flux f (with units of ergs per second and per square centimeter in some chosen band of frequencies) is defined to be

$$m = -2.5 \log f + \text{constant}. \tag{3.10}$$

Here and throughout, the base for $\log x$ is ten; natural logarithms are written $\ln x$. The normalizing constant is defined at the end of this section. The curious choice of basis for the logarithm, and the convention that the magnitude decreases with increasing energy flux, comes from the historical use of magnitudes as a way to rank the visual brightnesses of the stars. A way to remember it is to note that a change of five magnitudes represents a factor of 100 difference in flux. The absolute magnitude, M, of an object is related to its intrinsic luminosity, L, by the analogous relation,

$$M = -2.5 \log L + \text{constant}. \tag{3.11}$$

This constant is defined so that an object at 10 pc distance has apparent magnitude equal to its absolute magnitude (when there is no obscuration, for example by absorption and scattering by interstellar dust). Thus we have from the inverse square law (3.7) that the difference between the apparent and absolute magnitudes of an object at distance r is

$$m - M = 5 \log r_{\text{Mpc}} + 25. \tag{3.12}$$

The distance is measured in units of megaparsecs, so that at $r = 10$ pc this equation says $m - M = 0$, which is the definition of the absolute magnitude M. The magnitude difference $m - M$ is called the distance modulus.

From equation (3.10) the energy flux is

$$f \propto 10^{-0.4m}, \tag{3.13}$$

so Hubble's count law (3.9) for a spatially homogeneous mean distribution of objects is

$$N(<m) \propto 10^{0.6m}. \tag{3.14}$$

The differential counts have the same form,

$$\frac{dN(m)}{dm} \propto 10^{0.6m}. \tag{3.15}$$

In the 1920s it was known that star counts vary less rapidly than $10^{0.6m}$, indicating the Milky Way is inhomogeneous and the star counts probe the edges of the system. The test is readily applied to galaxy counts. Already in 1926 Hubble could report that the galaxy counts available in the literature agree with the $f^{-3/2}$ power law, consistent with the idea that the galaxy distribution is close to homogeneous in the large-scale average (Hubble 1926b). The deepest counts Hubble used amount to five galaxies per square degree in a half-magnitude interval in apparent magnitude. The direct estimates of the limiting magnitude in these old data are not reliable, but we can scale from the modern count-distance relations in figures 3.2 and 5.6. They indicate the counts reached blue magnitude $m \sim 17$, corresponding to a distance about two hundred times that of the Andromeda Nebula and about ten times that of the Virgo cluster (which is seen at the top of figure 3.1 and in figures 3.4 and 3.7 below). In the next decade, Hubble (1934) pushed the count test deeper, to limiting magnitude $m \sim 21$ in the magnitude system in figure 3.2. In the notation in equation (1.1) and in (3.18) below (and developed in section 5), Hubble's limiting depth is about $600h^{-1}$ Mpc, or 20% of the Hubble length in equation (1.2). The fact that Hubble encountered no pronounced evidence of an edge to the galaxy distribution was an impressive first extensive and quantitative test of Einstein's homogeneity postulate.

Hubble's test is powerful because the detection and measurement of the flux f is relatively easy even for very distant galaxies. The method is unlikely ever to yield a precision measure of modest fluctuations away from homogeneity on large scales, however, because the light from galaxies at great distance is shifted toward the red part of the spectrum, as will be discussed in section 5, and unless the redshift is known for each galaxy it is difficult to correct for the effect on the received energy flux measured in a fixed band of frequencies. Also, even at the depth of Hubble's 1934 survey the galaxies are distant enough so that in the standard model they are seen when they and the universe were significantly younger (due to the light travel time), so the time evolution of the galaxies and spacetime can cause an appreciable departure from Hubble's relation (3.15).

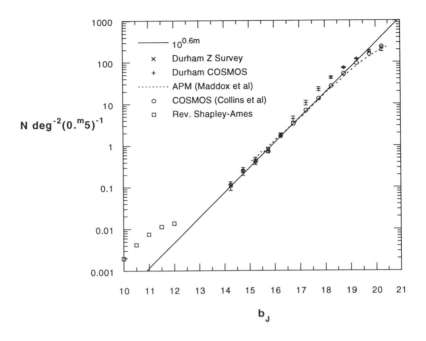

Figure 3.2. Galaxy counts (Shanks 1991). The horizontal axis is the apparent magnitude. The solid line is the Hubble law (3.15).

An example of the galaxy counts at what now are considered intermediate depths is shown in figure 3.2. The graph was made by Shanks (1991); the methods and observational surveys are reviewed in Shanks (1990). The solid line is the $10^{0.6m}$ Hubble law. The counts at magnitudes $m \lesssim 13$ are high because of the local concentration of galaxies in and around the Virgo cluster. In the relativistic Einstein-de Sitter model defined in equations (5.18) to (5.21), assuming the luminosities of the galaxies are not changing with time, the predicted counts corrected for the redshift and normalized at $m \sim 15$ are a factor of two below the observations at $m \sim 20$. From there to $m \sim 25$ the counts are well fitted by a shallower power law, $N \sim 10^{0.45m}$. This is presumed to be the combined effect of evolution of the galaxies and spacetime. There have been dramatic recent advances in the ability to measure galaxy redshifts at apparent magnitudes fainter than $m \sim 20$. The untangling of the effects of galaxy evolution and the departures from Minkowski spacetime on galaxy counts as a function of redshift and apparent magnitude are likely to be major topics in cosmology in the 1990s.

The apparent magnitudes on the horizontal axis of figure 3.2 are measured in a blue band at $\lambda \sim 4000$ to 5500 Å, in the Durham $m = b_J$ system (Shanks et al. 1984).[2] In section 5 we will need the normalization of Hubble's galaxy count law to estimate the mean space number density of galaxies. The solid line in figure 3.2 is

$$\frac{dN}{dm} = 10^{-5.70\pm0.10+0.6m} \text{ ster}^{-1} \text{ mag}^{-1}, \text{ for } 14 \lesssim m \lesssim 18. \tag{3.16}$$

Mapping the Galaxy Distribution

Figure 3.1 shows that the nearby galaxies are distributed in a decidedly inhomogeneous way. An understanding of the clustering of the galaxies is important as a measure of how closely the large-scale mean distribution can approach the homogeneity postulated in the cosmological principle. Equally important, the tendency for galaxies to appear in groups and clusters and superclusters is just a scaled-up version of the tendency of stars to appear in the concentrations we call galaxies. The phenomena surely are related, and an interpretation ought to teach us something important about the physics of the universe. Some of the attempts to rise to the challenge are chronicled in section 25.

We will be considering the nature of the galaxy distribution around us in a sequence of increasing distance. Since direct distance estimates are available for relatively few galaxies, we will use the result to be discussed in section 5, that the redshift of the spectrum of a galaxy is the sum of a cosmological part proportional to the galaxy distance r and a Doppler shift due to the line of sight component of the motion of the galaxy relative to the mean flow,

$$c(\lambda_o/\lambda_e - 1) \equiv cz \cong H_o r + v. \tag{3.17}$$

Here λ_o/λ_e is the ratio of observed wavelengths, λ_o, of features in the spectrum to the corresponding laboratory wavelengths, λ_e, that are presumed to be the wavelengths at emission as measured by an observer at rest in the observed galaxy. The second expression is the redshift, z, defined so $z = 0$ if there is no shift in the spectrum. The line-of-sight component of the motion of the galaxy relative to the mean, which is called the peculiar velocity, is v. The peculiar velocity produces an ordinary first-order Doppler shift, $\delta\lambda/\lambda = v/c$. The constant of proportionality in the cosmological redshift term $H_o r$ is Hubble's constant. This linear relation

---

[2] It is striking that there still is not a standard set of magnitudes based on fixed passbands and zero points (the constant in the relation between energy flux f and apparent magnitude m in eq. [3.10]), because each advance in detectors suggests new conventions. For the order-of-magnitude estimates presented in this book, the specific optical magnitude system does not much matter, and the choice will be based on convenience. In more detailed calculations one must of course use due caution in mixing magnitude systems.

between redshift and distance applies when $H_o r$ and v are small compared to the velocity of light; the relativistic corrections are dealt with in part 2. The value of Hubble's constant still is subject to debate, so the standard practice, going back at least to Kiang (1961), is to write it as

$$H_o = 100h \, \text{km s}^{-1} \, \text{Mpc}^{-1}, \qquad 0.5 \lesssim h \lesssim 0.85. \qquad (3.18)$$

As indicated, the dimensionless parameter h is believed to be between 0.5 and 0.85. In this section we attempt to reduce confusion by quoting distances in units of velocity. Thus, if the peculiar velocity in equation (3.17) is negligible, the distance in velocity units is

$$H_o r = cz = c\delta\lambda/\lambda, \qquad (3.19)$$

which of course is independent of the value of H_o.

As we have noted, the nearest large neighboring galaxy is the Andromeda Nebula, M31, at distance

$$D_{\text{M31}} = 770 \pm 30 \, \text{kpc}. \qquad (3.20)$$

This is about 20% of the mean distance between large galaxies (eq. [5.145] below). The spectrum of the Andromeda nebula is blueshifted, corresponding to a rate of approach of the centers of the Milky Way and M31 of about 100 km s^{-1}. This is a rare exception to the rule that galaxy spectra are redshifted. The blueshift and the relatively small separation both suggest the two galaxies are orbiting in a gravitationally bound system, called the Local Group. The other members of this group are the satellites of the Milky Way and of M31, such as the Magellanic Clouds and the Triangulum Nebula M33, along with a few outlying dwarf galaxies at distances $r \lesssim 1 \, \text{Mpc}$. A helpful guide to the Local Group is given by van den Bergh (1968). A model for its dynamics is presented in section 20.

Figure 3.3 shows the distribution of the known galaxies in our immediate neighborhood. The data for this figure come from a recent version of Huchra's (1991) compilation of measured redshifts, ZCAT.

The maps show two projections of the galaxy distribution, in a box of width $H_o r = 800 \, \text{km s}^{-1}$ centered on us. The width of the box is 0.3% of the Hubble length c/H_o at which the cosmological redshift in equation (3.17) extrapolates to the velocity of light. In distance units, the box width is $8h^{-1} \, \text{Mpc}$, an order of magnitude larger than the distance to the Andromeda Nebula. The coordinate axes are the de Vaucouleurs (1953) supergalactic coordinates. The Y axis points roughly in the direction of the Virgo cluster of galaxies, which is at distance $cz \sim 1200 \, \text{km s}^{-1}$, and the XY plane is oriented to include the plane of the local concentration of galaxies. The zone of avoidance caused by the obscuration by

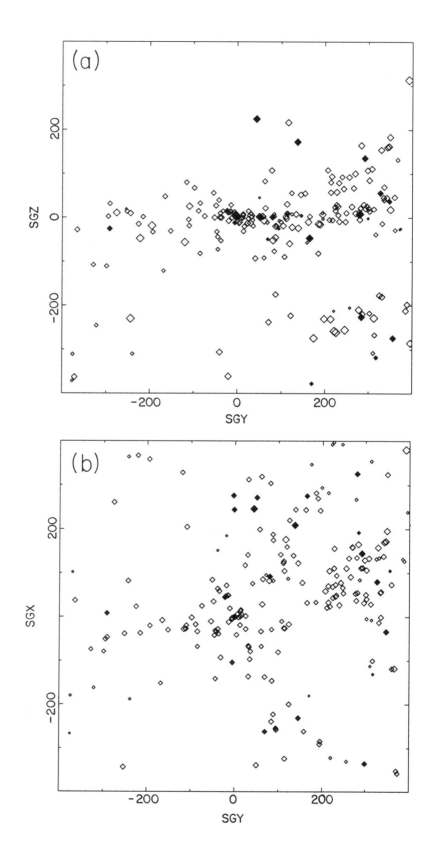

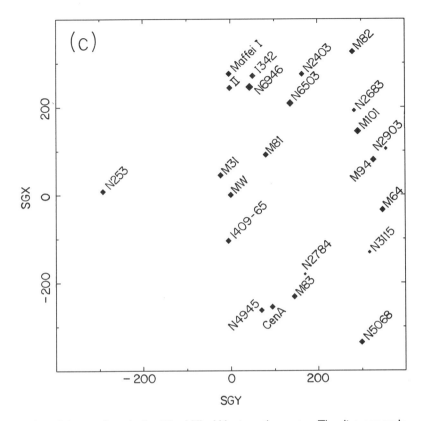

Figure 3.3. Map of the nearby galaxies. The Milky Way is at the center. The distance scale is the redshift, cz. The galaxies with luminosities greater than about 30% of the Milky Way are shown as filled diamonds, the fainter dwarf galaxies as the open symbols. The positions are plotted in the de Vaucouleurs supergalactic coordinates defined at the end of this section. The symbol size increases with increasing value of the coordinate along the projected direction. Panels (a) and (b) are orthogonal projections of the distribution. Panel (c) labels the more luminous galaxies.

dust in our galaxy is roughly in the XZ plane. This makes the map incomplete at $|Y| \lesssim 100\,\mathrm{km\,s^{-1}}$.

There are two cautionary remarks to make about this figure. First, it seems likely that all the bright galaxies within the volume of the box have been identified, but at the faintest absolute magnitudes and lowest surface brightnesses the fraction of the galaxies within this volume that have been detected and represented in the map has to vary across the sky, a function of the degree of interference by interstellar matter in the Milky Way and of the detail with which different parts of the sky have been searched. Also, there still is some confusion about measured values of the redshifts for a few of the faintest galaxies, and even about the identifications of some of them. This is not likely to have much effect on the general appearance of the map, unless there is an important class of galaxies that has been almost entirely missed.

Second, the radial distance coordinate, with origin centered on the Milky Way, is the redshift, cz. Galaxy motions v relative to the ideal Hubble flow in equation (3.17) (the peculiar motions relative to the peculiar velocity of the Local Group) tend to spread the apparent distribution of distances along the line of sight. The evidence is that the effect usually is relatively small, even at the small cosmological redshifts in this map, as is illustrated in figure 5.5 below, but there are important exceptions. Most prominent is the nearby Virgo cluster of galaxies (which appears at the top of figure 3.1). The recession velocity of this cluster is not much larger than the random velocities of the galaxies in the cluster, so the spectra of some of the galaxies in the cluster are shifted to the blue.[3] This effect has been suppressed by discarding all galaxies at angular positions in a circle of 6° radius centered on the Virgo cluster. We will note below some other examples of galaxies that are known to be in more distant groups and are in this map because they have peculiar motions toward us. Within the Local Group, the Andromeda Nebula and its more prominent satellites have been placed at about the right positions relative to the Milky Way, but the outlying group members are plotted at their redshift distances, which are considerably distorted by motions within the group. The map thus must be treated with caution, but it does give a useful first impression of where things are.

Panel (c) shows common names for the brighter galaxies. The galaxies M81, M82 and N2403 (where N, or more commonly NGC, stands for the New General Catalog of galaxies) are thought to be at a nearly common distance, members of the M81 group. The motions within the group would produce the dispersion in redshift distances. The giant spiral galaxy M101 is the dominant member of a group just outside the box, this galaxy having a small redshift distance because its peculiar motion is toward us. The Maffei group of galaxies, which is close to the X axis at $X \sim 400\,\text{km s}^{-1}$, is named after its discoverer (Maffei 1968); it was identified only fairly recently because it is in the zone of obscuration. The spiral N6946 appears close to the Maffei group in projection in panel (c), but one sees in panel (a) that it is quite isolated. The concentration of galaxies around N253 is called the Sculptor Group. The unusual galaxy N5128, or Centaurus A, looks like an elliptical crossed by a broad band of dust. Beautiful images of

---

[3] The only known galaxies with blueshifts are members of the Local Group and a few galaxies in the direction of the Virgo cluster that are presumed to be cluster members with velocities far enough into the tail of the distribution so that the peculiar velocity within the cluster is larger than the cosmological redshift and they have a net motion of approach toward us.

this galaxy and many of the others named in panel (c) are in *The Hubble Atlas of Galaxies* (Sandage 1961). Elegant maps of the galaxy distribution are in the *Nearby Galaxies Atlas* (Tully and Fisher 1987).

There are several notable features of the distribution in figure 3.3. First, one sees regions well away from the zone of avoidance at $Y \sim 0$ where the galaxy density is markedly low. Kirshner et al. (1981) named these regions voids. Second, there are many more dwarf galaxies than giants such as the Milky Way. Third, the distributions of giants and dwarfs are quite similar: the voids defined by the giants are avoided by the dwarfs. And fourth, the galaxies tend to lie on a plane, as evidenced by the strong concentration at small $|Z|$ in panel (a) compared to the spread in X and Y in panel (b). De Vaucouleurs identified this plane as part of the Local Supercluster, along with the concentration of galaxies around the Virgo cluster.

The local plane of galaxies defines a structure that extends to remarkable distances. It can be seen as a linear concentration in the sky map of the galaxies at distances $cz \lesssim 3000 \, \text{km s}^{-1}$ (Lahav 1987). As illustrated in figure 3.7 below, this plane contains several of the nearest clusters of galaxies. Particularly striking is the observation that most of the radio galaxies at distances $cz \lesssim 5000 \, \text{km s}^{-1}$ are concentrated in the same plane, though at this distance the mean angular distribution of the galaxies is close to isotropic (Shaver 1991). Section 26 contains a few thoughts on what this might be telling us.

Figure 3.4 shows the angular distribution of the galaxies brighter than apparent magnitude $m \sim 14.5$ and at redshifts $cz < 3000 \, \text{km s}^{-1}$, four times the width of the box in figure 3.3. The galaxies in this map were catalogued by Zwicky et al. (1961–68). The redshifts are from the first of the large-scale systematic redshift surveys, by Davis et al. (1982). The two panels show separately the high luminosity and low luminosity halves of the sample. The maps are centered on the north galactic pole, perpendicular to the plane of the Milky Way. The circles are at galactic latitudes $b = 70°$, $50°$, and $30°$ (that is, polar angles $20°$, $40°$, and $60°$). The survey boundary is $b = 40°$; at lower latitudes few galaxies are visible because they are obscured by the dust in the plane of the Milky Way, in the zone of avoidance. The dashed lines show the right ascension α and declination δ in the polar coordinate system centered on the Earth (and defined in a little more detail at the end of this section). The empty region to the lower left is the catalog boundary at $\delta = 0$.

The concentration of galaxies at declination $\delta \sim 13°$ and right ascension $\alpha \sim 12^h 25^m$ is the Virgo cluster. The cluster is about equally prominent in the distributions of the high and low luminosity galaxies shown in the two panels. That is, we see again that to quite a good approximation the less luminous dwarf galaxies cluster with the more luminous giants.

Another way to classify galaxies distinguishes the gas-rich late types, including spiral galaxies such as the Milky Way and irregulars such as the Magellanic Clouds, and relatively gas- and dust-free early types. High surface brightness

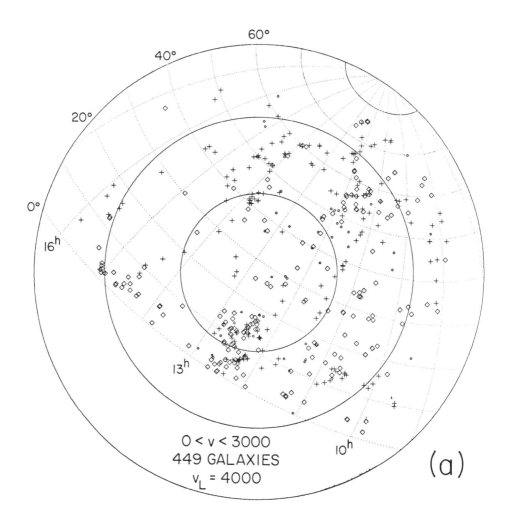

0 < v < 3000
449 GALAXIES
v_L = 4000

(a)

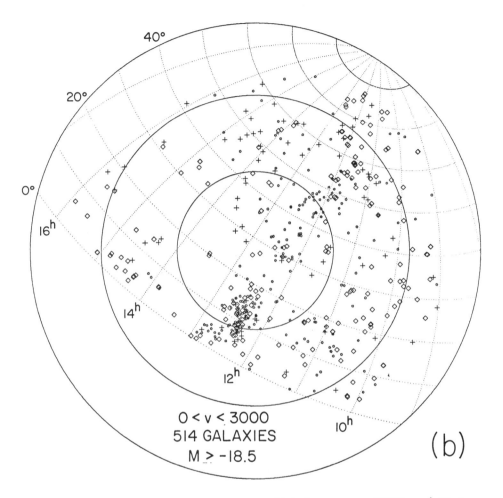

Figure 3.4. Angular distribution of galaxies at redshift distances $cz < 3000\,\mathrm{km\,s^{-1}}$ (Davis et al. 1982). The more luminous galaxies, with absolute magnitude $M \lesssim -18.5$, are shown in panel (a), the less luminous ones in panel (b). The pentagons represent galaxies at $v < 1000\,\mathrm{km\,s^{-1}}$, diamonds $1000 < v < 2000\,\mathrm{km\,s^{-1}}$, and plus signs $v > 2000\,\mathrm{km\,s^{-1}}$.

early-type galaxies are classified as elliptical and S0, according to their shapes.[4] Most of the brighter of the galaxies in our neighborhood shown in figure 3.3 are late types. Within $6°$ from the center of the Virgo cluster there are more nearly equal numbers of early- and late-type galaxies, and the early types are more strongly concentrated toward the cluster center. Following Baade and Spitzer (1951) and Gunn and Gott (1972), it is thought that collisions between young galaxies in the dense central parts of a developing cluster, or the ram pressure due to the motion of the young galaxies through the intracluster gas, have tended to inhibit the accumulation of the interstellar gas and dust characteristic of late-type galaxies, producing a higher concentration of early types in clusters.

The strong concentrations of galaxies in such clusters, and the usually prominent excess of early types toward the cluster centers, indicate that the clusters are localized concentrations in space, that is, that the cluster members have a common cosmological redshift. Thus the spread of values of the redshifts of the cluster members is interpreted as a distribution of the line-of-sight peculiar velocities in equation (3.17). In the Virgo cluster, the line-of-sight velocity dispersion (rms deviation from the mean) for the early-type galaxies is $\sigma_e \sim$ $500 \, \mathrm{km \, s^{-1}}$. The later types have a distinctly broader velocity distribution, $\sigma_s \sim$ $800 \, \mathrm{km \, s^{-1}}$ (de Vaucouleurs and de Vaucouleurs 1973; Huchra 1985; Binggeli, Tammann, and Sandage 1987). The cosmological redshift of the cluster is $1200 \pm$ $100 \, \mathrm{km \, s^{-1}}$, the main uncertainty being the correction for our motion toward the cluster caused by its gravitational attraction (as discussed in section 20).

Figure 3.5 shows the distribution of galaxies in the neighborhood of the Coma cluster, at distance $cz = 7000 \, \mathrm{km \, s^{-1}}$. Plotted are all galaxies in the field with measured redshifts in the range $4000 \, \mathrm{km \, s^{-1}}$ to $10000 \, \mathrm{km \, s^{-1}}$, from Huchra's (1991) ZCAT redshift catalog. The coordinate axes, in degrees of declination and hours of right ascension, are scaled to give an undistorted image. The symbol size decreases with increasing distance. The galaxies in the central parts of this cluster are predominantly early types; spirals are found only on the outskirts.

Abell (1958, 1965) catalogued the several thousand nearest and richest clusters in the northern hemisphere, to a limiting depth of about $60,000 \, \mathrm{km \, s^{-1}}$. Thus the Coma cluster also is known as A1656, after its place in the Abell catalog. The mean space number density of Abell clusters is $n_{\mathrm{cl}} \simeq 1 \times 10^{-5} h^3 \, \mathrm{Mpc^{-3}}$ (Bahcall 1988), and the mean distance between Abell clusters is $H_o n_{\mathrm{cl}}^{-1/3} = 5000 \, \mathrm{km \, s^{-1}}$. This is comparable to our distance from the Coma cluster. The ratio of distances of the Coma and Virgo clusters is

---

[4] The origins of such curious names are understood only by astronomers; Kormendy (1982), Kormendy and Djorgovski (1989), and van den Bergh (1990a) give helpful guides. At one time it was thought that early-type ellipticals might evolve into late-type spirals. In sections 25 and 26 we will take note of the arguments for and against the idea that the evolution might go the other way.

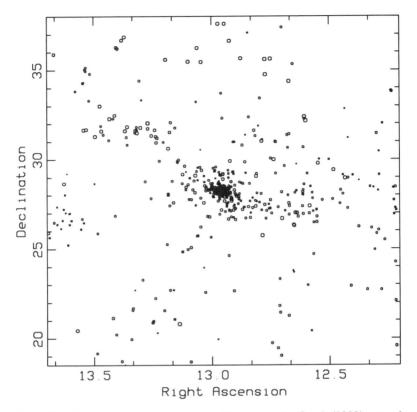

Figure 3.5. The Coma cluster of galaxies. This map is by Groth (1992), out of Huchra's ZCAT, after an earlier version by Huchra.

$$\frac{R_{\text{Coma}}}{R_{\text{Virgo}}} = 5.5 \pm 0.5 \tag{3.21}$$

(Aaronson et al. 1986; Dressler et al. 1987; Sandage and Tammann 1990). As discussed in section 5, this number is useful for establishing the distance scale.

Along with a greater concentration of galaxies than in the Virgo cluster, the Coma cluster has a larger line-of-sight velocity dispersion. The central value is $\sigma = 1200 \, \text{km s}^{-1}$, and the dispersion drops to $\sigma \sim 700 \, \text{km s}^{-1}$ at projected distance $H_o r \sim 1 \, \text{Mpc}$ from the cluster center (Kent and Gunn 1982). This decrease of σ with increasing projected radius is a common feature of rich clusters.

Clusters also tend to contain X-ray emitting plasma with space distribution similar to that of the galaxies. In the Coma cluster the plasma temperature derived from the X-ray spectrum is $kT = 8.5 \pm 0.3 \, \text{keV}$ (Hughes 1989). The mean particle mass in the plasma, allowing for the helium, is $m = 0.6$ times the proton mass, so the one-dimensional thermal velocity dispersion is

$$\sigma_p = (kT/m)^{1/2} = 1200 \, \text{km s}^{-1} \, . \tag{3.22}$$

This is close to the central galaxy velocity dispersion, consistent with the similar space distributions of plasma and galaxies. A measure of the core radius of the plasma distribution is that the surface brightness of the X-ray emission drops to half its central value at about 7 arc minutes from the cluster center, which translates to a projected radius $r = 150h^{-1}$ kpc. Measures of the mass distribution in rich clusters are discussed in section 20.

Figure 3.5 shows a prominent cluster core, which contains the bulk of the X-ray emitting plasma, and around that an irregular concentration that blends into the generally clumpy distribution of galaxies in the field. There is no clearly defined edge to the cluster. Zwicky (1957) noted that the concentration of galaxies around a rich cluster extends well beyond the central core, and the effect was clearly demonstrated in the first redshift survey around the Coma cluster (Chincarini and Rood 1975). The mean run of the space number density of galaxies with distance r from the cluster center, averaged over a sample of clusters to eliminate the fluctuations in the individual systems, is

$$\frac{\langle n(r) \rangle}{n_b} \sim 1 + \left(\frac{r_{cg}}{r}\right)^2 , \qquad r_{cg} \sim 15h^{-1} \, \text{Mpc} \tag{3.23}$$

(Seldner and Peebles 1977; Lilje and Efstathiou 1988), where n_b is the background mean galaxy number density. This is discussed further in section 19.

Figure 3.6 shows maps of the galaxy distribution in three adjacent strips of the sky, each $6°$ deep and about $100°$ wide, at the declination ranges indicated in the figure, and extending to a redshift slightly greater than that of the Coma cluster. The galaxies are from the catalog of Zwicky et al. (1961–68), at a limiting depth about one magnitude fainter than in figure 3.4. The radial coordinate is the redshift. The Coma cluster is the dense patch in panel (a) at right ascension $\alpha \sim 13^h$. The dispersion of peculiar velocities of the galaxies in the cluster causes it to appear in this redshift map as an elongated stripe pointing to us, an effect familiarly called the "Finger of God." It is presumed that outside the rich clusters the peculiar velocities of the galaxies do not greatly distort this redshift map from the true space distribution, because one sees neither the elongations along the line of sight to be expected from random motions nor the effect of collapse, which would make galaxy concentrations tend to appear flattened along the line of sight.

A striking feature of these maps is the tendency of the galaxies to trace linear sheetlike distributions in a larger-scale version of what is seen in figure 3.3. On the basis of the first of the maps in figure 3.6, in panel (a), de Lapparent, Geller, and Huchra (1986) argued that the linear structures are cuts through sheets of galaxies, rather than filaments, because one of these thin slices would not be likely to contain many long segments of filaments. They predicted therefore that the linear structures in the first map would continue to the neighboring slices, in agreement with what is seen in the second and third panels in figure 3.6. Other pioneering discussions of the tendency of galaxies to trace linear structures were given by Jôeveer, Einasto, and Tago (1978), Oort (1983), and Haynes and Giovanelli (1986).

In judging the physical significance of patterns, one has to take account of the tendency of the eye to see patterns even in noise. (It must be evolutionarily advantageous that the eye is very good at picking out the pattern of a leopard among the trees, and, if mistakes are made, that the eye should occasionally see phantom leopards that are not really there in preference to missing the occasional real one.) The linearity of the structures seen in figures 3.3 and 3.6 is manifest. The Great Wall is the high-density feature in figure 3.6 that runs from $cz \sim 10,000 \, \mathrm{km \, s^{-1}}$ at $\alpha = 17^h$ to $cz \sim 7000 \, \mathrm{km \, s^{-1}}$ at $\alpha = 8^h$. Is this a single physical entity more than $100h^{-1} \, \mathrm{Mpc}$ long, or the accidental juxtaposition of several smaller sheets? The test will come from what is found in still larger surveys.

One way to survey the character of the galaxy distribution on larger scales is to select a sparse sample for redshift measurements. The sparse sample of galaxies detected in the NASA IRAS (Infrared Astronomical Satellite) survey at 60 and 100 microns wavelength (discussed by Soifer, Houck, and Neugebauer 1987) are bright in the far infrared because they tend to contain unusually large amounts of dust heated by unusually large numbers of bright young stars. The production of both, which seems to be triggered by the disturbance of a close neighbor, also produces strong optical emission lines from the gas that make the measurement of the redshift relatively easy. Even more important, the IRAS galaxies are selected by their flux densities in the far infrared, which are not greatly affected by absorption by dust in our galaxy. That means the survey can cover a much larger area of the sky than an optically selected catalog.

Figure 3.7 shows the angular distributions of the 2700 nearest and brightest IRAS galaxies (with selection criteria discussed by Strauss et al. 1990), and figure 3.8 shows a sparser sample of some 2200 IRAS galaxies at redshifts $6000 < cz < 20,000 \, \mathrm{km \, s^{-1}}$. Galactic coordinates are used in these whole sky maps, as in figure 3.1, but in a slightly different projection and with the zero of longitude closer to what is now known to be the center of the Milky Way. The horizontal centerline is the plane of the Milky Way. The IRAS samples do not include galaxies in directions near the plane, because of interference by sources in the Milky Way, which causes the empty band running through the center of figure 3.8. Other narrow unsampled stripes are indicated in the map of a still deeper IRAS redshift sample of Saunders et al. (1991).

The galaxy concentrations in rich clusters of galaxies, such as the one illustrated in figure 3.5, are underrepresented in these IRAS maps because gas fuels the high infrared luminosities of most IRAS galaxies, and cluster members tend to be gas-poor. However, as we see in figure 3.5, most of the galaxies associated with a rich cluster are on the outskirts, where most of the brighter galaxies are spi-

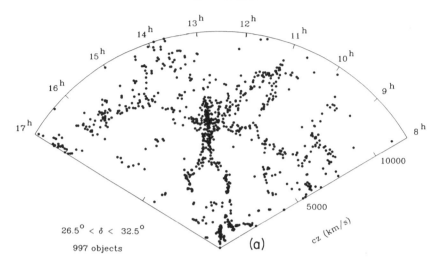

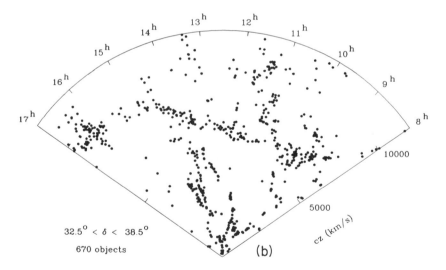

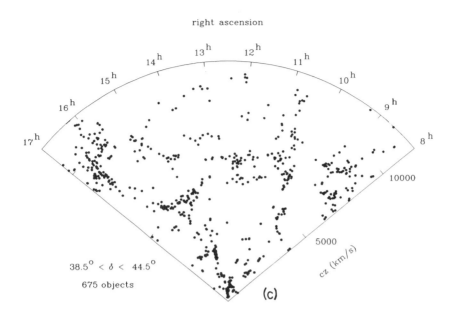

right ascension

Figure 3.6. The galaxy distribution at redshift distance $cz < 12,000 \, \text{km s}^{-1}$ and limiting apparent magnitude $m \lesssim 15.5$ (de Lapparent, Geller, and Huchra 1986; Geller and Huchra 1988, 1989). The radial coordinate is the redshift. The angular coordinate is the angular position of the galaxy along a strip of the sky, $6°$ deep in declination and $\sim 100°$ wide along the direction of right ascension. The three panels represent three adjacent strips, as indicated by the declination ranges.

rals, so the bias is not thought to be large outside the dense central parts of the clusters (Strauss et al. 1992b). The positions of the two most prominent clusters in figure 3.7a, Virgo and Ursa Major (which are named after the constellations in which they appear), are indicated by the dashed circles in figure 3.7c, and the solid circles show the positions of four clusters at $3000 < cz < 6000 \, \text{km s}^{-1}$ in figure 3.7b. The inner curve in figure 3.7c is the plane of the Local Supercluster, at $Z = 0$ in figure 3.3a. The local concentration of galaxies in the sheet in figure 3.3 is visible in figure 3.7a as the stripe running down to the right from the Virgo cluster. The general distribution of galaxies in the more distant slice in figure 3.7b shows no strong preference for the plane of the Local Supercluster, but it is striking that the clusters in this distance range are at low supergalactic latitude ($|Z|$ small compared to $|X|$ or $|Y|$). The Great Attractor discussed in section 5, where the peculiar velocity field seems to be unusually large, perhaps because of a particularly large mass concentration, is in the direction of the Centaurus cluster.

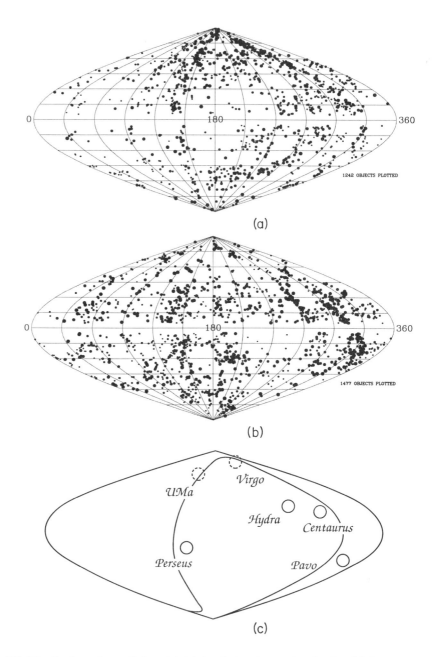

Figure 3.7. Distribution of IRAS (infrared bright) galaxies, in two bands of redshift, $cz <$ 3000 km s^{-1} and $3000 < cz < 6000$ km s^{-1} (Strauss 1992; Strauss et al. 1992a). The symbol size decreases with increasing redshift within each band. The two clusters of galaxies in the nearer band in panel (a) are indicated by the dashed circles in panel (c). The clusters in the more distant sample in panel (b) are indicated by the solid circles. The inner curve in panel (c) is the plane of the de Vaucouleurs Local Supercluster.

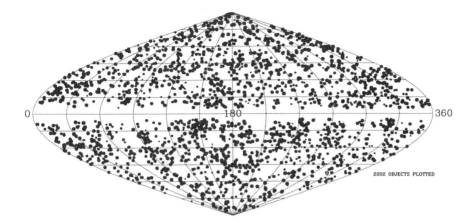

Figure 3.8. Distribution of more distant IRAS (infrared bright) galaxies (Strauss 1992; Strauss et al. 1992a). The symbol size decreases with increasing redshift in the range 6000 km s^{-1} to 20,000 km s^{-1}.

In the region of space covered by the map in figure 3.8 there are many galaxy concentrations similar to the Virgo cluster that is so prominent in figure 3.4. Such concentrations are difficult to see here because the depth is five times that of figure 3.4, so clusters are seen overlapping in projection, and each is very sparsely sampled. This means figure 3.8 is not useful for finding clusters, but it does give us a good picture of the fluctuations in the galaxy distribution on the scale of the depth of the map, $cz \sim 10,000\,\mathrm{km\,s^{-1}}$. The fact that there is no noticeable gradient in the density of galaxies across the sky means that this deeper sample has not revealed a new larger level in the hierarchy of clustering of galaxies. That is, in this figure we at last see an approach to the spatial homogeneity of the cosmological principle. We consider shortly some still deeper samples where the large-scale isotropy is even more evident. The following statistical test of the departure from homogeneity is discussed in more detail in sections 7 and 19.

As a measure of the departures from homogeneity, consider the counts of galaxies that are found within a cube of given width, l. The mean value of the count of galaxies found in the box placed at random within the region of the sample is $\langle N \rangle$. The count at a chosen position within the sample deviates from the mean by the amount $N - \langle N \rangle$. The mean square value of the fluctuation (the second central moment, or variance, or square of the standard deviation) is $\delta N^2 = \langle (N - \langle N \rangle)^2 \rangle = \langle N^2 \rangle - \langle N \rangle^2$. For the IRAS galaxy space distribution, the rms fractional fluctuation is found to be

$$\frac{\delta N}{\langle N \rangle} = 0.5 \pm 0.1\,, \qquad \text{at box width } l = 30h^{-1}\,\mathrm{Mpc} \qquad \textbf{(3.24)}$$

(Saunders et al. 1991; Efstathiou 1991). The methods of obtaining this measure are discussed in sections 7 and 19. The result means that the length $l \sim 30h^{-1}$ Mpc, or $H_o l \sim 3000$ km s^{-1}, marks a characteristic scale at which the galaxy distribution makes the transition from the large fluctuations seen on small scales to a nearly smooth large-scale distribution. This measure is discussed further in sections 7 and 19.

It might be noted that equation (3.24) does not conflict with the existence of systems considerably longer than $30h^{-1}$ Mpc; it only requires that they not have a large effect on the rms fluctuations in galaxy counts. An example is the Great Wall running across the three panels in figure 3.6. If this is a single entity, it is at least $100h^{-1}$ Mpc long, and a fascinating puzzle for theories of galaxy formation. However, the existence of this structure would conflict with the standard cosmological model only if the predicted gravitational effects of the mass concentrations in this system and others like it violated observed galaxy peculiar velocities or the isotropy of the cosmic background radiation. This is discussed in section 21, where it will be argued that the large-scale fluctuations in the mass distribution in systems such as the Great Wall, if fairly traced by the galaxies, produce large-scale peculiar velocity fields that are at least roughly in line with what is observed.

Figure 3.9 shows the angular distribution of optically selected galaxies at a limiting depth about twice that of the IRAS sample in figure 3.8. The field radius is 50° centered on the north pole of our galaxy, where optical counts are least affected by obscuration. These are counts to a limiting apparent magnitude, $m \sim 18.5$, rather than a limiting redshift, as in previous maps. Because the galaxy luminosity function (the frequency distribution dn/dL of galaxy luminosities) has a fairly abrupt upper cutoff, the map is similar to what one would see if the cutoff were at a limiting distance. The galaxy counts are shown in cells of size 10 by 10 arc minutes. The mean count is 1.5 galaxies per cell, for a total of about 4×10^5 galaxies in this map. The data were taken by Shane and Wirtanen (1967) at the Lick Observatory from a visual survey of photographic plates, as part of a larger sky survey that counted about one million galaxies. This was done before the days of high-speed computers; it is a tribute to the dedication and skill of these astronomers that their procedures gave ample control for the later statistical analyses that yielded, among other things, the first believable result for the statistic $\delta N/N$ in equation (3.24). Quite similar values for this number are inferred from the fluctuations in the counts in angular catalogs such as Lick (by the methods discussed in sections 7 and 19) and space maps such as IRAS.

Because the Lick map counts all optically bright galaxies, it is a considerably denser sample than the IRAS maps, so the "fine structure" on scales of a few tens of megaparsecs is much more visible. The prominent bright spot near the center is the Coma cluster shown in figure 3.5, here seen hanging in the foreground. In retrospect one might imagine that the linear features one sees in this map are edge-on sheets of galaxies, but we should bear in mind the tendency of the eye to

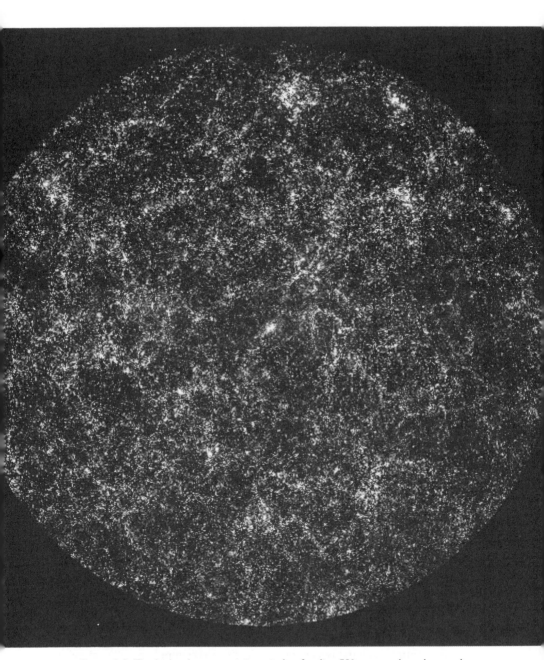

Figure 3.9. The Lick galaxy counts in a circle of radius 50° centered on the north galactic pole (Seldner et al. 1977).

see patterns in noise. The linear structures only became manifest with the redshift surveys.

In a modern and considerably improved version of the Lick survey, Maddox et al. (1990) used an automatic plate measuring system (APM) to catalog angular positions and apparent magnitudes of the two million galaxies at $m < 20.5$ in a field of the sky covering 1.3 steradians near the south galactic pole. The depth set by the limiting apparent magnitude is about three times that of the Lick catalog, and about 20% of the Hubble length c/H_o. It is essential that this machine-selected catalog does have much tighter control on accidental and systematic variations in limiting magnitude across the field than in the manually selected Lick catalog, for at the greater depth of the APM catalog fluctuations in the galaxy distribution across the sky due to structures such as those illustrated in figures 3.3 to 3.6 are quite small. At the depth of this survey a great many structures are seen overlapping in projection, leaving an exceedingly smooth mean distribution.

For our purpose, the central result from the APM survey is that the statistic $\delta N/N$ in equation (3.24), again inferred from the angular fluctuations in the counts, is within the errors of the measurements consistent with what is estimated in the shallower Lick catalog (Clutton-Brock and Peebles 1981) and the still shallower IRAS samples (Saunders et al. 1991; Strauss et al. 1992a). This gives two essential checks of the test of the cosmological principle and the measures of deviations from it. First, the three samples were obtained, and the statistic $\delta N/N$ estimated, in quite different ways, which one expects would tend to introduce different systematic errors. Hence, the consistency makes it seem quite unlikely that the results are seriously biased. Second, if there were appreciable gradients in the galaxy space number density on the scales of the distances sampled in IRAS and Lick, one would have expected to have seen the effect in progressively larger values of $\delta N/N$ at fixed cell size with increasing sample depth. As discussed in section 7, this is contrary to what is observed, that $\delta N/N$ at fixed cell size is independent of the depth of the sample.

Figure 3.10 shows Condon's (1991) map of the 30, 821 brightest radio sources in the 6 cm catalog of Gregory and Condon (1991). The right ascension increases in a clockwise direction, from $\alpha = 0$ at the top. The map radius is $r = (1 - \sin \delta)^{1/2}$, where δ is the declination, so equal areas of the sky are mapped into equal map areas. The circular hole at the center is a catalog boundary set by the range of the radio telescope. The outer boundary of the catalog is at declination $\delta = 0$. Interference by the Sun caused the ragged boundary at the lower left-hand edge. The small holes just above the central one are the result of interference by bright sources in the plane of the Milky Way. Other sources in the Milky Way produce the concentration of sources in the arc at the left side of the map. Apart from these regions, we have a good view of the extragalactic radio sky.

This is a flux-limited sample, as in the Lick map in figure 3.9. Identifications with optically detected objects indicate the bulk of these sources are galaxies and quasars at distances comparable to the Hubble length c/H_o (at which the redshift

Figure 3.10. Angular distribution of the $\sim 31,000$ brightest 6 cm radio sources (Gregory and Condon 1991.)

z in eq. [3.19] extrapolates to unity). At this very sparse sampling and great depth the "fine structure" in the distribution of extragalactic objects is quite lost. The central message from this map is that the distribution of objects smoothed over the scale of the Hubble distance fails to reveal substantial density gradients.

Finally, figure 3.11 shows the X-ray sky brightness at ~ 2 to 20 keV. As in figures 3.7 and 3.8, the map covers the whole sky, in galactic coordinates, but here galactic longitude increases to the left, and the map is centered at $l = 0$, on the center of the Milky Way. The contrast has been adjusted to emphasize the fluctuations, as indicated in the linear scale at the top of the map. Most of the objects visible in this map are compact sources whose sizes are set by the angular resolution of the detector. Local sources are concentrated along the plane of the

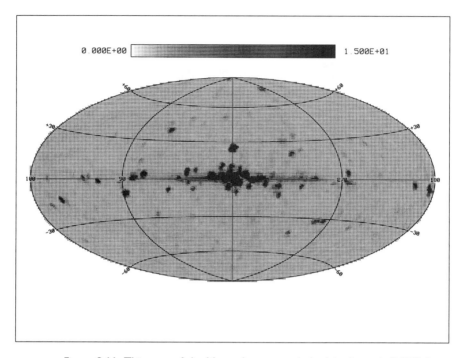

Figure 3.11. This map of the X-ray sky was made by Jahoda et al. (1992) from data obtained with the A2 instrument on the HEAO-1 satellite and provided by the NASA/Goddard Space Flight Center X-ray Branch.

Milky Way and toward $l = 0$, at the center of the map. The Large Magellanic Cloud is the dark patch toward the lower right, at longitude $l = 280°$ and latitude $b = -33°$, and the Small Magellanic Cloud is further down and to the left. The source near the top of the map at $l = 283°$ and $b = +75°$ is the plasma cloud around the galaxy M87 in the Virgo cluster. A few other nearby galaxies appear as dark patches. Between the patches there is a diffuse background (the sum of many distant sources) that is smooth to 3% within 3° by 3° fields, and smooth to one part in 10^3 in the average over 90°.

At the X-ray energies in this map the Milky Way is nearly transparent, and a typical path through the universe to the Hubble length c/H_o almost certainly is unaffected by absorption of X-rays. This means the X-ray flux density in any direction is proportional to the integrated column density of sources along the line of sight to the Hubble distance. (The integral is given in eq. [5.162] below.) Thus, the isotropy of the X-ray background tells us that the integrated column density of matter within 3° by 3° fields fluctuates by no more than 3% , and the density of sources averaged over one steradian of the sky and integrated to the Hubble depth fluctuates by no more than the large-scale anisotropy of the X-ray background, no more than about one part in 10^3.

As discussed in sections 6 and 21, the departure from isotropy of the radiation background at wavelengths on the order of one centimeter is about one part in 10^5 on angular scales greater than about $10°$. This adds to the evidence in figures 3.8 to 3.11, all of which show that, in the average over scales comparable to the Hubble length, our universe is quite close to isotropic, in striking agreement with what Einstein imagined. These measures do allow a universe that is spherically symmetric, with a radial density gradient, if the gradient is shallow enough to have avoided detection in the galaxy counts as a function of redshift, but then we would have to be very close to the center of symmetry. This seems unreasonable, for there are many distant galaxies that would appear to be equally good homes for observers, with the one difference that in this picture almost all would present observers with an anisotropic universe. The best direct constraint on a radial density gradient is the stability of the statistic $\delta N/N$ in equation (3.24) under changes in the depth of the sample. This test is discussed further in section 7.

Galaxies

The properties of galaxies are known in considerably greater detail than for groups and clusters of galaxies. Presented here are some results that seem memorable and simple enough that they might be expected to figure in the development of physical cosmology. Much more complete surveys are given in Kormendy (1982) and Kormendy and Djorgovski (1989).

Hubble (1926b) defined the two standard classes of giant galaxies, late-type spirals and early-type ellipticals, later adding the transitional S0 galaxies (as described by Sandage 1961). The three main components of a spiral galaxy are a thin disk, a spheroid, and a dark halo. The thin disk contains gas with young stars as well as old. The generally older stars in the spheroid are named for their space distribution, with surfaces of constant luminosity density approximated by ellipsoids, of the form

$$\frac{x^2}{a^2} + \frac{y^2}{b^2} + \frac{z^2}{c^2} = 1. \tag{3.25}$$

The ellipsoid component in a spiral galaxy approximates an oblate spheroid of revolution, with semiaxes $a = b > c$. Hubble classified spiral galaxies as Sa, Sb, Sc, and Sd, in a sequence of increasingly open spiral arms in the disk. The spheroid tends to be decreasingly prominent relative to the disk in the sequence from Sa to Sd. The Milky Way is usually classified as Sb.

Elliptical galaxies have at most inconspicuous disks, and spheroidal components that can have triaxial shapes, with the ratio of the lengths of the smallest to largest axis in the range $0.3 \lesssim c/a \lesssim 1$. Ellipticals contain relatively little gas, though they can have halos of hot X-ray emitting plasma. (The giant elliptical galaxy M87 in the Virgo cluster is the source at the top of the X-ray map in figure 3.11.) The spheroidal components in spirals and ellipticals follow quite similar relations between the luminosity and star velocity dispersion (eqs. [3.34] and [3.35] below). This could be read to mean that in the beginning there were spheroids, some of which later became adorned with disks (sections 22 and 25).

In luminous S0 galaxies the spheroidal component has a lenticular shape that may come close to the appearance of a spiral galaxy with a smooth disk rather than one having spiral arms. It has not escaped notice that if the gas were swept from a spiral galaxy and the young stars that outline the spiral arms were allowed to fade away, the result would resemble the more flattened S0 galaxies. There is no consensus on whether this really is what happened, however.

The following deals with some of the systematic properties of the regular giant spiral and elliptical galaxies. This oversimplifies the situation, for we see in figure 3.3 that dwarfs are much more common, but a more careful treatment would require a book on astronomy.

Figure 3.12 shows examples of the evidence for dark massive halos around spiral galaxies. The top panels show the optical surface brightness as a function of radius. The vertical axis is logarithmic, so the nearly straight curves mean the surface brightness in the disk approximates an exponential function of radius r,

$$i(r) = i_o e^{-\alpha r} . \tag{3.26}$$

The disk scale lengths for these two galaxies are $\alpha^{-1} \sim 2h^{-1}$ kpc; in giant spirals such as the Milky Way, the scale length may be twice that (Freeman 1970). In the lower panels the symbols with error flags are the measured circular speed of rotation as a function of distance from the center of the galaxy. This is the rotation curve $v_c(r)$. In these examples the rotation curve is derived from the Doppler shift, relative to the mean, in the 21 cm line of the atomic hydrogen in a disk that extends well beyond the optically bright part. The solid lines in the lower panels are the rotation curves predicted under the assumption that the mass has the same distribution as the light (with the addition of the observed mass of atomic hydrogen). This leaves one parameter, the mass-to-light ratio, which has been adjusted to fit the inner part of the rotation curve. The fit gives

$$\frac{\mathcal{M}}{L} = 2.5\,h\frac{\mathcal{M}_\odot}{L_\odot} \quad \text{and} \quad 5.3\,h\frac{\mathcal{M}_\odot}{L_\odot} \tag{3.27}$$

for NGC 2403 and NGC 3198. As in equation (3.1), these numbers are similar to what is observed for the matter in our neighborhood, and comparable to what might be expected for the mix of star masses that reproduces the spectra of the galaxies. The result is a reasonably close fit to the rotation curves in the central parts of the galaxies, meaning it is plausible to assume that the mass in the inner parts is dominated by ordinary stars. However, in the outskirts the luminosity

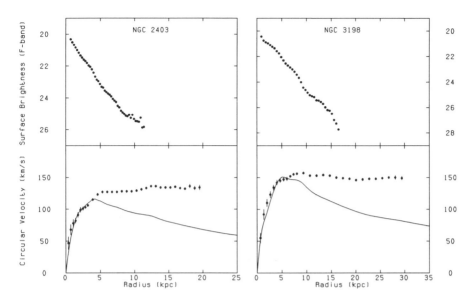

Figure 3.12. Luminosity profiles and rotation curves for spiral galaxies (van Albada and Sancisi 1986). The numerical value of the radius is based on the Hubble parameter $h = 0.75$.

in the exponential disk of the galaxy has converged, so if starlight traced mass we would have expected to see the rotation curves approach the Kepler form $v_c \propto r^{-1/2}$, as in the solid curves in the figure. This is quite different from what is observed.

The discrepancy means either that Newtonian gravitational physics fails or that the mass in the outer parts of the galaxies is dominated by low-luminosity material, a dark halo. Newtonian mechanics does give a reasonable description of the central parts, for the rotation curve is what would be expected if the mass were dominated by stars with astrophysically sensible mass-to-light ratios. As discussed in section 20, general relativity theory provides a good description of the gravitational lensing of background galaxy images by the mass concentrations in great clusters of galaxies, on scales ~ 100 kpc. Thus the reasonable interpretation to consider first is that the low velocity limit of general relativity theory, Newtonian mechanics, is a useful approximation to dynamics on the scale of galaxies, and that the material in the outer parts of the galaxies in figure 3.12 happens not to be readily visible. That certainly is easy to imagine: perhaps the mix of star masses changes with increasing radius, to favor lower mass stars with much higher mass-to-light ratios. Another popular idea is that the dark mass is some new exotic form of matter; some ideas are discussed in section 18. But since this subject is still being explored, it is well to bear in mind the alternative that we are not using the right physics.

Under Newtonian mechanics, figure 3.12 indicates that the nearly flat part of the rotation curve is dominated by the seen stellar mass at small radii, and by the dark halo in the outer parts, with the distributions of light and dark components just such as to keep the rotation curve close to flat. This "conspiracy" seems to be a common feature of spiral galaxies (van Albada and Sancisi 1986).

We can translate the rotation curve into a reasonable first approximation to the mass distribution by recalling that in Newtonian mechanics the gravitational acceleration g of a centrally concentrated mass distribution is well approximated by the expression for a spherically symmetric distribution,

$$g = \frac{G\mathcal{M}(r)}{r^2} = \frac{v_c^2}{r}, \tag{3.28}$$

where $\mathcal{M}(r)$ is the mass within radius r, and $v_c(r)$ is the circular rotation velocity as a function of radius. This gives

$$\mathcal{M}(r) = \frac{v_c^2 r}{G}, \qquad \rho(r) = \frac{v_c^2}{4\pi G r^2}, \tag{3.29}$$

where $\rho(r)$ is the mean mass density at radius r. Since the rotation curve is close to flat, the density varies as $\rho \propto 1/r^2$.

Another way to arrive at this mass model is to approximate the dark halo as a spherically symmetric isothermal ideal gas sphere. The "atoms" might be stars or exotic dark-matter particles. We will assume the velocity distribution is isotropic and the velocity dispersion is independent of radius, that is, it is isothermal. (A more general model is given in eq. [22.39]). Then the gas pressure is

$$p = \rho(r)\sigma^2, \tag{3.30}$$

where the velocity dispersion (rms deviation from the mean) in one direction is σ. The equation of hydrostatic equilibrium balances the pressure force per unit volume, $-dp/dr$, and the gravitational force $g\rho$ per unit volume,

$$-\frac{dp}{dr} = \frac{G\mathcal{M}(r)}{r^2}\rho(r), \qquad \mathcal{M}(r) = 4\pi \int_0^r \rho(r)\, r^2 dr. \tag{3.31}$$

The solution to these equations (with the boundary condition that the mass distribution is smooth at the origin, $d\rho/dr = 0$ at $r = 0$), is tabulated by Emden (1907; the differential equation is in eq. [18.70]). Outside the core the density run approaches the power law form $\rho \propto 1/r^2$. This limiting form in equation (3.31) gives

$$\mathcal{M}(r) = \frac{2\sigma^2 r}{G}, \qquad \rho = \frac{\sigma^2}{2\pi r^2 G}, \tag{3.32}$$

in agreement with equation (3.29) with

$$v_c = 2^{1/2}\sigma. \tag{3.33}$$

This is the relation between the circular velocity and the one-dimensional line-of-sight velocity dispersion when the rotation curve is close to flat. A model with an anisotropic velocity distribution is discussed in section 22 (eq. [22.39]).

An isolated spiral galaxy usually has a nearly flat rotation curve outside a central core, and v_c in the flat part is correlated with the galaxy luminosity, L. This is called the Tully-Fisher (or Fisher-Tully) relation after its discoverers, Tully and Fisher (1977). In the infrared 2.2 micron K band the relation is

$$v_c = 220(L/L_*)^{0.22} \text{ km s}^{-1} \tag{3.34}$$

(Aaronson et al. 1986), where L_* is a characteristic galaxy luminosity (eq. [5.141] below). The scatter in v_c at given L is about 5% . Since the mass within a fixed radius varies as $\mathcal{M}(r) \propto v_c^2$, at given luminosity the mass within a fixed radius is predicted to about 10% relative accuracy, a remarkable regularity.

The character of the mass distribution within elliptical galaxies is less well explored because these galaxies usually do not have material moving in circular orbits from which one easily gets the gravitational acceleration v_c^2/r. One can use the line-of-sight star velocity dispersion σ (derived from the Doppler spread of features in the integrated stellar spectrum). If the star motions are isotropically distributed, and the dispersion does not vary rapidly with radius, the mass within a fixed radius is well approximated by equation (3.32).

Faber and Jackson (1976) showed that the velocity dispersion in the central part of an elliptical galaxy is correlated with its luminosity, as is illustrated in figure 3.13. The Faber-Jackson relation is

$$v_c = 2^{1/2}\sigma = 220(L/L_*)^{0.25} \text{ km s}^{-1}. \tag{3.35}$$

This has been expressed in terms of the velocity of a particle that is in a circular orbit (eq. [3.33]); the mean streaming motion of rotation of the stars generally is well below v_c. The power law indices here and in equation (3.34) depend somewhat on the wavelength at which the luminosities are measured. Also, one should bear in mind that the Tully-Fisher velocities in equation (3.34) refer to the dark halo, while the Faber-Jackson relation refers to the velocity dispersion of the stars in the central parts of the galaxy. A close comparison thus must be taken with caution, but the general similarity of the relations does suggest we are seeing similar physical relations.

As in the case of the Tully-Fisher relation, there is not much scatter in σ at fixed L. The scatter is remarkably small in the "fundamental plane" for ellipticals,

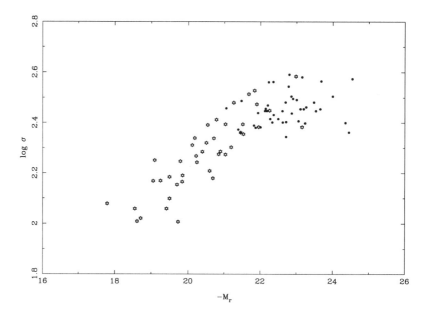

Figure 3.13. The correlation of elliptical galaxy velocity dispersion with luminosity (Oegerle and Hoessel 1991). The stars are galaxies in the Virgo and Coma clusters, the filled circles the brightest members of rich clusters.

where the galaxy is characterized by its velocity dispersion σ; a characteristic radius, such as the radius r_e of the circle that contains half the light; and the mean surface brightness i_e within r_e. Within the measurement errors, elliptical galaxies occupy a two-dimensional sheet in the three-dimensional space of these parameters. Important steps in the discovery of this effect were taken by Dressler et al. (1987) and Djorgovski and Davis (1987); the history of the discovery is detailed by Kormendy and Djorgovski (1989). In the example in figure 3.14, from Oegerle and Hoessel (1991), the equation for the fundamental plane is

$$r_e \propto \sigma^{1.33} i_e^{-0.83} . \tag{3.36}$$

The surface brightness i_e is the received energy flux per steradian averaged over the solid angle subtended by the galaxy. The standard way to express it is in units of apparent magnitude per square arc second,

$$\begin{aligned} \langle \mu \rangle &= m + 2.5 \log[2\pi(\theta_e'')^2] \\ &= -2.5 \log[L/r_e^2] + \text{constant} . \end{aligned} \tag{3.37}$$

The factor of two in the logarithm in the first line converts the galaxy apparent

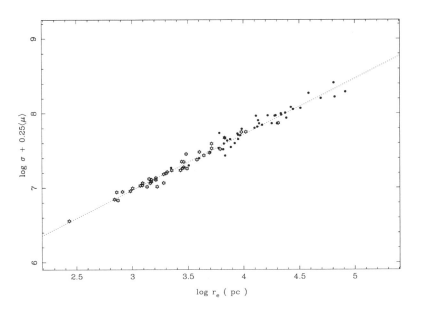

Figure 3.14. The fundamental plane for elliptical galaxies (eq. [3.36]) is viewed here in an edge-on projection (Oegerle and Hoessel 1991). The symbols are the same as in figure 3.13.

magnitude, m, to the half of the galaxy luminosity L that is within r_e. The factor $\pi\theta_e^2$, with the angle expressed in arc seconds, converts this to the apparent magnitude belonging to the mean energy flux per square arc second. As indicated in the second line, the argument of the logarithm is proportional to the luminosity per linear area of the galaxy (ergs $cm^{-2}\,s^{-1}$), which is proportional to the the the energy flux per steradian, which is the surface brightness.

The combination of velocity dispersion and surface brightness in the vertical axis in figure 3.14 is arranged to display the edge of the fundamental plane in equation (3.36). The startlingly small scatter surely betokens a tight regularity in the way these galaxies are made, but there is no agreement yet on the specific lesson.

Since the surface brightness i_e is proportional to L/r_e^2, we can replace i_e in equation (3.36) with the galaxy luminosity L, to get

$$\sigma \propto L^{0.62} r_e^{-0.50} . \tag{3.38}$$

The mass within the half-light radius scales as $\mathcal{M} \propto \sigma^2 r_e$ (eq. [3.32]). From equation (3.38) this is $\mathcal{M} \propto L^{1.25}$, almost independent of r_e. Thus the mass-to-light ratio scales as

$$\frac{\mathcal{M}}{L} \propto L^{0.25} \propto \mathcal{M}^{0.2}. \tag{3.39}$$

In the space of σ, r_e and i_e there is a second family of early-type galaxies that occupies a second plane at lower surface brightness. Galaxies in this family commonly are called dwarfs, though the range of luminosities overlaps the luminosities of the galaxies in the fundamental plane of the giant ellipticals (Kormendy 1987).

The most luminous ellipticals are classified as D, and the extreme cases cD, meaning the run of surface brightness with radius extends to an unusually broad low surface brightness envelope. In Morgan's (1958) original classification the optical image of a D galaxy is close to circularly symmetric. The notation cD came into common use with the discovery that galaxies with high radio luminosities tend to be giant ellipticals that are not very eccentric in their inner parts and very luminous (Matthews, Morgan, and Schmidt 1964; Morgan and Lesh 1965).

The broad envelopes of cD galaxies are at least in part a result of the fact that the brightest member of a cluster tends to occupy a central position, so it smoothly joins the diffuse light spread through the cluster. Hausman and Ostriker (1978) show that the broad envelope could have grown by gravitational capture of stars from passing galaxies by the giant elliptical that happens to stay closest to the cluster center. As discussed in section 24, the plasma in a rich cluster is losing energy by radiation, and in the central regions the plasma must either be receiving heat or contracting to something that is nearly dissipationless, such as stars. This cooling flow could produce part or all of a centrally placed cD galaxy (Fabian, Nulsen, and Canizares 1991). A pronounced example of the broad run of surface brightness with radius in a cD galaxy is given by Uson, Boughn, and Kuhn (1990). The central parts of a cD galaxy are not readily distinguished from a giant elliptical galaxy.

Galaxies have many sizes, with observed luminosities that span some five orders of magnitude, depending on what one chooses to call a galaxy. There is a rather distinct upper bound, however. Hubble and Humason (1931) found that the brightest members of rich clusters of galaxies tend to have a nearly standard luminosity (defined as the energy flux measured within a fixed surface brightness or radius chosen to eliminate the diffuse intracluster light around the cD galaxies). The effect is illustrated in figure 3.13, where one sees that the brightest cluster members represented in the figure by the filled circles cluster at the end of the plot, with little scatter in luminosity or velocity dispersion. The rms scatter in absolute magnitudes of brightest cluster members is about 0.3 mag, or 30% in luminosity (Sandage 1972a). Because the distribution of light within these galaxies has a nearly standard form, the small scatter in luminosity translates to a small scatter in the radius at which the surface brightness in a first-ranked cluster member drops to half the central value (Sandage 1972b). An indication of the strikingly hard upper cutoff in the distribution of velocity dispersions in elliptical galaxies is that in the Faber et al. (1989) sample, the central velocity dispersion exceeds $\sigma = 300\,\mathrm{km\ s^{-1}}$ in just 18 of 468 galaxies, and the largest value in the sample is 380 km s^{-1}. A similar effect is seen in large spirals. A typical circular velocity is 220 km s^{-1}, that of the Milky Way. While there are many dwarfs with considerably lower velocities, not many spirals have substantially larger v_c. The current record is $v_c = 500\,\mathrm{km\ s^{-1}}$, in the galaxy UGC 12591 (Giovanelli et al. 1986), about equivalent to the largest values of σ observed in ellipticals.

Since the mass concentrated within a fixed radius in a galaxy is proportional to the square of the circular velocity or velocity dispersion, the cutoff in the circular velocity in spirals translates into a bound on the mass concentration. At the fixed radius $r = 10h^{-1}\,\mathrm{kpc}$, about our distance from the center of the Milky Way, and for the circular velocity $v_c = 220\,\mathrm{km\ s^{-1}}$ of the Milky Way, which is typical for large galaxies, the mass contained is

$$\mathcal{M}(r < 10h^{-1}\,\mathrm{kpc}) = v_c^2 r/G = 1.1 \times 10^{11}\,h^{-1}\mathcal{M}_\odot. \qquad (3.40)$$

The largest known value for a spiral galaxy translates to a mass just four times this value.

Much larger mass concentrations are known in some giant elliptical and cD galaxies, but even here the central concentration is not much greater than the bound for known spirals. The giant elliptical galaxy M87 in the Virgo cluster is surrounded by a cloud of plasma at temperature $kT = 4\,\mathrm{keV}$ (detected by the X-ray emission, and unambiguously identified as thermal bremsstrahlung by the detection of X-ray lines from iron ions in the plasma). The symmetry of the cloud indicates it is gravitationally bound to the galaxy. The estimate of the pressure gradient from the plasma temperature and density yields a gravitational acceleration at 200 kpc radius that implies the mass within this radius is about $3 \times 10^{13}\mathcal{M}_\odot$ (Fabricant and Gorenstein 1983), three hundred times the mass in equation (3.40). However, the mass of M87 within the radius in equation (3.40) is more modest, about ten times the mass in equation (3.40) (Sargent et al. 1978; Huchra and Brodie 1987).

The conclusion is that Nature is adept at bringing together mass concentrations of the amount in equation (3.40), to make common bright galaxies, but proves to be quite reluctant to gather just four times this amount in the same volume. At lower mass concentrations we see the striking regularities in figures 3.13 and 3.14, and in equations (3.34) and (3.35), that show that the process which collected the material that produces the luminosity of a galaxy collected mass with a tight control on the relation between the final concentrations of light and mass. The lesson is that galaxies are built on a closely regulated plan, either in the process that collects the matter or in some feedback mechanism that limits the amount that can be assembled. The prospects for identifying the process are surveyed in sections 25 and 26.

Coordinate and Magnitude Systems

The galaxy maps shown in this section use three different coordinate systems. The relations among these coordinates are defined here, and this is a convenient place to record the normalization of the magnitude system used in some of the maps.

The equatorial or celestial coordinates used in figures 3.5 and 3.6 have polar axis along the Earth's north axis of rotation. The declination is

$$\delta = 90° - \theta, \tag{3.41}$$

where θ is the polar angle measured from the north pole. Thus, the Earth's equator is at declination $\delta = 0$, the north pole is at $\delta = 90°$, and the south pole is at $\delta = -90°$. The azimuthal angle in this coordinate system is the right ascension, α, usually measured in units of hours, minutes, and seconds of time, with one hour equal to $15°$. The right ascension α increases going to the east in the sky. The analogs for positions on the Earth are latitude and longitude, with the difference that the longitude increases going to the west (perhaps because one system is useful for looking down, the other for up). Since the axis of rotation of the Earth is precessing, the celestial coordinates are referred to the orientation of the Earth's axis at a given epoch, usually the year 1950, though it is increasingly common and timely to use the year 2000. The difference is irrelevant for our purposes.

The galactic spherical coordinates used in figures 3.4 and 3.7 are oriented so the pole is normal to the plane of the Milky Way and in the north celestial hemisphere. The galactic longitude (azimuthal angle) is l, with the origin $l = 0$ pointing to the center of the galaxy. The galactic latitude is $b = 90° - \theta$, where θ is the polar angle in these coordinates, so the plane of the Milky Way is the line $b = 0$. This is the center line in figures 3.7 and 3.8. A similar pole is used in figure 3.1, but the zero of longitude is different (because obscuration by dust made it difficult to see where the center of the galaxy is).

Figure 3.4 illustrates the relation between galactic and equatorial coordinates. The conversion uses the spherical triangle in figure 3.15. Recall that a side of the spherical triangle is the great circle defined by the intersection of the sphere with a plane passing through the center of the sphere. The length α of a side is the angle subtended by the arc at the center, so α is the length of the arc on the unit sphere. The angle A at the vertex facing the side α is the angle between the two planes that define the great circles β and γ on the other two sides of the triangle. The angle opposite the side β similarly is B, and the angle opposite γ is C. The law of sines for spherical triangles is

$$\frac{\sin \alpha}{\sin A} = \frac{\sin \beta}{\sin B}, \tag{3.42}$$

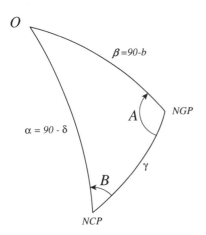

Figure 3.15. Spherical triangle for the conversion between equatorial and galactic coordinates.

and the law of cosines is

$$\cos \gamma = \cos \alpha \cos \beta + \sin \alpha \sin \beta \cos C. \tag{3.43}$$

One way to derive these relations is shown in section 12.

In the triangle in figure 3.15 the north celestial pole is in the direction NCP, the north pole of the galaxy is in the direction NGP, and $\mathcal{O}$ is the direction of an object with equatorial coordinates α, δ and galactic coordinates l, b. Since the declination is measured from Earth's equator, the angular distance from NCP to the object is $\alpha = 90° - \delta$. The distance from NGP to the object is $\beta = 90° - b$. The declination of the north pole of the galaxy is $\delta_p = 27.4°$, so the angular distance between NCP and NGP is $\gamma = 90° - \delta_p = 62.6°$. The right ascension of the north galactic pole is $\alpha_p = 192.25°$. The right ascension of the object $\mathcal{O}$ is α, so the angle facing side β is $B = \alpha - 192.25°$. The zero of galactic longitude is conveniently defined so the angle facing side α is $A = 123° - l$ for an object at galactic longitude l. The law of sines gives

$$\cos b \sin(123° - l) = \cos \delta \sin(\alpha - 192.25°). \tag{3.44}$$

The law of cosines gives

$$\sin b = \sin \delta \cos 62.6° + \cos \delta \sin 62.6° \cos(\alpha - 192.25°), \tag{3.45}$$

and

$$\sin \delta = \sin b \cos 62.6° + \cos b \sin 62.6° \cos(123° - l). \tag{3.46}$$

The result of substituting equation (3.45) into equation (3.46) and rearranging is

$$\cos b \cos(123° - l) = \sin \delta \sin 62.6°$$
$$- \cos \delta \cos 62.6° \cos(\alpha - 192.25°).$$

(3.47)

Equation (3.45) gives b and equations (3.44) and (3.47) give l (and both are needed to fix the sign of l).

Supergalactic coordinates SGL and SGB are used in figure 3.3 to show the plane of the local distribution of galaxies, and one sees in figure 3.5 that the nearby clusters of galaxies lie on the same plane. Following the definition by de Vaucouleurs, de Vaucouleurs, and Corwin (1976), let us put the pole of the galaxy at the vertex with angle B in figure 3.15, the supergalactic pole at vertex A, and the object $\mathcal{O}$ at C. The supergalactic pole is at galactic latitude $b_s = 6.32°$, so its distance from the galactic pole is $\gamma = 90° - b_s = 83.68°$. The longitude of the supergalactic pole is $l_s = 47.37°$, so the angle between the supergalactic pole and an object at longitude l is $B = l - 47.37°$. The origin of supergalactic longitude, at $SGL = SGB = 0$, is at $l_o = 137.29°$, $b_o = 0$. That means the triangle with base running from the galactic pole to the supergalactic pole and apex at $b = SGB = 0$ has two sides with length 90°, base with length 83.68°, and angles at the base equal to $l_o - l_s = 89.92°$. Thus, the angle at the supergalactic pole between the line to the galactic pole along the base γ and the line to an object at supergalactic longitude SGL is $A = 89.92° - SGL$. That fixes the spherical triangle, and the above operations give

$$\cos(SGB) \sin(89.92° - SGL) = \cos b \sin(l - 47.37°),$$
$$\cos(SGB) \cos(89.92° - SGL) = \sin b \sin 83.68°$$
$$- \cos b \cos 83.68° \cos(l - 47.37°),$$
$$\sin(SGB) = \sin b \cos 83.68° + \cos b \sin 83.68° \cos(l - 47.37°),$$

(3.48)

with SGL and SGB the supergalactic longitude and latitude. The inner curve in figure 3.7c is the line $SGB = 0$. The coordinates in figure 3.3 for a galaxy at distance R are $X = R \cos(SGB) \cos(SGL)$, $Y = R \cos(SGB) \sin(SGL)$, and $Z = R \sin(SGB)$.

Finally, this is a convenient place to list the zero points for the magnitude system defined in equations (3.10) to (3.12). The source for these and many other numbers is Allen's (1973) *Astrophysical Quantities*. Standard magnitude systems are V and B, the latter being roughly equivalent to the b_J system used in figure 3.2. Since the magnitude pass bands have nonzero width, the central wavelength depends on the spectrum. For spectra similar to that of the Sun, the central pass bands are

$$\lambda_V = 5500 \, \text{Å}, \qquad \lambda_B = 4450 \, \text{Å}.$$

(3.49)

The absolute magnitude of the Sun is

$$M(V)_\odot = 4.83 , \qquad M(B)_\odot = 5.48 . \tag{3.50}$$

The luminosity of the Sun in the wavelength range λ to $\lambda + \delta\lambda$ is $L_\lambda \delta\lambda$ (erg s^{-1}), and λL_λ is the energy radiated per unit time and logarithmic interval of wavelength (as in eq. [5.160]). The solar luminosity is

$$\lambda L_\lambda = 2.86 \times 10^{33} \text{ erg s}^{-1} \quad \text{at } V , $$
$$= 2.36 \times 10^{33} \text{ erg s}^{-1} \quad \text{at } B . \tag{3.51}$$

The energy flux received from a star at apparent magnitude m is

$$\lambda f_\lambda = \frac{\lambda L_\lambda}{4\pi (10 \text{ pc})^2} 10^{0.4(M_\odot - m)} , \tag{3.52}$$

where λL_λ is given by equation (3.51), and

$$1 \text{ parsec} = 3.086 \times 10^{18} \text{ cm} . \tag{3.53}$$

The origin of this length unit is described in the footnote for equation (3.5). Recall that the apparent magnitude of a star at 10 parsecs distance is its absolute magnitude, $m = M$. Thus, the first factor in the right-hand side of equation (3.52) is the energy flux received from a star with apparent magnitude $m = M_\odot$, and the second factor says that five magnitudes increase in the apparent magnitude represents a factor of 100 decrease in the received energy flux. The numerical values work out to

$$\lambda f_\lambda = 2.04 \times 10^{-5} 10^{-0.4m} \text{ erg cm}^{-2} \text{ s}^{-1} \quad \text{at } V , $$
$$= 3.07 \times 10^{-5} 10^{-0.4m} \text{ erg cm}^{-2} \text{ s}^{-1} \quad \text{at } B . \tag{3.54}$$

The total (bolometric) solar luminosity is

$$L_\odot = \int L_\lambda d\lambda = 3.83 \times 10^{33} \text{ erg s}^{-1} , \tag{3.55}$$

close to the numbers in equation (3.51) because the B and V pass bands are close to the peak of the spectrum. The mass of the Sun is

$$\mathcal{M}_\odot = 1.989 \times 10^{33} \text{ g} , \tag{3.56}$$

so the ratio of the luminosity and mass is

$$\frac{L_\odot}{\mathcal{M}_\odot} = 1.92 \text{ erg s}^{-1} \text{ g}^{-1} . \tag{3.57}$$

4. Einstein's World Model

When Einstein introduced his relativistic world model, almost nothing was known of the lore presented in the last section, so it hardly seems surprising that the model no longer is considered viable. It is remarkable that Einstein's model does predict the Hubble count law, $dN/dm \propto 10^{0.6m}$, and the isotropy of the deep maps shown in figures 3.8 to 3.11, on the basis of the cosmological principle. What is more, it is an easy step from Einstein's static model to the standard expanding cosmology. It is a measure of how revolutionary this last step is, at least within physical science, that Einstein did not even mention that his condition that spacetime is static on average is an assumption. It is now seen that a static world picture has conceptual problems (quite apart from the evidence that the universe is expanding), for if energy and entropy are conserved it cannot agree with the limited lifetimes of stars and stellar systems, and if general relativity theory is valid Einstein's static mass distribution is gravitationally unstable, so that even the present large-scale approximation to homogeneity could not persist very much longer than the lifetime of a solar mass star. These problems are resolved, or at least greatly adjusted, by going to the expanding world model to be discussed in the next section.

Einstein's solution still is well worth considering as a guide to conditions on an acceptable world model, and to the physics of the expanding version. To begin, let us consider the form for the line element ds^2 in equation (2.1) implied by the cosmological principle.

Geometry

Recall that the expression $ds = |g_{ij} dx^i dx^j|^{1/2}$ has an invariant meaning in terms of a proper (in principle measurable) length or time interval between the neighboring events with coordinate separation dx^i in the spacetime with metric tensor $g_{ij}(x)$. Given three neighboring events with coordinate intervals dx^i and dy^j connecting one event and the other two, another invariant expression is the four-dimensional scalar product,

$$dx \cdot dy = g_{ij} dx^i dy^j . \tag{4.1}$$

In the locally Minkowski coordinate system of equation (2.3), this quantity vanishes if dx^i and dy^j are directed along orthogonal coordinate axes (because the Minkowski tensor $g_{ij} = \eta_{ij}$ is diagonal). Since the quantity $dx \cdot dy$ is independent of the choice of coordinates, the invariant condition that dx^i and dy^j are orthogonal is $dx \cdot dy = 0$. We will use this as a guide to interpreting a convenient coordinate labeling of a spatially homogeneous and isotropic universe.

Imagine space is filled with a dense set of observers, each at rest relative to the mean motion of the nearby matter. Since the gravitational attraction of the mass concentrations in groups and clusters causes matter to move, we must imagine that each of these comoving observers moves with the mean flow of the matter averaged over scales large enough to remove the local fluctuations from homogeneity. Each observer is equipped with a clock, and each is assigned three numbers, x^α, $\alpha = 1$, 2, and 3, with neighboring observers assigned neighboring numbers. Then we can take the four coordinate labels x^i of an event in spacetime to be the three numbers x^α assigned to the observer who passes through the event and the observer's clock reading t at the time of event, $x^0 = t$.

Neighboring events along the world line of one of the comoving observers are separated by $dx^\alpha = 0$, because the observer has fixed spatial coordinates, and by $dx^0 = dt$ equal to the proper time interval read from the clock. The invariant interval connecting the events is then $dt^2 \equiv ds^2 = g_{ij} dx^i dx^j = g_{00} dt^2$, the first step following because ds is the proper time interval. We then have

$$g_{00} = 1 . \tag{4.2}$$

It has to be possible to synchronize the clocks to satisfy the three conditions,

$$g_{0\alpha} = 0 , \quad \text{for } \alpha = 1, 2, 3 , \tag{4.3}$$

because in a homogeneous and isotropic universe there is no preferred direction for $g_{0\alpha}$ to point. To see what this means, consider two intervals, $dx^i = (0, dx^\alpha)$ and $dy^j = (dt, 0)$ running from an event in spacetime. The first connects comoving observers at the same world time, t, and the second connects two events along the path of one of the comoving observers. Equation (4.3) says

$$dx \cdot dy = g_{0\alpha} dt dx^\alpha = 0 . \tag{4.4}$$

Since $dx \cdot dy$ is an invariant, it vanishes in a locally Minkowski coordinate system. This means each observer sees that the clocks of all the neighboring observers are synchronized with the observer's own clock. The cosmological principle says this construction is always possible, for isotropy allows synchronization of neighboring clocks, and homogeneity carries the synchronization through all space.

The conclusion is that the cosmological principle allows us to write the line element in the form

$$\begin{aligned} ds^2 &= dt^2 + g_{\alpha\beta} dx^\alpha dx^\beta \\ &= dt^2 - dl^2 . \end{aligned} \tag{4.5}$$

In the second line dl^2 is the proper spatial separation between events at the same world time, t.[5]

---

[5] In section 9 we will see that it is always possible to choose coordinates so that in a region of spacetime the line element has the time-orthogonal form (or synchronous gauge) of equation (4.5). The coordinates also can be comoving if the motion of the matter is force-free and irrotational.

We arrive at the form of the spatial part dl^2 in equation (4.5) as follows. Picture the spatial part of this model as a three-sphere, the three-dimensional analog of the two-dimensional surface of a balloon. We can imagine this three-sphere is embedded in a flat four-dimensional space with Cartesian (orthogonal) position coordinates x, y, z and w, and with the usual expression for the distance dl between neighboring points in the space,

$$dl^2 = dx^2 + dy^2 + dz^2 + dw^2 . \tag{4.6}$$

The three-sphere of Einstein's space is the set of points (x, y, z, w) at fixed distance R from the origin:

$$R^2 = x^2 + y^2 + z^2 + w^2 . \tag{4.7}$$

This condition leaves the wanted three independent space variables x, y, z. The fourth variable, w, is given by the equation

$$w^2 = R^2 - r^2 , \tag{4.8}$$

where

$$r^2 = x^2 + y^2 + z^2 . \tag{4.9}$$

The differential of equation (4.8) gives

$$dw = \frac{r\,dr}{w} = \frac{r\,dr}{(R^2 - r^2)^{1/2}} , \tag{4.10}$$

so the spatial line element dl^2 in equation (4.6) is

$$dl^2 = dx^2 + dy^2 + dz^2 + \frac{r^2 dr^2}{R^2 - r^2} . \tag{4.11}$$

With the change of coordinates from x, y, z to polar coordinates r, θ, ϕ, with the usual relations

$$z = r \cos\theta ,$$
$$x = r \sin\theta \, \cos\phi , \tag{4.12}$$
$$y = r \sin\theta \, \sin\phi ,$$

equation (4.11) becomes

$$dl^2 = dr^2 + r^2(d\theta^2 + \sin^2\theta \, d\phi^2) + \frac{r^2 \, dr^2}{R^2 - r^2}$$

$$= \frac{dr^2}{1 - r^2/R^2} + r^2(d\theta^2 + \sin^2\theta \, d\phi^2). \tag{4.13}$$

This is the space part of the line element of the Einstein world model. On adding the orthogonal time part in equation (4.5), we get the wanted four-dimensional line element,

$$ds^2 = dt^2 - \frac{dr^2}{1 - r^2/R^2} - r^2(d\theta^2 + \sin^2\theta \, d\phi^2). \tag{4.14}$$

The coordinate change

$$r = R \sin \chi \tag{4.15}$$

gives another useful form for the line element,

$$ds^2 = dt^2 - R^2[d\chi^2 + \sin^2\chi \, (d\theta^2 + \sin^2\theta \, d\phi^2)]. \tag{4.16}$$

This generalizes the familiar two-dimensional line element in polar coordinates to three dimensions. The length R in these expressions is a constant for the model, because spacetime is assumed to be static.

At $r \ll R$, the Einstein line element (4.14) approaches the Minkowski form in equation (2.4). When r is comparable to R, the departure from Euclidean spatial geometry follows the familiar behavior of the surface of a sphere. The proper length of the arc subtended by the angle $d\theta$ at fixed t, χ and ϕ (that is, the interval $(0, 0, d\theta, 0)$ in the coordinates of eq. [4.16]) is $dl = |ds| = R \sin \chi \, d\theta$. Thus the angular size $\delta = d\theta$ of a spiral galaxy of proper physical diameter $D = dl$ at coordinate position χ relative to an observer at $\chi = 0$ is

$$\delta = \frac{D}{R \sin \chi}. \tag{4.17}$$

This reaches the minimum value $\delta = D/R$ at $\chi = \pi/2$, when the galaxy is at the equator of the sphere with the observer at the pole. The angular size diverges to fill the sky as the distance to the galaxy approaches the antipodal point $\chi = \pi$.

The parallax of a distant object is the shift ϵ in its angular position when the observer is shifted by a distance A normal to the direction to the object.[6] It is an

---

[6] Recall that in the standard astronomical convention the parallax ϵ is one second of arc when the perpendicular shift A is one astronomical unit, the mean radius of the Earth's orbit around the Sun, and the object is at a distance of one parsec, as in equation (3.5).

interesting exercise to show that the parallax of an object at coordinate distance χ is

$$\epsilon = (A/R)\cot\chi. \tag{4.18}$$

The easy way is to put the object at the origin of the coordinate system in equation (4.16), and consider the paths of two light rays each running along lines of constant t and constant polar angles, one pointing at $\theta = \phi = 0$, the other at $\theta = \delta\theta$, $\phi = 0$. If $\chi = \pi/2$ the parallax vanishes, as one can visualize by imagining the observer is at the equator and the object is at the pole. And at $\chi > \pi/2$ the sign of the parallax is opposite to what we are used to.

Already in 1900 K. Schwarzschild had considered the possibility that the geometry of space is that of a closed three-sphere, with the the length R in the line element just large enough to contain the Milky Way galaxy. In the last of his three papers on general relativity theory, de Sitter (1917) was led to reconsider the idea in terms of Einstein's relativistic world model. In general relativity theory, a spatially homogeneous geometry requires a spatially homogeneous mass distribution. Since the stars in the Milky Way are known to be decidedly concentrated in space, Schwarzschild's picture would have required that the mass of the universe be dominated by spatially homogeneous "world matter," or what would now be called "dark matter." De Sitter preferred to assume rather that R is much larger than the Milky Way, and that the mass distribution is homogeneous on average because the spiral nebulae are other systems of stars, like the Milky Way, on average uniformly distributed through space. He used equation (4.17) to find a minimum value for R from the minimum angular sizes of known spiral nebulae and the assumption that the typical diameter D of one of these objects is comparable to the size of the Milky Way. He compared that to an upper bound on R derived from an estimate of the mean mass density in galaxies, as discussed next.

Dynamics

The dynamical properties of Einstein's model use some elements of general relativity theory that will be obtained in part 2 and are quoted here, as follows.

1. The differential equation that relates the metric tensor $g_{ij}(x)$ and the matter distribution is Einstein's field equation,

$$R_{ij} - \frac{1}{2}g_{ij}R = 8\pi G T_{ij}. \tag{4.19}$$

Here $R_{ij}(x)$ and $R(x)$ are functions of $g_{ij}(x)$ and its first two derivatives (and not to be confused with the length R in eq. [4.16]), Newton's gravitational constant is G, and T_{ij} is the stress-energy tensor that measures the relevant properties of the matter in the universe.

2. An ideal fluid is characterized by the mass per unit volume, $\rho(x)$, and the pressure, $p(x)$, both measured by an observer in the rest frame of the fluid, and by the fluid velocity. The stress-energy tensor of the fluid in the locally Minkowski coordinate system in which the fluid is instantaneously at rest is

$$T_{ij} = \begin{bmatrix} \rho & 0 & 0 & 0 \\ 0 & p & 0 & 0 \\ 0 & 0 & p & 0 \\ 0 & 0 & 0 & p \end{bmatrix}. \tag{4.20}$$

3. If coordinates can be chosen in a region of spacetime so matter velocities are small, then Einstein's field equation applied to a small enough region of spacetime predicts motions equivalent to the Newtonian gravity theory. This is a reasonable condition for any acceptable gravity theory, because Newtonian physics is known to be an excellent approximation at relatively small distances and velocities. A generalization of this result is that if the pressure is high the source for gravity changes from the mass density ρ to $\rho + 3p$, where p is the pressure. Thus Poisson's equation for the Newtonian gravitational acceleration **g** in a small region generalizes to

$$\nabla \cdot \mathbf{g} = -4\pi G(\rho + 3p). \tag{4.21}$$

This equation says the active gravitational mass density, which acts as the source for the gravitational acceleration, is $\rho_g = \rho + 3p$.[7]

4. Birkhoff's (1923) theorem says that for a spherically symmetric distribution of matter, Einstein's field equations have a unique solution (apart from the usual freedom of coordinate transformations). If space is empty ($T_{ij} = 0$) in some region that includes the point of symmetry, the solution in this empty hole is the flat spacetime of special relativity, with line element that can be written as (eq. [2.4])

$$\begin{aligned} ds^2 &= dt^2 - dx^2 - dy^2 - dz^2 \\ &= dt^2 - dr^2 - r^2(d\theta^2 + \sin^2\theta \, d\phi^2). \end{aligned} \tag{4.22}$$

[7] If the pressure results from particle motions in a gas, then when p is comparable to ρ the particle velocities are comparable to the velocity of light, so the conditions for the validity of Newtonian mechanics fail. However, Newtonian mechanics still describes the gravitational response of slowly moving particles, and the mean motion of the gas, as given by the gravitational acceleration from equation (4.21).

Newton's iron sphere theorem says the Newtonian gravitational acceleration inside a hollow spherical mass vanishes. The relativistic generalization is that spacetime is flat in a hole centered inside a spherically symmetric distribution of matter.

5. Now we are in a position to find the relation between the mass density, ρ, and the local rate of expansion or contraction of the material. We are considering a spatially homogeneous and isotropic mass distribution. Suppose the matter in the space within a sphere of radius l is removed and set to one side. Then result (4) says spacetime is flat within the sphere. Now replace the matter. If l is small enough, we have placed a small amount of material into flat space-time. Therefore result (3) says we can use Newtonian mechanics with equation (4.21) to describe the gravitational acceleration of the material. The active gravitational mass within the sphere of radius l is

$$M_g = \rho_g V = \frac{4}{3}\pi(\rho + 3p)l^3 . \tag{4.23}$$

Using the familiar inverse square law solution to Poisson's equation (4.21), we see that the gravitational acceleration at the surface of the sphere is

$$\ddot{l} = -\frac{GM_g}{l^2} = -\frac{4}{3}\pi G(\rho + 3p)l . \tag{4.24}$$

As usual, $\dot{l}$ means the first time derivative of $l(t)$, and $\ddot{l} = d^2l/dt^2$. This is the equation in general relativity theory for the evolution of a homogeneous isotropic mass distribution. It was first derived by the Newtonian limit by McCrea and Milne (1934). We can get a first integral of equation (4.24) by using energy conservation. Since ρ is the mass per unit volume, and mass is equivalent to energy, the net energy within the sphere is $U = \rho V$.[8] When material moves so as to change the sphere radius l that contains it, the energy U contained by the sphere changes because of the pressure work on the surface:

$$\begin{aligned} dU &= -p\, dV \\ &= \rho\, dV + V\, d\rho . \end{aligned} \tag{4.25}$$

The second line follows by differentiating out the product $U = \rho V$. On rearranging the second line, we get the energy equation for an ideal fluid,

---

[8] This neglects the gravitational energy of the mass within l. As can be readily checked, the gravitational energy is negligibly small compared to U when l is small enough.

$$\dot{\rho} = -(\rho+p)\frac{\dot{V}}{V} = -3(\rho+p)\frac{\dot{l}}{l}, \qquad (4.26)$$

where the volume of the sphere is $V \propto l^3$. The result of eliminating the pressure p from equations (4.24) and (4.26) is

$$\ddot{l} = \frac{8}{3}\pi G\rho l + \frac{4}{3}\pi G\dot{\rho}\frac{l^2}{\dot{l}}. \qquad (4.27)$$

This expression multiplied by $\dot{l}$ is a perfect differential, with the integral

$$\dot{l}^2 = \frac{8}{3}\pi G\rho l^2 + K, \qquad (4.28)$$

where the constant of integration is K.

6. The last result needed for this discussion is that the constant of integration in equation (4.28) for a static solution, where l is constant, is $K = -l^2/R^2$, where R is the radius of curvature in the line element of equation (4.16). This is derived in section 11 (eq. [11.56]). It means equations (4.24) and (4.28) in a static universe are

$$\frac{4}{3}\pi G(\rho_e + 3p_e) = 0, \qquad \frac{8}{3}\pi G\rho_e - \frac{1}{R^2} = 0. \qquad (4.29)$$

The subscript e indicates these are the mean mass density and pressure, averaged over local fluctuations, and chosen so as to satisfy the condition that the universe is in static equilibrium with l constant.

The first part of equation (4.29) shows that if the mean mass density ρ_e is positive, the pressure of the world matter has to be negative, $p_e = -\rho_e/3$, which is impossible for a collection of gas or stars or galaxies. To deal with this, Einstein proposed changing the field equation (4.19) to

$$R_{ij} - \frac{1}{2}g_{ij}R - \Lambda g_{ij} = 8\pi G T_{ij}, \qquad (4.30)$$

where Λ is called the cosmological constant (with units of reciprocal time squared). Einstein viewed this step as a modification of the field equations. The recent tendency is to move the cosmological constant term to the right-hand side of the equation, so it appears as a contribution to the stress-energy tensor (Zel'dovich 1968; Zel'dovich and Novikov 1983). On comparing equations (2.3) for the metric tensor g_{ij} and (4.20) for the stress-energy tensor T_{ij} for an ideal fluid in locally Minkowski coordinates, we see that the cosmological constant acts like a fluid with effective mass density and pressure

$$\rho_\Lambda = \frac{\Lambda}{8\pi G}, \qquad p_\Lambda = -\rho_\Lambda. \qquad (4.31)$$

These expressions for the effective pressure and energy density in the local energy conservation equation (4.26) say $\dot{\rho}_\Lambda = 0$, which is consistent. Other curious properties of the cosmological constant modeled as a fluid are discussed in section 18.

The quantities ρ_Λ and p_Λ are to be added to the density and pressure of ordinary matter in equation (4.29). To simplify the discussion, let us suppose the pressure of ordinary matter is negligible, $p_b \ll \rho_b$, as would be the case if the mass were dominated by ordinary stars moving at nonrelativistic speeds. Then equations (4.29) and (4.31) become

$$4\pi G\rho_b = \Lambda = 1/R^2 . \tag{4.32}$$

These are the relativistic relations between the mean background mass density, ρ_b, in ordinary matter like gas and stars, the cosmological constant, Λ, and the radius of curvature, R, of space sections in Einstein's static world model with the line element of equation (4.16).

This result was obtained by a computation within general relativity theory that exploits a convenient symmetry of the system, and not in a "quasi-Newtonian" model. In Newtonian mechanics the treatment of a homogeneous universe is problematic, because the gravitational potential energy of an infinite homogeneous mass distribution diverges.[9] This was recognized by Newton and others; for example, Seeliger (1895) noted that the potential energy per unit mass in a homogeneous universe would be finite if the potential at distance r from a mass element M were changed from the Newtonian form GM/r to what would now be called the Yukawa form, $(GM/r)e^{-r/r_o}$, with r_o a constant. Einstein (1917) also noted this possibility. However, Einstein's cosmological constant Λ acts not as a long-range cutoff of the gravitational interaction between particles, but as an effective negative active gravitational mass density that counters the mean gravitational attraction of ordinary matter.

De Sitter (1917) was the first to discuss the observational significance of Einstein's world model. He considered the possibility that the appropriate cosmological mass density ρ_b is the average from the spiral nebulae assumed to be star systems comparable to the Milky Way, and he used a rough estimate of ρ_b in the Einstein cosmological relations (4.32) to check that the radius of curvature R does not violate the lower bound from equation (4.17) and the minimum angular sizes of the spiral nebulae. De Sitter's value for the mass density is some four orders of magnitude larger than recent estimates, not unreasonably far off considering that this was the first very preliminary exploration. A decade later, Hubble (1926b)

---

[9] In flat space the volume at distance r to $r + \delta r$ from an observer is $\delta V = 4\pi r^2 \delta r$. With a homogeneous mean mass density ρ the mass within this volume element is $\delta M = 4\pi\rho r^2 \delta r$, and the gravitational potential energy of the observer due to this mass is $\delta U = G\delta M/r = 4\pi G\rho r \, \delta r$. The integral U of δU diverges as r^2 at $r \to \infty$.

used the considerably improved measures of galaxy distances discussed in the last section to arrive at a mean mass density

$$\rho_b \sim 10^{-31} \text{g cm}^{-3}, \tag{4.33}$$

due to the galaxies, not far from the present estimates discussed in sections 5 and 20.

Gravitational Instability

De Sitter did not comment on a curious feature of Einstein's relations (4.32)—that a variable, the mean mass density ρ_b, is set equal to a physical constant, $\Lambda/4\pi G$. What would happen if a perturbation raised or lowered the mean mass density? The answer is seen in equation (4.24): the Einstein universe is at a point of unstable equilibrium. If the mean mass density in ordinary matter were slightly larger than the equilibrium value ρ_b, then $\ddot{l}$ would be negative, the matter would contract, the contraction would further increase the mass density, and that would increase the rate of contraction. If the mass density were lower than the critical value, the universe would expand.

The same gravitational instablity applies to local fluctuations in the mass distribution, as one sees by considering a roughly spherical fluctuation large enough so that pressure gradient forces can be neglected. This is because Birkhoff's theorem says the acceleration equation (4.24) (with the source terms ρ_Λ and p_Λ in eq. [4.31] for the cosmological constant added to the density of ordinary matter) gives the gravitational acceleration within the density fluctuation no matter what is outside the perturbed region. To get an expression for the rate of change of the mass density within a perturbed region, let us keep the assumption that the pressure of ordinary matter is negligibly small, and write the mean density in ordinary matter within the patch as

$$\rho(t) = \rho_b \left(\frac{l_e}{l}\right)^3, \tag{4.34}$$

where $l = l(t)$ is the patch radius. Since we are assuming the matter pressure is negligible, the energy conservation equation (4.26) says the mass density within the patch varies inversely as the volume $\propto l^{-3}$. The normalization in equation (4.34) says the patch was slightly perturbed from the mean density when its radius was close to l_e. Equation (4.24) for $\ddot{l}$ with equations (4.31), (4.32), and (4.34) for the mass density is

$$\frac{\ddot{l}}{l} = -\frac{4}{3}\pi G \rho_b \left[\left(\frac{l_e}{l}\right)^3 - 1\right]. \tag{4.35}$$

On writing

$$l(t) = l_e(1 - \epsilon),\tag{4.36}$$

we see that for small fractional perturbations ϵ from homogeneity equation (4.35) is

$$\ddot{\epsilon} = 4\pi G \rho_b \epsilon,\tag{4.37}$$

with the solutions

$$\epsilon \propto e^{\pm\sqrt{4\pi G \rho_b}\, t}.\tag{4.38}$$

That is, the mass distribution is exponentially unstable, with characteristic growth time

$$\tau_e = (4\pi G \rho_b)^{-1/2} \sim 10^{11}\,\mathrm{y}.\tag{4.39}$$

The numerical value follows from Hubble's mass density (4.33). At some multiple of this timescale, depending on how homogeneous the universe was to begin with, the mass distribution in this model has become strongly clumpy, underdense regions expanding into ever larger voids, overdense patches collapsing. The collapse of a region of moderate size would be stopped by the growth of the nonradial motions that are ignored in equation (4.35), but density fluctuations on the scale of the length R, which have to be present at some level because our universe is not exactly homogeneous, are large enough to collapse to black holes, eventually leaving a decidedly inhomogeneous world.

Weyl (1922) and Eddington (1924) were among the first to have asked what would happen if the mass density were not identically equal to ρ_b in Einstein's cosmological equation (4.32). Eddington (1930) posed the problem to his research student, G. C. McVittie. News of Lemaître's (1927) study of the expanding world model presented in the next section led Eddington to the answer that the Einstein universe is at a point of unstable equilibrium.

Ages of Stars and Stellar Systems

In retrospect, now that we have reasonably convincing evidence that the universe really is expanding, it is easy to find reasons why a static universe is problematic. The one just discussed is that a relativistic universe would not remain close to homogeneous very much longer than the characteristic time $\sim (G \rho_b)^{-1/2}$. A second is the finite ages of stars and of stellar systems such as the Milky Way.

Suppose the Milky Way were an isolated system of stars, with total mass $\mathcal{M}$ and radius $\sim R$, in asymptotically empty space. The Newtonian gravitational binding energy per unit mass in the galaxy is $B \sim G\mathcal{M}/R$. Gravitational inter-actions among the stars would tend to populate the high energy tail of the Boltz-mann distribution of energies, promoting some to escape velocity at energy $\epsilon > B$. By this process, the Milky Way would slowly evaporate stars, leaving the remain-der increasingly tightly bound (to conserve the net energy of bound stars plus escapees). Schwarzschild (1900) suggested this evaporation problem might be avoided if the universe were closed, with the line element of equation (4.16), and with the radius R comparable to the size of the Milky Way. Einstein (1917) sug-gested that evaporation might be avoided by adopting a homogeneous universe. However, in either picture conventional physics says a massive localized bound system such as the Milky Way would evolve to a state in which an increasing part of the mass has positive energy, in the escaped stars, and the rest contracts to increasingly high binding energy (and in general relativity theory eventually produces a black hole). That is, conventional physics says the Milky Way as it is now cannot be a permanent object with an arbitrarily great age.

If energy is conserved locally, the stars in the Milky Way cannot be permanent either, for they must eventually exhaust their fuel supply. Nuclear burning con-verts a mass fraction $\epsilon \sim 0.007$ of hydrogen to radiation energy. The ratio of the luminosity of the Sun to its mass is (eq. [3.57])

$$\frac{L_\odot}{\mathcal{M}_\odot} \sim 2\,\mathrm{cm}^2\,\mathrm{s}^{-3} \sim 2 \times 10^{-21} c^2\,\mathrm{s}^{-1}. \qquad (4.40)$$

Thus, the lifetime of the sun is limited to

$$t \sim \frac{\epsilon \mathcal{M}_\odot c^2}{L_\odot} \sim 3 \times 10^{18}\,\mathrm{s} \sim 10^{11}\,\mathrm{y}. \qquad (4.41)$$

It is an interesting and perhaps suggestive coincidence that this is comparable to the instability time in equation (4.39).

If energy conservation as usually understood were violated within stars, so they could shine forever, what would become of the starlight? If the material universe were a finite island in asymptotically flat space, the starlight could stream away. In a homogeneous static universe, however, the mean energy per unit volume in starlight, u, would have to grow with time as $du/dt = j$, where j is the mean net rate of production of starlight energy per unit volume. The energy density in starlight presumably would continue to grow until the radiation temperature reached the internal temperature of the material within the stars, making it im-possible for the material to release any more energy unless the stars got still hot-ter, and of course making space quite uninhabitable for us. This effect, which is known as Olbers' paradox, is discussed further in section 5. As described by Jaki (1967) and Harrison (1987, 1990), the effect has been more or less understood since at least the seventeenth century, though people have not always been eager to face up to it, as we see from the fact that Einstein did not take note of this problem with his world model. One way out postulates that the energy spontaneously created in the stars spontaneously decays away when in the form of starlight, but this seems ridiculously ad hoc. In the steady-state model (section 7) hydrogen is spontaneously created at a steady mean rate to replace what is being burned in stars, and the dead stars and the starlight are swept away because space is expanding at a steady rate.

In standard physics, stars have a limited supply of energy, and in astronomically reasonable models for galaxy evolution most of the supply of hydrogen for new stars has by now been consumed. It is not surprising that in the standard expanding world model to be discussed next the expansion timescale is on the order of the instability timescale in equation (4.39), $\sim (G\rho_b)^{-1/2}$, because this is the only available expression with the right units. Also, it is not unreasonable that the evolutionary age (4.41) for a star like the Sun is comparable to the expansion timescale, for if star formation commenced when the universe was young most of the more massive short-lived stars would already have died. One can imagine that these considerations could have led people to the idea of an expanding universe. The connection was made by a different path, however, as discussed next.

5. The Expanding Universe

The discovery of the expansion of the universe was a deep surprise, though as so often happens in physical science one can see the connection with what came before. We noted that if one accepts the law of increase of entropy one must conclude that a static universe in the state we observe cannot exist forever. We will see that general relativity theory offers a way out. The theory says that a universe with the observed large mass and low relative velocity dispersion cannot be static: it has to expand or contract. The classical steady-state model offers another way to resolve the problem with entropy, but it is ruled out by the well-established evidence that the material content of the universe is evolving on a timescale comparable to the expansion time (as will be discussed in section 7). The picture that fits traces the expansion of the universe back to a dense state. We know this must offer another surprise, the way out of the formal singularity at $a \to 0$. Perhaps we will be able to demonstrate that the resolution already is available, in the inflation scenario to be presented in section 17, or we may find we are led in other directions. But we need not await the resolution to this puzzle. The remainder of this book reviews the broad variety of lines of research in progress or awaiting attention within the part of the framework for cosmology that does seem to work.

In this section we consider the meaning of the expansion, how the concept was discovered, and some of the consequences and observational tests.

Expansion Law

The expansion of the universe means that the proper physical distance between a pair of well-separated galaxies is increasing with time, that is, the galaxies are receding from each other. A gravitationally bound system such as the Local Group is not expanding, and we will see that gravitational instability tends to collect galaxies into increasingly more massive systems that break away from the general expansion to form a hierarchy of clusters. The homogeneous expansion law refers to galaxies far enough apart for these local irregularities to be ignored.

Neglecting the peculiar motions caused by local irregularities, the expansion follows the scaling law illustrated in figure 5.1. (This argument seems to have been first given by Lemaître 1931a.) The three galaxies define a triangle. At a later time they define a larger triangle, as indicated by the dashed lines, because these galaxies are moving apart. The cosmological principle says the motion must be homogeneous and isotropic, so the new and old triangles have to be similar, with the same angles and the length of each side scaled up by the same factor. That is, the proper physical distance $l(t)$ between a pair of well-separated galaxies scales with time as

$$l(t) = l_o a(t), \tag{5.1}$$

where l_o is constant for the pair and $a(t)$ is a universal expansion factor. The time derivative of this expression is the rate of recession of one galaxy as measured by an observer on the other,

$$v = \dot{l} = l_o \dot{a} = l \frac{\dot{a}}{a} \equiv Hl. \tag{5.2}$$

The recession causes a redshift in the spectrum of the light from one galaxy received by an observer on the other. For small recession speed v this is the ordinary first-order Doppler shift, where the observed wavelength λ_o differs from the wavelength λ_e at emission (as measured by an observer at rest at the source) by the fractional amount

$$z \equiv \frac{\lambda_o}{\lambda_e} - 1 = \frac{v}{c} = \frac{Hl}{c}. \tag{5.3}$$

This is Hubble's law in equation (3.17): the redshift z of a galaxy is proportional to its distance. We see from equation (5.2) that the coefficient in the velocity-distance relation, Hubble's parameter H, is fixed by the rate of change of the expansion parameter,

$$H = \frac{\dot{a}}{a}. \tag{5.4}$$

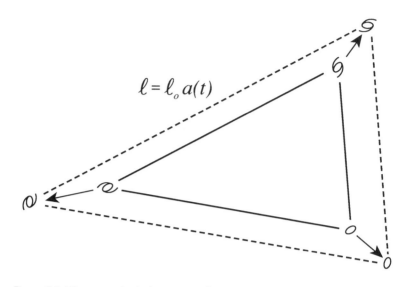

Figure 5.1. The cosmological expansion law.

In general H is a function of time. The standard convention is to write the present value as H_o.

As will be discussed below, equations (5.1) and (5.2) can be given an operational meaning (though not one that could be applied in practice) even when v is large. The definition of the redshift z in equation (5.3) also applies when the ratio λ_o/λ_e is quite different from unity. The application of the Doppler shift relation in the last parts of equation (5.3) assumes v is small, so $z \ll 1$. The meaning of the redshift when z is large is discussed below (eq. [5.45]).

Equation (5.2) says that if the recession velocity v for a given pair of galaxies were independent of time the expansion would trace back to a singular state, with zero distance between the galaxies, at the Hubble time,

$$T_H = l/v = H^{-1} . \tag{5.5}$$

Consistent with homogeneity, this characteristic timescale is independent of l, that is, it is a universal number. A more accurate calculation of the time from the start of expansion (as in eqs. [5.21], [5.63], and [13.9]) would take account of the fact that v in general is a function of time, but the Hubble time still sets the timescale, as one would expect from dimensional analysis. It is equivalent to the characteristic Hubble length,

$$L_H = c/H , \tag{5.6}$$

the distance at which the linear recession law in equation (5.2) extrapolates to the velocity of light. An object at a distance comparable to the Hubble length is seen as it was at a time in the past on the order of the expansion timescale, the Hubble time.

In terms of the galaxy distance modulus in equations (3.10) and (3.12), the linear redshift-distance relation (5.3) is

$$\log z = 0.2(m - M) - 5 - \log cH_o^{-1}, \tag{5.7}$$

where the Hubble length in the last term is measured in megaparsecs.

Line Element

It is natural to ask what the universe is expanding into. One can imagine the space we experience is expanding into a space with more dimensions, but that is not necessarily meaningful or even helpful. Instead it is best to note that the question does not arise in a metric description of spacetime, where physical distances and times are represented by the line element (2.1). In the homogeneous and isotropic world model of the cosmological principle, the convenient coordinate labeling for the line element is the comoving time-orthogonal construction used in the last section (eq. [4.5]). In this construction one imagines a set of observers, each equipped with a clock synchronized relative to the neighboring observers, and each comoving with the mean motion of the material averaged over a neighborhood large enough to remove the local fluctuations away from homogeneity. An event in spacetime is labeled by the three spatial coordinates x^α attached to the observer who passes through the event and the observer's clock reading $t = x^0$ at the time of the event. Homogeneity and isotropy require that the mean mass density and pressure are functions only of this world time t. Apart from the effects of local deviations from homogeneity, the galaxies have to be at rest relative to the comoving observers, at fixed spatial coordinate positions x^α.

The line element in these coordinates is the form given in equation (4.5),

$$\begin{aligned} ds^2 &= dt^2 - dl^2 \\ &= dt^2 - a(t)^2 g_{\alpha\beta}^o \, dx^\alpha dx^\beta . \end{aligned} \tag{5.8}$$

The construction has eliminated the off-diagonal terms $g_{0\alpha}$ that connect space and time coordinates (eq. [4.3]). Homogeneity and isotropy require that the spatial part of the metric tensor in this form can evolve only through a universal function of time, written here as $a(t)^2$. Each galaxy has fixed spatial coordinates x^α, apart from the usual local irregularities not represented by this line element. Thus equation (5.8) says the proper physical distance dl between a pair of comoving galaxies scales with time as $l(t) \propto a(t)$. This is equation (5.1).

Since the spatial geometry at a fixed instant of world time t is assumed to be homogeneous, we know the spatial part of the line element at fixed t can be

written in the form of equation (4.13). The full line element thus generalizes from
equations (4.14) and (4.16) to the time-dependent forms

$$ds^2 = dt^2 - a(t)^2 \left[\frac{dr^2}{1 - r^2/R^2} + r^2(d\theta^2 + \sin^2\theta \, d\phi^2) \right] \tag{5.9}$$

$$= dt^2 - a(t)^2 R^2 [d\chi^2 + \sin^2\chi \, d\Omega].$$

The factor R is a constant, as before, but the proper value belonging to this length
(the proper radius of the three-sphere in eq. [4.7]) is $a(t)R$, where $a(t)$ is the
expansion parameter in equations (5.1) and (5.8). In the second line of equation
(5.9) the radial coordinate is $r = R \sin \chi$, as in equation (4.15). Here and below the
expression for the line element has been shortened by writing the angular part as

$$d\Omega = d\theta^2 + \sin^2\theta \, d\phi^2. \tag{5.10}$$

As usual, the velocity of light is set to unity.

As an exercise, let us compute the volume of space at a fixed world time t in the
spacetime described by the line element (5.9) . The second line of equation (5.9)
says the proper area of the two-dimensional spherical shell of fixed coordinate
radius χ at fixed world time t is $A = 4\pi a(t)^2 R^2 \sin^2\chi$. (The integral over angles θ
and ϕ gives the usual 4π steradians.) The proper distance between the shells at
radii χ and $\chi + d\chi$ at fixed t, θ and ϕ is $dl = a(t)R d\chi$. The volume is then

$$V = \int A dl = 4\pi(aR)^3 \int_0^\pi d\chi \sin^2\chi = 2\pi^2(aR)^3. \tag{5.11}$$

The limits of integration cover the allowed range of w in equation (4.8). That is,
this is the total volume of space at fixed world time t in the model.

The model in equation (5.9) is said to be spatially closed, because the volume
of space at fixed time is finite. An open solution to Einstein's field equations, for
a spatially homogeneous universe in which the volume of space can be arbitrarily
large, is obtained by continuing equation (5.9) to $\chi \to i\chi$ and $R \to -iR$. This
brings the line element to

$$ds^2 = dt^2 - a(t)^2 R^2 [d\chi^2 + \sinh^2\chi \, d\Omega]. \tag{5.12}$$

Here the variable χ can range to arbitrarily large values, so the volume of the universe can be unbounded. Even if this open solution were a good approximation to
the part of the universe we can observe, however, the universe need not necessarily be infinite: the homogeneity assumption could fail at some large value of χ,
as is suggested by the inflation scenario in section 17, or one could even imagine
the universe is periodic.

Equations (5.9) and (5.12) are called the Robertson-Walker line element, after the demonstration by Robertson (1935) and Walker (1936) that they are the

most general form for the line element in a spatially homogeneous and isotropic spacetime, independent of general relativity theory.

The cosmological equations for the time evolution of the expansion parameter $a(t)$ follow from the dynamical arguments in the last section. Consider a comoving sphere, that is, a sphere expanding with the mean flow of matter so that there is no net flux of material through any part of its surface (apart from the small-scale fluctuations from homogeneity). A fixed point on the surface has fixed comoving spatial coordinates χ, θ, and ϕ in the line elements (5.9) or (5.12), so the proper physical radius of the sphere varies as

$$l(t) = l_o a(t), \tag{5.13}$$

with l_o a constant, as in equation (5.1). The acceleration of the radius of the sphere is given by the Newtonian equation (4.24), because Birkhoff's theorem says the material outside the sphere cannot have any gravitational effect on the behavior of what is inside. Since equation (4.24) is linear in l, the constant factor l_o divides out, leaving

$$\frac{\ddot{a}}{a} = -\frac{4}{3}\pi G(\rho + 3p). \tag{5.14}$$

This differential equation for $a(t)$ had to be independent of l_o, because the choice of sphere radius is arbitrary (as long as it is small enough that the weak-field limit in eq. [4.21] applies).

When the density and pressure are written as the sums of the mean values $\rho_b(t)$ and $p_b(t)$ in ordinary material (such as stars, gas and radiation) and the cosmological constant (eq. [4.31]), equation (5.14) becomes

$$\frac{\ddot{a}}{a} = -\frac{4}{3}\pi G(\rho_b + 3p_b) + \frac{\Lambda}{3}. \tag{5.15}$$

This is the standard relativistic form for the acceleration of the cosmological expansion.

In the energy conservation equation (4.26) the constant l_o in equation (5.13) again drops out, leaving

$$\dot{\rho}_b = -3(\rho_b + p_b)\frac{\dot{a}}{a}. \tag{5.16}$$

As in equation (5.15), this relation had to be independent of the arbitrary choice of the radius of the sphere.

Following the derivation in equations (4.27) and (4.28), one arrives at a first integral of equation (5.15) by using (5.16) to eliminate the pressure. This yields the second of the cosmological equations,

$$\left(\frac{\dot{a}}{a}\right)^2 = \frac{8}{3}\pi G \rho_b + \frac{K}{a^2} + \frac{\Lambda}{3}, \tag{5.17}$$

where K is the constant of integration.

It will be shown in section 11 (eqs. [11.51] to [11.56]) that the constant R in the line elements (5.9) and (5.12) is related to the constant of integration K. The cosmological equation (5.17) in the standard form is

$$H^2 = \left(\frac{\dot{a}}{a}\right)^2 = \frac{8}{3}\pi G \rho_b \pm \frac{1}{a^2 R^2} + \frac{\Lambda}{3}. \tag{5.18}$$

The sign in front of the curvature term is negative in the closed line element (5.9), positive in the open line element (5.12). As indicated in the first expression, this is an equation for Hubble's parameter H (eq. [5.4]).

Solutions to equations (5.16) and (5.18) are discussed in section 13; here we will need only the special Einstein-de Sitter limiting case, where the matter pressure is small, $p_b \ll \rho_b$, and the space curvature and cosmological constant Λ in equation (5.18) both are negligibly small compared to the mass density term. At zero pressure the energy conservation equation is $\dot{\rho}_b = -3\rho_b \dot{a}/a$, with the solution

$$\rho_b(t) \propto \frac{1}{a(t)^3}. \tag{5.19}$$

Since the proper volume of the comoving sphere in equation (5.13) varies with time as $a(t)^3$ (eq. [5.11]), this just says the mass per unit volume, ρ_b, varies inversely as the volume, as it must since mass is conserved here. When the mass density is the dominant term, the expansion rate equation (5.18) is

$$\left(\frac{\dot{a}}{a}\right)^2 = \frac{8}{3}\pi G \rho_b. \tag{5.20}$$

It is easy to check that the expanding solution to equations (5.19) and (5.20) is

$$a \propto t^{2/3}, \qquad t = \frac{2}{3H} = \frac{1}{(6\pi G \rho_b)^{1/2}}. \tag{5.21}$$

The zero of world time has been set to the formal singularity at $a \to 0$, where $\rho \to \infty$.

When the curvature term ($\propto a^{-2}$ in eq. [5.18]) dominates and is positive (the geometry is open), the solution is $a = (t - k)/R$, with k a constant. If the cosmological constant Λ dominates, the solutions are hyperbolic sines or cosines, as in the de Sitter solution to be discussed next.

The mean mass density in matter, $\rho_b(t)$, varies as a higher power of the expansion parameter $a(t)$ than the curvature and Λ terms in the right-hand side of equation (5.18). This means that if the expansion traces back to small enough values of $a(t)$ it must reach epochs where the mass term dominates and the limiting Einstein-de Sitter solution in equation (5.21) applies. That is, the solution traces back at a finite time in the past to the formal singularity at $a \to 0$. It is generally believed that this singularity is not real, but rather an indication of new physics that comes into play at small $a(t)$ and large energy density, and might modify the expansion rate equation or the classical picture for the geometry of spacetime.

To summarize, equations (5.2) to (5.4) express the law of general recession of the nebulae in the standard expanding cosmological model; equation (5.18) is the cosmological relation between the expansion rate $H(t)$, the mean mass density $\rho_b(t)$, the radius of curvature of space, $a(t)R$, and the cosmological constant, Λ; and equations (5.9) and (5.12) describe the spacetime geometry. Observational tests of the redshift-distance relation, and a more careful treatment of the redshift at large distances, where z approaches or exceeds unity, are discussed below, after we consider how this expanding world model was discovered.

Steps in the Discovery

The discovery of the expansion of the universe came from an interplay between the observations, which indicate the galaxies are moving away from us, and the theory, which suggests this behavior might make sense. An important step in the theory appeared in the same paper in which de Sitter (1917) first explored the possible astronomical implications of Einstein's static world model. De Sitter pointed out that one can find another solution to Einstein's field equations for a universe that is homogeneous, isotropic, and static. De Sitter's solution has negligibly small values for the mass density and pressure in ordinary matter. Since isotropic solutions are unique, de Sitter's solution has to satisfy equation (5.15) with $\rho_b = 0 = p_b$. In this limit the solution is a hyperbolic sine or cosine. A simple case is

$$a(t) = e^{H_\Lambda t}, \tag{5.22}$$

where

$$H_\Lambda = (\Lambda/3)^{1/2}. \tag{5.23}$$

Here Hubble's parameter is $H = H_\Lambda$ (eq. [5.4]), independent of time. Equation (5.18) with equation (5.22) and $\rho_b = 0$ says $R^{-2} = 0$ for this solution, so the line element (5.9) is

$$ds^2 = dt^2 - e^{2H_\Lambda t}[dr^2 + r^2 d\Omega]. \tag{5.24}$$

This is a standard form for the de Sitter solution (which should not be confused with the Einstein-de Sitter solution in eq. [5.21]). The form will reappear as the line element for the steady-state cosmology, and as a close approximation to the line element in some versions of the inflation scenario. It is discussed in more detail in section 12.

De Sitter's original form for the line element is obtained from equation (5.24) by changing the time and radial space variables from t and r to $\hat{t}$ and $\hat{r}$, with

$$
\begin{aligned}
e^{H_\Lambda t} &= \cos \hat{r} \, e^{H_\Lambda \hat{t}} \\
H_\Lambda r &= \tan \hat{r} \, e^{-H_\Lambda \hat{t}} .
\end{aligned}
\tag{5.25}
$$

It is left as an exercise to check that the result of using these equations to express dt and dr in equation (5.24) in terms of $d\hat{t}$ and $d\hat{r}$ is de Sitter's form for the line element,

$$
ds^2 = \cos^2 \hat{r} \, d\hat{t}^2 - H_\Lambda^{-2}[d\hat{r}^2 + \sin^2 \hat{r} \, d\Omega] .
\tag{5.26}
$$

This describes the same spacetime as equation (5.24), with a different coordinate label.

It is apparent from the form of the line element in equation (5.26) that de Sitter's solution represents a time-independent geometry, as one might have expected from the fact that the spacetime is determined by one fixed parameter, Λ. The source term in the Einstein gravitational field equation (4.30) is the effective stress-energy tensor associated with the cosmological constant, $T_{ij}(\Lambda) = \Lambda g_{ij}/(8\pi G)$. This is proportional to the metric tensor, which is invariant under a Lorentz velocity transformation, so it may not be surprising that the de Sitter solution is invariant under a velocity transformation (which is given in section 12). That is, in this spacetime there is nothing to define a preferred velocity (though inertial locally Minkowski frames are of course defined by the metric tensor, as discussed in section 2).

The construction of the time-orthogonal coordinates in equation (5.8) tells us that a test particle with fixed spatial coordinates r, θ and ϕ in equation (5.24) is moving freely, that is, it is not accelerated in a locally inertial reference frame. (In a homogeneous matter-filled universe the matter has to be moving freely because there can be no nongravitational force—which way could it point?—and the comoving observers at fixed x^α therefore move freely. This remains true in the limit $\rho \to 0$.) The second line in equation (5.25) thus indicates that a freely moving particle with constant coordinate position r is accelerated in the direction of increasing $\hat{r}$ in the static coordinate system of equation (5.26), and that the acceleration increases with increasing distance. That is, freely moving particles "scatter," or accelerate away from one another. We can interpret this as a result of the negative active gravitational mass density ρ_g (eqs. [4.21] and [4.31]) from the cosmological constant.

At the time of de Sitter's 1917 paper it was known from the work of Slipher and others that the spectra of the light from some of the nearer spiral nebulae are shifted toward the red, as if the light were Doppler shifted by motion of the galaxies away from us. (Prominent exceptions that at first confused the issue are the Andromeda Nebula M31 and its companion M33, both of which are approaching the Milky Way at $\sim 100\,\mathrm{km\,s^{-1}}$. As we noted in section 3, and will discuss in more detail in section 20, this is interpreted as the result of the gravitational attraction of these close neighbors in the Local Group.) During the 1920s Slipher considerably enlarged the list of measured spectra of galaxies and found that, with the exception of Local Group members, all are redshifted. This could be interpreted as the de Sitter scattering effect. There is the problem, however, that the de Sitter solution does not define a preferred velocity, so a prediction of the cosmological redshift depends on how one assigns initial conditions for the galaxy positions and velocities. Variety of possibilities were discussed (e.g., Silberstein 1924 and earlier references therein). Weyl (1923), Lemaître (1925), and Robertson (1928) all hit on the prescription consistent with the cosmological principle: if initial conditions are assigned such that galaxies move on geodesics with fixed spatial positions r, θ, ϕ in the coordinates of equation (5.24), then the pattern of relative velocities of the galaxies is independent of the galaxy to which the distances and velocities are referred (as we see from the fact that eq. [5.24] is a special case of the homogeneous Robertson-Walker line element). In this prescription the redshift of a galaxy is proportional to its distance, as we saw from the general argument in equation (5.2). Robertson (1928) remarked that this linear relation is not inconsistent with Slipher's redshifts and the galaxy distances Hubble was obtaining, and Robertson's estimate of the constant of proportionality was close to what Hubble gave the following year in his announcement of the linear relation. However, this prediction does depend on the prescription; there is nothing in the geometry of the de Sitter solution to give preferred velocities to the galaxies. The situation is different in the matter-filled evolving solution, where the prescription is replaced by the condition that the streaming motion of the matter has to yield a stress-energy tensor consistent with the assumption that the spacetime geometry is spatially homogeneous and isotropic.

Wirtz (1924) seems to have been the first to take note of astronomical evidence that galaxy redshifts tend to increase with increasing distance, and he remarked on the possible connection with the scattering effect in de Sitter's solution. Lundmark (1925) found a similar trend in galaxy redshifts and relative distances estimated from relative angular sizes. As we have noted, Robertson's (1928) consideration of homogeneous motion in the de Sitter solution led him to suggest Hubble's distance estimates may indicate a linear relation with redshift. Hubble's first paper on the observational evidence for a linear redshift-distance relation, based on his new measures of galaxy distances, appeared in 1929.

Kuhn (1962) writes of crises in which a scientific community is ready for a change of the paradigm or accepted framework within which questions are posed and answers sought. This describes the state of cosmology in the early part of 1930. The evidence that the spiral nebulae are galaxies of stars broadly spread through space made it clear that the universe is far from empty. That would argue for Einstein's matter-filled static solution, rather than de Sitter's empty static solution. The galaxy redshifts, on the other hand, do not fit with a static mass distribution, but could be neatly interpreted within de Sitter's solution. Eddington and de Sitter were aware of the conundrum, and in 1930 Lemaître informed them of the now accepted resolution, the expanding matter-filled solution. The solution was discovered by Friedmann (1922) and rediscovered by Lemaître (1927), who saw the possible connection to the galaxy redshifts. In a note to de Sitter (kindly provided by H. van der Laan) Eddington wrote that "it was the report of your remarks and mine at the [Royal Astronomical Society] which caused Lemaître to write to me about it. . . . A research student [G. C.] McVittie and I had been worrying at the problem and made considerable progress; so it was a blow to us to find it done much more completely by Lemaître (a blow softened, as far as I am concerned, by the fact that Lemaître was a student of mine)." Eddington (1930, 1931a) and de Sitter (1930) soon published papers that drew attention to Lemaître's work, and Eddington arranged for a shortened translation to be published in England (Lemaître 1931a). With these endorsements the expanding universe became the new paradigm.

The commentary in figure 5.2 on the discovery of the new world picture was drawn by an unidentified hand following an interview of de Sitter published in a Dutch newspaper. The sketch was found in the archives at Leiden Observatory, and kindly provided by H. van der Laan. De Sitter's comment referred to the special initial conditions in Lemaître's (1927) solution, in which the universe expands away from the unstable Einstein model. In this case, the universe is expanding because of the repulsion of a positive cosmological constant, as de Sitter said.

Lemaître turned to the idea that the universe expanded from a dense state, the primeval atom, but he kept a positive cosmological constant, at least in part because it increases the age of the universe over the relatively short value $\sim H_o^{-1}$ to be expected if the universe expanded from a dense initial state with $\Lambda = 0$ (eq. [5.63]). The advantage was at least in part only apparent, however, for the value of H_o had been overestimated by a factor of about five. As for the singular start of the expansion, Eddington (1931b) felt that "philosophically, the notion of a beginning to the present order of Nature is repugnant," while Lemaître (1931c) welcomed the idea for the new physics it surely will teach us.[10]

If the expansion does trace back to a dense state, the expansion rate at high redshift and high density is quite unaffected by the cosmological constant, and

[10] Since Lemaître was ordained a Roman Catholic priest in 1923, two years before the first of his great papers on cosmology, one can wonder how his physical cosmology interacted with his religion. There is no easy way to tell; apart from some comments in the earliest papers about the elegance of the physical universe as it seemed to be revealed by general relativity theory, Lemaître rigidly separated his public lives in religion and physical science.

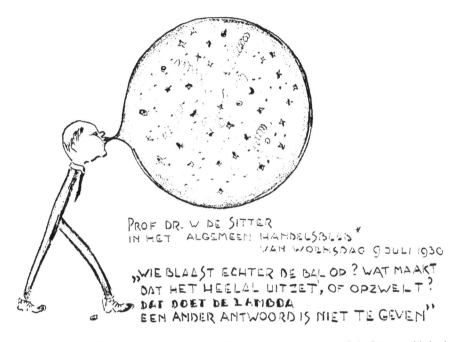

Figure 5.2. This sketch appeared following an interview of de Sitter published in a Dutch newspaper. The quote is translated by van der Laan as: "What, however, blows up the ball? What makes the universe expand or swell up? That is done by the Lambda. Another answer cannot be given."

we need some other resolution to the mystery of what caused the universe to expand from whatever it was like before it was expanding, whatever that means. It is notable that in the most popular of the proposed solutions to the mystery, the inflation scenario discussed in section 17, the expansion of the very early universe was driven by a large effective cosmological constant, just as pictured in figure 5.2. The balloon analogy remains a standard device for explaining what the expansion of the universe means, and, in the inflation picture, even a reasonable approximation to what might have caused it.

The abrupt change in the accepted picture for the physical universe, from static to expanding, was a quiet revolution, for once the concept of expansion was generally recognized it was accepted with little organized resistance. Who discovered the expansion of the universe? We have a wide choice. Einstein gave us the cosmological principle and the gravity theory that together require expansion or contraction, with a linear redshift-distance relation. Weyl was the first to take advantage of the cosmological principle, in the special case of the de Sitter solution, to obtain the linear relation between redshift and distance. Friedmann gave us the general expanding matter-filled solution. His first paper (Friedmann 1922) dealt with the closed geometry of Einstein's model; a second (Friedmann 1924) presented the open case (in a coordinate labeling that is quite different from eq. [5.12]; it is discussed in section 12). Einstein at first resisted Friedmann's generalization of his static solution, but in a subsequent note admitted that Friedmann's solutions are mathematically correct (Einstein 1923). The resolution of the conundrum thus was in the literature and even advertised as a result of Einstein's comments, but it became widely recognized only after Lemaître's discussion caught the general attention of the community. Lemaître (1927) showed how the redshift phenomenon could be related to the expansion of a matter-filled relativistic universe. This was the connection between theory and phenomena that set cosmology on the road to a mature physical science. It is not clear how much Lemaître knew about the astronomical evidence for a linear relation between redshift and distance. In his 1927 paper he indicated that available distance estimates seemed to him to be inadequate for an actual test. He instead assumed linearity and estimated the constant of proportionality by using Hubble's value for the typical luminosity of a galaxy along with measured redshifts and apparent magnitudes. Like Robertson's (1928) estimate, his result, $H_o = 630\,\mathrm{km\,s^{-1}\,Mpc^{-1}}$, is quite close to Hubble's 1929 value.[11] Wirtz was the first to note the evidence for a linear redshift-distance relation, and Hubble put the relation on a quantitative footing. It seems doubtful that Hubble was influenced by the theoretical arguments for a linear relation, for although he indicated that the redshifts may be related to the scattering effect in the de Sitter solution, he correctly noted that in this solution the redshift relation is model dependent. He cautioned also that "the linear relation found in the present discussion is a first approximation representing a restricted range in distance" (Hubble 1929). As will now be discussed, Hubble's relation has proved to be remarkably successful.

Observational Basis for Hubble's Law

Hubble's (1929) paper announcing the discovery of the linear relation between distance and redshift, what is now called Hubble's law, was based on individual distance estimates for some twenty-four relatively bright galaxies at redshifts $v \lesssim 1000\,\mathrm{km\,s^{-1}}$. Figures 3.3 and 3.6 show that at this depth the galaxy distribution is far from homogeneous, so the argument from the cosmological principle certainly does not require that the linear redshift-distance relation be a good approximation. Hubble's relation nevertheless was rapidly accepted as a useful working relation, in part surely because it agrees with the expanding universe picture, which in turn won quick approval, but certainly also because

---

[11] It is curious that the English translation of this paper, in Lemaître (1931a), omits the crucial paragraph explaining Lemaître's assessment of the test of the redshift-distance relation, and his method of estimating H_o.

it soon became apparent that, as predicted, the redshifts of much fainter and presumably more distant galaxies are much higher.

Precision tests of Hubble's law use estimates of distances from the inverse square law for galaxy apparent magnitudes (received energy flux densities). There are three ways to deal with the considerable spread in absolute magnitudes (intrinsic luminosities) of galaxies. First, one can identify a class whose luminosities have a nearly standard value. Second, one can use a predictor of absolute magnitude based on distance-independent features in the spectrum or surface brightness or appearance of the galaxy. And third, one can deal with the scatter of luminosities in a statistical way. In any of these approaches it proves considerably easier to compare ratios of distances than it is to find an absolute distance measure, so the linearity of the redshift-distance relation has been established to much better precision than the value of the constant of proportionality, Hubble's constant.

In the decade after Hubble's discovery he and Humason followed the first of the above approaches, establishing that the apparent magnitudes of the brighter galaxies in rich clusters correlate quite closely with redshift, indicating that these galaxies have nearly standard luminosities (Hubble and Humason 1931). This is an aspect of the galaxy systematics discussed in section 3, that there is a hard upper cutoff in the internal velocity dispersion σ and the correlated galaxy luminosity. The brightest galaxies in clusters tend to have luminosities (measured within a fixed surface brightness contour) close to the cutoff. The effect is illustrated by the distribution of filled circles in figure 3.13.

By 1936 Hubble could show that the relation between $\log z$ and apparent magnitude for brighter cluster members follows the predicted slope 0.2 in equation (5.7) out to redshift $z \sim 0.1$. In a modern version of this approach, using the brightest member of each cluster rather than Hubble's (1936) fifth-ranked galaxy, Sandage (1972a, b) extended the test to a sample of eighty-two rich clusters that reaches redshift $z \sim 0.5$. The redshift-magnitude Hubble diagram for these galaxies follows the trend predicted by Hubble's law, with an rms scatter $\sigma_m \sim 0.3$ magnitudes. If the galaxies had identical absolute magnitudes, this would mean the redshifts scatter by $\sim 15\%$ around the Hubble law prediction, but it is thought to be much more likely that most of the scatter is in the intrinsic galaxy luminosities. Sandages's test reaches one hundred times the depth of Hubble's (1929) discovery sample. The success of Hubble's linear relation is striking.

Galaxies with high intrinsic radio luminosities (such as those plotted in figure 3.10) have been observed to considerably greater distances than for cluster members, because the radio emission makes them relatively easy to find in the sky. Lilly and Longair (1982) took the important step of measuring the Hubble diagram of apparent magnitude as a function of redshift for radio galaxies in the infrared (mainly the K band centered on 2.2μ wavelength). Though the scatter is larger than for first-ranked cluster members, Lilly and Longair found that radio galaxies do exhibit a close correlation between redshift and magnitude, indicating that at a given redshift a galaxy with high radio luminosity has a close to standard

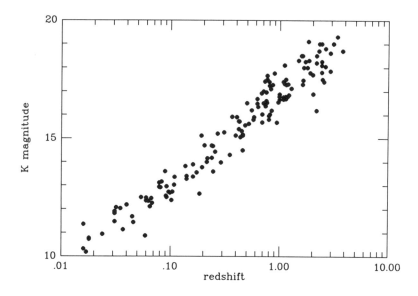

Figure 5.3. Redshift-magnitude relation for radio galaxies (McCarthy 1992).

optical luminosity (though it is thought that evolution makes the standard mean luminosity vary with redshift). Spinrad and Djorgovski (1987) extended the radio galaxy infrared Hubble diagram to still higher redshifts.

Figure 5.3 shows McCarthy's (1992) survey of the state of the measurements of the infrared K-band Hubble diagram for radio galaxies. The slope at $z \lesssim 1$ is again consistent with the Hubble law in equation (5.7). The hint of curvature at $z \gtrsim 1$ is presumed to be the result of two effects. As discussed in section 13, one expects to see relativistic corrections to the redshift-distance relation. If the relativistic effect could be isolated and measured, it would be a valuable test of the cosmology. That will be difficult, however, because the galaxies observed at high redshifts are seen as they were when they were younger and so very likely more luminous than the brightest cluster members observed at low redshifts.

The scatter in the Hubble diagram is smaller in the infrared than in optical bands of observation. This is presumed to be because optical radiation received from galaxies at $z \sim 1$ is emitted in the ultraviolet, where the light is dominated by short-lived massive stars whose total luminosity is sensitive to recent episodes of star formation, and possibly also radiation from the active nuclei of these galaxies. The K-band light tends to come from stars evolving off the hydrogen-burning main sequence and from less massive longer-lived stars whose net luminosities might be expected to be more stable. It is nonetheless startling to see how little scatter there is in the Hubble diagram for these radio-bright galaxies. The lesson for galaxy formation theories is yet to be interpreted. For our present purpose the main point is that the redshift is observed to increase with distance, at least roughly in accordance with Hubble's law, to redshifts in excess of unity.

The second way to test Hubble's law depends on the identification of a distance-independent predictor of the absolute magnitude of a galaxy (just as the period of a Cepheid variable star is a good predictor of its mean luminosity). The most successful so far use variants of the Faber-Jackson (1976) and Tully-Fisher (1977) relations between the intrinsic luminosity and internal velocity of a galaxy (eqs. [3.34] and [3.35]). For spiral galaxies, the preferred measure is the rotation velocity derived from the width δv_{21} of the 21-cm radio line emitted by atomic hydrogen in the disk of the galaxy. Aaronson, Huchra, and Mould (1979) proposed that the Tully-Fisher relation might be tighter at longer wavelengths, where there is less absorption by dust between us and the observed galaxy. The relation between the intrinsic luminosity and δv_{21} is established up to a multiplicative factor by the relation between line widths and apparent magnitudes observed in the galaxies in a cluster, since all are at the same (not well known) distance. The uncertainty in the magnitude calibration leads directly to an uncertainty in the value of Hubble's constant, but this does not affect the test of linearity of the redshift-distance relation.

Figure 5.4 shows near infrared Tully-Fisher distances to clusters. The error flags are the standard deviations in the means based on the internal scatter of distance estimates for the cluster members with useful Tully-Fisher distances. The uncertainties in the cluster redshifts are much smaller. The redshifts have been corrected for our motion relative to the thermal cosmic background radiation (CBR). As will be discussed in the next section, this radiation has a dipole ($\propto \cos\theta$) anisotropy across the sky, at $\delta T/T \sim 0.1\%$ of the mean value, which is interpreted to be the result of our peculiar motion. (The Doppler shift makes the radiation hotter in the direction toward which we are moving, and cooler in the opposite direction.) The motion of the Local Group relative to the preferred frame defined by the CBR is $\sim 600\,\text{km s}^{-1}$. The correction to the redshifts that would be measured by an observer moving so the background radiation has no dipole anisotropy lowers the scatter in the redshift-distance relation, consistent with this interpretation of the dipole anisotropy.

The filled symbols represent clusters in the neighborhood of the Great Attractor, a region in the direction of the Centaurus cluster in figure 3.7, and some distance behind it. Lynden-Bell et al. (1988) discovered that galaxy peculiar velocities in this region are unusually large, perhaps because of the presence there of an unusually large mass concentration, and perhaps consistent with the fact that the peculiar streaming motion in our neighborhood is toward this Great Attractor. (The effect is discussed further in Faber and Burstein 1988. Our peculiar motion relative to the reference frame defined by the cosmic background radiation is discussed in the next section; the interpretation as the result of the gravitational acceleration of large-scale fluctuations in the mass distribution is discussed in sections 20 and 21.) We see that something is happening in the neighborhood

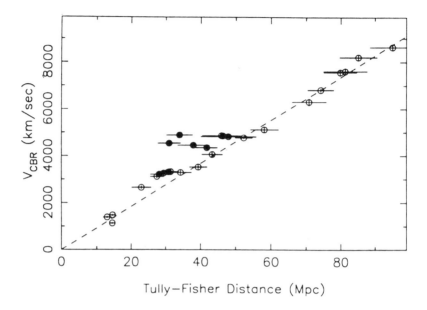

Figure 5.4. Test of Hubble's law using Tully-Fisher distances. Each symbol represents the mean of the distance estimates for the galaxies in a cluster (Mould et al. 1991).

of the Centaurus cluster, because the Tully-Fisher distances for the clusters in this region are distinctly off the general trend of redshift with distance.

The distances in figure 5.4 are expressed in megaparsecs, but this is based on a still somewhat controversial calibration of the absolute magnitude-δv_{21} relation. The straight line through the origin in this linear plot is Hubble's law. We see that, even with the anomaly in the direction of Centaurus, Hubble's law is quite a good description of the redshift-distance relation.

Figure 5.5 shows the redshift-distance relation for individual nearby galaxies, at distances comparable to the map of nearby galaxies in figure 3.3. The infrared H-band magnitudes and linewidths come from the catalog of Aaronson et al. (1982). They are converted to predicted Tully-Fisher galaxy distances, normalized to an earlier version of the redshift–Tully-Fisher distance relation in figure 5.4 (Aaronson et al. 1986). The distances are expressed in units of the predicted recession velocity, under the assumption of Hubble's law. Since the predicted redshifts in figure 5.5 are scaled from the observed redshifts of more distant clusters, they are independent of the parameter H_o. The observed redshifts plotted on the vertical axis have been corrected for our motion relative to the mean for the Local Group. (This motion is discussed in section 20 and illustrated in fig. 20.2.) If the redshifts had been corrected instead to what would be

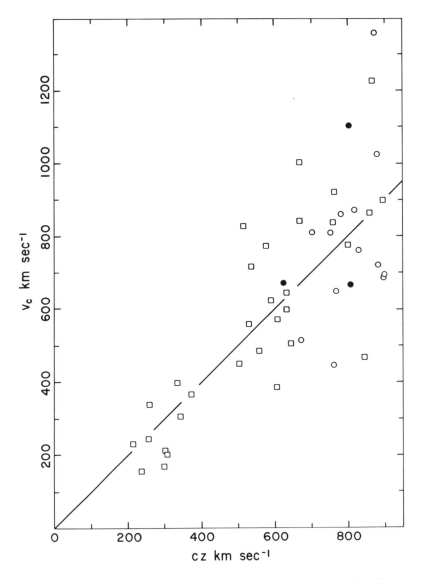

Figure 5.5. Test of Hubble's law using infrared Tully-Fisher distances for nearby galaxies (Peebles 1988a). The vertical axis is the observed redshift corrected for our motion within the Local Group. The horizontal axis is the Tully-Fisher prediction calibrated to the observed redshifts of more distant clusters.

measured by an observer in the rest frame defined by the CBR, as was done in figure 5.4, it would have introduced a considerable scatter. The galaxies plotted as boxes are at distances $|H_o Z| < 250\,\mathrm{km\,s^{-1}}$ from the sheet of galaxies shown in figure 3.3; galaxies at $H_o Z > 250\,\mathrm{km\,s^{-1}}$ are plotted as filled circles, and galaxies at $H_o Z < -250\,\mathrm{km\,s^{-1}}$ as open circles. Further details are given in Peebles (1988a).

The standard interpretation of these results is that the mean motion of the galaxies on scales $\sim 100h^{-1}\,\mathrm{Mpc}$, as defined by the clusters in figure 5.4, is close to the frame defined by the CBR, to the uncertainty of several hundred kilometers per second in the measurements. The Local Group has a peculiar velocity $\sim 600\,\mathrm{km\,s^{-1}}$ relative to this frame, and figure 5.5 shows that the nearby galaxies share this motion. This indicates that the galaxy peculiar velocity field has a coherence length considerably broader than the depth of the sample in figure 5.5. This is discussed further in section 6 (table 6.1) and section 21. The details of the large-scale galaxy flow and what it teaches us about the character of the mass distribution that is thought to drive it by gravitational acceleration are still under discussion.

Apart from the decision to use velocities relative to the Local Group rather than the CBR, there are no adjustable parameters in the comparison of predicted and observed redshifts in figure 5.5, for the predicted distances have been calibrated to the observed redshifts in the deeper sample in figure 5.4. The close correlation of observed and predicted redshifts means the local expansion rate is remarkably close to that observed on the larger scales in figures 5.3 and 5.4.

The peculiar velocity of the Local Group is comparable to the recession velocities of the galaxies in Hubble's (1929) original sample. On the above interpretation, he found a linear relation because the peculiar velocity coherence length is so large.

Let us consider finally a statistical test. Figure 5.6 shows a scatter plot of galaxy redshifts and apparent magnitudes, compiled by Shanks (1991) from a sequence of redshift surveys. There are abrupt changes in numbers of data points as a function of apparent magnitude, because quite different solid angles of the sky are sampled at different magnitudes. The broad spread in the observed redshifts at a given apparent magnitude results from the spread in intrinsic luminosities of galaxies. Since the sample size is large, one can average the data to see whether the mean trend of redshift with apparent magnitude agrees with Hubble's law, but of course one must take care to avoid a biased average. The distribution of absolute magnitudes of the galaxies seen at a given apparent magnitude is biased in favor of galaxies with higher intrinsic luminosities, because they can be seen to greater distances. In a catalog of galaxies complete to a limiting apparent magnitude, the bias increases with increasing redshift, because at greater distances only the brighter galaxies can be within the magnitude limit of the catalog. The result is that the mean apparent magnitude $\langle m \rangle$ increases with increasing z more slowly than the Hubble law (5.7), and at high z asymptotically approaches the magni-

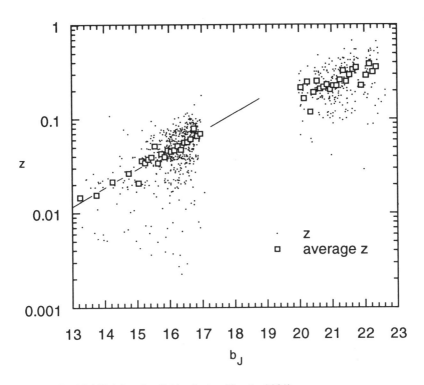

Figure 5.6. Hubble's law for field galaxies (Shanks 1991).

tude limit of the catalog. This is called the Malmquist or Scott (1957) bias.[12] If the space distribution of galaxies is homogeneous and one can neglect the effect on the apparent magnitude caused by the shift of the light toward the red, one avoids the bias by considering the distribution of redshifts in fixed bands of apparent magnitude. This was pointed out by Gunn and Oke (1975). Soneira (1979), who was commenting on the Malmquist bias in the analysis described by Segal (1976), showed how the approach can be used to find a test of Hubble's law, as follows.

Following Segal (1976), suppose the redshift varies as a power of the distance r to the galaxy,

$$z = Ar^p , \tag{5.27}$$

[12] The effect is familiar in the analysis of star counts. A catalog of stars selected by apparent magnitude contains a higher fraction of highly luminous stars than is found among the stars in a fixed volume of space, because more luminous stars are visible to greater distances, and therefore are counted within larger volumes of space.

with A and $p > 0$ constants. Let $\phi(M)$ be the galaxy luminosity function, that is, the mean number of galaxies per unit volume and unit increment of absolute magnitude M. It will be assumed that the galaxy distribution is homogeneous on average, so $\phi(M)$ is independent of distance. Also, this simplified analysis will assume the redshifts are small enough for us to ignore space curvature and deviations from the inverse square law for the apparent magnitude. Then the joint probability distribution in the redshifts and apparent magnitudes of the galaxies observed in a field of the sky with solid angle Ω is

$$\frac{d^2N}{dz\,dm} = \Omega \int r^2 dr\, \phi(M) dM\, \delta(m - M - 5\log r - C)\delta(z - Ar^p). \quad \textbf{(5.28)}$$

The first delta function is the relation between apparent and absolute magnitudes in equation (3.12), with C a constant, and the second delta function is the assumed redshift relation (5.27). The result of evaluating the integrals over distance and absolute magnitude is

$$\frac{d^2N}{dz\,dm} = \Omega\, z^{3/p-1} F(z/z_e(m)), \quad \textbf{(5.29)}$$

where F depends on the redshift only through the ratio $z/z_e(m)$, and z_e is a function of the apparent magnitude, m,

$$z_e(m) \propto 10^{0.2pm}. \quad \textbf{(5.30)}$$

That is, in this model the frequency distribution of redshifts at a given apparent magnitude m is a fixed functional form with characteristic width that varies with m by the scaling law in equation (5.30). In particular, the mean value of the redshift scales as

$$\langle z \rangle \propto 10^{0.2pm} \quad \textbf{(5.31)}$$

(Soneira 1979). This assumes a homogeneous space distribution of galaxies, and it assumes $z \ll 1$, so the relativistic corrections to the inverse square law and spatial geometry can be neglected.

The large symbols in figure 5.6 show the average values of the redshift in bins of apparent magnitude. The solid line is Hubble's law, which is equations (5.27) and (5.31) with $p = 1$. The redshifts in the faint galaxy sample are a factor of about two below the Hubble law prediction. This is in the expected direction, because the redshift brings the frequency band of the observations to the shorter wavelength part of the spectrum at emission, where galaxies typically are fainter.

Since galaxy spectra are reasonably well characterized, one can correct for the effect of the redshift of the spectrum on the apparent magnitude. (This is the K correction in eq. [13.59].) The curve in figure 5.7 shows Shanks' (1991) predicted

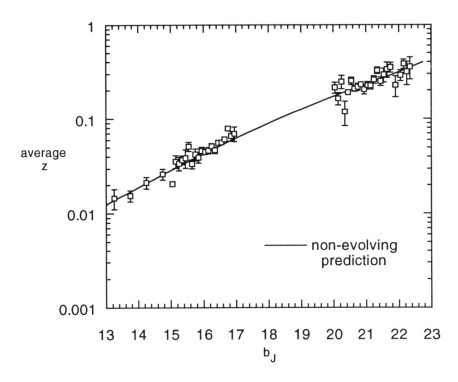

Figure 5.7. Hubble's law for field galaxies (Shanks 1991). The boxes are mean values of the redshifts in apparent magnitude bins, from figure 5.6. The curve is Hubble's law, corrected for the effect of the redshift on the apparent magnitude.

mean redshift as a function of apparent magnitude, taking account of the effect of the redshift and assuming Hubble's law ($p = 1$ in eq. [5.27]). The prediction is in good agreement with the observed mean redshifts taken from figure 5.6.

At $m \lesssim 17$, where the redshift correction is small, the mean value of the redshift as a function of the apparent magnitude is well approximated as

$$\langle z \rangle = 10^{0.2m - 4.53 \pm 0.05} . \tag{5.32}$$

This will be used in the normalization of the galaxy luminosity function.

The next subsection shows that the standard interpretation of the redshift as an effect of the expansion of the universe predicts that the same redshift factor applies to the observed rates of occurrence of distant events. Thus, if a supernova observed at redshift z were understood well enough so that the luminosity as a function of time t at the supernova could be predicted to be the function $L(t)$, then the received energy flux f would be predicted to vary with time as $f \propto L(t/(1+z))$.

That is, the timescale is predicted to be dilated by the redshift, which may be a testable effect.

The same time dilation effect contributes to the predicted dimming of the surface brightnesses of objects at high redshift (eqs. [6.43] and [9.53]). In the simplest case, one considers the surface brightness i integrated over frequencies (the energy flux per steradian) of resolved objects. If the redshift is the result of expansion, the observed surface brightness for objects with fixed intrinsic surface brightness is predicted to vary with the redshift as $i \propto (1+z)^{-4}$, two powers of $1+z$ coming from aberration, that opens the beam solid angle, one from the time dilation in the rate of reception of photons, and one from the loss of energy per photon. In a tired light cosmology, where spacetime is static and the redshift is a result of the loss of energy of each photon, only the last effect operates (Geller and Peebles 1972). The use of the $i \propto (1+z)^{-4}$ relation as a test for the expansion of the universe was proposed by Tolman (1930) and Hubble and Tolman (1935), revived by Geller and Peebles (1972) and Petrosian (1976), and most recently applied by Sandage and Perelmuter (1991). The most precise demonstration of the effect is the preservation of the thermal spectrum of the cosmic background radiation, as discussed in the next two sections.

Hubble's law carries with it the proposition that the redshift is independent of the wavelength of the spectral feature used to measure the ratio in equation (5.3). The best test is the comparison of redshifts of optical lines and the hyperfine 21-cm radio line from the coupling of the electron and proton spins in atomic hydrogen. An elegant presentation of the consistency of optical and 21-cm redshifts at $cz \lesssim 4000 \, \text{km s}^{-1}$ is given by Roberts (1972), and precision tests at higher redshifts are analyzed by Tubbs and Wolfe (1980) and Wolfe et al. (1985).

Wolfe et al. (1985) discuss the quasar PKS 0458-02. (These are round values for the position, with the right ascension in hours and minutes and the declination in degrees.) The quasar emission line redshift is $z_q = 2.29$. Its optical spectrum has a prominent feature identified as absorption by the Lyman α resonance line of atomic hydrogen, at rest wavelength $\lambda = 1216$ Å, in a gas cloud along the line of sight between us and the quasar. This is confirmed by the identification at the same redshift of absorption lines of the heavy elements commonly seen in interstellar gas. (These clouds are discussed in section 23.) The redshift of the cloud, fitted to the Lα and heavy element lines, is

$$z_\alpha = 2.0385 \,. \tag{5.33}$$

In the radio spectrum of the quasar, Wolfe et al. find two closely spaced absorption lines. If these are the 21-cm line, the redshifts are

$$z_{21} = 2.03937 \,, \quad \text{and} \quad 2.03953 \,. \tag{5.34}$$

These two radio redshifts presumably come from absorption in two subcondensations of hydrogen within the cloud. The redshift difference in equation (5.34) corresponds to a relative line of sight velocity of $15 \, \text{km s}^{-1}$ for the two subcondensations (eq. [5.49]).

The redshifts of the ultraviolet Lα line and the radio 21 cm lines agree within reasonable values for motions within the cloud. The radio line has a wavelength 10^6 times that of the Lα line. The close agreement is an impressive check on the predicted independence of the redshift on the wavelength at which it is measured.

This test does assume the ratios of wavelengths at the source are the same as the values measured in our laboratories. Since it would be absurd to imagine that variations of the ratios cancel a frequency-dependent cosmological redshift, we can conclude that we have a tight constraint on the variation of the physical constants with position and time (e.g., Savedoff 1956; Bahcall and Schmidt 1967; Wolfe, Brown, and Roberts 1976).

Estimates of the constant of proportionality, H_o, in the redshift-distance relation are discussed below in the subsection on the scales for distance and time.

Theoretical Significance of the Expansion

The cosmological expansion law is illustrated in figure 5.1. Here we discuss some aspects of the effect that may be confusing, and give a more careful discussion of what the expansion law means when the separations of the galaxies are large.

We have assumed the universe is homogeneous, but the Hubble law $v = H_o l$ would seem to mean that the universe is expanding away from us, at a special point at the center of the expansion. To see why this is not so, note that in a homogeneous isotropic universe the recession law is a vector equation,

$$\mathbf{v} = H_o \mathbf{l}, \tag{5.35}$$

where $\mathbf{l}$ and $\mathbf{v}$ are the vector position and velocity of the observed galaxy relative to our position. Consider the velocity of this galaxy measured by another observer who is in the galaxy at position $\mathbf{l}'$. The new observer has velocity $\mathbf{v}' = H_o \mathbf{l}'$ relative to us. Since we are assuming the velocities are small, the usual nonrelativistic velocity addition law says the velocity $\hat{\mathbf{v}}$ measured by this new observer is

$$\hat{\mathbf{v}} = \mathbf{v} - \mathbf{v}' = H_o(\mathbf{l} - \mathbf{l}') = H_o \hat{\mathbf{l}}, \tag{5.36}$$

where $\hat{\mathbf{l}}$ is the position of the galaxy relative to the new observer. This is Hubble's law. That is, the linear relation has the special property that it is measured by all comoving observers (that move with the mean galaxy flow). This agrees with the cosmological principle, which says there is no preferred site for measuring the expansion. The effect is illustrated in figures 5.1 and 5.2.

Milne (1935) emphasized that Hubble's relation follows from the condition that all comoving observers see the same recession law (together with the usual law of composition of nonrelativistic velocities), independent of the gravity law.

Another way to put it is that in a metric theory, in which length and time measurements are defined by a metric tensor $g_{ij}(x)$, the line element for a homogeneous and isotropic universe has to be the Robertson-Walker form, from which the linear relation follows by the scaling behavior of the line element in equation (5.8).

Since Einstein was led to the cosmological principle from the idea that the material content of the universe determines the preferred inertial motion, one might be inclined to think that the recession of the enormous amount of matter at great distances causes the matter in the Milky Way to tend to expand, the outer parts drifting away from us. That is not so in general relativity theory because, as Lemaître (1931b) pointed out, Birkhoff's (1923) theorem says the gravitational effect of the distant matter on local motions is limited to the tidal field, which is almost entirely eliminated by the spherical symmetry of the large-scale mass distribution. This is the basis for equation (4.21) for the local relative gravitational acceleration.

We have been assuming that the recession velocities are small, $z \ll 1$. As a first step in the discussion of large redshifts, let us state more carefully what is meant by the proper distance between a pair of galaxies in a metric theory. As discussed in section 2, when the line element ds^2 connecting neighboring events in spacetime is negative, the magnitude $|ds|$ is the proper distance between the events measured by an observer who moves so they are seen to be simultaneous. In the Robertson-Walker line element (eqs. [5.9] and [5.12]), a comoving observer halfway between the particles at fixed coordinate positions χ and $\chi + d\chi$ along a radial line ($d\theta = d\phi = 0$) finds that the proper distance between the particles at world time t is

$$dl = a(t)R d\chi . \tag{5.37}$$

We can imagine, in principle, placing a sequence of comoving observers along the radial line from a galaxy at $\chi = 0$ to one at comoving coordinate position χ. The sum of the distances between neighboring observers, all measured at the same world time t, is the proper physical distance between the galaxies,

$$l = a(t)R\chi , \tag{5.38}$$

as in equation (5.1).

The effect of the expansion of the universe is seen in the redshift of the radiation from distant objects. To understand the cosmological redshift effect, consider a packet of radiation with definite wavelength that is moving in a definite direction. At time t the packet passes a comoving observer who samples the radiation and measures the wavelength to be $\lambda(t)$. At world time $t + \delta t$ the packet passes a second comoving observer a proper distance $l = \delta t$ away from the first. (Recall that we are using units with the velocity of light equal to unity.) According to the expansion law in equation (5.2), the second observer is moving away from the first at speed $v = Hl = (\dot{a}/a)\delta t$. Therefore the frequency measured by the second observer is lowered by the first-order Doppler shift, and the wavelength is increased, to

$$\lambda(t + \delta t) = \lambda(t)(1 + v) = \lambda(t)[1 + (\dot{a}/a)\delta t]. \qquad (5.39)$$

On expanding the left side to first order in δt, we get the differential equation

$$\frac{\dot{\lambda}}{\lambda} = \frac{\dot{a}}{a}, \qquad (5.40)$$

with the solution

$$\lambda(t) \propto a(t). \qquad (5.41)$$

This is the evolution of the wavelength of the radiation as measured by the comoving observers it passes.

We can use the same method to get the time evolution of the momentum of a freely moving particle. Suppose observer $\mathcal{O}$ sees a particle with energy E, momentum p, and velocity p/E. A second observer $\mathcal{O}'$ who is moving with speed v along the direction of the momentum of the particle sees that the particle momentum is

$$p' = \frac{(p - vE)}{(1 - v^2)^{1/2}}. \qquad (5.42)$$

This flat spacetime Lorentz transformation equation applies because the observers are supposed to be close together relative to the curvature of spacetime. It is left as an interesting exercise to use these relations with the above argument to show that for a freely moving particle the momentum $p(t)$ measured by the comoving observer the particle passes at time t satisfies

$$p(t) \propto 1/a(t). \qquad (5.43)$$

If the particle proper peculiar velocity (the velocity relative to and measured by the comoving observers the particle is passing) is small, the momentum is $p = mv$, where m is the particle rest mass, and equation (5.43) says the proper peculiar velocity varies as

$$v \propto 1/a(t). \qquad (5.44)$$

This slowing of the peculiar motion of a freely moving particle should not be thought of as a dissipative drag; it simply results from the fact that the particle keeps overtaking observers who are moving away from it.

The momentum per photon in a packet of radiation is inversely proportional to the wavelength. Here equation (5.43) is equivalent to the redshift equation (5.41). More generally, the particle de Broglie wavelength λ is inversely proportional to the momentum p, so equation (5.43) says the de Broglie wavelength scales as $a(t)$. For yet another way to understand this result, suppose the Schrödinger wave equation for the particle is decomposed into normal spatial modes of oscillation. In a closed universe the mode satisfies the boundary condition that the wave completes an integral number of oscillations in going around the universe, as in the one-dimensional analog sketched in figure 5.8. In an open universe one gets discrete modes by imposing comoving boundary conditions at some unobservably large scale. In either case, the wavelength of a mode stretches as $\lambda \propto a(t)$. At wavelengths of interest the frequency of oscillation of the mode is much larger than the expansion rate $H = \dot{a}/a$, so adiabaticity says the occupation number in the mode is conserved in the absence of interactions with other fields. That is, the expansion of the universe stretches the de Broglie wavelength of a freely moving particle as $\lambda \propto a(t)$, which is equation (5.43).

The cosmological redshift z is defined in equation (5.3) by the ratio of observed to emitted wavelengths in the spectrum of the radiation from the galaxy. Since the wavelength scales as the expansion parameter, the cosmological redshift also satisfies

$$1 + z = \frac{\lambda_o}{\lambda_e} = \frac{a_o}{a_e}, \tag{5.45}$$

where a_e is the value of the expansion parameter at the time of emission of the radiation and a_o is the value at the time of observation. In terms of emitted and observed frequencies ν_e and ν_o, the relation is

$$\frac{\nu_o}{\nu_e} = \frac{a_e}{a_o} = \frac{1}{1+z}. \tag{5.46}$$

This also applies to proper rates of events, as one sees by the same application of a sequence of Lorentz time-dilation factors.

It is standard practice to label an epoch of the universe by the expansion factor, defined by the redshift z in equation (5.45), even when the epoch is so early the redshift cannot be observed in detectable radiation.

So far we have ignored the proper peculiar motions of source and observer relative to the comoving frame defined by the mean motion of the matter. Suppose the observer is moving at peculiar velocity v_o at observed angle θ_o from the line of sight to the source. It is left as an exercise in Lorentz transformations to check that the relation between the observed photon momentum p_o and the momentum p_c measured by a local comoving observer is

$$p_c = p_o(1 - v_o \cos \theta_o)/(1 - v_o^2)^{1/2}, \tag{5.47}$$

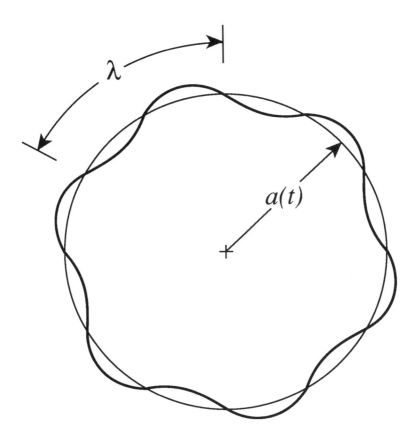

Figure 5.8. One-dimensional model for a mode of oscillation of a wave. The closed spatial part of the universe is modeled as the circle with radius $a(t)$. The oscillating line is the spatial part of a wave. The wavelength of this mode is proportional to the expansion parameter, $\lambda \propto a(t)$.

so that the observed redshift z_o is related to the redshift z_c measured by a comoving observer by the equation

$$1 + z_o = (1 + z_c)(1 - v_o \cos \theta_o)/(1 - v_o^2)^{1/2} . \tag{5.48}$$

Peculiar velocities of galaxies are small, so the expansion to first order in the velocity is adequate for practical purposes. To this order, the relation between the observed redshift of the spectrum, $1 + z_o = \lambda_o/\lambda_e$, and the cosmological redshift, $1 + z = a_o/a_e$, when source and observer are moving at speeds v_s and v_o at angles θ_s and θ_o to the line to the source, is

$$\frac{z_o - z}{1 + z} = v_s \cos \theta_s - v_o \cos \theta_o .$$ (5.49)

The left-hand side is just the fractional wavelength shift, $\delta\lambda_o/\lambda_o$, caused by the peculiar motions of the source and observer.

Horizons

The application of the redshift-distance relation at large distances or large differences between a_e and a_o raises problems, some only apparent, some very real. Let us begin with an apparent problem.

Since equation (5.38) for the proper distance l between two objects is valid whatever the coordinate separation, we can apply it to a pair of galaxies with separation greater than the Hubble length (eq. [5.6]). Here the rate of change of the proper separation, $\dot{l} = Hl$, is greater than the velocity of light. This is not a violation of special relativity; the superrelativistic velocity is just the sum of small relative velocities of closely spaced observers placed along the line connecting the galaxies, each of whom sees that special relativity describes what is happening in the neighborhood.

Suppose a packet of light is emitted from a galaxy and aimed toward another that is further than the Hubble length at the time of emission. Because the proper distance between the galaxies is increasing faster than the velocity of light, the proper distance between the packet and the galaxy toward which it is aimed actually is increasing. Depending on the parameters of the cosmological model, the packet may eventually start to move closer and eventually overtake the second galaxy.

To analyze this, choose coordinates so the galaxy that emitted the light packet is at $\chi = 0$ and the second galaxy toward which the packet is directed is at radial position χ in the coordinate system of equation (5.9). The symmetry of the model says the packet moves along a line of constant polar angles θ and ϕ. In the world time interval dt the packet moves a proper distance $dl = dt = a(t)Rd\chi$. (This makes $ds = 0$, as is appropriate for the null path of a light ray.) This says the rate of change of coordinate position of the light packet is

$$R\frac{d\chi}{dt} = \frac{1}{a(t)} .$$ (5.50)

Thus the coordinate position at time t_o of a light packet that left the origin $\chi = 0$ at time t_e is

$$R\chi = \int_{t_e}^{t_o} \frac{dt}{a} .$$ (5.51)

Let us evaluate the integral for the Einstein-de Sitter solution in equation (5.21), where the expansion rate is dominated by pressureless matter. (Here $r = R \sin \chi \to R\chi$ in eq. [5.9], because the curvature term in the expansion rate equation is supposed to be small, so R is large and χ is small.) Since the expansion parameter varies as $t^{2/3}$ in this solution, equation (5.51) is

$$a_o r = 3t_o[1 - (t_e/t_o)^{1/3}] = 3t_o[1 - (1 + z_e)^{-1/2}]. \qquad (5.52)$$

Here a_o is the present value of the expansion factor, at time t_o, and $1 + z_e = a_o/a_e$ is the expansion factor from the time of emission to detection. In this model the maximum net coordinate displacement r at t_o of a light packet that starts at an arbitrarily high redshift is $a_o r = 3t_o = 2/H_o$. Galaxies at greater coordinate separation are not causally connected at time t_o subsequent to the singularity at $t \rightarrow 0$, and are said to be outside each other's particle horizons. That is, in this model there are galaxies we cannot observe even in principle, though these galaxies will be visible to future observers in the Milky Way.

It might be noted that this result requires a little care with the limit $a \rightarrow 0$, for in the limit all comoving observers are arbitrarily close to each other, the physical separations $a(t)r$ approaching zero. Different observers cannot communicate in the limit because they are moving apart arbitrarily rapidly.

This particle horizon exists if the expansion parameter approaches zero in the past less rapidly than $a \propto t$, so the integral in equation (5.51) converges at $a \rightarrow 0$. This is equivalent to the condition that the acceleration $\ddot{a}$ is negative, which means $\rho + 3p$ is positive (eq. [5.14]). In the inflation scenario in section 17 one imagines that in the very early universe the stress-energy tensor is dominated by a term that behaves like a positive cosmological constant, making the effective value of $\rho + 3p$ negative. An example of the resulting line element is the de Sitter form (5.24). In this spacetime the integral for the coordinate displacement in equation (5.51) does not converge going back in time to $t_e \rightarrow -\infty$, so in the distant past there is causal connection between the parts of the universe we observe, and one can at least in principle imagine that there is a causal explanation for the striking large-scale homogeneity displayed in figures 3.8 to 3.11.

It is left as an exercise to check that in the de Sitter line element (5.24) applied into the indefinite future there is an event horizon, that is, there are events in spacetime that never become visible (even in principle) to an observer in the Milky Way. The inflation scenario finesses both particle and event horizons by arranging that the expansion parameter varies more rapidly than $a \propto t$ during the inflation epoch, less rapidly after inflation.

Parameters

The Friedmann-Lemaître model is characterized by dimensional parameters such as Hubble's constant H_o, and dimensionless parameters such as the product $H_o t_o$ of Hubble's constant and the age t_o of the universe computed from very high redshift. The main standard and useful parameters are listed here. The relations to the classical cosmological tests are discussed in section 13.

Since the expansion rate, $H = \dot{a}/a$, in general is a function of time (eqs. [5.2] and [5.4]), the present value is written as Hubble's constant, H_o, with $H(t)$ the expansion rate at time t. The universe at redshifts $z \lesssim 1000$ is thought to be dominated by matter such as stars and gas with pressure small compared to the mass density. In this case the mean mass density varies as $\rho_b \propto a(t)^{-3}$ (eq. [5.19]), and we can write the cosmological equation (5.18) for the expansion rate H as

$$H^2 = \left(\frac{\dot{a}}{a}\right)^2 = \left(\frac{\dot{z}}{1+z}\right)^2 = H_o^2[\Omega(1+z)^3 + \Omega_R(1+z)^2 + \Omega_\Lambda], \qquad (5.53)$$

and we can write the cosmological acceleration equation (5.15) as

$$\frac{\ddot{a}}{a} = H_o^2[\Omega_\Lambda - \Omega(1+z)^3/2], \qquad (5.54)$$

where Ω, Ω_R, and Ω_Λ are constants. The density parameter is

$$\Omega = \frac{8\pi G \rho_o}{3H_o^2}, \qquad (5.55)$$

where ρ_o is the present value of the mean mass density in material such as galaxies and intergalactic gas and stars. The redshift factor multiplying Ω in equation (5.53) says the mean mass density varies as $\rho_b \propto a^{-3} \propto (1+z)^3$. The curvature parameter is

$$\Omega_R = \frac{1}{(a_o H_o R)^2}. \qquad (5.56)$$

This is positive (R is real) for the open solution (eq. [5.12]), and negative (R is imaginary) if space is closed. The contribution of space curvature to H^2 varies as $a^{-2} \propto (1+z)^2$. The parameter associated with the cosmological constant is

$$\Omega_\Lambda = \frac{\Lambda}{3H_o^2}. \qquad (5.57)$$

The expressions for the variation with redshift of the expansion rate (eq. [5.53]) and acceleration (eq. [5.54]) assume the material pressure is negligible. The definitions of the three parameters in equations (5.55) to (5.57) still apply if pressure is important, and they give the relative contributions to the present expansion rate H_o, with

$$\Omega + \Omega_R + \Omega_\Lambda = 1. \qquad (5.58)$$

Opinions on sensible values for these parameters have evolved. In Lemaître's (1927) first expanding matter-filled world model, the parameters Ω and Ω_R are arranged so the expansion traces back to a finite redshift at which the expansion rate and acceleration in equations (5.53) and (5.54) approach zero, at the static Einstein model. Lemaître (1931d) soon moved to the idea that the universe expanded from a dense state at $a \to 0$, which he called the Primeval Atom, but he kept the two independent parameters. Einstein's position was that he had introduced the cosmological constant Λ for the purpose of making the universe static, and having learned that the universe is expanding he saw no logical need for Λ (Einstein 1945). Einstein and de Sitter (1932) argued that, pending the development of the astronomical techniques that would be needed to measure the two independent dimensionless parameters, it would be best to concentrate on the simplest reasonable case. The only source term in the expansion rate equation that we know is not negligibly small is the mass density, so in the simplest possibly realistic case $\Omega = 1$, and space curvature and the cosmological constant are unimportant. In this Einstein-de Sitter model the time evolution of the expansion parameter is given by equation (5.21), with $a \propto t^{2/3}$ and $H_o t_o = 2/3$, and the line element (5.9) is

$$ds^2 = dt^2 - a(t)^2[dr^2 + r^2 d\Omega].\tag{5.59}$$

The part in brackets represents ordinary flat space, and this spacetime accordingly is said to be cosmologically flat (despite the fact that the spacetime in general is curved; it is only the form multiplying $a(t)^2$ that is Euclidean). During the 1980s it was rather commonly believed that the Einstein-de Sitter case is the only sensible possibility. As discussed in sections 15 and 18, this was partly a result of the growing influence of Dicke's (1970) argument that it would be a remarkable coincidence if more than one of the parameters Ω, Ω_R, and Ω_Λ were appreciable at the present epoch (section 15), partly because inflation predicts the universe is cosmologically flat (section 17) and particle physicists could see no merit in an astronomically interesting value for the cosmological constant, and certainly also because the Einstein-de Sitter model is particularly simple. The current revival of interest in the possibility that Ω_Λ might be appreciable is driven by problems in fitting the Einstein-de Sitter model to the observations of the mass density and the expansion timescale (as will be discussed in the summary comments in section 26).

The expansion of the universe reverses to contraction in the past or future if the right-hand side of equation (5.53) has a zero (except for the Einstein case, where $\dot{a}$ and $\ddot{a}$ both approach zero). For example, if $\Lambda = 0$ and the universe is closed, so $\Omega_R < 0$, then $\dot{a}$ has a zero in the future, after which the universe collapses back to a "big crunch." The character of the solutions $a(t)$ and the conditions for zeros of $\dot{a}$ are discussed in section 13 and by Felten and Isaacman (1986). In the standard hot big bang cosmological model, the parameters are such that the expansion of the universe traces back to very high redshift (where new physics is presumed to save us from the singularity at $a \to 0$). That allows the material contents of the universe to reach a density high enough to relax to the thermal equilibrium that produces the cosmic background radiation discussed in sections 6 and 24.

When the redshift is large the dominant term in the expansion rate equation (5.53) is the mass density, because it appears with the highest power of $1 + z$, and the expansion rate approaches the Einstein-de Sitter limit,

$$H = \frac{\dot{a}}{a} = \frac{\dot{z}}{1+z} = H_o \Omega^{1/2}(1+z)^{3/2}, \tag{5.60}$$

with the solution

$$t = \frac{2}{3} \frac{1}{H_o \Omega^{1/2}(1+z)^{3/2}} \tag{5.61}$$

independent of space curvature or the cosmological constant. This Einstein-de Sitter limit applies when the first term in the brackets in equation (5.53) dominates, at redshift

$$1 + z \gg |\Omega^{-1} - 1|, \quad \text{when } |\Omega_\Lambda| \ll 1,$$
$$1 + z \gg |\Omega^{-1} - 1|^{1/3}, \quad \text{when } |\Omega_R| \ll 1. \tag{5.62}$$

The first limit assumes that the cosmological constant is negligible, the second that space curvature is negligible. In the latter case the cube root makes the transition to the Einstein-de Sitter limit with increasing redshift particularly rapid.

The present age, t_o, of the model universe computed from high redshift follows from equation (5.53),

$$H_o t_o = H_o \int_0^{a_o} \frac{da}{\dot{a}} = \int_1^\infty \frac{dy}{y[\Omega y^3 + \Omega_R y^2 + \Omega_\Lambda]^{1/2}}, \tag{5.63}$$

with $y = 1 + z = a_o/a$. This expression is plotted in figure 13.1. In the standard model the expansion traces back to high redshift (that is, a does not have a zero in the past). Since the universe is not empty (Ω is greater than zero), the integral converges at $y \to \infty$. The age of the expanding universe thus is set by the Hubble time H_o^{-1} with the dimensionless parameters Ω and Ω_R. The convergence of the integral does not have to mean that the physical universe came into being at a finite time in the past; the standard hope is that something new happened in the very early universe. The correction to the standard model would have to set in at exceedingly high densities, however, when objects like stars could not have existed, so it is appropriate to compare t_o in equation (5.63) to stellar evolution ages and the ages of the heavy elements determined from the decay of the unstable long-lived isotopes. This is discussed in the next subsection.

As in section 3, our standard practice is to write Hubble's constant as

$$H_o = 100h \text{ km s}^{-1} \text{ Mpc}^{-1}. \tag{5.64}$$

In some analyses, such as the dynamical estimates of the density parameter Ω, the value of the dimensionless parameter h cancels (section 20), while in other cases we can at least indicate how the results depend on the still somewhat uncertain distance scale. The reciprocal of Hubble's constant is the Hubble time,

$$H_o^{-1} = 0.98 \times 10^{10} h^{-1} \text{y}, \tag{5.65}$$

and the Hubble length,

$$c/H_o = 3000 h^{-1} \text{ Mpc}. \tag{5.66}$$

Equation (5.55) relates the mean mass density to Hubble's constant and the density parameter Ω. Numerical values are

$$\rho_o = 3\Omega H_o^2/(8\pi G) = \Omega \rho_{\text{crit}} = 2.78 \times 10^{11} \Omega h^2 \, \mathcal{M}_\odot \, \text{Mpc}^{-3}$$
$$= 1.88 \times 10^{-29} \Omega h^2 \text{g cm}^{-3}. \tag{5.67}$$

The equivalent number of hydrogen atoms per unit volume is

$$n_{\text{equiv}} = \rho_o/m_p = 1.124 \times 10^{-5} \Omega h^2 \text{ protons cm}^{-3}, \tag{5.68}$$

or on the order of one proton per cubic meter.

One of the lessons from the search for the nature of the dark matter illustrated in figure 3.12 and discussed in section 18 is that the bulk of the mass of the universe need not be in readily observable forms of protons and other baryons, or even in baryonic matter. As will be discussed in the last parts of this section, the mean mass density contributed by the baryonic material in the bright parts of galaxies is well below $\Omega = 1$, and we will see in the next section that the successful theory of light element production in the early universe requires that the net baryon density is well below this critical value. Opinion is divided on whether there has to be enough nonbaryonic matter to make $\Omega = 1$; the discussion of this situation begins in section 15. Section 20 details the dynamical estimates of the density parameter, and section 21 deals with the relation between the mean mass density and the large-scale fluctuations in the mass distribution. Section 26 presents an assessment of the current state of the constraints on the value of Ω.

Time and Length Scales

Equation (5.63) says that in the Einstein-de Sitter model, where $\Omega_R = \Omega_\Lambda = 0$, the product of Hubble's constant and the age of the universe reckoned from high redshift is $H_o t_o = 2/3$ (eq. [5.21]). If the magnitude of the pressure is small compared to the energy density (including the effective contributions from a component that acts like a comological constant), then $\ddot{a} < 0$ (eq. [5.54]) so $\dot{a} > \dot{a}_o$ at $t < t_o$, and it follows from the second expression in equation (5.63) that

$H_o t_o < 1$. If it were shown that $H_o t_o$ exceeds unity, it would mean the expansion rate $\dot{a}$ has been accelerating rather than decreasing, and within the standard model that would mean the active gravitational mass density is negative, as would happen if there were an appreciable positive cosmological constant. The measurement of this product thus is of considerable interest. Some elements of the state of the art are reviewed here.

Radioactive decay ages give a reliable measure of the age of the Solar System, and very useful constraints on the age of the galaxy. The methods are illustrated by the decay of the uranium isotopes ^{235}U and ^{238}U. By 1930 it was known that ^{238}U is the progenitor of the radium series of activities ending in the lead isotope ^{206}Pb, and that the thorium isotope ^{252}Th decays to ^{208}Pb. It was inferred that there must be a uranium isotope ^{235}U that decays through the actinium series to ^{207}Pb. Rutherford (1929) showed how to use these decay series to find the age of the elements, as follows.

Most of the decays from natural uranium go through the radium series, so the net decay rate gives a reasonable approximation to the decay rate for ^{238}U. Using this number Rutherford could estimate the ages of uranium ores from the accumulated decay product ^{206}Pb. Knowing the accumulated amounts of ^{207}Pb in different ores, Rutherford could estimate the decay rate of ^{235}U and its present abundance relative to ^{238}U. It seems reasonable to expect that whatever made the elements produced the two isotopes of uranium in about equal amounts, so the difference between the decay rates fixes the time at which the abundance ratio drops to the present value. Rutherford found $t_g \sim 3 \times 10^9 y = 3$ Gy.

Patterson (1956) used an adaptation of Rutherford's method in the first precision measurement of the age of the Solar System. Suppose the Earth and each of the meteorites formed as separate isolated and closed systems all at the same time, and the Earth and meteorites condensed from well-mixed material with uniform lead and uranium isotope ratios. Chemical fractionation gave different meteorites different abundances of uranium and lead, and uranium decay after isolation changed the lead isotope ratios in each meteorite. The isotope ^{204}Pb has no long-lived parents, so its abundance is whatever was present at the formation of the Solar System. The present isotope ratios by number in a meteorite thus satisfy the equations,

$$R(235) = \frac{^{207}Pb}{^{204}Pb} = \frac{^{207}Pb(i)}{^{204}Pb} + \frac{^{235}U}{^{204}Pb} \left[e^{\lambda(235)t_e} - 1 \right],$$

$$R(238) = \frac{^{206}Pb}{^{204}Pb} = \frac{^{206}Pb(i)}{^{204}Pb} + \frac{^{238}U}{^{204}Pb} \left[e^{\lambda(238)t_e} - 1 \right]. \tag{5.69}$$

Here t_e is the time since the chemical isolation of the meteorite, and the argument i means the isotope ratio at the time of isolation. The measured mean lives are

$$\lambda(235)^{-1} = 1.015 \, \text{Gy}, \qquad \lambda(238)^{-1} = 6.45 \, \text{Gy}. \tag{5.70}$$

Suppose the ratios $R(235)$ and $R(238)$ are measured in two meteorites, a and b. The difference between the ratios for the two meteorites eliminates the initial abundance ratios, which are assumed to have a common value. The ratio of the differences is

$$\frac{R(235)_a - R(235)_b}{R(238)_a - R(238)_b} = \frac{^{235}U}{^{238}U} \frac{e^{\lambda(235)t_e} - 1}{e^{\lambda(238)t_e} - 1} . \tag{5.71}$$

The present abundance ratio is

$$\frac{^{235}U}{^{238}U} = 0.00725 . \tag{5.72}$$

The remaining unknown in equation (5.71) is the age t_e. Patterson showed that a single value of t_e gives a close fit to the isotope ratios in the meteorites and in oceanic sediments, consistent with the assumption that the meteorites and the Earth were isolated at the same time with common isotope ratios. A standard value for the age is

$$t_e = 4.6 \pm 0.1 \, \text{Gy} \tag{5.73}$$

(Wasserberg et al. 1977), quite close to Patterson's number. Consistent results are obtained from the other long-lived isotopes, including $^{40}K \rightarrow {}^{40}A$ and $^{87}Rb \rightarrow {}^{87}Sr$.

To find the radioactive decay age of the galaxy, we need a model for the time history of the production of the elements. As discussed in section 25, the evidence is that the disk was assembled in a relatively short time interval out of material already enriched in heavy elements. A useful model introduced by Dicke (1962) assumes that a fraction F of the lead and uranium in the Solar System was produced in a short burst at time t_g measured back from the present, the remaining fraction $1 - F$ produced at a steady rate in the time interval $t_g - t_e$ between the initial burst and the isolation of the Solar System. It is an easy exercise to check that the numbers of each of the uranium isotopes satisfy

$$^{\alpha}U = ({}^{\alpha}U)_p e^{-t_g \lambda_\alpha} \left[F + \frac{1 - F}{\lambda_\alpha(t_g - t_e)} \left(e^{\lambda_\alpha(t_g - t_e)} - 1 \right) \right] . \tag{5.74}$$

The index α means 235 or 238. The subscript p means the amount produced. An estimate of the production ratio, based on the relative number of progenitor isotopes in an exploding star that can decay to ^{235}U or ^{238}U without being lost by fission, is

$$\frac{^{235}U}{^{238}U} = 1.4 \pm 0.2 \tag{5.75}$$

(Cowan, Thielemann, and Truran 1991). The above numbers give

$$t_g = \begin{cases} 6.3 \pm 0.2 \,\text{Gy} & \text{for } P = 1 \\ 8.0 \pm 0.6 \,\text{Gy} & \text{for } P = 0.5 \\ 12 \pm 2 \,\text{Gy} & \text{for } P = 0 \end{cases} \qquad (5.76)$$

A more complete analysis constrains the model for the time history of element production by seeking a concordant fit to the abundances of two or more long-lived isotopes. An example based on the above method is given by Fowler (1989); more detailed models for the production history are described in Meyer and Schramm (1986) and Cowan, Thielemann, and Truran (1991). The arguments reviewed in section 25 suggest the time interval from high redshift to the peak of formation of the elements in the initial burst might amount to 1 Gy. The conclusion is that in the standard model the age of the universe is not less than $t_o = 7$ Gy, and the age of the material in the disk of our galaxy is not likely to be greater than about 13 Gy.

The measurement of stellar evolution ages is an art too subtle to be summarized here; let us simply note that the practitioners report that a reasonable working number for the oldest globular star clusters in the halo of our galaxy is

$$t_h = 16 \pm 2 \,\text{Gy} \qquad (5.77)$$

(Demarque, Deliyannis, and Sarajedini 1991), and that it is considered difficult to reconcile theory and observation with an age less than about 14 Gy. The evidence is that the oldest star clusters in the disk are younger than the oldest globular clusters. The cooling ages of the white dwarf in the disk are about 10 Gy (section 18), at least roughly in line with the radioactive decay ages.

As we noted, one would like to compare these numbers to the Hubble time. This depends on the extragalactic distance scale, through the relation $H_o = cz/R$ for an object at cosmological redshift z and physical distance R. The observational tests of the linearity of the relation, as illustrated in figures 5.3 to 5.6, are considerably easier than the measurement of the distance scale. The difficulty with the latter is that galaxies that are close enough for a convincing measurement of the distance by the identification of objects with reasonably well known intrinsic luminosities have such small cosmological redshifts that there is no reliable way to separate the cosmological part from the peculiar motion.

Hubble's original analysis of the extragalactic distance scale indicated $h \sim$ 5, making the Hubble time, $H_o^{-1} \sim 2 \times 10^9$ y, uncomfortably close to the then current estimates of the radioactive decay age of the Earth (and well below the present numbers). Baade (1956) gives a beautiful description of the first of the two major steps to the present scale, the calibration of the classical Cepheid variable stars.

Baade had shown that in our neighborhood there are two broad classes of stars, the old metal-poor halo population II stars, and the metal-rich disk population I stars that have a broader range of ages. There are variable stars in each population; for our purpose we can focus on the population II RR Lyrae or cluster-type variables, so called because they are found in globular star clusters, and the classical population I Cepheid variables. Shapley found a calibration of the period-luminosity relation, from which he could estimate the distances to the globular clusters from their RR Lyrae variables and the distance to the Magellanic Clouds from their Cepheid variables. He used the assumption one would try first—that the variables follow a common continuous period-luminosity relation; later work showed that the population I Cepheids are about four times brighter. The distances to the RR Lyrae stars were the more reliable ones, because they are found at higher galactic latitudes where there is less obscuration by dust, and they have higher velocities from which one can estimate distances by the comparison of angular and radial velocities. This meant that the luminosities of the classical Cepheids were underestimated by a factor of about four, and the distances to the Magellanic Clouds and M31 based on the apparent magnitudes of the Cepheids were underestimated by a factor of two. A symptom of the error was known in the 1930s: at the accepted distance the globular clusters in M31 are fainter than those in the galaxy. The inconsistency became manifest when Baade used the newly commissioned 200-inch telescope at Palomar Observatory to "study the two kinds of variables side by side, so to speak," in M31. He observed the difference between the luminosities of the population I Cepheids and the RR Lyrae stars, and concluded that the distances to M31 and the Magellanic Clouds are about twice the old numbers. Since distances outside the Local Group are scaled from these galaxies, this reduced H_o by a factor of two.

The second large step was Sandage's (1958) discovery that the objects Hubble had identified as stars in more distant galaxies are actually star clusters, or tight knots of hot stars surrounded by plasma ionized by the stars. This was established by the use of red-sensitive plates on which Sandage could identify the hydrogen Hα recombination line from the plasma. The individual resolved bright stars are a factor of about five fainter than these star clusters, meaning Hubble's distance estimates relative to Local Group members must be reduced by a factor of about two. Sandage concluded that the Hubble parameter is between $H_o = 50$ and $100 \text{ km s}^{-1} \text{ Mpc}^{-1}$.

It is interesting to read Bondi's (1960) reaction in the second edition of his book, *Cosmology*: "The dominating feature of recent observational work has undoubtedly been the revision of the distance scale, and with it of Hubble's constant, by Baade and Sandage. It is not easy to appreciate now the extent to which for more than fifteen years all work in cosmology was affected and indeed oppressed by the short value of T (1.8×10^9 years) so confidently claimed to have been established observationally."

Most of the current estimates of the distance scale follow Hubble in using a distance ladder, in which the luminosities of nearby galaxies are determined by the apparent magnitudes of stars or star systems whose absolute magnitudes one hopes one knows, and the distances to galaxies at higher redshifts are found by comparing their apparent magnitudes to the absolute magnitudes of the nearby calibrator galaxies. This difficult art is surveyed by Rowan-Robinson (1985). Here we consider some elements of a case that is simple and useful.

As indicated in equation (3.21), the ratio of distances to the Virgo and Coma clusters of galaxies is known, to perhaps 10% accuracy, from the comparison of the luminosity functions of the elliptical galaxies, the predictors of relative absolute magnitude based on the galaxy internal velocity and surface brightness, as indicated in equations (3.34) and (3.36), and the other measures reviewed by Huchra (1992) and Jacoby et al. (1992). The redshift of the Coma cluster is $cz = 7000$ km s^{-1}. This is large enough so that one can expect the correction for peculiar motion to be small (according to the estimates discussed in section 21), and one can hope the remaining error is reduced in the average over the several clusters at similar redshifts for which there are useful estimates of the distance ratio. The distance to the Virgo cluster is estimated by comparing the cluster members to nearby calibrator galaxies, or by the identification of bright stars and star clusters in this relatively nearby cluster. At the time this is written, the flow of preprints is favoring a value of about 16 ± 2 Mpc. With equation (3.21) for the distance ratio, this would say the distance to the Coma cluster is 90 Mpc, and Hubble's constant is

$$H_o = 80 \pm 15 \, \text{km s}^{-1} \, \text{Mpc}^{-1} \, . \tag{5.78}$$

The uncertainty might rank as two standard deviations, in the sense that, at the time this is written, many of the practitioners of this art would be mildly surprised if the true value were outside the indicated range.

The ratio of the stellar evolution age (5.77) to the Hubble time is

$$H_o t_o \sim (1.3 \pm 0.3)h \, . \tag{5.79}$$

The Einstein-de Sitter model, which predicts $H_o t_o = 2/3$, is well into the surprising range, but the history of the measurement of this number teaches us to treat it with caution. Section 26 continues this discussion.

Gravitational Instability

As we see in figures 3.3 to 3.11, the universe is only approximately homogeneous on the average over large enough scales. The topic here is the gravitational instability of the mass distribution in the expanding Friedmann-Lemaître model. This instability does not cause the same problem as for the static Einstein model, because the expanding universe has a limited lifetime anyway, but it does tell us that the early universe had to have been remarkably close to homogeneous to have ended up as close to uniform as it is now. The instability also is a first step to an understanding of where the structure in the galaxy distribution came from: it grew by gravity out of smaller structures that existed earlier. Attempts to turn this gravitational instability picture into a theory for the origin of observed structures is discussed in part 3. Here we consider some general features of the instability.

It is convenient to write the mass density in the form

$$\rho(\mathbf{x}, t) = \rho_b(t)[1 + \delta(\mathbf{x}, t)]. \qquad (5.80)$$

Here $\rho_b(t)$ is the mean background mass density; it is a function only of the world time t. The density contrast, or fractional departure of the local mass density from the mean, is $\delta(\mathbf{x}, t)$. This contrast is a function of world time and the spatial position $\mathbf{x}$.

The spatial coordinates $\mathbf{x}$ are comoving or expanding with an ideal homogeneous background model. When the coherence length (the length scale over which the density contrast changes appreciably) is larger than the Hubble length, this definition, and the density contrast, require a convention, for what does it mean to say that parts of the universe at separations greater than the particle horizon are overdense by the same or different amounts? The prescription used here follows the construction of the time-orthogonal coordinates in equations (4.5) and (5.8). One imagines space is filled with freely moving observers, each of whom keeps a record of the local mass density as a function of the observer's proper time. The records eventually can be collected (assuming there is no event horizon) and compared. The spatial coordinate $\mathbf{x}$ in equation (5.80) is the label of one of these observers, t is the time kept by the observer, and $\rho(\mathbf{x}, t)$ is the record of densities kept by the observer with label $\mathbf{x}$.

The following discussion of the evolution of the density contrast assumes the material pressure is small compared to the mass density, so the mean density varies as $\rho_b \propto a(t)^{-3}$ (eq. [5.19]). We begin with a linear perturbation calculation that assumes the density fluctuations are small, $|\delta| \ll 1$, and that nongravitational forces on the material can be neglected.

As discussed in section 4, Birkhoff's theorem says we can imagine that different parts of the universe evolve as independent homogeneous universes, the local expansion parameter measured by the observer at $\mathbf{x}$ being

$$a = a_b(t)[1 - \epsilon(\mathbf{x}, t)]. \qquad (5.81)$$

Since pressure is negligible, we can take ρa^3 to be constant, so the fractional perturbations to the mass density and expansion parameter are related by the equation

$$\delta = 3\epsilon. \qquad (5.82)$$

Suppose we have a family of solutions to equation (5.18) for the cosmological expansion rate. The solutions are of the form $a(t, \alpha)$, where α is a parameter labeling the solution. Then equation (5.82) says a solution for the time evolution of the density contrast in linear perturbation theory is

$$\delta = -3\frac{\delta\alpha}{a}\frac{\partial a}{\partial\alpha}, \tag{5.83}$$

because this is the fractional difference in expansion parameters in neighboring solutions whose parameters differ by the constant amount $\delta\alpha$.

Equation (5.18) for the expansion parameter as a function of time t can be written as

$$t = \int^a \frac{da}{X^{1/2}} + \delta t_c, \tag{5.84}$$

where δt_c is a constant of integration, and

$$X = \dot{a}^2 = \frac{8}{3}\pi G\rho_b a^2 + R^{-2} + \frac{1}{3}\Lambda a^2. \tag{5.85}$$

Here $a(t)$ and $\rho_b(t)$ refer to the homogeneous background model. The constant R^{-2} can be negative as well as positive. The results of differentiating equation (5.84) with respect to the two parameters R^{-2} and δt at fixed t are

$$0 = \frac{1}{X^{1/2}}\frac{\partial a}{\partial(R^{-2})} - \frac{1}{2}\int^a \frac{da}{X^{3/2}}, \qquad 0 = \frac{1}{X^{1/2}}\frac{\partial a}{\partial(\delta t_c)} + 1. \tag{5.86}$$

Then, with equation (5.83), the two solutions are

$$\delta_1(t) = -\frac{3X^{1/2}\delta R^{-2}}{2a}\int^a \frac{da}{X^{3/2}}, \qquad \delta_2(t) = 3\frac{X^{1/2}\delta t_c}{a}. \tag{5.87}$$

Other approaches to this result are given in sections 10 (eq. [10.123]), 11 (eq. [11.68]), and 13 (fig. 13.13).

Consider first the Einstein-de Sitter model, where Λ and R^{-2} in equation (5.85) both are negligibly small compared to the mass density term, so the mean expansion rate is

$$H^2 = \left(\frac{\dot{a}}{a}\right)^2 = \frac{8}{3}\pi G\rho_b. \tag{5.88}$$

Here $X \propto a(t)^{-1}$, and the first part of equation (5.87) is the growing perturbation,

$$\delta_1 = -\frac{9}{40\pi}\frac{\delta R^{-2}}{G\rho_b a^2} \propto a \propto t^{2/3}. \tag{5.89}$$

This result assumes that pressure can be neglected.

It is an interesting exercise to check that if X at high redshift is dominated by radiation with the equation of state $p = \rho/3$, and the coherence length is large compared to the Hubble length so the pressure gradient force can be neglected, then the energy conservation equation (5.16) says $\rho \propto a^{-4}$, this in equation (5.88) gives $a \propto t^{1/2}$, and equation (5.89) changes to

$$\delta_1 \propto a^2 \propto t. \tag{5.90}$$

The second part of equation (5.87) in the Einstein-de Sitter model is the decaying solution,

$$\delta_2 \sim (G\rho_b)^{1/2} \delta t_c \sim \delta t_c / t. \tag{5.91}$$

This has the simple interpretation that the different patches of the universe are expanding out of synchronization by the time δt_c. The effect of the time shift decays as $\delta t_c / t$.

At low redshifts the curvature term in equation (5.85) may dominate, making X nearly constant. In this case the solutions (5.87) are

$$\delta_1 \sim \text{constant}, \qquad \delta_2 \propto 1/a \propto 1/t. \tag{5.92}$$

If the cosmological constant term dominates, both δ_1 and δ_2 approach constant values at large times.

The solution in equation (5.89) for the growing density perturbation in an Einstein-de Sitter model had to vary as a power of the world time, because the model has no fixed characteristic time to enter a function that is not a power law, such as the exponential in equation (4.38). In the open model in equation (5.92), the density fluctuations in effect are frozen when the characteristic time for the growth of density fluctuations, $\sim (G\rho_b)^{-1/2}$, becomes much longer than the expansion time, when the latter becomes dominated by space curvature.

It is worth pausing to consider another method of analysis, based on Newtonian mechanics, that is better suited to the study of the development of structures such as galaxies and clusters of galaxies. This computation requires that we be able to isolate a region small enough for the Newtonian gravitational potential energy and the relative particle velocities within the region to be small (nonrelativistic). We will establish orders of magnitude by considering first a homogeneous mass distribution, where the gravitational potential energy belonging to the mass M contained within a sphere of proper radius R is

$$\Phi \sim \frac{GM}{R} \sim G\rho_b R^2 \sim (HR)^2. \tag{5.93}$$

The last step assumes $\Omega \sim 1$, so the expansion rate is dominated by the mass density term, which is true within an order of magnitude at the present epoch.

If the sizes R of the structures to be studied are small compared to the Hubble length H^{-1}, the Hubble velocities are nonrelativistic. If the density contrast is small, this also guarantees that the gravitational potential is nonrelativistic. Where the contrast δ is large, it increases the gravitational potential; but for objects such as galaxies we know R can be chosen so the potential is small, because galaxies are nonrelativistic objects (apart from what is happening in the nucleus). That is, we can satisfy the conditions for the Newtonian limit that led to equation (5.14) for the acceleration of the expansion rate. The same limiting case applies to the dynamics of the observed clustering of matter.

We will treat the matter as an ideal pressureless fluid. Recall that in Newtonian mechanics the velocity field $\mathbf{u}$ is referred to an inertial coordinate system with proper position vector $\mathbf{r}$ measured in physical length units. The velocity field and the mass density ρ considered as functions of the proper coordinates $\mathbf{r}$ and time t satisfy the mass conservation equation

$$\left(\frac{\partial \rho}{\partial t}\right)_{\mathbf{r}} + \nabla_{\mathbf{r}} \cdot (\rho \mathbf{u}) = 0. \tag{5.94}$$

Since we are ignoring the pressure gradient force, the Euler equation of motion is

$$\left(\frac{\partial \mathbf{u}}{\partial t}\right)_{\mathbf{r}} + (\mathbf{u} \cdot \nabla_{\mathbf{r}})\mathbf{u} = -\nabla_{\mathbf{r}}\Phi, \tag{5.95}$$

where Poisson's equation for the gravitational potential Φ is

$$\nabla_{\mathbf{r}}^{2}\Phi = 4\pi G\rho. \tag{5.96}$$

The subscripts indicate that the independent variables are t and $\mathbf{r}$.

These equations apply equally well in the expanding Friedmann-Lemaître cosmological model, on scales small compared to the Hubble length and well away from black holes. In the standard cosmological model the convenient position variables are the coordinates $\mathbf{x}$ comoving with the mean rate of expansion,[13]

$$\mathbf{x} = \mathbf{r}/a(t), \tag{5.97}$$

where $a(t)$ is the expansion parameter. In these expanding coordinates the velocity field can be written as

$$\mathbf{u} = \dot{a}\mathbf{x} + \mathbf{v}(\mathbf{x}, t). \tag{5.98}$$

---

[13] The comoving coordinates $\mathbf{x}$ we are using here, which move with the general expansion of the universe, have the same name as the coordinates in equation (5.80) that are attached to freely moving observers, but they are equivalent only in the limit of a nearly homogeneous universe.

The Hubble flow is $Hr = \dot{a}x$, and the peculiar velocity relative to the general expansion is $v(x, t) = a\dot{x}$. Thus a particle at fixed comoving position x is moving away from the origin with the general expansion of the universe, with velocity $u = \dot{a}x$, so the particle has zero peculiar velocity. We will write the mass density as before as

$$\rho = \rho_b(t)[1 + \delta(x, t)] . \tag{5.99}$$

The pressure is assumed to be small, so the mass is in the rest mass of particles, and the mean density therefore varies with time as

$$\rho_b \propto 1/a(t)^3 . \tag{5.100}$$

With the change of position variables to the comoving coordinates in equation (5.97), the time derivative at fixed r of a function $f = f(t, x = r/a)$ is

$$\left(\frac{\partial f}{\partial t}\right)_r = \left(\frac{\partial f}{\partial t}\right)_x - \frac{\dot{a}}{a}x \cdot \nabla f , \tag{5.101}$$

where the gradient with respect to x at fixed time is

$$\nabla = a\nabla_r . \tag{5.102}$$

The mass conservation equation (5.94) in comoving coordinates is then

$$\left(\frac{\partial}{\partial t} - \frac{\dot{a}}{a}x \cdot \nabla\right) [\rho_b(t)(1 + \delta)] + \frac{\rho_b}{a}\nabla \cdot [(1 + \delta)(\dot{a}x + v)] = 0 . \tag{5.103}$$

On expanding out the derivatives, and recalling that $\dot{\rho}_b = -3\rho_b\dot{a}/a$ (eq. [5.100]), we get

$$\frac{\partial \delta}{\partial t} + \frac{1}{a}\nabla \cdot [(1 + \delta)v] = 0 . \tag{5.104}$$

Here and below it will be understood that the derivatives are with respect to the comoving position coordinates x and t.

Poisson's equation (5.96) is

$$-\frac{1}{a}\nabla \cdot g = \frac{1}{a^2}\nabla^2\Phi = 4\pi G\rho_b(1 + \delta) - \Lambda , \tag{5.105}$$

where the last term is the cosmological constant, as in equations (4.21) and (4.31). We can remove the unperturbed part of this equation by writing

$$\Phi = \phi(x, t) + \frac{2}{3}\pi G\rho_b a^2 x^2 - \frac{1}{6}\Lambda a^2 x^2 . \tag{5.106}$$

This brings Poisson's equation to

$$\nabla^2 \phi = 4\pi G \rho_b a^2 \delta . \tag{5.107}$$

The final step is to use these expressions in the Euler equation of motion (5.95), with equation (5.98) for the fluid velocity, equation (5.101) for the time derivative, and equation (5.106) for the gravitational potential. This gives

$$\left(\frac{\partial}{\partial t} - \frac{\dot{a}}{a} \mathbf{x} \cdot \nabla \right) (\dot{a}\mathbf{x} + \mathbf{v}) + \frac{1}{a}(\dot{a}\mathbf{x} + \mathbf{v}) \cdot \nabla(\dot{a}\mathbf{x} + \mathbf{v})$$
$$= -\frac{1}{a}\nabla\phi - \frac{4}{3}\pi G \rho_b a \mathbf{x} + \frac{1}{3}\Lambda a \mathbf{x} . \tag{5.108}$$

The terms in this expression that do not contain $\mathbf{v}$ or ϕ cancel (with eq. [5.15] with $p_b = 0$ for $\ddot{a}$). That just says the equations allow an unperturbed universe with $\mathbf{v} = 0 = \phi$. The remainder works out to

$$\frac{\partial \mathbf{v}}{\partial t} + \frac{\dot{a}}{a}\mathbf{v} + \frac{1}{a}(\mathbf{v} \cdot \nabla)\mathbf{v} = -\frac{1}{a}\nabla\phi . \tag{5.109}$$

The term $\mathbf{v}\dot{a}/a$ reflects the fact that the expansion of the universe tends to make the peculiar velocity decay as $\mathbf{v} \propto 1/a(t)$, as in equation (5.44).

Equations (5.104), (5.107), and (5.109) describe the evolution of mass fluctuations in an expanding universe in the approximation of a pressureless ideal fluid. The treatment of the gravitational response to the departures from homogeneity assumes that the peculiar velocities and the gravitational potential ϕ are small compared to unity, which is the Newtonian limit, but the motions and density fluctuations can be nonlinear, as in a galaxy. It is easy to add a pressure term to the equations of motion (eq. [5.121] below). One can add models for heating and cooling of the material, the effects of star formation and explosions, and other elements of astrophysics, though it is not easy to do this in a realistic and useful approximation.

We arrive at a simple problem by averaging over the nonlinear concentrations of material on small scales, leaving small amplitude fluctuations in the mass density and small streaming motions.[14] Then δ and $\mathbf{v}$ can be computed in linear perturbation theory, where terms of order $v\delta$ or v^2 are dropped. In this limit, equations (5.104) and (5.109) are

---

[14] This linear approximation for the large-scale fluctuations in the mass distribution works because the heavily nonlinear interactions within small-scale mass concentrations do not affect the motions of centers of mass of bound systems. They respond to the large-scale gradients in the gravitational potential. This is discussed further in section 22 (eq. [22.3]).

$$\frac{\partial \delta}{\partial t} + \frac{1}{a} \nabla \cdot \mathbf{v} = 0 ,$$

$$\frac{\partial \mathbf{v}}{\partial t} + \frac{\dot{a}}{a} \mathbf{v} + \frac{1}{a} \nabla \phi = 0 . \tag{5.110}$$

The peculiar velocity is eliminated by subtracting the time derivative of the first equation from the divergence of the second. With Poisson's equation (5.107) for $\nabla^2 \phi$, the result is

$$\frac{\partial^2 \delta}{\partial t^2} + 2 \frac{\dot{a}}{a} \frac{\partial \delta}{\partial t} = 4\pi G \rho_b \delta . \tag{5.111}$$

This is the time evolution equation for the mass density contrast $\delta = \delta\rho/\rho$ in the mass distribution, modeled as a pressureless fluid, in linear perturbation theory.

In the Einstein-de Sitter model, where space curvature and the cosmological constant are negligibly small, the expansion parameter varies as $a \propto t^{2/3}$ (eq. [5.21]), and equation (5.111) becomes

$$\frac{\partial^2 \delta}{\partial t^2} + \frac{4}{3t} \frac{\partial \delta}{\partial t} = \frac{2}{3t^2} \delta , \tag{5.112}$$

with the solution

$$\delta = A t^{2/3} + B t^{-1} , \tag{5.113}$$

where A and B are constants. This agrees with equations (5.89) and (5.91). Solutions for more general cases are given in LSS, §§ 11 to 13; numerical solutions are shown in figure 13.13.

The velocity field in linear perturbation theory is obtained as follows. The solution to equation (5.111) is of the form $\delta = A(\mathbf{x})D_1(t) + B(\mathbf{x})D_2(t)$, where D_1 and D_2 are linearly independent. Suppose one solution dominates, as the growing mode, so we can write $\delta = A(\mathbf{x})D(t)$. Then the linear mass conservation equation (5.110) is

$$\nabla \cdot \mathbf{v} = -a \frac{\partial \delta}{\partial t} = -a\delta \frac{\dot{D}}{D} . \tag{5.114}$$

We can write the velocity field as the sum of a part with no divergence and an irrotational part. The former does not contribute to the density contrast in the first line of equation (5.110), and we see from the second line that this component of the velocity decays as $a(t)^{-1}$. Equation (5.114) is a Poisson equation for the irrotational part. The solution familiar from electrostatics is

$$\mathbf{v}(\mathbf{x}) = a \frac{fH}{4\pi} \int \frac{\mathbf{y} - \mathbf{x}}{|\mathbf{y} - \mathbf{x}|^3} \delta(\mathbf{y}) d^3 y , \tag{5.115}$$

where the dimensionless velocity factor is

$$f = \frac{a\,\dot{D}}{\dot{a}\,D} = \frac{1}{H}\frac{\dot{D}}{D}\,. \tag{5.116}$$

Comparing equation (5.114) to Poisson's equation (5.107) for the peculiar gravitational acceleration $\mathbf{g} = -\nabla\phi/a$, we see the peculiar velocity also can be written as

$$\mathbf{v} = \frac{fH}{4\pi G\rho_b}\mathbf{g} = \frac{2}{3}\frac{f}{\Omega H}\mathbf{g}\,, \tag{5.117}$$

where the density parameter is Ω (eq. [5.55]). In the Einstein–de Sitter model, where $\Omega = 1 = f$ and $H = 2/(3t)$, the peculiar velocity has the simple form

$$\mathbf{v} = \mathbf{g}t\,. \tag{5.118}$$

For a spherical mass fluctuation, equation (5.115) is

$$v(x) = -a\frac{fH}{x^2}\int_0^x y^2\,dy\delta(y) = -\frac{1}{3}fHax\bar{\delta}\,, \tag{5.119}$$

where $\bar{\delta}(x)$ is the mass density contrast averaged within radius x.

Numerical values for the factor f are shown in figure 13.14. If the cosmological constant is negligibly small or space curvature is negligible, the Friedmann-Lemaître model leaves us with one parameter, Ω, and a useful approximation to the velocity factor is

$$f \sim \Omega^{0.6}\,. \tag{5.120}$$

Let us consider next how fluid pressure affects the solution. The pressure force per unit volume is

$$\mathbf{F} = -\frac{1}{a}\nabla p = -\frac{dp}{d\rho}\frac{\nabla\rho}{a} = -c_s^2\rho_b\frac{\nabla\delta}{a}\,. \tag{5.121}$$

The second step assumes the pressure p is a function of the density alone. The last step defines the velocity of sound, $c_s = (dp/d\rho)^{1/2}$, and uses equation (5.99) for the density contrast. The equation of motion (5.109) becomes

$$\frac{\partial\mathbf{v}}{\partial t} + \frac{\dot{a}}{a}\mathbf{v} + \frac{1}{a}(\mathbf{v}\cdot\nabla)\mathbf{v} = -\frac{1}{a}\nabla\phi - \frac{c_s^2}{a}\nabla\delta\,. \tag{5.122}$$

This brings the density perturbation equation (5.111) to

$$\frac{\partial^2\delta}{\partial t^2} + 2\frac{\dot{a}}{a}\frac{\partial\delta}{\partial t} = 4\pi G\rho_b\delta + \frac{c_s^2}{a^2}\nabla^2\delta\,. \tag{5.123}$$

To interpret the pressure term in this equation, write the density contrast as a Fourier series, $\delta = \sum \delta_k \exp i\mathbf{k} \cdot \mathbf{x}$. Then the amplitude $\delta_k(t)$ belonging to wave number $\mathbf{k}$ satisfies

$$\frac{d^2\delta_k}{dt^2} + 2\frac{\dot{a}}{a}\frac{d\delta_k}{dt} = \left(4\pi G\rho_b - \frac{k^2 c_s^2}{a^2}\right)\delta_k. \tag{5.124}$$

The source term on the right-hand side vanishes at wavelength

$$\lambda_J = \frac{2\pi a}{k_J} = \left(\frac{\pi c_s^2}{G\rho_b}\right)^{1/2}. \tag{5.125}$$

At this critical Jeans length the competing pressure and gravitational forces cancel (Jeans 1928). Another way to put it is that the characteristic gravitational growth time, on the order of $(G\rho_b)^{-1/2}$, is comparable to the time $\sim \lambda_J/c_s$ for a pressure wave to move across the Jeans length. At wavelengths long compared to the Jeans length, $k \ll k_J$, the pressure term in equation (5.123) is unimportant because the response time for the pressure wave is long compared to the growth time for the density contrast, and the zero pressure solutions apply. At wavelengths shorter than λ_J, the contrast oscillates as a sound wave. These effects are illustrated in figure 6.9 below.

Finally, let us return to equation (5.89) for the relation between the growing density perturbation in an Einstein-de Sitter universe and the local value of the space curvature. In this model the unperturbed background has negligible space curvature, so the radius of curvature of the space within a perturbed patch with parameter δR^{-2} is $aR_1 = a|\delta R^{-2}|^{-1/2}$. If the coordinate radius of the perturbed patch is x, the distortion to the geometry within the patch is characterized by the dimensionless and constant ratio of the comoving patch size to the comoving radius of curvature of the space within the patch,

$$\left(\frac{x}{R_1}\right)^2 = \frac{40\pi}{9}(ax)^2 G\rho_b|\delta_1| = \frac{5}{3}(Hl)^2|\delta_1| = \frac{10}{3}\frac{G|\delta M|}{l}. \tag{5.126}$$

The second step uses the Einstein-de Sitter equation (5.21) for the expansion rate. The proper radius of the patch at time t is $l = a(t)x$, and Hl is the cosmological expansion velocity of the patch. In the last step, $\delta M \sim \rho_b \delta_1 l^3$ is the mass in excess of homogeneity within the perturbed region. This says the distortion $(x/R_1)^2$ to the geometry is on the order of the Newtonian gravitational potential belonging to the density fluctuation.

When the size of the perturbed patch is comparable to the radius of curvature of the space within it, $x \sim R_1$, the perturbation to the geometry can be compared to a knob, in which the ratio of the circumference of a circle to its radius can be quite different from 2π. This is the condition for relativistic collapse to a black hole, as is the condition $G\delta M \sim l$ in equation (5.126), and it means we really should not have been treating the problem in perturbation theory. When $x \ll R_1$ the distortion to the geometry is small, to be compared to a low-amplitude ripple. In this case we see from equation (5.126) that the fluctuation in the mass distribution develops into a strong mass concentration, with $\delta_1 \sim 1$, that can break away from the general expansion to form a bound system, only when the perturbed patch is well within the Hubble length, $Hl \ll 1$. This is the appropriate case for observed mass concentrations in galaxies and clusters of galaxies. That is, in the search for a theory of the origin of large-scale structure, we are looking for something that caused ripples in the geometry of the universe.

The density contrast at the epoch when the size of a perturbed region is on the order of the Hubble length[15] is of particular interest because that marks the epoch at which free-streaming radiation can leave the region, perhaps carrying information to us about the amplitude of the perturbation in the anisotropy of the thermal cosmic background radiation to be discussed in the next section. When $Hl \sim 1$ the density contrast in equation (5.126) is

$$\delta_h \sim (x_h/R_1)^2 . \tag{5.127}$$

As we have remarked, structures like galaxies and clusters of galaxies must have formed as bound systems when they were much smaller than the Hubble length. This means that at the redshift when the region of a protogalaxy or protocluster was comparable in size to the Hubble length the density fluctuation in it satisfied $\delta_h \ll 1$.

The linear perturbation solutions for the density contrast and peculiar velocity in general relativity theory, with the result $\delta \propto t^{2/3}$ in the pressureless Einstein-de Sitter model, were first obtained by Lifshitz (1946). The spherically symmetric case is easier and captures the essential elements of the instability. This was first analyzed by Lemaître (1931b, 1933), Tolman (1934), and Bonnor (1957), who found the Jeans criterion (5.125). From the fact that the density contrast grows only as a power of time (eq. [5.113]) rather than by the familiar exponential growth for an unstable system, as for equation (4.38) in the Einstein universe or for the growth of velocity fluctuations at the onset of turbulent flow of fluid in a pipe, Lifshitz, Bonnor and others were led to conclude that the expanding universe could not produce galaxies by gravitational instability. The more recent fashion has been to argue that, since there is no known reason why the power-law evolution of $\delta\rho/\rho$ should not trace back to exceedingly high redshifts, that is, very small values of the world time t, density fluctuations nevertheless can

---

[15] It often is said that the patch is "coming into the horizon" when its physical size l is on the order of the Hubble length $\sim H^{-1}$. That is true if the expansion traces back, with positive pressure, to a singularity, for then the particle horizon is on the order of H^{-1}, as in equation (5.52). Of course, the particle horizon is much larger if the expansion traces back to an inflation epoch. Everyone knows what is meant by "the horizon," even if the patch is not really entering it, but we will attempt to minimize confusion by saying that the patch size has become comparable to the Hubble length.

grow by a large factor. We see from equation (5.127) that density fluctuations that can grow by gravity into interesting objects such as galaxies and clusters of galaxies are characterized in the very early universe by values of the space curvature parameter $(x_h/R_1)^2$ that are constant and small but nonzero (Novikov 1964; Peebles 1967). As will be discussed in part 3, an interesting value for this parameter is $(x_h/R_1)^2 \sim 10^{-4}$.

Part 3 also deals with what has become one of the central themes in the study of the physics of the very early universe: the search for a believable theory for the origin of the ripples in the geometry that grew into the large-scale structures we observe. Debate over this important issue should not obscure a simpler fundamental point, however. There are growing and decaying modes of perturbation from homogeneity. The decaying mode means a chaotic universe can grow homogeneous, but that would require precisely balanced initial conditions. The more reasonable presumption is that a growing mode also is present and, in time, dominates. That is, the standard model is gravitationally unstable, and the fluctuations away from homogeneity thus surely are growing with time, the matter becoming more strongly clustered. The present roughly homogeneous state of the universe cannot have grown out of primeval chaos. Rather, the very early universe must have been exceedingly close to homogeneous to have produced what is seen in figures 3.8 to 3.11.

The Galaxy Luminosity Function and Luminosity Density

Since observational cosmology is based on galaxies as fundamental building blocks, their characteristic properties are of particular interest. This subsection explores some orders of magnitude for galaxy luminosities and number densities and their contributions to the mean luminosity density and mass density. The simplified analysis presented here allows us to see how the characteristic numbers follow from the observations. More complete discussions of more efficient methods of estimating the luminosity function are given by Binggeli, Sandage, and Tammann (1988) and Efstathiou, Ellis, and Peterson (1988).

The galaxy luminosity function, ϕ, is defined by the mean number of galaxies per unit volume with luminosity in the range L to $L+dL$,

$$dn = \phi(L/L_*)dL/L_* , \qquad (5.128)$$

where L_* is a characteristic galaxy luminosity. The problem of finding the luminosity function at the present epoch conveniently separates into three parts: find the shape of the function $\phi(y)$, find its normalization, and find the luminosity scale set by L_*.

For the shape, we know that in our neighborhood in figure 3.3 there are many galaxies much less luminous than the Milky Way, few much more luminous. Zwicky (1942) and Abell (1962) noted this effect in clusters. Abell's model for the luminosity function is a broken power law, $\phi \sim L^{-1.5}$ at $L < L_*$, bending to $\phi \sim L^{-3}$ at the bright end. Schechter (1976) introduced a nearly equivalent more elegant form,

$$\phi(y) = \phi_* y^\alpha e^{-y}, \tag{5.129}$$

where ϕ_* and α are constant parameters. The power law index for the faint end slope is

$$\alpha = -1.07 \pm 0.05 \tag{5.130}$$

(Efstathiou, Ellis, and Peterson 1988). The formal divergence in the number density integral at $L \ll L_*$ is harmless because the more interesting integrals are dominated by the luminosity function at $L \sim L_*$. One can do better by a direct estimate of the luminosity function rather than a fit to parameters in a functional form, and by fitting separate luminosity functions to separate morphological types, as described by Binggeli, Sandage, and Tammann (1988). For the purpose of understanding the orders of magnitude, however, the Schechter function has proved to be useful.

Having fixed the shape of the luminosity function, we are left with two unknowns, L_* and ϕ_*, which we can get from the two measured coefficients in the mean redshift and mean galaxy count as functions of apparent magnitude (eqs. [3.16] and [5.32]). From L_* and ϕ_* we have the mean luminosity per unit volume,

$$j = \int_0^\infty L\phi(L/L_*)dL/L_* = L_* \int_0^\infty y\phi(y)\,dy$$
$$= L_*\phi_* \int_0^\infty y^{1+\alpha}e^{-y}\,dy = (1+\alpha)!\,\phi_* L_*. \tag{5.131}$$

The second line uses Schechter's form, and the gamma function is $n! = \Gamma(n+1)$.

The joint distribution in galaxy redshifts z and energy flux densities f observed in a region of the sky with unit solid angle is

$$\frac{d^2N}{dz\,df} = \int \phi(L/L_*)(dL/L_*)\,r^2dr\,\delta(z - H_o r/c)\delta(f - L/4\pi r^2). \tag{5.132}$$

This ignores the effects of space curvature and redshift on f and the space volume element. As in equation (5.28), the delta functions fix the wanted differential counts. The result of working the integrals is

$$\frac{d^2N}{dz\,df} = \frac{4\pi}{L_*}\left(\frac{c}{H_o}\right)^5 z^4\phi(\kappa z^2), \qquad \kappa = \frac{4\pi f c^2}{H_o^2 L_*}. \tag{5.133}$$

The redshift distribution at a fixed apparent magnitude (fixed f) is $dN/dz \propto z^4\phi(\kappa z^2)$. The effect of the prefactor z^4 is to suppress the number of low redshift galaxies. This just says there is less volume at low z from which to draw the galaxies. The result is that the distribution of luminosities of galaxies selected by apparent magnitude approximates a bell-shaped curve, despite the fact that there are many more dwarfs than giant galaxies per unit volume. This is a form of Malmquist bias, applied to galaxies rather than stars, that equation (5.31) is designed to avoid. The dimensionless factor κ sets the scale of the redshift distribution; it varies with the flux density as $z \propto \kappa^{-1/2} \propto f^{-1/2}$, as in equation (5.31) with $p = 1$.

The mean value of the redshifts of galaxies with given flux density f is found by multiplying equation (5.133) by z, integrating over redshift, and dividing by the normalizing integral. The result is

$$\langle z \rangle = \left(\frac{H_o^2 L_*}{4\pi c^2 f} \right)^{1/2} \frac{\int dy \, y^2 \phi(y)}{\int dy \, y^{3/2} \phi(y)}$$

$$= \left(\frac{H_o^2 L_*}{4\pi c^2 f} \right)^{1/2} \frac{(2+\alpha)!}{(3/2+\alpha)!} .$$

(5.134)

The second line uses the Schechter function (5.129). The result of integrating equation (5.133) over redshifts is the mean galaxy count per steradian as a function of f,

$$f \frac{dN}{df} = \frac{1}{2} \left(\frac{L_*}{4\pi f} \right)^{3/2} \int dy \, y^{3/2} \phi(y)$$

$$= \frac{\phi_*}{2} \left(\frac{L_*}{4\pi f} \right)^{3/2} (3/2+\alpha)! .$$

(5.135)

This is equation (3.9) expressed as an integral.

To convert these expressions to magnitudes, let M_* be the absolute magnitude belonging to the characteristic galaxy luminosity L_*. Then the flux f belonging to apparent magnitude m is

$$f = \frac{L_*}{4\pi(10\,\mathrm{pc})^2} 10^{0.4(M_*-m)} .$$

(5.136)

(Recall that the distance modulus is normalized to $m - M = 0$ at a distance of 10 parsecs, as in eqs. [3.10] and [3.12].) This expression in equation (5.134) for the mean redshift is

$$\langle z \rangle = \frac{h}{3 \times 10^8} \frac{(2+\alpha)!}{(3/2+\alpha)!} 10^{0.2(m-M_*)} .$$

(5.137)

The numerical factor is the ratio of the normalizing length, 10 parsecs, to the Hubble length. Equation (5.135) for the galaxy count per steradian, with equation (5.136) to change to magnitudes and equation (5.134) to eliminate L_*, is

$$\frac{1}{\langle z \rangle^3} \frac{dN}{dm} = 0.2 \ln 10\, \phi_* \left(\frac{c}{H_o}\right)^3 \frac{(3/2+\alpha)!^4}{(2+\alpha)!^3} . \qquad (5.138)$$

The expression for the mean redshift in equation (5.137), with $\alpha = -1.07 \pm 0.05$, and equation (5.32) for the observed mean redshift as a function of magnitude, gives

$$M_* = -19.53 \pm 0.25 + 5 \log h . \qquad (5.139)$$

This is based on the b_J magnitude system in figures 3.2 and 5.6. The absolute magnitude of the Sun at this band of wavelengths is (eq. [3.50])

$$M_\odot = 5.48 , \qquad (5.140)$$

so the characteristic luminosity belonging to M_* is

$$\begin{aligned} L_* &= 10^{0.4(M_\odot - M_*)} L_\odot \\ &= 1.0 \times 10^{10} e^{\pm 0.23} h^{-2} L_\odot . \end{aligned} \qquad (5.141)$$

Equation (5.138) with equations (3.16) and (5.32) for dN/dm and $\langle z \rangle$ gives

$$\phi_* = 0.010 e^{\pm 0.4} h^3 \, \text{Mpc}^{-3} . \qquad (5.142)$$

The uncertainties in L_* and ϕ_* might be counted as two standard deviations, for it would be only mildly surprising if the true results within this method were somewhat outside the indicated error range. The errors are best entered as multiplicative factors, because that is the way the power laws for $\langle z \rangle$ and dN/dm are calibrated, and in the products the arguments of the error exponentials add in quadrature. More detailed estimates of these parameters are given by Efstathiou, Ellis, and Peterson (1988), and Loveday et al. (1992).

Equation (5.131) for the mean luminosity per unit volume is

$$j = 1.0 \times 10^8 e^{\pm 0.26} h\, L_\odot \, \text{Mpc}^{-3} . \qquad (5.143)$$

(The error is smaller than in eq. [5.142] for ϕ_*, because j is proportional to $\langle z \rangle^{-1} dN/dm$.) This luminosity density counts only the light within the relatively high surface brightness parts of the galaxies luminous enough to be detected in the galaxy counts. In the next subsection the light of the night sky is used to bound the possible contribution from diffuse light or very small galaxies.

The ratio of j to L_* is a characteristic mean number of galaxies per unit volume,

$$n_* \equiv j/L_* = (1+\alpha)! \phi_* = 0.010 e^{\pm 0.4} h^3 \, \text{Mpc}^{-3} . \qquad (5.144)$$

A characteristic distance between galaxies is the width of a cube that on average contains one galaxy at number density n_*,

$$d_* = n_*^{-1/3} = 4.7 e^{\pm 0.13} h^{-1} \text{ Mpc}. \tag{5.145}$$

The luminosities of the Milky Way and our nearest large neighbor, M31, both are comparable to L_*. The distance to M31, 770 kpc, is well below d_*. This is one of the reasons for assigning these galaxies to the same group (fig. 3.3).

How many galaxies does the universe contain? A characteristic number within the Hubble length is

$$n_* (c/H_o)^3 \sim 3 \times 10^8 \text{ galaxies}. \tag{5.146}$$

If the universe is closed, the total characteristic number is (eqs. [5.11] and [5.56])

$$N_* = 2\pi^2 |\Omega_R|^{-3/2} (c/H_o)^3 n_* \sim 5 \times 10^9 |\Omega_R|^{-3/2} \text{ galaxies}. \tag{5.147}$$

What is the mean mass density contributed by the bright parts of galaxies? As indicated in equation (3.39), the mass-to-light ratio is only weakly sensitive to the galaxy luminosity, and since most of the luminosity density comes from the galaxies near L_* we can use a typical value at this luminosity. The survey of Faber and Gallagher (1979) indicates

$$\mathcal{M}/L = 12 \, e^{\pm 0.2} h, \tag{5.148}$$

in solar units, for the mass and luminosity within the Holmberg radius (where the surface brightness drops to 26.5 magnitudes per square arc second) in field spirals. The product of $\mathcal{M}/L$ and the mean luminosity density in equation (5.143) is the mean mass density contributed by the mass in the central parts of galaxies,

$$\begin{aligned}
\rho_* &= 1.2 \times 10^9 e^{\pm 0.3} h^2 \mathcal{M}_\odot \text{ Mpc}^{-3} \\
&= 8 \times 10^{-32} h^2 \text{g cm}^{-3} \\
&= 5 \times 10^{-8} h^2 \text{ protons cm}^{-3}.
\end{aligned} \tag{5.149}$$

The ratio to the critical mass density (eq. [5.67]) is the density parameter,

$$\Omega_* = 0.004 e^{\pm 0.3}. \tag{5.150}$$

The value of Hubble's constant scales out of this number.

The critical mass-to-light ratio in the Einstein-de Sitter model is the ratio of the critical mass density in equation (5.67) to the luminosity density in equation (5.143),

$$(\mathcal{M}/L)_{\text{crit}} = 3000e^{\pm 0.3}h(\mathcal{M}/L)_{\odot}. \tag{5.151}$$

The mass density ρ_* in equation (5.149) counts what is in the central parts of the galaxies. To get the total we have to add the contribution from the dark halos illustrated in figure 3.12, and the contribution from whatever is in less readily detected matter between the galaxies. We consider next the bound on the contribution to the light of the night sky by galaxies or star clusters that might have been missed because they have low surface brightness or low luminosity or are unusually compact. Dynamical measures for this "missing mass" are discussed starting in section 18.

Olbers' Paradox and the Light of the Night Sky

As we saw in Einstein's world model, the assumption of a homogeneous unchanging universe leads to a double bind: we have to postulate an eternal source of material for new stars and galaxies, and we need some provision for the disposal of the debris. In the absence of the latter the energy density of starlight would build up until it became intense enough to suppress energy production within stars. This is Olbers' paradox; its interpretation was a good deal more subtle before people became convinced of the conservation of energy and of the entropy conservation law that forbids the conversion of the excess starlight back into matter for the next generation of stars. The history of these ideas is described by Jaki (1967) and Harrison (1987, 1990).

One could imagine that the universe as we know it is eternal and the accumulation of starlight is avoided because matter occupies a vanishingly small fraction of space, as in the fractal model to be discussed in section 7. Another possibility is the steady-state cosmology, in which the universe is homogeneous, continuous spontaneous creation of matter provides a steady source of material for new generations of galaxies, and the expansion of the universe sweeps away the debris. It will be shown in section 7 that both ideas lead to observational problems.

In the standard cosmological model the material for stars and galaxies was present at very high redshift (and originated by a process not yet convincingly identified). As discussed in the next section, the universe at redshifts $z \gtrsim 10^{10}$ was hot enough to assure that photodissociation eliminated any complex nuclei and the radiation relaxed to a thermal blackbody spectrum. At much lower redshifts some of the matter was cycled through massive stars that burned the hydrogen and helium to carbon, nitrogen, oxygen, and the other heavy elements, and stellar winds and supernovae spread the elements into the interstellar material out of which terrestrial-type planets and we were made. Much of the primeval material now is locked up in slowly evolving low mass stars and other now unobtrusive objects, and much is in stellar remnants such as white dwarfs, neutron stars, and perhaps black holes. Measurements of the integrated background of starlight produced by all this activity are of fundamental interest for two purposes. First, this is a probe of the history of the universe, for galaxies had to have been more luminous in the past to have produced the heavy elements present in them now. Second, the background light tests the possibility that equation (5.143) for the density of luminous matter might have missed a good deal of light from unobtrusive systems. Studies of the interpretation of the background light as a probe of galaxy evolution trace back at least to Whitrow and Yallop (1964) and Partridge and Peebles (1967b), and as a constraint on the luminosity density to PC. An excellent survey of the present situation is given in Bowyer and Leinert (1990).

If the galaxies radiate at mean luminosity density j for an expansion time $\sim H_o^{-1}$, they produce a net energy density $u \sim j H_o^{-1}$. The equivalent surface brightness i is the energy flux per unit area and solid angle,

$$i = \frac{cu}{4\pi} \sim \frac{j}{4\pi} \frac{c}{H_o}. \tag{5.152}$$

A more detailed estimate integrates the rate of production of the energy density u by sources and the rate of loss caused by the expansion of the universe and perhaps also by absorption. Let the mean number of photons per unit volume with frequency in the range ν to $\nu + \delta\nu$ at time t be

$$\delta n = n(\nu, t)\delta\nu, \tag{5.153}$$

as measured by comoving observers. At time $t + dt$ each of these photons has been redshifted by the amount

$$\nu \rightarrow \nu a(t)/a(t+dt) = \nu(1 - dt\dot{a}/a), \tag{5.154}$$

to order dt. If photons are not created or destroyed, the distribution function at time $t + dt$ is

$$\delta n = n(\nu(1 - dt\dot{a}/a), t+dt)(1 - dt\dot{a}/a)\delta\nu$$
$$= (1 - 3dt\dot{a}/a)n(\nu, t)\delta\nu. \tag{5.155}$$

In the first line the argument of the photon distribution function is the new frequency in equation (5.154), with the new bandwidth. The first factor in the second line says the number density is lower because the volume of the universe has expanded by the factor $[a(t+dt)/a(t)]^3 = 1 + 3dt\dot{a}/a$. This gives

$$\frac{\partial n}{\partial t} = \nu \frac{\dot{a}}{a} \frac{\partial n}{\partial \nu} - 2\frac{\dot{a}}{a}n. \tag{5.156}$$

We get the energy per unit volume and frequency interval by multiplying the number density by the energy per photon: $u(\nu, t) = h\nu n(\nu, t)$. The extra factor of ν changes the factor 2 in equation (5.156) to 3 in the differential equation for u. If in addition there is an energy source with mean luminosity $j(\nu, t)$ per unit volume and frequency interval, the differential equation for the radiation energy density becomes

$$\frac{\partial u}{\partial t} = \nu \frac{\dot{a}}{a} \frac{\partial u}{\partial \nu} - 3 \frac{\dot{a}}{a} u + j(\nu, t). \qquad (5.157)$$

On changing to the surface brightness defined in equation (5.152) and writing a time-dependent frequency as $\nu(t) \propto a(t)^{-1}$, we have from equation (5.157)

$$\frac{d}{dt} i(\nu(t), t) = \frac{\partial i}{\partial t} - \nu(t) \frac{\dot{a}}{a} \frac{\partial i}{\partial \nu} = -3 \frac{\dot{a}}{a} i(\nu(t), t) + \frac{c}{4\pi} j(\nu(t), t). \qquad (5.158)$$

The integral of this expression gives the surface brightness at the present time, t_o, with expansion parameter $a_o = a(t_o)$, and at observed frequency ν_o,

$$i(\nu_o, t_o) = \frac{c}{4\pi} \int_0^{t_o} dt \left[\frac{a(t)}{a_o} \right]^3 j(\nu_o a_o / a(t), t). \qquad (5.159)$$

This is the surface brightness computed as an integral of the luminosity density along the line of sight. It differs from the usual expression for radiative transfer because the redshift makes the frequency in the argument of the luminosity density a function of time, and the factor $(a/a_o)^3$ converts the integrand to the luminosity per unit comoving volume. A more general treatment takes account of absorption; an example is given in section 7.

It simplifies the units to consider the energy per increment of the natural logarithm of the frequency or wavelength,

$$\frac{di}{d \ln \nu} = \frac{di}{d \ln \lambda} = \nu i_\nu = \lambda i_\lambda, \qquad (5.160)$$

where i_ν and i_λ are the energies per unit increment of frequency and wavelength. Then we can write the surface brightness integrated over a band of frequencies or wavelengths as

$$i = \int \nu i_\nu d \ln \nu = \int \lambda i_\lambda d \ln \lambda, \qquad (5.161)$$

where $i_\nu = i(\nu, t)$ is the surface brightness per frequency interval and i_λ is the surface brightness per increment of wavelength. That is, the integrated surface brightness is the area under the curve in a graph of νi_ν as a function of $\log \nu$ or $\log \lambda$. In this convention, equation (5.159) is

$$\nu_o i(\nu_o, t_o) = \frac{c}{4\pi} \int_0^{t_o} dt \left[\frac{a(t)}{a_o} \right]^4 \nu(t) j(\nu(t), t), \qquad (5.162)$$

where $\nu(t) = \nu_o a_o / a(t)$. The integrated energy density in the absence of sources and absorption thus scales as $u \propto a(t)^{-4}$. This figures in the discussion of the cosmic background radiation in the next section.

In normal present-day L_* galaxies, the luminosity L_λ is close to constant from $\lambda \sim 4500$ Å to well into the infrared. In the approximation that L_λ is flat, and neglecting the evolution of the luminosities of the galaxies, the luminosity density scales as

$$j(\nu, t) \propto a(t)^{-3} \nu^{-2}, \tag{5.163}$$

and equation (5.159) is

$$i(\nu_0, t_0) = \frac{c j(\nu_0, t_0)}{4\pi} \int_0^{t_o} \left(\frac{a}{a_o}\right)^2 dt. \tag{5.164}$$

At observed wavelengths ~ 5000 Å it may be a reasonable approximation to neglect the evolution of the galaxies (unless light from a luminous blue phase in young galaxies is redshifted into this band) because when L_λ is flat or decreasing to the blue the integral in equation (5.164) converges before the redshift is very large. For the same reason, the integral is insensitive to the density parameter. For example, in the Einstein-de Sitter model, where $a \propto t^{2/3}$, the integral is $H_o^{-1}/3.5$, while in a very low density open model with negligible cosmological constant, where $a \propto t$, the integral is $H_o^{-1}/3$.

The luminosity density in equation (5.143) is expressed in units of the solar luminosity at the blue B band centered at about 4400 Å, where the luminosity of the Sun is (eq. [3.51])

$$\nu_0 L_{\nu_o} = 2.4 \times 10^{33} \text{ erg s}^{-1}. \tag{5.165}$$

With this conversion factor, equation (5.143) for j in equation (5.164) for the surface brightness is

$$\nu_0 i_{\nu_o} = 2 \times 10^{-6} e^{\pm 0.3} \text{ erg cm}^{-2} \text{ s}^{-1} \text{ ster}^{-1}. \tag{5.166}$$

A more detailed analysis showing the sensitivity to the galaxy spectra is given by Code and Welch (1982).

As Felten (1966) pointed out, it is no surprise that equation (5.166) is independent of Hubble's constant, because it represents the sum of observed galaxy flux densities, with an extrapolation to higher redshifts. The direct approach sums the contribution to $\nu_0 i_{\nu_o}$ from the observed counts of galaxies as a function of apparent magnitude. In Tyson's (1990) sum over observed galaxy counts, the dominant contribution to the sum is at $m \sim 24$, the counts extend to $m \sim 27$, and the sum is

$$\nu_0 i_{\nu_o} = 3.0^{+0.6}_{-0.3} \times 10^{-6} \text{ erg cm}^{-2} \text{ s}^{-1} \text{ ster}^{-1}, \tag{5.167}$$

at $\lambda = 4500$ Å. Within the uncertainties this is consistent with equation (5.166), but one might expect that Tyson's sum is higher than what follows from the method in equation (5.166), which ignores the galaxy evolution that is expected to have made galaxies brighter and bluer in the past.

As in equation (5.166), the surface brightness from galaxy counts in equation (5.167) represent the luminous matter in objects compact and bright enough to be detected as galaxies. The test for light between the galaxies is to compare these results to the difference between the mean net sky brightness and what can be accounted for in local sources (in the atmosphere, the Solar System, and the Milky Way). The upper bounds on the extragalactic part are surveyed by Mattila (1990); the bound at $\lambda \sim 5000$ Å is

$$\nu_0 i_{\nu_0} \lesssim 2 \times 10^{-5} \, \mathrm{erg \, cm^{-2} \, s^{-1} \, ster^{-1}} \,. \tag{5.168}$$

Comparing with equations (5.166) and (5.167), we conclude that the net mean luminosity density could be as much as ten times the value in equation (5.143) for the light from the bright parts of normal galaxies. If this material between the galaxies had the same mass-to-light ratio as for galaxies, it would increase the density parameter in equation (5.150) to $\Omega_* \sim 0.04$, still well below unity. That is, the Einstein-de Sitter model requires that the mass-to-light ratio of the matter outside the bright parts of galaxies is $\mathcal{M}/L \gtrsim 100h$ for consistency with the bound from the light of the night sky.

It is thought that galaxies were more luminous in the past, when they were younger, because the present rate of production of heavy elements in normal galaxies such as the Milky Way is not large enough to yield the observed element abundances in the Hubble time. We get a measure of the integrated luminosity needed to make the heavy elements by noting that the mass fraction released as radiation (subtracting the part lost in neutrinos) in burning hydrogen to helium is

$$\epsilon = 0.007 \,. \tag{5.169}$$

The extra energy released in burning to heavier elements is not large. The net energy released in a galaxy with baryonic mass $\mathcal{M}$ to convert a fraction Z of the hydrogen to heavy elements is

$$H_o^{-1} \langle L \rangle \sim \int L \, dt = \epsilon \mathcal{M} c^2 Z \,. \tag{5.170}$$

If $\mathcal{M}/\langle L \rangle$ is the observed characteristic mass-to-light ratio in equation (5.148), then equation (5.170) (with $L_\odot / \mathcal{M}_\odot = 1.9 \, \mathrm{erg \, s^{-1} \, g^{-1}}$, as in eq. [3.57]) is

$$Z = 0.007 h^{-2} \,. \tag{5.171}$$

The actual production at the present rate is considerably less than this, because much of the luminosity in present-day L_* galaxies comes from stars with masses close to that of the sun, which sequester most of their ashes in white dwarfs and neutron stars. Since the present heavy element abundance is $Z \sim 0.03$, the conclusion is that galaxies had to have passed through a significantly brighter phase when more massive stars were producing the heavy elements.

If this phase of element production is centered on redshift z_e, the contribution to the radiation background is

$$\nu_o i_{\nu_o} \sim \frac{c}{4\pi} \frac{\rho_* Z \epsilon c^2}{1 + z_e}$$

$$= 8 \times 10^{-3} \Omega_* h^2 (1 + z_e)^{-1} \text{ erg cm}^{-2} \text{s}^{-1} \text{ ster}^{-1}. \tag{5.172}$$

This assumes the mass fraction $Z = 0.03$ of baryonic matter with mean mass density $\rho_* = \Omega_* \rho_{\text{crit}}$ has burned from hydrogen to heavier elements, releasing a fraction $\epsilon = 0.007$ as radiation energy. The present energy density in this radiation is reduced by the factor $1 + z_e$, because the redshift reduces the energy $h\nu$ of each photon by this factor. (This energy loss effect is discussed in the next section.) The spectrum of this background radiation depends on the types of stars producing the elements, the effect of dust that may absorb the starlight and reradiate it at longer wavelengths, and the range of cosmological redshifts. Unless such effects spread the energy over many decades in wavelength, we get close to the right value for i_ν at the peak of the spectrum by considering the energy per logarithmic frequency interval.

A production phase at $z_e \sim 10$ would shift the radiation of the massive stars that are efficient producers of heavy elements from emitted wavelengths ~ 1000 to 2000 Å to the infrared at $\lambda_o \sim 1$ to 2 microns. The zodiacal light (a combination of scattered sunlight and sunlight that has been absorbed and reradiated by interplanetary dust grains) has a minimum near 3 microns, with

$$\nu_o i_{\nu_o} \sim 3 \times 10^{-5} \text{ erg cm}^{-2} \text{s}^{-1} \text{ ster}^{-1}. \tag{5.173}$$

Matsumoto (1990) shows that measurements of the extragalactic background at this level are feasible. With $1 + z_e \sim 10$ this would bound the mass density of the heavy element producing material at $\Omega_* \lesssim 0.03 h^{-2}$, larger than the density of luminous matter (eq. [5.150]), but close to the baryon density in the model for helium production to be discussed in the next section.

There is evidence that a significant fraction of the heavy elements was produced at redshifts well below $z = 10$. Songaila, Cowie, and Lilly (1990) note that many of the galaxies seen at redshifts on the order of unity have spectra L_ν (luminosity per frequency interval) that are close to flat, the form that might be expected from the light from a young star population. This flat spectrum would be expected to be suppressed at frequencies larger than the Lyman limit at $\lambda_L = c/\nu_L = 912$ Å, because at shorter wavelengths photoionization in the stellar atmosphere strongly absorbs the radiation. Thus a reasonable model for the spectrum of one of these galaxies is

$$L_\nu = L/\nu_L, \quad \nu < \nu_L,$$
$$= 0, \quad \nu > \nu_L. \tag{5.174}$$

The mass density scales as $n \propto a(t)^{-3}$, so equation (5.159) is

$$i_{\nu_o} = \frac{c}{4\pi\nu_L} \int \mathcal{L} \, dt . \tag{5.175}$$

The integrand is the mean luminosity per unit volume, $\mathcal{L}$, in these objects, scaled to the present epoch. It is integrated over the luminosity history, if the observations are at long enough wavelengths so that the integral is not affected by the frequency cutoff in equation (5.174). This integral is

$$\int \mathcal{L} \, dt = \rho_1 Z \epsilon c^2 , \tag{5.176}$$

if the fraction Z of the present mean mass density ρ_1 is converted to heavy elements. From the counts of these objects, Songaila, Cowie, and Lilly (1990) estimate that their contribution to the mean surface brightness of the sky at one micron wavelength is

$$\nu_o i_{\nu_o} = 4 \times 10^{-7} \, \text{erg cm}^{-2} \text{s}^{-1} \text{ster}^{-1} . \tag{5.177}$$

Collecting, we see that the mean mass density of the heavy elements produced by these objects is

$$\rho_1 Z = 3 \times 10^{-34} \text{g cm}^{-3} . \tag{5.178}$$

By comparison, the mass density in heavy elements in the bright parts of galaxies, with equation (5.149) and $Z = 0.03$, amounts to

$$\rho_* Z = 3 \times 10^{-33} \text{g cm}^{-3} . \tag{5.179}$$

The observed flat-spectrum galaxies approximated by equation (5.174) cannot have redshifts greater than about $z \sim 3$, because the Lyman limit is not seen redshifted into the visible (Guhathakurta, Tyson, and Majewski 1990). And we see that these relatively low redshift galaxies make an appreciable contribution to the observed amount of heavy elements. It is not clear whether the heavy elements from these low redshift sources end up in the luminous galaxies that are counted in ρ_*, or in dwarf galaxies, or perhaps even in dark halos. Resolving the issue, and establishing the timetable for production of the heavy elements in L_* galaxies, is a matter of current debate, some parts of which are reviewed in part 3.

6.   The Thermal Cosmic Background Radiation

The 2.7K Background Radiation

At wavelengths in the range of millimeters to centimeters, the extraterrestrial electromagnetic radiation background is dominated by an isotropic component, the cosmic background radiation, or CBR. The isotropy suggests the CBR is a sea of radiation that uniformly fills space. This would mean an observer in any other galaxy would see the same intensity of radiation, equally bright in all directions, consistent with the cosmological principle. The spectrum is very close to a thermal Planck form at a temperature near 3 K, suggesting the radiation has almost completely relaxed to thermodynamic equilibrium. This could not have happened in the universe as it is now, because the universe is optically thin to radio radiation, as evidenced by the observation of radio sources at redshifts well above unity (as in fig. 5.3). That is, the CBR moves across the Hubble length of the present universe with little change apart from that caused by the expansion. Therefore, the standard interpretation is that the CBR is left over from early epochs when the expanding universe was dense and hot enough to have relaxed to thermal equilibrium, filling space with a sea of blackbody radiation. We will see that when the interaction with matter is negligible a homogeneous expansion of the universe causes the radiation to cool, as in an adiabatic expansion process, preserving a thermal spectrum. That is, a thermal radiation spectrum remains thermal even in the absence of the speck of dust one usually invokes to make the expansion reversible. The spectrum tends to remain close to blackbody when matter and radiation interact, because the heat capacity of the radiation is very much larger than that of the matter. Thus the expected signature of this blackbody radiation left over from the early hot and dense phase of the expanding universe is its nearly thermal spectrum.

Figure 6.1 shows the background radiation spectrum at wavelengths of 500 microns to 5 mm, from the COBE satellite measurement (Mather et al. 1990). The line through the points is a thermal Planck blackbody spectrum. A Planck function has just one free parameter, the temperature T_0. The best fit to this parameter from the COBE measurements and from a rocket measurement made at very nearly the same time (Gush, Halpern, and Wishnow 1990) is the quoted temperature from the rocket experiment,

$$T_o = 2.736 \pm 0.017 \, \text{K} . \qquad (6.1)$$

Measurements at longer wavelengths, from a summary by Wilkinson (1992), are

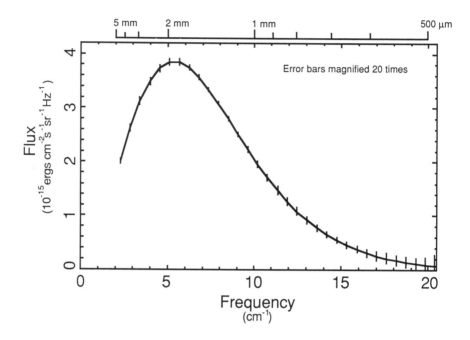

Figure 6.1. Spectrum of the cosmic background radiation (courtesy of the COBE Science Working Group). The frequency is expressed in terms of the reciprocal wavelength. The midpoints of the vertical bars show a spectrum measured by the Far Infrared Spectrometer (FIRAS) on the NASA Cosmic Background Explorer (COBE) satellite. The data are from observations of a region of particularly low emission from galactic dust ("Baade's Hole"), for a total exposure of 100 minutes. Only data from one particular channel ("Left/Low") and mirror mechanism scan mode ("Short/Slow") were used. The sizes of the bars are ±20 standard deviations, as 1σ error flags would be invisible. The solid curve shows the expected flux from a pure blackbody spectrum, with temperature 2.730 K. The precision of the absolute temperature measurement, 0.06 K, is limited by the current understanding of possible systematic errors in the FIRAS thermometry.

shown in figure 6.2. As in the last figure, the thin line is the blackbody spectrum at the temperature in equation (6.1). Within the uncertainties, all these measurements agree with a pure thermal blackbody spectrum. This beautiful result comes out of some three decades of development of methods of measurement. A guide to the methods is given by Uson and Wilkinson (1988).

The CBR is remarkably close to isotropic. The main feature is a dipole ($\propto \cos\theta$) variation in the thermodynamic temperature as a function of position across the

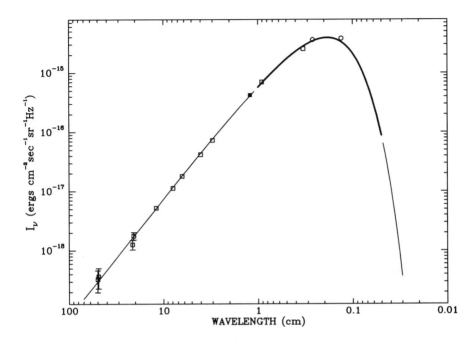

Figure 6.2. Spectrum of the cosmic background radiation (Wilkinson 1992). The scale is chosen to emphasize the measurements in the long wavelength Rayleigh-Jeans part of the spectrum. The COBE measurement from figure 6.1 is shown as the solid curve near the peak of the spectrum, the more precise of the Dicke radiometer measurements are plotted as the squares, and the excitation temperatures of the first two excited levels of the interstellar molecule cyanogen as circles. The thin line is the Planck blackbody spectrum. Where error flags are not shown they are comparable to or smaller than the sizes of the line or symbol. The measurements at long and short wavelengths are limited by emission from the galaxy, which is larger than the extragalactic part at wavelengths longer than 30 cm and shorter than 400 μ. Some elements of these measurements are discussed below.

sky, at an amplitude $\sim 0.1\%$ of the mean. The evidence is that this anisotropy is dominated by the effect of our motion relative to the rest frame defined by the radiation and by the mean redshifts of distant galaxies (table 6.1 later in this section). The quadrupole anisotropy, also measured by COBE, is about one part in 10^5. As discussed in sections 21 and 25, this is in the range expected from the gravitational perturbation to the CBR by the large-scale fluctuations in the mass distribution. At smaller angular scales down to about 20 seconds of arc, the CBR is isotropic to better than one part in 10^4.

The discovery of the cosmic background radiation twenty-five years ago had a profound effect on the directions and pace of research in physical cosmology. Its presence makes the evolving Friedmann-Lemaître picture considerably more credible, because it is difficult to see how the distinctive thermal spectrum could have been produced in the universe as it is at the present epoch. That is, the CBR is considered almost tangible evidence that the universe really did expand from a dense state. This is what led most people to abandon the steady-state cosmology (as will be discussed in section 7). The knowledge of the present radiation temperature of the universe makes the exploration of scenarios for its evolution a good deal more definite. In particular, the straightforward computation of the thermal history of matter as it expands and cools through temperatures $kT \sim 1\,\text{MeV}$ indicates that about three quarters of the baryons end up as hydrogen, with most of the rest becoming helium and small but significant traces of other light isotopes. This agrees with the observed light element abundances as they now are understood, accounting for the otherwise puzzling observation that the helium abundance has a nearly universal value, quite unlike the large variations observed in the heavy elements that are thought to have come from stars.

The CBR also gives us an exceedingly important probe of the history of structure formation in the evolving universe, through the effects of structure on the spectrum and isotropy of the radiation. The perturbation to the CBR by the plasma in a cluster of galaxies has been observed, and will be discussed in section 24. The closely thermal and isotropic nature of the CBR is not inconsistent with the existence of galaxies and the large-scale structure in their distribution, as far as is known, but it certainly constrains the possibilities, and the observed departures from isotropy are an invaluable if still enigmatic clue to how structures formed. A survey of the state of understanding of the lessons commences in section 21.

Blackbody Radiation in an Expanding Universe

The easiest way to see why the expansion of the universe preserves a thermal spectrum is to consider the decomposition of the electromagnetic field into normal modes of oscillation, as illustrated in figure 5.8. At temperature T the occupation number, or mean number $\langle N \rangle = \mathcal{N}$ of photons per mode, is given by the Planck function,

$$
\begin{aligned}
\mathcal{N} &= \frac{1}{e^{\hbar\omega/kT} - 1} \\
&= \frac{1}{e^{hc/kT\lambda} - 1}.
\end{aligned}
\tag{6.2}
$$

Here, Boltzmann's constant is k, Planck's constant is $h = 2\pi\hbar$, the angular frequency is $\omega = 2\pi\nu$, and the wavelength of the mode is $\lambda = c/\nu$. At the observed range of wavelengths the frequency of the CBR is very much larger than the expansion rate H of the universe, so we know from adiabaticity that in the absence of interactions with other fields the quantum number for the mode is conserved, that is, $\mathcal{N}$ is independent of time. As illustrated in figure 5.8, the expansion of the universe stretches the wavelength of the mode as $\lambda \propto a(t)$, where $a(t)$ is the expansion parameter. Thus, when $\mathcal{N}$ is constant the temperature associated with the mode has to scale as

$$T \propto 1/a(t). \tag{6.3}$$

Since this is independent of the wavelength, we see that if the radiation initially is in thermal equilibrium, so that the temperature is the same for all modes, then the mode temperature remains independent of wavelength and the spectrum remains thermal.

The conclusion is that the expansion of the universe makes the CBR temperature scale with redshift as

$$T(t) = T_o a_o / a(t) = T_o(1 + z). \tag{6.4}$$

The redshift is z, where $1 + z = a_o/a$ is the expansion factor (eq. [5.45]). This temperature scaling applies, with minor corrections for the energy taken up in making thermal particle pairs as the temperature gets high enough, back to whatever produced the radiation.

For a gas of nonrelativistic pointlike particles, such as hydrogen atoms, the occupation number is given by the Boltzmann expression,

$$\mathcal{N}_m \propto e^{-p^2/2mkT_m}. \tag{6.5}$$

Here the matter temperature is T_m and the particle mass is m. If the gas is noninteracting, or free, the same mode-stretching argument says the particle momenta vary as $p \propto 1/a(t)$ (eq. [5.43]), so the kinetic temperature of a free gas varies with cosmic time as

$$T_m \propto a(t)^{-2}, \tag{6.6}$$

preserving the thermal Boltzmann form. More generally, the expression for the occupation number of an ideal gas of fermions or bosons is

$$\mathcal{N} = \frac{1}{e^{(\epsilon - \mu)/kT} \pm 1}. \tag{6.7}$$

At thermal equilibrium, the temperature T and chemical potential μ are the same in all modes. In the relativistic limit the particle energy ϵ is proportional to the momentum. Since the expansion of the universe stretches the de Broglie wavelength as $p^{-1} \propto a(t)$, we see that when $\mathcal{N}$ is conserved the temperature and chemical potential in a relativistic gas both scale as $\mu \propto T \propto 1/a(t)$. In the nonrelativistic limit we can subtract the particle rest mass from ϵ and μ. The remaining energy scales as $p^2 \propto a(t)^{-2}$, so when $\mathcal{N}$ is conserved the temperature and the remaining part of the chemical potential scale as a^{-2}.

Central to the interpretation of the CBR is the result that the expansion of the universe keeps a thermal blackbody spectrum thermal. This is a consequence of the fact that the energy is proportional to a power of the momentum. In a gas of diatomic molecules with negligible collisions, the temperature belonging to translational kinetic energy scales as $T \propto 1/a(t)^2$, while the spin temperature remains constant. A second more relevant example is a gas of massive neutrinos, with energy $\epsilon = (p^2 c^2 + m^2 c^4)^{1/2}$. If this gas were thermalized at a temperature well above the neutrino rest mass, that is at $kT \gg mc^2$, and then freely expanded to nonrelativistic velocities, the momentum distribution would remain that of a relativistic gas, and the energy distribution therefore would be markedly different from Maxwell-Boltzmann. This is discussed further in section 18, in connection with dark matter candidates. The expansion of the universe tends to break thermal equilibrium between blackbody radiation and a gas of nonrelativistic particles, because the cooling laws in equations (6.4) and (6.6) are different. The CBR dominates because the heat capacity in the radiation is so much higher (eq. [6.16] below).

Let us recall how the expression for the photon occupation number in equation (6.2) translates into the blackbody radiation spectrum. A convenient way is to use periodic boundary conditions in a box of volume $V = L^3$, with the side L taken to be large compared to wavelengths of interest and small compared to the radius of curvature of space. The electromagnetic field in a single mode of oscillation in the box is proportional to the real part of the plane wave $\exp i\mathbf{k} \cdot \mathbf{r}$. The periodic boundary condition implies that the propagation vector is of the form $\mathbf{k} = 2\pi\mathbf{n}/L$, where $\mathbf{n} = (n_x, n_y, n_z)$ is a triplet of positive or negative integers n_α. The number of plane waves with propagation vector in the range d^3k is then

$$d^3N = \frac{L^3}{(2\pi)^3} d^3k = \frac{V}{(2\pi)^3} k^2 dk d\Omega = \frac{V}{(2\pi c)^3} \omega^2 d\omega d\Omega. \tag{6.8}$$

The second step expresses the element d^3k in polar coordinates, where k is the magnitude of the propagation vector and $d\Omega = \sin\theta d\theta d\phi$ is the element of solid angle for the direction θ, ϕ of $\mathbf{k}$. The last step uses the dispersion relation $\omega = kc$ for radiation, with c the speed of light. (And we will rely on context to distinguish the identical symbols for the propagation vector and Boltzmann's constant.)

The number of photons in the quantization volume V with wave number $\mathbf{k}$ in the range d^3k is the product of the occupation number per mode (eq. [6.2]) and the number of modes, $2 d^3N$, where the factor of two takes account of the two polarization states. The number of photons per unit volume with frequency ω in the range $d\omega$ and moving in any direction (the solid angle $d\Omega$ integrated to 4π) is then

$$n(\omega)d\omega = \frac{1}{\pi^2 c^3} \frac{\omega^2 d\omega}{e^{\hbar\omega/kT} - 1}. \tag{6.9}$$

It is worth pausing to note that this expression offers yet another way to see that the expansion of the universe preserves the thermal spectrum of noninteracting radiation. Suppose at the epoch labeled by expansion parameter a the universe contains a uniform sea of blackbody radiation at temperature T. The number of photons per unit volume with frequency ω in the range $d\omega$ is $dN = n(\omega)d\omega$. At the later epoch labeled by a', each of these photons has been redshifted to frequency $\omega' = \omega a/a'$. We are assuming the radiation is free, and hence photons are conserved, so the number of photons per unit volume varies inversely as the volume, $\propto a(t)^{-3}$. Thus, at epoch a' the number $n'(\omega')$ of photons per unit volume and frequency interval satisfies

$$dN = n'(\omega')d\omega' = n(\omega)d\omega(a/a')^3, \qquad \omega' = \omega a/a'. \tag{6.10}$$

If $n(\omega)$ is given by the blackbody function (6.9), this equation says $n'(\omega')$ also is a blackbody function, with the redshifted temperature $T' = Ta/a'$.

The blackbody radiation energy $u(\omega)$ per unit volume and unit frequency interval is the product of the photon number density (6.9) and the energy $\hbar\omega$ per photon,

$$u(\omega)d\omega = \frac{\hbar}{\pi^2 c^3} \frac{\omega^3 d\omega}{e^{\hbar\omega/kT} - 1}. \tag{6.11}$$

This is the Planck blackbody radiation spectrum. Another way to express it is as the spectral surface brightness, $i(\nu)$, which is the energy flux per unit area, solid angle, and frequency interval ν. This follows by dividing equation (6.11) by 4π, to get the energy per steradian, and multiplying by c to get the energy flux (as in eq. [5.152]). On changing to the circular frequency $\nu = \omega/(2\pi)$ we get

$$i(\nu)d\nu = \frac{2h}{c^2} \frac{\nu^3 d\nu}{e^{h\nu/kT} - 1}. \tag{6.12}$$

The limit $h\nu \ll kT$ is the Rayleigh-Jeans law

$$i(\nu) = 2kT\nu^2/c^2. \tag{6.13}$$

The integral of equation (6.11) over frequencies is the blackbody radiation energy per unit volume,

$$u = \int_0^\infty u(\omega)d\omega = \frac{\pi^2}{15} \frac{(kT)^4}{(\hbar c)^3} = a_B T^4, \tag{6.14}$$

$$a_B = 7.56 \times 10^{-15} \text{ erg cm}^{-3} \text{ K}^{-4}.$$

The second step follows from the change of variables $x = \hbar\omega/kT$, with the dimensionless integral

$$I_- = \int_0^\infty \frac{x^3\, dx}{e^x - 1} = \frac{\pi^4}{15}. \tag{6.15}$$

Equation (6.14) is the Stefan-Boltzmann law. We will use the standard symbol, a, for Stefan's constant when it is easily distinguished from the expansion parameter.

The heat capacity of the CBR at fixed volume is $4aT^3$. If the matter consists of atomic hydrogen, the ratio of its heat capacity to that of the CBR is

$$\frac{1.5 n_B k}{4aT^3} = 4 \times 10^{-9} \Omega h^2, \tag{6.16}$$

where n_B is the mean number density of hydrogen atoms (eq. [5.68]). This ratio is independent of redshift. Its small value explains why the CBR might be expected to have a closely thermal spectrum: at high redshifts, where the interaction between matter and radiation is appreciable, the matter relaxes to the radiation temperature, because the radiation has by far the higher heat capacity, and at thermal equilibrium the radiation spectrum remains thermal no matter how strong the interaction.

The ratio of the mean mass density in matter (eq. [5.67]) to the mass density in the CBR at the temperature in equation (6.1) is

$$\frac{\rho_b c^2}{aT^4} = 4.0 \times 10^4 \, \Omega h^2 (1+z)^{-1}. \tag{6.17}$$

The redshift dependence follows because the energy density in the CBR varies as $(1+z)^4$ (because $T \propto 1/a(t)$ in eq. [6.4]), one power faster than for the nonrelativistic matter. With the lower bound on the mass density parameter Ω in equation (5.150), we see that at the present epoch the energy density in the radiation is a small fraction of the total. It follows that when the redshift is not too large the energy available from annihilation of mass by nuclear burning (or perhaps by the more efficient process of accretion by black holes) is sufficient to produce an appreciable perturbation to the radiation temperature. Whether this can have happend depends on whether there is a way to transfer the energy to the CBR while keeping the spectrum close to thermal. Some details on how this might happen are discussed in the next section and in section 24.

Two features in the standard interpretation of the CBR tend to be confusing. We have already noted in section 1 that the name for the standard model, the hot big bang, is misleading, for a bang suggests a localized explosion. In the standard picture the source of the CBR is not localized; the radiation is uniformly and isotropically distributed throughout the space we can observe. This agrees with the fact that the radiation is equally bright in all directions. The number density of photons is decreasing with time as $a(t)^{-3}$, not because photons are leaving the universe—there is nowhere else to go—but because the volume of space is increasing as $a(t)^3$.

The second confusing point is the nature of energy balance in the CBR. Since the energy density in blackbody radiation varies as the fourth power of the temperature, the expansion of the universe causes the radiation energy density to evolve as $\rho_\gamma \propto T^4 \propto a(t)^{-4}$. As indicated in equation (6.17), this is faster by one power of the expansion parameter than for the mass density in a gas of nonrelativistic particles such as baryons (eq. [5.19]). The number density of photons varies as $a(t)^{-3}$, as for baryons, but there is an extra factor of $1/a(t)$ for the redshift of the mean energy per photon. Another way to get the cooling law is to recall that the pressure of the radiation is $p_\gamma = \rho_\gamma/3$. With this equation of state, the local energy conservation equation (5.16) is

$$\frac{d\rho_\gamma}{dt} = -3(\rho_\gamma + p_\gamma)\frac{\dot{a}}{a}$$

$$= -4\rho_\gamma \frac{\dot{a}}{a}.$$

(6.18)

The solution is

$$\rho_\gamma \propto a(t)^{-4},$$

(6.19)

consistent with the Stefan-Boltzmann law (6.14) and the redshift law (6.4) for the radiation temperature. We see that the faster decrease of ρ_γ compared to the mass density of a nonrelativistic gas is the result of the pressure work done by the expanding radiation. However, since the volume of the universe varies as $a(t)^3$, the net radiation energy in a closed universe decreases as $1/a(t)$ as the universe expands. Where does the lost energy go? Since there is no pressure gradient in the homogeneously distributed radiation, the pressure does not act to accelerate the expansion of the universe. (The active gravitational mass due to the pressure has the opposite effect, slowing the rate of expansion, as indicated in eq. [5.15]). The resolution of this apparent paradox is that while energy conservation is a good local concept, as in equation (6.18), and can be defined more generally in the special case of an isolated system in asymptotically flat space, there is not a general global energy conservation law in general relativity theory.

Discovery

The history of the discovery and interpretation of the CBR is worth considering as an example of the curious paths progress in science can take.

Lemaître was the first to speculate on the physics and possible observable remnants of the very early stages of expansion of the universe. He imagined a hot beginning: "The evolution of the world can be compared to a display of fireworks that has just ended: some few red wisps, ashes and smoke. Standing on a well-chilled cinder, we see the slow fading of the suns, and we try to recall the vanished brilliance of the origin of the worlds" (Lemaître 1931e). This is readily interpreted as a description of the standard hot big bang model, but Lemaître's candidate for the remnant radiation from high redshifts was cosmic rays rather than a millimeter-wavelength thermal background.

Tolman introduced the idea of the thermodynamic history of an expanding universe. He showed that the expansion of the universe cools blackbody radiation while keeping the spectrum thermal (Tolman 1934, § 171).

The search for the origin of the chemical elements led people to consider the possibility that matter passed through a phase dense and hot enough to have promoted nuclear reactions that could have built up the elements. It was considered that this might have happened in stars, but there also were discussions of the possibility that the nuclear reactions occurred during the early dense epochs of an expanding universe. Chandrasekhar and Henrich (1942) concluded that if matter had relaxed to thermal equilibrium at a density $\sim 10^7 \, \mathrm{g \, cm^{-3}}$ and temperature $\sim 10^{10} \, \mathrm{K}$, and if the abundances had been frozen in at that point because of the rapid expansion and cooling of the universe, then the relative abundances of the lighter elements would agree reasonably well with cosmic abundances. Chandrasekhar and Henrich noted that their solution does considerably underestimate the abundances of the heavier elements and that, for that matter, the theory is not very satisfactory because it seems unrealistic to suppose the material in an expanding universe makes a discontinuous break from thermal equilibrium at one epoch to a frozen set of abundances immediately thereafter.

Gamow (1942, 1946) emphasized that the thermal equilibrium model is questionable because the high mass density in the early universe causes a rapid rate of expansion (eq. [5.18]). He argued that an analysis of the element abundances that would be left over from the early universe really involves a dynamic rather than equilibrium calculation, taking account of reaction rates in rapidly expanding and cooling material. The first step in this direction was the "$\alpha\beta\gamma$" paper (Alpher, Bethe, and Gamow 1948).[16] The events leading up to and following this paper are described in PG and by Alpher and Herman (1988). In the $\alpha\beta\gamma$ paper Alpher and Gamow suggested that the elements were built up by rapid capture of neutrons (with relaxation by beta decay). In this process, the final abundance of an element would depend on the competition between loss by absorption of neutrons and production by neutron capture by the next lighter nucleus. The analysis used the neutron capture cross sections at energies $\sim 1 \, \mathrm{Mev}$ that had been made public at the end of the Second World War. They found that the general trend of cross section with atomic mass would lead to a reasonable-looking pattern of rel-

---

[16] Bethe's name was added to complete the symmetry.

ative abundances in their building-up process. That still is the case, but the site has been moved from the early universe to exploding stars. It is now thought that element formation in the early universe did not go far beyond helium.

The $\alpha\beta\gamma$ paper ignored the mass density associated with blackbody radiation. Alpher (1948) and Gamow (1948a,b) noted that at the temperatures of interest for element formation, where thermal energies kT are on the order of 1 MeV, the universe would be filled with blackbody radiation (the CBR) with mass density considerably larger than the baryon mass density. Since the density determines the expansion rate (eq. [5.18]), this means that at the nucleosynthesis epoch the radiation temperature could determine the rate of expansion and cooling of the universe. Gamow (1948a,b) gave the now standard order-of-magnitude considerations by which we understand that the present radiation temperature of our universe ought to be a few degrees Kelvin, to allow consistency with the cosmic abundances of the lightest chemical elements. The argument, with the added refinement of the processes that fix the neutron-to-proton abundance ratio, goes as follows. (Some further details are given in the last part of this section.)

Let us extrapolate the expansion of the universe back to redshift $z \sim 10^{10}$, when the temperature was $T \sim 3 \times 10^{10}$ K, and the characteristic photon energy was $\sim kT \sim 3$ MeV. At this epoch, the CBR photons are hard enough to photodissociate complex nuclei, leaving free neutrons and protons. Hayashi (1950) pointed out that the CBR also is hot enough to produce a sea of electron-positron pairs, and neutrino pairs, and that these pairs interact with the neutrons and protons, through neutrino reactions such as

$$e^- + p \leftrightarrow \nu + n, \qquad (6.20)$$

which are fast enough at temperatures $T \sim 10^{10}$ K to produce a thermal abundance ratio, $n/p = \exp -Q/kT$, where Q is the neutron-proton mass difference. The neutrons and protons have a rapid rate for radiative capture,

$$n + p \rightarrow d + \gamma. \qquad (6.21)$$

At high temperatures the reverse reaction breaks up the deuterons as fast as they form. When the temperature has fallen to $T \sim 10^9$ K, there are too few photons hard enough to photodissociate the deuterons, and deuterium can accumulate. Once there is an appreciable deuterium abundance, particle exchange reactions can rapidly burn the deuterium to helium.

Gamow (1948a,b) recognized that a key to the element buildup process is the reaction in equation (6.21). The probability per neutron for completion of the reaction as the universe passes through the epoch $T \sim 10^9$ K should be appreciable, so isotopes heavier than deuterium form in significant amounts, but not very large, so there are significant remnants of deuterium and other light isotopes, consistent with the observations. Thus the Gamow condition is

$$\langle \sigma v \rangle nt \sim 1, \tag{6.22}$$

at the epoch of deuterium formation, $T \sim 10^9$ K. Here σ is the cross section for the reaction (6.21) at relative velocity v, and the brackets indicate a thermal average (which is easy because the product σv is close to constant). The baryon number density is n, and the expansion timescale for the universe is t.

There are three contributions to the expansion rate equation (5.18), representing mass density, space curvature, and Λ. If the parameters are such that the expansion does trace back to a dense state, then the situation at high redshift simplifies because the mass density grows with redshift faster than the other terms, and the mass density in radiation grows faster than the density in nonrelativistic matter (eq. [6.17]). Thus, at $z \gg 10^4$ the expansion rate is well approximated by a radiation-dominated universe with negligible space curvature and Λ,

$$\left(\frac{\dot{a}}{a} \right)^2 = \frac{8\pi G a_B T^4}{3c^2}. \tag{6.23}$$

Since $T \propto 1/a(t)$, the solution is

$$t = \left(\frac{3c^2}{32\pi G a_B T^4} \right)^{1/2}. \tag{6.24}$$

This says the age of the universe at $T = 10^9$ K is $t \sim 200$ s.

With this expression for the time as a function of temperature, and a theoretical estimate of the neutron radiative capture rate coefficient $\langle \sigma v \rangle$, Gamow found that at $T = 10^9$ K the baryon number density ought to be $n \sim 10^{18}$ cm^{-3}. He did not estimate the present temperature, but that is an easy exercise. The temperature and baryon number density scale with the expansion of the universe as $T \propto 1/a(t) \propto n^{1/3}$, so Gamow's estimate gives $T_o \sim 4$ K at the present epoch, $n \sim 10^{-7}$ cm^{-3} (eq. [5.149]). This is as close as could be expected to the observed value in equation (6.1).

Alpher and Herman (1948) corrected some numerical errors in Gamow's calculation and used the result to compute the present temperature predicted by this theory. Their value, $T_o \sim 5$ K, is the first numerical estimate of the present CBR temperature, and again quite close to what is observed.

It is now believed that the element production reactions essentially stop at helium, with trace amounts beyond that, and that the bulk of the heavier elements are produced in stars. This is consistent both with the theory of element production in the early universe and the observation that the oldest stars have low heavy element abundances. There is no known source in stars for the observed abundances of the isotopes of helium or the heavier isotopes of hydrogen; they are thought to have been produced in the hot big bang. The Gamow condition in equation (6.22) still is considered to be one of the keys to the formation of these light elements. Further details of the physics are given in the last part of this section (and summarized in table 6.2).

In the 1950s and early 1960s the Gamow-Alpher hot universe picture certainly was not forgotten. An example is Gamow's 1956 article on the "Importance of Thermal Radiation in Cosmology." However, it was not widely discussed, and attention shifted to element formation in stars. This was in part because, as Alpher (1948) had foreseen, the $\alpha\beta\gamma$ neutron capture process gets hung up at ^{4}He by the instability of ^{5}He, and certainly also because in the steady-state cosmology there never was a time when the universe was hot and dense, so one had to show that the elements formed in the only other hot place one can think of, the stars.

There were some notable early clues to the validity of the Gamow-Alpher picture. One was the growing recognition that helium abundances in stars and gaseous nebulae are surprisingly uniform, suggesting the bulk of the helium was present before the stars formed, and perhaps originated in the hot big bang (Osterbrock and Rogerson 1961; O'Dell, Peimbert, and Kinman 1964). Another was the observed excitation temperature of interstellar cyanogen.

Adams (1941) detected absorption lines of the diatomic molecule cyanogen, CN, in interstellar clouds along the lines of sight to stars, and found that there are lines from the rotationally excited state of CN as well as the ground state. McKellar (1941) remarked that the relative abundance in the ground state and each first rotationally excited state defines an effective excitation temperature T by the Boltzmann equation

$$\frac{n_1}{n_0} = e^{-E/kT} . \tag{6.25}$$

The energy difference between the states is E. If the rotational excitation of the CN molecules were in statistical equilibrium with the background radiation field at the resonance for the transition between ground and first rotationally excited states, $\lambda = hc/E = 2.6$ mm, the parameter T would be the effective background radiation temperature. One expects that the excitation temperature in equation (6.25) in diffuse molecular clouds is fixed by energy exchange with the radiation background, rather than by particle collisions, because the CN molecule has a relatively high dipole moment for emission and absorption of radiation. McKellar found a temperature $T = 2.3$ K, in the range for the Gamow-Alpher hot universe. However, the connection was recognized independently by N. J. Woolf and George B. Field only after the proposed interpretation of the centimeter wavelength background as the CBR. The results of recent applications of this measure of the CBR spectrum are shown as the circles in figure 6.2.

O'Dell, Peimbert, and Kinman (1964) measured element abundances in a planetary nebula in the globular star cluster M15. This cluster is a representative of the oldest known star population. The oxygen abundance in the nebula is markedly lower than the present cosmic mean found in relatively young objects like the

Sun, consistent with the idea that the material in old systems like M15 has not been as much enriched by heavy element production in earlier generations of stars. But the helium abundance in M15 is close to the present cosmic mean. O'Dell et al. noted that the material in a planetary nebula is ejected from a star that is nearing the endpoint of its evolution, and it is conceivable that the helium was produced within the evolving star. However, it would be a surprising coincidence if the star produced just the present cosmic helium abundance. Hoyle and Tayler (1964) reviewed the evidence for a nearly universal helium abundance in young stars and old. They noted that this universal abundance is not what one would expect if helium were formed in the stars that made the heavy elements. Hoyle and Tayler considered two possibilities: the helium was produced in the Gamow-Alpher hot big bang, or in an early generation of supermassive stars. The latter "little bang" picture could produce element abundances similar to those coming out of a hot big bang, because a supermassive star is unstable and tends to expand so rapidly that element formation need not proceed much beyond helium.

The direct evidence of a hot big bang would be a detection of the blackbody radiation. This was recognized independently and at about the same time by Robert Dicke in Princeton and by the group around Yakov Zel'dovich in Moscow. The latter work was summarized in a review paper by Zel'dovich (1965). Smirnov (1964) had rediscussed the Gamow condition; he arrived at $T_o \sim 1$ K to 10 K at baryon density $\rho_B = 1 \times 10^{-30}$ g cm^{-3}, about the present value. Doroshkevich and Novikov (1964) pointed out that one could find a very useful bound on the present radiation temperature from published reports of a Bell Telephone Laboratories radio telescope in Holmdel, New Jersey (Ohm 1961).

The Holmdel telescope was used in the Echo satellite experiment as part of the first trials of communication by means of satellites. Detailed studies of the noise properties of the telescope plus receiver yielded a noise excess that was taken to be ground radiation picked up by the back lobes of the antenna. Ohm's report indicated that an excess system temperature "not otherwise accounted for" is "somewhat larger than the calculated temperature expected from back lobes measured on a similar antenna." D. T. Wilkinson's judgment is that a radio astronomer looking for evidence for an isotropic radiation background such as the CBR might have considered this an encouraging sign, because it is hard to distinguish radiation entering the antenna by the back lobes from an isotropic background. Doroshkevich and Novikov interpreted the description of the noise properties of the telescope to mean that the cosmic background temperature could be no greater than about ~ 1 K. At this value for the present CBR temperature, the computed helium abundance seemed unreasonably high. That led Zel'dovich (1965) to turn briefly to a cold big bang cosmological model. It is now recognized that the excess noise in the Holmdel telescope is extraterrestrial—the thermal cosmic background radiation.

The Bell Telephone Laboratories instrument is a radiometer, a device Dicke invented as part of research during the Second World War at the MIT Radiation

Laboratory. Dicke (1946) used this device to measure centimeter wavelength radiation from atmospheric water vapor and from the Sun and Moon, and he placed an upper limit $\sim 20 \, K$ on the temperature of isotropic "radiation from cosmic matter" (Dicke et al. 1946). In 1964 Dicke proposed the construction of a radiometer that might be capable of detecting the blackbody radiation in a hot big bang universe. By this time he had forgotten about his 1946 measurement. And he was unaware of the problem with excess noise in a Dicke radiometer in Holmdel, some 30 miles away.

Dicke was motivated not by element production in the Gamow-Alpher theory (about which he had also forgotten), but by element destruction in an oscillating universe. In a closed model with negligible cosmological constant, the expansion of the universe eventually is stopped (by the gravitational deceleration of the matter), and the universe collapses back to a dense state. Many have speculated that physics outside the standard model causes a bounce producing a new expanding phase; Lemaître (1933) considered this an attractive picture that recalled to him the legendary phoenix. Dicke noted that if our universe is expanding from a bounce from a collapsing phase in which much of the hydrogen had been burned to heavier elements in stars, then there must be some provision for the conversion of the heavy elements back to hydrogen for a fresh generation. When hydrogen in a star is burned to heavier elements, the energy released is radiated as starlight, with photon energies on the order of one electron volt. The heavy elements from the last cycle would be produced when the universe was at a low density, perhaps on the order of the present value. As we have seen, the expansion of the universe lowers the energies of these photons. The same process during a subsequent collapse of the universe increases the photon energies, so that when the collapse brings the mean density back to the present value there is again a sea of photons with energies comparable to starlight, $\epsilon \sim 1 \, eV$. Further collapse eventually brings the typical photon energy to $\epsilon \sim 1 \, MeV$. At this point, the photons are hard enough to evaporate stars and photodissociate heavy elements, producing fresh hydrogen for the next cycle. This uses only a small fraction of the radiation, for the burning of each proton to heavier elements releases a few MeV of binding energy as some 10^6 starlight $\sim 1 \, eV$ photons. In the photodissociation phase only a few of these photons, now blueshifted by the contraction of the universe to gamma ray energies, are needed to release the proton again. The net gain is some $\sim 10^6$ gamma ray photons per proton.[17] In other words, the burning and photodissociation of the hydrogen by irreversible processes has created entropy. It is this entropy, in thermalized radiation and cooled by the expansion of the universe after the bounce, that Dicke proposed looking for.

At Dicke's suggestion Peter Roll and David Wilkinson at Princeton University

---

[17] There are 10^8 to 10^9 photons per baryon in our universe, as shown in equation (6.58). We see that if the bounce conserved baryons and photons, this number of photons could accumulate in a few hundred oscillations of the universe.

built a radiometer capable of detecting this radiation, and he assigned to me the task of thinking of the possible significance of a detection or nondetection. This led to a rediscovery of the Gamow condition for light element production, and to the physics of galaxy formation. I was given leave to present a colloquium on this work at The Johns Hopkins Applied Physics Laboratory in Baltimore, it being reckoned that there was no chance for the development of competition with the Roll-Wilkinson experiment already under construction.

A few points from this talk are of some historical interest. Figure 6.3, which was prepared for the talk, shows bounds on the extragalactic radiation background. The vertical axis is the surface brightness per logarithmic frequency interval, $\nu i(\nu)$ (eq. [5.160]). Thus if the graph were replotted with a linear vertical scale, keeping the logarithmic frequency axis, the area under the curve would be proportional to the net energy density. The large rectangle is placed in the range from millimeter to micron wavelengths, which was the largest gap in the measured surface brightness of the sky. The area under the rectangle is the critical Einstein-de Sitter mass density (eq. [5.67]). A Planck spectrum at $T = 40\,\mathrm{K}$ would have the same mass density, but a thermal spectrum this hot is ruled out by the Dicke measurement from 1946 at 6-cm wavelength, which is plotted as a circle with down-pointing arrow. Even the tighter spectrum indicated by the rectangle could be excluded, because it was known that the cosmic ray background extends to events at energies greater than $10^{19}\,\mathrm{eV}$ that are thought to be protons. As will be discussed in more detail below, a proton at this energy is not appreciably deflected by the magnetic field in the Milky Way. Since these particles move in nearly straight lines through the galaxy, and their arrival directions show no preference for the plane of the Milky Way, they are thought to be extragalactic. In the rest frame of a $10^{19}\,\mathrm{eV}$ proton, the background photons in the right-hand side of the rectangle are energetic enough for photopion production,

$$p + \gamma \rightarrow p + \pi^0,\ n + \pi^+, \tag{6.26}$$

and at the background level indicated by the rectangle, the mean free path of a $10^{19}\,\mathrm{eV}$ cosmic ray proton for this reaction is less than the size of the galaxy. Thus it seemed clear that there is no reasonable chance that the cosmic background radiation makes a cosmologically interesting contribution to the present mean mass density. (Mean free paths for energetic particles in the observed thermal cosmic background radiation are discussed later in this section.)

If the expanding universe started out cold, with vanishing lepton number, the matter would start out as neutrons, because in the dense early universe the degeneracy energy of the electrons would force them onto the protons. As the neutrons decayed, the protons would capture other neutrons, leaving a hydrogen-poor universe, which is unacceptable. This could be avoided by postulating a degenerate neutrino sea, or else a hot universe. I argued that in the latter case the background temperature could be as high as $T_o \sim 10\,\mathrm{K}$, which would be well within reach of

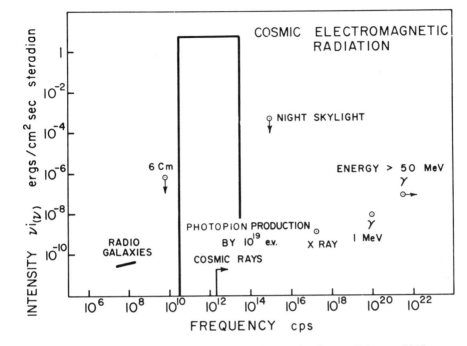

Figure 6.3. Spectrum of the cosmic background radiation, February 1965.

the Roll-Wilkinson experiment. Word of the experiment spread from Ken Turner, who was at the Baltimore colloquium, to Bernie Burke, who with Turner was at the Carnegie Institution of Washington (then a center for radio astronomy), and from there to Arno Penzias and Bob Wilson at the Bell Telephone Laboratories in Holmdel. They were planning to use the Holmdel telescope for radio astronomy but were unable to understand the excess noise known for some time to be present in this instrument. A meeting of the Holmdel and Princeton groups led to the conclusion that the noise excess very likely is extraterrestrial. Since the noise excess was known to be close to isotropic, it at least meets the first condition for the thermal background in a hot big bang cosmology. Thus the "excess antenna temperature," $T_o = 3.5 \pm 1.0\,$K at 7-cm wavelength, was reported (Penzias and Wilson 1965), and interpreted as the "cosmic blackbody radiation" (Dicke et al. 1965).

The antenna temperature is a conventional way to express the background radiation surface brightness $i(\nu)$ through the Rayleigh-Jeans law (eq. [6.13]). The key signature of the hot big bang radiation is that $i(\nu)$ is close to a thermal blackbody spectrum. Figure 6.4 shows the situation shortly after the discovery, in an illustration I prepared for a talk at the American Physical Society. The straight line on the left is the extrapolated contribution of known radio sources, with slope similar to the diffuse synchrotron radiation from our galaxy. The discovery point of Pen-

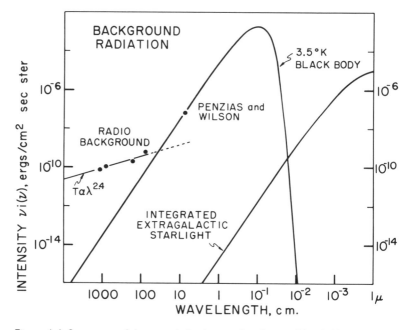

Figure 6.4. Spectrum of the cosmic background radiation, March 1965.

zias and Wilson at 7.35-cm wavelength is above this extrapolation and well above the expected integrated contribution from starlight from galaxies. That is, it is a distinctly new component of the background. Since a thermal spectrum is fixed by a single parameter, the temperature, the Penzias and Wilson measurement predicts the background surface brightness as a function of wavelength, within the limits set by the accuracy of the measurement, if the CBR interpretation is correct. The Roll-Wilkinson radiometer detects radiation at 3.2-cm wavelength. Their result, obtained a few months later, was $T_o = 3.0 \pm 0.5$ K. The consistency with the Penzias-Wilson measurement, at roughly twice the wavelength, was a dramatic positive test of the hot big bang interpretation (Roll and Wilkinson 1966). The present state of the tests shown in figures 6.1 and 6.2 is the product of a long history of difficult innovative measurements.

When it was proposed that the radio background at centimeter wavelengths is blackbody radiation left over from the dense early stages of expansion of the universe, and the spectrum was not well known, it was natural to consider other possible interpretations. The CBR energy density can be compared to that of other local fields. The luminosity of the Milky Way is $L = 10^{10} L_\odot$, so at our position near the outskirts, at distance $r \sim 8$ kpc from the center, the energy density is

$u_* \sim L/(4\pi r^2 c) \equiv aT_*^4$. This defines an effective temperature for the starlight; the result is $T_* = 2\,\mathrm{K}$. The interstellar magnetic field is $B \sim 10^{-6}$ Gauss. If we again define an effective temperature by the energy density, $B^2/8\pi = aT_B^4$, we get $T_B = 2\,\mathrm{K}$. The curious coincidences of these two numbers with the CBR temperature might have suggested a very local origin for the background radiation. However, we know the Milky Way is transparent at centimeter wavelengths, because extragalactic objects are visible in all parts of the sky. Since we are located toward the edge of the Milky Way it is hard to imagine how radiation originating within it could be even remotely close to isotropic, while the CBR is strikingly smooth.

When the measurements of the CBR were mainly at long wavelengths, where the thermal spectrum is the Rayleigh-Jeans form $i(\nu) \propto \nu^2$, it was natural to ask whether the extragalactic radio sources might add up to this power law behavior (Narlikar and Wickramasinghe 1968). However, it seems quite unreasonable to imagine that the integrated background from radio sources could conspire to produce the remarkably close approximation to a thermal spectrum shown in figures 6.1 and 6.2.

At low redshifts there is enough energy from nuclear burning of matter in stars to produce the CBR (eq. [6.17]). However, to convert starlight to a thermal spectrum by statistical relaxation, the universe would have to be optically thick at the Hubble distance, which is inconsistent with the observations of high redshift galaxies shown in the redshift-magnitude diagram in figure 5.3. A numerical example of this point is worked out in the next section for the steady-state cosmology. It is considerably easier to imagine that the CBR was made out of starlight if one goes to an evolving initially cold universe, where the radiation is produced and thermalized at moderately high redshifts (Layzer and Hively 1973; Rees 1978). This eliminates the problem with the transparency of the universe at $z \sim 1$, and the higher density and shorter CBR wavelengths ease the problem of finding candidate dust grains that could make the universe opaque. The epoch at which the CBR could have been produced out of starlight is bounded by the energy available in the matter (eq. [6.17]); a favored redshift is $z \sim 100$. The most recent computations prior to the precision satellite and rocket measurements illustrated in figure 6.1 predicted that in this picture there would be departures from a thermal spectrum at a level that now is ruled out (Hawkins and Wright 1988). It would be useful to see whether there is any related way the cold big bang picture can be arranged to fit the new spectrum measurements.

As we have discussed, another path to the idea that the early universe was hot comes from the problem of accounting for the abundances of the light elements. The first detailed computations of element production in the hot big bang, following the discovery of the CBR, are in Peebles (1966) and Wagoner, Fowler, and Hoyle (1967). The present state of the standard model for light element production is illustrated in figure 6.5. The computation, which is described in the last

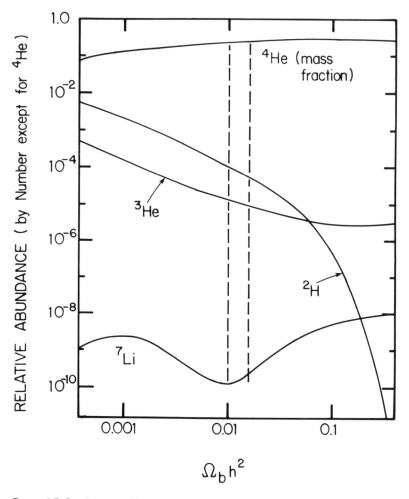

Figure 6.5. Production of light elements in the hot big bang (Schramm 1991).

part of this section, assumes that the distributions of baryons and radiation are close to homogeneous at the epoch of light element production, and it assumes there are negligibly small lepton numbers (meaning the neutrinos are not degenerate). Since the present CBR temperature is accurately known, the free parameters are the present mean baryon number density and the net mass density in relativistic components such as neutrinos, gravitational radiation, and the CBR. This relativistic part dominates the mass density and hence the expansion rate during light element production. The abundances in the figure are computed under the assumption that the additional relativistic energy density is that of the three families of neutrinos associated with the known three charged leptons, e, μ, and τ. The horizontal axis is proportional to the present mean number density of baryons. The vertical dashed lines indicate the range of baryon densities for which the predicted abundances agree with the observations (in some cases with mild corrections for additional production in stars). The range is

$$n_B = 1.4 \pm 0.3 \times 10^{-7} \text{ nucleons cm}^{-3},$$
$$\Omega_B = 0.013 \pm 0.003h^{-2}$$

(6.27)

(Walker et al. 1991). The wanted baryon density is greater than what is observed in the bright stellar matter in galaxies (eq. [5.150]), but it is easy to imagine the rest is in low mass stars in dark halos, or in gas. For example, there is a high density of plasma in rich clusters. Thus the fact that the wanted value of Ω_B is in a reasonable range is a nontrivial success. The present evidence is that this picture gives an observationally successful account of the otherwise puzzling systematics of the light element abundances. This important result, from the group at Chicago that produced figure 6.5 and from other observational and theoretical groups, is a central step in the establishment of the hot expanding relativistic cosmology as the standard model.

One might draw many lessons from the history of these discoveries. Perhaps the most significant is that the physical world presented us with a considerable variety of hints to a hot evolving universe that eventually were recognized. The abundance of hints is one of the things that encourages belief that the hot universe really is a useful approximation to reality.

Aether Drift: Observation

Blackbody radiation can appear isotropic only in one frame of motion. An observer moving relative to this frame finds that the Doppler shift makes the radiation hotter than average in the direction of motion, cooler in the backward direction. That means the CBR acts as an aether, giving a local definition for preferred motion. This does not violate relativity; it always is possible to define motion relative to something, in this case the homogeneous sea of radiation. The pattern of galaxy redshifts similarly can only be seen to be isotropic relative to a preferred frame of motion.

The analysis in the next subsection shows that an observer moving at speed v relative to the preferred frame in which blackbody radiation is isotropic sees that the thermodynamic temperature of the radiation is a function of direction relative to the motion. To first order in v/c the anisotropy is dipole,

$$T(\theta) = T_o(1 + v \cos \theta).$$

(6.28)

Here θ is the angle between the line of sight and the direction of motion. The CBR has a dipole anisotropy with this property. The effect was first convincingly seen by Conklin (1969) and Henry (1971). The present result, in table 6.1, is an average over many elegant experiments, as summarized by Smoot et al. (1991 and 1992).

The standard interpretation is that the CBR dipole anisotropy is the result of our peculiar motion caused by the gravitational field of the irregularities in the mass distribution. There is the possibility that the anisotropy instead is the effect of a large-scale inhomogeneity in the distribution of the radiation. Fluctuations in the mass density on scales large compared to the Hubble length would produce a quadrupole anisotropy, rather than dipole, because a constant gradient in the hypersurface of constant density is eliminated by a velocity transformation that tilts the surfaces of constant time, eliminating the dipole and leaving the second derivatives that produce a quadrupole (Grishchuk and Zel'dovich 1978). If the total mass density were homogeneous, a large-scale gradient in the entropy per baryon (or whatever dominates the mass of the universe now) would produce an anisotropy dominated by the linear term in the power series expansion of the temperature as a function of position, producing a dipole anisotropy, which could fit the observations (Paczyński and Piran 1990; Turner 1991).

In the standard interpretation, the same preferred comoving rest frame is defined by the CBR, the redshift-distance relation for galaxies, and the X-ray background. (For X rays, the dipole anisotropy is called the Compton-Getting effect, after the effect of motion on the distribution of cosmic rays; Compton and Getting 1935.) If, as is usually assumed, our motion relative to this preferred frame is caused by gravity, our motion ought to agree with the gravitational field produced by the fluctuations in the mass distribution. If the CBR and X-ray dipoles were caused by large-scale inhomogeneities in the radiation distribution and sources, one would look for quite different effects in these differently defined frames. The evidence is that the frames are consistent to perhaps 300 km s^{-1} (Aaronson et al. 1986; Shafer and Fabian 1983; Rubin and Coyne 1988). Still tighter tests will be of considerable interest.

The motion of the Solar System relative to the frame in which the cosmic background radiation is isotropic is

$$v_\odot - v_{CBR} = 370 \pm 10 \, \text{km s}^{-1} \quad \text{to}$$
$$\alpha = 11.2^h, \quad \delta = -7°; \quad\quad\quad\quad (6.29)$$
$$l = 264.7 \pm 0.8°, \quad b = 48.2 \pm 0.5°.$$

The conventional correction for the solar motion relative to the Local Group is 300 km s^{-1} to $l = 90°$, $b = 0$. (This is close to the mean motion defined by the Local Group members, as discussed by Yahil, Tammann, and Sandage 1977, and to the velocity that minimizes the scatter in the local redshift-distance relation in figure 5.5.) With this correction, the velocity of the Local Group relative to the CBR is 600 km s^{-1} toward $\alpha = 10.5^h$, $\delta = -26°$ ($l = 268°$, $b = 27°$). This velocity is considerably larger than the scatter in the local redshift-distance relation in

Table 6.1
Velocity of the Local Group

	V_x	V_y	V_z
CBR	-16 ± 8	-540 ± 15	275 ± 15
Rubin et al.	-420 ± 120	420 ± 120	-80 ± 120
Aaronson et al.	-190 ± 220	-720 ± 190	240 ± 180
Great Attractor	340	-450	90
IRAS gravity $\mathbf{v}/f(\Omega)$	-100 ± 100	-300 ± 100	400 ± 100

figure 5.5. That means the peculiar velocity field has to have a broad coherence length, such that the galaxies in our neighborhood are moving with us at nearly the same peculiar velocity.

The properties of the galaxy peculiar velocity field still are under discussion. An indication of the evolution of the state of understanding is given in table 6.1, which lists some estimates of the the velocity of the Local Group relative to several standards. The x component is the velocity toward the center of the Milky Way ($l = 0$, $b = 0$), with y the component in the disk in the direction of rotation ($l = 90°$, $b = 0$), and z the component normal to the disk ($b = 90°$).

The top line in the table is the velocity of the Local Group relative to the CBR, in kilometers per second. The first comparisons of our motion relative to the CBR and relative to the mean defined by the galaxies were by Sciama (1967), who used de Vaucouleurs' earlier analysis of our motion within the Local Supercluster, and by de Vaucouleurs and Peters (1968). Rubin et al. (1976) obtained the first all-sky galaxy sample designed to probe the large-scale velocity field. Their result is based on 96 Sc galaxies at distances $3500 < cz < 6500$ km s^{-1}. The velocity of the Local Group relative to this sample is listed in the second line of the table. The third line is the velocity of the Local Group relative to the clusters with Tully-Fisher distances that are represented in figure 5.4 (Aaronson et al. 1986). Lynden-Bell et al. (1988) found a peculiar velocity field that seems to converge toward a Great Attractor in the direction of Centaurus, $l = 307°$, $b = 9°$, at distance $cz \gtrsim 5000$ km s^{-1}, the flow having peculiar velocity 570 ± 60 km s^{-1} at the Local Group. The unusually large peculiar velocity field in the region of the Centaurus cluster is seen in the fluctuation in the redshift-distance relation in figure 5.4. The Great Attractor effect was found in a sample of elliptical galaxies and, in an important check, was then seen to be indicated also in an independent sample of spiral galaxies (as well as the clusters in figure 5.4). An estimate of the Great Attractor flow at the Local Group is entered in the fourth line of the table. Finally, if large-scale fluctuations in the galaxy distribution trace the mass density fluctuations $\delta = \delta\rho/\rho$, one can use the linear perturbation equation (5.115) to predict the peculiar velocity of the Local Group, up to the factor $f(\Omega)$ that depends on the parameters of the cosmological model. The result in the table is from an analysis of the IRAS galaxy distribution at distances $cz \lesssim 8000$ km s^{-1} by Strauss and Davis (1988). Other examples of analyses of the peculiar velocity field are in Rubin and Coyne (1988).

The numbers in table 6.1 are meant to represent a process rather than an established result for a very difficult measurement. The velocity found by Rubin et al. was at first resisted because it does not point in the same direction as the CBR dipole, and because it was possible to imagine systematic errors in the sample (though none has been convincingly demonstrated, and indeed the latter objection is true of every sample in the table). One might have thought that the velocities in the table are unreasonably large, because they are based on samples at distances at which the fluctuations in galaxy counts are starting to average out, according to the measure in equation (3.24). One must bear in mind, however, that the coherence length of the gravitational field is broader than that of the mass fluctuations that produce it, because gravity is a long-range force. We will see in section 21 that the large-scale fluctuations in the galaxy distribution, if reflected in the mass distribution, produce velocities comparable to those shown in the table (Clutton-Brock and Peebles 1981).

We stand to learn several things from a resolution of the very difficult issue of the nature and origin of the large-scale peculiar velocity field. If the velocity interpretation of the CBR dipole anisotropy really is correct, the velocity frames defined by galaxy redshifts and by the X-ray background ought to converge to the CBR frame. If the peculiar velocity field is produced by the gravity of mass fluctuations that are usefully traced by the galaxy distribution, we ought to find that as the depths of galaxy samples increase the motion defined by their gravity converges and stays at the CBR frame. If that happens, it fixes the velocity function $f(\Omega)$, which is a very useful constraint on the parameters of the cosmological model (as indicated in fig. 13.14). And finally, the gravitational field measures the large-scale fluctuations in the mass distribution, a critical datum for the theories discussed in part 3 for the origin of the mass fluctuations.

Aether Drift: Theory

This subsection gives a derivation of the Lorentz transformation law for radiation surface brightness. The "one line" derivation in section 9 uses the brightness theorem (along with the many lines leading up to it). The method presented here is a lengthy but strengthening exercise in Lorentz transformations (Peebles and Wilkinson 1968).

Suppose two observers, $\mathcal{O}$ and $\mathcal{O}'$, are in a homogeneous sea of radiation, with observer $\mathcal{O}'$ moving relative to $\mathcal{O}$ at velocity v to the right along the x axis. $\mathcal{O}'$ has a detector, at rest relative to $\mathcal{O}'$, that collects photons with energy p' in the range

dp' and approaching within a cone of directions at angle θ' to the x' axis, and in the solid angle

$$d\Omega' = d\cos\theta'd\phi. \tag{6.30}$$

We will suppose the face of the detector is perpendicular to the parallel x and x' axes. The face area of the detector is A_o. Then in the frame of $\mathcal{O}'$ the collecting area A' seen by the incident radiation is

$$A' = A_o\cos\theta'. \tag{6.31}$$

That is, $\mathcal{O}'$ sees that the number of photons detected in the time dt' is

$$dN' = f'(p',\theta')\,dp'd\Omega'dt'A_o\cos\theta', \tag{6.32}$$

where $f'(p',\theta')$ is the number of photons per unit volume, solid angle, and energy interval.

The corresponding quantities measured by $\mathcal{O}$ are written without primes. Thus, in the frame of $\mathcal{O}$ the number of photons per unit volume, solid angle, and energy interval is $f(p,\theta)$. As sketched in figure 6.6, $\mathcal{O}$ sees that in the time interval dt the detector attached to $\mathcal{O}'$ moves to the right along the x axis by distance vdt, and the photons to which the detector is sensitive move to the left by the amount $dt\cos\theta$ (for we are taking the velocity of light to be unity). The observers agree that the face area of the detector, which is normal to the axis of the velocity

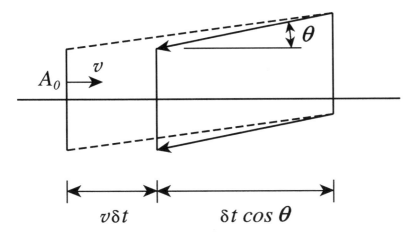

Figure 6.6. Lorentz transformation for radiation surface brightness. The photons within the dashed lines are detected in the time interval δt.

transformation, is A_o, so $\mathcal{O}$ determines that the volume occupied by the photons swept up by the detector in time dt is

$$dV = A_o(v + \cos\theta)\, dt \,. \tag{6.33}$$

$\mathcal{O}$ therefore reckons that the number of photons detected is

$$dN = f(p, \theta)dp\, d\Omega\, dt\, A_o(v + \cos\theta) \,. \tag{6.34}$$

Of course, $\mathcal{O}$ and $\mathcal{O}'$ agree on the number detected, $dN = dN'$.

The final step is to write down the Lorentz transformations relating primed and unprimed quantities. Recall that the Lorentz transformation relations are

$$\begin{aligned}
x &= \gamma(x' + vt') & t &= \gamma(t' + vx') \,, \\
x' &= \gamma(x - vt) & t' &= \gamma(t - vx) \,,
\end{aligned} \tag{6.35}$$

with $\gamma = (1 - v^2)^{-1/2}$ and x the distance measured along the direction of motion of $\mathcal{O}'$ relative to $\mathcal{O}$ (and the sign fixed by the prescription that $\mathcal{O}'$ at fixed x' is moving in the direction of increasing x). The two events pictured in figure 6.6 are separated by $dx' = 0$, so $\mathcal{O}$ measures the time between the events as

$$dt = \gamma\, dt' \,. \tag{6.36}$$

The detected photons have momentum in the x direction $p_x' = -p'\cos\theta'$, so the Lorentz transformations (6.35) applied to the photon energy-momentum four-vector say $\mathcal{O}$ sees that the photon energy and momentum along the x axis are

$$p = \gamma p'(1 - v\cos\theta') \qquad p_x = -\gamma p'(\cos\theta' - v) \,. \tag{6.37}$$

Thus $\mathcal{O}$ sees that the detected photons approach at an angle to the x axis given by

$$\cos\theta = -\frac{p_x}{p} = \frac{\cos\theta' - v}{1 - v\cos\theta'} \,. \tag{6.38}$$

The solid angle of the cone of detected photon directions follows by differentiating this expression (eq. [6.30]), to get

$$d\Omega = \frac{d\Omega'}{\gamma^2(1 - v\cos\theta')^2} \,. \tag{6.39}$$

These relations in equations (6.32) and (6.34), with $dN = dN'$, give

$$f'(p', \theta') = \frac{f(p, \theta)}{\gamma^2(1 - v\cos\theta')^2} \,. \tag{6.40}$$

This is the wanted Lorentz transformation law for the photon distribution defined in equation (6.32). In terms of the momentum relation in equation (6.37), the transformation law is

$$f'(p', \theta') = (p'/p)^2 f(p, \theta).$$ (6.41)

The radiation surface brightness is the energy flux per unit area, solid angle, and frequency interval,

$$i(p, \theta) = pf(p, \theta).$$ (6.42)

Equation (6.41) says the surface brightness measured by $\mathcal{O}'$ is

$$i'(p', \theta') = (p'/p)^3 i(p, \theta) = (1 + z)^{-3} i(p, \theta).$$ (6.43)

The second step represents the frequency shift as the redshift factor, $1 + z = p/p'$. The result of integrating this expression over frequency p' is the net surface brightness,

$$i'(\theta') = (1 + z)^{-4} i(\theta),$$ (6.44)

where i is the integrated surface brightness observed by $\mathcal{O}$.

Suppose now the radiation observed by $\mathcal{O}$ has the thermal spectrum in equation (6.9). Then the distribution function f defined in equation (6.32) is of the form

$$f(p) = p^2 F(p/kT),$$ (6.45)

where F is a function of the single variable $\hbar\omega/kT = p/kT$. With equation (6.37) for p', equation (6.41) says the spectrum measured by $\mathcal{O}'$ is

$$f'(p', \theta') = (p')^2 F(\gamma p'[1 - v \cos \theta']/kT).$$ (6.46)

This is the same form as the blackbody function in equation (6.45), with temperature

$$T' = \frac{T}{\gamma(1 - v \cos \theta')}$$
$$= T[1 + v \cos \theta' + v^2(\cos^2 \theta' - 1/2) + \cdots].$$ (6.47)

The conclusion is that the moving observer sees in any direction a thermal blackbody spectrum, but with a temperature that is a function of the angle θ' between the direction of observation and the direction of motion relative to the preferred frame in which the radiation is isotropic. The leading term is the dipole in equation (6.28); the term of order v^2 is a quadrupole anisotropy.

The relation between redshift and surface brightness in equation (6.43) applies to the observations of the surface brightnesses of high redshift galaxies, because the redshift and the surface brightness shift can be represented as the results of a sequence of velocity transformations between neighboring observers along the light path, as in equation (5.39). The same argument applies to the evolution of the surface brightness of the CBR. Thus we have as a special case yet another demonstration that in a homogeneous and isotropic universe an initially black-body CBR spectrum remains blackbody. In an anisotropic or inhomogeneous universe the redshift can be different along different lines of sight, producing a corresponding anisotropy in the CBR temperature. Since the redshift factor is the same as the expansion factor along the line of sight when scattering can be neglected, the fact that the CBR is isotropic to better than one part in 10^4 requires that the expansion of the universe measured back to last scattering is isotropic to like accuracy. This remarkably tight constraint is formalized in the Sachs-Wolfe relation to be discussed in section 21.

Characteristic Quantities for the CBR

The cosmic background radiation leads to a variety of characteristic quantities for cosmology. We begin with some numbers for the CBR spectrum.

In blackbody radiation at temperature T the number of photons per unit volume, from the integral of equation (6.9) over frequencies, is

$$n_\gamma = \frac{2\zeta(3)}{\pi^2} \left(\frac{kT}{\hbar c}\right)^3 . \tag{6.48}$$

The dimensionless integral is

$$J_- = \int_0^\infty \frac{x^2\, dx}{e^x - 1} = 2\zeta(3) = 2.404, \tag{6.49}$$

where ζ is the Riemann ζ-function.

To get the entropy in blackbody radiation, recall from classical thermodynamics that the reversible addition of heat dU to a system at temperature T and fixed volume increases the entropy by the amount $dS = dU/T$. Thus, with the Stefan-Boltzmann law $u_\gamma = aT^4$ (eq. [6.14]), the entropy per unit volume satisfies

$$ds_\gamma = du_\gamma/T = 4aT^2\, dT , \tag{6.50}$$

and the integral is

$$s_\gamma = \frac{4}{3}aT^3 = \frac{4\pi^2 k}{45} \left(\frac{kT}{\hbar c}\right)^3 . \tag{6.51}$$

This is the entropy per unit volume in blackbody radiation at temperature T. The ratio of the entropy to the photon number (6.48) is

$$\frac{s_\gamma}{n_\gamma k} = \frac{2\pi^4}{45\zeta(3)} = 3.6. \tag{6.52}$$

Since a reversible adiabatic expansion of the radiation conserves both entropy and photons, it is no surprise that this ratio is independent of temperature.

The frequency ω_m at the maximum of the blackbody function for the energy u_ω per frequency interval (eq. [6.11]) satisfies

$$\frac{\hbar\omega_m}{kT} = \frac{h\nu_m}{kT} = \frac{hc}{kT\lambda_m} = 2.82. \tag{6.53}$$

The frequency ω_h at the half-energy point in the spectrum (such that half the integrated energy is at lower frequency) is

$$\frac{\hbar\omega_h}{kT} = \frac{h\nu_h}{kT} = \frac{hc}{kT\lambda_h} = 3.50. \tag{6.54}$$

Up to numerical factors, the number density of photons is $n_\gamma \sim \lambda_h^{-3}$ (eq. [6.48]), the characteristic photon energy is $\epsilon_\gamma = h\nu_h = hc/\lambda_h \sim kT$, and the energy density (eq. [6.14]) is $u_\gamma \sim \epsilon_\gamma n_\gamma$.

With the present CBR temperature in equation (6.1), the wavelength at half energy is

$$\lambda_h = 1.50(1+z)^{-1} \, \text{mm}. \tag{6.55}$$

The redshift dependence applies back to $z \sim 10^{10}$, where the thermal electron pairs annihilate (as discussed below). The characteristic energy belonging to this wavelength is

$$\epsilon_h = hc/\lambda_h = 3.5kT = 1.32 \times 10^{-15}(1+z) \, \text{erg}$$
$$= 8.2 \times 10^{-4}(1+z) \, \text{eV}. \tag{6.56}$$

The photon number density (eq. [6.48]) is

$$n_\gamma = 420(1+z)^3 \, \text{cm}^{-3}. \tag{6.57}$$

The ratio to the number density of baryons (eq. [5.68]) is

$$\eta = \frac{n_B}{n_\gamma} = 2.7 \times 10^{-8}\Omega_B h^2. \tag{6.58}$$

Apart from numerical factors, this is the same as the baryon number per dimensionless unit of entropy (eq. [6.52]), and the ratio of the heat capacities in baryons and radiation (eq. [6.16]). The small value of η means the universe has an enormous entropy compared to its matter content. One way to put it is that adiabatic compression back to redshift $z \sim 10^{10}$, where the baryon number density is comparable to that in ordinary solid material, brings the CBR temperature to $T \sim 10^{10}$ K.

The energy density in the background radiation is

$$
\begin{aligned}
u_\gamma &= 4.2 \times 10^{-13}(1+z)^4 \, \text{erg cm}^{-3} \\
&= 0.26(1+z)^4 \, \text{eV cm}^{-3} .
\end{aligned}
\tag{6.59}
$$

We noted above that this is comparable to the energy density in starlight near the edge of the Milky Way. An observer placed at random, rather than near a galaxy, would see on average the extragalactic starlight surface brightness νi_ν in equation (5.166), with energy density $u = 4\pi \nu i_\nu / c$ about three orders of magnitude below the CBR energy density u_γ in equation (6.59). The upper bound in equation (5.168) is two orders of magnitude below u_γ. That means it would take 100 to 1000 cycles of an oscillating universe that conserves baryons and entropy to produce the CBR out of starlight.

The rms values of the electric and magnetic fields, B, in the CBR satisfy $u_\gamma = B^2/4\pi$, giving

$$
B_\gamma = 2 \times 10^{-6}(1+z)^2 \, \text{gauss} ,
\tag{6.60}
$$

six orders of magnitude smaller than the magnetic field on the surface of the Earth, and as we noted above comparable to the static magnetic field in the interstellar medium in the Milky Way.

Relict Neutrinos

At high redshifts, the CBR photons are energetic enough to produce an equilibrium abundance of neutrinos. This is of interest because the mass density in thermal neutrinos at high redshift is appreciable, and if one of the families has nonzero rest mass the relict neutrinos could make an important contribution to the present mass density. In the standard model there are three neutrino families, each with two states for a given momentum. Below we will check that the neutrinos decouple from thermal equilibrium with the CBR at redshift $z \sim 10^{10}$; here we consider some properties of the relict neutrinos after decoupling under the assumptions that they are not degenerate and that all three neutrino rest masses are well below the typical photon energy kT at decoupling. (As discussed in section 18, this gives an acceptable present mass density in neutrinos.)

Since neutrinos are fermions, and behave in an excellent approximation to an ideal gas, the equilibrium occupation number at temperature T and chemical potential μ in a mode with energy ϵ is

$$\mathcal{N} = \frac{1}{e^{(\epsilon-\mu)/kT}+1} \,. \tag{6.61}$$

The negative sign in the denominator of equation (6.2) for bosons is replaced by the positive sign that makes $\mathcal{N} \leq 1$ for fermions. Recall from classical thermo-dynamics that the Gibbs function for a mixture of particles, with the N_i of type i having chemical potential μ_i, is $G = \sum N_i \mu_i$. If the particle species are in thermal equilibrium through the reactions

$$a+b \leftrightarrow c \,, \tag{6.62}$$

then G relaxes to an extremum such that $\delta G = 0$ if the reaction shifts the abun-dances by $\delta N_a = \delta N_b = -\delta N_c$. That means the particle chemical potentials in the reaction (6.62) satisfy

$$\mu_a + \mu_b - \mu_c = 0 \,. \tag{6.63}$$

Photons can be absorbed, $a+\gamma \rightarrow a$, so $\mu = 0$ for photons at thermodynamic equi-librium. A neutrino and its partner (antiparticle or opposite spin state) can annihi-late, so we know that at thermal equilibrium

$$\mu(\nu) = -\mu(\bar{\nu}) \,. \tag{6.64}$$

The parameter $\mu(\nu)$ thus fixes the difference between the number densities of neutrinos and their partners.

We are assuming the neutrino rest masses and chemical potentials are small compared to kT. The latter means the fractional difference between the number densities of particles and antiparticles is small, that is, the thermal neutrino sea is far from degenerate. Under these assumptions we can drop the chemical potential μ from the occupation number $\mathcal{N}$ in equation (6.61), and we can set the momen-tum equal to $p = \epsilon/c$. Then the energy density in a single family of neutrinos is

$$u_\nu = \frac{2}{(2\pi\hbar)^3} \int_0^\infty \frac{4\pi p^2 \, dp \, pc}{e^{pc/kT_\nu}+1} \,. \tag{6.65}$$

The first factors are the sum over phase space in equation (6.8), with two spin states (for neutrino and antineutrino). On changing variables of integration, we get

$$u_\nu = \frac{I_+}{\pi^2} \frac{(kT_\nu)^4}{(\hbar c)^3} \,, \tag{6.66}$$

where the dimensionless integral is

$$I_+ = \int_0^\infty \frac{x^3\, dx}{e^x + 1}. \tag{6.67}$$

We can evaluate it by using the identity

$$\frac{1}{e^x - 1} - \frac{1}{e^x + 1} = \frac{2}{e^{2x} - 1}. \tag{6.68}$$

The result of multiplying this expression by x^3, integrating, and changing variables on the right-hand side to $y = 2x$, is

$$I_+ = \frac{7}{8} I_-, \tag{6.69}$$

where I_- is given by equation (6.15). The energy density in a single family is then

$$u_\nu = \frac{7}{8} a T_\nu^4. \tag{6.70}$$

That is, the energy density is $7/8$ times the energy density in electromagnetic radiation at the same temperature. A similar calculation shows that the number density of neutrinos plus their partners in one family is

$$n_\nu = \frac{3\zeta(3)}{2\pi^2} \left(\frac{kT_\nu}{\hbar c} \right)^3, \tag{6.71}$$

which is $3/4$ of the number density of photons at the same temperature (eq. [6.48]).

At $kT \gtrsim m_e c^2$ the reactions

$$\gamma + \gamma \leftrightarrow e^+ + e^- \tag{6.72}$$

produce a sea of electron-positron pairs. Since the pairs are far more abundant than the extra electrons belonging to the protons, we can neglect the electron chemical potential, so the number density of pairs is given by an expression similar to equation (6.71) for neutrinos. We will see at the end of this subsection that the neutrinos decouple from thermal equilibrium with the radiation before the sea of electron pairs annihilates and dumps its entropy into the radiation. The result is that the CBR temperature T_o is larger than the present value of the parameter T_ν in the above equations.

To find T_ν, note that at kT large compared to the electron rest mass the energy density in electromagnetic radiation plus the sea of relativistic electron pairs is

$$u = aT^4 \left[1 + 2\frac{I_+}{I_-} \right] = \frac{11}{4} aT^4 . \qquad (6.73)$$

The first term in the brackets represents the electromagnetic radiation energy density. The second term for the electron pairs is twice equation (6.70), because there are four states for given momentum (two spin states for the electron and positron). Since u varies as T^4, the entropy density is $s = 4u/3T$ (eq. [6.51]), and the entropy in a volume V expanding with the general expansion of the universe is

$$S = \frac{11}{3} a(T')^3 V' = \frac{4}{3} aT^3 V . \qquad (6.74)$$

The primed variables apply to the relativistic sea of radiation and electron pairs in equation (6.73), the unprimed variables to the radiation after the pairs have annihilated. The annihilation is reversible, so the entropy in the volume expanding with the universe is unchanged. The temperature parameter T_ν for the neutrinos scales with the expansion parameter in the usual way for free particles,

$$(T'_\nu)^3 V' = T_\nu^3 V . \qquad (6.75)$$

Well before the electron pairs are annihilated, the neutrinos are in thermal equilibrium with the radiation, $T'_\nu = T'$, so equations (6.74) and (6.75) say that well after annihilation the neutrino temperature parameter is

$$T_\nu = \left(\frac{4}{11} \right)^{1/3} T . \qquad (6.76)$$

With equation (6.1) for T_o, the present value is

$$T_\nu = 1.95 \, \text{K} . \qquad (6.77)$$

It is best to call T_ν a parameter rather than a temperature, for as we noted at the beginning of this section the expansion preserves a thermal neutrino distribution only when kT_ν is large compared to the neutrino rest mass. Independent of the neutrino mass, the neutrino number density varies as T_ν^3, so equations (6.57), (6.71) and (6.77) give the present number density of neutrinos plus antineutrinos in a single family,

$$n_\nu = \frac{3}{11} n_\gamma = 113 \, \text{cm}^{-3} . \qquad (6.78)$$

After electron pair annihilation, the net energy density in electromagnetic radiation plus the relict relativistic neutrinos is (eq. [6.70] and [6.76])

$$u_r = aT^4 \left[1 + \frac{7}{8} \left(\frac{T_\nu}{T} \right)^4 N_\nu \right] = 1.68 a T^4 . \tag{6.79}$$

The numerical value assumes that $N_\nu = 3$ relativistic neutrino families are left over from high redshift, and it assumes that the neutrino masses are small compared to the CBR temperature. Following the definition of the density parameter (eq. [5.55]), we can write the fractional contribution Ω_r to the present expansion rate by radiation and the neutrinos as

$$\Omega_r = \frac{8\pi}{3} \frac{G u_r}{H_o^2 c^2} = 4.22 \times 10^{-5} h^{-2} . \tag{6.80}$$

The fractional contribution to the present expansion rate by nonrelativistic matter is estimated to be $\Omega \gtrsim 0.05$ (table 20.1). Because u_r varies with redshift as $(1+z)^4$, the mass densities in relativistic and nonrelativistic matter are equal at redshift

$$1 + z_{eq} = \Omega/\Omega_r = 2.37 \times 10^4 \Omega h^2 . \tag{6.81}$$

This number is of interest because the gravitational growth of clustering of free nonrelativistic material is suppressed when the expansion rate is dominated by relativistic matter. As we will discuss, z_{eq} is comparable to the redshift at which the plasma combines to neutral atomic hydrogen and the matter is released from the radiation drag (eq. [6.96]).

At the epoch of formation of light elements represented in figure 6.5, $10^{10} \gtrsim z \gtrsim 10^9$, the expansion rate is dominated by the energy density in relativistic matter. It would be easy to imagine that there is an appreciable contribution to the energy density from gravitational radiation, and perhaps from other very low mass fields yet to be discovered, but if these contributions were appreciable they would increase the expansion rate and spoil the concordance shown in figure 6.5. Either this concordance is an unlucky accident or we have a reasonably complete catalog of the mass density when our universe was a few seconds old, which is a remarkable concept.

Let us consider finally the thermal decoupling of the neutrinos from the matter and radiation. Near decoupling the main thermalizing reaction is

$$\nu + \bar{\nu} \leftrightarrow e^+ + e^- , \tag{6.82}$$

with cross section

$$\sigma \sim G_F^2 E_\nu^2 \sim 4 \times 10^{-44} T_{10}^2 \, \text{cm}^2 \tag{6.83}$$

(Weinberg 1975). The cross section varies as the square of the energy, because that fixes the volume in phase space for the final state, as in equation (6.178) below. At thermal equilibrium the energy is $E \sim kT$. In equation (6.83) the temperature is expressed in units of 10^{10} K. The neutrino number density (eq. [6.78]) for three families is

$$n_\nu = 1.6 \times 10^{31} T_{10}{}^3 \, \text{cm}^{-3} . \tag{6.84}$$

At the redshifts of interest here the expansion rate is dominated by the relativistic mass density u_r, and the solution to equation (5.18) for the age of the universe is

$$t = \left(\frac{3c^2}{32\pi G u_r} \right)^{1/2} = 2 T_{10}{}^{-2} \, \text{s} . \tag{6.85}$$

The mean number of times a neutrino suffers the reaction (6.82) in an expansion time is

$$\sigma n_\nu c t \sim 0.04 \, T_{10}{}^3 . \tag{6.86}$$

When this number is large the neutrinos are in close thermal contact with the radiation, when it is small the neutrinos are free, and thermal contact is broken when $\sigma n_\nu c t \sim 1$. This is at $T_{10} \sim 3$, or $kT / m_e c^2 \sim 5$. Equation (6.76) for T_ν assumes this number is large. The approximation thus is reasonable but not abundantly satisfied.

Thermal Ionization

At high redshifts the CBR keeps the matter fully ionized, and the radiation drag on the free electrons prevents the formation of a gravitationally bound system such as a protogalaxy. At redshift $z_{\text{dec}} \sim 1400$ the primeval plasma combines to neutral atomic hydrogen, leaving a small but possibly interesting ionized fraction. This is called the decoupling epoch, for, as we will see, the nearly neutral matter is able to move through the CBR to form gas clouds that can produce the stars and active galactic nuclei that later reionize much of the remaining diffuse matter (as will be discussed in section 23).

We consider here the thermal equilibrium ionization, and in the next subsection we will deal with the recombination rate. The calculation is simplified without eliminating any of the interesting physics by taking the matter to be pure hydrogen, ignoring the helium.

The photoionization reaction is

$$e + p \leftrightarrow H + \gamma . \tag{6.87}$$

The photons have no chemical potential, so the equilibrium condition (6.63) is

$$\mu(e) + \mu(p) = \mu(H) . \tag{6.88}$$

We can get the chemical potentials from the expression for the occupation number, $\mathcal{N}$. Since these particles behave as classical gases with $\mathcal{N} \ll 1$, equation (6.61) for $\mathcal{N}$ becomes

$$\mathcal{N} = e^{(\mu-\epsilon)/kT} . \tag{6.89}$$

The number of particles per unit volume, n, is the sum over phase space in equation (6.8),

$$n = \frac{g}{(2\pi\hbar)^3} \int d^3p \, e^{(\mu-\epsilon)/kT} , \tag{6.90}$$

where g is the number of spin states. The energy is the sum of the annihilation and kinetic energies, $\epsilon = mc^2 + p^2/2m$. The integral works out to

$$n = g \frac{(2\pi mkT)^{3/2}}{(2\pi\hbar)^3} e^{(\mu-mc^2)/kT} . \tag{6.91}$$

On using this expression for the chemical potential in the equilibrium equation (6.88) and recalling that the masses of the electron and proton exceed the hydrogen atom mass by the binding energy,

$$(m_e + m_p - m_h)c^2 = B = 13.60 \, \text{eV} , \tag{6.92}$$

we get the Saha thermal ionization equilibrium equation

$$\frac{n_e n_p}{n_h n} = \frac{x^2}{1 - x} = \frac{(2\pi m_e kT)^{3/2}}{n(2\pi\hbar)^3} e^{-B/kT} . \tag{6.93}$$

The factor $g_e g_p / g_h$ is unity, because the electron and proton each have $g = 2$ spin states and the hydrogen atom has four spin states in the ground energy levels within the hyperfine structure. The difference between the mass of the proton and hydrogen atom has been ignored. The total number density of protons, free and in atoms, is $n_p + n_h = n$. The fractional ionization is

$$n_e = n_p = xn . \tag{6.94}$$

Let us write $T = 2.736(1 + z) \, \text{K}$ and $n = 1.12 \times 10^{-5} \Omega_B h^2 \, (1 + z)^3 \, \text{cm}^{-3}$ (eq. [5.68]), ignoring the helium. This gives

$$\log[x^2/(1 - x)] = 20.99 - \log[\Omega_B h^2 (1+z)^{3/2}] - 25050/(1+z). \tag{6.95}$$

This equilibrium ionization is shown as the steeply falling curves in figure 6.8 below. At the baryon density that is wanted for light element nucleosynthesis (eq. [6.27]), the equilibrium fractional ionization is $x = 0.5$ at

$$z_{\text{dec}} = 1360, \quad T_{\text{dec}} = T_o z_{\text{dec}} = 3700 \, \text{K}, \quad \text{for } \Omega_B h^2 = 0.013. \tag{6.96}$$

A liberal upper bound on the baryon density is $\Omega h^2 = 1$. This increases the redshift very slightly, to $z_{\text{dec}} = 1520$.

This equilibrium ionization calculation assumes homogeneous distributions of matter and radiation. The radiation has to be close to homogeneous, because diffusion through the matter is quite effective at dissipating small-scale irregularities (as discussed in section 25), and the isotropy of the CBR strongly bounds large-scale fluctuations. The matter might be irregularly distributed in a smooth sea of radiation. In this case, the mean residual ionization at $x \ll 1$ is $\langle n_e \rangle \propto \langle n^{1/2} \rangle < \langle n \rangle^{1/2}$, so an irregular matter distribution lowers the mean density of ions.

The conclusion is that the primeval plasma is free to combine to atomic hydrogen at redshift $z_{\text{dec}} \sim 1400$, when the matter density is

$$n_{\text{dec}} = 3 \times 10^4 \Omega_B h^2 \, \text{cm}^{-3}, \tag{6.97}$$

and the cosmic time is (eq. [5.61])

$$\begin{aligned}
t_{\text{dec}} &= 4 \times 10^{12} \Omega^{-1/2} h^{-1} \, \text{s} \\
&= 1.2 \times 10^5 \Omega^{-1/2} h^{-1} \text{y}.
\end{aligned} \tag{6.98}$$

Under these conditions, the rate of capture of the electrons by protons is much faster than the rate of expansion and cooling of the universe, but the resulting recombination rate depends on some subtle effects, to be considered next.

Recombination Rate

When an electron is captured direct to the ground state in atomic hydrogen, it produces a photon that with high probability ionizes another atom, leaving no net change. When an electron is captured to an excited state, the allowed decay to the ground state produces a resonant Lyman series photon. These photons have large cross sections for capture by a hydrogen atom, putting the atom in a high energy state that is easily photoionized again. That leaves two main routes to the production of atomic hydrogen. One is two-photon decay from the metastable $2s$ level to the ground state, at the rate

$$\Lambda = 8.23 \, \text{s}^{-1}. \tag{6.99}$$

The second is the loss of the Lα resonance photons by the cosmological redshift. The resulting recombination history is worked out in Peebles (1968) and

Zel'dovich, Kurt, and Sunyaev (1969). Some of the subtleties are resolved by Krolik (1990).

The approximation one uses to find the residual ionization is based on the short mean free time for the scattering of Lα resonance photons. The scattering cross section near resonance (eqs. [23.43] and [23.97]) is

$$\sigma = \frac{3\lambda_\alpha^2}{8\pi} \frac{\gamma^2}{(\omega - \omega_\alpha)^2 + \gamma^2/4}. \tag{6.100}$$

The photon frequency is ω, the resonance is $\omega_\alpha = 2\pi c/\lambda_\alpha$, with wavelength

$$\lambda_\alpha = 1216 \,\text{Å}, \tag{6.101}$$

and the decay rate from a $2p$ state is

$$\gamma = 6.25 \times 10^8 \,\text{s}^{-1}. \tag{6.102}$$

The thermal motion of the atoms broadens the resonance seen by the radiation by the amount

$$\frac{\delta\omega}{\omega_\alpha} = \frac{v}{c} = \left(\frac{3kT}{m_p c^2}\right)^{1/2} = 3 \times 10^{-5} \tag{6.103}$$

at T_{dec}. Since $\gamma/\omega_\alpha \sim 10^{-7}$ is considerably smaller than $\delta\omega/\omega_\alpha$, equation (6.100) is broadened to the thermal width, bringing the mean cross section at maximum to

$$\bar{\sigma} \sim \frac{3\lambda_\alpha^2}{8\pi} \frac{\gamma^2}{\omega_\alpha^2} \frac{m_p c^2}{3kT} = 3 \times 10^{-17} \,\text{cm}^2. \tag{6.104}$$

If the neutral fraction is large, the mean free time of a Lα photon at recombination at density n_{dec} (eq. [6.97]) is

$$(\bar{\sigma} n_{\text{dec}} c)^{-1} \sim 30 \,\Omega_B^{-1} h^{-2} \,\text{s}. \tag{6.105}$$

The conclusion is that the mean free time for the Lα resonance photons is very much shorter than the expansion time (6.98). That means the abundance ratio of atoms in the $1s$ and $2p$ states and the shape of the radiation spectrum within the thermally broadened resonance relax to statistical equilibrium.

A recombination followed by the allowed transition to the ground state adds a photon to the radiation within the Lα resonance. Scattering by moving atoms rearranges the radiation within the resonance. On a much longer timescale, the cosmological redshift pulls photons away from the red side of the resonance and adds them on the blue side from the unperturbed background radiation spectrum.

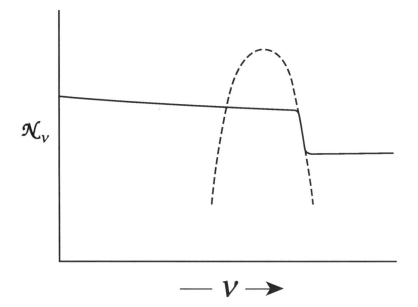

Figure 6.7. Radiation spectrum near the $L\alpha$ resonance. Within the dashed re-gion the mean free path is much shorter than the expansion time, so the spec-trum in this region is in statistical equilibrium with the atoms. The production of resonance photons by formation of hydrogen atoms produces the step in the spectrum.

The result is a step at the short wavelength side of the resonance, as illustrated in figure 6.7. The spectrum of the photons across the relatively narrow resonance has to be very close to flat, because the photon energy distribution is in statistical equilibrium with the atoms. (A more formal demonstration is given by Krolik 1990.) This is a moving step in the spectrum, which leaves a slight excess in the short wavelength side of the CBR spectrum from the smeared-out distribution of recombination photons. The effect is small, however, and not likely to be observable.

In terms of the photon occupation number $\mathcal{N}$ as a function of photon momen-tum p, the step in the spectrum satisfies the relation

$$R = \frac{2 \cdot 4\pi p^2}{h^3} \frac{p\dot{a}}{a} \left(\mathcal{N}_\alpha - e^{-h\nu_\alpha/kT} \right) . \qquad (6.106)$$

In the parentheses, $\mathcal{N}_\alpha$ is the occupation number across the resonance, and join-ing smoothly to the spectrum on the red side. The unperturbed occupation number on the blue side is the Planck function; it can be approximated as the exponential because $\mathcal{N}$ is much less than unity. The first prefactor converts the occupation number to the number of photons per unit volume and momentum interval. That multiplied by the mean energy drift from the cosmological redshift gives the rate at which photons are leaving the resonance at the red side. With

$$K \equiv \frac{\lambda_\alpha^3}{8\pi} \frac{a}{\dot{a}},$$ (6.107)

this equation is

$$KR = \mathcal{N}_\alpha - e^{-h\nu_\alpha/kT}.$$ (6.108)

This gives the net rate R of production of hydrogen atoms per unit volume by recombination followed by allowed $L\alpha$ emission and with the elimination of one $L\alpha$ photon on average by the cosmological redshift.

Now let us write out the expression for the net rate of production of hydrogen atoms. The rate per unit volume is

$$\alpha_e n_e^2 - \beta_e n_{2s} = R + \Lambda \left(n_{2s} - n_{1s} e^{-h\nu_\alpha/kT} \right).$$ (6.109)

The first term on the left-hand side is the rate of recombinations to excited states of the atom, ignoring recombination direct to the ground state. The recombination coefficient is $\alpha_e = \langle \sigma v \rangle$, and n_e is the number density of free electrons and of free protons. The second term on the left is the rate of ionization from excited states of the atom. This ionization rate depends on the numbers of atoms in each excited state. The relative distribution of the excited states is close to what is given by the thermal Boltzmann factors, because absorption and emission of radiation at the relatively low energy differences among excited states rapidly rearranges the population, relaxing the distribution to equilibrium with the background radiation. The ionization rate thus is proportional to the number density of atoms in the $2s$ state. The constant of proportionality, β_e, follows from the condition that the rates of ionization and recombination are such as to tend to bring n_{2s}/n_e^2 to thermal equilibrium. The Saha equation for the thermal value of this ratio is given by equation (6.93) with the binding energy changed to the binding energy $B_2 = 3.4$ eV in the $n = 2$ energy level,

$$\beta_e = \alpha_e \frac{(2\pi m_e kT)^{3/2}}{(2\pi\hbar)^3} e^{-B_2/kT}.$$ (6.110)

The difference of recombination and ionization rates on the left side of equation (6.109) is the net rate of production of hydrogen atoms. As indicated in the right-hand side of the equation, this net rate is the sum of the rate R for the allowed $L\alpha$ transition and the rate for two-photon emission from the $2s$ state. The latter is the difference between the decay rate, with Λ given by equation (6.99), and the reverse rate for absorption of two photons from the CBR. Since these photons are softer than $L\alpha$ they are present in great abundance and very close to thermal equilibrium, so the rate of excitation is just such as to tend to bring n_{2s}/n_{1s} to the thermal equilibrium value $\exp -h\nu_\alpha/kT$.

Next, let us consider the ratio of the numbers of atoms in the ground and excited states. The tight statistical equilibrium between the atoms and the radiation within the $L\alpha$ resonance pictured in figure 6.7 fixes the ratio of number densities n_{2p}/n_{1s} in terms of the intensity of the radiation. The ratio n_{2p}/n_{2s} is fixed by the relaxation of the population of the excited states to the background radiation temperature. Thus we know that the ratio of numbers of atoms in the zero orbital angular momentum ground and first excited levels is

$$\frac{n_{2s}}{n_{1s}} = \mathcal{N}_\alpha, \tag{6.111}$$

because at thermal equilibrium both would be equal to the Boltzmann factor $\exp -h\nu_\alpha/kT$.

The final steps are to use equation (6.111) to eliminate n_{2s} from equation (6.109), use equation (6.108) to eliminate R from equation (6.109), solve the resulting equation for the occupation number $\mathcal{N}_\alpha$, and finally use the expression for $\mathcal{N}_\alpha$ to eliminate it from the net recombination rate in the left side of equation (6.109). This algebra gives

$$-\frac{d}{dt}\frac{n_e}{n} = \left(\frac{\alpha_e n_e^2}{n} - \frac{\beta_e n_{1s}}{n} e^{-h\nu_\alpha/kT} \right) C, \tag{6.112}$$

where

$$C = \frac{1 + K\Lambda n_{1s}}{1 + K(\Lambda + \beta_e)n_{1s}}. \tag{6.113}$$

In the left side of equation (6.112), n_e is the free electron number density and n is the total baryon number density, free and in hydrogen atoms. The ratio n_e/n is unaffected by the expansion of the universe. The exponential factor in the second term on the right-hand side brings β_e to the Saha form (6.93) for the ground state. The expression in parentheses in equation (6.112) thus is the recombination rate one would use if one could ignore the $L\alpha$ resonance photons. These photons reduce the rate by the factor C.

The ratio of the rates of formation of hydrogen atoms by the allowed and forbidden transitions to the ground state is, from equations (6.108) and (6.109),

$$\frac{L\alpha \text{ rate}}{\text{two-photon rate}} = \frac{R}{\Lambda\left(n_{2s} - n_{1s}e^{-h\nu_\alpha/kT}\right)} = \frac{1}{K\Lambda n_{1s}} \sim \frac{0.01}{1-x}\frac{\Omega^{1/2}}{\Omega_B h}. \tag{6.114}$$

The last entry is evaluated at z_{dec}. In this expression, the expansion rate $\dot{a}/a$ (eq. [5.60]) allows for the possibility that the net density parameter, Ω, is larger than the density Ω_B in baryons. Unless the latter is considerably smaller, the ratio in equation (6.114) is smaller than unity, indicating that most of the hydrogen is produced by two-photon decay. In this case, the coefficient in equation (6.113) is

$$C \sim \frac{\Lambda}{\Lambda + \beta_e}. \tag{6.115}$$

As one would expect, the suppression factor is the ratio of the rates of decay and ionization from the excited states.

The recombination rate is discussed in section 23 (eq. [23.26]). The rate coefficient $\alpha_e = \langle \sigma v \rangle$ for recombinations to excited states is reasonably well approximated by the fitting formula

$$\alpha_e = 2.6 \times 10^{-13} T_4^{-0.8} \, \text{cm}^3 \, \text{s}^{-1}, \tag{6.116}$$

with $T = 10^4 T_4$ K. The ratio of the ionization and decay rates (eqs. [6.99] and [6.110]) is

$$\beta_e/\Lambda = 8 \times 10^7 T_4^{0.7} e^{-3.95/T_4}. \tag{6.117}$$

This ratio is large at T_{dec} but drops rapidly, reaching unity at $T_c = 2300$ K, redshift $z_c \sim 850$. That is, not long after the plasma is allowed to combine, the ionization rate β_e from the excited state drops below the decay rate Λ, and the inhibition factor C approaches unity.

Figure 6.8 shows the result of numerical integration of the recombination equation. For comparison, the steeply falling lines show the equilibrium ionization in equation (6.95). The calculation assumes that there are no sources of ionizing radiation (as from the early stars in the isocurvature scenarios discussed in section 25), that the baryons are nearly homogeneously distributed, and that the density parameter is large enough for the expansion rate to be dominated by nonrelativistic matter rather than radiation (eq. [6.81]). Then the relevant parameters are Ωh^2 and $\Omega_B h^2$ for the densities in nonrelativistic matter and in the baryons (because Ωh^2 fixes the expansion rate in the Einstein–de Sitter limit in eq. [5.61], and $\Omega_B h^2$ fixes the baryon number density in eq. [5.68]). The solid and dashed curves in the figure use the same value for the baryon density, $\Omega_B h^2 = 0.1$, so the equilibrium ionization is the same. The solid curves assume that the baryons dominate the mass, $\Omega = \Omega_B = 0.1$, and the dashed curve assumes that the total mass density is ten times larger, $\Omega h^2 = 1$. The more rapid expansion rate in the latter case increases the residual ionization. In the dot-dash curves $\Omega h^2 = 1$ and the baryon density is lowered to $\Omega_B h^2 = 0.01$. That slightly increases the equilibrium ionization, and considerably increases the residual ionized fraction.

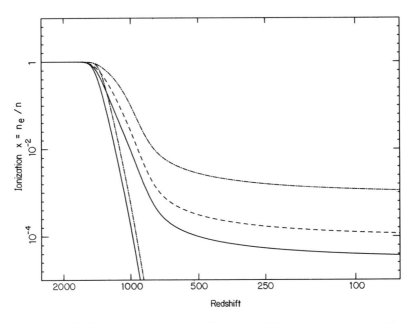

Figure 6.8. Evolution of the ionization. The steeply falling curves are the equilibrium ionization in equation (6.95). The flatter curves show $x = n_e/n$ computed from equations (6.112) and (6.113). For the solid curves the density parameters are $\Omega h^2 = \Omega_B h^2 = 0.1$, for the dashed curve $\Omega h^2 = 1$ and $\Omega_B h^2 = 0.1$, and for the dot-dashed curves $\Omega h^2 = 1$ and $\Omega_B h^2 = 0.01$.

The effect of the factor C in equation (6.112) is to keep the ionization at $z \gtrsim 800$ considerably larger than it would have been if C were unity. At lower redshifts β_e/Λ is small, $C \sim 1$, and the capture probability per electron per expansion time is

$$-\frac{t}{n_e}\frac{dn_e}{dt} = \alpha_e xnt \sim 400 h \Omega_B \Omega^{-1/2}(1+z)^{0.7}x . \qquad (6.118)$$

The expansion time is $t \propto \Omega^{-1/2}h^{-1}$ (eq. [5.61]) and the baryon number density is $n \propto \Omega_B h^2$. Thus, the residual ionization at low redshifts (in the absence of inhomogeneities or sources of ionizing radiation) scales as $x \propto \Omega^{1/2}/(\Omega_B h)$. The fit to the computed ionization is

$$x = 1.2 \times 10^{-5}\Omega^{1/2}/(h\Omega_B) \qquad (6.119)$$

at $z = 100$.

As a first application, let us consider the optical depth of the universe for Thomson scattering by free electrons. The cross section is

$$\sigma_t = 6.65 \times 10^{-25} \, \text{cm}^2 \,. \tag{6.120}$$

The characteristic number for the scattering probability at redshift z is the optical depth

$$\tau = \sigma_t n_e ct = 0.046 x (1+z)^{3/2} \Omega_B \Omega^{-1/2} h \,. \tag{6.121}$$

With the residual ionization in equation (6.119), the optical depth at $z = 800$ is

$$\tau_{\text{dec}} \sim 0.01 \,. \tag{6.122}$$

This calculation shows that if there are no sources of ionization at decoupling (from very early stars, active protogalactic nuclei, or matter-antimatter annihilation), the decrease in ionization leads to an abrupt transition from high optical depth to near transparency to the CBR. As will be discussed in sections 24 and 25, the universe could become opaque again at $z \gtrsim 30$, depending on the epoch of formation of the first sources of ionization.

The residual free electrons and ions serve as catalysts for the formation of molecular hydrogen, through the reactions

$$H + e \leftrightarrow H^- + \gamma \,.$$
$$H^- + H \leftrightarrow H_2 + e \,, \tag{6.123}$$

and

$$H + p \leftrightarrow H_2^+ + \gamma \,.$$
$$H_2^+ + H \leftrightarrow H_2 + p \tag{6.124}$$

(Saslaw and Zipoy 1967; Peebles and Dicke 1968). The molecular hydrogen may be the main source of cooling in the first generation of gravitationally bound gas clouds.

The sequence in equation (6.124) produces molecular hydrogen at a higher redshift than for (6.123), because the binding energy of the proton in the molecular hydrogen ion H_2^+ in the intermediate state is $B_+ = 2.65 \, \text{eV}$, considerably larger than the binding energy $B_- = 0.75 \, \text{eV}$ of the electron in the negative hydrogen ion H^-. The amount of molecular hydrogen that forms depends on the evolution of the gas density. Lepp and Shull (1984) have analyzed in detail the case where the hydrogen expands with the general expansion of the universe. We can understand the order of magnitude of the molecular hydrogen production by the method used to estimate the residual ionization in equation (6.118). The Saha equilibrium abundance ratio (6.93) applied to the number density of molecular hydrogen ions relative to free protons is

$$\frac{n_{2+}}{n_p} = \frac{n(2\pi\hbar)^3}{(2\pi m_p kT)^{3/2}} e^{B_+/kT} \tag{6.125}$$

$$= \exp\left[-59.6 + \ln[\Omega_B h^2 (1+z)^{3/2}] + 11200/(1+z)\right] ,$$

where n is the number density of hydrogen atoms. This equilibrium ratio rises sharply through unity at redshift

$$z_f \sim 200 . \tag{6.126}$$

Most of the H_2 forms at $z \sim z_f$, because at higher redshift photodissociation by the CBR suppresses the abundance of H_2^+, and at lower redshifts the lower density suppresses molecule formation, if the gas is expanding with the universe. The rate coefficient for the formation of H_2^+ in equation (6.124) is

$$\alpha = \langle \sigma v \rangle = 1.4 \times 10^{-18} \, \text{cm}^3 \, \text{s}^{-1} \tag{6.127}$$

(Lepp and Shull 1984). This coefficient is close to constant at low energy because the distribution of final states in phase space is determined by the energy released in the reaction, which is almost independent of the incident energy. The particle exchange reaction in equation (6.124) is faster than radiative capture, so we can take it that all the H_2^+ that is produced ends up as H_2. The final abundance of molecular hydrogen relative to atoms is then

$$n_2/n \sim \alpha x n t \sim 1 \times 10^{-7} . \tag{6.128}$$

The ionization x is from equation (6.119), and the number density n of atoms and the expansion time t are evaluated at the redshift z_f in equation (6.126).

This calculation assumes the gas is expanding. In a gas cloud that forms at high redshift, the production of molecular hydrogen is considerably increased because the density stays higher, and when the background temperature is cool enough production of molecular hydrogen in a gas cloud may be dominated by the negative atomic hydrogen ions produced by the more mobile electrons. The role of the molecules as an energy sink depends on the temperature history of the cloud. The massive first-generation clouds in the adiabatic hot dark matter picture (section 25) are hot enough to ionize the matter, and the plasma dissipates energy by thermal bremsstrahlung. In the adiabatic cold dark matter and isocurvature baryonic dark matter pictures, low mass gas clouds form at high redshifts and collapse as a result of radiation by collisionally excited H_2.

Coupling of Matter and the CBR

The thermal cosmic background radiation plays a key role in the early evolution of structure, setting the epoch at decoupling (eq. [6.98]) when matter becomes free to move through the radiation to form the first generation of bound systems, and fixing the matter temperature that determines the Jeans length for the minimum size of the first generation. It is prudent to pay careful attention to the possibility that remnants of these effects are to be found in the present observed structures, though it is of course also quite possible that the evidence has been thoroughly erased by later generations of structure formation. This discussion deals with the role of radiation drag on the free electrons in determining the matter temperature and motions. Further details are in sections 24 and 25.

Consider an electron moving at nonrelativistic speed $v \ll c$ through the CBR. In the electron rest frame, the CBR temperature measured at angle θ from the direction of motion is (eq. [6.47])

$$T(\theta) = T \left[1 + \frac{v}{c} \cos \theta \right] . \tag{6.129}$$

The radiation energy per unit volume moving into the element of solid angle $d\Omega$ at angle θ to the direction of motion is

$$du = aT(\theta)^4 d \cos \theta \, d\phi / (4\pi) . \tag{6.130}$$

We get the energy flux by multiplying this by c, and the momentum flux by dividing the energy flux by c. The momentum flux multiplied by the Thomson scattering cross section (6.120) is the rate of transfer of momentum to the particle. The component of the momentum transfer along the direction of motion, and integrated over all directions of the radiation, is the net drag force on an electron moving at speed v,

$$\begin{aligned}
F &= \int \sigma_t aT^4 \cos \theta \frac{d \cos \theta d\phi}{4\pi} \left(1 + 4 \frac{v}{c} \cos \theta \right) \\
&= \frac{4}{3} \frac{\sigma_t aT^4 v}{c}
\end{aligned} \tag{6.131}$$

(Peebles 1965). We will use this result to find the rate of relaxation of the matter temperature to that of the radiation, and the rate of slowing of moving matter.

Consider the heat transfer rate in a gas of protons, electrons, and hydrogen atoms at rest at temperature T_e in a sea of blackbody radiation at temperature T. The mean energy per electron in the plasma is

$$u = \frac{3}{2} kT_e = \frac{1}{2} m_e \langle v^2 \rangle . \tag{6.132}$$

The rate at which the electron is doing work against the radiation drag force is Fv, so the plasma transfers energy to the radiation at the mean rate, per free electron,

$$-\frac{du}{dt} = \langle Fv \rangle = \frac{4}{3} \frac{\sigma_t a T^4}{c} \langle v^2 \rangle . \qquad (6.133)$$

This is proportional to the plasma temperature, T_e. At thermal equilibrium, when $T_e = T$, this rate has to be balanced by the rate at which the fluctuating force of the photons scattering off the electrons increases the matter energy. (This detailed balance result is derived in eq. [24.45].) Thus the time rate of change of the matter temperature T_e due to the interaction with the CBR at temperature T is

$$\frac{dT_e}{dt} = \frac{x}{1+x} \frac{8\sigma_t a T^4}{3m_e c} (T - T_e), \qquad (6.134)$$

Where x is the fractional ionization (eq. [6.94]). The factor $x/(1+x)$ takes account of the fact that when a hydrogen atom is ionized it produces two particles to share the thermal energy. The electrons exchange energy with the CBR; collisions among the particles keeps them all at the same temperature.

With equation (5.61) for the expansion timescale, the characteristic number for the approach to temperature equilibrium in equation (6.134) works out to

$$\frac{t}{T_e} \frac{dT_e}{dt} = \frac{0.0028}{\Omega^{1/2} h} \left(\frac{T}{T_e} - 1 \right) \frac{2x}{1+x} (1+z)^{5/2} . \qquad (6.135)$$

If the gas is fully ionized, $x = 1$, the coefficient of the fractional temperature difference in equation (6.135) is unity at the thermalization redshift

$$1 + z_t = 10.5(\Omega h^2)^{1/5} . \qquad (6.136)$$

With increasing redshift at $z > z_t$ it becomes increasingly difficult to keep optically thin plasma hotter than the CBR. It is interesting that the critical redshift z_t for fully ionized matter is comparable to the highest redshifts of observed objects, now $1 + z \sim 6$.

Before the first energy sources perturb the matter, it has the residual ionization after decoupling gives in equation (6.119). With this ionization, the characteristic number for relaxation of the matter temperature is

$$\frac{t}{T_e} \frac{dT_e}{dt} \sim \frac{10^{-7}}{\Omega_B h^2} \left(\frac{T}{T_e} - 1 \right) (1+z)^{5/2} . \qquad (6.137)$$

The coefficient of the fractional temperature difference reaches unity at redshift

$$1 + z_t \sim 1000(\Omega_B h^2)^{2/5} . \qquad (6.138)$$

At the baryon density favored for light element production, $\Omega_B h^2 \sim 0.015$, this is $z_t \sim 100$. That is, the residual ionization can be enough to keep the matter in temperature equilibrium with the radiation well after the universe becomes transparent, at $z \sim 1400$. Of course, all this neglects mass density fluctuations and the heat sources they might produce.

The formation of the first gravitationally bound systems is limited by radiation drag. The drag force F per free electron is given by equation (6.131), where now v is the mean streaming velocity relative to the CBR. The drag force per baryon is xF, where x is the fractional ionization (and we are ignoring the minor correction for helium). The mean drag force divided by the mass m_p of a hydrogen atom is the deceleration of the streaming motion,

$$\frac{1}{v}\frac{dv}{dt} = -\frac{4}{3}\frac{\sigma_t a T^4 x}{m_p c} . \tag{6.139}$$

The ratio of the expansion time to the dissipation time for the streaming motion is

$$-\frac{t}{v}\frac{dv}{dt} = \frac{1.5 \times 10^{-6}}{(\Omega^{1/2}h)}(1+z)^{5/2}x . \tag{6.140}$$

This number is unity at the velocity dissipation redshift

$$1+z_v = 210(\Omega h^2)^{1/5}x^{-2/5} . \tag{6.141}$$

The critical redshift z_v for velocity dissipation is higher than the redshift z_t in equation (6.136) for thermal coupling of the plasma to the CBR, because to transfer heat the CBR only has to slow the electrons, while to transfer momentum it has to slow the heavier baryons.

For the residual ionization in equation (6.119), equation (6.140) is

$$-\frac{t}{v}\frac{dv}{dt} \sim \frac{2 \times 10^{-11}(1+z)^{5/2}}{\Omega_B h^2} \sim \frac{0.0003}{\Omega_B h^2} \quad \text{for } z \sim 800 . \tag{6.142}$$

With the usual estimate for the baryon density, $\Omega_B h^2 \gtrsim 0.015$, this says the residual ionization is too low to dissipate peculiar motions. Since the characteristic timescale for the gravitational growth of mass density fluctuations in linear perturbation theory is on the order of the expansion timescale, this means mass density fluctuations become free to grow at decoupling (unless some source of ionizing radiation prevents recombination).

If in the primeval baryon distribution there are regions with density contrast $\delta\rho/\rho$ greater than unity, as in the isocurvature scenarios discussed in section 25, gravitationally bound systems can form prior to decoupling (Hogan 1978). In a region of comoving size w and mass density contrast $\delta = \delta\rho/\rho$, the peculiar gravitational acceleration on the periphery is just balanced by the radiation drag, assuming fully ionized matter, when the peculiar velocity has relaxed to the terminal velocity given by the equation

$$g = \frac{4}{3}\pi G\rho_b a(t) \, w \, \delta = \frac{dv}{dt} = \frac{4}{3} \frac{\sigma_t a T^4 v}{m_p c} . \tag{6.143}$$

The fractional change in the radius of the concentration in an expansion timescale is

$$\frac{vt}{aw} \sim 0.005 \frac{\Omega_B h}{\Omega^{1/2}} \left[\frac{1000}{1+z} \right]^{5/2} \delta . \tag{6.144}$$

This says that immediately before decoupling, radiation drag allows collapse of an overdense region if the contrast is $\delta \gtrsim 300\Omega^{1/2}/(\Omega_B H)$. Since this critical contrast increases rapidly with increasing redshift, z_{dec} marks the effective bound for the formation of structures out of baryons.

It might be noted that gravity also can overcome the radiation drag force if the density concentration is larger than the Jeans length (5.125) for the matter and radiation treated as a single fluid. However, z_{dec} is comparable to z_{eq} (eq. [6.81]), so in this case the gravitational potential belonging to an upward mass fluctuation would be large enough to tend to draw the region to relativistic collapse to a black hole. This could not have been a common event, for the mass density at $z > z_{eq}$ varies as $\rho \propto (1+z)^{-4}$, while the mass density in radiation captured in black holes varies as $\rho \propto (1+z)^{-3}$, so a small mass fraction captured at high redshift could seriously overproduce dark matter in black holes at the present epoch.

If the fluctuations in the baryon distribution at decoupling are linear, $|\delta| \lesssim 1$, the Jeans length sets a lower bound on the size and mass of the first generation of gravitationally bound objects. At $z \gtrsim z_t$ in equation (6.138), radiation drag keeps the matter temperature close to uniform at the CBR temperature, so the velocity of sound is $c_s^2 = dp/d\rho \sim kT/m$, and the critical Jeans length (5.125) is

$$\lambda_J = \left(\frac{\pi kT}{G\rho_b m_p} \right)^{1/2} \tag{6.145}$$

(Peebles 1965). At $z \gtrsim z_t$ the matter temperature is $T \sim a(t)^{-1}$, and the mean mass density varies as $\rho_b \propto a(t)^{-3}$, so the Jeans length scales as $a(t)$, and it defines a time-independent Jeans mass,

$$M_J \sim \frac{\pi}{6}\rho_b \lambda_J^3 \sim 10^5 (\Omega_B h^2)^{-1/2} \mathcal{M}_\odot . \tag{6.146}$$

To understand the evolution of the baryon distribution $\delta(\mathbf{x}, t)$ through decoupling, let us consider the representation of δ as a sum of Fourier components

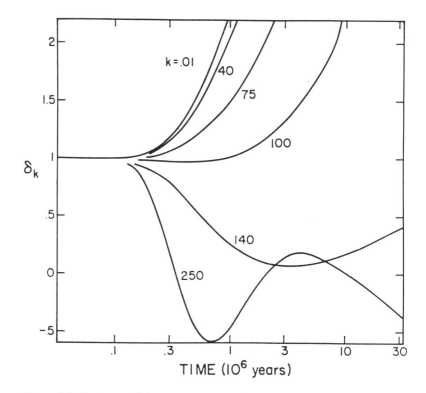

Figure 6.9. Evolution of the amplitude of the baryon distribution at fixed co-moving wave number (Peebles 1969b). The wave number is expressed in units of $10^{-20}\,\mathrm{cm}^{-1}$ at the epoch $T = 10,000\,\mathrm{K}$.

$\propto \exp i\mathbf{k} \cdot \mathbf{x}$. In linear perturbation theory each Fourier amplitude $\delta_k(t)$ with fixed comoving wave number $\mathbf{k}$ evolves independently, according to equation (5.124) with the addition of the drag force (6.131). An example of the solutions is shown in figure (6.9). At redshift $z > z_{\mathrm{dec}}$, at the left side of the figure, radiation drag holds the amplitudes constant. At small k, where the wavelength is much larger than the Jeans length, the growth of the amplitude after decoupling is almost independent of k. At large k, short wavelength, the matter pressure is important and the amplitude tends to oscillate as a sound wave. The oscillation commences when the wave is moving in a medium with dissipation time comparable to the oscillation time, for that is what allows the amplitude to change, so the oscillation is close to critically damped. This heavily suppresses the final amplitudes at wavelengths shorter than the Jeans length. The result is that the smallest mass for the first generation is set by the Jeans mass in equation (6.146).

It may be a suggestive coincidence that the Jeans mass in the primeval matter is comparable to the typical mass of a globular star cluster (Peebles and Dicke 1968). One can identify processes that would produce globular clusters in the interstellar medium in a young galaxy (Fall and Rees 1987), and consistent with that are the observations of what look like young globular clusters in the Magellanic Clouds. Also, the space distribution of the globular clusters around a galaxy is more extended than that of the spheroid stars, but tighter than that of the dark matter, which is most directly interpreted by assuming the clusters formed out of the material in the protogalaxy. On the other hand, it is striking to see how similar are the gross properties of the globular clusters found in environments as different as giant elliptical galaxies and the extreme dwarf satellites of our galaxy, as might be expected if the protoglobular gas clouds were assigned their masses before the protogalaxies formed. Some other aspects of the baryon Jeans mass are discussed in sections 23 and 25.

Interaction with Relativistic Particles

The radiation drag force in equation (6.131) applies to particles with nonrelativistic peculiar velocities. When the cosmic background radiation was discovered it was recognized that the interaction of relativistic particles with the CBR also would be of considerable interest. Felten (1965) discussed the slowing of electrons from cosmic radio sources by scattering by the CBR, and Hoyle (1965) noted that the upscattered CBR photons can be an appreciable X-ray flux. This is equivalent to the synchrotron radiation from electrons accelerated in magnetic fields in cosmic radio sources, but here operating on the magnetic field of the CBR (eq. [6.60]). The universe is predicted to be opaque to energetic gamma rays through pair production with CBR photons,

$$\gamma + \gamma_b \to e^+ + e^- \tag{6.147}$$

(Gould and Schréder 1966). The photopion production reaction,

$$p + \gamma_b \to p + \pi^0, \; n + \pi^+, \tag{6.148}$$

slows energetic cosmic ray protons. This effect, which was discussed by Peebles and Dicke (1965), Greisen (1966), and Zatsepin and Kuz'min (1966), is of particular interest because cosmic ray events are observed at the threshold energy for pion production, so we have the possibility of a quantitative test of the process from the shape of the cosmic ray energy spectrum. This has been analyzed by Hill, Schramm, and Walker (1986).

Sciama (1990a) gives a thorough review of the effects of interactions of high energy particles with the CBR. Here we consider some examples.

The center of mass energy Q in the reactions in equations (6.147) and (6.148) follows from the square of the sum of the four-momenta p_a^i and p_b^i of the two initial particles,

$$Q^2 = (\epsilon_a + \epsilon_b)^2 - (\mathbf{p}_a + \mathbf{p}_b)^2$$
$$= m_a{}^2 + m_b{}^2 + 2(\epsilon_a\epsilon_b - \mathbf{p}_a \cdot \mathbf{p}_b), \quad (6.149)$$

because in the center of mass frame the spatial part of the total momentum vanishes. In the pair production reaction (6.147) the photons are massless, so if the background photon has energy ϵ_b the threshold at $Q = 2m_e$ is at gamma ray energy

$$\epsilon = m_e{}^2/\epsilon_b. \quad (6.150)$$

At the characteristic CBR energy in equation (6.56), this is

$$\epsilon \sim 3 \times 10^{14} \, \text{eV} = 300 \, \text{TeV}. \quad (6.151)$$

Near the threshold, the pair production cross section is close to the Thomson cross section in equation (6.120), $\sigma_{\gamma\gamma} \sim 10^{-25} \, \text{cm}^2$. The CBR photon number density is $n_\gamma \sim 400 \, \text{cm}^{-3}$ (eq. [6.57]), so the mean free path for pair production near the threshold center of mass energy is

$$l_\gamma = (\sigma_{\gamma\gamma} n_\gamma)^{-1} \sim 10 \, \text{kpc}, \quad (6.152)$$

comparable to the distance to the center of the Milky Way. At somewhat lower gamma ray energies the mean free path is longer because there are fewer background photons above the threshold.

The same pair production process operating on background radiation at shorter wavelengths limits the mean free path at lower gamma ray energies. The sky brightness normal to the plane of the Milky Way at wavelength $\lambda = 100 \, \mu$ is $\nu i_\nu \sim 1 \times 10^{-4} \, \text{erg s}^{-1} \text{cm}^{-2} \, \text{ster}^{-1}$ (Hauser et al. 1991). This corresponds to photon number density

$$n_\gamma = \frac{4\pi}{c} \frac{\nu i_\nu}{h\nu} \sim 1 \, \text{photon cm}^{-3}. \quad (6.153)$$

If an appreciable part of this flux is extragalactic, the mean free path for pair production is

$$l_\gamma \sim 1 \, \text{Mpc} \quad \text{at} \quad \epsilon \sim 30 \, \text{TeV}, \quad (6.154)$$

comparable to the distance across the Local Group. At wavelength $\lambda = 5000 \, \text{Å}$ the background energy density in starlight from the galaxies (eq. [5.166]) translates to a mean photon number density $n_\gamma = 10^{-4}\,\mathrm{cm}^{-3}$. That brings the mean free path to

$$l_\gamma \sim 10^4\,\mathrm{Mpc} \quad \text{at } \epsilon \sim 100\,\mathrm{GeV}. \tag{6.155}$$

This is longer than the Hubble length, so at $\epsilon \sim 100\,\mathrm{Gev}$ the universe may be transparent back to redshifts in excess of unity. The optical depth for photon annihilation is discussed in more detail, and at lower energies and higher redshifts, by Babul, Paczyński, and Spergel (1987).

Ultrahigh-energy cosmic ray events are detected by the air shower that the primary produces in the atmosphere. At the highest observed energies, $\epsilon \sim 10^{20}\,\mathrm{eV}$, the primaries are thought to be individual protons or neutrons, because starlight can photodissociate complex nuclei. A proton with momentum p normal to a magnetic field B moves in a circle at the Larmor radius $R = pc/Be$. At the interstellar field $B \sim 10^{-6}$ gauss and proton energy $\epsilon = pc = 10^{20}\,\mathrm{eV}$, the Larmor radius is $R \sim 100\,\mathrm{kpc}$. Since this is an order of magnitude larger than the distance to the center of the galaxy, locally produced cosmic ray events at this energy ought to point back to their sources. Since the cosmic rays show no preference for the plane of the Milky Way, the indication is that the ultrahigh-energy primaries are mainly extragalactic. The distance they can travel is limited by the loss of energy by e^+e^- and pion production by the CBR photons that appear as gamma rays in the rest frame of a $10^{20}\,\mathrm{eV}$ proton. Let us estimate the mean free path for pion production.

Since the center of mass energy Q at the threshold is small compared to the proton rest mass, equation (6.149) is

$$Q = m_p + (1 - \cos\theta)\epsilon_p\epsilon_b/m_p, \tag{6.156}$$

where the proton and CBR photon have energies ϵ_p and ϵ_b at initial relative angle θ. Near the threshold for photoproduction of pions, the cross section peaks at $\sigma_{\gamma\pi} = 500\,\mu\mathrm{b} = 5 \times 10^{-28}\,\mathrm{cm}^2$ at the Δ resonance at energy $M_\Delta = 300\,\mathrm{MeV}$. For the characteristic CBR photon energy in equation (6.56), the nucleon energy at the resonance is $\epsilon_p = 1.5 \times 10^{20}\,\mathrm{eV}$, and the mean free path is

$$l_p = (\sigma_{\gamma\pi}n_\gamma)^{-1} \sim 2\,\mathrm{Mpc}, \tag{6.157}$$

again comparable to the distance across the Local Group of galaxies. Nucleons coming from greater distances should be slowed (for the pions and electron pairs are emitted in the forward direction in the comoving frame) and accumulate at energies just below threshold for the CBR photons near the peak of the distribution. The cosmic ray spectrum thus would be expected to be the sum from local sources and an extragalactic component with a peak and cutoff at energy slightly below $10^{20}\,\mathrm{eV}$ (Hill, Schramm, and Walker 1986). Unambiguous detection of this peak is of considerable interest as a check of this picture, and, it appears, within reach. Whether nearby extragalactic sources could be detected from the distribution of arrival directions depends on the intergalactic magnetic field, which again is something we would like to know a good deal more about.

Helium Production

In the standard hot expanding cosmological model, helium and trace but important amounts of other light elements are produced as the universe expands and cools through temperatures kT on the order of 1 MeV. This process is exceedingly interesting as a test of the standard cosmological model applied when the radius of the universe would have been ten orders of magnitude smaller than at the present epoch. The main result is the comparison of the predicted and observed abundances shown in figure 6.5. We begin the discussion with a survey of the main relevant orders of magnitude, and then work through some details of the computation.

At redshifts $z \gtrsim 10^{10}$, where $kT \gtrsim 3$ MeV, photodissociation assures the elimination of any complex nuclei, and the free neutrons and protons are thermally coupled to the sea of electron and neutrino pairs by the reactions

$$
\begin{aligned}
e^- + p &\leftrightarrow n + \nu, \\
\bar{\nu} + p &\leftrightarrow n + e^+, \\
n &\leftrightarrow p + e^- + \bar{\nu},
\end{aligned}
\tag{6.158}
$$

at rates fast enough to keep the relative abundance of neutrons and protons close to the thermal equilibrium Boltzmann ratio,

$$
n/p = e^{-Q/kT}, \qquad Q = (m_n - m_p)c^2 = 1.2934 \text{ MeV}.
\tag{6.159}
$$

There is a high rate for radiative capture of the neutrons to form deuterium,

$$
n + p \leftrightarrow d + \gamma,
\tag{6.160}
$$

but the reverse photodissociation reaction keeps the deuterium abundance quite low until the temperature has fallen to $kT \sim 100$ keV and there no longer is a significant abundance of thermal photons more energetic than the deuteron binding energy $B = 2.2$ MeV. At this point the deuterium abundance grows large enough for the deuterons to burn to helium, the fastest ways being the particle exchange reactions

Table 6.2
Helium Production

t (sec)	$T(10^{10}\,\mathrm{K})$	aT	λt	$\hat{\lambda} t$	$\ln(n_d n/n_p n_n)$	σnvt	$n_n/(n_n + n_p)$
0.0020	22.5	1.0000	4700	5000	−25.2	8400	0.483
0.0058	13.1	1.0000	900	1010	−26.0	4900	0.471
0.0170	7.6	1.0001	170	208	−26.6	2800	0.451
0.050	4.45	1.0004	31.0	43	−27.2	1700	0.418
0.148	2.59	1.0012	5.4	9.7	−27.6	970	0.363
0.44	1.51	1.0035	0.85	2.3	−27.7	570	0.292
1.29	0.887	1.0102	0.11	0.60	−27.3	340	0.225
3.78	0.526	1.0282	0.010	0.181	−26.1	200	0.185
11.0	0.319	1.0720	—	0.071	−23.6	130	0.166
32	0.201	1.1579	—	0.056	−19.6	97	0.156
92	0.129	1.2805	—	0.111	−13.1	75	0.144
267	0.081	1.3790	—	0.302	−1.94	53	0.118
780	0.048	1.4006	—	0.87	19.3	32	0.066
2280	0.0279	1.4010	—	2.6	57	18	0.012
6730	0.0163	1.4010	—	7.6	122	11	0.000

$$d + d \leftrightarrow t + p\,,$$
$$d + d \leftrightarrow {}^3\mathrm{He} + n\,,$$
$$t + d \leftrightarrow {}^4\mathrm{He} + n\,, \tag{6.161}$$
$$^3\mathrm{He} + d \leftrightarrow {}^4\mathrm{He} + p\,.$$

Table 6.2 summarizes the characteristic numbers from the computations to be described below. The first column is the expansion time computed from very high temperature. The second column is the radiation temperature in units of $10^{10}\,\mathrm{K}$ (with $kT_{10} = 0.86\,\mathrm{MeV}$). The third column is the product of the expansion parameter $a(t)$ and the radiation temperature. The annihilation of the thermal electron pairs increases $a(t)T(t)$ by the factor $(11/4)^{1/3}$ (eq. [6.76]). The neutrinos decouple from the radiation at $T_{10} \sim 3$ (eq. [6.86]), before electron pair annihilation has appreciably increased $a(t)T(t)$, so we can take it that the neutrino temperature scales as $T_\nu \propto 1/a(t)$ from $T_\nu = T$ at the start of the table.

In the next two columns, λ and $\hat{\lambda}$ are the rates for the conversion between neutrons and protons in the reactions in equation (6.158),

$$\frac{dn_n}{dt} = \lambda n_p - \hat{\lambda} n_n\,. \tag{6.162}$$

When the products λt and $\hat{\lambda} t$ are large, the neutron abundance relaxes to the equilibrium value in equation (6.159). We see that the neutron abundance becomes frozen at $T_{10} \sim 1$, and in the absence of the reactions that lock the neutrons in atomic nuclei the neutron abundance would again relax to the equilibrium value (with no neutrons) at $T_{10} \sim 0.03$.

The Saha expression in equation (6.187) below for the thermal abundance ratio of deuterons relative to neutrons and protons under the reactions (6.160) is tabulated in the sixth column. There is a sharply defined switch from the high-temperature limit, where the greater phase space strongly prefers free neutrons and protons, to a strong preference for deuterium at low temperature. That means there is essentially no accumulation of complex nuclei prior to the epoch at which the ratio passes through unity, and thereafter photodissociation of deuterium can be ignored.

The characteristic number $\sigma n v t$ from the Gamow criterion in equation (6.22) is listed in column 7 for $\Omega_b h^2 = 0.015$. Since this number is greater than unity, the radiative capture of neutrons by protons is fast enough to produce deuterium when thermal equilibrium allows it. Almost all the deuterium ends up in ^{4}He by the particle exchange reactions in equation (6.161), because the coulomb barriers are more than offset by the advantage of not having to create a photon.

Helium forms well after it is thermodynamically favored over free nucleons, because the reactions that are fast enough to produce it go through the less strongly bound intermediate deuteron. The production of molecular hydrogen after z_{dec} similarly occurs after it is thermodynamically favored, through the less tightly bound molecular hydrogen ion or negative hydrogen ion (eqs. [6.123] and [6.124]).

The production of an astronomically reasonable abundance of the light elements depends on several numerical coincidences. First, the neutron abundance freezes at a temperature just below the neutron-proton mass difference Q, causing about 10% of the baryons to end up as neutrons. If the freezing temperature, which is set by the weak interactions, had been much higher, there would have been nearly equal numbers of neutrons and protons, and they would have combined to place most of the baryons in ^{4}He. If the freezing temperature had been below the temperature at which deuterium can accumulate, it would have eliminated most of the neutrons, and with it any interesting amount of element production. Second, the binding energy of the deuteron is such that it can accumulate and burn to helium while the neutron abundance is frozen. If the binding energy were significantly higher, helium production would have commenced when the neutron-proton abundance ratio was close to unity, again eliminating most of the hydrogen, while a much lower binding energy would have eliminated most of the light element production because the neutrons would have decayed. Third, the rate coefficient $\sigma n v t$ when deuterium production is allowed has to be greater than about unity for appreciable light element production. We see that the condition is satisfied, but not abundantly.

Dicke (1961) points out that some coincidences in cosmology really are consistency conditions. For example, we could not be in a universe with a timescale from the start of expansion to the final collapse that is much shorter than 10^{10} years, for it takes about that long to make us. In a universe in which most of the baryons come out of the big bang as helium, the lifetimes of stars like the Sun would be considerably shorter, with a possibly deleterious effect on the prospects for slowly evolving life forms such as us. It is difficult to see the evolutionary disadvantage of a universe with negligible light element production at high redshift, save for the loss of a very educational phenomenon.

There are no subtleties in this calculation, and only the one practical problem that the reactions (6.161) and (6.162) are numerically unstable when the rates are high, but one knows how to deal with that from the much more complex processes in stellar evolution. The extrapolation to a state of the universe when it was only one second old is daring, but easy for a theorist. The truly remarkable thing is that the calculation yields a pattern of light element abundances that agrees with the observations as now understood. If these processes did not occur in the early universe, what produced the strikingly uniform and high abundance of helium?

The computation goes through so simply in part because of some important constraints from the physics, and in part with the help of some assumptions that seem reasonable. The radiation drag force (eq. [6.131]) prevents matter from moving through the radiation to produce complicated objects such as stars. A mass density fluctuation with length scale larger than the radiation Jeans length can bind radiation as well as matter, but we have noted that this is unacceptable because the fluctuations would tend to collapse to black holes, storing an unacceptable amount of mass. Thus we are left with the simple case of nearly homogeneous expansion.

The baryons can be distributed in an inhomogeneous way if that is how they were placed when they were created or last found themselves locked to the radiation. In the inflation scenario discussed in section 17, the baryons have to be created at or after the end of the inflation epoch, and the simplest possibility is that the process created a universal value for the baryon number per unit of entropy. That gives the nearly homogeneous distribution assumed in the standard calculation to be described here. There are alternatives: the phase transition as quarks collect to form nucleons could leave small-scale irregularities in the matter distribution (Applegate and Hogan 1985; Iso, Kodama, and Sato 1986), and one can imagine there were large-scale fluctuations in the efficiency of conversion of entropy to baryons, or even that inflation never happened. We consider here only the homogeneous case, because it is simple, but we should bear in mind that the primeval baryon distribution may have been inhomogeneous.

The rate of expansion of the universe during light element production is determined by the mass density in the relativistic components, which includes the radiation and neutrinos, and perhaps also gravitational radiation and other massless fields. The net contribution is parametrized by the effective number N_ν of neutrino families, in equation (6.174) below. This is one of the two free parameters in the standard homogeneous computation, the other being the number density of baryons relative to photons. The best present fit to the observations is with $N_\nu \sim 3$, which allows for the known neutrino families and little else (Walker et al. 1991).

The standard calculation also assumes that the chemical potentials of the neutrino families can be neglected in comparison to the temperature, meaning the numbers of particles and their partners differ by a small fraction of the number of thermal pairs. This has to be the case for the electrons, because charge neutrality demands that the difference between the numbers of electrons and positrons is equal to the relatively small proton number, and it may be natural to require that the neutrino numbers be similarly small. The other possibly natural choice makes the neutrino numbers comparable to the number of photons, which would mean the chemical potentials are appreciable. This adds another free parameter, μ/kT for the neutrinos that exchange protons and neutrons, to the parameter for the density of other neutrinos and other types of relativistic matter (Fowler 1970; Yahil and Beaudet 1976).

There is the possibility that primeval magnetic fields affect the reactions in equation (6.158) (Greenstein 1969; Matese and O'Connell 1970), and there is some interest developing in the idea that magnetic fields might have been produced in the early universe (as discussed by Ratra 1992). To see what is involved for light element production, note that an electron of mass m_e and charge e moves at speed v in magnetic field B in a circle of radius a, with force $Bev/c = m_e v^2/a$. We can get the Landau levels for the motion transverse to B by setting the angular momentum $m_e va$ to a multiple of Planck's constant. That gives the energy level splitting relative to the CBR temperature,

$$\frac{\epsilon}{kT} \sim \frac{e\hbar}{m_e c} \frac{B_o}{kT_o}(1+z). \tag{6.163}$$

Here B_o is the present value of the field, and the redshift dependence follows because homogeneous expansion that preserves magnetic flux changes the field as $B \propto (1+z)^2$. If the magnetic field were to play a significant role in fixing the neutron abundance, ϵ/kT would have to be greater than about unity at $z \sim 10^{10}$, which gives

$$B_o \gtrsim 10^{-6}\,\text{gauss}. \tag{6.164}$$

If this field were isotropically compressed from the cosmic mean baryon density to what is found within a galaxy it would be increased by some four orders of magnitude to $B \sim 10^{-2}$ gauss, which is about four orders of magnitude larger than the observed value. That is, primeval magnetic fields could affect light element production only if the field strength at high redshifts were well above the usual estimates.

There is considerable room for new physics. This includes processes within the standard cosmological model, such as particles with decay or annihilation radiation that can photodissociate nuclei (Dicus et al. 1978; Miyama and Sato 1978), and adjustments to the physics that determine the expansion and reaction rates (Dicke 1968). The results could be relatively minor adjustments to the predictions of the standard model, or an entire revision of the picture. The former would become of intense interest if the observations revealed discrepancies with the standard computation, though this has not happened so far. The latter is to be borne in mind, but it would mean the apparent success so far for the standard computation is a remarkably misleading coincidence.

Let us consider now some elements of the computation. This discussion emphasizes the physics and uses simple numerical integrations to establish orders of magnitudes. A more detailed guide to how these orders of magnitude come about is given by Bernstein, Brown, and Feinberg (1989). We will concentrate on the helium abundance, because it is the most robust prediction. Results of the computations and observational tests for the other light elements are surveyed in Kolb and Turner (1990).

The first step is to get the effect of the thermal electron pair annihilation on the radiation temperature. Since this is a reversible process we can compute the temperature from the condition that the entropy of radiation plus electron pairs is conserved.

To see that we can approximate the electron pairs as an ideal gas, note that at kT larger than the electron mass the mean distance between electrons is $\lambda \sim T^{-1}$, in units where $\hbar$, c, and k are unity (eq. [6.54]). Thus the typical electrostatic energy is e^2/λ, and the ratio of electrostatic to thermal energy kT is the fine structure constant, $e^2 = 1/137$, which is a reasonably small perturbation to the free gas picture.

To get the entropy in the electron pairs, recall that the free energy is $F = -kT \ln Z$, where the partition function Z is the sum of $\exp -E/kT$ over states. For electrons the exclusion principle allows a single particle mode two energies, $E = 0$ and $E = \epsilon$, so the partition function for the mode is $Z = 1 + \exp -\epsilon/kT$. The free energy summed over all modes for the electron pairs is

$$F = -kT \int \frac{4V d^3 p}{(2\pi\hbar)^3} \ln \left(1 + e^{-\epsilon/kT}\right) = U - TS. \qquad (6.165)$$

The first factor in the integral is the usual sum over states in equation (6.8), allowing two spin states each for electrons and positrons. The energy is

$$U = \int \frac{4V d^3 p}{(2\pi\hbar)^3} \frac{\epsilon}{e^{\epsilon/kT} + 1}, \qquad (6.166)$$

and the entropy is given by the difference of these equations. The next step is to eliminate the logarithm in equation (6.165) by integrating by parts. Then a short calculation leads to the entropy per unit volume in the electron pairs,

$$s_e = \frac{S}{V} = \frac{1}{T} \int \frac{4 \, d^3p}{(2\pi\hbar)^3} \frac{1}{\epsilon} \frac{\epsilon^2 + p^2/3}{e^{\epsilon/kT} + 1}. \tag{6.167}$$

Here and below the velocity of light is set to unity, so the electron momentum satisfies

$$\epsilon^2 = p^2 + m_e^2. \tag{6.168}$$

The entropy in the radiation is given by equation (6.51). With equation (6.14) for Stefan's constant a_B, we can write the entropy density of radiation plus electron pairs as

$$s = \frac{4}{3} a_B T^3 \left[1 + \frac{45}{2(\pi kT)^4} \int_0^\infty \frac{p^2 dp}{\epsilon} \frac{\epsilon^2 + p^2/3}{e^{\epsilon/kT} + 1} \right]. \tag{6.169}$$

Since the annihilation is reversible, the entropy $s(T(t))a(t)^3$ is constant. The result of differentiating this expression is

$$\frac{1}{s} \frac{ds}{dT} \frac{dT}{da} = -\frac{3}{a}. \tag{6.170}$$

This gives a differential equation for T as a function of the expansion parameter, a. The result of differentiating equation (6.169) for the entropy with respect to T, and rearranging, is

$$\frac{d \ln aT}{d \ln a} = \frac{I_n}{3 + I_d}, \tag{6.171}$$

with

$$I = \frac{45}{2(\pi kT)^4} \int_0^\infty \frac{dp}{e^{\epsilon/kT} + 1} J, \tag{6.172}$$

and $J_n = \epsilon^3 - p^4/\epsilon$ and $J_d = \epsilon^3 + 3p^2\epsilon$ for the numerator and denominator. Numerical integration of these equations gives the third column in table 6.2.

The expansion rate is given by the mass density, which is dominated by the energy density u in radiation, particle pairs, and any other relativistic components. We will write the energy density as

$$u = a_B T_\nu^4 E, \tag{6.173}$$

with

$$E = \left(\frac{T}{T_\nu}\right)^4 + \frac{7}{8}N_\nu + \frac{15}{\pi^2} \frac{\hbar^3}{(kT_\nu)^4} \int \frac{4 \, d^3p}{(2\pi\hbar)^3} \frac{\epsilon}{e^{\epsilon/kT} + 1}. \tag{6.174}$$

The neutrino temperature T_ν varies inversely with the expansion parameter, with $T = T_\nu$ at $kT \gg m_e$. The first term in F is the energy density in radiation. The second term represents the contribution from relativistic weakly interacting matter with energy density that varies as $a(t)^{-4}$. It is written as the effective number N_ν of neutrino families (eq. [6.70]), but it includes all the relativistic forms of energy that dominate the mass density at high redshift. The last term is the energy density in electron pairs (eq. [6.166]). If the expansion of the universe does trace back to the densities of interest here, the contributions of space curvature and the cosmological constant to the expansion rate equation (5.18) are quite negligible, so we have

$$\left(\frac{\dot{T}_\nu}{T_\nu}\right)^2 = \frac{8}{3}\pi G a_B T_\nu^{\,4} E = 0.047(T_{\nu,10})^4 E \text{ s}^{-2}. \tag{6.175}$$

We see that the expansion timescale is on the order of one second at the thermal energies for element production. The variation of the radiation temperature T with time in table 6.2 comes from numerical integration of equations (6.174) and (6.175) to get T_ν, with equations (6.171) and (6.172) for T/T_ν.

Next, we need the rates for the weak reactions (6.158) that convert the baryons between neutrons and protons. The best normalization still comes from the neutron half-life, as follows. Recall that in linear perturbation theory the transition probability per unit time to a definite final state is proportional to the delta functions of energy and momentum conservation. Thus we can write the probability per unit time for decay of a neutron, $n \rightarrow p + e^- + \nu$, in a box with volume V and periodic boundary conditions, as

$$\begin{aligned}
\lambda_d &= \kappa \int \frac{V d^3 p_\nu}{(2\pi\hbar)^3} \frac{2V d^3 p_e}{(2\pi\hbar)^3} \delta(Q - \epsilon_e - p_\nu) \\
&= \frac{\kappa V^2}{2\pi^4 \hbar^6} m_e^5 f,
\end{aligned} \tag{6.176}$$

where the factor κ represents the matrix element and the integral is

$$\int_{m_e}^{Q} p_e \epsilon_e d\epsilon_e (Q - \epsilon_e)^2 = m_e^5 f, \qquad f = 1.636, \tag{6.177}$$

with the neutron-proton mass difference in equation (6.159). Here and below we are ignoring the coulomb correction to the electron wave function.

Suppose the box contains a sea of electron pairs at temperature T and neutrino pairs at temperature T_ν. Then the probability per unit time that a proton in the box converts to a neutron by the capture of an electron, $p + e^- \rightarrow n + \nu$, is

$$\lambda_1 = \kappa \int \frac{2V d^3 p_e}{(2\pi\hbar)^3} \frac{1}{e^{\epsilon_e/kT} + 1} \cdot \frac{V d^3 p_\nu}{(2\pi\hbar)^3} \frac{\delta(\epsilon_e - p_\nu - Q)}{1 + e^{-p_\nu/kT_\nu}}. \tag{6.178}$$

The delta function with the integral over final neutrino momentum is the expression for the transition probability from a given incident electron momentum. The first factor is the sum over the thermal distribution of incident electron momenta. Since the decay can only leave the neutrino in a state that is not already filled, we have to multiply the integrand by the probability that the final neutrino state is empty,

$$1 - \frac{1}{e^{p_\nu/kT_\nu} + 1} = \frac{1}{1 + e^{-p_\nu/kT_\nu}} \, . \tag{6.179}$$

This is the last factor in equation (6.178). On dividing this equation by equation (6.176) for the free neutron decay rate, we get

$$\frac{\lambda_1}{\lambda_d} = \frac{1}{m_e^5 f} \int_0^\infty \frac{p_e \epsilon_e p_\nu^2 dp_\nu}{(e^{\epsilon_e/kT} + 1)(1 + e^{-p_\nu/kT_\nu})}, \qquad \epsilon_e = p_\nu + Q \, . \tag{6.180}$$

The rate for the reverse process, $n + \nu \to p + e^-$, similarly is

$$\frac{\hat{\lambda}_1}{\lambda_d} = \frac{1}{m_e^5 f} \int_0^\infty \frac{p_e \epsilon_e p_\nu^2 dp_\nu}{(e^{p_\nu/kT_\nu} + 1)(1 + e^{-\epsilon_e/kT})}, \qquad \epsilon_e = p_\nu + Q \, , \tag{6.181}$$

the rate for $p + \bar{\nu} \to n + e^+$ is

$$\frac{\lambda_2}{\lambda_d} = \frac{1}{m_e^5 f} \int_0^\infty \frac{p_\nu^2 p_e^2 dp_e}{(e^{p_\nu/kT_\nu} + 1)(1 + e^{-\epsilon_e/kT})}, \qquad p_\nu = \epsilon_e + Q \, , \tag{6.182}$$

and the rate for the reverse process $n + e^+ \to p + \bar{\nu}$ is

$$\frac{\hat{\lambda}_2}{\lambda_d} = \frac{1}{m_e^5 f} \int_0^\infty \frac{p_\nu^2 p_e^2 dp_e}{(e^{\epsilon_e/kT} + 1)(1 + e^{-p_\nu/kT_\nu})}, \qquad p_\nu = \epsilon_e + Q \, . \tag{6.183}$$

All these rates represent sums over the final spin state of the nucleon and averages or sums over the spins of the electron and neutrino, so we automatically have taken account of the spin-dependence of the matrix element.

We can ignore the blocking of $n \to p + e^- + \bar{\nu}$ by the pairs already present, and the reverse of this process, because by the time free decay is important the pairs are too cool to matter. The net rate coefficients for equation (6.162) are then

$$\lambda = \lambda_1 + \lambda_2 \, , \qquad \hat{\lambda} = \hat{\lambda}_1 + \hat{\lambda}_2 + \lambda_d \, . \tag{6.184}$$

At thermal equilibrium, the rate equation has to have the stationary thermal solution $n_n/n_p = \exp{-Q/kT}$, and consistent with this it is easy to check that when $T_e = T_\nu$ these coefficients satisfy $\lambda/\hat{\lambda} = \exp{-Q/kT}$ (apart from the term λ_d, which we have not treated in full). The general argument for this equilibrium abundance ratio uses the condition $\mu_n = \mu_p$, for μ_e is small and we are assuming the chemical potential for the neutrinos is small also.

The reaction rates are normalized to the free neutron decay rate. The value recommended by Olive et al. (1990) is

$$\lambda_d = (1.124 \pm 0.006) \times 10^{-3}\, \text{s}^{-1}\,. \tag{6.185}$$

The numbers for λt and $\hat{\lambda} t$ in table 6.2 are from numerical evaluation of the integrals in equations (6.180) to (6.183).

The only problem in integrating equation (6.162) for the neutron abundance is the numerical instability, which is eliminated by rewriting the equation as

$$\frac{n_n}{n_n + n_p} \equiv x(t) = x_o e^{-\int_{t_o}^{t}(\lambda + \hat{\lambda})\, dt'} + \int_{t_o}^{t} \frac{\lambda}{\lambda + \hat{\lambda}} de^{-\int_{t'}^{t}(\lambda + \hat{\lambda})\, dt''}\,. \tag{6.186}$$

Numerical integration of this equation with $N_\nu = 3$ gives the last column in the table.

The reactions $n + p \leftrightarrow d + \gamma$ tend to relax the abundance of deuterium relative to neutrons and protons to the thermal equilibrium value given by the Saha equation. This is based on (6.91) for the chemical potentials with the condition $\mu_p + \mu_n = \mu_d$, as in equation (6.88) for atomic hydrogen. The neutron and proton have two spin states each, and the deuteron has three states for spin one, so the equilibrium abundance ratio is

$$\frac{n_p n_n}{n_d n} = \frac{4}{3}\left(\frac{m_p m_n}{m_d}\right)^{3/2} \frac{(2\pi kT)^{3/2}}{(2\pi \hbar)^3 n} e^{-B/kT}$$

$$= \exp\left[25.82 - \ln \Omega_B h^2 T_{10}^{3/2} - 2.58/T_{10}\right]\,. \tag{6.187}$$

The present baryon number density is $n = 1.124 \times 10^{-5} \Omega_B h^2\, \text{cm}^{-3}$, and the density has been scaled as the cube of the radiation temperature. This is not quite right because the radiation is receiving entropy from the electron pairs, but the correction is small. The deuteron binding energy is

$$B = 2.225\, \text{MeV}\,. \tag{6.188}$$

In column 6 in table 6.2 this equilibrium deuterium abundance is evaluated for $\Omega_B h^2 = 0.015$.

The rate coefficient for the reaction $n + p \rightarrow d + \gamma$ is

$$\langle \sigma v \rangle \sim 4.6 \times 10^{-20}\, \text{cm}^3\, \text{s}^{-1}\,. \tag{6.189}$$

As for equation (6.127), σv is nearly independent of energy because the energy released in the reaction dominates the final state energy.

Figure 6.10 shows the results of a numerical integration of equation (6.186) for the evolution of the neutron abundance $x = n_n/(n_n + n_p)$ as a function of the radiation temperature, ignoring the sequestering of neutrons in deuterium and heavier nuclei. The parameter is the effective number N_ν of neutrino families in equation (6.174). Increasing N_ν increases the expansion rate (6.175), which freezes the neutron abundance at a higher temperature and higher neutron abundance. The short vertical lines mark the critical temperature at which the equilibrium deuteron abundance ratio in equation (6.187) passes through unity. The line at the higher temperature assumes $\Omega_B h^2 = 1$, and the line at the lower temperature assumes the value $\Omega_B h^2 = 0.013$ in equation (6.27) that gives the best present fit to the observations. One sees in the sixth column of table 6.2 that the equilibrium swings quite sharply from free nucleons at high temperature to deuterons at late times, so in effect deuterium production commences at this critical temperature. Because $\sigma n v t$ is larger than unity, as shown in the seventh column of the table, most of the neutrons present at the critical temperature are promptly incorporated in deuterons, most of which end up in helium.

Since ^{4}He contains equal numbers of protons and neutrons, the final helium abundance by mass is $Y = 2x$, where $x = n_n/(n_n + n_p)$ is the abundance of neutrons when the thermal equilibrium swings to favor deuterons. At $N_\nu = 3$ and $\Omega_B h^2 = 0.013$, the curve in the figure cuts the critical vertical line at $x = 0.11$, giving helium abundance $Y = 0.22$. This is considered to be close to the maximum allowed value for the primeval abundance consistent with observations of helium in systems with low heavy element abundances (Pagel 1991; Olive, Steigman, and Walker 1991). Increasing the baryon density parameter to $\Omega_B h^2 = 1$ increases Y to about 0.25, which seems to be observationally unacceptable, and increasing the relativistic energy density parameter to $N_\nu = 4$ is almost as bad.

A more complete computation follows the evolution of the abundances of deuterium and the other light nuclei. This has little effect on the computed ^{4}He abundance, for the reasons we have discussed. The most important addition is the predicted residual deuterium abundance in figure 6.5. Lowering the baryon density lowers the efficiency of burning the deuterium, along with the intermediate tritium and ^{3}He, leaving higher residual abundances of these intermediate isotopes. The interpretation of these and the other trace elements requires fine arguments that can be contentious, because these nuclei are readily destroyed in stars and it always is conceivable that Nature has found some clever way to produce them.

The ^{4}He abundance is of special interest because lowering the present value from what came out of the early universe would require even more unusual circumstances than what would be needed to cause appreciable changes in the abundances of the other light elements. What are likely responses to a believable observation of a helium abundance in another galaxy at a value significantly below $Y \sim 0.22$? Current ideas include photodissociation of helium by radiation from decaying or annihilating exotic particles; thermal destruction in early generations of very massive stars; small-scale fluctuations in the primeval baryon-to-entropy

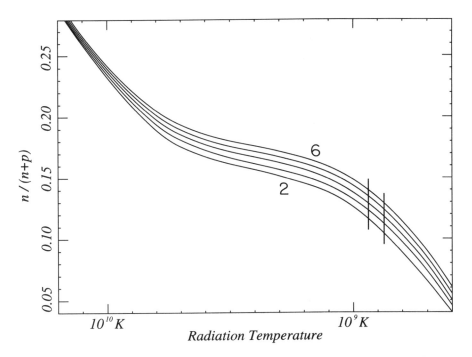

Figure 6.10. Neutron abundance as a function of radiation temperature. The parameter is the effective number N_ν of neutrino families in equation (6.174). The vertical lines mark the critical temperature where deuterium can accumulate and burn to helium, for baryon density parameters $\Omega_B h^2 = 1$ and 0.013.

ratio, perhaps caused by fluctuations at the quark-hadron phase transition; and large-scale variations in the primeval baryon-to-entropy ratio, with the prediction that where the ratio is low, to make a low helium abundance, the deuterium is very high. Since none of these pictures seem particularly attractive, the difficult art of securing precise and believable helium abundances will be followed with particular attention.

The baryon density parameter for the best fit to helium and the other light element abundances shown in figure 6.5, $\Omega_B h^2 \sim 0.013$, is between the lower bound from the seen baryons in the bright parts of galaxies (eq. [5.150]) and the dynamical measures of the total mass density discussed in section 20, $\Omega \gtrsim 0.1$. This is a comforting consistency check, though we must note that the check is loose because the best choice for $\Omega_B h^2$ is not close to either observational measure. That is, if all these numbers are correct, we must invoke hypothetical baryonic and nonbaryonic dark matter. Both are entirely possible, and the former is even observationally reasonable, but still it will be interesting to see whether the situation might simplify.

The standard model bounds the possible mass density in gravitational radiation or other massless quanta in the very early universe, through the effect on the expansion rate through the epoch of light element production (Peebles 1966; Shvartsman 1969). The fit to the observed abundances requires $N_\nu < 4$, meaning the standard model requires that there was little beyond the three families of neutrinos that are known to exist and are the limit allowed by collider measurements (Walker et al. 1991). The consistency is a dramatic success for the standard cosmological model.

7. Alternative Cosmologies

Many of the elements of the standard hot big bang cosmological model were in place six decades ago. Lemaître (1927) gave a clear explanation of the physics of an expanding universe within general relativity theory. A more standard and complete reference for cosmological models was the review article by Robertson (1933). In the 1930s Robertson, Tolman, and others worked out the classical cosmological tests for the predicted effect of the expansion of the universe on observations of distant galaxies, and work began on the 200-inch telescope on Mount Palomar, which, it was hoped, might allow the application of some of these tests. In the 1930s Lemaître explored what were to become central elements of physical cosmology, including the gravitational instability of the expanding universe as a possible explanation for the observed tendency of matter to be concentrated in galaxies and clusters of galaxies, and the physics of the violent early stages of expansion that might have left behind observable remnants. Tolman had explained the behavior of a particularly important remnant, blackbody radiation. The evolution of all these ideas into what now might be called a mature physical science has been so slow because it has depended on a wide range of developments in technology and an enormous effort in the application. As one example, it has only recently been possible to show that galaxies observed at redshifts $z \gtrsim 1$, when in the standard model they and the universe were young, really do look something like the younger cousins of the galaxies in our neighborhood.

It is natural to ask whether such a slowly moving subject really can be promising, whether there might not be other more likely cosmologies. People have considered a series of candidates that arguably are at least no less plausible a priori than the hot big bang, and the debate over their merits has been an important stimulus to progress in research and in the development of concepts. At a time when it was not possible to make very meaningful tests of the relativistic Friedmann-Lemaître picture, the discovery of the steady-state cosmology was particularly important as a stimulus for research (and even for enraging the participants). Alternative cosmologies give us a way to gauge the significance of observational tests. For example, Milne explained that the linear redshift-distance relation follows from the assumptions of homogeneity and the nonrelativistic law of addition of velocities for neighboring observers (eq. [5.2]). Since we have precision laboratory tests of Lorentz invariance, and as discussed in section 3 the tests for large-scale homogeneity seem quite convincing, it is likely that any interesting cosmology will contain good approximations to these symmetries. Thus it is reasonable to count the linear redshift-distance relation as an element in the net of evidence for the homogeneous large-scale structure of the universe, but not for the Friedmann-Lemaître picture in particular.

The darker sides to the tradition of dissent in cosmology are the overreactions of those who consider themselves the guardians of the true and canonical faith (as presented in this book), and the tendency for dissent to slip into pathological science. There are documented examples of frivolous criticism of dissenting arguments, which is irrational and destructive. On the other hand, it is doubtful that the establishment view has been overly reactionary on average, for we observe that most cosmologists would be delighted to abandon the standard model for something new to think about if only the alternatives looked reasonably promising. Minor examples are the bursts of activity produced by what are at best moderately interesting ideas within the standard model. We also observe that there has not been a high rate of production of ideas in cosmology that have proved to be of lasting interest. This has two implications: the old ideas have proved to be durable, and the standard model remains schematic, with considerable room for improvements and even revolutionary changes.

Those who would seek a revolution in cosmology must bear in mind that the days are gone when it was easy to think of viable alternatives. Now any serious attempt at a revision of the main elements of the standard world picture would involve a survey of a considerable (though certainly limited) store of observational and laboratory constraints. An excellent picture of the state of the art is given in the proceedings of the conference "Theory and Observational Limits in Cosmology" (Stoeger 1987). There are two lessons: nature is quite capable of surprising us, but in a mature science surprises are rare. The major surprise in the century since Maxwell's unification of electromagnetism is the quantum principle, but the expansion of the universe is startling enough, and we do know that the standard cosmological model has to be pointing to another surprise of some sort, because the model applied in a straightforward way traces back to a singularity.

Despite the fact that surprises must lurk there is good reason for the tabu against the postulate of new physics to solve new problems, for in the silly limit one invents new physics for every new phenomenon. We become convinced that a physical science is a believable approximation to reality in the opposite limit, where new results are predicted or could have been predicted from the accepted physics. For example, superconductivity was a surprise only because quantum mechanics is so subtle. Is the missing mass puzzle in extragalactic astronomy (section 18) a result of the application of the wrong gravity physics, or only of our limited imagination in the telling of the masses? The record suggests we ought to put most of our money on the latter possibility, but save a little for the former.

The steady-state theory is discussed here as a part of the central lore of cosmology. The other examples are selected for their use in clarifying the significance of various cosmological tests. Ideas on the physics of the very early universe are placed in part 3, because they have not been subject to serious tests.

Milne

E. A. Milne was one of the most interesting figures in cosmology in the 1930s. In earlier work he had contributed to the theory of stellar structure, with emphasis on the analysis of radiative transport in stellar photospheres. His kinematic cosmology now receives little attention, but his methods, and his critique of the Friedmann-Lemaître cosmology, have had a lasting influence. They are summarized in his book, *Relativity, Gravitation and World Structure* (Milne 1935).

Milne pointed out that it is trivial to find a model for Hubble's law, the linear relation between redshift and distance. Consider a collection of noninteracting particles in a bounded region in flat spacetime. Suppose the particles suddenly are given velocities, perhaps by an explosion. We will imagine that the velocities are drawn from an isotropic and broad distribution. Then as the particles move freely, the fastest move farthest away, leading asymptotically to a linear relation between recession velocity v and distance r from the explosion,

$$v \to r/t, \tag{7.1}$$

where t is the time since the explosion. This velocity sorting effect reappears in the plasma universe discussed below.

Milne showed that Hubble's law follows from the assumptions that the universe is homogeneous, that it is expanding in the sense that the proper distances between neighboring comoving observers are increasing, and that we can apply the usual law of vector addition of relative velocities. This is the argument in equation (5.2).

Milne (1934) and McCrea and Milne (1934) showed how the Friedmann-Lemaître equation (5.17) for the expansion rate can be derived using elements of Newtonian mechanics (as in eqs. [4.21] to [4.28] and [5.14] to [5.17]). As for so much of the fundamental physics of the standard model, the point was first made by Lemaître (1931e), but in a popular article that was not much noticed by cosmologists.

Within the standard Friedmann-Lemaître cosmology, the Newtonian computation is a simple way to apply Einstein's field equation (though more work is required to make the connection between the constant of integration in eq. [5.17] and the radius of curvature of space sections of fixed world time). However, the derivation really only requires that Newtonian gravity apply in the weak-field low-velocity limit, as well as some arrangement to prevent the enormous mass of distant matter from affecting the local expansion rate. The former is a condition for any reasonable gravity physics, because the Newtonian theory works so well on scales ranging from the Solar System to the inner parts of galaxies. In general relativity theory, the latter is Birkhoff's theorem.

In Milne's cosmology spacetime is flat, as in special relativity. He showed that under the cosmological principle observers can be assigned fixed spatial coordinates r, θ, and ϕ, with t the proper time kept by the comoving observers, and that the line element is

$$ds^2 = dt^2 - t^2 \left(\frac{dr^2}{1+r^2} + r^2 d\Omega \right) . \tag{7.2}$$

It is an interesting exercise to show how one knows (from the theorems of Birkhoff and of Robertson and Walker) that this must be a coordinate labeling of flat spacetime, and to find the coordinate transformation that brings it to the standard Minkowski form.

We saw in section 5 that if the standard relativistic model with pressure $p \geq 0$ is applied all the way to the singularity, $a \to 0$, it has a particle horizon (eq. [5.52]). This means we can observe galaxies that have not been in causal contact with each other subsequent to the singularity, and so arguably never have been in contact. How can these galaxies know how to look so similar? Milne considered the particle horizon an argument against relativistic cosmology. The recent tendency is to assume this embarrassment can be resolved by inflation or some other adjustment of the physics of the very early universe.

Milne introduced the phrase "Einstein's cosmological principle," or for short the "cosmological principle," for the assumption that the universe is spatially homogeneous and isotropic in the large-scale average. His attempt to base a complete world picture on this principle is no longer considered interesting, but it did make clear the power of the homogeneity assumption, and his use of it foreshadowed the steady-state cosmology that became the major competitor to the Friedmann-Lemaître model in the two decades following the Second World War.

The Steady-State Cosmology

An excellent description of the state of thinking in cosmology in the late 1940s is given by Bondi (1960). Bondi and others emphasized the remarkable observation that, as far as can be deduced from the broadband spectra and the arrangement of the spectral lines in the light from distant galaxies, the gas and stars in these remote objects are governed by the physics we observe operating nearby. An example discussed in section 5 is the close agreement of redshifts derived from the radio 21-cm line and the optical lines in the spectra of objects with redshifts well in excess of unity. It is natural to ask whether this remarkable stability of the physics might be related to the fact that the mass distribution is close to homogeneous, for Einstein did hit on the cosmological principle through the idea that inertia is determined by the distribution and motion of matter. Galaxies at redshifts greater than unity are seen as they were in the past, when the expanding model would say the universe was denser. We have other glimpses of the past from the pattern of decay products of long-lived radioactive nuclei in meteorites and minerals. Could such observations reveal evidence of evolution of the dimensionless constants of physics induced by the evolution of the universe? The question was addressed by Dicke (1970); the search has revealed fascinating hints but so far no pronounced indication that physics is evolving. The usual interpretation is that there apparently is not a tight connection between microscopic physics and the state of the universe. The alternative considered in the steady-state cosmology is that the universe is not evolving. Bondi and Gold (1948) expressed this idea as the perfect cosmological principle. Hoyle (1948) adopted the same steady-state picture, but arrived at it by writing down a generalization of general relativity theory. The approach from the perfect cosmological principle will be discussed here.

According to this concept, the universe is spatially homogeneous and isotropic on average, that is, unchanged by translations or rotations in space, and it is unchanged also under time translations, that is, it is in a steady state. To provide fresh hydrogen to keep a steady supply of young stars, it is assumed that there is spontaneous creation of matter, at a steady mean rate. To remove the evolved stars and the radiation they have produced, the universe has to be expanding. It is assumed that the geometry of spacetime is described by a metric tensor. Since the universe is spatially homogeneous and isotropic, we can write the line element in the general Robertson-Walker form of equation (5.9). Since the radius of curvature of constant time sections, $a(t)R$, is in principle observable, and observable quantities describing the structure of the universe are supposed to be independent of time, R^{-1} has to vanish. The rate of expansion of the universe is (eq. [5.2])

$$\dot{a}/a = H_o . \tag{7.3}$$

Since H_o must be constant, the expansion parameter is an exponential function of time. Thus the line element (5.9) has to reduce to

$$ds^2 = dt^2 - e^{2H_o t}(dx^2 + x^2 d\Omega) \tag{7.4}$$

in the steady-state theory. This is the de Sitter solution (eq. [5.24]), but here the form is dictated by symmetry, not by general relativity theory.

The mean number density n of baryons satisfies the equation

$$\dot{n} = -3n\dot{a}/a + C . \tag{7.5}$$

The first term on the right-hand side is the effect of the expansion of the universe, as in equation (5.16). In a steady state $\dot{n} = 0$, so the mean rate of continual creation of baryons in the second term is

$$C = 3H_o n. \tag{7.6}$$

With the numbers in equations (5.65), (5.68), and (5.150), the mean creation rate is

$$C \sim 10^{-24} \text{ baryons cm}^{-3} \text{ s}^{-1}. \tag{7.7}$$

This would be detectable only if the creation were preferentially near preexisting matter (such as us).

Since the stars in galaxies eventually die, and the galaxies are being swept away by the general expansion of the universe, new galaxies have to be forming out of the newly created matter, at the rate $C_g = 3H_o n_g$ to maintain the mean number density of galaxies at n_g. At a given epoch, t_o, the mean number of galaxies in the volume V that have ages in the range t to $t + dt$ is the mean number that were created in the same comoving region at the size it had at the epoch $t_o - t$:

$$dN = dt\, C_g V [a(t_o - t)/a(t_o)]^3. \tag{7.8}$$

The mean distribution in galaxy ages t at any epoch is then

$$\frac{dn_g}{dt} = 3H_o n_g e^{-3H_o t}. \tag{7.9}$$

It follows that the mean age of a galaxy is

$$\langle t \rangle = 1/(3H_o). \tag{7.10}$$

When the steady-state cosmology was discovered, the accepted value of Hubble's constant was the original estimate from fifteen years earlier, $h \sim 5$, and the Hubble time, $H_o^{-1} \sim 2 \times 10^9$ y (eq. [5.65]), was uncomfortably close to the estimates of the radioactive decay age of the Earth. This was a problem for the Friedmann-Lemaître model, but certainly not difficult for the steady-state theory because some galaxies are considerably older than the mean. Section 5 describes the correction to the timescale by Baade and Sandage, which contributed to a shift of attention to pioneering work on the galaxies much younger than the Hubble age H_o^{-1} to be expected in the steady-state picture (Burbidge, Burbidge, and Hoyle 1963). As we will discuss shortly, the evidence for youthful-looking galaxies at low redshifts continues to accumulate, but it now seems clear that the abundance of these immature galaxies increases with increasing redshift, contrary to what would be expected in the steady-state picture.

The cosmological tests to be discussed in sections 13 and 14 are based on the counts and observed properties of distant objects compared to nearby ones. Since the line element in the steady-state cosmology is fixed (eq. [7.4]), and in this theory distant objects are on average the same as the ones observed nearby, the theory gives definite predictions for testable relations such as the angular sizes of galaxies as a function of redshift, or the counts of extragalactic radio sources as a function of flux density. The conclusion is that the perfect cosmological principle gives an elegant path to interesting and testable predictions in cosmology. Evidence of problems with the predictions has surfaced only fairly recently. We begin with the evolution of galaxies.

The prediction that galaxies at all redshifts are drawn from the same distribution is tested by comparing morphologies and spectra of galaxies at different redshifts. Stebbins and Whitford (1948) found indications that the colors of galaxies correlate with z, but Whitford and others later concluded that this was a systematic error (Oke and Sandage 1968). Color evolution reappeared, with the opposite sign, in the discovery of the Butcher-Oemler (1978) effect, that at redshifts $z \gtrsim 0.3$ the galaxies in rich clusters tend to be bluer than is typical of cluster members at lower redshifts. This agrees with the observation that cluster members at $z \sim 0.4$ to 0.6 tend to have spectra indicating an unusually high population of main-sequence A stars, whose lifetimes are on the order of 10^9 years (Dressler and Gunn 1983; Gunn and Dressler 1988). The conclusion is that these cluster members have passed through recent bursts of star formation, as would befit their youth. Dressler (1987) estimates the rate of occurrence of this phenomenon at the present epoch is down by at least an order of magnitude from what is observed at $z \sim 0.5$.

Another example of galaxy evolution is given by the radio galaxies shown in the redshift-magnitude diagram in figure 5.3. The radio and optical appearances of these galaxies show a distinct trend with redshift. At $z \lesssim 0.6$ the radio lobes exhibit some preference for alignment along the minor axes of the optical images, while at larger redshifts there is a distinct preference for alignment of radio and optical axes, and for increasingly prominent clouds of gas (identified by the emission lines) and increasingly irregular images in the optical (ultraviolet at the source) (Chambers and McCarthy 1990). The continuity and small scatter in the redshift-magnitude relation in figure 5.3 suggest the objects at $z > 1$ are drawn from the same physical population as the low redshift galaxies, but as they were when they were young.

Faint galaxy counts, at redshifts $z \sim 0.5$, are larger at galaxy rest frame wavelengths in the ultraviolet than in the red. That is, field galaxies were considerably bluer at $z \sim 0.5$, and there likely was a larger number density of galaxies with ultraviolet luminosities comparable to L_* (Cowie 1991).

The purpose of this lengthy sampler is to emphasize that there is abundant and unambiguous evidence that galaxy properties correlate with redshift. The indicated evolution is in the direction expected in the standard model, where high redshift galaxies are seen as they were when they were young. And these observations are inconsistent with the classical steady-state cosmology.

As it happened, most people were led to abandon the steady-state picture before galaxy evolution was unambiguously observed. The deciding event was the difficulty of understanding the thermal cosmic background radiation and the light element abundances within this cosmology and, as discussed in the last section, the apparent ease of understanding them within the Friedmann-Lemaître picture. With the recent dramatic advances in the measurement of the CBR spectrum, it is useful to consider in more detail the problem of accounting for the CBR if the universe has always been as it is now. The steady-state cosmology is ideal for this purpose because, as we have noted, it makes it easy to arrive at cleanly defined models.

It would be reasonable to suppose radiation is produced along with the continuous creation of baryons, but absurd to imagine the created radiation is just such that the sum over redshifts adds up to a blackbody spectrum. Let us suppose therefore that the spontaneously produced radiation is absorbed and reradiated by intergalactic dust grains at a fixed grain temperature, T. Let κ be the opacity per unit path length through this dust. To keep the discussion simple, let us assume κ is independent of wavelength in the CBR band.

Recall that in a static spacetime the radiative transfer equation for the rate of change of the surface brightness $i(\nu)$ (energy flux per unit area, solid angle and frequency interval) with respect to displacement along the path of the radiation is

$$\frac{di}{dl} = \frac{1}{c}\frac{di}{dt} = \frac{j}{4\pi} - \kappa i, \tag{7.11}$$

where j is the luminosity density (energy radiated per unit time, volume, and frequency interval), and κi is the loss of surface brightness by absorption per unit path length. For blackbody radiation, $i(\nu)$ is the Planck function (6.12),

$$i \to P(T,\nu) = \frac{2h\nu^3}{c^2}\frac{1}{e^{h\nu/kT} - 1}. \tag{7.12}$$

If the radiation were blackbody at the same temperature as the dust, the right-hand side of equation (7.11) would have to vanish, absorption balancing emission. This means that if the dust grains are at temperature T their luminosity density is

$$j(\nu) = 4\pi\kappa P(T,\nu). \tag{7.13}$$

In the expanding steady-state universe, the radiation emitted at time t in the past is redshifted by the factor

$$(1+z) = \frac{\lambda_o}{\lambda_e} = \frac{a(t_o)}{a(t_o - t)} = e^{H_o t}. \tag{7.14}$$

The contribution to the sky brightness by the radiation produced in the time interval t to $t + dt$ in the past starts out as $\kappa P(T, \nu) c \, dt$, by equations (7.11) and (7.13). Absorption by dust attenuates this contribution by the factor $\exp -\kappa c t$. The sequence of velocity transformations leading to equation (6.45) shows that the temperature parameter for this radiation at the detector is shifted to $T/(1+z)$. The detected integrated surface brightness of the radiation is then

$$i(\nu) = \int_0^\infty \kappa c \, dt \, \frac{2h\nu^3}{c^2} \frac{1}{e^{h\nu y/kT} - 1} e^{-\kappa c t}, \tag{7.15}$$

with $y = e^{H_o t}$. This is equation (5.159), with the addition of the absorption term, and equation (7.13) for the luminosity density.

Absorption by the dust decreases the observed flux density from a distant radio source. This is described by equation (7.11) with $j = 0$, giving

$$f_o/f = e^{-\kappa c t} = (1 + z)^{-\kappa c/H_o}, \tag{7.16}$$

where the observed flux density from a source at redshift z is f_o, and f is the flux density expected in the absence of absorption.

The dashed line in figure 7.1 shows the spectrum predicted by equation (7.15), with the temperature T adjusted to get the best fit to the COBE observations shown as boxes (Mather et al. 1990), and $\kappa c/H_o = 3.3$. With this opacity, the optical depth for absorption is unity at redshift $z = 0.3$ ($f_o/f = e^{-1}$ in eq. [7.16]). This theoretical spectrum departs from the observations by many times the measurement errors, which are well below the box size in the figure. With

$$\kappa c/H_o = 10, \tag{7.17}$$

the predicted spectrum still is convincingly excluded by the CBR spectrum measurements. With this opacity, the extinction (7.16) at redshift $z = 2$ is close to five orders of magnitude. This certainly does not agree with the observations of radio galaxies in figure 5.3 at redshifts greater than two and with apparently normal ratios of radio to optical luminosities.

The point of this calculation is that if the universe were postulated to be opaque enough at radio wavelengths to have caused the radiation background to relax to the observed very nearly thermal spectrum of the CBR, space would be predicted to have been too opaque to have allowed the observations of distant radio sources (Peebles et al. 1991; Wright 1992).

It is an interesting exercise to consider the properties of the dust grains that would be needed to produce the opacity in equation (7.17). Following Wright (1982), who was considering the idea that the CBR is starlight thermalized at red-

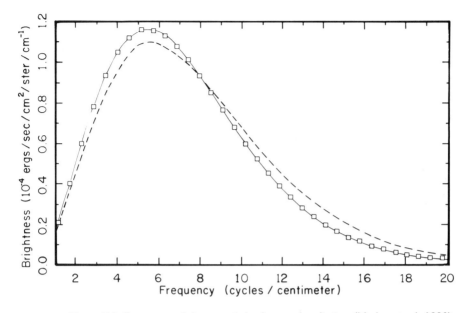

Figure 7.1. Spectrum of the cosmic background radiation (Mather et al. 1990). This is an earlier version of the measurement shown in figure 6.1. The dashed line shows the steady-state prediction in equation (7.15) with opacity $\kappa c/H_o =$ 3.3 (Peebles et al. 1991).

shift $z \sim 100$ in an initially cold Friedmann-Lemaître universe, suppose the grains consist of straight needles all with the same diameter d and length l, which is small compared to CBR wavelengths $\lambda \sim 1$ cm, and made of material with conductivity σ. That is, we are modeling the grains as wire segments with resistance $R \sim l/(\sigma d^2)$. Consider one of these needles placed in a radiation field with the electric field vector $E = E_o \cos \omega t$ pointing along the needle. If the current is limited by the resistance of the wire segment, rather than its electrostatic capacity, the current is $i = \sigma d^2 E$. Then the charge on the wire ends is $Q \sim i/\omega$, and the condition that the current is not limited by the capacity of the wire segment is

$$Q/l^2 \sim \sigma d^2 E/(\omega l^2) \lesssim E .\tag{7.18}$$

Thus the condition on the radiation frequency for resistance-limited current is

$$\omega \gtrsim \sigma(d/l)^2 .\tag{7.19}$$

Still following Wright (1982), let us suppose the grains are graphite, with conductivity

$$\sigma \sim 10^{14}\,\mathrm{s}^{-1}\,. \tag{7.20}$$

At 1-cm wavelength, the frequency is $\omega \sim 10^{11}\,\mathrm{s}^{-1}$. Then the condition (7.19) is satisfied if the aspect ratio for the grains is $l/d \gtrsim 30$, which is not unreasonable for graphite whiskers.

In this simple resistance-limited case, the current density within a grain is σE, so the mean rate of dissipation of energy per unit volume is σE^2, and the rate of dissipation of energy per unit volume of space is

$$\mathcal{P} = n_g v_g \sigma \langle E^2 \rangle / 3 = \kappa c \langle E^2 \rangle / 4\pi\,. \tag{7.21}$$

In the second expression v_g is the volume of a grain, n_g is the number density of grains, and the factor of three corrects to the mean square component of the electric field along the grain. In the third expression κ is the resulting opacity, or optical depth per unit distance, defined in equation (7.11). Let us take the mass density within a grain to be that of graphite, $\rho_g \sim 2\,\mathrm{g\,cm}^{-3}$. Then, if the opacity of space is the number in equation (7.17), the mean mass density in these grains is

$$\langle \rho \rangle = n_g v_g \rho_g = (10 H_o)(3\rho_g / 4\pi\sigma) \sim 10^{-31}\,\mathrm{g\,cm}^{-3}\,. \tag{7.22}$$

If the conductivity were degraded by the ionizing radiation in interstellar space, or if the aspect ratio in the grains were less favorable than in equation (7.19), it would increase the needed mass density.

The abundance of carbon observed in stars and gaseous nebulae is less than one percent by mass relative to hydrogen, and it would not be plausible to imagine that most of it is in these grains. Since the total mass density in galaxies (eq. [5.149]) is comparable to the wanted mass density in grains in equation (7.22), the conclusion is that grains are not likely to make the universe as it is now opaque at radio wavelengths. That is a good thing, for it agrees with the fact that high redshift radio galaxies are observed.

We noted in the last section that the dust picture is considerably less problematic if the CBR is thermalized at redshift $z \sim 100$ in an evolving Friedmann-Lemaître universe, because the universe is denser, the CBR wavelengths shorter and more readily coupled to the dust, and there is no reason to think the universe has to be transparent (Wright 1982). But of course this is not allowed by the perfect cosmological principle.

Hoyle has consistently emphasized that the perfect cosmological principle could only apply in the sense of a time average, just as the cosmological principle applies in the sense of an average over large enough spatial volumes. Independent of the discovery of the CBR, Hoyle and Narlikar (1966) had considered extending this point to the idea that matter creation occurs in bursts between which the universe expands as a conventional evolving world model. If the interval between bursts were long enough, and the bursts correspondingly intense, the results could be observationally indistinguishable from the standard model.

The steady-state cosmology, the inflation scenario to be discussed in section 17, and the standard cosmological model each were found by a bold *ansatz* that had no direct empirical basis, but led an elegant world picture. The classical steady-state cosmology has floundered, first on the CBR and light element abundances, more recently on the observed evolution of galaxies. The observation of evolution is a success for the standard model, and commonly considered to be a general philosophical problem as well, because it generally is believed that the universe ought not to have a beginning or end. It is interesting to note the similarity of the Hoyle-Narlikar resolution of this conundrum and versions of the inflation scenario in section 17.

Plasma Universe

The cosmology of Klein (1971) and Alfvén follows another attractive and sensible path that keeps to standard physics and minimal extrapolations from what is observed of the universe around us. But as we will discuss, there is no way the results can be consistent with the isotropy of the CBR and X-ray backgrounds.

In his book, *Worlds-Antiworlds*, Alfvén (1966) raises an interesting question. Given that the galaxies are observed to be moving apart, and that the natural interpretation is that the density averaged over the observed system of galaxies is growing lower, why extrapolate this motion all the way back to the enormous densities of a big bang? We cannot observe the transverse components of the motions of distant galaxies, but it is not difficult to imagine they are present. If so, one might expect that the transverse motions were larger in the past, and that there was a time and a finite mean density at which transverse and radial motions were comparable. Prior to that, if galaxies already existed, they would have been moving together, in a general contraction of the material universe.

The beginning assumption of the plasma universe is that the material in the galaxies we observe originated as a dilute, slowly contracting cloud of matter and antimatter. That goes considerably beyond what is observed, but the big bang is by far the greater extrapolation. As the cloud contracts, matter and antimatter would start to annihilate. The pressure of the annihilation radiation eventually would become high enough to reverse the contraction and turn it into violent expansion. One can imagine that instabilities in this explosion are capable of piling matter into clumps such as galaxies, and perhaps also capable of segregating matter and antimatter well enough to avoid violating the limits on the present rate of annihilation from bounds on the gamma ray luminosities of well-mixed systems such as the plasma in rich clusters of galaxies (Steigman 1976; eq. [18.114] below). The Milne velocity sorting process would ensure that the galaxy motions approach Hubble's law well after the explosion (eq. [7.1]).

The size of the cloud of gas and galaxies in this picture is limited by the condition that during the collapse phase the gravitational potential energy per unit mass, ϕ, cannot exceed the critical value for relativistic collapse of the bulk of the material to a black hole, for the remnant would be hard to miss. On writing the present mean mass density within the system in terms of the density parameter in equation (5.55), and the present radius of the system as $r \sim cz_r/H_o$, where z_r is the redshift at the edge, we have that the present potential energy per unit mass is

$$\phi \sim \frac{GM}{rc^2} \sim \frac{1}{2}\Omega z_r^2 . \tag{7.23}$$

The dynamical estimates to be discussed in section 20 indicate that the density parameter is not less than about $\Omega \sim 0.1$. Galaxies are seen in abundance in redshift surveys to at least $z \sim 0.5$. (Many more galaxies are observed at higher redshifts, but let us suppose this high velocity tail carries relatively little net mass.) The present value of the dimensionless potential ϕ is then at least $\phi \sim 0.01$. This means that, if the cloud radius at maximum collapse were less than a few percent of the present value, the conventional physics of general relativity theory says the cloud would have suffered relativistic collapse, however violent the explosion. More important, if z_r were much larger than 0.5 it would leave no room for collapse and reexpansion. That is, the edge of the system has to be at a modest redshift, and the edge therefore can be probed by available observations. The powerful test is the angular distribution of radiation from distant matter.

It would be hard to imagine that the explosion produced a spherically symmetric expanding system of galaxies, and if that happened it would be quite implausible to imagine we are in one of the few galaxies at the center of symmetry. Therefore, the prediction of this picture surely is that the surface density of material integrated along a line of sight to the edge of the system of galaxies is substantially different in different directions. That is not seen in the radio source map in figure 3.10, despite the fact that many of these sources are known to be at $z \gtrsim 1$, as one sees in figure 5.3 for radio galaxies. The X-ray background in figure 3.11 is isotropic to $\delta i/i \lesssim 3\%$ on angular scales $\sim 3°$, and $\delta i/i \lesssim 0.001$ on an angular scale of one radian. As in equation (7.11), the X-ray background is a line integral of the source density along the line of sight. Since individual X-ray sources are observed at redshifts greater than unity, the isotropy of the integrated background could not be due to scattering of the radiation. Rather, it is telling us once again that the integrated mass per unit area out to the Hubble length is quite close to isotropic.

The centimeter to submillimeter cosmic background radiation also could not have been made isotropic by scattering, for high redshift objects are observed at radio wavelengths. One might imagine the CBR was present to begin with, with a thermal spectrum, along with the original dilute cloud of matter and antimatter. Since the CBR moves freely through the present expanding system of galaxies, as does the radiation from observed radio sources, the dipole anisotropy of the

CBR says the velocity of the Local Group relative to the center of mass of the primeval cloud is only a few hundred kilometers per second (eq. [6.29]). Since the redshifts of most galaxies correspond to at least mildly relativistic speeds of recession away from us, our low velocity relative to the CBR would have to be a truly remarkable coincidence.

The standard cosmological model steps from the observed roughly homogeneous distribution and expansion of the nearby galaxies to a universe that is close to homogeneous within the Hubble length. That is a considerable extrapolation, but one that we see has received impressive observational support from the isotropy of the radiation backgrounds. The alternative is that we are very close to the center of a bounded spherically symmetric system, but that seems highly unlikely, for there appears to be nothing special about our galaxy. In the inflation scenario, the nearly homogeneous universe we observe is bounded, but the redshift at the boundary can be enormous, adjusted to avoid any problem with the observations.

Fractal Universe and Large-Scale Departures from Homogeneity

The idea that the galaxy space distribution might be a pure scale-invariant fractal, or clustering hierarchy, traces back to Charlier's map in figure 3.1. Mandelbrot (1975a) gives a fascinating survey of still earlier arguments. The geometrical picture is elegant, but since it has not been translated into a physical model we cannot discuss some of the precision cosmological tests. In particular, it is not clear how to deal with such observations as the isotropic radiation backgrounds, the thermal spectrum of the CBR, or the cosmological redshift. It is a useful exercise, however, to confine the discussion to a single question: From the observations of the galaxies, can we find a convincing test for or against the idea that the galaxy space distribution is a fractal extending to very large scales? The conclusion from the analysis to be presented here is that, if a single fractal dimension is adjusted to fit the observations of galaxy clustering on small scales, the predicted large-scale distribution strongly violates well-established observations of galaxy counts as a function of apparent magnitude (Sandage, Tammann, and Hardy 1972), and counts as a function of direction at fixed depth (LSS, § 62). If the fractal dimension (eq. [7.24]) for the large-scale distribution were close to $d = 3$, it would remove the problem with the counts. We will see that to get consistency with the large-scale isotropy of the X-ray background, either we live at the center of a spherically symmetric universe or d differs from homogeneity, $d = 3$, by no more than about one part in 10^3 (eq. [7.70]).

The analysis presented here is lengthy but useful for a broader purpose, for it yields a successful positive test of the assumption that the galaxy space distribution is a stationary random process, and it yields a numerical measure of the length scale on which the fluctuations from homogeneity become small.

In astronomy the original name for a fractal distribution is a clustering hierarchy. In a scale-invariant clustering hierarchy, observations at any length scale show that the mass tends to be concentrated in clusters. Observations at a larger scale would show that these concentrations tend to appear in superclusters, and observations at a smaller scale that the concentrations break up into subclusters. For a more explicit prototype, place a stick of length l. At each end place sticks of length l/λ, with random orientations. On each end of each stick place sticks of length l/λ^2, randomly oriented. Continue, shrinking the length by successive powers of $\lambda > 1$, until there are enough sticks that we can place a galaxy at the ends of each of the shortest ones. The stick we started with could be on one end of a stick of length $l\lambda$, and so on. In the extreme, one could imagine that all observable galaxies are on the ends of the shortest sticks in one hierarchy.[18]

The fractal dimension of the galaxy distribution is based on the mean value $N(<r)$ of the number of galaxies found within distance r of a galaxy, averaged over counts centered on a fair sample of galaxies. In a scale-invariant fractal, with all the galaxies in the same hierarchy, this statistic varies as a power of the distance

$$N(<r)=Ar^d,\qquad(7.24)$$

where A is a constant and the index d is Mandelbrot's fractal dimension. In a spatially uniform distribution, $N(<r)$ would be proportional to the volume enclosed, so d would be the dimension of the space, here three. To get d for the stick construction, note that if r is increased to λr it increases N by a factor of two, because on average we have doubled the number of levels of the hierarchy within the distance to which we are counting. This gives

$$N(<\lambda r)=2Ar^d=A(\lambda r)^d,\qquad(7.25)$$

so the fractal dimension, d, is given by the equation

$$\lambda^d=2.\qquad(7.26)$$

In this model, the mean number density of galaxies within distance r of a galaxy is

$$n(r)\propto N(<r)/r^3\propto r^{-\gamma},\qquad\gamma=3-d.\qquad(7.27)$$

---

[18] Figures 19.1 and 19.2 show model galaxy maps based on a variant of this construction, with finite clustering hierarchies built of sticks, and started at random places so as to satisfy the cosmological principle. This gives a reasonable approximation to the observed galaxy distribution, but it is not a scale-invariant fractal because the density fluctuations average to zero on scales large compared to the mean distance between independent clustering hierarchies.

In three spatial dimensions a fractal dimension $d < 3$ has zero mean density, in the sense that the average in equation (7.27) approaches zero in the limit of arbitrarily large averaging length r, yet each observer is surrounded by galaxies in a distribution that is statistically independent of the observer's galaxy. Mandelbrot points out that the gravitational potential energy per galaxy belonging to a typical mass concentration scales with the radius r of the concentration as

$$\phi \sim \frac{GNm^2}{rc^2} \propto r^{d-1} = r^{2-\gamma}, \tag{7.28}$$

where m is a characteristic galaxy mass. The fractal dimension $d = 1$, or $\gamma = 2$, has the attractive feature that the potential does not diverge to relativistic velocities on large or small scales. In a static fractal universe, in which it is postulated that stars live forever, the mean integrated surface brightness of the sky is proportional to the mean mass per unit area reckoned from a galaxy,

$$i \propto \int_0^\infty n(r) dr. \tag{7.29}$$

For γ greater than unity the integral converges at large r, so in this static universe there would be no Olbers' paradox even if it were assumed that stars shine forever.

This fractal picture is elegant enough to motivate a cosmology, and there are even observations that seem to support it. As Charlier (1908, 1922) noted, the strong clustering of the extragalactic nebulae in his map reproduced in figure 3.1 certainly looks like a hierarchical clustering pattern. The conspicuous empty band across the middle of the figure is now known to be an artifact of the obscuration by dust in the plane of the Milky Way, but the larger number of galaxies in the top half of the map is real, the concentration of galaxies in the Virgo cluster and the surrounding Local Supercluster. In the 1930s Shapley directed galaxy surveys at shallower depths and in larger fields than Hubble (1936) was using for his test of the $10^{0.6m}$ law for a homogeneous universe (eq. [3.15]). The surveys revealed concentrations of galaxies much larger than the Virgo cluster (Shapley 1934). Abell (1958) found that in his catalog of the richest clusters of galaxies there is again a tendency for the clusters to appear in concentrations, which he called superclusters.

With all this evidence, it is natural to follow Shapley (1938), de Vaucouleurs (1970), Mandelbrot (1975a,b), Coleman and Pietronero (1992), and others in asking whether the clustering hierarchy might continue to very large scales, perhaps even indefinitely. As it has happened, however, deeper galaxy counts do not support this picture, at least for a scale-invariant fractal.

The following analysis compares two models for the galaxy space distribution, a scale-invariant fractal, and a distribution that is a statistically homogeneous and isotropic (stationary) random process. The latter follows the cosmological principle in assuming that the galaxy space distribution is homogeneous on average. The fractal will be a mathematically convenient variant of the above construction of a clustering hierarchy. The analysis is lengthy but useful as a way to measure the fluctuations in the mass distribution within the standard model. This aspect of the discussion is continued in sections 19 to 21.

The idea behind the test of the pure clustering hierarchy model is that a scale-invariant fractal space distribution of galaxies with $d < 2$ would look the same, in a statistical sense, when viewed at any depth. This follows from the fact that there is no characteristic length in the distribution to set the depth at which it could start to look smooth. The effect is vividly illustrated in Mandelbrot's (1975a) examples. And it is not observed in the maps in figures 3.3 to 3.11. We will see in equation (7.70) below that this requires that the fractal dimension of the large-scale distribution of matter differs from the value $d = 3$ required by the cosmological principle by no more that about one part in 10^3.

The measures used to test these models will be two-point correlation functions of the galaxy distributions in space and projected on the sky. Since the projected distributions give an indirect test of the fractal model, it may seem surprising that there is not more emphasis on the direct test of the three-dimensional galaxy distribution based on redshift maps. The reason in part is that only recently have we had redshift surveys deep enough compared to the known galaxy clustering length for useful tests for clustering on larger scales. (These are based on the IRAS sources, as in figs. 3.7 and 3.8.) There also are practical limitations to the use of redshift surveys to measure galaxy clustering. As we noted in connection with figure 3.6, galaxy peculiar motions bias the small-scale clustering seen in redshift maps. Until galaxy distance measures are a good deal better than any we have now, statistical measures of spatial clustering on small scales will have to be based on projected distributions, following the methods outlined here. An example, for the analysis of the CfA redshift survey in figure 3.4, is given by Davis and Peebles (1983). Peculiar motions are thought not to affect redshift maps of the large-scale galaxy distribution, but to use these maps to test for radial variations of the galaxy density we must be sure the efficiency of detection of the galaxies does not vary with distance. Since distant galaxies are seen as they were when they and the universe were young, we need some way to separate the effects of evolution of galaxies from a possible radial gradient in the galaxy number density. The angular distributions compare galaxies in different directions at the same depth, and so presumably at the same state of evolution. This allows a much more precise test of large-scale density gradients, always assuming Nature has not placed us at the center of an inhomogeneous but spherically symmetric world.

The statistically homogeneous model for the space distribution of the galaxies is characterized by statistics that are functions only of relative position. Thus, the probability that a galaxy is found to be centered in the randomly placed volume element dV is proportional to the size of the element,

$$dP = n dV .$$ (7.30)

The constant of proportionality, the galaxy number density n, is a universal quantity, reflecting the assumption that the process is stationary. The joint probability of finding galaxies centered within the two volume elements dV_1 and dV_2 at separation r is proportional to the sizes of each of the elements,

$$dP = n^2[1 + \xi(r/r_o)] \, dV_1 dV_2 .$$ (7.31)

In a stationary random Poisson process (galaxies placed independently at random with spatially homogeneous probability), the joint probability would be the first term in this expression, the product of the one-point probabilities from equation (7.30). The dimensionless function $\xi(r/r_o)$ is the reduced two-point correlation function for the distribution. Since we are assuming a stationary process, ξ can only be a function of the distance between the volume elements. We will assume that ξ is small when r is large compared to a characteristic clustering length r_o, and that r_o is small compared to the Hubble length c/H_o. The observations indicate that for galaxies r_o is about one-tenth of one percent of the Hubble length.

Analyses of galaxy distributions within the framework of this model, along the lines to be presented below, show that at small separations the reduced two-point correlation function for galaxies is well approximated as a power law,

$$\xi = (r_o/r)^\gamma ,$$ (7.32)

where the parameters are

$$\gamma = 1.77 \pm 0.04 , \qquad r_0 = 5.4 \pm 1 \, h^{-1} \, \text{Mpc} .$$ (7.33)

Since $\xi(r)$ at $r < r_o$ is proportional to the mean density of galaxies as a function of distance r from a galaxy, this says the galaxy distribution on small scales has fractal dimension $d = 1.23 \pm 0.04$. The slope of the correlation function steepens at $r \gtrsim 2r_o$, and at much larger separations we have only upper bounds on ξ (Groth and Peebles 1977, 1986; Maddox et al. 1990).

Higher-order correlation functions generalize the definition in equation (7.31) to the three-point correlation function, $\zeta(r_a, r_b, r_c)$, which the stationary process allows to be a function only of the sides of the triangle defined by the three points, and so on. As described in LSS and section 19 below, the estimates of the higher-order functions are consistent with the picture that on scales less than r_o the galaxy distribution is a clustering hierarchy.

For the alternative fractal model, we will use an elegant construction, the Rayleigh-Lévy random walk, which Mandelbrot (1975b) introduced in cosmology. It proceeds as follows.

Starting from a galaxy, place the next galaxy in a randomly chosen direction at distance l drawn from the distribution

$$P(>l) = \begin{cases} (l_o/l)^\alpha & \text{for } l \geq l_o \\ 1 & \text{for } l < l_o \end{cases} \tag{7.34}$$

Repeat from the new galaxy position, using a new displacement drawn from this distribution, and a new randomly chosen direction. This continues indefinitely. It is an easy exercise in statistics (LSS, § 62) to check that if $\alpha > 2$ this is equivalent to the more familiar random walk with fixed steps, producing a Gaussian distribution in the net displacement after a fixed number of steps. If $0 < \alpha < 2$, the probability that after any number of steps starting from a chosen galaxy another is placed in the volume element d^3r at distance $r \gg l_o$ from the initial one is

$$dP = Cr^{-\gamma}d^3r, \qquad \gamma = 3 - \alpha. \tag{7.35}$$

The constant C is fixed by l_o and α.

The statistics in equations (7.32) and (7.35) are power laws, and by taking the same value for the index γ in the two models we can get the same mean numbers of close neighbors. The higher-order n-point correlation functions derived from the random walk model also have forms quite similar to the measured functions for galaxies (LSS, § 62); that is, on small scales the two models are quite similar. The issue is whether the fractal model might be a valid description of the galaxy distribution on large scales as well as small.

We have to assign galaxy luminosities. In both models, we will assume they are drawn independently from a universal distribution, and we will use the Schechter form in equation (5.129) as a convenient and realistic approximation.

In the computation of galaxy counts as a function of apparent magnitude, we will assume Euclidean geometry, so the energy flux density f from a galaxy of luminosity L at distance r is

$$f = \frac{L}{4\pi r^2}. \tag{7.36}$$

Then a galaxy at distance r is bright enough to be seen at flux density f if its luminosity exceeds $4\pi r^2 f$. That is, the probability, P, that a galaxy at distance r is included in a catalog with limiting flux density f is

$$P \propto \psi(4\pi r^2 f / L_*) = \int_{4\pi r^2 f/L_*}^{\infty} \phi(w)\, dw. \tag{7.37}$$

where $\phi(L/L_*)$ is the luminosity function (5.128). In the Schechter model,

$$\psi(x) = \int_x^\infty w^\alpha e^{-w}\, dw, \quad \alpha = -1.07 \pm 0.05, \quad x = L/L_* = 4\pi r^2 f/L_* \,. \quad (7.38)$$

The selection function, ψ, and the geometry of the galaxy distribution determine the distribution of distances of galaxies selected by apparent magnitude.

In the homogeneous random process model, the mean number of galaxies per steradian brighter than f is

$$\mathcal{N}(>f) = \int_0^\infty n r^2 dr\, \psi(4\pi r^2 f/L_*). \quad (7.39)$$

The first factor is the constant one-point probability in equation (7.30), and ψ gives the probability that a galaxy at distance r is bright enough to be in the sample. The mean space number density, n, has be chosen to agree with the normalization of ψ.

The change of variables $4\pi r^2 f = L_* y^2$ in equation (7.39) gives

$$\mathcal{N} = n \left(\frac{L_*}{4\pi f}\right)^{3/2} \int_0^\infty y^2\, dy\, \psi(y^2). \quad (7.40)$$

This is the $f^{-3/2}$ law of equation (3.9). The Schechter function (7.38) gives

$$\mathcal{N} = n \frac{(3/2 + \alpha)!}{3} \left(\frac{L_*}{4\pi f}\right)^{3/2}. \quad (7.41)$$

In the fractal model, the probability of finding a galaxy in the volume element $d^3 r$ at distance r from a galaxy is given by the power law in equation (7.35). The expectation value for the number of galaxies per steradian brighter than f counted by an observer in a galaxy is

$$\mathcal{N} = 2C \int_0^\infty r^{2-\gamma} dr\, \psi(4\pi r^2 f/L_*)$$

$$= 2C \left(\frac{L_*}{4\pi f}\right)^{(3-\gamma)/2} \int_0^\infty y^{2-\gamma}\, dy\, \psi(y^2) \quad (7.42)$$

$$= 2C \frac{(3/2 - \gamma/2 + \alpha)!}{3 - \gamma} \left(\frac{L_*}{4\pi f}\right)^{(3-\gamma)/2}.$$

In the first line there is a factor of two because the observer's galaxy could have been placed before or after the one at distance r. This expression for the mean number density represents an ensemble average over observations from a fair sample of galaxies. It is a useful approximation to what is observed from a typical galaxy, though as we will see the fluctuations are large.

The prediction in equation (7.42) for the fractal model, expressed in apparent magnitudes, is

$$\frac{d\mathcal{N}}{dm} \propto 10^{\beta m}, \qquad \beta = 0.2(3 - \gamma). \tag{7.43}$$

With the fractal dimension adjusted to fit the distribution of neighbors observed at small scales, $\gamma \sim 1.8$, the slope is $\beta \sim 0.24$. The first crisis for the fractal picture is that this is well below the observed slope of the galaxy counts (Sandage, Tammann, and Hardy 1972). The situation is illustrated in figure 3.2. At the bright end the slope is shallow because of the concentration of galaxies in the Local Supercluster. At $m \sim 13$ the slope of the counts steepens to a reasonable approximation to the $10^{0.6m}$ Hubble relation, and well above the prediction of the fractal model with $\gamma \sim 1.8$.

An even more manifest problem for the fractal model comes from the predicted fluctuations in counts of galaxies as a function of angular position at a fixed apparent magnitude. Consider the joint probability that an observer in a galaxy finds that galaxies brighter than f are centered in both of the elements of solid angle $d\Omega_1$ and $d\Omega_2$ at angular separation θ in the sky. Following the definition of the spatial two-point correlation function in equation (7.31), this probability can be written as

$$dP = \mathcal{N}^2 d\Omega_1 d\Omega_2 [1 + w(\theta)]. \tag{7.44}$$

In the fractal model this again represents an average over observations from an ensemble of galaxies. In the statistically homogeneous model, observations at depths large compared to the clustering length r_o would be very nearly the same viewed from any galaxy, or any point in space.

In the homogeneous model the probability in equation (7.44) for the angular distribution is an integral over the spatial probability in equation (7.31),

$$dP = n^2 d\Omega_1 d\Omega_2 \int r_1^2 dr_1 r_2^2 dr_2 [1 + \xi(r_{12}/r_o)] \psi_1 \psi_2. \tag{7.45}$$

The last factors are the selection functions, $\psi_i = \psi(4\pi r_i^2 f / L_*)$, as in equation (7.39). The distance between the points of integration along the two lines of sight is

$$r_{12} = (r_1^2 + r_2^2 - 2r_1 r_2 \cos\theta)^{1/2}. \tag{7.46}$$

With equation (7.39) for the mean number per steradian, $\mathcal{N}$, we see from equations (7.44) and (7.45) that the angular two-point correlation function in the homogeneous model is

$$w(\theta) = \frac{\int (r_1 r_2)^2 dr_1 dr_2 \psi_1 \psi_2 \xi(r_{12}/r_o)}{[\int r^2 dr \, \psi]^2}. \tag{7.47}$$

This relation between the spatial and angular two-point correlation functions was first written down by Limber (1953) and Rubin (1954).

With the change of variables used in equation (7.40), and the power law model for the spatial function ξ in equation (7.32), equation (7.47) becomes

$$w_h(\theta) = \left(\frac{4\pi r_o^2 f}{L_*}\right)^{\gamma/2} \frac{\int (y_1 y_2)^2 \, dy_1 dy_2 \psi(y_1^2) \psi(y_2^2) \, y_{12}^{-\gamma}}{[\int y^2 \, dy \, \psi(y^2)]^2}. \tag{7.48}$$

The subscript on this statistic stands for the homogeneous model. This equation says that $w_h(\theta)$ at fixed θ decreases with increasing sample depth, which is decreasing f, as $f^{\gamma/2}$. This is because we are assuming the galaxy distribution is homogeneous in the large-scale average, so the angular fluctuations are averaged away as the depth of the sample increases.

In the fractal model, the probability that an observer in a galaxy finds that galaxies are centered in the volume elements $d^3 r_1$ and $d^3 r_2$ at positions $\mathbf{r}_1$ and $\mathbf{r}_2$ relative to the observer is, from equation (7.35),

$$dP = 2C^2 d^3 r_1 d^3 r_2 \left[(r_1 r_2)^{-\gamma} + (r_1 r_{12})^{-\gamma} + (r_{12} r_2)^{-\gamma}\right]. \tag{7.49}$$

The distance between $\mathbf{r}_1$ and $\mathbf{r}_2$ is r_{12}. The three terms represent the three ways the random walk can connect the galaxies. In the first term, the walk connects $\mathbf{r}_1$ to the observer to $\mathbf{r}_2$; in the second, it connects the observer to $\mathbf{r}_1$ to $\mathbf{r}_2$. The factor of two allows for the two directions the walk can go. On repeating the calculation that led to equation (7.48), we get in the fractal model

$$\begin{aligned} w_f(\theta) + \frac{1}{2} &= \frac{\int r_1^{2-\gamma} r_2^2 r_{12}^{-\gamma} dr_1 dr_2 \psi_1 \psi_2}{[\int r^{2-\gamma} dr \, \psi]^2} \\ &= \frac{\int y_1^{2-\gamma} y_2^2 y_{12}^{-\gamma} \, dy_1 dy_2 \psi(y_1^2) \psi(y_2^2)}{[\int y^{2-\gamma} \, dy \, \psi(y^2)]^2}. \end{aligned} \tag{7.50}$$

In this pure fractal model, the change of variables $4\pi r^2 f = L_* y^2$ shows that the angular two-point correlation function is independent of the limiting magnitude m or flux density f of the sample. This had to be so, because the fractal model is scale-free: the limiting flux density cannot appear in the expression for w_f

because there is nothing to match its units. In the homogeneous clustering model, the depth scale is set by the clustering length r_o with the characteristic galaxy luminosity L_* (eq. [7.48]).

For another way to understand why w_f is independent of the sample depth in the scale-invariant fractal model, recall that the appearance of a scale-invariant or renormalizable fractal is unchanged by a change of the scale at which it is displayed, as Mandelbrot (1975a) shows. The prediction is then that in the scale-invariant fractal model for the galaxy distribution, an increase in the sample depth reveals ever larger voids and clusters that on average subtend the same solid angle. This means that at any limiting magnitude the angular fluctuations in the counts across the sky have the same character, with the same amplitude $\delta \mathcal{N}/\mathcal{N}$ in the fractional fluctuation in galaxy counts across the sky.

The observed fluctuations in galaxy counts at a fixed apparent magnitude show a distinct decrease in amplitude with increasing apparent magnitude, as is illustrated in the comparison of figures 3.4 and 3.7 to 3.10. This is quite contrary to the prediction of the fractal model. To make the test quantitative, we will consider two cases. If the fractal dimension is $d < 2$, we can use a convenient approximation to the integrals for the angular correlation functions $w_h(\theta)$ and $w_f(\theta)$ when the angular separation θ is small compared to one radian. The case $d \geq 2$ is considered in equation (7.69).

At $\theta \ll 1$ radian, and for the homogeneous model or the fractal model with $d < 2$ ($\gamma > 1$), the integrals in the numerators of equations (7.48) and (7.50) are dominated by the small value of the denominator $y_{12}{}^\gamma$ when y_1 is close to y_2. That means we can approximate the integrals by changing the variables of integration to $y = y_1$ and $s = y_2 - y_1$, and in the integrands set $y_2 = y$ everywhere except in the denominator, where we can write the separation in equation (7.46) as

$$y_{12} = (s^2 + \theta^2 y^2)^{1/2} . \tag{7.51}$$

This is the small-angle approximation.

In the homogeneous model, the small-angle approximation, with the changes of variable $u = r_2 - r_1 = r_o v$ and $4\pi r^2 f = L_* y^2$, brings the expression in equation (7.47) for the angular correlation function to

$$w_h(\theta) = \left(\frac{4\pi r_o^2 f}{L_*} \right)^{1/2} \frac{\int y^4 \, dy \, \psi(y^2)^2 \int_{-\infty}^{\infty} dv \, \xi(A)}{[\int y^2 \, dy \, \psi(y^2)]^2} . \tag{7.52}$$

The dimensionless argument of the spatial function is

$$A^2 = v^2 + \theta^2 y^2 L_* / (4\pi r_o^2 f) . \tag{7.53}$$

This expression for the angular function is of the form

$$w_h(\theta) = \frac{r_o}{D_*} W(\theta D_* / r_o) . \tag{7.54}$$

The characteristic sample depth, D_*, is the distance at which a galaxy with luminosity L_* appears at the limiting flux density f,

$$D_* = (L_*/4\pi f)^{1/2}. \tag{7.55}$$

The scaling law in equation (7.54) is immediately understandable. The scaled angular separation $\theta D_*/r_o$ brings the argument of the function W to a given fraction of the clustering length. The prefactor is the suppression in the angular clustering due to the number $\sim D_*/r_o$ of independent clustering patches seen overlapping in projection along the line of sight.

With the power law model in equation (7.32) for the spatial function $\xi(A)$, equation (7.52) for the angular function is

$$w_h(\theta) = \theta^{1-\gamma} H_\gamma \left(\frac{4\pi r_0^2 f}{L_*} \right)^{\gamma/2} \frac{\int y^{5-\gamma} \, dy \, \psi(y^2)^2}{[\int y^2 \, dy \, \psi(y^2)]^2}. \tag{7.56}$$

The dimensionless integral over the difference in distances v along the line of sight is

$$H_\gamma = \int_{-\infty}^{\infty} \frac{dx}{(1+x^2)^{\gamma/2}} = \frac{(-1/2)!(\gamma/2 - 3/2)!}{(\gamma/2 - 1)!}. \tag{7.57}$$

If $\gamma \sim 2$ this factor is on the order of unity, as is the other dimensionless integral in equation (7.56), so the correlation function is approximately

$$w_h \sim \theta^{1-\gamma}(r_0/D_*)^\gamma. \tag{7.58}$$

That is, when the characteristic depth D_* in equation (7.55) is much larger than the clustering length r_o, the value of the correlation function approaches unity, $w_h \sim 1$, at angular separation θ much less than one radian.

Tests of the depth scaling law in equation (7.54) are shown in figures 7.2 and 7.3. In figure 7.2 (Groth and Peebles 1977, 1986), the triangles in panel (a) are from the Zwicky catalog (Zwicky et al. 1961–68). Redshift samples from this catalog are shown in figures 3.4 and 3.6, but here we are using the angular distribution of galaxies brighter than a fixed limiting apparent magnitude, $m = 15$. The circles are from the deeper Lick catalog in figure 3.9 (Shane and Wirtanen 1967), and the boxes from the still deeper Jagellonian field (Rudnicki et al. 1973). The relative limiting magnitudes in these catalogs are not well calibrated, so the ratio of characteristic sample depths D_* is based on the mean densities $\mathcal{N}$ of galaxies per unit area on the sky, and uses the relation $\mathcal{N} \propto D_*^3$ appropriate to the homogeneous model (eqs. [7.40] and [7.55]). The scaling relation (7.54) says that if in the logarithmic plot in panel (a) the correlation function for one catalog

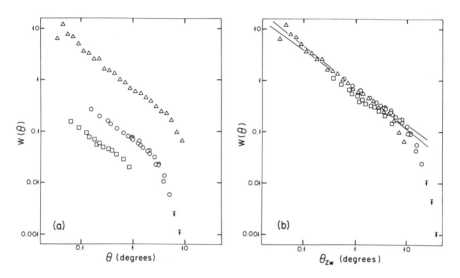

Figure 7.2. Scaling test for the galaxy angular two-point correlation function. Panel (b) shows the result of applying the homogeneous scaling law in equation (7.54) to the correlation function estimates in panel (a) from catalogs at three different depths.

is shifted up and to the right by the logarithm of the depth relative to a second catalog,

$$\frac{D_1}{D_2} = \left(\frac{\mathcal{N}_1}{\mathcal{N}_2}\right)^{1/3}, \tag{7.59}$$

the scaled functions ought to coincide. The result in panel (b) shows that the scaling law is quite successful.

Figure 7.3 shows a test of the homogeneous scaling law (7.54) based on a deeper catalog obtained with an automatic plate-measuring machine (APM), in a version provided by S. J. Maddox from the analysis of Maddox et al. (1990). Panel (a) shows the angular correlation functions in successive slices $\Delta m = 0.5$ in the magnitude range $17.5 < m < 20.5$. The deeper the slice, the smaller $w(\theta)$. Panel (b) shows the result of scaling these estimates to the depth of the Lick sample shown as the open circles in figure 7.2, and here as open diamonds. At larger angular separations, where $w \lesssim 0.02$ in the Lick catalog, the APM estimates are systematically higher, but this is not a significant disagreement because at the small values of the angular correlation function where this difference appears, the estimates are prey to systematic error. At smaller θ, where the correlation function

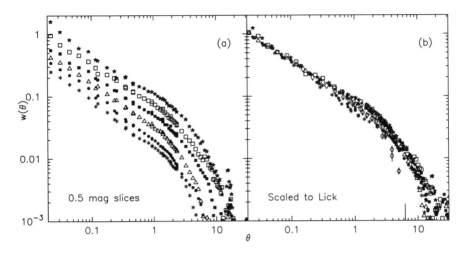

Figure 7.3. Scaling test in the APM galaxy catalog (Maddox et al. 1990).

estimates are larger and thought to be reliable in both catalogs, the results are quite close to the scaling prediction of the statistically homogeneous model.

The conclusion is that the correlation function analyses have yielded a new and positive test of the assumption that the galaxy space distribution is a stationary (statistically homogeneous) random process. And this result is a serious failure of the fractal model with $d < 2$, where $w(\theta)$ is predicted to be statistically independent of the catalog depth.

We can push the test of the fractal model a little further by considering the magnitude of the predicted correlation function $w_f(\theta)$. Equation (7.50) for the angular function in the fractal model, evaluated in the small angle limit, is

$$w_f(\theta) = \theta^{1-\gamma} H_\gamma \frac{\int y^{5-2\gamma} \, dy \, \psi(y^2)^2}{[\int y^{2-\gamma} \, dy \, \psi(y^2)]^2} . \qquad (7.60)$$

If the parameters γ and α are close to the values in equations (7.33) and (7.38), the integrals in this expression converge. (If they did not, the redshift distributions in figure 5.6 would not scale properly with the sample depth f.) With $\gamma = 1.8$ and $\alpha = -1.07$, the numerical result is

$$w_f = 1.2 \, \theta^{-0.8} . \qquad (7.61)$$

That is, independent of the sample depth, the ensemble average correlation function at θ equal to one radian is on the order of unity, as we knew it had to be by dimensional analysis. That certainly is not what is seen in the Lick and APM samples.

To pursue the point still further, let us consider in more detail the meaning of the two-point correlation function.

Suppose an observer in a galaxy counts the galaxies brighter than f that appear in two circular fields, each of radius θ, with centers separated by angular distance $\Theta \gg \theta$. We will compute the mean square difference in counts in the two fields.

An easy way to compute is to imagine that each field is divided into elements of solid angle, $\delta\Omega_i$. Since the elements are small, the number of galaxies in an element is $n_i = 0$ or $n_i = 1$. The expectation values of n_i and $n_i{}^2$ thus are the same,

$$\langle n_i \rangle = \langle n_i{}^2 \rangle = \mathcal{N}\,\delta\Omega_i \,. \tag{7.62}$$

Here $\mathcal{N}$ is the mean count per steradian in equation (7.41) or (7.42). For disjoint elements $d\Omega_i$ and $d\Omega_j$, we have from the equation (7.44) that defines the angular correlation function

$$\langle n_i n_j \rangle = \mathcal{N}^2 \delta\Omega_i \delta\Omega_j [1 + w(\theta_{ij})] \,. \tag{7.63}$$

The number of galaxies in one of the fields is

$$N = \sum n_i \,. \tag{7.64}$$

The expectation value of this sum is

$$\langle N \rangle = \mathcal{N} \int d\Omega \,, \tag{7.65}$$

as usual. The mean square value of the count in the field is

$$\langle N^2 \rangle = \sum \langle n_i^2 \rangle + \sum \langle n_i n_j \rangle$$
$$= \langle N \rangle + \langle N \rangle^2 + \mathcal{N}^2 \int d\Omega_i d\Omega_j w(\theta_{ij}) \,. \tag{7.66}$$

The square and cross terms have been treated separately, using equations (7.62) and (7.63). The mean value of the product $N_1 N_2$ of the counts in the two circular fields is given by a similar expression, but without the shot noise term, $\langle N \rangle$, because $d\Omega_1$ and $d\Omega_2$ are integrals in separate fields instead of the same one.

Ignoring the shot noise, the mean square difference between the counts in the two fields is

$$\langle (N_1 - N_2)^2 \rangle / 2 = \langle N_1^2 \rangle - \langle N_1 N_2 \rangle$$
$$= \langle N \rangle^2 \left[\int_{11} \frac{d\Omega_i d\Omega_j}{\Omega^2} w(\theta_{ij}) - \int_{12} \frac{d\Omega_i d\Omega_j}{\Omega^2} w(\theta_{ij}) \right] \,. \tag{7.67}$$

In the second line the subscripts indicate that both elements of solid angle in the first integral are in one field, while in the second integral the two elements are in separate fields at separation Θ. Because the integrals converge it is easy to write down the order of magnitude in the fractal case in equation (7.61):

$$\langle (N_1 - N_2)^2 \rangle \sim \langle N \rangle^2 (\theta^{-0.8} - \Theta^{-0.8}). \tag{7.68}$$

This means that in a catalog at any depth the galaxy counts in fields much smaller than one radian scatter across the sky by more than 100% of the mean value. This applies in the sense of an ensemble average, but galaxies cataloged in different directions, and those cataloged at appreciably different apparent magnitudes, sample different parts of space, in effect giving us different samples from the ensemble. And as we have seen in figures 3.8 to 3.11, the predicted large fluctuations do not appear in the deep catalogs.

If the large-scale galaxy distribution were a fractal with dimension $d > 2$, so $\gamma < 1$, the small-angle approximation in equation (7.60) would fail, and the correlation function $w_f(\theta)$ would be a slowly varying function of θ. Here the most direct test of the fractal model is the large-scale angular distribution of counts of galaxies in deep samples, as in figure 3.9 for optically selected galaxies, figure 3.10 for radio-luminous galaxies, and figure 3.11 for the X-ray background. We will use the last of these samples.

An appreciable part of the X-ray background is known to come from active galaxies (quasars and radio galaxies) at redshifts on the order of unity. This would be expected in a fractal model with dimension $d > 2$, where the column density of sources integrated to distance R diverges as $f \propto R^{d-2}$ (eq. [7.29]), and the large-scale cutoff is provided by the cosmological redshift which suppresses the contribution to the background by sources at redshifts well in excess of unity.

To find the relation between the fractal dimension and the large-scale anisotropy of the X-ray background, note that in a clustering hierarchy model the mean density within distance r of a galaxy scales as $\bar{\rho} \propto r^{d-3}$ (eq. [7.27]). If we are not at the center of a spherically symmetric universe, the same scaling applies to the mean density measured from galaxies at the edge of the region of size R that supplies us with X rays, so there will be two hemispheres of our sky in which the means of the observed X-ray flux densities differ by the fractional amount

$$\frac{\delta f}{f} \sim 2^{3-d} - 1 \lesssim 10^{-3}. \tag{7.69}$$

The last number is the bound on the dipole and quadrupole anisotropies (Boldt 1987). Thus, the bound on the fractal dimension of the large-scale space distribution of X-ray galaxies is

$$3 - d \lesssim 0.001. \tag{7.70}$$

If we accept general relativity theory, we can improve this constraint by another order of magnitude by using the Sachs-Wolfe effect on the anisotropy of the thermal cosmic background radiation, as discussed in section 21.

The computation of the second moments of counts of galaxies in cells applies as well to the fluctuations around the mean space distribution in the standard cosmological model, and gives a useful measure of the scale of clustering of galaxies. Equation (7.66) applied to the spatial correlation function $\xi(r)$ for a stationary random process gives the mean square fluctuation in the count of galaxies found within a sphere of radius r,

$$
\left(\frac{\delta N}{N}\right)^2 \equiv \frac{\langle (N - \langle N \rangle)^2 \rangle}{\langle N \rangle^2} = \frac{1}{\langle N \rangle} + \int_r \frac{d^2V}{V^2} \xi(r_{12}).
\tag{7.71}
$$

The angular brackets mean averages over counts in spheres placed at random across a fair sample of the process. The shot noise term in the last expression is unimportant on large scales. In the power law model for $\xi(r)$ in equations (7.32) and (7.33), the integral is

$$
C = \left(\frac{3}{4\pi}\right)^2 \int_0^1 \frac{d^3x_1 d^3x_2}{x_{12}^\gamma} = \frac{72}{2^\gamma(3-\gamma)(4-\gamma)(6-\gamma)}.
\tag{7.72}
$$

This works out to

$$
\frac{\delta N}{N} = 1.35 \left(\frac{r_o}{r}\right)^{\gamma/2} = 0.4 \quad \text{for } r = 20h^{-1} \text{ Mpc}.
\tag{7.73}
$$

The similar result in equation (3.24) for a square cell is based on fluctuations in the counts in the redshift map of IRAS galaxies. Either statistic says that the rms fluctuation in the count of galaxies found within a randomly placed cell of diameter $30h^{-1}$ Mpc is about 50%. By moving the cell around it is not hard to find a 2σ fluctuation, where $\delta N/N \sim -1$. That is, these statistics indicate that voids of size $\sim 30h^{-1}$ Mpc are not uncommon, consistent with what we see in figure 3.6 and in figure 19.2 below.

The lesson from this analysis is that we have indirect but powerful measures of the statistics of the large-scale galaxy space distribution. The results rule out the elegant fractal picture, unless the fractal dimension on large scales is exceedingly close to $d = 3$ (eq. [7.70]), which is not what is observed in the small-scale fractal galaxy space distribution (and discussed further in section 19). The successful prediction of the scaling of the two-point correlation function with sample depth, in figures 7.2 and 7.3, is an additional piece of evidence for the cosmological principle.

Noncosmological Redshifts

The observed redshifts and blueshifts of the nearest galaxies almost certainly are dominated by the Doppler shifts from their motions. At greater distances this sorts out into the pattern of increasing redshift with increasing distance that in the conventional interpretation is the result of the general expansion of the universe. In a tired light model, photons moving through apparently free space lose energy at the rate

$$\frac{d\nu}{dl} = -H_o\nu\,, \tag{7.74}$$

where H_o is a constant and dl is the proper displacement along the path of the light in a static world. This picture seems to have been first discussed by Zwicky (1929), who considered the idea that gravitational interactions cause photons to lose momentum and hence frequency at a steady rate. The most immediate problem for the picture is the cosmic background radiation spectrum shown in figures 6.1 and 6.2. Suppose at some given time space is filled with blackbody radiation, with the spectrum in equation (6.12). Then after a Hubble time H_o^{-1} the number density of the original photons would be unchanged, but each would have 30% of its original energy, giving a distinctly nonthermal spectrum. One might remedy this by introducing a thermalizing process, but that would lead to the problems we detailed for the steady-state cosmology. As discussed in section 5, the same effect implies that in a static tired light world model the surface brightnesses of distant galaxies would vary as $i \propto (1+z)^{-1}$ rather than the $i \propto (1+z)^{-4}$ law to be expected if the redshift is caused by expansion (eq. [6.44]). This is because in a tired light model the only effect on the surface brightness is the diminishing energy of each photon, leading to one power of $1+z$, while in the expanding case there is another power of $1+z$ from the rate of reception of photons, and two more from the aberration that opens up the solid angle of a cone of rays. Tolman (1930) and Hubble and Tolman (1935) pointed out that the $i \propto (1+z)^{-4}$ law tests the expanding universe picture, Geller and Peebles (1972) noted that it tests the tired light model, and Sandage and Perelmuter (1991) present the most recent assessment of the test for the redshift-dependence of galaxy surface brightnesses.

Differences of gravitational potential energy produce redshifts and blueshifts. Ellis, Maartens, and Nel (1978) point out that a static world model in which the mass density is a function of position can contain background radiation in a steady, thermally relaxed state, with a spectrum that is everywhere thermal and isotropic, but with a temperature that varies with position. If stationary observer a sees that the radiation from stationary observer b is gravitationally redshifted by the factor $1+z$, then the ratio of the background radiation temperatures measured by the two observers is $T_b/T_a = 1+z$. This follows by the argument in equation

(6.45): the gravitational frequency and temperature shifts between observers are equivalent to the effect of a sequence of velocity shifts between a sequence of freely moving observers. For the same reason, the surface brightness of an object at a different potential would vary with its redshift as $i \propto (1+z)^{-4}$. This is not a cosmology, however, for it is not known how one could get a reasonable redshift-distance relation from a stable static mass distribution, or what provision one would make for the apparently finite lifetimes of stars and galaxies.

If the redshifts of quasars did not follow the redshift-distance relation observed for galaxies, it would show we have missed something very significant. The debate on this point has a long history (Field, Arp, and Bahcall 1973; Arp 1987; Arp et al. 1990), which to most participants has been settled by observational tests such as the clustering of galaxies around low redshift quasars and the gravitational lensing of high redshift quasars by galaxies that have to be well in front of the lensed quasars. Examples of the latter effect are the luminous arcs discussed in section 20 (eq. [20.46]). The literature on these and other tests of quasar distances is reviewed in Dar (1991) and Peebles et al. (1991).

Is There More?

It is sensible and prudent that people should continue to think about alternatives to the standard model, because the evidence is not all that abundant. The examples presented here do show that the observational constraints are far from negligible, however. In the 1930s one could have made a reasonable case for the scale-invariant fractal model, on the basis of elegance and even on the available observational evidence, but that is not the way subsequent observations have gone. In the 1950s the steady-state model proved to be a viable alternative to the Friedmann-Lemaître cosmology, but the classical steady-state model is now convincingly ruled out by the CBR and by the observed evolution of galaxies with redshift. The moral is that the invention of a credible alternative to the standard cosmological model would require consultation of a considerable suite of evidence. It is equally essential that the standard model be subject to scrutiny at a still closer level than the alternatives, for it takes only one well-established failure to rule out a model, but many successes to make a convincing case that a cosmology really is on the right track. That is one of the purposes of the cosmological tests in part 2 and of the search chronicled in part 3 for an understanding of details such as the origin of the galaxies.

In section 26 we return to the survey of possibilities for surprises in cosmology.

II. General Relativity and Cosmology

General relativity theory was one of the keys to the discovery of the expansion of the universe: Einstein's interpretation of Mach's principle led to the cosmological principle, and the instability of a relativistic world model consistent with this principle led to the expansion law that agrees with the observed variation of galaxy redshifts with distances. The cosmology that developed out of this is observationally and experimentally rich, but much of it uses only the simplest of elements of general relativity, as summarized in section 1 and used throughout the previous sections. This is changing, however, for observations now reach gravitationally lensed objects, and objects at redshifts so high that they are viewed through the gravitational lens of the universe itself in a way that is not trivially computed by symmetry arguments. Out of this is coming a fuller understanding of how mass is distributed and evolving, and we may hope eventually it will yield meaningful and detailed tests of the relativity theory underlying the standard cosmological model.

The purpose of this part is to present the elements of general relativity theory that figure in the theoretical interpretation of the observations and in some of the ideas discussed in part 3 for how to fit what is observed into a believable picture for the physics of the very early universe. Sections 8 through 12 present the formal development of the theory, along with some observationally useful results, such as the brightness theorem. The cosmological tests in an ideal homogeneous and isotropic universe are discussed in section 13. The complications of cosmology in our lumpy universe are likely to be of increasing interest as the ability to apply the tests improve. This is considered in section 14.

8. General Covariance

We noted in section 2 that the coordinates x^i of an event or point in spacetime have no absolute significance; they are just labels. A quantity that is measurable (even if only in principle) has to have a value that is independent of the choice of coordinate labels. An object that satisfies this condition — that the value at a point in spacetime is unchanged by a change of coordinates — is said to be a scalar. In writing down field equations we will be considering vectors, such as the four-velocity of a fluid, and tensors, such as the metric tensor that determines physical lengths and times. These fields transform by definite rules under a change of the coordinate labels. Since the components of a vector thus depend on the coordinate labeling, they have no absolute significance; observables are scalars constructed out of vectors and tensors. Thus a measurement of a velocity field really means the measurements of scalar products of the field with vectors that define a frame of reference.

General relativity is a metric theory, which means length and time intervals are assumed to be independent of the choice of the physical rod or clock used to measure them, and are described by a metric tensor field. We will see in the next two sections how this leads to the equivalence principle, that the local physics seen by a freely falling observer is unaffected by gravitational fields. Here we take the first step of defining the coordinate transformation laws for vectors and tensors, and the ways to make observable scalars out of them. We will then consider the special role of the metric tensor in defining scalars and prescribing how to integrate and differentiate vectors to get field equations.

Scalars, Vectors, and Tensors

In a coordinate transformation, the four coordinates x^i, with $i = 0$, 1, 2, and 3, that label events in spacetime are changed to new labels $\bar{x}^j$ that are functions of the old ones,

$$\bar{x}^j = \bar{x}^j(x). \tag{8.1}$$

The $\bar{x}^j(x)$ are four differentiable functions of the x^i.

A physical quantity that has a definite value at each point in spacetime can be represented by a function $\phi(x)$ of the position labels x^i. The condition that the value of the quantity at a definite point is independent of the choice of coordinates means that the function $\bar{\phi}(\bar{x})$ representing the quantity in the new coordinate system of equation (8.1) is related to the original function by the equation

$$\bar{\phi}(\bar{x}) = \phi(x). \tag{8.2}$$

This is the coordinate transformation law for a scalar field.

By equation (8.1), the gradient of the scalar field $\phi(x)$ in equation (8.2) is

$$\frac{\partial \bar{\phi}(\bar{x})}{\partial \bar{x}^i} = \sum_j \frac{\partial \phi(x)}{\partial x^j} \frac{\partial x^j}{\partial \bar{x}^i}. \tag{8.3}$$

We will use the comma notation for partial derivatives,

$$\phi_{,j} = \frac{\partial \phi(x)}{\partial x^j}, \tag{8.4}$$

and we will use Einstein's summation convention — that a pair of repeated indices, such as j in equation (8.3), is summed unless otherwise stated. For Latin indices such as i and j, which refer to components in four-dimensional spacetime, the sum is understood to run from 0 to 3. With these conventions, the transformation law in equation (8.3) is

$$\bar{\phi}_{,i} = \phi_{,j}\frac{\partial x^j}{\partial \bar{x}^i} \, . \tag{8.5}$$

The indices in equation (8.5) are arranged in a helpful pattern. Let us agree that an index that appears as a superscript in the denominator in a derivative, as in equation (8.4), counts as a subscript in the expression, as $\phi_{,j}$. Then, in a pair of repeated and summed indices, as j in equation (8.5), one index is a superscript and one is a subscript. The unsummed indices match on the two sides of the equation, as the subscript i in equation (8.5).

Now we will define the two classes of vectors, called covariant and contravariant (for reasons that are lost in the mists of time). The four components of a covariant four-vector field A_i are defined to transform under a change of coordinates by the transformation law in equation (8.5):

$$\bar{A}_i = A_j\frac{\partial x^j}{\partial \bar{x}^i} \, . \tag{8.6}$$

Both sides of this equation refer to the same point in spacetime, so the argument of A_j is x and the argument of the new vector field $\bar{A}_i$ on the left side of the equation is $\bar{x} = \bar{x}(x)$.

Another vector transformation law follows by writing out the differential of equation (8.1):

$$d\bar{x}^j = \frac{\partial \bar{x}^j}{\partial x^i}\, dx^i \, . \tag{8.7}$$

The four components of a contravariant vector B^j are defined to transform according to this law:

$$\bar{B}^j(\bar{x}) = B^i(x)\frac{\partial \bar{x}^j}{\partial x^i} \, . \tag{8.8}$$

Consistent with the rules for placing indices, the index in this vector is a superscript, so the summed index appears as a subscript-superscript pair, leaving superscripts j on both sides of the equation.

We get an identity for the transformation coefficients $\partial x^j/\partial \bar{x}^i$ by differentiating equation (8.1):

$$\frac{\partial \bar{x}^i}{\partial \bar{x}^j} = \delta^i_j = \frac{\partial \bar{x}^i}{\partial x^k}\frac{\partial x^k}{\partial \bar{x}^j} \, , \tag{8.9}$$

where the Kronecker delta function is defined as

$$\delta_j^i = \begin{cases} 1 & \text{if } i = j, \\ 0 & \text{if } i \neq j. \end{cases} \tag{8.10}$$

Note again the elegant balancing of indices. On multiplying equation (8.6) by $\partial \bar{x}^i / \partial x^k$, and using the identity (8.9), we get

$$A_k = \bar{A}_i \frac{\partial \bar{x}^i}{\partial x^k}, \tag{8.11}$$

as expected since the x and $\bar{x}$ are interchangeable.

In a similar way, the identity (8.9) with the covariant and contravariant transformation laws in equations (8.6) and (8.8) gives

$$\bar{A}_i \bar{B}^i = A_j \frac{\partial x^j}{\partial \bar{x}^i} \frac{\partial \bar{x}^i}{\partial x^k} B^k = A_j \delta_k^j B^k = A_j B^j. \tag{8.12}$$

That is, the covariant and contravariant vectors A_i and B^j yield a scalar expression $A_i B^i = \bar{A}_j \bar{B}^j$, which might represent a measurable quantity. For example, the transformation laws (8.5) and (8.7) for $\phi_{,i}$ and dx^i show that the differential of the scalar field $\phi(x)$,

$$\phi_{,i} \, dx^i = d\phi, \tag{8.13}$$

is a scalar product. This is just the invariant difference between the values of the scalar ϕ at neighboring given points in spacetime.

A tensor has more than one index, and it has transformation laws that follow equations (8.6) and (8.8). For example, the transformation law for a tensor T_{ij} with two covariant indices is

$$\bar{T}_{ij} = T_{kl} \frac{\partial x^k}{\partial \bar{x}^i} \frac{\partial x^l}{\partial \bar{x}^j}. \tag{8.14}$$

The two covariant vectors A_i and B_j thus define the tensor

$$U_{ij} = A_i B_j. \tag{8.15}$$

The transformation law for a tensor $T^j{}_i$ with mixed indices is

$$\bar{T}^j{}_i = T^l{}_k \frac{\partial x^k}{\partial \bar{x}^i} \frac{\partial \bar{x}^j}{\partial x^l}. \tag{8.16}$$

It is an easy exercise, following equation (8.12), to check that $T_{ij} B^i$ is a covariant vector, and $T_{ij} B^i C^j$ is a scalar.

Finally, let us note that the identity in equation (8.9) for the transformation coefficients can be rewritten as

$$\delta^i_j = \delta^k_l \frac{\partial \bar{x}^i}{\partial x^k} \frac{\partial x^l}{\partial \bar{x}^j} . \tag{8.17}$$

Thus the Kronecker delta function is a special tensor, one that is the same in any coordinate system.

The Metric Tensor

The metric tensor $g_{ij}(x)$ plays the central role in general relativity theory of defining the geometry of spacetime and, as discussed below, giving the prescription for integrals and derivatives. The metric tensor is symmetric,

$$g_{ij} = g_{ji} , \tag{8.18}$$

and it satisfies the usual tensor transformation equation

$$\bar{g}_{ij} = \frac{\partial x^k}{\partial \bar{x}^i} \frac{\partial x^l}{\partial \bar{x}^j} g_{kl} . \tag{8.19}$$

The metric tensor defines the line element (eq. [2.1]),

$$ds^2 = g_{ij} \, dx^i dx^j , \tag{8.20}$$

connecting neighboring events x^i and $x^i + dx^i$ in spacetime. Since ds^2 is a scalar, we are allowed to assign it a meaning as a quantity that is in principle measurable. As discussed in section 2, the definition is that if ds^2 is positive, ds is the time interval measured by an observer on a world line that passes between the events; and if ds^2 is negative, $|ds|$ is the distance between the events measured by an observer who is moving such that they are seen to happen simultaneously.

We will see shortly that it is always possible to assign coordinates such that at a chosen point in spacetime the components of the metric tensor bring the line element to the special relativity form (eq. [2.4]),

$$ds^2 = dt^2 - dx^2 - dy^2 - dz^2 . \tag{8.21}$$

In this coordinate system at the chosen event, the intervals dt and dx are proper — and in principle measurable — time and length intervals.

In analogy to the line element, the scalar

$$A^2 = g_{ij} A^i A^j \tag{8.22}$$

can be considered the square of the invariant length A of the vector A^i, for that is its value in the coordinates of equation (8.21). Also,

$$A \cdot B = g_{ij} A^i B^j \tag{8.23}$$

is the invariant scalar product of the two vectors.

We can construct the corresponding expressions for covariant vectors by using the contravariant or reciprocal metric tensor g^{jk} defined by the expression

$$g_{ij} g^{jk} = \delta_i^k . \tag{8.24}$$

This just says that g^{jk} is the matrix inverse of the matrix g_{ij}. Recall that the inverse matrix element g^{kl} can be computed from the determinants of the matrix g_{ij} with row k and column l removed. (The operation is shown in eq. [8.63] below.) The inverse is symmetric, $g^{kl} = g^{lk}$, and its existence requires that the determinant g of g_{ij} does not vanish. Further reasons for this condition are indicated below (eq. [8.37]). Since the Kronecker delta function on the right-hand side of equation (8.24) is a tensor (eq. [8.17]), and g_{ij} on the left side is a covariant tensor, g^{jk} has to transform as the components of a contravariant tensor.

As in equation (8.22), we can define the invariant length C of the vector C_i by the equation

$$C^2 = g^{ij} C_i C_j . \tag{8.25}$$

This expression is made more compact by using the metric tensors g_{ij} and g^{kl} to raise and lower indices. Thus the covariant vector C_i defines the contravariant vector

$$C^i = g^{ij} C_j , \tag{8.26}$$

so we can write the length of the vector as

$$C^2 = C_i C^i . \tag{8.27}$$

Equation (8.23) for the scalar product of two vectors similarly is

$$A \cdot B = A_j B^j , \qquad A_j = g_{ji} A^i . \tag{8.28}$$

The same operations apply to tensors, for example,

$$R^i{}_{klm} = g^{ij} R_{jklm} . \tag{8.29}$$

The Minkowski form η_{ij} for the metric tensor in equation (2.3), and the corresponding line element in equation (8.21), are of particular interest because they approximate the proper orthogonal coordinates an observer would set up in a neighborhood of spacetime. Let us check that it is possible to choose a coordinate transformation to bring the metric tensor to this locally Minkowski form in the neighborhood of a point in spacetime.

The coordinate transformation for g_{ij} in equation (8.19) has sixteen transformation coefficients $\partial x^i / \partial \bar{x}^j$. Since the symmetric tensor g_{ij} has ten components, we have enough free parameters to transform the metric tensor to the Minkowski form η_{ij} at a point in spacetime. (We always exclude bad things such as a metric tensor with vanishing determinant.) The first derivatives $g_{ij,k}$ of the metric tensor at the chosen point consist of $4 \times 10 = 40$ components. The transformation law for $g_{ij,k}$ involves the sixteen transformation coefficients $\partial x^i / \partial \bar{x}^j$ and their forty derivatives $\partial^2 x^i / \partial \bar{x}^j \partial \bar{x}^k$. That is, we have fifty-six numbers available to satisfy the fifty conditions that g_{ij} is equal to the Minkowski form η_{ij} and $g_{ij,k} = 0$. This leaves six free parameters for three free spatial rotations and three Lorentz velocity transformations.

There are one hundred second derivatives $g_{ij,kl}$, but only eighty second derivatives $\partial^3 x^i / \partial \bar{x}^j \partial \bar{x}^k \partial \bar{x}^l$ of the transformation coefficients. This means it is not in general possible to eliminate the second derivatives of the metric tensor.

The conclusion is that we can choose the coordinate labels so the ordinary line element of special relativity applies in the neighborhood of a chosen point, with the metric tensor g_{ij} equal to the Minkowski form η_{ij}, and the first derivatives of the metric tensor equal to zero. We will see in the next section that this represents the locally orthogonal coordinate system that a freely falling observer would set up using proper physical rods and clocks. If all the second derivatives of the metric tensor cannot be eliminated, they describe the tidal field that causes the relative gravitational acceleration of neighboring freely moving test particles, and one says spacetime is curved.

To summarize, we have prescribed the way scalars, tensors, and vectors transform under a change of the coordinate labels of spacetime (eqs. [8.2], [8.6], and [8.14]). The metric tensor and its inverse are used to raise and lower indices (eqs. [8.26] and [8.28]). We can reduce the number of indices on a tensor by summing over an upper and lower index. Thus the contraction $R^i{}_{kil} = R_{kl}$ is a two-index tensor, and the contractions T^i_i and $A_i B^i$ are scalars. Quantities that are measurable in principle are scalars, which might be constructed out of vectors and tensors.

Permutation Symbol

We have seen that the Kronecker delta function is a special constant tensor (eq. [8.17]). Another special tensor constructed out of the metric tensor and the four-component permutation symbol ϵ_{ijkl} will be useful later.

The permutation symbol has the value $\epsilon_{0123} = 1$, and it is antisymmetric under exchange of any two indices, so

$$\epsilon_{1023} = -1, \qquad \epsilon_{1123} = 0, \tag{8.30}$$

and so on. The permutation symbol defines the determinant A of the four-by-four square matrix A_{ij} by the equation

$$A = \sum_{stuv} \epsilon_{stuv} A_{0s} A_{1t} A_{2u} A_{3v}, \tag{8.31}$$

for this gives the usual sums and differences of the products of the components of the matrix. The rule for balancing a superscript with a subscript does not apply here; this deals with any square matrix and its determinant. Equation (8.31) generalizes to

$$\epsilon_{ijkl} A = \sum_{stuv} \epsilon_{stuv} A_{is} A_{jt} A_{ku} A_{lv}, \tag{8.32}$$

because this agrees with equation (8.31) when $ijkl = 0123$, and it has the right antisymmetry when the free indices are exchanged. It follows that the determinant of the matrix product $C_{ik} = A_{ij} B_{jk}$ is

$$\begin{aligned} C &= \sum \epsilon_{ijkl} A_{0s} B_{si} A_{1t} B_{tj} A_{2u} B_{uk} A_{3v} B_{vl} \\ &= B \sum \epsilon_{stuv} A_{0s} A_{1t} A_{2u} A_{3v} \\ &= AB. \end{aligned} \tag{8.33}$$

That is, the determinant of a product of square matrices is the product of the determinants.

The coordinate transformation law (8.19) for the metric tensor is a triple matrix product, so the determinant of this transformation law is

$$\bar{g} = g \left| \frac{\partial(x)}{\partial(\bar{x})} \right|^2. \tag{8.34}$$

Here g and $\bar{g}$ are the determinants of the metric tensors g_{ij} and $\bar{g}_{jk}$, and $|\partial(x)/\partial(\bar{x})|$ is the determinant of the matrix $\partial x^i / \partial \bar{x}^j$. Shortly we will use the fact that this is the Jacobian of the coordinate transformation $\bar{x} = \bar{x}(x)$ in equation (8.1). Since the determinant of the Minkowski form η_{ij} in equation (8.21) is $\eta = -1$, equation (8.34) says the determinant of g has to be nonzero and negative in any nonsingular coordinate system.

On setting $A_{ij} = \partial x^j / \partial \bar{x}^i$ in equation (8.32) for the determinant, and using equation (8.34), we get

$$\sqrt{-\bar{g}} \, \epsilon_{ijkl} = \sqrt{-g} \, \epsilon_{stuv} \frac{\partial x^s}{\partial \bar{x}^i} \frac{\partial x^t}{\partial \bar{x}^j} \frac{\partial x^u}{\partial \bar{x}^k} \frac{\partial x^v}{\partial \bar{x}^l}. \tag{8.35}$$

This says $\sqrt{-g}\,\epsilon_{stuv}$ transforms as the components of a four-index covariant tensor. The determinant g is entered with a minus sign, because g has to be negative. It is left as an exercise to check that $\epsilon_{stuv}/\sqrt{-g}$ transforms as the components of a contravariant tensor. (It would be better therefore to write it as $\epsilon^{stuv}/\sqrt{-g}$, where ϵ^{stuv} is the same as ϵ_{stuv}.)

Integrals and Derivatives

The next step is to set up coordinate-independent integrals to represent relations such as charge conservation, and the covariant generalizations of derivatives of vectors and tensors, so we can write down field equations.

Recall that the way to change the volume element in an integral when the variables of integration are changed by the coordinate transformation (8.1) is

$$d^4x = \frac{\partial(x)}{\partial(\bar{x})}d^4\bar{x}, \tag{8.36}$$

where $\partial(x)/\partial(\bar{x})$ is the Jacobian of the change of variables, that is, it is the determinant of the matrix of partial derivatives $\partial x^i/\partial \bar{x}^j$. This is the determinant that appears in equation (8.34), so equation (8.36) can be rewritten as

$$\sqrt{-g}\,d^4x = \sqrt{-\bar{g}}\,d^4\bar{x}. \tag{8.37}$$

This is a scalar relation. Thus, if $\phi(x)$ is a scalar field, then $\sqrt{-g}\,d^4x\phi(x)$ is a scalar, and the expression

$$\int \sqrt{-g}\,d^4x\,\phi(x), \tag{8.38}$$

integrated to boundaries fixed in spacetime, is independent of the choice of coordinates, because it is a sum of scalars.

Now let us consider derivatives of fields. Equation (8.5) says that the derivative of the scalar $\phi(x)$ is a vector. However, the derivative of the transformation law in equation (8.6) for a vector gives

$$\bar{A}_{i,k} = A_{j,m}\frac{\partial x^j}{\partial \bar{x}^i}\frac{\partial x^m}{\partial \bar{x}^k} + A_j\frac{\partial^2 x^j}{\partial \bar{x}^i\partial \bar{x}^k}. \tag{8.39}$$

To bring this to the canonical form for the transformation law for a tensor, we have to get rid of the second term, which comes from the derivative of the transformation coefficients. This is done by introducing a set of functions, the Christoffel symbols, designed so their transformation law is

$$\bar{\Gamma}^l_{ik} = \Gamma^n_{jm}\frac{\partial x^j}{\partial \bar{x}^i}\frac{\partial x^m}{\partial \bar{x}^k}\frac{\partial \bar{x}^l}{\partial x^n} + \frac{\partial^2 x^j}{\partial \bar{x}^i\partial \bar{x}^k}\frac{\partial \bar{x}^l}{\partial x^j}. \tag{8.40}$$

This defines the transformation law for the symbols; we will see that they can be constructed out of the components of the metric tensor (eq. [8.57]). In the first term on the right-hand side, the transformation coefficients are arranged as for a tensor with two covariant indices and one contravariant index. With the transformation law (8.6) for A_i, we have

$$\bar{\Gamma}^l_{ik}\bar{A}_l = \Gamma^n_{jm}A_n\frac{\partial x^j}{\partial \bar{x}^i}\frac{\partial x^m}{\partial \bar{x}^k} + A_j\frac{\partial^2 x^j}{\partial \bar{x}^i\partial \bar{x}^k}\,. \tag{8.41}$$

The result of subtracting this expression from equation (8.39) for the partial derivative of the vector gives

$$\bar{A}_{i,k} - \bar{\Gamma}^l_{ik}\bar{A}_l = (A_{j,m} - \Gamma^n_{jm}A_n)\frac{\partial x^j}{\partial \bar{x}^i}\frac{\partial x^m}{\partial \bar{x}^k}\,, \tag{8.42}$$

which is the standard transformation law for a tensor.

The conclusion is that the expression

$$A_{i;k} = A_{i,k} - \Gamma^j_{ik}A_j \tag{8.43}$$

transforms as a covariant tensor. The first term on the right-hand side is the ordinary partial derivative of the vector field, and the second term has been arranged to eliminate the derivative of the transformation coefficient. The semicolon on the left side represents the covariant derivative defined by this equation.

The covariant derivative of a tensor is

$$A_{ik;l} = A_{ik,l} - \Gamma^j_{il}A_{jk} - \Gamma^j_{kl}A_{ij}\,. \tag{8.44}$$

It is a somewhat more tedious but still straightforward exercise to follow the approach in equations (8.41) and (8.42) to check that the covariant derivative $A_{ik;l}$ transforms as a three-index covariant tensor. Since the product of two vectors is a tensor, we have from equation (8.44)

$$(A_iB_k)_{;l} = A_{i;l}B_k + A_iB_{k;l}\,. \tag{8.45}$$

That is, the ordinary rule for differentiating a product applies to covariant derivatives.

Since A_iB^i is a scalar, its derivative is a vector (eq. [8.5]). The result of expanding out the derivative, and using equation (8.43) to replace the derivative of A_i with the covariant derivative, is

$$\begin{aligned}(A_iB^i)_{,k} &= A_{i,k}B^i + A_iB^i{}_{,k}\\ &= (A_{i;k} + \Gamma^l_{ik}A_l)B^i + A_iB^i{}_{,k}\\ &= A_{i;k}B^i + A_l(B^l{}_{,k} + \Gamma^l_{mk}B^m)\,.\end{aligned} \tag{8.46}$$

The expression in the last line is a vector, and so is the first term, so the second term is a vector. Since this is true for any A_i, the expression in parentheses in the last line has to be a tensor, which we will write as

$$B^l{}_{;k} = B^l{}_{,k} + \Gamma^l_{mk} B^m .\tag{8.47}$$

This defines the covariant derivative of a contravariant tensor. With this definition, equation (8.46) is

$$(A_i B^i){}_{,k} = A_{i;k} B^i + A_i B^i{}_{;k} .\tag{8.48}$$

As in equation (8.45), this agrees with the usual rule for the derivative of a product.

The covariant derivative of a tensor T^{ij} similarly works out to

$$T^{ij}{}_{;k} = T^{ij}{}_{,k} + \Gamma^i_{lk} T^{lj} + \Gamma^j_{lk} T^{il} ,\tag{8.49}$$

and the covariant derivative of a mixed tensor is

$$T^i{}_{j;k} = T^i{}_{j,k} + \Gamma^i_{lk} T^l{}_j - \Gamma^l_{jk} T^i{}_l .\tag{8.50}$$

The same pattern applies to higher order tensors.

The Γ^i_{jk} are constructed out of the metric tensor. To see how this goes, note first that the antisymmetric expression

$$F_{ik} = A_{i,k} - A_{k,i}\tag{8.51}$$

is a tensor, because the subtraction eliminates the second term on the right-hand side of equation (8.39), for it is symmetric in i and k. The definition of the covariant derivative in equation (8.43) gives

$$A_{i;k} - A_{k;i} = A_{i,k} - A_{k,i} - \left(\Gamma^j_{ik} - \Gamma^j_{ki}\right) A_j .\tag{8.52}$$

The left side is a tensor, and so is the first part of the right-hand side. The last term is eliminated by demanding that the Christoffel symbol is symmetric in its lower indices,

$$\Gamma^j_{ik} = \Gamma^j_{ki} .\tag{8.53}$$

Next, we will use the *ansatz* that the covariant derivative of the metric tensor vanishes,

$$g_{ij;k} = 0 .\tag{8.54}$$

This expression written out is (eq. [8.44])

$$g_{ij,k} - \Gamma^l_{ik} g_{lj} - \Gamma^l_{jk} g_{il} = 0 . \tag{8.55}$$

The results of cyclically permuting the three indices are

$$g_{jk,i} - \Gamma^l_{ji} g_{lk} - \Gamma^l_{ki} g_{jl} = 0 ,$$
$$g_{ki,j} - \Gamma^l_{kj} g_{li} - \Gamma^l_{ij} g_{kl} = 0 . \tag{8.56}$$

Adding these two equations and subtracting equation (8.55) gives

$$\Gamma_{kij} \equiv g_{kl} \Gamma^l_{ij} = \frac{1}{2} (g_{ki,j} + g_{kj,i} - g_{ij,k}) . \tag{8.57}$$

The first step defines the Christoffel symbol Γ_{kij} with all three indices down. The form with the first symbol up is then

$$\Gamma^m_{ij} = g^{mk} \Gamma_{kij} . \tag{8.58}$$

These are the expressions for the Christoffel symbols in terms of the metric tensor g_{ij}. They uniquely follow from the condition in equation (8.53) that the Christoffel symbol is symmetric.

Below we will use the relation from equation (8.55) for the derivative of the metric tensor,

$$g_{ij,k} = \Gamma_{ijk} + \Gamma_{jik} . \tag{8.59}$$

We will also need the expression for the result of contracting a pair of indices in the Christoffel symbol,

$$\Gamma^j_{ji} = \frac{1}{2} g^{jk} (g_{ki,j} + g_{kj,i} - g_{ij,k})$$
$$= \frac{1}{2} g^{jk} g_{jk,i} , \tag{8.60}$$

where the second line follows because the first term in the parentheses cancels the last.

Equation (8.60) for Γ^j_{ji} is further simplified by using the expression for the derivative of a determinant A of a matrix A_{ij}. The determinant of a four-by-four matrix is given in equation (8.31). The determinant of the three-by-three matrix obtained by eliminating row 0 and column i of the four-by-four matrix A_{ij} is (up to a sign)

$$B_{0i} = \sum_{tuv} \epsilon_{ituv} A_{1t} A_{2u} A_{3v} , \tag{8.61}$$

for the permutation symbol ϵ_{ituv} eliminates the terms where t, u, or v are equal to i. This gives

$$\sum_i B_{0i} A_{ji} = A \delta_{0j} , \tag{8.62}$$

for if $j = 0$ this gives back the expression for the determinant A in equation (8.31), and if $j \neq 0$ the antisymmetry of the permutation symbol makes the sum from equation (8.61) vanish. The conclusion from equation (8.62) is that the matrix inverse of the matrix A_{ij} is the matrix

$$A^{-1}{}_{ij} = B_{ji}/A . \tag{8.63}$$

Suppose now the elements of A_{mn} are changed by the infinitesimal amounts δA_{mn}. Then the change in the sum in equation (8.31) for the determinant A is, with equations (8.61) and (8.63),

$$\delta A = \sum_{mn} \delta A_{mn} B_{mn} = A \sum_{mn} A^{-1}{}_{nm} \delta A_{mn} . \tag{8.64}$$

This gives the derivative of a determinant with respect to its matrix elements. Since g^{ij} is the inverse of the metric tensor g_{jk}, the derivative of the determinant g of g_{jk} is

$$g_{,i} = g g^{kj} g_{jk,i} . \tag{8.65}$$

Here the indices are properly placed for the sum convention. This result in equation (8.60) gives the wanted identity,

$$\Gamma^j_{ji} = \frac{1}{2} \frac{g_{,i}}{g} = \frac{\partial}{\partial x^i} \ln \sqrt{-g} . \tag{8.66}$$

An example of why this identity is useful is the equation of charge conservation. In Newtonian mechanics, the conservation equation is

$$\frac{\partial \rho}{\partial t} + \nabla \cdot \mathbf{j} = 0 , \tag{8.67}$$

where ρ is the charge per unit volume and $\mathbf{j}$ is the current density (the flow of charge per unit time and per unit area normal to $\mathbf{j}$). This generalizes to the covariant expression

$$j^i{}_{;i} = 0 . \tag{8.68}$$

In a locally Minkowski coordinate system, where the Christoffel symbols vanish, this is equation (8.67), and the covariant divergence gives a scalar that can be set equal to zero independent of the coordinate system. On writing out the covariant derivative, we see that the conservation equation is

$$0 = j^i{}_{;i} = j^i{}_{,i} + \Gamma^i_{ki} j^k = \frac{(\sqrt{-g}\, j^k)_{,k}}{\sqrt{-g}}, \tag{8.69}$$

where the last step uses the identity in equation (8.66). Since $j^i{}_{;i}$ is a scalar, we can integrate over it, as in equation (8.38), to get

$$0 = \int \sqrt{-g}\, j^i{}_{;i}\, d^4x = \int (\sqrt{-g}\, j^k)_{,k}\, d^4x. \tag{8.70}$$

The last expression is an ordinary divergence, so Gauss's theorem converts it to a surface integral. Suppose the charges are confined to some bounded region of space, and the integral in the spatial directions covers all of the region where j^i is nonzero. If the integral is bounded by two three-dimensional hypersurfaces at fixed values of $x^0 = t$, Gauss's theorem says the three-dimensional integral over x^1, x^2, and x^3 at fixed x^0,

$$Q = \int d^3x \sqrt{-g}\, j^0, \tag{8.71}$$

is independent of time. This is the conserved total charge.

As indicated in equation (8.54), the Christoffel symbols are designed so the covariant derivative of the metric tensor g_{ij} vanishes. To see that the covariant derivative of the inverse tensor g^{kl} also vanishes, consider the covariant derivative of the defining equation (8.24) for g^{kl}. The Kronecker delta function on the right-hand side is a tensor, and one sees from equation (8.50) that its covariant derivative vanishes. The covariant derivative of the product $g_{ij}g^{jk}$ is the sum of the covariant derivatives of g_{ij} and g^{jk}, as in equation (8.48). Since the covariant derivative of g_{ij} vanishes, the covariant derivative of the inverse tensor has to vanish:

$$g^{ij}{}_{;k} = 0. \tag{8.72}$$

To summarize, equation (8.38) prescribes an integral of a scalar function of position in spacetime. The integral is independent of the choice of coordinates and therefore can be assigned a physical meaning. This cannot be extended to integrals over vectors or tensors, because there is no way to eliminate the transformation coefficients $\partial x^i / \partial \bar{x}^j$ that in general are functions of position. The covariant derivative of a scalar field is the usual partial derivative (eq. [8.5]). The prescription for the covariant derivative of a vector or tensor adds to the usual partial derivative one term containing a Christoffel symbol for each contravariant index and subtracts one term for each covariant index (eqs. [8.43] and [8.47]). The rule for expanding out the covariant derivative of a product is the same as the rule for ordinary derivatives (eq. [8.45]). The Christoffel symbols are constructed out of the metric tensor, as shown in equations (8.57) and (8.58).

The covariant derivatives provide a guide to the generalization of Lagrangian densities from special relativity theory to curved spacetime. Because covariant derivatives transform as tensors, we know how to make the Lagrangian density a scalar, and we know that a wave equation derived from the Lagrangian will transform in a covariant way. The next section deals with the simplest example, the action for a freely moving test particle, and section 10 treats field equations.

The Curvature Tensor

We now have the machinery to construct the curvature tensor from the metric tensor and its first and second derivatives. We have seen that it is always possible to choose coordinates so that at a point in spacetime the metric tensor is equal to the Minkowski form $g_{ij} = \eta_{ij}$ and the first derivatives vanish, $g_{ij,k} = 0$, but that it need not be possible to eliminate the second derivatives, $g_{ij,kl}$. If the second derivatives can all be eliminated, the curvature tensor to be defined here vanishes, because it is constructed out of these derivatives. Since the coordinate transformation law is linear in the tensor, if the curvature tensor vanishes when expressed in one coordinate system it vanishes in any coordinate system. In particular, in the flat spacetime of special relativity, coordinates can be assigned so $g_{ij} = \eta_{ij}$ everywhere, so the curvature tensor vanishes everywhere, and it therefore has to vanish when expressed in any coordinate system. If on the other hand the curvature tensor is nonzero at a point in spacetime, then the second derivatives of g_{ij} cannot all be eliminated in locally Minkowski coordinates, and spacetime in the neighborhood of the point is curved.

The curvature tensor is used in the next section to describe the relative motions of neighboring freely moving test particles, and in section 10 in the Einstein gravitational field equation.

Suppose A_i is a vector field, so the covariant derivative $A_{i;k}$ is a tensor. The game is to write down the covariant derivative of this field, $A_{i;k;l}$, and then the result of taking the derivatives in the opposite order, $A_{i;l;k}$. In the difference of these two tensors the partial derivatives of the A_i cancel, leaving an expression proportional to A_i. The part multiplying A_i, which is a function of the metric tensor and its first and second derivatives, is the wanted curvature tensor.

The covariant derivative of the vector field is as usual

$$A_{i;k} = A_{i,k} - \Gamma_{ik}^m A_m, \qquad (8.73)$$

and the covariant derivative of this expression is

$$A_{i;k;l} = (A_{i,k} - \Gamma_{ik}^m A_m)_{,l} - \Gamma_{il}^m A_{m;k} - \Gamma_{kl}^m A_{i;m}$$
$$= A_{i,kl} - (\Gamma_{ik}^m)_{,l} A_m - \Gamma_{ik}^m A_{m,l} - \Gamma_{il}^m (A_{m,k} - \Gamma_{mk}^u A_u) \qquad \textbf{(8.74)}$$
$$- \Gamma_{kl}^m (A_{i,m} - \Gamma_{im}^u A_u) .$$

The expression for the derivatives in the opposite order is

$$A_{i;l;k} = A_{i,lk} - (\Gamma_{il}^m)_{,k} A_m - \Gamma_{il}^m A_{m,k} - \Gamma_{ik}^m (A_{m,l} - \Gamma_{ml}^u A_u)$$
$$- \Gamma_{lk}^m (A_{i,m} - \Gamma_{im}^u A_u) . \qquad \textbf{(8.75)}$$

The difference is

$$A_{i;k;l} - A_{i;l;k} = A_m R^m{}_{ikl} , \qquad \textbf{(8.76)}$$

where the curvature tensor is

$$R^m{}_{ikl} = (\Gamma_{il}^m)_{,k} - (\Gamma_{ik}^m)_{,l} + \Gamma_{il}^n \Gamma_{nk}^m - \Gamma_{ik}^n \Gamma_{nl}^m . \qquad \textbf{(8.77)}$$

Since equation (8.76) is a tensor equation with an arbitrary vector field A_i, $R^m{}_{ikl}$ is a tensor.

We will need the curvature tensor with all indices down:

$$R_{iklm} = g_{ij} R^j{}_{klm} . \qquad \textbf{(8.78)}$$

A somewhat simpler expression for this latter tensor uses the derivative of the inverse metric tensor. The result of differentiating out equation (8.24) and rearranging is

$$g^{ij}{}_{,k} = -g^{il} g_{lm,k} g^{mj} . \qquad \textbf{(8.79)}$$

With this expression, and equation (8.57) for the Christoffel symbol, it is a straightforward if somewhat tedious exercise to check that equation (8.77) for the curvature tensor becomes

$$R_{iklm} = \frac{1}{2}(g_{kl,im} + g_{im,kl} - g_{km,il} - g_{il,km})$$
$$+ g_{st}(\Gamma_{im}^s \Gamma_{kl}^t - \Gamma_{il}^s \Gamma_{km}^t) . \qquad \textbf{(8.80)}$$

This is useful in the weak field approximation to Einstein's gravitational field equation, where the derivatives of g_{ij} are small and the squared terms in the last parentheses are even smaller, and more generally in locally Minkowski coordinates, where the Christoffel symbols vanish. Also, we read from this expression that the curvature tensor satisfies the identities

$$
\begin{aligned}
R_{iklm} &= -R_{ikml} \\
&= -R_{kilm} \\
&= R_{lmik} \, .
\end{aligned}
\tag{8.81}
$$

We also need the tensors that can be constructed by contracting indices in R_{iklm}. The Ricci tensor is the result of contracting two of the indices,

$$
R_{km} = R^i{}_{kim} = (\Gamma^i_{km}),_i - (\Gamma^i_{ki}),_m + \Gamma^i_{ni}\Gamma^n_{km} - \Gamma^i_{mn}\Gamma^n_{ki} \, .
\tag{8.82}
$$

This tensor is symmetric, $R_{km} = R_{mk}$ (as follows from the last line of eq. [8.81], or by noting that Γ^i_{ki} is the total derivative in eq. [8.66]). The identities in equation (8.81) tell us that the contraction of any other pair of indices in R_{iklm} either vanishes or is equal to the Ricci tensor up to a minus sign. The scalar curvature is the result of contracting the indices in the Ricci tensor,

$$
R = g^{km}R_{km} \, .
\tag{8.83}
$$

This appears in the gravitational part of the action for the Einstein field equation.

The traceless part of the tensor R_{ijkl} is defined by the equation

$$
\begin{aligned}
R_{ijkl} = {}&(g_{ik}R_{jl} - g_{il}R_{jk} - g_{jk}R_{il} + g_{jl}R_{ik})/2 \\
&- (g_{ik}g_{jl} - g_{il}g_{jk})R/6 \\
&+ C_{ijkl} \, .
\end{aligned}
\tag{8.84}
$$

The expressions in the first two lines have the symmetries of the curvature tensor in equation (8.81), so the Weyl tensor C_{ijkl} in the last line has the same symmetries. The coefficients in the first two lines are chosen so the contraction of the Weyl tensor vanishes,

$$
C^i{}_{jil} = 0 \, ,
\tag{8.85}
$$

as can be checked by working out the contractions of the first two expressions. The Weyl tensor will be helpful in simplifying the optical equation for the distortion in the image of a distant galaxy in an inhomogeneous universe (eq. [14.21]).

Finally, the covariant derivative of the Ricci tensor satisfies an interesting identity. At a point in spacetime where the coordinates are locally Minkowski, the Christoffel symbols vanish, so the derivative of equation (8.80) is

$$
R_{iklm,n} = \frac{1}{2}(g_{kl,imn} + g_{im,kln} - g_{km,iln} - g_{il,kmn}) \, .
\tag{8.86}
$$

The result of cyclically permuting the last three indices and adding is

$$R_{iklm,n} + R_{ikmn,l} + R_{iknl,m} = 0 . \tag{8.87}$$

In a general coordinate system this is the Bianchi identity

$$R_{iklm;n} + R_{ikmn;l} + R_{iknl;m} = 0 . \tag{8.88}$$

The result of contracting the indices and rearranging a little is the identity

$$(R^{ij} - g^{ij}R/2)_{;j} = 0 . \tag{8.89}$$

The expression in parentheses is the gravitational field part of the Einstein equation (10.62). The identity shows that the source term in the field equation, the stress-energy tensor T^{ij}, has zero divergence. As we will discuss, this is the expression for the local conservation of energy and momentum (eq. [10.32]).

9. Motions of Free Test Particles

The previous section presents the elements of a metric theory of spacetime, in which the tensor g_{ij} describes geometry through the prescription for measuring lengths and times. The next step is to consider how particles and fields behave in a given spacetime geometry. This section deals with particles.

Geodesic Equation of Motion

Consider a test particle with rest mass m so small it has no appreciable effect on the behavior of spacetime. The particle moves freely, with an equation of motion that follows from a generalization of the action in flat spacetime. In Newtonian mechanics, the Lagrangian for a free particle is the kinetic energy, so the action is

$$S = \int dt \, \frac{1}{2} m \dot{\mathbf{x}}^2 , \tag{9.1}$$

where $\dot{\mathbf{x}} = d\mathbf{x}/dt$ is the velocity. In special relativity this generalizes to

$$S = -m \int dt (1 - \dot{\mathbf{x}}^2)^{1/2} = -m \int ds , \tag{9.2}$$

where ds is the line element in equation (8.21). This expression is unchanged by a Lorentz transformation, as required, because ds is a Lorentz scalar. In the nonrelativistic limit, $\dot{x} \ll 1$, the Lagrangian differs from that of equation (9.1) by the constant term, m, which does not affect the equation of motion. The generalization to a metric theory of spacetime is to write ds as the line element in equation (8.20) defined by the metric tensor,

$$S = -m \int ds = -m \int (g_{ij} dx^i dx^j)^{1/2} . \tag{9.3}$$

This expression is independent of the choice of coordinates because ds is a scalar. In locally Minkowski coordinates, it just says the action is equation (9.2) for a particle that moves in a locally straight line.

Let us derive the equation of motion from this action by the traditional methods of classical mechanics. For this purpose let $x^0 = t$ and $\dot{x}^i = dx^i/dt$, so the action in equation (9.3) is

$$S = \int L \, dt , \qquad L = -m(g_{ij}\dot{x}^i\dot{x}^j)^{1/2} . \tag{9.4}$$

The three coordinates are $x^\alpha(t)$, with $\alpha = 1$, 2, and 3. In what follows, repeated Greek indices will be understood to be summed from 1 to 3.

The canonical momenta are the derivatives of the Lagrangian with respect to the coordinate velocities,

$$p_\alpha = \frac{\partial L}{\partial \dot{x}^\alpha} = -\frac{m g_{\alpha i}\dot{x}^i}{(g_{jk}\dot{x}^j\dot{x}^k)^{1/2}} . \tag{9.5}$$

The denominator is

$$(g_{jk}\dot{x}^j\dot{x}^k)^{1/2} = ds/dt , \tag{9.6}$$

so the momenta are

$$p_\alpha = -m g_{\alpha i}u^i , \tag{9.7}$$

where the four-velocity is

$$u^i = \frac{dx^i}{ds} . \tag{9.8}$$

This is a vector, for dx^i is a vector (eq. [8.7]) and ds is a scalar. The length of the four-velocity is unity, $g_{ij}u^i u^j = g_{ij} dx^i dx^j/ds^2 = 1$.

Equation (9.7) says the canonical momenta are the three spatial components of the covariant version of the four-vector

$$p^i = -m u^i . \tag{9.9}$$

Apart from the sign, this is the usual energy-momentum four-vector for a particle of mass m. The rate of change of coordinate position with respect to coordinate time is the velocity

$$\dot{x}^i = \frac{dx^i}{dt} = \frac{p^i}{p^0}.$$

(9.10)

The Euler-Lagrange equation of motion is

$$\frac{dp_\alpha}{dt} = \frac{\partial L}{\partial x^\alpha} = -\frac{1}{2}\frac{mg_{jk,\alpha}\dot{x}^j\dot{x}^k}{(g_{lm}\dot{x}^l\dot{x}^m)^{1/2}}.$$

(9.11)

With equation (9.7) for the momentum and equation (9.6) to simplify the right-hand side, we get

$$\frac{d}{dt}g_{\alpha i}u^i = \frac{1}{2}g_{jk,\alpha}u^j\frac{dx^k}{dt}.$$

(9.12)

Equation (9.12) has three components, sufficient to determine the three functions $x^\alpha(t)$ that define the path of the particle. To write the equation of motion in a manifestly covariant form we need a fourth component, which has to be an identity derived from the three components of the Euler-Lagrange equation. The easy way to find it is to recall that the Hamiltonian gives an energy equation. The Hamiltonian is

$$H = p_\alpha \dot{x}^\alpha - L$$
$$= -mg_{\alpha i}\frac{dx^i}{ds}\frac{dx^\alpha}{dt} + m\frac{ds}{dt}.$$

(9.13)

In the first term on the right-hand side the index α is summed only over the space coordinates, so the difference of the two terms is

$$H = mg_{0i}u^i = -p_0.$$

(9.14)

As might be expected, this is the fourth component of the energy-momentum four-vector in equations (9.7) and (9.9).

The energy equation from classical mechanics is

$$\frac{dH}{dt} = \frac{\partial H}{\partial t},$$

(9.15)

where the partial derivative on the right-hand side is computed at fixed coordinates x^α and momenta p_β. We have from equation (9.9) for the four-momentum the identity

$$m^2 = g^{jk}p_jp_k = g^{\alpha\beta}p_\alpha p_\beta + 2g^{\alpha 0}p_\alpha p_0 + g^{00}p_0^2.$$

(9.16)

The derivative of this expression with respect to t at fixed x^α and p_β is

$$0 = \dot{g}^{jk} p_j p_k + 2 g^{i0} p_i \dot{p}_0 = -\dot{g}_{jk} p^j p^k + 2 p^0 \dot{p}_0 \,. \tag{9.17}$$

The last step follows from equation (8.79) for the derivative of the reciprocal tensor g^{ij}. The energy equation is then

$$\frac{dH}{dt} = -\frac{dp_0}{dt} = -\dot{p}_0 = -\frac{1}{2} g_{jk,0} p^j \frac{p^k}{p^0} \,. \tag{9.18}$$

The last factor is the coordinate velocity (9.10), so the energy equation is

$$\frac{d}{dt} g_{0i} u^i = \frac{1}{2} g_{jk,0} u^j \frac{dx^k}{dt} \,. \tag{9.19}$$

Equation (9.19) has the form of the fourth component of the equation of motion (9.12), so the four-component equation is

$$\frac{d}{dt} g_{ij} u^j = \frac{1}{2} g_{jk,i} u^j \frac{dx^k}{dt} \,. \tag{9.20}$$

The product with $u^0 = dt/ds$ gives a more elegant form,

$$\frac{d}{ds} g_{ij} u^j = \frac{1}{2} g_{jk,i} u^j u^k \,. \tag{9.21}$$

On differentiating out the left side, and using $dg_{ij}/ds = g_{ij,k} \, dx^k/ds$, we get

$$\begin{aligned}
g_{ij} \frac{du^j}{ds} &= -g_{ij,k} u^j u^k + \frac{1}{2} g_{jk,i} u^j u^k \\
&= -\frac{1}{2} (g_{ij,k} + g_{ik,j} - g_{jk,i}) u^j u^k \\
&= -\Gamma_{ijk} u^j u^k \,.
\end{aligned} \tag{9.22}$$

The expression in parentheses in the second line is the Christoffel symbol (8.57). Thus on raising the index we can write the equation of motion as

$$\frac{du^i}{ds} + \Gamma^i_{jk} u^j u^k = 0 \,. \tag{9.23}$$

For yet another form, write the rate of change of the four-velocity u^i with respect to the proper time s as

$$\frac{du^i}{ds} = \frac{\partial u^i}{\partial x^k} \frac{dx^k}{ds} \,. \tag{9.24}$$

Here we are imagining a sea of freely moving test particles — in a region of spacetime where the orbits do not cross — that define a smooth velocity field $u^i(x)$. The product of derivatives on the right-hand side just gives the rate of change of the velocity field at the position of a chosen particle. This product in equation (9.23) gives

$$0 = \left(\frac{\partial u^i}{\partial x^k} + \Gamma^i_{jk} u^j \right) \frac{dx^k}{ds}. \tag{9.25}$$

The expression in parentheses is the covariant derivative of the velocity field (eq. [8.47]), so the equation of motion for the field of test particles is the still more elegant form

$$u^i_{;k} u^k = 0. \tag{9.26}$$

We can interpret equation (9.26) as the expression for the parallel transport of a vector field. In the flat spacetime of special relativity, with a Minkowski coordinate system, the vectors $A^i(x)$ at the positions x^i and $x^i + dx^i$ are parallel if the components are the same. This condition is

$$dA^i = A^i_{,j} dx^j = 0. \tag{9.27}$$

If we use polar coordinates instead of orthogonal Cartesian ones, we find that the components of parallel vectors at different positions are different, and we can take that into account by rewriting equation (9.27) in the covariant form

$$DA^i = A^i_{;j} dx^j = 0, \tag{9.28}$$

because we know that this is the covariant expression that reduces to equation (9.27) when the coordinates are orthogonal. Since the same applies to locally Minkowski coordinates in curved spacetime, the general condition that a local observer sees that the vector field $A^i(x)$ has been parallel transported from x^i to $x^i + dx^i$ is $DA^i = 0$. The geodesic equation (9.26) is

$$\frac{Du^i}{ds} = 0. \tag{9.29}$$

This says the velocity of a freely moving test particle is parallel transported along the path the particle is following. A local freely moving observer would just say the particle is moving without acceleration.

Equations (9.20), (9.21), (9.23), and (9.26) are equivalent versions of the geodesic equation of motion for free test particles, and a solution is said to be the geodesic path traced out by the particle.

The action $-m \int ds$ (eq. [9.3]) is proportional to the proper time elapsed in moving along the geodesic, as measured by a clock attached to the particle. The action principle says the time interval is an extremum at the geodesic and a minimum for local variations of the path away from the geodesic. The minus sign in the action tells us that the proper time measured by observers who move from given initial to final positions in spacetime is longest for the freely moving observer. (To be complete, we should note that there may be another geodesic path that connects the given initial and final points in spacetime with a longer elapsed proper time, for the principle only says the action is minimum under local variations of the path.) This sometimes is called the twins paradox, though it is not paradoxical and the effect is observed.

Suppose coordinates are chosen so the test particle instantaneously is at rest, $\dot{x}^\alpha(t) = 0$, at a point in spacetime where the metric tensor is static and equal to the Minkowski form, with $g_{00} = 0$, $g_{\alpha\beta} = -\delta_{\alpha\beta}$, and $\dot{g}_{\alpha 0} = 0$ (as we will see can always be arranged). Then $u^0 = 1$, and the geodesic equation (9.12) is

$$\frac{d^2 x^\alpha}{dt^2} = -\frac{1}{2} g_{00,\alpha} \, . \tag{9.30}$$

In Newtonian mechanics, the acceleration is

$$\frac{d^2 x^\alpha}{dt^2} = -\phi_{,\alpha} \, , \tag{9.31}$$

where ϕ is the gravitational potential. This means the first-order correction to the time part of the metric tensor in the neighborhood of the particle has to be

$$g_{00} = 1 + 2\phi \, . \tag{9.32}$$

Geodesic and Synchronous Coordinates

It was shown in the last section that it is always possible to choose locally Minkowski coordinates such that at a chosen point the metric tensor is equal to the Minkowski form η_{ij} and all the first derivatives of the metric tensor vanish. Here we check that this can be extended to locally Minkowski coordinates along the path of a freely moving observer.

Suppose that in a coordinate labeling of spacetime the metric tensor is $g_{ij}(x)$. We will choose new coordinates in which the observer is at a fixed spatial position $\bar{x}^\alpha$, so the new four-velocity of the observer is $\bar{u}^i = \delta^i_0 \bar{u}^0$. In these new coordinates the equation of motion (9.20) is

$$\frac{d}{dt} \bar{g}_{i0} \bar{u}^0 = \frac{1}{2} \bar{g}_{00,i} \bar{u}^0 \, . \tag{9.33}$$

Next, let us seek coordinates to satisfy the ten equations $\bar{g}_{ij} = \eta_{ij}$ along the path of the observer, and the thirty equations $g_{ij,\alpha} = 0$ along the path. Equation (9.33) gives us three of these conditions, leaving thirty-seven to be satisfied by the choice of the coordinate transformation. The transformation coefficients we have available to do this are[1]

$$\frac{\partial \bar{t}}{\partial t}, \quad \frac{\partial \bar{x}^i}{\partial x^\alpha}, \quad \frac{\partial^2 \bar{x}^i}{\partial x^\alpha \partial x^\beta}. \qquad (9.34)$$

These are thirty-seven free functions, enough to satisfy the conditions for locally Minkowski coordinates along the path of a freely moving observer.

These locally Minkowski coordinates are the locally orthogonal coordinates a freely falling observer could set up using physical measuring rods and clocks. At the position of the observer, test particles made of any material move with no gravitational acceleration. Test particles some distance away may drift away from or toward the observer, with acceleration proportional to the distance. This tidal effect comes from the second derivatives of the metric tensor; it is described by the geodesic deviation equation to be presented below.

Another version of this prescription yields the time-orthogonal or synchronous coordinates that were used in section 4 (eq. [4.5]) and will be useful in following sections. Choose a three-dimensional hypersurface with unit normal n^i that is everywhere timelike ($n^i n^j g_{ij} = +1$). Fill spacetime with a collection (congruence) of observers, each of whom moves as a free test particle, and with masses that have negligible effect on the spacetime. The observers' initial velocities at the hypersurface are normal to it, $u^i = n^i$. Each observer is equipped with a clock, and the clocks are synchronized to a common value at the hypersurface. Each observer is assigned three numbers x^α, with neighboring observers having neighboring numbers. Then the coordinates of an event in spacetime are the clock reading t of the observer who passes through the event and the numbers x^α belonging to the observer. Orbits may eventually cross, producing a coordinate (unphysical) singularity. Where this happens one has to choose a new starting hypersurface.

With this prescription, events with coordinate separation $\delta x^\alpha = 0$ are along the path of the same observer, so their proper separation is

$$\delta t^2 = \delta s^2 = g_{ij} \delta x^i \delta x^j = g_{00} \delta t^2, \qquad (9.35)$$

because the proper separation is the clock time. This means $g_{00} = 1$. The four-velocity of each observer is then

$$u^i = \delta^i_0. \qquad (9.36)$$

---

[1] The condition $\bar{g}_{ij} = \eta_{ij}$ along the path gives the final set of derivatives, $\bar{g}_{ij,0} = 0$. Similarly, the condition that $\bar{x}^\alpha(x^\beta(t), t)$ is constant along the path $x^\beta(t)$ of the particle in the original coordinate system, with $\partial \bar{x}^\alpha / \partial x^\beta$, fixes $\partial \bar{x}^\alpha / \partial t$.

At the initial hypersurface the clocks are synchronized, so an interval δx_a^i with $\delta x_a^0 = 0$ is in the hypersurface. An interval δx_b^i with $\delta x_b^\alpha = 0$ points along the path of an observer, perpendicular to the hypersurface. The intervals therefore are orthogonal, $\delta x_a{}^i \delta x_{bi} = 0$. This says $g_{\alpha 0} = 0$ at the hypersurface. With the four-velocity in equation (9.36), the geodesic equation (9.33) says $dg_{\alpha 0}/dt = 0$. Thus $g_{\alpha 0}$ vanishes everywhere, and the line element is

$$ds^2 = dt^2 + g_{\alpha\beta}\, dx^\alpha dx^\beta \,. \tag{9.37}$$

This is the time-orthogonal coordinate system used in equations (4.5) and (5.8) for the Einstein and Friedmann-Lemaître cosmological models.

One might want to choose time-orthogonal coordinates in a cosmological model in such a way that each observer at fixed x^α sees that the material in the neighborhood has no streaming velocity. If this is possible, it means that in these time-orthogonal coordinates the material defines a four-velocity field that satisfies equation (9.36), with all the derivatives vanishing. We can express this condition in a covariant form by the identity

$$w_{ij} \equiv u_{i;j} - u_{j;i} = u_{i,j} - u_{j,i} \,, \tag{9.38}$$

where the difference of the covariant derivatives leaves ordinary partial derivatives, because the Christoffel symbol Γ_{lm}^k is symmetric in its lower indices (as in eq. [8.51]). If we can find a time-orthogonal coordinate system in which the material velocity field has the form of equation (9.36), then in this coordinate system w_{ij} vanishes, and since it is a tensor w_{ij} vanishes in any coordinate system. That is, the conditions that time-orthogonal coordinates can be assigned to the streaming motion of the material is that the matter moves along geodesics with an irrotational velocity field, $w_{ij} = 0$.

The construction of the time-orthogonal coordinates in equation (9.37) has reduced the number of fields to the six spatial components $g_{\alpha\beta}$. As will be seen in the next section, there are six independent gravitational field equations to propagate these six fields from initial values on the hypersurface we started with. There remains the freedom to adjust the initial hypersurface and the coordinate assignments x^α on the surface. This is most easily examined for infinitesimal changes. With

$$\bar{t} = t + \psi \,, \qquad \bar{x}^\alpha = x^\alpha + d^\alpha \,, \tag{9.39}$$

the coordinate transformation to a new time-orthogonal coordinate system gives, to first order in ψ and d^α,

$$g_{00} = 1 = 1 + 2\dot\psi ,$$

$$g_{0\alpha} = 0 = \psi_{,\alpha} + g_{\alpha\beta}\dot d^\beta ,$$ (9.40)

$$g_{\alpha\beta}(x, t) = \bar g_{\alpha\beta}(\bar x, \bar t) + d^\gamma_{,\alpha}g_{\gamma\beta} + d^\gamma_{,\beta}g_{\alpha\gamma} .$$

The first equation says $\psi = \psi(\mathbf{x})$ is a function of the space coordinates alone, independent of time. It describes the shift in the hypersurface used to construct the coordinates. The second equation says

$$d^\gamma = \chi^\gamma(\mathbf{x}) - \psi_{,\alpha}\int^t g^{\alpha\gamma}dt .$$ (9.41)

The time-independent functions χ^α describe a reassignment of position coordinates on the hypersurface. The four free functions ψ and χ^α leave two physically significant functions on the hypersurface to be propagated forward in time by the field equations.

Massless Particles

We need special treatment of the geodesic equation of motion for massless test particles, such as photons, because the displacement dx^i along the path of a massless particle which moves at the velocity of light is a null vector, $(g_{ij}\,dx^i dx^j)^{1/2} = ds = 0$. One way is to start from particles with nonzero rest mass m, where the components of the energy-momentum four-vector satisfy the equations

$$\frac{dp_i}{dt} = \frac{1}{2}g_{jk,i}p^j\dot x^k .$$ (9.42)

The three space components $i = \alpha$ are the Euler-Lagrange equation (9.12) for the canonical momenta (9.7) and the part $i = 0$ is the energy equation (9.19). Since the energy-momentum four-vector is $-p^i = mu^i$, the coordinate velocity is, as before,

$$\dot x^i = \frac{p^i}{p^0} .$$ (9.43)

In the limit $m \to 0$ and $ds \to 0$ equations (9.42) and (9.43) remain well defined, and describe the motions of particles such as photons with zero rest mass. From the equation $g_{ij}p^i p^j = m^2 g_{ij}u^i u^j = m^2$ we have the condition

$$g_{ij}p^i p^j = 0$$ (9.44)

for particles such as photons with negligible rest mass. It is a useful exercise to work through the argument that shows that, in time-orthogonal coordinates, $-p^0$ is the photon energy measured by an observer at fixed coordinate position x^α.

We get back to equations that look like those for particles with nonzero mass by defining an affine parameter λ by the equation

$$p^0 = \frac{dt}{d\lambda} = \frac{dx^0}{d\lambda}. \tag{9.45}$$

Points along the path of the particle are labeled by the value of the parameter λ one gets by integrating this equation along the path. This definition of λ gives

$$\frac{dx^i}{d\lambda} = \frac{dx^i}{dt}\frac{dt}{d\lambda} = p^i, \tag{9.46}$$

by equation (9.43). Thus the geodesic equation (9.42) becomes

$$\frac{d}{d\lambda}g_{ij}\frac{dx^j}{d\lambda} = \frac{1}{2}g_{jk,i}\frac{dx^j}{d\lambda}\frac{dx^k}{d\lambda}, \tag{9.47}$$

which is the same form as equation (9.21) with the parameter λ replacing the proper time s along the path, and with the condition from equation (9.44)

$$g_{ij}\frac{dx^i}{d\lambda}\frac{dx^j}{d\lambda} = 0. \tag{9.48}$$

The equivalent of equation (9.23) is

$$\frac{d^2 x^i}{d\lambda^2} + \Gamma^i_{jk}\frac{dx^j}{d\lambda}\frac{dx^k}{d\lambda} = 0. \tag{9.49}$$

The p_α in the Euler-Lagrange equation (9.42) are the canonical momenta conjugate to the coordinates x^α (eq. [9.7]). Since these coordinates and momenta satisfy Hamilton's equations of motion, Liouville's theorem tells us we can use the x^α and p_α to define a six-dimensional single-particle phase space, in which the distribution function f for a gas of test particles is defined by the number of particles found in the volume element $dx^1 dx^2 dx^3$ and $dp_1 dp_2 dp_3$,

$$dN = f(x, p, t)\, d^3 p\, d^3 x. \tag{9.50}$$

Liouville's theorem says that in the absence of collisions f is constant along the path of a particle in phase space. Since this is true independent of the coordinate system, the distribution function f has to be a scalar.

This result applied to radiation is known as the brightness theorem. The surface brightness $i(\nu)$ of a beam of radiation is the constant of proportionality in the expression for the energy δu in the bandwidth $\delta\nu$ observed to move normally through the element of area δA into the element of solid angle $\delta\Omega$ in the time interval δt,

$$\delta u = i(\nu)\, \delta\nu\, \delta A\, \delta\Omega\, \delta t. \tag{9.51}$$

Equation (9.50) in a locally Minkowski coordinate system says the photon number flux density per unit interval of p and solid angle is $p^2 f$, and since the photon energy is p the surface brightness is

$$i(p) = p^3 f .$$
(9.52)

Consider radiation that leaves a galaxy with surface brightness $i_e(p_e)$, as measured by an observer at rest at the galaxy. A distant observer finds that photons emitted at energy p_e are received at energy $p_o = p_e/(1+z)$, where $1+z$ is the redshift factor. Since f is unchanged, the observed surface brightness (9.52) is

$$i_o(p_o) = i_e(p_e)/(1+z)^3 .$$
(9.53)

The surface brightness integrated over all frequencies is then

$$i_o = i_e/(1+z)^4 .$$
(9.54)

Equation (9.53) is equivalent to equation (6.43), and the $(1+z)^{-4}$ surface brightness law is equation (6.44).

Geodesic Deviation

The equation of motion of a freely moving test particle can be rewritten in a form that is helpful in interpreting the gravitational field equation (section 10), and in the analysis of the effect of an inhomogeneous mass distribution on the appearance of distant galaxies (section 14). The game is to write down the equation for the rate of change of the physical separation of neighboring geodesics of freely moving test particles. It will be seen that this introduces the curvature tensor. We will discuss particles with nonzero rest mass. For photons, $d\lambda$ replaces ds, as in equation (9.49).

Consider a four-vector field $A^i(x)$. In locally Minkowski coordinates, the rate of change of the field components with respect to the path label s at the position of an observer moving with four-velocity $u^i = dx^i/ds$ is

$$\frac{dA^i}{ds} = A^i_{,j} \frac{dx^j}{ds} = A^i_{,j} u^j .$$
(9.55)

The covariant expression in a general coordinate system follows by replacing the partial derivative with a covariant derivative, giving

$$\frac{DA^i}{ds} = A^i_{;j} u^j = (A^i_{,j} + \Gamma^i_{jk} A^k) u^j$$

$$= \frac{dA^i}{ds} + \Gamma^i_{jk} A^k u^j .$$
(9.56)

If the observer sees that the field is parallel, this equation is $DA^i/ds = 0$, as in the geodesic equation (9.29).

Since DA^i/ds is a vector, we can write the covariant form for the rate of change of DA^i/ds with respect to the path parameter s as

$$\frac{D^2A^i}{ds^2} = \left(\frac{DA^i}{ds}\right)_{,l}u^l + \Gamma^i_{lm}\frac{DA^m}{ds}u^l$$

$$= \left(\frac{dA^i}{ds} + \Gamma^i_{jk}A^ku^j\right)_{,l}u^l + \left(\frac{dA^m}{ds} + \Gamma^m_{uv}A^uu^v\right)\Gamma^i_{lm}u^l. \tag{9.57}$$

We can eliminate the derivative of the velocity field in this equation by using the geodesic equation (9.23),

$$u^k_{,l}u^l = -\Gamma^k_{mn}u^mu^n. \tag{9.58}$$

Then the result of multiplying out equation (9.57) is

$$\frac{D^2A^i}{ds^2} = \frac{d^2A^i}{ds^2} + (\Gamma^i_{jk})_{,l}A^ku^ju^l + \Gamma^i_{jk}A^k_{,l}u^ju^l - \Gamma^i_{jk}A^k\Gamma^j_{lm}u^lu^m$$

$$+ \Gamma^i_{jk}A^j_{,l}u^lu^k + \Gamma^i_{jk}\Gamma^j_{lm}u^kA^lu^m. \tag{9.59}$$

Now let us apply this equation to two geodesic paths traced by neighboring test particles,

$$x^i(s) \quad \text{and} \quad x^i(s) + \xi^i(s). \tag{9.60}$$

We will compute to terms of first order in the separation ξ^i. The geodesic equations (9.23) for the two paths are

$$\frac{d^2x^i}{ds^2} + \Gamma^i_{jk}(x)\frac{dx^j}{ds}\frac{dx^k}{ds} = 0,$$

$$\frac{d^2(x^i + \xi^i)}{ds^2} + \Gamma^i_{jk}(x + \xi)\frac{d(x^j + \xi^j)}{ds}\frac{d(x^k + \xi^k)}{ds} = 0. \tag{9.61}$$

The difference between these two equations, to first order in the separation ξ, is

$$\frac{d^2\xi^i}{ds^2} + (\Gamma^i_{jk})_{,l}\xi^l\frac{dx^j}{ds}\frac{dx^k}{ds} + 2\Gamma^i_{jk}\frac{d\xi^j}{ds}\frac{dx^k}{ds} = 0. \tag{9.62}$$

The final step is to use equation (9.59) to write out the equation for the covariant derivative $D^2\xi^i/ds^2$ of the separation between the paths, with $A^i = \xi^i$, and the part $d^2\xi^i/ds^2$ given by equation (9.62). The expression, which is best worked through in a quiet place, is

$$\frac{D^2\xi^i}{ds^2} = -(\Gamma^i_{jk}),_l\xi^l u^j u^k - 2\Gamma^i_{jk}\frac{d\xi^j}{ds}u^k$$

$$+ (\Gamma^i_{jk}),_l\xi^j u^k u^l + \Gamma^i_{jk}\frac{d\xi^j}{ds}u^k - \Gamma^i_{jk}\xi^j\Gamma^k_{mn}u^m u^n \qquad (9.63)$$

$$+ \Gamma^i_{lm}u^l\frac{d\xi^m}{ds} + \Gamma^i_{lm}\Gamma^m_{uv}\xi^u u^v u^l .$$

The terms containing $d\xi^l/ds$ cancel, leaving

$$\frac{D^2\xi^i}{ds^2} = \xi^j u^k u^l \left[(\Gamma^i_{jk}),_l - (\Gamma^i_{kl}),_j + \Gamma^i_{lm}\Gamma^m_{jk} - \Gamma^i_{jm}\Gamma^m_{kl}\right] . \qquad (9.64)$$

The expression in parentheses is the curvature tensor (8.77), so we have

$$\frac{D^2\xi^i}{ds^2} = R^i_{klj}u^k u^l\xi^j . \qquad (9.65)$$

This is the equation of geodesic deviation.

In flat spacetime, the curvature tensor vanishes and equation (9.65) is $D^2\xi/ds^2 = 0$. This just says that if the neighboring geodesics are parallel to begin with, they remain parallel. In curved spacetime, a freely moving observer sees a relative acceleration of neighboring freely moving test particles. We will see in the next section that this would be interpreted as the tidal effect of distant masses or the attraction of the distributed gravitating mass in the region (eq. [10.66]).

10. Field Equations

This section deals with Einstein's gravitational field equation, which relates the metric tensor to the material content of the universe. We start with the action principle for the dynamics of the matter and radiation. That leads to the stress-energy tensor, which is the source term for the gravitational field equation. The significance of the field equation is illustrated with the help of some linear approximations.

Scalar Field Equation

The field equations for matter and radiation, and Einstein's field equation for the metric tensor, are derived from an action, of the form

$$S = \int \sqrt{-g}\mathcal{L}\, d^4x . \qquad (10.1)$$

The action has to be independent of the choice of the spacetime coordinate labels, so the Lagrangian density $\mathcal{L}$ must be a scalar constructed out of the fields, as in equation (8.38). Since the fields transform as scalars or tensors or vectors, this prescription guarantees that the field equations are generally covariant. Equation (10.39) below shows how to write the action (9.3) for a freely moving test particle in the form of equation (10.1).

The way the game is played is illustrated by the term in the action for a free scalar field $\phi(x)$ with mass m. The Lagrangian density for this field is

$$\mathcal{L}_\phi = \frac{1}{2}(g^{ij}\phi_{,i}\phi_{,j} - m^2\phi^2). \tag{10.2}$$

This is a scalar, because $\phi_{,i}$ is a vector. The change in the action produced by a small change $\delta\phi$ in the field is

$$\begin{aligned}
\delta S_\phi &= \int d^4x\sqrt{-g}\,\delta\mathcal{L}_\phi \\
&= \int d^4x\sqrt{-g}\,[g^{ij}\phi_{,i}\delta\phi_{,j} - m^2\phi\delta\phi] \\
&= \int d^4x\delta\phi\left[-(\sqrt{-g}\,g^{ij}\phi_{,i})_{,j} - \sqrt{-g}\,m^2\phi\right].
\end{aligned} \tag{10.3}$$

The last step is the result of integrating by parts; we eliminate the surface term by agreeing to confine the variation of the field to some bounded region of spacetime. The factor multiplying $\delta\phi$ in the integrand is the variational derivative of S_ϕ with respect to the field ϕ. The action principle says $\delta S = 0$ for infinitesimal variations of the fields around a true solution, so the field equation is

$$(\sqrt{-g}\,g^{ij}\phi_{,i})_{,j} + \sqrt{-g}\,m^2\phi = 0. \tag{10.4}$$

The first term is proportional to the covariant divergence of the vector $g^{ij}\phi_{,j}$ (eq. [8.69]). On dividing through by $\sqrt{-g}$ we get the manifestly covariant form for the wave equation,

$$(g^{ij}\phi_{,i})_{;j} + m^2\phi = 0. \tag{10.5}$$

It is a useful exercise to check that when the wavelengths in $\phi(x)$ are short compared to the radius of curvature of spacetime, the wave equation (10.5) implies that wave packets move along the geodesic paths discussed in the last section. In this short-wavelength limit it is convenient to write the field as

$$\phi = Ae^{i\chi}, \tag{10.6}$$

where A and χ are scalar fields. The central assumption is that the wavelength of ϕ from the gradient of the phase χ is much shorter than the other lengths in the problem.

On substituting the expression for ϕ in equation (10.6) into the wave equation (10.5), and keeping only the largest term from the square of the gradient of the phase, we get

$$g^{ij}\chi_{,i}\chi_{,j} = m^2 . \tag{10.7}$$

We can define a momentum for the wave packet by the equation

$$p_i = \chi_{,i} . \tag{10.8}$$

This brings equation (10.7) to the usual relation for the momentum of a particle,

$$g^{ij}p_i p_j = m^2 . \tag{10.9}$$

In a region small compared to the scale over which the momentum changes, the function in equation (10.6) can be expanded as

$$\phi \propto \exp(i\chi_{,\alpha}x^\alpha + i\dot{\chi}t) = \exp(ip_\alpha x^\alpha + ip_0 t) . \tag{10.10}$$

A wave packet is a linear superposition of waves of this form with a slight spread of momenta p_α. These oscillating functions of position sum to negligibly small values except where the phase in equation (10.10) as a function of the p_α is an extremum at the mean momentum in the linear combination. That tells us the group velocity for the wave packet is

$$\frac{dx^\alpha}{dt} = v^\alpha = -\frac{\partial p_0}{\partial p_\alpha} . \tag{10.11}$$

The derivative is computed at fixed position and time, and equation (10.9) gives p_0 as a function of the p_α. The result of differentiating equation (10.9) is

$$0 = \frac{\partial}{\partial p_\alpha}\left(g^{\beta\gamma}p_\beta p_\gamma + 2g^{0\gamma}p_0 p_\gamma + g^{00}p_0{}^2\right)$$
$$= 2g^{\alpha\gamma}p_\gamma + 2g^{0\alpha}p_0 + 2\left(g^{0\gamma}p_\gamma + g^{00}p_0\right)\frac{\partial p_0}{\partial p_\alpha} . \tag{10.12}$$

Thus the group velocity (10.11) is

$$v^\alpha = \frac{g^{\alpha i}p_i}{g^{0i}p_i} = \frac{p^\alpha}{p^0} . \tag{10.13}$$

This agrees with equation (9.10) for a point particle.

With equation (10.13) for the group velocity, the time rate of change of the momentum (10.8) is

$$\frac{dp_i}{dt} = \frac{d\chi_{,i}}{dt} = \chi_{,ij}\frac{dx^j}{dt}$$

$$= \chi_{,ij}\frac{p^j}{p^0} = \chi_{,ij}g^{jk}\frac{\chi_{,k}}{p^0}\,. \tag{10.14}$$

The gradient of equation (10.7) is

$$0 = 2g^{jk}\chi_{,ji}\chi_{,k} + g^{jk}{}_{,i}\,\chi_{,j}\,\chi_{,k}\,. \tag{10.15}$$

This brings equation (10.14) to

$$p^0\frac{dp_i}{dt} = -\frac{1}{2}g^{jk}{}_{,i}p_jp_k$$

$$= \frac{1}{2}g_{jk,i}\,p^j\,p^k\,. \tag{10.16}$$

The second line uses equation (8.79) for the derivative of the reciprocal metric tensor. Using equation (10.13) for the group velocity once again, we get

$$\frac{dp_i}{dt} = \frac{1}{2}g_{jk,i}p^j\frac{dx^k}{dt}\,, \tag{10.17}$$

which is the Euler-Lagrange equation of motion (9.42). This approach is discussed further by Kulsrud and Loeb (1992).

Stress-Energy Tensor

The next step is to use the Lagrangian to derive the stress-energy tensor for matter and radiation, for use in the gravitational field equation. Let us recall first some properties of this tensor in special relativity.

The stress-energy tensor is symmetric, $T^{ij} = T^{ji}$, and in special relativity satisfies the conservation law

$$T^{ij}{}_{,j} = 0\,. \tag{10.18}$$

This is similar to the charge conservation law in equation (8.67), and the interpretation is the same. If we integrate equation (10.18) over a region of space with fixed boundaries and at fixed time $t = x^0$, we get

$$\frac{\partial}{\partial t}\int d^3x\,T^{00} = -\int da_\beta T^{0\beta}\,,$$

$$\frac{\partial}{\partial t}\int d^3x\,T^{\alpha 0} = -\int da_\beta T^{\alpha\beta}\,. \tag{10.19}$$

Gauss's theorem replaces the spatial divergence with the integral over the spatial boundary with surface element da_β. The component T^{00} is the mass (energy) density, and the mass in the region of the integral is $M = \int d^3x T^{00}$. The rate of change of the mass is the integral of the mass flux density $T^{0\beta}$ through the bounding surface of the volume. That is, $T^{0\beta}$ is the amount of mass (energy) that flows through unit area perpendicular to the direction β in unit time. The mass flux density is the momentum per unit volume, $T^{\alpha 0}$, and the volume integral in the left side of the second line is the total momentum within the region. The time rate of change of the momentum is the momentum flux density. Thus $T^{\alpha\beta}$ is the flux density of the α component of momentum in the direction β. Since a force is a rate of change of momentum, and pressure is the force per unit area, the pressure in the z direction is T^{33}, and the shear stress in the y direction on a surface with area da and normal pointing along the positive z axis is $T^{23}da$.

In the rest frame of an ideal fluid with mass density ρ and pressure p the stress-energy tensor has to be diagonal (where could the off-diagonal components point?) with $T^{00} = \rho$ and $T^{11} = T^{22} = T^{22} = p$. The Lorentz transformation to an ideal fluid with four-velocity $u^i = dx^i/ds$ is

$$T^{ij} = (\rho + p)u^i u^j - \eta^{ij} p, \qquad (10.20)$$

because this is the Lorentz covariant expression that reduces to the correct form when $u^\alpha = 0$ and $u^0 = 1$.

A mass concentration that represents a nearly pointlike particle has a stress-energy tensor that vanishes outside some small region of space. In special relativity the position of the particle is

$$X^\alpha = \frac{\int d^3x\, x^\alpha T^{00}}{\int d^3x\, T^{00}}. \qquad (10.21)$$

The time derivative of the position is the velocity

$$V^\alpha = \frac{dX^\alpha}{dt} = \frac{\int d^3x\, x^\alpha T^{00}{}_{,0}}{\int d^3x\, T^{00}} = -\frac{\int d^3x\, x^\alpha T^{0\beta}{}_{,\beta}}{\int d^3x\, T^{00}}. \qquad (10.22)$$

The denominator is constant, because the stress-energy tensor is assumed to vanish outside the small region of space occupied by the particle. The second step uses the conservation law (10.18). The result of integrating by parts is

$$V^\alpha = \frac{\int d^3x\, T^{0\alpha}}{\int d^3x\, T^{00}}. \qquad (10.23)$$

As expected, the momentum $\int d^3x T^{0\alpha}$ is the velocity multiplied by the mass $\int d^3x T^{00}$. The same operation shows the velocity V^α is constant. In a more general calculation we might suppose the divergence of the stress-energy tensor for the particle is nonzero because the particle is interacting with other fields. That would give the particle an acceleration.

In general relativity, the conservation law (10.18) is generalized to a covariant divergence, $T^{ij}{}_{;j} = 0$. The tensor that satisfies this condition is derived from the action, as follows.

Let the action for all matter and fields except the metric tensor be

$$S_m = \int \sqrt{-g}\, d^4x\, \mathcal{L}_m. \tag{10.24}$$

This quantity is independent of the choice of coordinates. Let us evaluate it in two different coordinate systems, x^i and $\bar{x}^i$, that differ by the expression

$$\bar{x}^i = x^i + \xi^i(x). \tag{10.25}$$

We will compute to first order in the small change ξ^i. Each field as a function of the variables of integration x^j is different in the two coordinate systems, but that cannot affect the value of S_m because the action principle says S_m is stationary under small changes of the fields. The metric tensor in the two coordinate systems also differs by terms of order ξ^i. Since the value of S_m is independent of the choice of coordinates, by considering the form of the derivative of S_m with respect to the g_{ij} we get an identity satisfied by the fields. This is the wanted stress-energy conservation law.

The transformation law for the metric tensor is

$$\bar{g}^{ij}(\bar{x}) = g^{kl}(x)\frac{\partial \bar{x}^i}{\partial x^k}\frac{\partial \bar{x}^j}{\partial x^l}. \tag{10.26}$$

With $\bar{x}^i = x^i + \xi^i$ we get, to first order in ξ^i,

$$\begin{aligned}
\delta g^{ij} &= \bar{g}^{ij}(x) - g^{ij}(x)\\
&= \xi^i{}_{,k}g^{kj} + \xi^j{}_{,k}g^{ik} - \xi^k g^{ij}{}_{,k}.
\end{aligned} \tag{10.27}$$

This is the change in the metric tensor as a function of the argument, x. Let us change the partial derivatives into covariant derivatives. We have to add to the first two terms the expression

$$\begin{aligned}
\Gamma^i_{kl}g^{kj}\xi^l + \Gamma^j_{kl}g^{ki}\xi^l &= \xi^l g^{im}g^{jk}(\Gamma_{mkl} + \Gamma_{kml})\\
&= \xi^l g^{im}g^{jk}g_{km,l} = -\xi^l g^{ij}{}_{,l}.
\end{aligned} \tag{10.28}$$

The second step uses equation (8.59) for the sum of the Christoffel symbols, and the last step uses the identity (8.79) for the derivative of g^{ij}. With this relation, equation (10.27) becomes

$$\delta g^{ij} = \xi^i{}_{;k}g^{kj} + \xi^j{}_{;k}g^{ki}. \tag{10.29}$$

The infinitesimal change in the coordinates changes the components of the metric tensor as a function of the variable of integration by the amount δg^{ij} in equation (10.29), and that changes the form for the action, giving

$$
0 = \delta S_m = \int d^4x \delta g^{ij} \left[\frac{\partial(\sqrt{-g}\,\mathcal{L}_m)}{\partial g^{ij}} - \frac{\partial}{\partial x^k} \frac{\partial(\sqrt{-g}\,\mathcal{L}_m)}{\partial g^{ij}{}_{,k}} \right]
$$
$$
\equiv \int d^4x \frac{1}{2} \delta g^{ij} \sqrt{-g}\, T_{ij}. \tag{10.30}
$$

The first line shows the elimination of the derivatives of δg^{ij} by the usual trick of integration by parts, using $\delta(g^{ij}{}_{,k}) = (\delta g^{ij})_{,k}$. This leaves the variational derivative of the action with respect to g^{ij}. The second line defines the stress-energy tensor. With equation (10.29) for δg^{ij}, we have

$$
0 = \delta S_m = \int d^4x \sqrt{-g}\, T_{ij} \xi^i{}_{;k} g^{kj}
$$
$$
= \int d^4x \sqrt{-g}\, [(\xi^i T_i{}^k)_{;k} - \xi^i T_i{}^k{}_{;k}] \tag{10.31}
$$
$$
= -\int d^4x \sqrt{-g}\, \xi^i T_i{}^k{}_{;k}.
$$

The second line uses the fact that the covariant derivative of a product may be expanded like an ordinary derivative of a product (eq. [8.45]). The third line follows from equation (8.70), which shows that we can change the integral of the covariant divergence $(\xi^i T_i{}^k)_{;k}$ to a surface integral that vanishes if the surface is placed outside the region where the coordinate shift ξ^i is nonzero. Since the expression in the last line has to vanish for arbitrary ξ^i, the stress-energy tensor has to satisfy the conservation law

$$
T_i{}^k{}_{;k} = 0 = T^{jk}{}_{;k}. \tag{10.32}
$$

The last expression follows because the covariant derivative of g^{ij} vanishes.

Equation (10.32) is the covariant generalization of the conservation equation (10.18). The definition of the stress-energy tensor in equation (10.30) is

$$
T_{ij} = \frac{2}{\sqrt{-g}} \left[\frac{\partial(\sqrt{-g}\,\mathcal{L}_m)}{\partial g^{ij}} - \frac{\partial}{\partial x^k} \frac{\partial(\sqrt{-g}\,\mathcal{L}_m)}{\partial g^{ij}{}_{,k}} \right]. \tag{10.33}
$$

Now let us consider some examples. The scalar field Lagrangian density in equation (10.2) gives

$$
\frac{\partial \mathcal{L}_\phi}{\partial g^{ij}} = \frac{1}{2} \phi_{,i} \phi_{,j}. \tag{10.34}
$$

Equation (8.65) for the derivative of the determinant g of g_{ij} with equation (8.79) for the derivative of the reciprocal tensor g^{ij} is

$$\delta g = g g^{ij} \delta g_{ij} = -g g_{ij} \delta g^{ij} . \qquad (10.35)$$

We have then[2]

$$\frac{\partial \sqrt{-g}}{\partial g^{ij}} = -\frac{1}{2} \sqrt{-g} \, g_{ij} . \qquad (10.36)$$

With these equations, the stress-energy tensor (10.33) for the scalar field Lagrangian density (10.2) is

$$T_{ij} = \phi_{,i} \phi_{,j} - \frac{1}{2} g_{ij} [g^{kl} \phi_{,k} \phi_{,l} - m^2 \phi^2] . \qquad (10.37)$$

It is a useful exercise to check that the field equation (10.5) for ϕ does guarantee that this expression satisfies $T^j_{i\,;j} = 0$. In locally Minkowski coordinates, the field energy density and pressure are

$$\rho = T^{00} = T_{00} = \frac{1}{2} [\dot{\phi}^2 + (\nabla \phi)^2 + m^2 \phi^2] ,$$

$$\qquad (10.38)$$

$$p_{11} = T^{11} = T_{11} = \left(\frac{\partial \phi}{\partial x^1}\right)^2 + \frac{1}{2} [\dot{\phi}^2 - (\nabla \phi)^2 - m^2 \phi^2] .$$

It is another interesting exercise to use the Lagrangian formalism in equations (9.5) to (9.13) to check that this expression for the energy density ρ agrees with the Hamiltonian density in these locally Minkowski coordinates.

As a second example, let us find the stress-energy tensor for a gas of pointlike particles with masses m_a at positions $x_a^\alpha(t)$. Equation (9.4) gives the action for a single particle as a one-dimensional integral along the world line. We can express the action as a four-dimensional integral by writing the Lagrangian density as a product of the time integral along the world line and a space integral over the three-dimensional Dirac delta function $\delta(\mathbf{x} - \mathbf{x}(t))$, where the three spatial components of the particle position as functions of time $t = x_0$ are $x^\alpha(t)$. This brings the particle action to

$$S_p = -m \int d^4x \, \delta(\mathbf{x} - \mathbf{x}(t)) (g_{ij} \dot{x}^i \dot{x}^j)^{1/2} . \qquad (10.39)$$

We know from the construction that this somewhat cumbersome expression is independent of the choice of coordinates.

---

[2] The notation here and in equation (10.33) may be confusing, for since g^{ij} is symmetric the off-diagonal terms satisfy $\delta g^{01} = \delta g^{10}$ and so on. As one sees in the derivation in equation (8.64) for the differential of a determinant, one is supposed to proceed as if δg^{01} and δg^{10} were independent.

264 10 FIELD EQUATIONS

The action for a gas of particles with masses m_a and paths $x_a^\alpha(t)$ is the sum

$$S_p = -\sum_a m_a \int d^4x\, \delta(\mathbf{x} - \mathbf{x}_a(t))(g_{ij}\dot{x}_a^i\dot{x}_a^j)^{1/2}.$$ (10.40)

With equation (8.79) to change the derivative with respect to g^{ij} to a derivative with respect to g_{ij}, equation (10.33) for the stress-energy tensor is

$$T^{ij} = \frac{2}{\sqrt{-g}}\frac{\partial}{\partial g_{ij}}\sum_a m_a\delta(\mathbf{x} - \mathbf{x}_a(t))(g_{kl}\dot{x}_a^k\dot{x}_a^l)^{1/2}$$

$$= \sum_a m_a \frac{\delta(\mathbf{x} - \mathbf{x}_a(t))}{\sqrt{-g}}\frac{dx_a^i}{ds}\frac{dx_a^j}{dt}.$$ (10.41)

Recall that $\dot{x}^i = dx^i/dt$, so $(g_{ij}\dot{x}^i\dot{x}^j)^{1/2}\,dt = ds$. That changes one of the coordinate velocities to the four-velocity $u^i = dx^i/ds$.

Equation (10.41) is the stress-energy tensor for a gas of pointlike particles. In locally Minkowski coordinates, the energy density in this expression is

$$T^{00} = \sum_a m_a u_a^0\delta(\mathbf{x} - \mathbf{x}_a(t)).$$ (10.42)

As expected, this is a sum over the particle energies $m_a u_a^0$. The energy flux density and momentum density are

$$T^{0\alpha} = \sum_a m_a u_a^0 v_a^\alpha\delta(\mathbf{x} - \mathbf{x}_a(t)),$$ (10.43)

where $\mathbf{v} = \dot{\mathbf{x}}$, again the expected form.

Suppose the locally Minkowski coordinates are chosen so the gas of particles has zero streaming velocity. This means the average value of $T^{0\alpha}$ vanishes:

$$\langle mu^0 v^\alpha\rangle = 0.$$ (10.44)

The brackets are an average over the particles in a region small enough for us to ignore the variation of mean values within the region, large enough to contain many particles. The energy density (10.42) in this rest frame for the gas is

$$\rho = \langle T^{00}\rangle = n\langle mu^0\rangle,$$ (10.45)

where n is the mean number of particles per unit volume measured by an observer in the gas rest frame. The pressure in the x direction is

$$p = \langle T^{11}\rangle = n\langle mu^0(v^1)^2\rangle.$$ (10.46)

If the random particle velocities in the gas rest frame are isotropic, the pressure is

$$p = \rho v^2 / 3 , \tag{10.47}$$

with v the mass-weighted rms particle velocity. This is the usual expression from kinetic theory. In the relativistic limit, the pressure is

$$p = n \langle m u^0 \rangle / 3 = \rho / 3 . \tag{10.48}$$

Again, this is the usual result — that for a gas of photons (or electromagnetic radiation) the pressure is one third the energy density.

In a general coordinate system an ideal fluid is characterized by the energy density ρ and isotropic pressure p as they would be measured by an observer in the rest frame of the fluid, and by the fluid four-velocity u^i. Its stress energy tensor,

$$T^{ij} = (\rho + p) u^i u^j - g^{ij} p , \tag{10.49}$$

is the covariant generalization of equation (10.20).

Einstein's Gravitational Field Equation

The complete form for the action in general relativity theory is

$$S = \int d^4 x \sqrt{-g} \left(\mathcal{L}_m - \frac{1}{16 \pi G} R \right) . \tag{10.50}$$

The first term is the action (10.24) for the matter and radiation fields. The scalar curvature R comes from contraction of the indices of the curvature tensor, which is constructed out of the metric tensor and its first and second derivatives. The first contraction is the Ricci tensor (8.82),

$$R_{kl} = R^i{}_{kil} , \tag{10.51}$$

and the contraction of this is the scalar curvature defined in equation (8.83),

$$R = g^{kl} R_{kl} . \tag{10.52}$$

In the action the prefactor of R contains Newton's constant, G, and the numerical factors are arranged to get the standard relations in the Newtonian limit, as will be discussed shortly.

In equation (10.50) the scalar curvature provides the Lagrangian density for the spacetime geometry represented by the metric tensor. This agrees with the condition that the Lagrangian density has to be a scalar density, and it is pleasant that after an integration by parts the derivative terms enter the action as a standard-looking quadratic form. But equation (10.50) is an assumption, to be justified by its successful predictions. Einstein arrived at the field equation (10.62) by a more circuitous route, as described by Pais (1982). Hilbert first wrote down the action in the complete form of equation (10.50). Einstein introduced the cosmological constant by adding to the Lagrangian density for the gravitational field a term proportional to $\Lambda\sqrt{-g}$. The recent tendency is to add any such term to the Lagrangian density $\mathcal{L}_m\sqrt{-g}$ for matter. Examples of field Lagrangian densities that act like a cosmological constant (which may be time-variable) are discussed in sections 17 and 18.

The field equations for matter and radiation come from varying the fields in $\mathcal{L}_m$. We get the gravitational field equations by varying the metric tensor g_{ij} in both terms in the action. The first term gives the the stress-energy tensor (eq. [10.33]). In the second term we have from equation (10.36)

$$\delta\sqrt{-g} = -\frac{1}{2}\sqrt{-g}\, g_{km}\delta g^{km}\,. \tag{10.53}$$

The result of varying g^{ij} in the integrand in the second term in the action (10.50) is then

$$\delta\int d^4x\sqrt{-g}\,R = \int d^4x\sqrt{-g}\,\left(R_{km}\delta g^{km} + g^{km}\delta R_{km} - \frac{1}{2}g_{km}R\,\delta g^{km}\right)\,. \tag{10.54}$$

The following argument shows that the middle term, containing δR_{km}, vanishes.

Let us consider the curvature tensor at a point where the coordinates are locally Minkowski, so all the first derivatives of the metric tensor vanish and the Christoffel symbols therefore vanish:

$$\Gamma^i_{jk} = 0\,. \tag{10.55}$$

Then the Ricci tensor in equations (8.82) and (10.51) is

$$R^i{}_{kim} = R_{km} = (\Gamma^i_{km}),_i - (\Gamma^i_{ki}),_m\,. \tag{10.56}$$

By suitable choice of coordinates we can limit the variations δg_{ij} to those whose first derivatives vanish, so they satisfy equation (10.55). Then the only nonzero terms in the variation of the Ricci tensor appearing in equation (10.54) are

$$\begin{aligned}
g^{km}\delta R_{km} &= g^{km}\left[\delta(\Gamma^i_{km}),_i - \delta(\Gamma^i_{ki}),_m\right] \\
&= (g^{km}\delta\Gamma^i_{km}),_i - (g^{km}\delta\Gamma^i_{ki}),_m\,.
\end{aligned} \tag{10.57}$$

The right-hand side is an ordinary divergence, in locally Minkowski coordinates. On the left side δR_{lm} is a the difference of two tensors (representing neighboring spacetimes), so the left side is a scalar. We make the right-hand side a scalar by changing it to a covariant divergence,

$$g^{km}\delta R_{km} = U^i{}_{;i},$$ (10.58)

where U^i is a vector. This is a covariant relation, so it has to be valid in any coordinate system. It is left as an exercise to check that the variation of the Christoffel symbol can be written as

$$\delta\Gamma^i_{km} = \frac{1}{2}g^{ij}(\delta g_{jk;m} + \delta g_{jm;k} - \delta g_{km;j}).$$ (10.59)

That is, this expression is a tensor, which gives the tensor U^i in equations (10.57) and (10.58).

We have

$$\int d^4x\sqrt{-g}\,g^{km}\delta R_{km} = \int d^4x\sqrt{-g}\,U^i{}_{;i} = 0.$$ (10.60)

As in equation (8.70), the integral of the divergence can be written as a surface integral that vanishes if we place the surface outside the region where we are varying the field g^{ij}. This shows that the term proportional to δR_{km} in equation (10.54) vanishes.

With equation (10.30) for the effect of the variation of g_{km} on the action for the material fields, and equation (10.54) for the gravitational part, the net variation in the action is

$$\delta S = \int d^4x\sqrt{-g}\,\delta g^{km}\left[\frac{1}{2}T_{km} - \frac{1}{16\pi G}\left(R_{km} - \frac{1}{2}g_{km}R\right)\right].$$ (10.61)

Since the action is stationary, Einstein's gravitational field equation is

$$R_{ij} - \frac{1}{2}g_{ij}R = 8\pi GT_{ij}.$$ (10.62)

With

$$g_{ij}g^{ij} = \delta^i_i = 4,$$ (10.63)

we see that the trace of the field equation is

$$R = -8\pi GT,$$ (10.64)

with $T = T^i_i$ the trace of the stress-energy tensor. This expression in the field equation brings it to the form

$$R_{ij} = 8\pi G \left(T_{ij} - \frac{1}{2} g_{ij} T \right) . \tag{10.65}$$

The field equation is symmetric in i and j, so there are ten components. The identity $T^k_{i;k} = 0$ in equation (10.32) means there are four constraints, leaving six independent components.

Small-Scale Limit

An immediate and useful application of the gravitational field equation is to the relative motion of neighboring freely moving test particles. Recall that a gravitational field can be "transformed away" by choosing locally Minkowski coordinates that approximate the orthogonal system that a small freely moving observer would use (eq. [9.34]). This expresses the fact that an observer in free fall experiences no gravitational acceleration, or any other difference from the physics of flat spacetime. However, the observer may see that neighboring test particles are accelerating, at a rate proportional to their distance. This effect of the tidal fields of distant masses, or of the gravitational attraction of smoothly distributed matter in the neighborhood, is described by the equation of geodesic deviation.

In coordinates that are locally Minkowski along the path of the freely moving observer, the observer's four-velocity is $u^i = \delta^i_0$, and the geodesic deviation equation (9.65) is

$$g^\alpha = \frac{D^2 \xi^\alpha}{ds^2} = R^\alpha_{\ 00\beta} \xi^\beta . \tag{10.66}$$

The two middle indices in the curvature tensor are set to $i = 0$, because that is the only nonzero component of the four-velocity, and the antisymmetry of the curvature tensor (eq. [8.81]) eliminates the time part of the last index. This equation gives the relative acceleration $\mathbf{g}$ of a freely moving particle at proper distance ξ^β from the observer. The divergence of the acceleration is

$$\nabla \cdot \mathbf{g} = R^\alpha_{\ 00\alpha} = -R_{00} = -8\pi G \left(T_{00} - \frac{1}{2} T \right) . \tag{10.67}$$

This follows because the antisymmetry of the curvature tensor in the first and last pair of indices means $R^0_{\ 000} = 0$, so we can extend the sum over α to a sum over all four values of the index, and we can exchange the last two indices, giving the Ricci tensor (10.51). Since we are in locally Minkowski coordinates, equation (10.67) with $T^{00} = \rho$ is

$$\nabla \cdot \mathbf{g} = -4\pi G (\rho + \sum_\alpha T^{\alpha\alpha}) . \tag{10.68}$$

In nonrelativistic material, the mass density ρ is much larger than the stresses (as in eqs. [10.45] and [10.46] with $v \ll 1$), and equation (10.68) is

$$\nabla \cdot \mathbf{g} = -4\pi G\rho, \tag{10.69}$$

which is Poisson's equation in Newtonian mechanics.

If the pressure is appreciable and the material is isotropic in the observer's frame, we can identify T^{11} as the isotropic pressure, p, as in equation (10.46). Then equation (10.68) is

$$\nabla \cdot \mathbf{g} = -4\pi G(\rho + 3p). \tag{10.70}$$

That is, the active gravitational mass density is $\rho_g = \rho + 3p$. This is the result used in equation (5.14) for the cosmological equation for the rate of change of the expansion rate.

Equation (10.68) is a linear expression for the relative gravitational acceleration in the neighborhood of a freely moving observer. It requires no special conditions on the curvature of spacetime (except of course that the curvature is finite), because the equivalence principle tells us it is always possible to transform away the gravitational acceleration along the world line of a freely moving particle. The relative gravitational acceleration has to be small in a small enough region around the path, allowing the linear expansion in equation (10.66). We consider next another linear limit, which assumes there are only small departures from the flat spacetime of special relativity.

Weak-Field Limit

Spacetime in the Local Supercluster in figure 3.3 seems to be quite close to flat, for there are no obvious distortions of images, and no high velocities. It makes sense therefore to describe the Local Supercluster in coordinates chosen so the metric tensor is of the form

$$g_{ij} = \eta_{ij} + h_{ij}, \tag{10.71}$$

where the h_{ij} are corrections to the Minkowski tensor η_{ij}, and to compute the gravitational field equations to first order in the h_{ij} (while allowing for nonlinear motion of the matter and radiation). This is the weak-field limit.

To first order in h_{ij}, equation (8.80) for the Ricci tensor (10.51) is

$$R_{km} = \frac{1}{2}\eta^{il}(h_{kl,im} + h_{im,kl} - h_{km,il} - h_{il,km}), \tag{10.72}$$

because the other terms contain products of Christoffel symbols, and so are of order h^2.

If motions in the material that dominate the source term T^{ij} are nonrelativistic, which again agrees with what is seen in the Local Supercluster, and we can ignore gravitational radiation, then the time derivatives of h_{ij} are small compared to the space derivatives. Keeping only the space derivatives in equation (10.72), we get

$$R_{00} = -\frac{1}{2}\eta^{il}h_{00,il} = \frac{1}{2}\nabla^2 h_{00}. \tag{10.73}$$

For nonrelativistic matter the dominant term in the stress-energy tensor is $T^{00} = \rho = T$, so the R_{00} part of the field equation (10.65) is

$$\frac{1}{2}\nabla^2 h_{00} = 8\pi G\left(T_{00} - \frac{1}{2}T\right) = 4\pi G\rho. \tag{10.74}$$

Comparing this with Poisson's equation (10.69), we see that $h_{00}/2$ is the Newtonian gravitational potential ϕ, with $\mathbf{g} = -\nabla\phi$. It follows that the time part of the metric tensor in this weak-field nonrelativistic limit is $g_{00} = 1 + 2\phi$. As indicated in equation (9.32), this makes the geodesic equation agree with the Newtonian gravitational acceleration equation.

For a more general expression for the gravitational field in linear perturbation theory, let us rearrange the terms in the linear equation (10.72) for the Ricci tensor to

$$R_{km} = \frac{1}{2}\left[\left(h^i_{k,i} - \frac{1}{2}h_{,k}\right)_{,m} + \frac{\partial}{\partial x^k}\left(h^i_{m,i} - \frac{1}{2}h_{,m}\right) - \Box h_{km}\right], \tag{10.75}$$

where

$$\Box\phi = \eta^{il}\phi_{,il}, \tag{10.76}$$

and the Minkowski tensor has been used to raise indices:

$$h^i_k = \eta^{il}h_{lk}, \qquad h = \eta^{il}h_{il}. \tag{10.77}$$

We can change the h^i_j by a coordinate transformation, $\bar{x}^i = x^i + \xi^i$. This gives four free functions, $\xi^i(x)$, which we can choose to satisfy the four conditions

$$h^i_{k,i} - \frac{1}{2}h_{,k} = 0. \tag{10.78}$$

This constraint brings equation (10.75) for the Ricci tensor to

$$R_{km} = -\frac{1}{2}\Box h_{km}. \tag{10.79}$$

Then the gravitational field equation (10.65) in the weak-field limit is

$$\Box h_{km} = -16\pi G(T_{km} - \eta_{km}T/2)$$

$$\Box \left(h_j^i - \frac{1}{2}\delta_j^i h \right) = -16\pi G T_j^i. \tag{10.80}$$

In this limit the equation of stress-energy conservation is $T_{j;i}^i = T_{j,i}^i = 0$, consistent with the constraint in equation (10.78) in the second form for the wave equation. The wave equation in the first line describes gravitational radiation as well as the response of the gravitational field to the source T_{ij}.

In stars and galaxies and cluster of galaxies the material is moving at speeds well below the velocity of light (with the likely exception of the nuclei of galaxies or other black holes, but this is difficult to detect). Thus in describing a galaxy or a cluster of galaxies we can assume that the dominant term in the stress-energy tensor is the mass density, $T^{00} = \rho(\mathbf{r})$. As in equation (10.73), in this nonrelativistic limit the characteristic time for the mass distribution to change is large compared to the characteristic distances over which the density varies, so the time derivatives of the fields h_{ij} produced by the mass distribution are negligible compared to the space derivatives. This brings the weak-field equation (10.80) to

$$\nabla^2 h_{00} = 8\pi G\rho, \qquad \nabla^2 h_{\alpha\beta} = 8\pi G\rho\delta_{\alpha\beta}, \tag{10.81}$$

and

$$\nabla^2 h_{0\alpha} = 0. \tag{10.82}$$

These are Poisson equations, like (10.69). Neglecting tidal fields, as in gravitational radiation, we can write the solutions in terms of the Newtonian potential ϕ,

$$h_{00} = 2\phi, \qquad h_{\alpha\beta} = 2\delta_{\alpha\beta}\phi, \tag{10.83}$$

with $h_{0\alpha} = 0$. As can be checked, this solution is consistent with the constraint equation (10.78) (because we are ignoring the time derivative of ϕ).

The conclusion is that the line element in the weak-field nonrelativistic limit can be written as

$$ds^2 = (1 + 2\phi)\,dt^2 - (1 - 2\phi)\delta_{\alpha\beta}\,dx^\alpha dx^\beta. \tag{10.84}$$

In this approximation we have dropped terms of order ϕ^2, we have assumed the stress-energy tensor is dominated by the mass density $T^{00} = \rho(\mathbf{r})$, we have neglected gravitational radiation, and we have assumed the time derivatives in the sources and metric tensor are negligibly small compared to the space derivatives, that is, the motions that rearrange the mass distribution are nonrelativistic.

Gravitational Deflection of Light

At this point we have completed the development of the wanted elements of general relativity theory, and we are ready to apply the theory to astronomy and cosmology. A first interesting example is the gravitational deflection of light, as observed near the Sun and in the gravitational lensing of background galaxies and quasars by the mass concentrations in galaxies and clusters of galaxies. The effect is derived here, and applied in sections 13, 14, 18, and 20.

A bounded quasistatic mass distribution $\rho(\mathbf{x})$ produces a Newtonian gravitational potential $\phi(\mathbf{x})$ that can be represented in the line element (10.84). The geodesic equation of motion for the four-momentum p_i of the photons in a wave packet is given by equation (9.42). In the coordinate system of the line element (10.84), the space components of the equation of motion are

$$-\frac{d}{dt}\left[(1-2\phi)p^\alpha\right] = \phi_{,\alpha}p^0\left[1+\delta_{\beta\gamma}p^\beta p^\gamma/(p^0)^2\right] . \tag{10.85}$$

The time part is

$$\frac{dp}{dt} = \frac{d}{dt}(1+2\phi)p^0 = 0 , \tag{10.86}$$

where $p \equiv (1+2\phi)p^0$. We are assuming that time derivatives are negligible compared to space derivatives, so the expression p is constant along the path. Then the space part (10.85) is, to first order in ϕ,

$$-\frac{d}{dt}\left[(1-4\phi)p^\alpha/p^0\right] = \phi_{,\alpha}\left[1+\delta_{\beta\gamma}p^\beta p^\gamma/(p^0)^2\right] , \tag{10.87}$$

We can simplify the right-hand side by taking account of the fact that p^i is a null vector:

$$0 = g_{ij}p^i p^j = (1+2\phi)(p^0)^2 - (1-2\phi)\delta_{\alpha\beta}p^\alpha p^\beta . \tag{10.88}$$

This shows that the expression in brackets on the right-hand side of equation (10.87) is equal to 2 to lowest order. On the left side, $p^\alpha/p^0 = v^\alpha$ is the coordinate velocity (eq. [9.43]). This brings the equation of motion (10.87) to

$$\frac{d}{dt}\left[(1-4\phi)\frac{dx^\alpha}{dt}\right] = -2\frac{\partial\phi}{\partial x^\alpha} . \tag{10.89}$$

Equation (10.89) gives the path of a light ray moving in the Newtonian gravitational potential $\phi(\mathbf{r})$ of a quasistatic mass distribution. The factor $(1-4\phi)$ keeps the coordinate velocity v consistent with the constraint in equation (10.88), to first order in ϕ. The right-hand side is twice the acceleration of a particle moving at low velocity.

The equation of motion (10.89) is equivalent to the equation for the path of a light ray in a medium with index of refraction $n = 1 - 2\phi$, to first order in ϕ. Fermat's principle says the time for a light pulse to travel the path from given initial to final positions,

$$\tau = \int n(\mathbf{r})(\delta_{\alpha\beta}\, dx^{\alpha} dx^{\beta})^{1/2}, \tag{10.90}$$

is an extremum along the true path. It is an interesting exercise to check that this reproduces equation (10.89).

To find the expression for the gravitational deflection of a light ray, suppose the ray is approaching in the y direction a bounded mass concentration with Newtonian potential $\phi(\mathbf{r})$. The rate of increase of the transverse velocity in the x direction is $-2\phi_{,x}$, so the angle of deflection in the x direction is

$$\alpha_x = -2 \int dy\, \phi_{,x}. \tag{10.91}$$

If the potential is that of a pointlike mass $\mathcal{M}$,

$$\phi = -G\mathcal{M}/r, \tag{10.92}$$

the deflection angle is

$$\alpha_x = -2G\mathcal{M}x \int_{-\infty}^{\infty} \frac{dy}{(x^2 + y^2)^{3/2}}, \tag{10.93}$$

where x is the distance of closest approach to $\mathcal{M}$. The negative sign means the ray is bent toward the mass. The integral works out to

$$\alpha = \frac{4G\mathcal{M}}{x}, \tag{10.94}$$

where α is the magnitude of α_x. This is the Einstein deflection angle for a light ray moving past a pointlike mass $\mathcal{M}$. It is twice the value one gets from Newtonian mechanics.

As illustrated in figure 3.12, the mass distribution around an isolated spiral galaxy is reasonably well approximated by an isothermal gas sphere with a small core radius, where the mass within radius r is $\mathcal{M}(r) = 2\sigma^2 r/G$, with σ the line-of-sight rms velocity dispersion (eq. [3.32]). The Newtonian potential for this mass distribution is

$$\phi(r) = \int^r dr\, G\mathcal{M}/r^2 = 2\sigma^2 \ln r .$$ (10.95)

This expression in equation (10.91) is

$$\alpha_x = -4\sigma^2 x \int_{-\infty}^{\infty} \frac{dy}{x^2 + y^2} ,$$ (10.96)

which works out to

$$\alpha = 4\pi(\sigma/c)^2 .$$ (10.97)

We have put back the velocity of light.

Equation (10.97) is the gravitational deflection angle for a light ray produced by a distributed mass, such as a galaxy, in which the mass density scales with radius as $\rho \sim 1/r^2$. If this mass distribution is supported by an isotropic distribution of velocities, the one-dimensional line-of-sight velocity dispersion is σ. In a giant elliptical galaxy, σ can reach 300 km s^{-1} (eq. [3.35]). That brings the angle between the two images of a background source lensed by a galaxy halfway between us and the source to $\alpha = 2.6$ arc sec, just larger than the angular resolution allowed by the seeing in our atmosphere. The geometry for lensing in a cosmological model is discussed in section 13, microlensing by low mass stars in section 18, and the luminous arcs caused by lensing by the mass concentrations in great clusters of galaxies in section 20.

Large-Scale Density Fluctuations in an Expanding Universe

As a final example of the weak-field limit, let us consider the evolution of mass density fluctuations in the early universe, when the energy density in the cosmic background radiation can make the space parts of the stress-energy tensor comparable to the time part. We will arrive at a relation between large-scale fluctuations in the mass distribution and the perturbation to the spacetime geometry, which can be compared to the more informal treatment in section 5.

We will consider density fluctuations on the comoving scale x_g characteristic of galaxies or clusters of galaxies. Since x_g is small compared to the allowed value for the mean radius of curvature of constant time sections, $a_o x_g \ll H_o^{-1} \lesssim a_o R$, it is a good approximation to take $(aR)^{-2} = 0$ in the Robertson-Walker line element (5.9). We will consider the behavior of the density fluctuation at high redshift, where the Hubble length $H(t)^{-1}$ is small compared to $a x_g$, making it a good approximation to ignore the pressure gradient force and the cosmological constant. The physical meaning of density fluctuations on scales broader than the Hubble length is discussed in section 5 (eq. [5.80]), and in more detail in chapter 5 of LSS and in Efstathiou (1990, sec. 3); the solutions are discussed in more detail by Ratra (1988).

We can simplify the problem by assigning time-orthogonal coordinates, such that the mean streaming velocity of the material vanishes on an initial hypersurface.[3] In these coordinates the streaming velocity of the material stays small until the physical size ax_g of the fluctuation approaches the Hubble length, and the pressure gradient force can start to accelerate the material relative to freely moving observers.

The line element in the time-orthogonal coordinates of equation (9.37) is

$$ds^2 = dt^2 - a(t)^2(\delta_{\alpha\beta} - h_{\alpha\beta})\,dx^\alpha dx^\beta \,. \tag{10.98}$$

As usual, the Greek indices indicate the space coordinates $\alpha = 1, 2$, and 3. We will compute to first order in the departures $h_{\alpha\beta}$ from homogeneity. To this order, the determinant of the metric tensor is

$$\sqrt{-g} = a^3(1 - h/2)\,, \tag{10.99}$$

where

$$h = \sum_\alpha h_{\alpha\alpha}\,, \tag{10.100}$$

and the components of the reciprocal tensor are

$$g^{00} = 1\,, \quad g^{0\alpha} = 0\,, \quad g^{\alpha\beta} = -a^{-2}(\delta_{\alpha\beta} + h_{\alpha\beta})\,. \tag{10.101}$$

We will model the material as an ideal fluid with mass density ρ, pressure p, and stress-energy tensor (eq. [10.49])

$$T^{ij} = (\rho + p)u^i u^j - g^{ij}p\,. \tag{10.102}$$

To simplify the expressions a little, let us take the equation of state to be

$$p = \nu\rho\,, \tag{10.103}$$

where ν is a constant. For electromagnetic radiation, or a gas of free relativistic particles, $\nu = p/\rho = 1/3$ (eq. [10.48]).

---

[3] This assumes the velocity is irrotational. A more complete analysis allows for streaming velocity on the hypersurfaces of constant time. As discussed in LSS, § 86, when the pressure is comparable to the energy density, irrotational streaming motion generates another more slowly growing solution for the mass density perturbation $\delta\rho/\rho$.

We are assigning coordinates so the fluid is at rest, with four-velocity $u^i = \delta^i_0$, so the components of the stress-energy tensor (10.102) are

$$T^{00} = \rho_b(1+\delta),$$
$$T^{0\beta} = 0,$$
$$T^{\alpha\beta} = (\nu\rho_b/a^2)(1+\delta)(\delta_{\alpha\beta} + h_{\alpha\beta}),$$
$$T_{\alpha\beta} = \nu\rho_b a^2(1+\delta)(\delta_{\alpha\beta} - h_{\alpha\beta}).$$

(10.104)

The mean background mass density is $\rho_b(t)$, and the mass density contrast is $\delta(\mathbf{x}, t) = \rho/\rho_b - 1$. Since $u_i u^i = 1$, the trace of the stress-energy tensor (10.102) is

$$T = T^i_i = \rho - 3p = \rho_b(1 - 3\nu)(1+\delta).$$

(10.105)

The stress-energy conservation law is $T^j_{i;j} = 0$ (eq. [10.32]). We can simplify this expression by the trick used in equation (8.69) for particle conservation. The covariant divergence is

$$T^j_{i;j} = \frac{\partial T^j_i}{\partial x^j} + \Gamma^j_{jk} T^k_i - \Gamma^k_{ij} T^j_k.$$

(10.106)

The factor Γ^j_{jk} is given by equation (8.66). Because T^{kj} is symmetric, the last term with equation (8.57) for the Christoffel symbol is $\Gamma_{kij} T^{jk} = g_{jk,i} T^{jk}/2$. The result of collecting these terms is the conservation law in the form

$$\frac{1}{\sqrt{-g}} \partial_j \sqrt{-g} T^j_i = \frac{1}{2} g_{jk,i} T^{jk}.$$

(10.107)

With the metric tensor in equations (10.98) to (10.100), and the stress-energy tensor (10.104), the $i = 0$ part of the conservation equation (10.107), carried to first order in the perturbation, is

$$\left(3\frac{\dot{a}}{a} - \frac{\dot{h}}{2}\right)\rho_b(1+\delta) + \dot{\rho}_b(1+\delta) + \rho_b\dot{\delta} = \frac{1}{2}\dot{g}_{jk}T^{jk}$$

$$= -a\dot{a}(\delta_{\alpha\beta} - h_{\alpha\beta})\frac{\nu\rho_b}{a^2}(1+\delta)(\delta_{\alpha\beta} + h_{\alpha\beta}) + \frac{1}{2}a^2\dot{h}_{\alpha\beta}\frac{\nu\rho_b}{a^2}\delta_{\alpha\beta} \quad (10.108)$$

$$= -3\frac{\dot{a}}{a}\nu\rho_b(1+\delta) + \frac{1}{2}\nu\rho_b\dot{h}.$$

The unperturbed part is

$$\dot{\rho}_b = -3\frac{\dot{a}}{a}(1+\nu)\rho_b.$$

(10.109)

This is the energy conservation equation (5.16). The first-order part is

$$\delta = (1 + \nu)\dot{h}/2. \tag{10.110}$$

This also expresses local energy conservation, for we see from equation (10.99) that the fractional perturbation to a proper volume element moving with the fluid is $\delta V / V = -h/2$.

The results of substituting the components of the line element in equation (10.98) into equation (8.60) for the Christoffel symbols, and keeping only terms of first order in the $h_{\alpha\beta}$, are

$$
\begin{aligned}
&\Gamma^0_{0i} = 0, \qquad \Gamma^\alpha_{00} = 0, \\
&\Gamma^0_{\alpha\beta} = a\dot{a}(\delta_{\alpha\beta} - h_{\alpha\beta}) - a^2 \dot{h}_{\alpha\beta}/2, \\
&\Gamma^\alpha_{0\beta} = \delta_{\alpha\beta}\dot{a}/a - \dot{h}_{\alpha\beta}/2, \\
&\Gamma^\gamma_{\alpha\beta} = -(h_{\alpha\gamma,\beta} + h_{\beta\gamma,\alpha} - h_{\alpha\beta,\gamma})/2.
\end{aligned} \tag{10.111}
$$

Now it is only a matter of labor to use these expressions in equation (8.82) to get the Ricci tensor, and then use this with the above forms for the components of the stress-energy tensor to get the field equations. The time-time part works out to

$$
\begin{aligned}
R_{00} &= -3\frac{\ddot{a}}{a} + \frac{1}{2}\ddot{h} + \frac{\dot{a}}{a}\dot{h} \\
&= 8\pi G(T_{00} - g_{00}T/2) \\
&= 4\pi G\rho_b(1 + 3\nu)(1 + \delta).
\end{aligned} \tag{10.112}
$$

We will also want the space-space part,

$$
\begin{aligned}
R_{\alpha\beta} &= a^2\left[\left(\frac{\ddot{a}}{a} + 2\frac{\dot{a}^2}{a^2}\right)(\delta_{\alpha\beta} - h_{\alpha\beta}) - \frac{3}{2}\frac{\dot{a}}{a}\dot{h}_{\alpha\beta} - \frac{1}{2}\frac{\dot{a}}{a}\dot{h}\delta_{\alpha\beta} - \frac{1}{2}\ddot{h}_{\alpha\beta}\right] \\
&\quad + (h_{\alpha\beta,\gamma\gamma} + h_{,\alpha\beta} - h_{\alpha\gamma,\gamma\beta} - h_{\beta\gamma,\gamma\alpha})/2 \\
&= 8\pi G(T_{\alpha\beta} - g_{\alpha\beta}T/2] \\
&= 4\pi G\rho_b a^2(1 - \nu)(1 + \delta)(\delta_{\alpha\beta} - h_{\alpha\beta}).
\end{aligned} \tag{10.113}
$$

The unperturbed parts of these two equations are

$$\frac{\ddot{a}}{a} = -\frac{4}{3}\pi G\rho_b(1 + 3\nu), \qquad \frac{\ddot{a}}{a} + 2\frac{\dot{a}^2}{a^2} = 4\pi G\rho_b(1 - \nu). \tag{10.114}$$

The first agrees with the cosmological equation (5.14). The difference is

$$\frac{\dot{a}^2}{a^2} = \frac{8}{3}\pi G\rho_b. \tag{10.115}$$

This agrees with the expansion rate equation (5.18) for the case we are considering, where space curvature and the cosmological constant may be neglected.

The first-order part of equation (10.112) is

$$\ddot{h} + 2\frac{\dot{a}}{a}\dot{h} = 8\pi G\rho_b(1+3\nu)\delta , \qquad (10.116)$$

and the first-order part of equation (10.113) is

$$-\frac{3}{2}\frac{\dot{a}}{a}\dot{h}_{\alpha\beta} - \frac{1}{2}\frac{\dot{a}}{a}\dot{h}\delta_{\alpha\beta} - \frac{1}{2}\ddot{h}_{\alpha\beta} + \frac{1}{2a^2}(h_{\alpha\beta,\gamma\gamma} + h_{,\alpha\beta} - h_{\alpha\gamma,\gamma\beta} - h_{\beta\gamma,\gamma\alpha}) \qquad (10.117)$$
$$= 4\pi G\rho_b(1-\nu)\delta\delta_{\alpha\beta} .$$

On multiplying equation (10.116) by $(1+\nu)/2$, and using the energy conservation equation (10.110), we get

$$\ddot{\delta} + 2\frac{\dot{a}}{a}\dot{\delta} = 4\pi G\rho_b(1+\nu)(1+3\nu)\delta . \qquad (10.118)$$

This generalizes equation (5.111) for the evolution of the mass density contrast in a nonrelativistic fluid.

Now we can write down some solutions. The energy conservation equation (10.109) says the mean mass density varies with the expansion parameter as

$$\rho_b \propto a^{-3(1+\nu)} . \qquad (10.119)$$

This expression in the expansion rate equation (10.115) gives a differential equation for the expansion parameter as a function of time. The solution is

$$a(t) \propto t^{2/(3+3\nu)} , \qquad (10.120)$$

with

$$t^{-2} = 6\pi G(1+\nu)^2\rho_b . \qquad (10.121)$$

With these expressions, the density contrast equation (10.118) is

$$\ddot{\delta} + \frac{4}{3(1+\nu)}\frac{\dot{\delta}}{t} = \frac{2}{3}\frac{1+3\nu}{1+\nu}\frac{\delta}{t^2} . \qquad (10.122)$$

This is homogeneous in the world time t, so we know the solutions are powers of time. The results of solving for the index n are

$$\delta_{(1)} \propto t^n , \quad n = \frac{2}{3}\frac{1+3\nu}{1+\nu} , \qquad (10.123)$$
$$\delta_{(2)} \propto 1/t .$$

The growing solution for nonrelativistic matter, with $\nu = 0$, agrees with equation (5.89), and the solution for radiation-dominated material, with $\nu = 1/3$, agrees with equation (5.90).

An important aspect of this problem, which requires the full machinery we have developed, is the curvature of spacetime that accompanies a given perturbation to the mass distribution. The effect is worked out for the special case of a spherical homogeneous perturbation in equation (5.126). The more general result obtained here is used in section 17 to find the density fluctuations at the present epoch produced by quantum fluctuations in the field that drives an inflation epoch.

Since the coefficients in the perturbation equations are independent of position, we can write the solutions as linear combinations of plane waves with independently evolving amplitudes. For a given plane wave, let us place the space coordinates with the propagation vector along the $x^3 = z$ axis, so the space part of the plane wave is $\propto \exp ikz$. Symmetry of rotation about the z axis says that in the inhomogeneous solution $h_{\alpha\beta}$ is diagonal, with $h_{11} = h_{22}$. Then the two independent components of equation (10.117) for the inhomogeneous solution (which ignores tidal fields) are

$$\frac{k^2}{a^2}h_{11} + \frac{\dot{a}}{a}(3\dot{h}_{11} + \dot{h}) + \ddot{h}_{11} = -8\pi G\rho_b(1-\nu)\delta,$$

$$\frac{k^2}{a^2}(h - h_{33}) + \frac{\dot{a}}{a}(3\dot{h}_{33} + \dot{h}) + \ddot{h}_{33} = -8\pi G\rho_b(1-\nu)\delta. \tag{10.124}$$

Since $h = 2h_{11} + h_{33}$, these are two equations for the two functions h and h_{11}. We can use equation (10.116) to eliminate $\ddot{h}$, and then rearrange the equations to get

$$\ddot{h}_{11} + 3\frac{\dot{a}}{a}\dot{h}_{11} = 8\pi G\rho_b\nu\delta, \tag{10.125}$$

and

$$\frac{k^2}{a^2}h_{11} + \frac{\dot{a}}{a}\dot{h} + 8\pi G\rho_b\delta = 0. \tag{10.126}$$

The second equation, with equation (10.120) for $a(t)$ and equation (10.110) and (10.123) for the growing density perturbation $\delta_{(1)}(t)$, gives

$$h_{11} = -\frac{4}{9}\frac{5+9\nu}{(1+\nu)^3}\left(\frac{a}{tk}\right)^2\delta_{(1)}. \tag{10.127}$$

The time variations of $\delta_{(1)}$ and a are such that this expression is independent of time. Equation (10.125) applied to the growing mode gives

$$\dot{h}_{11} = \frac{4\nu}{(3\nu+5)(\nu+1)}\frac{\delta_{(1)}}{t}, \tag{10.128}$$

which says $t\dot{h}_{11} \sim \delta_{(1)}$. This is a small correction to the constant term in equation (10.127), because we are computing in the limit where the spatial scale a/k of the density fluctuation is large compared to the Hubble length,

$$Hax_g \sim \frac{a}{tk} \gg 1 . \tag{10.129}$$

Thus the contribution from the time-varying part of h_{11} in equation (10.128) is down from the leading term in equation (10.127) by the factor $t\dot{h}_{11}/h_{11} \sim (Hax_g)^{-2} \ll 1$.

The field equations do not fix the value of h_{33}. That is because this component enters the line element (10.98) in the form $[1 - h_{33}]\,dz^2$, so we are free to change the leading constant value of h_{33} by a coordinate transformation from z to a function of z. The fact that the leading term in h_{11} is independent of time tells us that a growing mass density perturbation is associated with a fixed wrinkle in the spacetime of the very early universe. This is what we concluded from equation (5.126). Equation (10.127) extrapolated to the epoch at which the wavelength $\sim a/k$ of the density fluctuation is comparable to the Hubble length $\sim t$, is $h_{11} \sim \delta$. As in equation (5.127), this means that if the perturbation $\delta\rho/\rho$ to the mass distribution is small when the scale of the density fluctuation reaches the Hubble length, then the fluctuation has a small effect on the geometry of spacetime, and the density fluctuation stops expanding to form a bound system only when its width is small compared to the Hubble length.

11. Wall, String, and Spherical Solutions

In a situation with enough symmetry for the fields to be functions of only one variable, the field equations become ordinary differential equations that are easy to integrate numerically or even solve analytically. The three standard cases are plane symmetry, axial symmetry, and spherical symmetry. All play interesting roles in recent ideas in cosmology.

Walls

An idealized wall is a sheetlike distribution of matter that is invariant under translations in the two directions in the plane, rotations around the normal, and reflection $z \rightarrow -z$ through the midplane of the sheet at $z = 0$. Under these conditions, the line element can be written as

$$ds^2 = D\,dt^2 - A(dx^2 + dy^2) - C\,dz^2 . \tag{11.1}$$

We will consider the simple limiting case where the stress-energy tensor vanishes outside a region of width z_o chosen to satisfy the following condition. The mass per unit area in the sheet is $\sigma \sim \rho_o z_o$, where ρ_o is the mass density within the sheet. This defines a characteristic length, $l = (G\sigma)^{-1}$. We will assume $z_o \ll l$, and we will calculate in a region of space and time containing the wall and small compared to l. Let us check that in this limit the gravitational field can be described in linear perturbation theory.

The symmetry makes the derivatives $D' = dD/dz$ vanish at the midplane, and we can choose units of t and x such that at the midplane

$$D(0) = 1, \quad D'(0) = 0, \quad A(0) = 1, \quad A'(0) = 0. \tag{11.2}$$

In the Ricci tensor R_{ij} inside the wall, the dominant terms are the second derivatives of the metric tensor. The components of the field equation are of the form $D'' \sim G\rho_0$, and at the wall edge $D' \sim G\rho_0 z_0 \sim 1/l$. In the Ricci tensor the second derivatives appear in linear terms, the first derivatives in products (eq. [8.82]), so the ratio of nonlinear to linear terms is on the order of $(D')^2/D'' \sim z_0/l$, which we have arranged to be small. This means we can use linear perturbation theory to describe a small enough region of spacetime around a segment of a thin enough wall, whatever the mass per unit area in the wall.

In this limiting thin-wall case, we can choose the z coordinate label so $C = 1$ in equation (11.1), meaning the coordinate z is the proper perpendicular distance from the midplane of the mass sheet. The dominant terms in equations (8.82) and (10.72) are the second derivatives with respect to position normal to the wall, which gives

$$R_{00} = D''/2, \quad R_{11} = -A''/2, \quad R_{33} = -D''/2 - A'', \quad R = D'' + 2A''. \tag{11.3}$$

The gravitational field equations are

$$R^{00} - g^{00}R/2 = -A'' = 8\pi G\rho,$$
$$R^{11} - g^{11}R/2 = A''/2 + D''/2 = 8\pi GT^{11}, \tag{11.4}$$
$$R^{33} - g^{33}R/2 = 0 = 8\pi GT^{33}.$$

In the approximation of equation (11.3), the 33 component of the field equation vanishes, and the pressure in this direction has to be negligibly small because we are assuming that the wall is thin.

In a sheet of nonrelativistic material, the stress T^{11} in the sheet is small compared to the mass density $T^{00} = \rho$, so the second of the field equations (11.4) says $D'' = -A''$, and the first is

$$D'' = 8\pi G\rho. \tag{11.5}$$

The solution just outside the sheet is

$$D = 1 + 4\pi G\sigma|z|$$
$$A = 1 - 4\pi G\sigma|z|, \tag{11.6}$$

where the mass per unit area in the sheet is

$$\sigma = \int \rho \, dz. \tag{11.7}$$

We recall that $g_{00} = 1 + 2\phi + \cdots$ defines the Newtonian potential ϕ (eq. [9.32]), so equation (11.6) says the gravitational acceleration just outside the sheet is

$$g = 2\pi G\sigma, \tag{11.8}$$

directed toward the sheet. This Newtonian expression is valid for any surface mass density σ when the sheet is thin enough for the departure of D and A from unity at the edge of the sheet to be small, so the line element in a region of spacetime around a small enough section of the sheet is only slightly perturbed from Minkowski.

In the next chapter, we take note of domain walls that could have been left over from phase transitions in the early universe. These sheetlike structures are characterized by the energy per unit area, σ, in the wall, which is fixed by the potential energy term in the field equation for a single scalar field. This means σ has to be a fixed quantity, and therefore the surface tension has to be equal to σ, for that makes the work done in increasing the area of the domain wall equal to the increase of energy in the added area. Thus, the stress in a domain wall with normal along the x^3 axis is

$$T^{11} = T^{22} = -T^{00}. \tag{11.9}$$

That is, the field acts as if it had negative pressure in the two directions in the wall, with the magnitude of the pressure equal to the energy density.

For a plane domain wall the symmetry allows us to make the metric tensor diagonal, as in equation (11.1), and an easy way to find the field equations for a diagonal metric tensor is to use the general expressions Dingle (1933a) worked out and Tolman (1934) recorded. Vilenkin's (1983) solution is

$$ds^2 = (1 - \kappa|z|)^2 \, dt^2 - dz^2 - (1 - \kappa|z|)^2 e^{2\kappa t}(dx^2 + dy^2), \tag{11.10}$$

where $\kappa = 2\pi G\sigma$. We can understand the limiting properties of this elegant solution, in a neighborhood of the wall with size small compared to κ^{-1}, and for a time small compared to κ^{-1}, by using equation (11.4) for the line element in equation (11.1). That gives

$$D'' = A'' = -8\pi G\rho, \tag{11.11}$$

or

$$D = A = 1 - 4\pi G\sigma|z| \tag{11.12}$$

just outside the wall, as in equation (11.10). The sign of D'' is opposite to equation (11.5) for a nonrelativistic mass sheet. That means the gravitational acceleration just outside the domain wall is $g = 2\pi G\sigma$ directed away from the wall rather than toward it; that is, the wall acts as a negative gravitational mass.

The fact that the domain wall repels rather than attracts could have been anticipated from Poisson's equation (10.68) for the local relative gravitational acceleration, **g**, because the source is the sum of the components T^{00}, T^{11}, T^{22}, and T^{33} of the stress-energy tensor in locally Minkowski coordinates. That is, all these terms are active gravitational mass densities. The cosmological constant acts like an isotropic fluid with pressure equal to the negative of the mass density $\propto \Lambda$ (eq. [4.31]), and the de Sitter scattering effect for positive Λ arises because there are three negative pressure terms. In a domain wall, the diagonal space parts of T^{ij} in the plane of the wall are equal to the negative of T^{00}, and the wall repels because there are two negative pressure terms. In the cosmic string solution to be discussed next, only the one diagonal space part of T^{ij} along the string is equal to the negative of T^{00}. The two just cancel, so a static cosmic string has no active gravitational mass.

It is worth pausing to note that a negative active gravitational mass can behave in a curious way. Suppose we had a compact stable body with negative active mass, $-m$, and we placed it next to an ordinary body with positive active mass m, with both bodies initially at rest and freely moving. The ordinary body produces an attractive gravitational field. The equivalence principle says we can transform away the field locally by going to free-fall. The negative mass knows this, so it falls toward the positive mass. The same argument applied to the gravitational field of the negative mass shows that the positive mass falls away from it. The result is that the two accelerate away from a nearby inertial observer, keeping a constant separation relative to each other. If the two bodies were attached to the rim of a wheel and next to each other, they would cause the wheel to turn, from which we could get work. In conventional physics that does not happen; bodies with negative masses are not allowed.

One might have thought that a domain wall gives a counterexample. Suppose a domain wall is attached to a plane circular ring made of ordinary matter (which is strong enough to hold the wall stationary). We can imagine attaching an ordinary material wall to the ring with a spacer so the two sheets are separated and nearly parallel. We can stipulate that the material sheet and the domain wall interact with each other by gravity alone, and that they have the same mass per unit area. The first impression might be that the material sheet is repelled by the domain wall and the domain wall is attracted to the material sheet. That would say the ring in asymptotically flat spacetime is pushed in the direction of the material sheet, which certainly is contrary to general relativity. Vilenkin (1991) points out that the flaw is an inappropriate use of the equivalence principle.

We know it is always possible to choose coordinates that are Minkowski in the neighborhood of a compact test body (assuming spacetime has not become singular), and in these coordinates the test body has no acceleration, as asserted in the equivalence principle. But the domain wall is extended, and we have no guarantee the coordinates can be chosen to be Minkowski over the whole wall. An easy way to check is to use energy-momentum conservation in the form of equation (10.107). The result of integrating this expression across a range of z that might include one of the sheets, with normal in the z direction, is

$$
\dot{P} = -\frac{d}{dt} \int dz \sqrt{-g} T_z^{\ 0} = - \int dz \left(\sqrt{-g} T_z^{\ i}\right)_{,i}
$$
$$
= -\frac{1}{2} \int dz \sqrt{-g} T^{jk} g_{jk,z} .
$$

(11.13)

This is the rate of change of the momentum per unit area, and in linear perturbation theory the velocity of the sheet is $P/\int \rho\, dz$, as in equation (10.23). Now let us apply this to material sheets and domain walls.

An observer on the positive z side of an ordinary material sheet sees that the line element in the approximation of equation (11.6) is

$$
ds^2 = (1+\alpha z)dt^2 - (1-\alpha z)(dx^2+dy^2) - dz^2 ,
$$

(11.14)

with $\alpha = 4\pi G\sigma$. As a check on the expression for $\dot{P}$, consider a test sheet of ordinary material, where the only significant component of the stress-energy tensor is $T^{00} = \rho$, placed above the original sheet. We get from equations (11.13) and (11.14) that $\dot{P} = -\int \alpha\rho\, dz/2$ for this test sheet. Since the mass per unit area is $\int \rho\, dz$, the acceleration of the test sheet is $\alpha/2$ in the negative z direction, toward the other material sheet, in agreement with equation (11.8).

Consider next an observer placed on the negative z side of a domain wall. Here the line element (11.10) is, to first order,

$$
ds^2 = (1+\alpha z)dt^2 - (1+\alpha z)(dx^2+dy^2) - dz^2 .
$$

(11.15)

Here the same operation shows that $\dot{P}$ for a test sheet of ordinary matter is negative, indicating the particle is accelerated away, which again checks.

Now let us put the domain wall on the positive z side of the material sheet. For the material sheet below the domain wall, we get

$$
\dot{P}_m = -\frac{1}{2} \int \alpha\rho\, dz ,
$$

(11.16)

meaning the matter accelerates down, away from the domain wall, as we have just observed. For the rate of change of momentum in the domain wall above the ordinary mass sheet, we get

$$
\dot{P}_d = -\frac{1}{2} \int dz(\alpha\rho - 2\alpha\rho) = \frac{1}{2} \int \alpha\rho\, dz .
$$

(11.17)

The second term in the integrand comes from the terms $T^{11} = T^{22} = -p$ in equation (11.13). The result is that the domain wall accelerates toward positive z, which is away from the matter sheet. And we see that the sum of $\dot{P}_m$ and $\dot{P}_d$ vanishes, so an observer sees local conservation of momentum, which we knew had to happen because momentum is conserved locally in general relativity theory. In our thought experiment with the hoop, the material sheet and the domain wall would bow away from each other, as in a convex lens.

Cosmic Strings

The line element for a static system that is invariant under rotations around an axis and translations and reflections along the axis is

$$ds^2 = D(r)dt^2 - dr^2 - B(r)d\theta^2 - C(r)dz^2 . \tag{11.18}$$

We will consider only a cosmic string, in which the energy per unit length, μ, is a constant fixed by the potential energy function for a postulated cosmic field (as discussed in section 16). In a straight static cosmic string, the work done in increasing the length of string by the amount L is the increase in energy, μL. The tension therefore has to be μ. As we have noted, the negative active gravitational mass associated with the tension balances the positive gravitational mass of the energy density, so there is no gravitational acceleration, and $D = 1$. This leaves the one function $B(r)$ to describe the geometry.

The component T^{zz} of the stress tensor along the string is equal to the negative of $T^{00} = \rho(r)$, and other two components of T^{ij} can be taken to be negligibly small. With $D = 1$, Dingle's expressions for the rr and $\theta\theta$ components of Einstein field equation require that either B or C is constant. We can set $C = 1$ by a choice of coordinates, and this brings the zz and tt components of the field equation to the same form,

$$8\pi G T^0_0 = 8\pi G T^z_z = -\frac{B''}{2B} + \frac{(B')^2}{4B^2} . \tag{11.19}$$

With $B = b(r)^2$ this is

$$\frac{1}{b}\frac{d^2 b}{dr^2} = -8\pi G \rho(r) . \tag{11.20}$$

To get a definite expression for the solution, let us take $\rho(r)$ to be uniformly distributed across the string,

$$\rho(r) = \begin{cases} \rho_o & \text{for } r \leq r_o , \\ 0 & \text{for } r > r_o , \end{cases} \tag{11.21}$$

with r_o and ρ_o constants. Then the solution at $r < r_o$ is

$$b = \frac{\sin \kappa r}{\kappa}, \qquad \kappa = (8\pi G \rho_o)^{1/2}. \tag{11.22}$$

The normalization in the expression for $b(r)$ is fixed by the condition $b \to r$ at $r \to 0$, so the geometry along the axis of the string is regular. The solution at $r > r_o$ is $b(r) = c + rd$. The constants c and d are fixed by the condition that $b(r)$ and its first derivative are continuous at the surface of the string. That gives

$$b(r) = (r - r_o) \cos \kappa r_o + \frac{\sin \kappa r_o}{\kappa}. \tag{11.23}$$

In situations of interest $\kappa r_o \ll 1$. Then at $r \gg r_o$ the solution is

$$b = (1 - 4\pi G \rho_o r_o{}^2)r, \tag{11.24}$$

and the line element (11.18) is

$$ds^2 = dt^2 - dr^2 - (1 - 8G\mu)r^2 d\theta^2 - dz^2, \tag{11.25}$$

where the mass per unit length is

$$\mu = \pi r_o^2 \rho_o. \tag{11.26}$$

Vilenkin (1981a) found the solution in equation (11.25) in linear perturbation theory; Gott (1985) found the exact solution in equations (11.23) and (11.24).

The circumference of a circle at fixed r, z, and t in the line element in equation (11.25) is

$$C = (2\pi - 8\pi G\mu)r. \tag{11.27}$$

Vilenkin (1981a) noted that this represents the geometry of a conical space, in which a wedge of angle

$$\Phi = 8\pi G\mu \tag{11.28}$$

has been removed from locally flat space, as is illustrated in figure 11.1. Two points in space that face each other at the same radius and on opposite sides of the wedge are to be identified as the same point. The dashed and dotted lines in the figure are paths of freely moving particles. The lines are straight, because spacetime locally is flat, with no gravitational acceleration in the coordinates of equation (11.25). Where the path arrives at the edge of the wedge it is continued on the other side, and at the same angle to the edge. The result is that the two

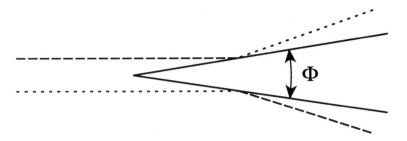

Figure 11.1. Geometry in the plane normal to a static cosmic string.

paths that on the left are parallel and normal to the string cross on the right-hand side, at the angle $\Phi = 8\pi G\mu$ in equation (11.28). The string thus acts as a gravitational lens, giving two undistorted images at angular separation Φ of a background object. Possible values for this angle are discussed in sections 16 and 25.

Schwarzschild Solution and Birkhoff's Theorem

Here we consider a spherically symmetric system, invariant under rotations about the center of symmetry. This is a useful model for a mass concentration and for the standard cosmological model, which is spherically symmetric about any point. We can write the line element as

$$ds^2 = D\,dt^2 - A\,dr^2 - B(d\theta^2 + \sin^2\theta d\phi^2) + 2E dr dt, \qquad (11.29)$$

where the coefficients A, B, D, and E are functions of r and t. The angular part is familiar from polar coordinates; it is the form that is symmetric under rotations. Cross terms between the polar angles and r and t are not allowed because that would spoil the rotational symmetry.

We can simplify the line element. First, the change in radial coordinate to $\bar{r} = B^{1/2}$ changes A, D, and E. Dropping the bars on the new radial coordinate, we have

$$ds^2 = D\,dt^2 - A\,dr^2 - r^2(d\theta^2 + \sin^2\theta d\phi^2) + 2E\,dr dt. \qquad (11.30)$$

Second, let us define a new time coordinate by the equation

$$d\bar{t} = \eta(E\,dr + D\,dt), \qquad (11.31)$$

where $\eta(r,t)$ is the integrating factor that makes the right-hand side a perfect differential. On squaring and rearranging, we get

$$D\,dt^2 + 2E\,dr\,dt = \frac{(d\bar{t})^2}{D\eta^2} - \frac{E^2\,dr^2}{D}\,. \tag{11.32}$$

This expression in the line element (11.30) eliminates the cross term $E\,drdt$.

The conclusion is that we can write the line element in the standard form (and the standard notation),

$$ds^2 = e^\nu\,dt^2 - e^\lambda\,dr^2 - r^2 d\Omega\,, \tag{11.33}$$

with

$$d\Omega = d\theta^2 + \sin^2\theta d\phi^2\,, \tag{11.34}$$

and ν and λ functions of r and t. This line element in Dingle's (1933a) expressions give the field equations

$$8\pi G T_0^0 = e^{-\lambda}\left(\frac{\lambda'}{r} - \frac{1}{r^2}\right) + \frac{1}{r^2}\,,$$

$$8\pi G T_1^1 = -e^{-\lambda}\left(\frac{\nu'}{r} + \frac{1}{r^2}\right) + \frac{1}{r^2}\,, \tag{11.35}$$

$$8\pi G T_0^1 = -e^{-\lambda}\dot{\lambda}/r\,.$$

The coordinates are $x^0 = t$ and $x^1 = r$, the prime means a derivative with respect to r, and the dot is a derivative with respect to time. The other nonzero components of the field equations are constructed out of these.

The Schwarzschild solution applies to an empty region (inside and outside of which there may be material). With $T_0^1 = 0$ the last of equations (11.35) says λ is a function of r alone. Then the solution to the first equation with $T_0^0 = 0$ is

$$e^{-\lambda} = 1 - r_o/r\,, \tag{11.36}$$

where r_o is the constant of integration. This in the last of the equations, with $T_1^1 = 0$, gives

$$\nu' = \frac{1}{r - r_o} - \frac{1}{r}\,, \tag{11.37}$$

with the solution

$$e^\nu = F(t)(1 - r_o/r)\,. \tag{11.38}$$

The constant of integration, F, is a free function of t, but we can eliminate it by writing $d\bar{t}^2 = F(t)\,dt^2$.

The result is the Schwarzschild line element,

$$ds^2 = \left(1 - \frac{r_o}{r}\right) dt^2 - \frac{dr^2}{1 - r_o/r} - r^2 d\Omega. \tag{11.39}$$

The change of radial coordinate defined by the equation

$$r = \left(1 + \frac{r_o}{4\bar{r}}\right)^2 \bar{r} \tag{11.40}$$

gives

$$1 - \frac{r_o}{r} = \left(\frac{1 - r_o/(4\bar{r})}{1 + r_o/(4\bar{r})}\right)^2, \quad dr = d\bar{r}\left[1 - \left(\frac{r_o}{4\bar{r}}\right)^2\right], \tag{11.41}$$

and the equivalent form for the line element,

$$ds^2 = \left(\frac{1 - r_o/(4\bar{r})}{1 + r_o/(4\bar{r})}\right)^2 dt^2 - \left(1 + \frac{r_o}{4\bar{r}}\right)^4 (d\bar{x}^2 + d\bar{y}^2 + d\bar{z}^2). \tag{11.42}$$

The line element (11.42) to first order in $\bar{r}/r_o$ is

$$ds^2 = (1 - r_o/\bar{r}) dt^2 - (1 + r_o/\bar{r})(d\bar{x}^2 + d\bar{y}^2 + d\bar{z}^2). \tag{11.43}$$

This is the form of the line element in equation (10.84), with the Newtonian potential $\phi = -G\mathcal{M}/\bar{r} = -r_o/2\bar{r}$ for an isolated mass $\mathcal{M}$, so we see that the Schwarzschild radius r_o can be written as

$$r_o = 2G\mathcal{M}. \tag{11.44}$$

That is, the Schwarzschild solution at $r \gg r_o$ acts like the Newtonian gravitational field of a mass concentration, $\mathcal{M}$, which might represent the external part of the gravitational field of a spherically symmetric star, or it might be the gravitational field of a black hole, in which equation (11.39) continues inside the Schwarzschild radius r_o to a singularity (from which one presumes we are saved by new physics).

Birkhoff (1923) remarked that in the above calculation we did not have to stipulate that the stress-energy tensor represents a static situation in the regions where it is nonzero. That means the Schwarzschild line element is the general spherically symmetric solution in a range of radii where T_i^j vanishes, independent of the time-variation of the stress-energy tensor inside or outside this region (as long as it remains spherically symmetric). This generalizes Newton's iron sphere theorem — that there is no gravitational acceleration inside a hollow spherical mass distribution. To see the analog in general relativity theory, note that the line element for a spherically symmetric mass distribution with a void (with $T_i^j = 0$) that contains the center of symmetry has to be regular at $r \to 0$, and the only way this can be consistent with the unique solution in equation (11.39) is that r_o has to vanish. That is, spacetime is flat within a void that contains the center of symmetry, independent of how the external mass is behaving.

Birkhoff's theorem is a very useful guide. For example, it tells us that in general relativity theory a supernova explosion, no matter how violent, can generate gravitational radiation (through the source term in the right-hand side of eq. [10.80] in linear perturbation theory) only if the explosion is not spherically symmetric. The enormous amount of mass in the universe at the Hubble distance, which is moving away from us at relativistic speeds, has negligible local effect, because the evidence in section 3 is that the distribution is very close to spherically symmetric about us (and in the standard model spherically symmetric about any galaxy we can see). This is the basis for the analysis in sections 4 and 5 of the dynamics of an expanding universe.

Spherical Lemaître-Tolman Solution

The spherical solution to be discussed here includes the Robertson-Walker line element of the standard cosmological model, and it describes the evolution of a zero pressure spherical fluctuation in the mass distribution.

We will start with the time-orthogonal coordinates of equation (9.37). This gives $g_{00} = 1$ and $g_{0\alpha} = 0$, so the spherically symmetric line element in equation (11.33) can be written as

$$ds^2 = dt^2 - e^{2\alpha} dr^2 - e^{2\beta} d\Omega, \tag{11.45}$$

with α and β functions of r and t. We will find solutions for these functions under the assumption that the stress-energy tensor is dominated by pressureless dust ($p = 0$ in eq. [10.49]), along with a cosmological constant.

Recall that in a time-orthogonal coordinate system a path with constant spatial coordinates x^γ is a geodesic normal to the hypersurface $t = $ constant. The spherical symmetry allows us to choose the hypersurface so its normal everywhere points along the four-velocity of the dust. Then the dust particles have four-velocity $u^i = \delta_0^i$, and the only nonzero component of the stress-energy tensor (10.49) is the mass density in the rest frame of the dust, $T^{00} = \rho(r, t)$, along with the cosmological constant.

Dingle's expressions give the field equations

$$8\pi G T_0^0 = 8\pi G \rho + \Lambda$$
$$= \dot{\beta}^2 + 2\dot{\alpha}\dot{\beta} + e^{-2\beta} - e^{-2\alpha}[2\beta'' + 3(\beta')^2 - 2\alpha'\beta'],$$
$$8\pi G T_1^1 = \Lambda = 2\ddot{\beta} + 3\dot{\beta}^2 + e^{-2\beta} - (\beta')^2 e^{-2\alpha}, \tag{11.46}$$
$$8\pi G T_1^0 = 0 = 2\dot{\beta}' + 2\dot{\beta}\beta' - 2\dot{\alpha}\beta'.$$

As before, dots and primes mean partial derivatives with respect to t and r. Because T_1^0 vanishes, the last of these equations is

$$\frac{\dot{\beta}'}{\beta'} = \frac{\partial}{\partial t} \log \beta' = \dot{\alpha} - \dot{\beta}. \tag{11.47}$$

The solution is $\beta' = g(r)e^{\alpha - \beta}$, where g is a function of r alone. This gives

$$e^{\beta} \beta' = \frac{\partial e^{\beta}}{\partial r} = g(r)e^{\alpha}. \tag{11.48}$$

We can get an expression that looks somewhat more familiar by defining the expansion parameter $a(r, t)$ by the equation

$$e^{\beta} \equiv ra(r, t), \tag{11.49}$$

and rewriting the function $g(r)$ as $g = [1 - r^2/R(r)^2]^{1/2}$. Then equation (11.48) is

$$e^{\alpha} = \frac{(ar)'}{[1 - r^2/R(r)^2]^{1/2}}. \tag{11.50}$$

This brings the line element (11.45) to

$$ds^2 = dt^2 - \frac{[(ar)']^2 dr^2}{1 - r^2/R(r)^2} - (ar)^2 d\Omega. \tag{11.51}$$

We see that if a and R are independent of r this is the Robertson-Walker line element for the standard cosmological model (eq. [5.9]). The function R^{-2} has been written as a square, to conform with the standard notation for the homogeneous model, but of course R^{-2} can be negative, as in an open homogeneous model.

Equations (11.49) for e^{β} and (11.50) for e^{α} in the first of the field equations (11.46) give

$$\left[\frac{8}{3}\pi G\rho(r, t) + \frac{\Lambda}{3} \right] \frac{\partial}{\partial r} (ar)^3 = \frac{\partial}{\partial r} (\dot{a}^2 ar^3 + ar^3/R^2), \tag{11.52}$$

and the second of the field equations gives

$$\Lambda = 2\frac{\ddot{a}}{a} + \frac{\dot{a}^2}{a^2} + \frac{1}{a^2 R^2}. \tag{11.53}$$

This last expression multiplied by $a^2 \dot{a}$ is a total time derivative, which we can integrate to get

$$\dot{a}^2 a + a/R^2 - \Lambda a^3/3 = F(r). \tag{11.54}$$

Suppose space is not too heavily curved, so the function $F(r)$ from the integration is positive. Then the change of the radial coordinates to

$$\bar{r} = r \left[\frac{F(r)}{A} \right]^{1/3} , \qquad \bar{a} = a \left[\frac{A}{F(r)} \right]^{1/3} , \qquad \bar{R} = R \left[\frac{F(r)}{A} \right]^{1/3} , \quad (11.55)$$

where A is a constant, leaves the form of the line element (11.51) unchanged, and it brings the right-hand side of equation (11.54) in the new coordinate system to a constant. Dropping the bar, equations (11.52) and (11.54) become

$$\dot{a}^2 a + a/R(r)^2 - \Lambda a^3/3 = A ,$$
$$\frac{8}{3} \pi G \rho(r,t) = \frac{A}{a^2[ra(r,t)]'} , \qquad (11.56)$$

where A is a constant.

In this solution, a mass shell is assigned a fixed value for the radial coordinate r, and it is assigned an expansion parameter $a(r,t)$ that measures the proper circumference of the shell, $C = 2\pi ar$. We see from the first line of equation (11.56) that the expansion parameter belonging to the mass shell evolves with time according to the standard homogeneous cosmological model in equation (5.18). The homogeneous Robertson-Walker model follows if we take the mass density ρ and expansion parameter a to be independent of radius.

Lemaître (1933), Dingle (1933b), and Tolman (1934) were the first to discuss the inhomogeneous spherical zero pressure solution in equations (11.51) and (11.56). They noted that, since the expansion parameters belonging to different mass shells can evolve according to different cosmological models, the expanding universe is gravitationally unstable.

The gravitational development of a mass concentration such as a cluster of galaxies, with coordinate size $r = l$, might be modeled by a solution in which R and a at $r > l$ are independent of the radial coordinate, to represent the homogeneous background model, and with the density $\rho(r,t)$ larger than the background at $r < l$, to model a growing mass concentration. We see from the second line in equation (11.56) that in this case $a(r,t)$ within the mass concentration is smaller than the background value, so a' is positive at $r \sim l$, and the density on the edges of the mass contrast has to be less than the background value. The density concentration is said to be compensated by a density minimum around it, at $r \sim l$, so that there is no perturbation to the expansion rate outside this region.

As will be discussed in section 19, there is no evidence that the mass concentration in a typical great cluster of galaxies is compensated. To see what this implies, note that the second line in equation (11.56) with ρ independent of r requires that $a^2(ar)'$ is independent of r. The solution to this equation is

$$a = (C + D/r^3)^{1/3} , \qquad (11.57)$$

where C and D are independent of r. The second term indicates that at large r the circumference ar of a mass shell is perturbed from the homogeneous case by a fractional amount that varies as r^{-2}. This is the ordinary Newtonian inverse square law, as discussed below (eq. [11.70]). A more detailed solution for an uncompensated mass concentration is shown in figure 22.2.

We have noted that a growing density perturbation that develops into a nonrelativistic mass concentration such as a cluster of galaxies produces a wrinkle in the curvature of spacetime, rather than a knob (eqs. [5.126] and [10.127]). It is easy to get the same result from the above solution, as follows.

We will drop the cosmological constant, and we will suppose the function $R(r)$ has the value $R(0)$ within a growing mass fluctuation with comoving radius $r \sim l$, with $R(r) \gg R(0)$ at larger radii. At very high redshift, the right-hand side of the first of equations (11.56) is dominated by the expansion rate term, $\dot{a}^2 \gg R(0)^{-2}$, and the mass fluctuation acts as a small perturbation. The matter at $r \lesssim l$ stops expanding, $\dot{a}(0, t_x) = 0$, when the expansion parameter reaches the value

$$a_x = a(0, t_x) = AR(0)^2 , \tag{11.58}$$

at time

$$t_x \sim a_x R(0) . \tag{11.59}$$

The proper physical size of the mass concentration when it stops expanding is

$$L_x \sim l a_x , \tag{11.60}$$

and the ratio of this proper size to the Hubble length when the mass concentration stops expanding is

$$\frac{L_x}{t_x} \sim \frac{l}{R(0)} . \tag{11.61}$$

This ratio is small in observed mass concentrations, which have stopped expanding well within the Hubble length. That means the denominator $1 - r^2/R(0)^2$ in the line element in equation (11.51) is close to unity, which means that the mass concentration is causing only a small perturbation to the geometry. If the coordinate size l of the mass concentration were comparable to $R(0)$, the geometry in equation (11.51) at $r \sim l$ would resemble that of a knob, that would tend to collapse to a black hole.

Gravitational Green's Function

In linear perturbation theory, and when pressure can be neglected, we can represent the effect on the metric tensor of an irregular mass distribution as a linear combination of elementary departures $m(t)$ from homogeneity, each confined to a small region of space, as in a Green's function solution. The gravitational response to a single mass element is given by the spherical model we have just considered.

Let us write the mass density, expansion parameter, and curvature function around $m(t)$ as

$$a(r,t) = a_b(t)[1 - \alpha(r,t)],$$
$$\rho(r,t) = \rho_b(t)[1 + \delta(r,t)], \qquad (11.62)$$
$$R^{-2}(r) = R_b^{-2} + \delta R^{-2}(r).$$

The subscript refers to the homogeneous background, in which the expansion parameter and mass density are functions of time alone and the unperturbed radius of curvature R_b is a constant. On substituting these expressions into the spherical solution (11.56), one finds the usual zero pressure equations for the background model,

$$\frac{\dot{a}_b^2}{a_b^2} + \frac{1}{a_b^2 R_b^2} - \frac{\Lambda}{3} = \frac{A}{a_b^3} = \frac{8}{3}\pi G\rho_b, \qquad (11.63)$$

and for the perturbation to the line element one gets

$$\delta R^{-2}(r) = 2\dot{\alpha}a_b\dot{a}_b + \alpha(3\dot{a}_b^2 + R_b^{-2} - \Lambda a_b^2), \qquad r^2\delta = (r^3\alpha)'. \qquad (11.64)$$

The result of differentiating the first part of equation (11.64) with respect to time, and using equation (11.63) to simplify, is

$$\ddot{\alpha} + 2\dot{\alpha}\dot{a}/a = 4\pi G\rho_b\alpha. \qquad (11.65)$$

Since $\delta \propto \alpha$ at fixed r (from the second part of eq. [11.64]), the density contrast satisfies the same equation. This linear perturbation equation for the evolution of the mass contrast was obtained another way in section 5 (eq. [5.111]).

To express the solution to equation (11.65) as an integral, differentiate equation (11.63) and rearrange, to get

$$2\frac{\ddot{a}_b}{a_b} + \frac{\dot{a}_b^2}{a_b^2} + \frac{1}{a_b^2 R_b^2} - \Lambda = 0, \qquad (11.66)$$

and use this to rewrite the first part of equation (11.64) as

$$\dot{\alpha} + \alpha\left(\frac{\dot{a}_b}{a_b} - \frac{\ddot{a}_b}{\dot{a}_b}\right) = \frac{\delta R^{-2}}{2a_b\dot{a}_b}. \qquad (11.67)$$

This is a linear first-order differential equation; the solution can be written as an integral, which works out to

$$\alpha = K \frac{\dot{a}_b}{a_b} + \frac{\delta R^{-2}}{2} \frac{\dot{a}_b}{a_b} \int_0^t \frac{dt}{\dot{a}_b^2} , \tag{11.68}$$

where K is the constant of integration. This is the solution in equation (5.87).

We are considering an uncompensated mass fluctuation, $\delta(r, t)$, that is confined to a small region of space, so the mass in excess of homogeneity in the fluctuation is

$$m(t) = 4\pi \rho_b a_b^3 \int r^2 \delta \, dr . \tag{11.69}$$

With the second part of equation (11.64) for δ, we have

$$\alpha(r, t) = \frac{m(t)}{4\pi \rho_b a_b^3 r^3} . \tag{11.70}$$

This gives the perturbation to the expansion parameter as a function of the angular size distance r from the mass element.

If the growing term dominates the solution in equation (11.68) for the density perturbation, then we can rewrite it as an expression for the perturbation to the curvature function,

$$\delta R^{-2} = 2\alpha \left(\frac{\dot{a}_b}{a_b} \int_0^t \frac{dt}{\dot{a}_b^2} \right)^{-1} . \tag{11.71}$$

With equation (11.70) for α, the line element (11.51) belonging to the mass element works out to

$$ds^2 = dt^2 - \frac{a_b^2(1 + 4\alpha) dr^2}{1 - r^2 R_b^{-2} - r^2 \delta R^{-2}(r)} - (1 - 2\alpha) a_b^2 r^2 d\Omega . \tag{11.72}$$

This linear perturbation result does not take account of the nonlinear behavior within a mass fluctuation, but it is the general solution for what happens well away from an isolated uncompensated mass concentration or deficit, independent of what is happening on small scales. The net mass concentration or deficit grows with time, as given by equations (11.68) and (11.70). In the Einstein–de Sitter model, the growth factor since decoupling is $\sim z_{\text{dec}} \sim 1000$ (eq. [5.113]), so a great cluster with present mass $10^{15} \mathcal{M}_\odot$ would have amounted to a mass excess of $\sim 10^{12} \mathcal{M}_\odot$, about the mass of a giant galaxy, at decoupling. This does not mean protoclusters existed at decoupling as nonlinear objects, but that the mass excess had to have been present if clusters formed by gravity (and as will be discussed in section 25 there is no known acceptable alternative).

The tidal field far away from an uncompensated mass concentration can be defined by the relative motions of neighboring observers at fixed coordinate positions. Let $l_r(t)$ be the proper distance between observers at r and $r + \delta r$ and the same θ and ϕ, and let l_θ be the proper separation at the same r and polar angle differing by $\delta\theta$,

$$l_\theta(t) = (1 - \alpha)a_b r \delta\theta, \qquad l_r(t) = \frac{(1 + 2\alpha)a_b \delta r}{[1 - r^2(R_b^{-2} + \delta R^{-2})]^{1/2}} . \qquad (11.73)$$

The relative velocities of the observers are

$$\frac{\dot{l}_\theta}{l_\theta} = \frac{\dot{a}_b}{a_b} - \dot{\alpha} \qquad \frac{\dot{l}_r}{l_r} = \frac{\dot{a}_b}{a_b} + 2\dot{\alpha} . \qquad (11.74)$$

Since there are two transverse directions the volume is unperturbed, as expected, because this is a tidal field. With equation (11.70) for α, the relative peculiar velocity in the radial direction is

$$v = \frac{l_r \dot{m}}{2\pi \rho_b a_b^3 r^3} . \qquad (11.75)$$

Following equation (5.116), let $\dot{m} = mHf(\Omega)$, where f is the velocity function plotted in figure 13.13. With equation (5.55) for the density parameter, we can rewrite equation (11.75) as

$$v = \frac{2}{3} \frac{f(\Omega)\delta g}{\Omega H} , \qquad (11.76)$$

where the peculiar gravitational acceleration has been defined as

$$g = \frac{Gm}{(a_b r)^2} , \qquad \delta g = 2\frac{Gml_r}{(a_b r)^3} . \qquad (11.77)$$

The tidal field between the observers is the difference of peculiar accelerations, δg . Equation (11.76) is the same as equation (5.117). It is interesting to note that the inverse square law defined in equation (11.77), $g \propto 1/(a_b r)^2$, applies at separations larger than the Hubble length. In a cosmologically flat model, the proper radial distance from m to the observer is $a_b r$. In an open model, the radial distance is $a_b R_b \chi$, where $r = R_b \sinh \chi$ (eq. [5.12]), so $g \propto (\sinh \chi)^{-2}$. In terms of the radial coordinate χ, the inverse square law has an exponential cutoff at $\chi \sim 1$.

Schwarzschild Solution in a Cosmological Model

Birkhoff's theorem tells us that in a homogeneous zero-pressure cosmological model we can evacuate a spherical region and replace the material with a compact mass $\mathcal{M}$ at the center, without affecting spacetime outside the region. Let us find the relation between the mass $\mathcal{M}$ and the rest mass evacuated from the cavity in this compensated rearrangement of the mass distribution. We will simplify the discussion by dropping the cosmological constant.

The Schwarzschild line element (11.39) in the cavity is

$$ds^2 = \left(1 - \frac{2G\mathcal{M}}{r}\right) dt^2 - \frac{dr^2}{1 - 2G\mathcal{M}/r} - r^2 d\Omega, \tag{11.78}$$

and the Robertson-Walker line element outside the cavity is

$$ds^2 = d\bar{t}^2 - a(\bar{t})^2 \left[\frac{d\bar{r}^2}{1 - \bar{r}^2/R^2} + \bar{r}^2 d\Omega\right]. \tag{11.79}$$

The edge of the cavity has fixed comoving radius $\bar{r}_e$ in the coordinates of equation (11.79). The radius $r_e(t)$ of the cavity in the coordinates of equation (11.78) is fixed by the condition that the circumference C of the cavity is the same measured either way,

$$C/2\pi = r_e(t) = a(\bar{t})\bar{r}_e. \tag{11.80}$$

The time derivative of this equation is

$$\frac{1}{a}\frac{da}{d\bar{t}} = \frac{1}{r_e}\frac{dr_e}{dt}\left(\frac{\partial t}{\partial \bar{t}}\right)_{\bar{r}_e}. \tag{11.81}$$

To get $\partial t/\partial \bar{t}$ for an observer at rest at the edge of the cavity, note that the relation between proper and coordinate time intervals at fixed $\bar{r}_e$ is

$$ds^2 = d\bar{t}^2 = \left[1 - \frac{2G\mathcal{M}}{r_e} - \left(1 - \frac{2G\mathcal{M}}{r_e}\right)^{-1}\left(\frac{dr_e}{dt}\right)^2\right] dt^2. \tag{11.82}$$

Also, since the components of the metric tensor are independent of time, the $i=0$ component of the geodesic equation of motion for this observer, expressed in Schwarzschild coordinates, is

$$\left(1 - \frac{2G\mathcal{M}}{r_e}\right)\frac{dt}{ds} = K^{1/2}, \tag{11.83}$$

where K is a constant. The result of eliminating dt/ds from these two equations is

$$\frac{dr_e}{dt} = \left(1 - \frac{2G\mathcal{M}}{r_e}\right)\left[1 - \frac{1}{K}\left(1 - \frac{2G\mathcal{M}}{r_e}\right)\right]^{1/2}. \tag{11.84}$$

This with equation (11.82) in equation (11.81) gives

$$\left(\frac{1}{a}\frac{da}{dt}\right)^2 = \frac{2GM}{(\bar{r}_e a)^3} + \frac{K-1}{(\bar{r}_e a)^2} . \tag{11.85}$$

The expansion rate in the homogeneous cosmological model is

$$\left(\frac{1}{a}\frac{da}{dt}\right)^2 = \frac{8}{3}\pi G\rho - \frac{1}{(aR)^2} . \tag{11.86}$$

Comparing these two equations, we have

$$M = \frac{4}{3}\pi\rho(\bar{r}_e a)^3 , \qquad R^{-2} = \frac{(1-K)}{\bar{r}_e^2} . \tag{11.87}$$

Equation (11.87) fixes the value of the compact mass M that has replaced what has been evacuated from the homogeneous solution to make the void. The latter amounts to

$$M' = 4\pi\rho a^3 \int_0^{\bar{r}_e} \frac{\bar{r}^2 d\bar{r}}{(1 - \bar{r}^2/R^2)^{1/2}} . \tag{11.88}$$

In a cosmologically flat model, where $R^{-2} = 0$, this agrees with the mass in equation (11.87). In a closed model, where $R^{-2} > 0$, the evacuated mass is larger, $M' > M$. In an informal way of describing this, the net gravitational mass M seen by an observer at $\bar{r}_e$ in the homogeneous case is the sum of the rest mass M' within $\bar{r}_e$, the positive kinetic energy of expansion, and the negative gravitational potential energy of the smoothly distributed matter. In the Einstein-de Sitter model the last two just cancel, while in a higher density model the larger magnitude of the gravitational potential energy term makes M smaller than the rest mass M'.

12. Robertson-Walker Geometry

This section deals with some general properties of the spatial geometry in the standard homogeneous world model, including some forms for the line element and the spherical trigonometry used to figure distances between well-separated objects.

Forms for the Line Element

As discussed in sections 4 and 5, we can write the line element in the time-orthogonal form

$$ds^2 = dt^2 - a(t)^2 dl^2 , \tag{12.1}$$

where the expansion parameter is $a(t)$. In a closed model the part dl^2 follows by imagining that the space sections of fixed world time t are embedded in a four-dimensional space with coordinates x, y, z, w, and line element

$$dl^2 = dx^2 + dy^2 + dz^2 + dw^2 . \tag{12.2}$$

Our three-dimensional space is the surface of the sphere

$$x^2 + y^2 + z^2 + w^2 = R^2 , \tag{12.3}$$

at fixed R. Spherical coordinates for this space are

$$
\begin{aligned}
w &= R \cos \chi , \\
z &= R \sin \chi \cos \theta , \\
x &= R \sin \chi \sin \theta \cos \phi , \\
y &= R \sin \chi \sin \theta \sin \phi .
\end{aligned}
\tag{12.4}
$$

The familiar transformation from Cartesian to polar coordinates in three dimensions easily generalizes to the extra dimension, bringing the line element (12.2) at fixed radius R to

$$
\begin{aligned}
dl^2 &= R^2[d\chi^2 + \sin^2\chi(d\theta^2 + \sin^2\theta d\phi^2)] \\
&= R^2[d\chi^2 + \sin^2\chi d\Omega] .
\end{aligned}
\tag{12.5}
$$

For the negative curvature case, the substitution

$$w \to iw , \quad R \to iR , \quad \chi \to -i\chi \tag{12.6}$$

brings equation (12.4) to $z = R \sinh \chi \cos \theta$, and so on, and it brings the line element to

$$dl^2 = R^2[d\chi^2 + \sinh^2\chi d\Omega] . \tag{12.7}$$

Except where noted, this relation between positive and negative curvature applies to all that follows.

Some other forms for the line element are worth recording. The change of radial variable

$$r = R \sin \chi \tag{12.8}$$

gives

$$dl^2 = \frac{dr^2}{1 - r^2/R^2} + r^2 d\Omega . \tag{12.9}$$

The expression is singular at $r = R$, or $\chi = \pi/2$, but we see from equation (12.5) that this is a failure of the coordinates, for we can continue the space to the other pole at $\chi = \pi$.

It is left as an exercise to check that the change of radial variable

$$r = \frac{\hat{r}}{1 + \hat{r}^2/4R^2} \tag{12.10}$$

gives

$$dl^2 = \frac{d\hat{r}^2 + \hat{r}^2 d\Omega}{(1 + \hat{r}^2/4R^2)^2} = \frac{d\hat{r}_1^2 + d\hat{r}_2^2 + d\hat{r}_3^2}{(1 + \hat{r}^2/4R^2)^2}. \tag{12.11}$$

In this coordinate system the line element differs from the flat Cartesian case, $dl^2 = d\hat{x}_1^2 + d\hat{x}_2^2 + d\hat{x}_3^2$, by the conformal factor $1 + \hat{r}^2/4R^2$.

The change in radial coordinate

$$\bar{r} = 2R \sin \chi/2 \tag{12.12}$$

in equation (12.5) gives

$$dl^2 = \frac{d\bar{r}^2}{1 - \bar{r}^2/4R^2} + \left(1 - \frac{\bar{r}^2}{4R^2}\right) \bar{r}^2 d\Omega. \tag{12.13}$$

A somewhat more lengthy change of variables that applies to the open (negative curvature) case goes as follows. We can use the three polar angles χ, θ and ϕ in the line element (12.7) for the open model to define three new variables,

$$y_1 = R \tanh \chi \sin \theta \cos \phi,$$
$$y_2 = R \tanh \chi \sin \theta \sin \phi,$$
$$y_3 = \frac{R}{\cosh \chi}. \tag{12.14}$$

We will also use an auxiliary variable,

$$y_4 = R \tanh \chi \cos \theta. \tag{12.15}$$

The four variables satisfy

$$y_1^2 + y_2^2 + y_3^2 + y_4^2 = R^2. \tag{12.16}$$

On using the familiar transformation from polar to Cartesian coordinates in three dimensions, we see that the first two of the coordinates (12.14) with the auxiliary coordinate (12.15) satisfy

$$dy_1{}^2 + dy_2{}^2 + dy_4{}^2 = R^2 \left[\frac{d\chi^2}{\cosh^4\chi} + \tanh^2\chi \, (d\theta^2 + \sin^2\theta \, d\phi^2) \right] . \quad \textbf{(12.17)}$$

The differential of the third variable is

$$dy_3{}^2 = R^2 \frac{\sinh^2\chi}{\cosh^4\chi} d\chi^2 . \quad \textbf{(12.18)}$$

Adding these two expressions, we get

$$\begin{aligned} (dy_1{}^2 + dy_2{}^2 + dy_3{}^2 + dy_4{}^2)\cosh^2\chi \\ = R^2 [d\chi^2 + \sinh^2\chi(d\theta^2 + \sin^2\theta \, d\phi^2)] = dl^2 . \end{aligned} \quad \textbf{(12.19)}$$

The second line of equation (12.19) is the form for the negative curvature line element in equation (12.7). On the left side, we can express the differential dy_4 in terms of the three independent variables y_1, y_2, and y_3. We have from equation (12.16)

$$y_4{}^2 = R^2 - y^2 , \qquad y^2 = y_1{}^2 + y_2{}^2 + y_3{}^2 . \quad \textbf{(12.20)}$$

The differential of this expression is

$$dy_4 = \frac{y \, dy}{y_4} = \frac{y \, dy}{(R^2 - y^2)^{1/2}} , \quad \textbf{(12.21)}$$

so the line element (12.19) is

$$dl^2 = \left(dy_1{}^2 + dy_2{}^2 + dy_3{}^2 + \frac{y^2}{R^2 - y^2} \, dy^2 \right) \cosh^2\chi . \quad \textbf{(12.22)}$$

Now let us define new polar coordinates by the usual equations

$$\begin{aligned} y_3 &= y \cos\bar\theta , \\ y_1 &= y \sin\bar\theta \cos\phi , \\ y_2 &= y \sin\bar\theta \sin\phi . \end{aligned} \quad \textbf{(12.23)}$$

This brings the line element (12.22) to

$$dl^2 = \left[\frac{dy^2}{1 - y^2/R^2} + y^2(d\bar\theta^2 + \sin^2\bar\theta d\phi^2) \right] \frac{R^2}{(y \cos\bar\theta)^2} . \quad \textbf{(12.24)}$$

The last factor results from using equation (12.14) to write $\cosh^2\chi$ in terms of the polar angles in equation (12.23). It will be noted that ϕ is the same angle in equations (12.14) and (12.23), while the new polar angle $\bar{\theta}$ is given by the equation

$$\tan\bar{\theta} = \sinh\chi \sin\theta . \tag{12.25}$$

The factor in brackets in equation (12.24) has the form of the line element for positive space curvature (though the complete line element really is the negative curvature case). We can bring this factor to a conformally flat expression by the trick in equation (12.10). With

$$y = \frac{w}{1 + w^2/4R^2} , \tag{12.26}$$

equation (12.24) is

$$dl^2 = R^2 \frac{dw^2 + w^2(d\bar{\theta}^2 + \sin^2\bar{\theta} \, d\phi^2)}{w^2 \cos^2\bar{\theta}} . \tag{12.27}$$

On expressing this in Cartesian coordinates, as in equation (12.23), we finally get the wanted result,

$$dl^2 = R^2 \frac{dw_1^2 + dw_2^2 + dw_3^2}{w_3^2} . \tag{12.28}$$

The line element (12.28) is of historical interest, because it is the form Friedmann (1924) used in the first discussion of a homogeneous evolving universe with negative space curvature. (In his first paper on an evolving universe with positive curvature, Friedmann 1922 used the line element [12.5]). As Bernstein and Feinberg (1986) note, it certainly is not immediately apparent that equation (12.28) represents a homogeneous space. It also is not obvious, and an interesting exercise to demonstrate, that the same trick cannot bring the positive curvature line element to a similar form.

Friedmann's line element for negative space curvature may also be of some practical interest, because it gives simple forms for geodesics. The path of a stationary stretched string in the three-dimensional space with line element

$$dl^2 = g_{\alpha\beta}^{(0)} dx^\alpha dx^\beta , \tag{12.29}$$

where the $g_{\alpha\beta}^{(0)}$ are functions of the three coordinates x^γ, is fixed by the condition that the proper length

$$l = \int dl \tag{12.30}$$

is stationary under small variations of the path with fixed end points. Following section 9, we see that this is equivalent to the three-dimensional geodesic equation [4]

$$\frac{d}{dl} g^{(0)}_{\alpha\beta} \frac{dx^\beta}{dl} = \frac{1}{2} g^{(0)}_{\beta\gamma,\alpha} \frac{dx^\beta}{dl} \frac{dx^\gamma}{dl}. \tag{12.31}$$

The metric tensor in equation (12.28) is diagonal, with the diagonal components equal to R^2/z^2 (with x, y, and z replacing the w^α). Since the x and y derivatives of $g^{(0)}_{\beta\gamma}$ vanish, the x and y components of the geodesic equation are

$$\frac{dx}{dl} = Az^2, \qquad \frac{dy}{dl} = Bz^2, \tag{12.32}$$

where A and B are constants. The third component of the geodesic equation (12.31) is

$$\frac{d}{dl} \frac{1}{z^2} \frac{dz}{dl} = -\frac{1}{z^3} \left[(A^2 + B^2)z^4 + \left(\frac{dz}{dl} \right)^2 \right], \tag{12.33}$$

or

$$\frac{d}{dl} \frac{1}{z} \frac{dz}{dl} + (A^2 + B^2)z^2 = 0. \tag{12.34}$$

On multiplying this by $(dz/dl)/z$, we get a total derivative, with integral

$$\frac{1}{z^2} \left(\frac{dz}{dl} \right)^2 = \frac{1}{R^2} - (A^2 + B^2)z^2. \tag{12.35}$$

The first term on the right is the constant of integration; its value is set by the condition $g^{(0)}_{\alpha\beta} dx^\alpha dx^\beta = dl^2$.

The integral of equation (12.35) is

$$\frac{1}{z} = R(A^2 + B^2)^{1/2} \cosh(l - l_o)/R, \tag{12.36}$$

---

[4] For another way to get this equation, note that the four-dimensional spacetime line element is $ds^2 = dt^2 - a(t)^2 g^{(0)}_{\alpha\beta} dx^\alpha dx^\beta$. The space components of the four-dimensional geodesic equation (9.21) reduce to the geodesic equation in three dimensions in equation (12.31) if we set $ds = Ka^2 dl$, with K a constant. Since $ds^2 = dt^2 - a^2 dl^2$, that says the proper peculiar velocity is $v = adl/dt = (1 + K^2 a^2)^{-1/2}$. This is the general expression for the peculiar velocity of a freely moving particle whose proper momentum varies as $p \propto 1/a(t)$.

with l_o a constant. This expression in equation (12.32) gives

$$x = x_o + \frac{A \tanh(l - l_o)/R}{R(A^2 + B^2)},$$

$$y = y_o + \frac{B \tanh(l - l_o)/R}{R(A^2 + B^2)}.$$

(12.37)

The constants of integration are x_o and y_o. The result of eliminating l from these equations is

$$\frac{(x - x_o)^2}{A^2} = \frac{(y - y_o)^2}{B^2} = \frac{1}{R^2(A^2 + B^2)^2} - \frac{z^2}{A^2 + B^2}.$$

(12.38)

Equations (12.36) to (12.38) are the geodesics in the coordinates of the line element (12.28).

Spherical Trigonometry

If two objects are observed at cosmological redshifts z_1 and z_2 and at angular separation θ in the sky, what is the distance between the objects in a given Robertson-Walker line element? The conversion from redshift to coordinate distance is discussed in the next section. Here we consider the relations among the coordinate distances and angles defined in figure 12.1. The observer at $\mathcal{O}$ in the figure sees an object $\mathcal{P}$ at coordinate distance χ_1. The line $\mathcal{O}\mathcal{P}$ is the spatial part of the geodesic of the light ray that travels from source to observer. The line $\mathcal{O}\mathcal{O}'$ is the geodesic running to another observer $\mathcal{O}'$ at coordinate distance χ_0. The observer $\mathcal{O}$ sees that the angular distance between the lines of sight to $\mathcal{P}$ and to $\mathcal{O}'$ is the polar angle θ. The observer $\mathcal{O}'$ sees that the object $\mathcal{P}$ is at coordinate distance χ_2 at polar angle θ' from the line $\mathcal{O}\mathcal{O}'$.

The relations among these lengths and angles are the same as for the usual spherical trigonometry on a two-dimensional sphere (eqs. [3.42] and [3.43]). This is because our three-dimensional space can be embedded in a four-dimensional flat space, as indicated in equation (12.3), and we can place the triangle in the hypersurface $z = 0$. Then equations (12.2) and (12.3) become identical to what we would write down for great circles on a two-dimensional sphere.

To be more specific, choose the Cartesian coordinate system x_1, y_1, z_1, and w_1 in equation (12.3) so $\mathcal{O}$ is at the pole of the angular coordinates defined in equation (12.4), at $\chi_1 = 0$ with $x_1 = y_1 = z_1 = 0$ and $w_1 = R$. We can orient the coordinates so the second observer $\mathcal{O}'$ at distance χ_0 is at polar angle $\theta_1 = 0$, with Cartesian position coordinates $x_1 = y_1 = 0$ and $z_1 = R \sin \chi_0$. The object $\mathcal{P}$ is at polar angles χ_1, θ and ϕ in this coordinate system.

Now introduce a second set of coordinates, x_2, y_2, z_2, and w_2, centered on $\mathcal{O}'$, so the Cartesian coordinates of $\mathcal{O}'$ are $x_2 = y_2 = z_2 = 0$. These new coordinates are

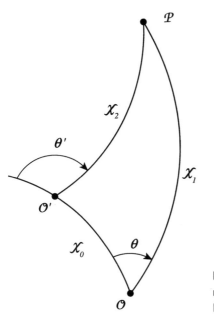

Figure 12.1. Spherical triangle for coordinate distances and angular separations in the Robertson-Walker line element.

translated along the z_1 axis from the original ones, so we can stipulate that the coordinates of any point satisfy

$$x_2 = x_1, \qquad y_2 = y_1. \tag{12.39}$$

In these new coordinates the polar angles of $\mathcal{P}$ are χ_2, θ' and ϕ. Equation (12.39) keeps the azimuthal angle ϕ the same. The polar angles θ and θ' are shown in figure 12.1.

The Cartesian coordinates are related by the rotation

$$
\begin{aligned}
w_2 &= w_1 \cos \chi_0 + z_1 \sin \chi_0, \\
z_2 &= z_1 \cos \chi_0 - w_1 \sin \chi_0,
\end{aligned}
\tag{12.40}
$$

with equations (12.39) for the x and y components. To check this, note first that these relations preserve equations (12.2) and (12.3). Second, $\mathcal{O}'$ is at $x_1 = y_1 = 0$, $z_1 = R \sin \chi_0$, $w_1 = R \cos \chi_0$, and one sees that equation (12.40) correctly brings these coordinates to $w_2 = R$ and $z_2 = 0$ for the position of $\mathcal{O}'$ in the second coordinate system.

Now we can write down the relations between the polar coordinates of $\mathcal{P}$ as observed by $\mathcal{O}$ and $\mathcal{O}'$. The equations $x_1 = x_2$ and $y_1 = y_2$ in equation (12.4) for the polar angles are

$$\sin \chi_1 \sin \theta = \sin \chi_2 \sin \theta' . \tag{12.41}$$

This is the law of sines in spherical trigonometry.

The first of equations (12.40) expressed in polar coordinates is

$$\cos \chi_2 = \cos \chi_1 \cos \chi_0 + \sin \chi_1 \sin \chi_0 \cos \theta . \tag{12.42}$$

This is the law of cosines for the sides of the spherical triangle shown in figure 12.1.

For negative curvature, the prescription in equation (12.6) brings the law of cosines to

$$\cosh \chi_2 = \cosh \chi_1 \cosh \chi_0 - \sinh \chi_1 \sinh \chi_0 \cos \theta , \tag{12.43}$$

and similarly for the law of sines.

Another way to derive equations (12.41) and (12.42) is to write out the solutions to the differential equation (12.31) for a geodesic path expressed in polar coordinates. This is left as an interesting if somewhat lengthy exercise.

If $\theta \ll 1$ in equation (12.42), and $\chi_0 = \chi_1 = \chi$, then the result of expanding $\cos \theta$ and $\cos \chi_2$ to second order in the small arguments is

$$\chi_2 = \theta \sin \chi . \tag{12.44}$$

This gives the angular size, θ, of an object with coordinate width χ_2 at coordinate distance χ.

To get the infinitesimal version of the coordinate transformation in figure 12.1, let $\chi_0 = \epsilon \ll 1$, and write $\chi_2 = \chi_1 + \delta \chi$. Then the law of cosines in equation (12.42) to first order in ϵ and $\delta \chi$ is

$$\delta \chi = \chi_2 - \chi_1 = -\epsilon \cos \theta . \tag{12.45}$$

This expression in the law of sines gives

$$\delta \theta = \theta' - \theta = \epsilon \sin \theta \cot \chi_1 . \tag{12.46}$$

With $\delta \phi = 0$, these are the components of a translation ξ^α of the coordinates.

In a general infinitesimal coordinate transformation,

$$\bar{x}^\alpha = x^\alpha - \xi^\alpha , \tag{12.47}$$

the change in the metric tensor as a function of position follows as in equation (10.27):

$$\bar{g}_{\alpha\beta}(x) = g_{\alpha\beta}(x) + g_{\alpha\beta,\gamma}\xi^\gamma + g_{\alpha\gamma}\xi^\gamma{}_{,\beta} + g_{\gamma\beta}\xi^\gamma{}_{,\alpha} . \tag{12.48}$$

A symmetry of the space under a rotation or translation means the vectors ξ^α can be chosen so the new metric tensor is the same function as the old one, $\bar{g}_{\alpha\beta}(x) = g_{\alpha\beta}(x)$. This gives the Killing equation

$$g_{\alpha\beta,\gamma}\xi^\gamma + g_{\alpha\gamma}\xi^\gamma_{,\beta} + g_{\gamma\beta}\xi^\gamma_{,\alpha} = 0. \tag{12.49}$$

It is left as an exercise to check that the Killing field in equations (12.45) and (12.46),

$$\xi^\chi = \cos\theta, \quad \xi^\theta = -\sin\theta\cot\chi, \quad \xi^\phi = 0, \tag{12.50}$$

satisfies the Killing equation (12.49). The existence of this field reflects the translational symmetry of the Robertson-Walker line element.

De Sitter Solution

In the de Sitter solution the material density and pressure are negligibly small, so the Einstein field equation is

$$R_{ij} - \frac{1}{2}g_{ij}R = \Lambda g_{ij}, \tag{12.51}$$

and the solution is spatially homogeneous and isotropic. The main purpose here is to check that the solution also is invariant under velocity transformations, as one would expect from the fact that the source term on the right-hand side of the equation is proportional to the metric tensor, which has the same symmetry. The easy way to do this shows some other elegant properties of the solution.

Since the de Sitter solution is spatially homogeneous and isotropic, the line element has to be the Robertson-Walker form in equation (12.9), and the solution in the last section (eq. [11.56]) shows that the expansion parameter satisfies the usual cosmological equations with zero mass density. To simplify the equations, let us choose the length unit so

$$\Lambda/3 = 1. \tag{12.52}$$

Then the cosmological equations (5.15) and (5.18) are

$$\frac{\ddot{a}}{a} = 1, \quad \frac{\dot{a}^2}{a^2} = 1 - \frac{1}{a^2 R^2}. \tag{12.53}$$

Different solutions to these equations are different coordinate labelings of the de Sitter model. The solution $a \propto \exp t$ for the first equation says $R^{-2} = 0$ in the second, so the line element is

$$ds^2 = dt^2 - e^{2t}(dr^2 + r^2\,d\Omega). \tag{12.54}$$

This is the form discussed in section 5 in connection with the discovery of the expansion of the universe (eq. [5.24]), and in section 7 for the steady-state cosmology. Another solution to the first part of equation (12.53) is $a = \cosh t$. This solution requires $R^{-2} = 1$ in the second part, so the line element is

$$ds^2 = dt^2 - \cosh^2 t \left(\frac{dr^2}{1 - r^2} + r^2 d\Omega \right)$$

$$= dt^2 - \cosh^2 t \, (d\chi^2 + \sin^2 \chi d\Omega),$$

$$(12.55)$$

with $r = \sinh \chi$. Yet another solution is $a = \sinh t$, which requires $R^{-2} = -1$. Since the space curvature is negative, the line element belonging to this solution is

$$ds^2 = dt^2 - \sinh^2 t \, (d\chi^2 + \sinh^2 \chi d\Omega). \qquad (12.56)$$

All these forms refer to parts of the same spacetime, which can be embedded in a five-dimensional space with Cartesian coordinates a, b, c, u, v, and line element

$$-ds^2 = da^2 + db^2 + dc^2 - du^2 + dv^2. \qquad (12.57)$$

The four-dimensional de Sitter spacetime is the surface

$$1 = a^2 + b^2 + c^2 - u^2 + v^2. \qquad (12.58)$$

In the plane $b = 0 = c$, we have $1 + u^2 = a^2 + v^2$, which is the equation for a two-dimensional hyperboloid of revolution about the u axis.

We can define polar coordinates for the five-dimensional space by the equations

$$
\begin{aligned}
u &= \sinh t, \\
v &= \cosh t \cos \chi, \\
c &= \cosh t \sin \chi \cos \theta, \\
a &= \cosh t \sin \chi \sin \theta \cos \phi, \\
b &= \cosh t \sin \chi \sin \theta \sin \phi.
\end{aligned}
$$

$$(12.59)$$

These points lie on the unit surface in equation (12.58), and they cover the whole surface with the range of arguments

$$
\begin{aligned}
-\infty &< t < \infty, \\
0 &< \chi < \pi, \\
0 &< \theta < \pi, \\
0 &< \phi < 2\pi.
\end{aligned}
$$

$$(12.60)$$

Following the usual method of converting the line element from Cartesian to polar coordinates, one sees that equation (12.57) in these new coordinates becomes

$$ds^2 = dt^2 - \cosh^2 t \left[d\chi^2 + \sin^2\chi \left(d\theta^2 + \sin^2\theta d\phi^2 \right) \right] . \tag{12.61}$$

This is the line element in equation (12.55).

Another set of coordinates is defined by the equations

$$u = \sinh t + r^2 e^t / 2 ,$$
$$v = \cosh t - r^2 e^t / 2 ,$$
$$a = xe^t , \tag{12.62}$$
$$b = ye^t ,$$
$$c = ze^t ,$$

with $r^2 = x^2 + y^2 + z^2$. It is easy to check that this places the points on the unit sphere in equation (12.58), and that it brings the line element (12.57) to the form

$$ds^2 = dt^2 - e^{2t} \left(dx^2 + dy^2 + dz^2 \right) , \tag{12.63}$$

which is equation (12.54). This coordinate labeling does not cover all of the four-dimensional surface in equation (12.58), for we see that it is restricted to the range $u + v = e^t \geq 0$.

It is an amusing exercise, if to your taste, to find the coordinates that yield the third of the forms for the line element in equation (12.56).

We can rotate the surface in equation (12.58) about the u axis to generate space translations, as in equation (12.40) for the general Robertson-Walker metric. Just as in a Lorentz transformation, we can rotate about other axes to generate velocity translations. The infinitesimal transformation

$$u' = u - \epsilon c , \qquad c' = c - \epsilon u , \tag{12.64}$$

with $|\epsilon| \ll 1$, and

$$a' = a , \qquad b' = b , \qquad v' = v , \tag{12.65}$$

keeps the points on the surface (12.58) and it preserves the line element (12.57), so it is an allowed rotation. In the coordinates of equation (12.59), the first of equations (12.64) is

$$\sinh t' = \sinh t - \epsilon \cosh t \sin \chi \cos \theta . \tag{12.66}$$

Since ϵ is small, this is

$$t' = t - \epsilon \sin \chi \cos \theta$$
$$= t - \epsilon c / \cosh t.$$

(12.67)

The condition $v' = v$ similarly gives

$$\chi' = \chi - \epsilon \tanh t \cos \chi \cos \theta,$$

(12.68)

and we get from the second part of equation (12.64)

$$\theta' = \theta + \epsilon \tanh t \sin \theta / \sin \chi.$$

(12.69)

At $|t| \ll 1$, equations (12.64) and (12.67) are

$$c' = c - \epsilon t, \qquad t' = t - \epsilon c.$$

(12.70)

This is an infinitesimal Lorentz transformation at velocity ϵ along the c axis.

The elegant symmetry in equations (12.57) and (12.58), which allows the rotation in equation (12.64), shows that a pure de Sitter solution does not define a preferred frame of motion. This is the cause of the dilemma discussed in section 5 — that one can define motions of test particles consistent with the cosmological principle, as was shown by Weyl (1923), Lemaître (1925), and Robertson (1928), but there really is nothing in the geometry to dictate this prescription. The general Robertson-Walker line element reduces the symmetry to the space translations and rotations allowed by the sphere in equation (12.3), thus defining a preferred motion for the material content of the universe. The classical steady-state cosmology has a pure de Sitter line element (eq. [7.4]). This means some new provision is needed to define the motion of the newly created matter, for the geometry does not do it. In the inflation scenario in section 17 the geometry during inflation is close to de Sitter, and it is the gradient of an inflaton field that tells the entropy created at the end of inflation how to move.

13.   Neoclassical Cosmological Tests

The purpose of the classical cosmological tests is to derive from observations such as counts and angular sizes of distant objects measures of the global cosmological parameters, such as the mean mass density and the space curvature of the cosmological model. If we knew the parameters well enough we could predict the future of the universe, whether it will collapse back to a big crunch or expand into an indefinitely prolonged future where bound systems of galaxies move arbitrarily far apart and the quiet is punctuated only by the occasional collapse of a bound system into a black hole. The concept is romantic, but the prediction will be believable only when the physics and the astronomy are a good deal better established. For that purpose, the interesting science is in the attempts to unravel what has happened from whatever fossils might be uncovered. An important element of the program is the search for independent and redundant measures of the parameters from which we can test the consistency of the standard relativistic expanding cosmological model.

In the cosmological tests to be discussed here the emphasis is on discrete objects such as galaxies and quasars, but the relations are relevant to features in diffuse radiation backgrounds, such as the angular scale of fluctuations of the thermal cosmic background radiation produced by scattering in plasma in galaxies at high redshift.

The basic ideas of the tests were recognized with the discovery of the expanding world model, and one of the original purposes for the 200-inch telescope on Mount Palomar, which was designed in the 1930s, was their application. Sandage (1987, 1988) reviews the history of the results. It is sobering to reflect that now, some six decades after the work started, the cosmological parameters still are only very loosely constrained.

Despite past experience there is growing interest in a new round of applications of the tests. This is in part because it is at last feasible to observe galaxies at redshifts well in excess of unity, where the effects of the cosmological model become quite significant, and in part because it has become possible to apply the tests in new and quite powerful ways, as in gravitational lensing and the superluminal motions in quasars. That is why this section refers to the tests as "neoclassical."

Along with the much greater power of the tests applied at higher redshifts there is the greater chance for systematic error from misinterpretations of how the objects are behaving. For example, the counts of faint galaxies are surprisingly high for an Einstein-de Sitter model. Is this telling us we should consider a cosmological model with larger space volume per unit increase in redshift, or that we have underestimated the abundances of luminous galaxies at high redshifts? It is easy to imagine how the latter could have happened: perhaps galaxies were brighter in the past, perhaps merging of galaxies is reducing their abundances. There are ways to decide, for an adjustment of free parameters to secure agreement with one set of observations may spoil another. Thus the physical state of low redshift galaxies offers constraints on their merger histories, as does the redshift-magnitude relation. A convincing application of the cosmological tests will yield redundant and concordant constraints on the cosmology and the picture for the evolution of the objects used in the tests. This section discusses elements of the rich suite of observational programs that are feasible now or perhaps in the reasonably well understood future. They are sure to teach us something of value; we will have to wait to see whether the lessons center on the nature of the large-scale structure of the universe and its cosmological parameters or on the nature of the galaxies.

The numerical relations presented in the figures in this section assume the material pressure in matter and radiation (excluding that of a cosmological constant) is small compared to the mass density. At observationally interesting values of the redshift the CBR pressure is small. The generalization to models with high material pressure at low redshifts, perhaps from a sea of degenerate low mass neutrinos, is not thought to be interesting because the pressure reduces the predicted age of the universe, which at the time this is written already is uncomfortably low unless the mass density is appreciably lower than the critical Einstein-de Sitter value. But this is a situation that has evolved, and if it does so again the generalization of the numerical relations in the figures to models with pressure will be an interesting exercise for the reader.

The analysis presented here treats the universe as homogeneous apart from the mass concentration in an observed object. The possible effects of density fluctuations along the line of sight are considered in the next section.

To establish notation, let us recall that the negative curvature line element is

$$
\begin{aligned}
ds^2 &= dt^2 - a(t)^2 \left[\frac{dr^2}{1+r^2/R^2} + r^2(d\theta^2 + \sin^2\theta\, d\phi^2) \right] \\
&= dt^2 - a^2 R^2 \left[d\chi^2 + \sinh^2\chi(d\theta^2 + \sin^2\theta\, d\phi^2) \right] ,
\end{aligned}
\tag{13.1}
$$

with $R \to iR$ and $\chi \to -i\chi$ for positive curvature. The cosmological equations for the expansion rate are (eqs. [5.15] and [5.18])

$$
\begin{aligned}
\frac{\ddot{a}}{a} &= -\frac{4}{3}\pi G(\rho_b + 3p_b) + \frac{\Lambda}{3} , \\
\left(\frac{\dot{a}}{a}\right)^2 &= \left(\frac{\dot{z}}{1+z}\right)^2 = \frac{8}{3}\pi G\rho_b + \frac{1}{a^2 R^2} + \frac{\Lambda}{3} .
\end{aligned}
\tag{13.2}
$$

If the mean mass density is dominated by nonrelativistic matter, so $p_b \ll \rho_b \propto a^{-3}$, we can rewrite these equations as (eq. [5.53])

$$
\begin{aligned}
\frac{\ddot{a}}{a} &= H_o{}^2 \left[\Omega_\Lambda - \Omega(1+z)^3/2 \right] , \\
\frac{\dot{a}}{a} &= H_o E(z) = H_o \left[\Omega(1+z)^3 + \Omega_R(1+z)^2 + \Omega_\Lambda \right]^{1/2} .
\end{aligned}
\tag{13.3}
$$

The redshift is $1 + z = a_o/a(t)$ (eq. [5.45]). The fractional contributions to the present value of Hubble's constant H_o by the present mean mass density ρ_o, the radius of curvature $a_o R$ of space sections at fixed world time, and the cosmological constant are

$$
\Omega = \frac{8\pi G\rho_o}{3H_o^2} , \qquad \Omega_R = \frac{1}{(H_o a_o R)^2} , \qquad \Omega_\Lambda = \frac{\Lambda}{3H_o^2} ,
\tag{13.4}
$$

with $\Omega + \Omega_R + \Omega_\Lambda = 1$. In this notation, the relation between the coordinate distances r and χ in equation (13.1) is

$$H_o a_o r = \Omega_R^{-1/2} \sinh \chi . \tag{13.5}$$

In a closed geometry, $\sinh \chi$ becomes $\sin \chi$.

As will be discussed in section 15, the Einstein-de Sitter model, in which space curvature and the cosmological constant both are negligibly small, is preferred on several grounds. First, it is the simplest case. Second, inflation as an explanation for the large-scale homogeneity of the observable universe requires that space curvature is negligibly small (section 17). If we accept this argument we still can consider a nonzero cosmological constant, though a value that would be interesting in cosmology is not considered plausible within the particle physics community (Carroll, Press, and Turner 1992). Within physical cosmology it nonetheless is sensible to consider measurements of all three of the parameters in equation (13.4) that might contribute to the expansion rate. If it could be shown that independent measurements of the parameters by means of different cosmological tests of the kind discussed here give consistent results, it would be an exceedingly important positive test of the standard model, as well as a believable demonstration of what the values of the parameters really are.

The Dicke coincidences argument offers a way to approach the tests. As discussed in section 15, the parameters Ω, Ω_R, and Ω_Λ are different functions of the world time at which they are evaluated. It would be a curious coincidence if two of the parameters were on the order of unity at the present epoch, and it would be an exceedingly special case if all three happened to be of order unity. This suggests that it is reasonable to consider first those models in which only two of the parameters are significantly different from zero (Peebles 1984c; Kofman and Starobinsky 1985; Carroll, Press, and Turner 1992). In the more commonly discussed case, the cosmological constant vanishes, and the one free parameter, Ω, fixes the mass density and space curvature. In the other case, motivated by inflation, space curvature is negligibly small, and the free parameter Ω fixes the term that is (or acts like) a cosmological constant. These are the cases, both with negligible material pressure, in the numerical examples in this section.

Lookback Time

At low redshifts, we can write the expansion parameter as a function of world time t as

$$a(t) = a_0 \left[1 - H_o(t_o - t) - q_o H_o^2 (t_o - t)^2 / 2 + \cdots \right] , \tag{13.6}$$

where $t_o - t$ is the lookback time (the time measured back from the present world time, t_o). Hubble's constant is $H_o = \dot{a}_o/a_o$ (eq. [5.4]), and the dimensionless acceleration (or deceleration) parameter is

$$q_o = -\frac{\ddot{a}_o a_o}{\dot{a}_o{}^2} = \frac{\Omega}{2} - \Omega_\Lambda. \tag{13.7}$$

The last step uses equation (13.3); it assumes the material pressure is small compared to the mass density. With $1 + z = a_o/a(t)$, we can rewrite equation (13.6) as an expression for the lookback time as a function of redshift,

$$H_o(t_o - t) = z - (1 + q_o/2)z^2 + \cdots. \tag{13.8}$$

The more general expression for the time evolution is

$$H_o t(z) = H_o \int_0^a \frac{da}{\dot{a}} = \int_0^a \frac{da}{aE(z)} = \int_z^\infty \frac{dz}{(1+z)E(z)}. \tag{13.9}$$

The dimensionless function $E(z)$ is given by equation (13.3). This expression is easy enough to integrate numerically; a useful array of results for the expansion parameter as a function of world time is given by Felten and Isaacman (1986).

Examples of the lookback time, $t_o - t$, as a function of redshift are shown in figure 13.1. Panel (a) assumes space curvature is negligible, $|\Omega_R| \ll 1$, so the one free parameter is the density parameter in matter, Ω, which fixes the cosmological constant. Panel (b) assumes the cosmological constant can be ignored, again leaving the one free parameter, Ω.

The horizontal lines on the right-hand axes are the limiting values of the time t_o computed from high redshift. If $\Lambda = 0$ and the matter pressure is not negative, then $H_o t_o \leq 1$; if space curvature is negligible, $H_o t_o$ can be somewhat larger than unity in the range of estimates of Ω discussed in section 20 from the local dynamical measures of the mean mass density. Measurements of H_o and t_o that fix the product $H_o t_o$ to 10% accuracy would provide an exceedingly valuable constraint on the cosmological models.

Some analytic results for the expansion parameter as a function of time are worth recording. This is the subject of the remainder of the subsection.

If the cosmological constant can be ignored, and the material pressure is small so the mass density varies as $\rho \propto a(t)^{-3}$, then the first line of equation (13.2) is $\ddot{a} = -\kappa/a^2$, with κ a constant. This is the equation for radial motion in an inverse square force; the solution can be written as the parametric equation for a cycloid, $a = A(\cosh \eta - 1)$ and $t = B(\sinh \eta - \eta)$, with A and B constants. The result of substituting in equation (13.3) and matching constants is

$$\frac{a(t)}{a_o} = \frac{\Omega}{2(1-\Omega)}(\cosh \eta - 1),$$

$$H_o t = \frac{\Omega}{2(1-\Omega)^{3/2}}(\sinh \eta - \eta), \tag{13.10}$$

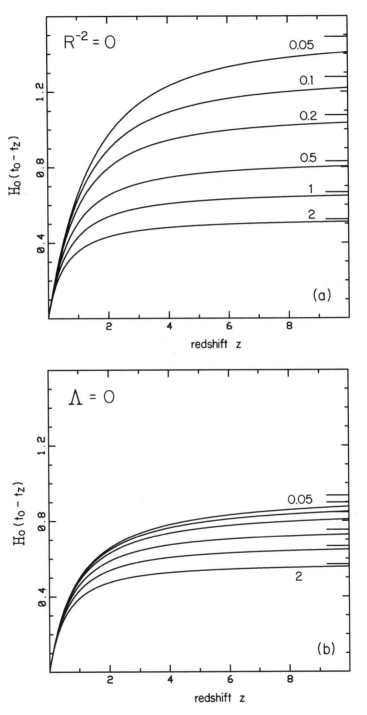

Figure 13.1. Lookback time as a function of redshift. The long dashes on the right-hand axis show the age t_o of the universe computed from $z \to \infty$. In panel (a) space curvature is negligible, and in panel (b) the cosmological constant, Λ, is negligibly small. The curves are labeled by the density parameter, Ω.

for $\Lambda = 0$. The present value of the parameter, at $a = a_o$, is

$$\cosh \eta_o = 2\Omega^{-1} - 1. \qquad (13.11)$$

This continues to the closed case where $\Omega > 1$ in the standard way. At redshift $1 + z = a_o/a \gg 2(\Omega^{-1} - 1)$ the parameter is $\eta \ll 1$. In this limit the lowest nontrivial expansion of the solution in powers of η yields the Einstein-de Sitter limit for the expansion time t as a function of redshift in equation (5.61).

In a model with negligible cosmological constant, and with the mass density dominated by radiation rather than nonrelativistic matter, equations (13.3) and (13.9) become

$$H_o t = \int_0^a \frac{da}{\left[\Omega a_o^4/a^2 + (1 - \Omega) a_o^2\right]^{1/2}}. \qquad (13.12)$$

The density parameter is multiplied by an extra factor of a_o/a because the mass density in radiation scales as $\rho \propto a^{-4}$. The integral is easy to evaluate (by multiplying numerator and denominator by a) to get

$$H_o t = \frac{\Omega^{1/2}}{1 - \Omega}\left[\left(1 + \frac{1 - \Omega}{\Omega}\frac{a^2}{a_o^2}\right)^{1/2} - 1\right]. \qquad (13.13)$$

This is equivalent to the parametric form

$$\frac{a}{a_o} = \frac{\sinh \eta}{(\Omega^{-1} - 1)^{1/2}}, \qquad H_o t = \frac{\Omega^{1/2}}{1 - \Omega}(\cosh \eta - 1), \qquad (13.14)$$

with $\Lambda = 0$. If the density parameter is close to unity, this radiation-dominated case gives $H_o t_o = 1/2$.

Let us consider next the situation in the standard model at redshift $100 \lesssim z \lesssim 10^9$, where the cosmological constant and space curvature both are negligibly small compared to the mass density, and the mass density is the sum of a nonrelativistic component from the baryons (and perhaps also exotic dark matter) and the relativistic contributions from the thermal cosmic background radiation and the relict neutrinos. Since the mass density in the relativistic components varies as a^{-4}, the relativistic components add to the expression in equation (13.3) for E^2 the term $\Omega_r(1 + z)^4$, where Ω_r is the present fractional contribution to the expansion rate $\dot{a}^2/a^2$. This brings equation (13.9) to

$$H_o t = \int_0^a \frac{da}{\left[\Omega a_o^3/a + \Omega_r a_o^4/a^2\right]^{1/2}}. \qquad (13.15)$$

The result of multiplying the numerator and denominator of equation (13.15) by a, working the integral, and rearranging, is

$$H_o t = \frac{1}{\Omega^{1/2}} \left(\frac{a_{eq}}{a_o} \right)^{3/2} \left[\frac{2}{3} \left(1 + \frac{a}{a_{eq}} \right)^{3/2} - 2 \left(1 + \frac{a}{a_{eq}} \right)^{1/2} + \frac{4}{3} \right]. \quad (13.16)$$

The redshift at equality of mass densities in matter and radiation is

$$1 + z_{eq} = \frac{a_o}{a_{eq}} = \frac{\Omega}{\Omega_r} = 2.4 \times 10^4 \Omega h^2. \quad (13.17)$$

The last expression, from equation (6.81) for Ω_r, assumes the relativistic component is that of the cosmic background radiation and three families of massless relict neutrinos. At $a \gg a_{eq}$ this solution approaches the Einstein–de Sitter limit in equation (5.61), because we are ignoring space curvature and Λ. At $a \ll a_{eq}$ the solution is

$$H_o t = \frac{1}{2\Omega^{1/2}} \frac{a^2}{a_o^{3/2} a_{eq}^{1/2}} = \frac{1}{2\Omega_r^{1/2}} \frac{a^2}{a_o^2}. \quad (13.18)$$

This is the equivalent of the Einstein–de Sitter limit for a radiation-dominated universe.

If space curvature can be neglected, and the expansion rate is dominated by a cosmological constant and nonrelativistic matter with density that varies as $\rho \propto a^{-3}$, the time equation (13.9) is

$$H_o t = \int_0^a \frac{da}{\left[\Omega a_o^3 / a + (1 - \Omega) a^2 \right]^{1/2}}. \quad (13.19)$$

When numerator and denominator are multiplied by $a^{1/2}$ this becomes a trigonometric integral,

$$H_o t = \frac{2}{3(1 - \Omega)^{1/2}} \sinh^{-1} \left[\left(\frac{1 - \Omega}{\Omega} \right)^{1/2} \left(\frac{a}{a_o} \right)^{3/2} \right]. \quad (13.20)$$

Let us consider finally some properties of the Eddington and Lemaître models. As discussed in section 5, the former expands away from the static Einstein solution. This is the model Lemaître (1927) first used in finding an interpretation of the cosmological redshift; it seems to have been named after Eddington because he at first resisted Lemaître's move to a universe that expands from a dense state. In this Eddington model the expansion traces back to a maximum redshift z_e at which $\dot{a}$ and $\ddot{a}$ both vanish, so if the pressure may be neglected we have from equation (13.3)

$$\Omega_\Lambda = \Omega(1+z_e)^3/2 \, ,$$
$$\Omega_R = 1 - \Omega - \Omega_\Lambda = -3\Omega^{2/3}\Omega_\Lambda^{1/3}/2^{2/3} \, , \tag{13.21}$$

which give

$$\Omega = \frac{2}{z_e^2(3+z_e)} \, . \tag{13.22}$$

The maximum redshift z_e in this model would have to be greater than about 5, the maximum observed for quasars, so Ω would have to be less than about 0.01, well below the dynamical estimates in section 20. Quite apart from this observational problem, the model is not now taken seriously because it is so difficult to see how the initial condition could have been arranged.

In the Lemaître model, the universe expands from a dense state, the big bang or what Lemaître called the primeval atom, and passes through a quasistatic or coasting phase where the expansion rate and acceleration are small because the cosmological parameters almost satisfy the conditions for the static Einstein solution. In the 1930s this had the apparent advantage that the quasistatic phase increases the expansion time, but as described in section 5 it is now recognized that the timescale problem was at least in large part the result of a considerable overestimate of the value of the Hubble parameter. Lemaître also liked the fact that gravitational instability during the quasistatic phase favors mass clustering at a characteristic density, which is fixed by the mean during this phase and might be compared to the densities of clusters of galaxies. As noted in section 19, however, there is no known evidence for a preferred mass density in the spectrum of galaxy clustering. Interest in the Lemaître model was revived by indications of a peak in the quasar redshift distribution at $z \sim 2$ that has now gone away, but could have resulted from a long path length through a Lemaître quasistatic phase (Petrosian, Salpeter, and Szekeres 1967; Shklovsky 1967). More recently, it has been noted that the enhanced gravitational growth of clustering during the quasistatic phase could help reconcile the observed large-scale clustering of galaxies with the small anisotropy of the cosmic background radiation (Sahni, Feldman, and Stebbins 1992). But following the Dicke coincidences argument, it would seem surprising if the primeval atom produced the exceedingly special initial conditions for a coasting phase.

Let us see how the expansion time depends on the parameters in the Lemaître model. Since the parameters are close to those in equation (13.21), it is convenient to write

$$\Omega_R = -3(1 - \epsilon)\Omega^{2/3}\Omega_\Lambda^{1/3}/2^{2/3} \, , \tag{13.23}$$

where ϵ is a small positive number, and

$$a = x a_o \left(\frac{\Omega}{2\Omega_\Lambda} \right)^{1/3} . \tag{13.24}$$

Then for nonrelativistic matter the expansion time is

$$H_o t = \frac{1}{\Omega_\Lambda^{1/2}} \int_0^x \frac{dx}{(2/x + x^2 - 3 + 3\epsilon)^{1/2}} . \tag{13.25}$$

We are interested in the effect of ϵ on the expansion time, so consider

$$H_o \frac{\partial t}{\partial \epsilon} = -\frac{3}{2\Omega_\Lambda^{1/2}} \int_0^x \frac{dx}{(2/x + x^2 - 3 + 3\epsilon)^{3/2}} . \tag{13.26}$$

At small ϵ this integral is dominated by the part near $x = 1$; with the lowest nontrivial terms in the series expansion of the denominator around $x = 1$, we get

$$H_o t = \frac{1}{(3\Omega_\Lambda)^{1/2}} \ln \epsilon^{-1} . \tag{13.27}$$

This shows that a dwell time much larger than the Hubble time requires a very small value of ϵ, meaning a very close coincidence of the cosmological parameters. A similar analysis for the gravitational growth of linear mass fluctuations is given in LSS, § 13.

Angular Size Distance

The coordinate distance $r = R \sinh \chi$ for an object, with the observer at the origin of the coordinates in equation (13.1), is called the angular size distance, because the observed angular size of the object is inversely proportional to r, as shown in equation (12.44). To get the angular size distance as a function of redshift, note that a light ray moves in the radial direction with $ds = 0$, or $dt = aR d\chi$, so the coordinate distance from emission of the radiation at the source, at time t_e, to detection at time t_o, is

$$R\chi = \int_{t_e}^{t_o} \frac{dt}{a} = \int_{a_e}^{a_o} \frac{da}{a\dot{a}} = \frac{1}{H_o} \int_{a_e}^{a_o} \frac{da}{a^2 E(z)} = \frac{1}{H_o a_o} \int_0^{z_e} \frac{dz}{E(z)} , \tag{13.28}$$

where E is given by equation (13.3). The angular size distance is then

$$y(z_e) \equiv H_o a_o r(z_e) = H_o a_o R \sinh \left[\frac{1}{H_o a_o R} \int_0^{z_e} \frac{dz}{E(z)} \right] . \tag{13.29}$$

If space curvature is positive, the function changes to a sine. If space curvature is negligible, the function is

$$y = H_o a_o r(z) = \begin{cases} \int_0^z dz/E(z) & \text{for } \Omega_R = 0, \\ 2\left[1 - (1+z)^{-1/2}\right] & \text{for } \Omega_R = \Omega_\Lambda = 0. \end{cases} \qquad (13.30)$$

In the series expansion in equations (13.6) and (13.8), the angular size distance in equation (13.29) is

$$y = H_o a_o r = z - (1 + q_o)z^2/2 + \cdots . \qquad (13.31)$$

The effect of space curvature appears only in order z^3, because χ and $\sinh \chi$ differ in order χ^3, so this is not an effect one would look for in observations at redshifts below unity. At $z \gtrsim 1$ the angular size distance is sensitive to the geometry, as is illustrated in figure 13.2.

If the cosmological constant is negligible, there is a useful analytic solution to the angular size distance as a function of redshift (Mattig 1958). With $\Lambda = 0$, we can rewrite the integrand in equation (13.28) as

$$\frac{dr^2}{1 + H_o^2 a_o^2(1 - \Omega)r^2} = \frac{dt^2}{a^2} = \frac{da^2}{H_o^2[\Omega a_o^3 a + (1 - \Omega)a_o^2 a^2]} . \qquad (13.32)$$

The first equation says the path of the light ray satisfies $ds = 0$ with $d\theta = d\phi = 0$ in the first line of equation (13.1). The expansion rate in the last expression is given by equation (13.3), and the parameter $\Omega_R = 1 - \Omega$ in equation (13.4) has been used to eliminate R^{-2}. On setting

$$H_o a_o (1 - \Omega)^{1/2} r = x , \qquad 2(\Omega^{-1} - 1)a/a_o = b - 1 , \qquad (13.33)$$

we get

$$\frac{dx^2}{1 + x^2} = \frac{db^2}{b^2 - 1} . \qquad (13.34)$$

With $x = \sinh \chi$ and $b = \cosh \beta$, this is $\chi = c - \beta$. The constant of integration is $c = \beta_o$, where β_o is the value of β in equation (13.33) at $a = a_o$. This satisfies the wanted boundary condition, that $x = 0$ at $a = a_o$ and x is positive at $a < a_o$. We have then

$$x = \sinh(\beta_o - \beta) = \sinh \beta_o \cosh \beta - \sinh \beta \cosh \beta_o$$
$$= b\sqrt{b_o^2 - 1} - b_o\sqrt{b^2 - 1} . \qquad (13.35)$$

The result of using equation (13.33) to replace x with r and b with a/a_o is

Mattig's relation for the angular size distance r of an object at redshift z in a model with $\Lambda = 0$,

$$y = H_o a_o r(z) = \frac{2[2 - \Omega + \Omega z - (2 - \Omega)(1 + \Omega z)^{1/2}]}{\Omega^2(1 + z)} \,. \tag{13.36}$$

The inverse relation is

$$1 + z = \frac{y + \Omega/2 - y\Omega + (1 - \Omega/2)[1 + y^2(1 - \Omega)]^{1/2}}{(1 - y\Omega/2)^2} \,. \tag{13.37}$$

At $\Omega \to 0$ the angular size distance in the $\Lambda = 0$ model is

$$y = H_o a_o r = \frac{z(1 + z/2)}{1 + z} \qquad 1 + z = y + \sqrt{1 + y^2} \,. \tag{13.38}$$

Figure 13.2 shows the angular size distance as a function of redshift for the two special cases shown in figure 13.1. In panel (b), for negligible Λ, the distance is given by equation (13.36). Panel (a), for negligible space curvature, is based on a numerical integration of equation (13.29).

The limiting value of the angular size distance in equation (13.36) at large redshift is

$$H_o a_o r = 2/\Omega, \quad \text{for } z \gg \Omega^{-1}, \Lambda = 0 \,. \tag{13.39}$$

This is shown as the dashed line in figure 13.3. The solid line shows the limiting value for negligible space curvature.

The Neoclassical Tests

In the standard cosmological model the predicted relations among observables for distant objects are determined by two functions of redshift, the angular size distance $r(z)$ in figure 13.2 and the lookback time function in figure 13.1. The latter is equivalent to the rate of proper displacement with redshift along the path of a light ray, where the proper displacement dl in world time dt is

$$dl = dt = \frac{da}{\dot{a}} = \frac{dz}{1 + z} \frac{a}{\dot{a}} \,. \tag{13.40}$$

With equation (13.3) for the expansion rate, this is the radial displacement function

$$\frac{dl}{dz} = \frac{H_o^{-1}}{(1 + z)E(z)} \,. \tag{13.41}$$

The lookback time is the integral of the radial displacement function dl/dz. In cosmological tests based on this function, as shown in figure 13.1, one compares

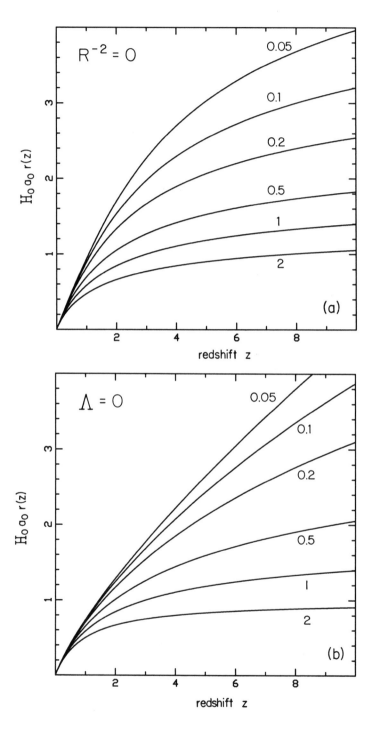

Figure 13.2. Angular size distance r as a function of redshift. The parameters are arranged as in figure 13.1.

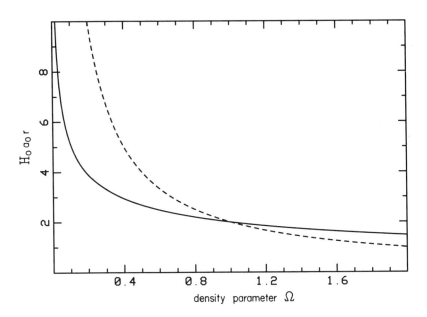

Figure 13.3. Angular size distance r at $z \to \infty$ as a function of the density parameter. The solid curve assumes negligible space curvature, the dashed curve, $\Lambda = 0$.

maximum observed physical ages, from stellar evolution and from the radioactive decay ages of the elements, to the lookback time at large z, and one compares the lookback times at lower redshifts to estimates of the rate of physical evolution of objects as a function of redshift.

The radial displacement function dl/dz also determines the probability $dP = \sigma n dl$ that a line of sight intersects an object at redshift z in the interval dz from a population with number density $n = n_o(1 + z)^3$ and cross section σ. The intersection probability per unit redshift interval is

$$\frac{dP}{dz} = \sigma n \frac{dl}{dz} = \sigma n_o H_o^{-1} \frac{(1 + z)^2}{E(z)} . \tag{13.42}$$

Figure 13.4 shows how dP/dz varies with redshift. At high z, where $E \to \Omega^{1/2}(1 + z)^{3/2}$ (eq. [13.3]), the optical depth for intersection of objects up to redshift z is

$$\tau_i(z) = \int^z dP = \frac{2}{3} \frac{\sigma n_o c H_o^{-1}}{\Omega^{1/2}} (1 + z)^{3/2} . \tag{13.43}$$

The convergence to this limit is much faster when space curvature is negligible, because the two terms in equation (13.3) for $E(z)$ differ by three powers of $1 + z$.

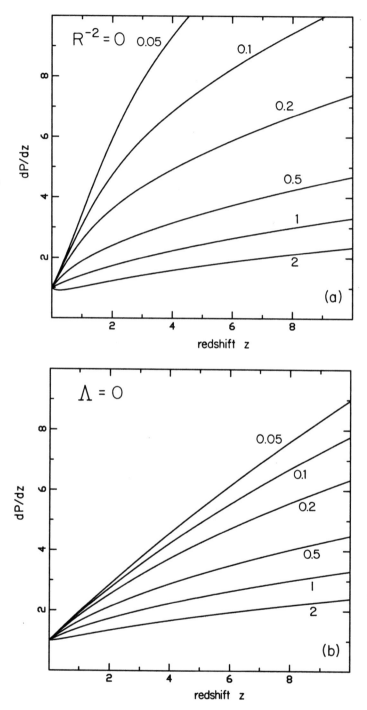

Figure 13.4. Probability for intersecting objects along a line of sight as a function of redshift. Plotted is the dimensionless part $(1+z)^2/E(z)$ of equation (13.42) for dP/dz. The arrangement of parameters is the same as in figure 13.1.

For ordinary galaxies, characteristic quantities discussed in sections 3 and 5 are

$$n_g \sim 0.02 h^3 \, \mathrm{Mpc}^{-3}, \qquad r_g \sim 10 h^{-1} \, \mathrm{kpc}, \qquad (13.44)$$

where the cross section for the bright part of the galaxy is $\sigma = \pi r_g^2$. With these numbers, the optical depth (13.43) for intersecting a galaxy within impact parameter r_g is

$$\tau_i \sim 0.01 (1+z)^{3/2} \Omega^{-1/2}. \qquad (13.45)$$

This predicts that at $z = 1$ the fraction of the sky covered by galaxies is $\tau_i \sim 0.04 \Omega^{-1/2}$, and that the sky is plated with galaxies, $\tau_i = 1$, at redshift $z_1 \sim 20 \Omega^{1/3}$. If young galaxies were more diffuse, or were in pieces that later merged, it would increase $\tau_i(z)$; if galaxies formed as luminous objects at relatively low redshift, it would decrease τ_i. The measurement and interpretation of τ_i is of considerable interest (Koo 1989; Tyson 1990; Lilly 1991; Cowie 1991). Some indications from the current results are discussed in section 26.

The intersection probability is discussed further in section 23 in connection with the redshift distribution of absorption lines in quasar spectra. This intersection probability depends on the cosmological model through the time evolution of the expansion parameter; given $a(t)$ the probability does not depend on the geometry. The same is true of the mean surface brightness of the sky due to the integrated emission from sources with a given mean luminosity density, j, as a function of time or redshift (for the surface brightness is the integral of $(1+z)^{-3} j \, dl$, as in eq. [5.159]). The same of course applies to timing measurements such as $H_o t_o$. We consider next tests based on the second function, the angular size distance.

The proper length $l_\perp$ subtended by angle θ at angular size distance $r = R \sinh \chi$ in equations (12.44) and (13.1) is

$$l_\perp = a r \theta. \qquad (13.46)$$

The angular diameter of an object of proper diameter $l_\perp$ observed at redshift z is then

$$\theta = H_o l_\perp F_\theta, \qquad F_\theta = \frac{(1+z)}{H_o a_o r}. \qquad (13.47)$$

The denominator is the dimensionless integral in equation (13.29). At high redshift the integral for the angular size distance r converges to a finite value, as shown in figure 13.3, and in this limit the angular size at fixed proper linear size varies as $\theta \propto (1+z)$. This magnification effect arises because, as we noted in section 5, the light from an object at high redshift was emitted when its proper distance $ar = a_o r / (1+z)$ was small. An object at very high redshift thus is predicted to have a large angular size because the light we detect was emitted when the object was looming over us. (Since the object was looming over us because the mass of the universe has strongly decelerated the expansion rate subsequent to the redshift when the received radiation left the object, one also can consider this an effect of magnification caused by the mass of the universe.)

Figure 13.5 shows the factor F_θ in the angular size equation (13.47). With the radius r_g in equation (13.44), the prefactor is

$$H_o r_g / c \sim 0.7 \text{ arc sec} . \tag{13.48}$$

That is, a galaxy at redshift $z \sim 1$ is expected to have an angular size on the order of one arc second, which coincidentally is comparable to the angular resolution, or seeing, permitted by our atmosphere. At $z \gtrsim 2$ the angular size at given linear size is predicted to increase with increasing redshift. Optical images of radio galaxies observed at $z \gtrsim 2$ tend to have relatively large angular sizes, but this need not be the magnification effect because the optical images are less regular than in low z galaxies and tend to line up with the radio source lobes, indicating strong evolution of the galaxies (Chambers and McCarthy 1990). Observations above the atmosphere may show that there are classes of elliptical or spiral galaxies at $z \gtrsim 1$ that look similar enough to their low redshift cousins so that there is a reasonable case for an estimate of their physical sizes relative to nearer galaxies. If so, the angular sizes will provide a very useful cosmological test. Kapahi (1989) discusses the prospects for using the angular sizes of the radio images of radio galaxies to test the relation between angular size and redshift.

The energy flux f received from a galaxy at a given redshift and luminosity is determined by the angular size distance, because the surface brightness (integrated over frequencies) varies according to the $(1+z)^{-4}$ law (eqs. [6.44] and [9.54]), and f is the integral over the angular distribution of the surface brightness. We can compute f by modeling the galaxy as a uniformly bright sphere of physical radius r_g and surface brightness i at the source (because any source distribution is a sum of little spheres). The energy flux through unit area of the surface is

$$f_g = \int_0^1 2\pi i \cos\theta \, d(\cos\theta) = \pi i , \tag{13.49}$$

and the luminosity is the product of f_g and the surface area of the sphere,

$$L = 4\pi^2 r_g^2 i . \tag{13.50}$$

The expansion of the universe brings the observed surface brightness to $i_o = i/(1+z)^4$. The angular radius of the galaxy at the observer is $\theta = r_g(1+z)/[a_o r(z)]$. The

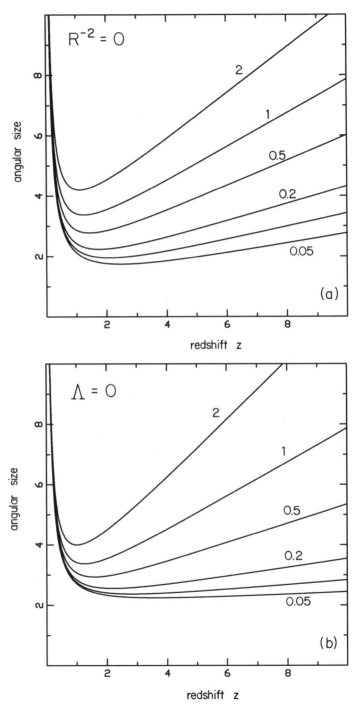

Figure 13.5. Angular size as a function of redshift. The vertical axis is the factor $F_\theta = (1+z)/H_o a_o r(z)$ in equation (13.47). The parameters are arranged as in figure 13.1.

observed energy flux is the product of the surface brightness and the solid angle the galaxy subtends, $f_o = \pi\theta^2 i_o$. Collecting, we have

$$f_o = \frac{L}{4\pi(a_o r)^2(1+z)^2} . \tag{13.51}$$

For another way to understand this relation, note that the line element (13.1) says the flux of light from a galaxy at coordinate distance r has spread over an area $4\pi(a_o r)^2$ at the present epoch. The energy flux per unit area is diminished by the factor $(1+z)^2$, because the energy of each photon is diminished by the redshift factor $1 + z$, and the rate of arrival of photons is diminished by the same factor, giving equation (13.51).

The energy flux usually is expressed in terms of the distance modulus $m - M$ (eq. [3.12]), where $m - M = 0$ for an object at 10 parsecs distance, and $m - M$ increases by 2.5 for each decade of decrease of the received energy flux relative to what would have been received if the object had been at the canonical 10 parsecs distance. The distance modulus for equation (13.51) is then

$$m - M = 25 + 5 \log \left[3000(1+z)H_o a_o r(z)\right] - 5 \log h . \tag{13.52}$$

The numerical factor in the logarithm is the present Hubble length, $c/H_o = 3000h^{-1}$ Mpc. The dimensionless factor $y = H_o a_o r$ is given by equation (13.29); it is a function of the density parameters, not of Hubble's constant.

With the series expansion for the angular size distance in equation (13.31), equations (13.51) and (13.52) are

$$f_o \propto z^{-2} \left[1 + (q_o - 1)z + \cdots\right] ,$$
$$m - M + 5 \log h = 42.38 + 5 \log z - 1.086(q_o - 1)z + \cdots . \tag{13.53}$$

Figure 13.6 shows some numerical values for the distance modulus.

In the Einstein-de Sitter model, the flux is predicted to deviate from the z^{-2} law by 25% at $z = 0.5$, so a useful application of the z-m test at this redshift would require that the galaxy luminosities be understood to better than 25%. In a classic analysis, Tinsley (1972) demonstrated how difficult this is. We get a rough idea of what is involved by taking account only of the luminosities of the stars burning hydrogen on the main sequence. For star masses with lifetimes comparable to the Hubble time, the luminosity scales with the mass as $L \propto \mathcal{M}^\alpha$, with $\alpha \sim 3$, and the mass distribution is approximated by the Salpeter (1955) function, $dN/d\mathcal{M} \propto \mathcal{M}^{-(x+1)}$, with $x \sim 1$. The amount of hydrogen a star can burn before its core collapses or the star otherwise dies is not very sensitive to the star mass, so the lifetime scales about as $t \propto \mathcal{M}^{-\alpha}$. That is, if all stars were created at high redshift, those dying at world time t would have mass $\mathcal{M} \propto t^{-1/\alpha}$,

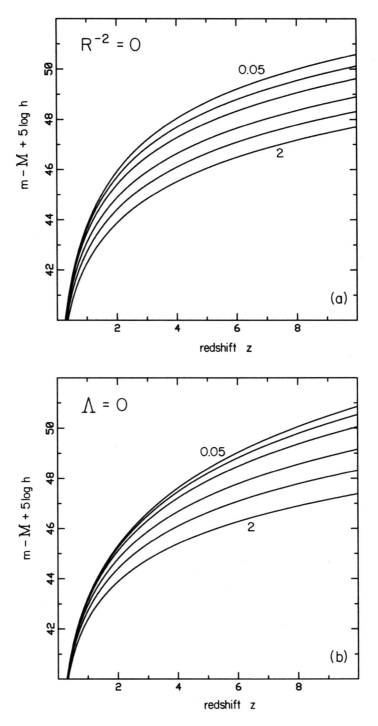

Figure 13.6.
Bolometric distance modulus
$m - M + 5 \log h$
as a function
of redshift. The
parameters are
arranged as in
figure 13.1.

and the net luminosity of the main sequence stars in a galaxy would scale with time as

$$L \propto \int^{\mathcal{M}} L(\mathcal{M})\mathcal{M}^{-(x+1)}d\mathcal{M} \propto \mathcal{M}^{\alpha-x} \propto t^{-(1-x/\alpha)} . \tag{13.54}$$

A more careful calculation takes account of the light from the luminous stars that have just evolved off the main sequence. Tinsley's result is

$$L \propto t^{-(1-0.2x)} . \tag{13.55}$$

For the Einstein-de Sitter model, where $t \propto (1+z)^{-3/2}$, and with $x = 1$, this is

$$L \sim 1 + 1.2z + \cdots . \tag{13.56}$$

The evolution correction in equation (13.56) is equivalent to a shift $\delta q_o \sim 1$ in the apparent acceleration parameter in equation (13.53). The theory of the effect of evolution has been studied at length, with the conclusion that the Tinsley relation (13.55) is considered likely to be a useful rough approximation to the size of the correction and that the correction is not reliably known. That means the z-m relation alone cannot distinguish between an Einstein-de Sitter universe and the low density open case (Ellis 1991). As we noted at the beginning of this section, the program instead must be to overconstrain the models by seeking a consistent picture from a variety of cosmological and astronomical tests.

Equations (13.51) and (13.52) are the bolometric relations for the energy flux integrated over all frequencies or wavelengths. The energy flux observed at frequency ν in the bandwidth $\delta\nu$ was radiated at frequency $\nu a_o/a$ in bandwidth $\delta\nu a_o/a$, so the observed energy flux density (flux per unit frequency interval) is

$$f_o(\nu) = \frac{L(\nu a_o/a)a_o/a}{4\pi(a_o r)^2(1+z)^2} . \tag{13.57}$$

The distance modulus belonging to this expression is

$$m - M = 25 + 5 \log \left[3000(1+z)H_o a_o r(z) \right] - 5 \log h + K , \tag{13.58}$$

where the last term, the K-correction, is

$$K = -2.5 \log[(1+z)L(\nu a_o/a)/L(\nu)] . \tag{13.59}$$

Thus the difference $m - M - K$ is the bolometric distance modulus in equation (13.52) and figure 13.6.

Figure 13.7 shows K-corrections in the blue B band (eq. [3.49]) for elliptical and spiral galaxies (Shanks 1990; King and Ellis 1985). The correction at $z \sim 1$ is

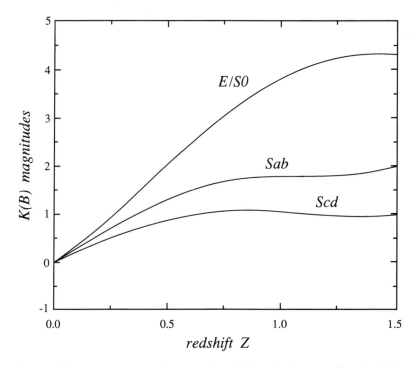

Figure 13.7. K-correction as a function of redshift and galaxy type (Shanks 1990; King and Ellis 1985).

large, particularly for ellipticals, because light detected in the B band is emitted in the ultraviolet, where galaxies are not very luminous. The K-corrections are smaller in the red where the spectrum L_λ tends to be close to flat, so $L_\nu \sim \nu^{-2}$. This gives $K \sim 2.5 \log(1 + z) \sim z$. In the best of all cases, where the spectrum and redshift are known, one eliminates the K-correction by integrating the spectrum $L(\nu a_o/a)$ at the source over the pass band of the detector.

A characteristic absolute magnitude for low redshift galaxies is $M_* \sim -19.5 + 5 \log h$ (eq. [5.139]). The Hubble parameter cancels from the predicted apparent magnitude m at given redshift (because M_* is calibrated on observed apparent magnitudes and redshifts). At $z = 1$ the blue-band K-correction for a spiral galaxy is $K \sim 2$, and the predicted apparent magnitude is $m \sim 25$, about one magnitude fainter than the present limit at which redshifts can be measured. The radio galaxies at $z \gtrsim 1$ in the redshift-magnitude diagram in figure 5.3 are inferred to be considerably more luminous than M_*, consistent with the evolution suggested by their youthful-appearing morphologies.

The count of galaxies as a function of redshift depends on both of the functions we have been considering. The angular size distance r determines the area at redshift z that is subtended by the field of solid angle $\delta\Omega$ within which the galaxies are counted, $\delta A = a^2 r^2 \delta\Omega$, and dl/dz in equation (13.41) gives the linear depth of the sample at redshift z to $z + dz$. The product is the proper volume element,

$$\delta V = \frac{H_o^{-1}\delta z}{(1+z)E(z)}\frac{(a_o r)^2 \delta\Omega}{(1+z)^2}. \tag{13.60}$$

If objects are conserved, the proper number density varies as $n = n_o(a_o/a)^3$, and the predicted count of objects per steradian and per unit increment in redshift is

$$\frac{d\mathcal{N}}{dz} = n_o H_o^{-3} F_n(z), \qquad F_n(z) = \frac{[H_o a_o r(z)]^2}{E(z)}. \tag{13.61}$$

Numerical results for the function $F_n(z)$ are shown in figure 13.8. One sees that for a given space number density, the predicted counts at low Ω and $0.5 \lesssim z \lesssim 4$ are considerably higher if space curvature is negligible. Loh and Spillar (1986) have demonstrated that $d\mathcal{N}/dz$ is a powerful test for the cosmological parameters.

Having a model for the galaxy count and apparent magnitude as functions of redshift, one can predict the count as a function of apparent magnitude. The application to the observations is even more difficult to interpret, however, because it is more sensitive to the galaxy luminosity function and its evolution with redshift than are the redshift-magnitude and redshift-count relations.

The two cosmological functions $r(z)$ and dl/dz appear in another combination in a test that seems to have been first discussed by Alcock and Paczyński (1979). Suppose quasars form in associations that are not gravitationally bound, but rather expand with the general expansion of the universe. In an expanding association with proper linear diameter δl at redshift z, the redshift difference δz across the association and its angular size $\delta\theta$ satisfy

$$\frac{dl}{dz}\delta z = \delta l = ar(z)\delta\theta. \tag{13.62}$$

With equation (13.41) this is

$$\frac{1}{z}\frac{\delta z}{\delta\theta} = \frac{H_o a_o r(z)E(z)}{z}. \tag{13.63}$$

Alcock and Paczyński note that this function does not change much with Ω when $\Lambda = 0$, but that it is sensitive to Λ. The latter is illustrated in figure 13.9 for zero space curvature.

To apply this test one need not assume that the individual associations are spherically symmetric if they are common enough to average over a fair sample. The central assumption is that one can identify associations of objects, perhaps

Figure 13.8.
Counts as a
function of red-
shift. The ver-
tical axis is the
dimensionless
function $F_n(z)$ in
$d\mathcal{N}/dz$ in equation (13.61).
The parameters
are arranged as
in figure 13.1.

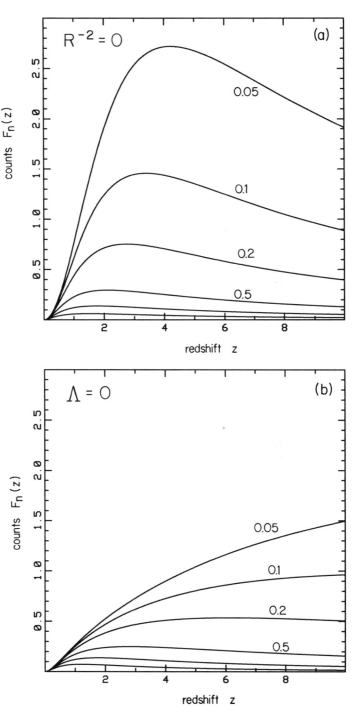

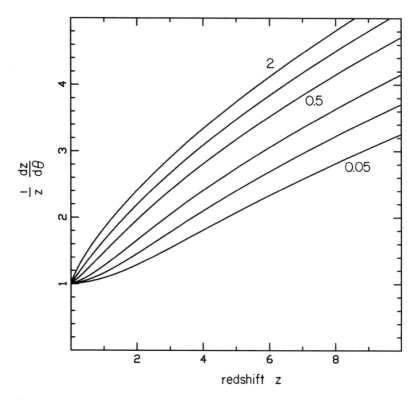

Figure 13.9. The ratio of redshift difference to angular size for an association of objects moving with the general expansion of the universe (eq. [13.63]). This assumes space curvature is negligible. The label is Ω.

quasars, perhaps superclusters of galaxies, that are expanding with the general expansion of the universe.

A charming example of the relations between observables at source and observer is given by the superluminal motions observed in radio sources in quasars. The angular motions of bright spots in a radio image of a quasar are observed to translate to a rate of change of the projected position normal to the line of sight at the quasar that exceeds the velocity of light. The standard model for this effect was introduced by Rees (1966), and is reviewed by Pearson and Zensus (1987). In the sketch in figure 13.10, a radiating component leaves the central quasar engine at time t_1 and moves at fixed speed v at angle θ to our direction along the line of sight. At time $t_2 = t_1 + \delta t$, the component has moved distance $\delta l_\perp = v \sin \theta \delta t$ normal to the line of sight, and distance $v \cos \theta \delta t$ along the line of sight. A packet of radiation that left the engine with the moving component has moved distance δt toward us at time t_2, so a second packet of radiation that leaves the moving

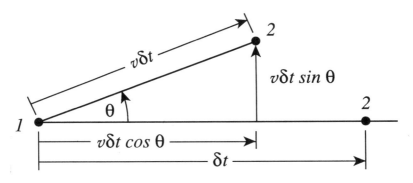

Figure 13.10. Model for superluminal motion.

component at time t_2 is behind the first packet by the distance $\delta t_e = \delta t(1 - v \cos \theta)$ along the line of sight. Thus an observer along our line of sight, and near the quasar and at rest relative to it, would see that the image of the component is moving normal to the line of sight at the rate

$$\frac{\delta l_\perp}{\delta t_e} = \frac{v \sin \theta}{1 - v \cos \theta} . \tag{13.64}$$

The maximum value, at $\cos \theta = v$, is

$$\frac{\delta l_\perp}{\delta t_e} = \gamma v , \tag{13.65}$$

with $\gamma = (1 - v^2/c^2)^{-1/2}$. We see that the apparent transverse velocity $dl_\perp/dt_e$ exceeds the velocity of light when v is greater than $c/2^{1/2}$.

The time interval δt_e measured by an observer at the quasar at redshift z translates to the interval $\delta t = (1 + z)\delta t_e$ observed by us. The displacement $\delta l_\perp$ is an observed angular displacement $\delta \theta = \delta l_\perp/ar(z)$. The observed angular proper motion is then

$$\mu = \frac{d\theta}{dt} = \frac{1}{a_o r(z)} \frac{v \sin \theta}{1 - v \cos \theta} , \tag{13.66}$$

and the maximum in equation (13.65) for given γ is

$$\mu = \frac{\gamma v}{a_o r(z)} . \tag{13.67}$$

Since the dimensionless cosmological parameters fix $H_o a_o r$ (eq. [13.29]), the maximum proper motion in a given cosmological model is determined by the product $H_o\gamma v$. Cohen *et al.* (1988) show that $\gamma = 9h^{-1}$ gives a reasonable fit to the observations. With the present data the test does not strongly constrain Ω, but it does show striking discrepancies in some alternative models for quasar redshifts.

Fukugita, Futamase, and Kasai (1990) and Turner (1990) have shown that yet another application of the relations in the classical tests, for the predicted frequency of gravitational lensing events, is quite sensitive to the cosmological parameters. The application of this test is discussed further in Fukugita and Turner (1991).

In a lensing event a mass concentration along the line of sight causes a gravitational deflection of light rays large enough to produce more than one observable image of a background source, or an appreciable perturbation to the energy flux received in a single image. The geometry in an otherwise homogeneous cosmological model is illustrated in figure 13.11. The lens and source are at angular size distances r_{ol} and r_{os} in a coordinate system with the origin at the observer. The angular size distance between lens and source is r_{ls} in a coordinate system with the origin at the source. An image of the source seen by the observer is deflected by angular distance θ from the position in the sky it would have had in the absence of the gravitational lens. The deflection angle of the line of sight at the lens is α. We get the relation between these angles by noting that the comoving coordinate length s at the source subtended by the angles θ at the observer and α at the lens is (eq. [13.46])

$$s = \theta r_{os} = \alpha r_{ls} . \tag{13.68}$$

Thus the comoving distance d at the lens subtended by the angle θ at the observer is

$$d = \theta r_{ol} = \alpha r_{ol} r_{ls} / r_{os} . \tag{13.69}$$

In a cosmologically flat model, angular size distances add: $r_{os} = r_{ol} + r_{ls}$. If space curvature is important, the polar angle χ adds: $\chi_{os} = \chi_{ol} + \chi_{ls}$ (as we see from the geodesic equation for a radial light ray, $R d\chi = a^{-1} dt$). Since $r = R \sinh \chi$, and the curvature parameter is $\Omega_R = (H_o a_o R)^{-2}$ (eq. [13.4]), we have (as in eq. [13.35])

$$\begin{aligned} y_{ls} &= H_o a_o r_{ls} = \Omega_R^{-1/2} \sinh(\chi_{os} - \chi_{ol}) \\ &= y_{os}\left(1 + y_{ol}^2 \Omega_R\right)^{1/2} - y_{ol}\left(1 + y_{os}^2 \Omega_R\right)^{1/2} . \end{aligned} \tag{13.70}$$

The second line, which gives the angular size distance between lens and source in terms of the angular size distances of lens and source from the observer, also applies in positive curvature models, where $\Omega_R < 0$.

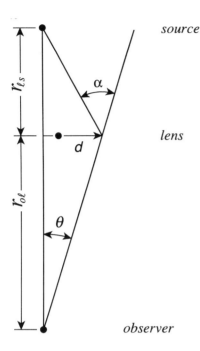

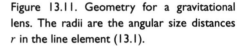

Figure 13.11. Geometry for a gravitational lens. The radii are the angular size distances r in the line element (13.1).

To simplify the analysis let us suppose that the mass distribution in the lens varies inversely as the square of the radius, as in an isothermal sphere with small core radius. Then, independent of the impact parameter, the bending angle at the lens is $\alpha = 4\pi\sigma^2$, where the line-of-sight velocity dispersion is σ (eq. [10.97]). This model is a useful first approximation to the mass distribution in a galaxy or in a rich cluster of galaxies.

When α is independent of the impact parameter, the condition for a lensing event that produces two images is that the perpendicular distance of the lens from the undeflected line of sight to the source is less than the comoving distance given by equation (13.69) for d. The probability that a line of sight passes within comoving distance d of a gravitational lens at redshift z to $z + dz$ is given by equation (13.42) with cross section $\sigma_{gl} = \pi a^2 d^2$. Collecting all these relations, we have that the lensing probability per unit increment of redshift is

$$\frac{dP}{dz} = \frac{16\pi^3}{H_o^3} \frac{\sum \left[n(i)\sigma_i^4\right]}{(1+z)^3 E(z)} \left(\frac{y_{ol}y_{ls}}{y_{os}}\right)^2 , \qquad (13.71)$$

where $n(i)$ is the number density of lensing systems with line-of-sight velocity dispersion σ_i at redshift z.

A useful summary statistic is the integral of the lensing probability over all lens redshifts between $z = 0$ and the redshift z_s of the source. This integral is the gravitational lensing optical depth (Turner, Ostriker, and Gott 1984). If lenses are conserved, meaning $n(i) = n_o(i)(1 + z)^3$, we have

$$\tau_{gl} = \int dP = H_o^{-3} \sum \left[n_o(i) \sigma_i^4 \right] F_{gl}(z_s),$$

$$F_{gl}(z_s) = 16\pi^3 \int_0^{z_s} \frac{dz}{E(z)} \left(\frac{y_{ol} y_{ls}}{y_{os}} \right)^2. \tag{13.72}$$

If space curvature may be neglected, the angular size distances add, $y_{ls} = y_{os} - y_{ol}$, and $dy_{ol} = dz_l/E(z_l)$ (eq. [13.30]). In this case the integral is

$$F_{gl} = 8\pi^3 y_{os}^3/15, \tag{13.73}$$

and the lensing probability per increment of y_{ol} is maximum at $y_{ol} = y_{os}/2$ (Gott, Park, and Lee 1989).

The function $F_{gl}(z_s)$ in equation (13.72) for the lensing optical depth is shown in figure 13.12. It is striking that a logarithmic scale is needed to show how this statistic varies with the density parameter in the cosmologically flat case. If Λ may be ignored, the lensing probability is less sensitive to Ω.

Finally, let us use the Einstein-de Sitter model to see the orders of magnitude for the probability of lensing a quasar by the mass concentrations in galaxies. At the most likely distance for a lensing event, $y_{ol} = y_{os}/2$, the angular separation of the images is

$$\phi = \alpha = 4\pi(\sigma/c)^2 = 2 \text{ arc sec} \quad \text{at} \quad \sigma = 250 \text{ km s}^{-1}. \tag{13.74}$$

The velocity dispersion is characteristic of giant elliptical galaxies, and the angle is near the lower end of the range of observed lensing events. With $\Omega = 1$ and $\Lambda = 0$, the angular size distance to a quasar at redshift z_s is $y_{os} = 2[1 - (1 + z_s)^{-1/2}]$ (eq. [13.30]). For quasar redshift $z_s = 3$ this is $y_{os} = 1$, and the maximum lensing probability is at $y_{ol} = 0.5$ and $z_l = 0.8$. The impact parameter D of the line of sight at the galaxy at this distance is

$$D \lesssim \frac{\alpha}{2} \frac{c y_{ol}}{H_o(1 + z_l)} = 4h^{-1} \text{ kpc}. \tag{13.75}$$

This is in the range of radii where the mass model used here, with a flat rotation curve, is thought to be reasonable. In the survey of Faber et al. (1989), the number density of elliptical galaxies with velocity dispersions greater than 250 km s^{-1} is $n_o \sim 0.001h^3$ Mpc^{-3}. We really want the sum of $n\sigma^4$, but at this value of σ

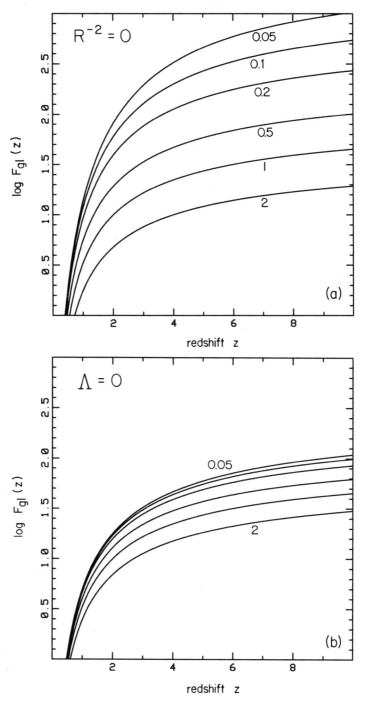

Figure 13.12. Optical depth for gravitational lensing as a function of the redshift of the source (eq. [13.72]). The parameters are arranged as in figure 13.1.

the density is dropping so rapidly with increasing σ that we make little error by setting σ equal to the value at the cutoff. Then the above numbers give

$$P \sim 0.001 \left[1 - (1 + z_s)^{-1/2}\right]^3 . \tag{13.76}$$

This is in the range of the observed rate of lensing.

A more careful treatment uses more realistic mass models for the lens, including the contributions from the clusters in which elliptical galaxies tend to be found, corrects for events that may be lost by obscuration by dust in spiral galaxies, and tests the models against the observed frequency distribution of the lens redshifts (eq. [13.71]) and the distribution of angular separations of the images. The applications of tests based on the lensing effect are under active study at the time this is written; the results will be followed with considerable interest.

To summarize, the tests of the standard model are based on two relations that are independent of the cosmological parameters (within general relativity theory): the surface brightness varies as $i_o \propto (1 + z)^{-4}$, and the time dilation is the same as the redshift, $\delta t_o = (1 + z)\delta t_e$; and the tests depend on two model-dependent expressions, the expansion parameter $a(t)$ as a function of time, and the spacetime geometry as fixed by the parameter R^{-2} in the Robertson-Walker line element (13.1). These model-dependent parts appear in the cosmological tests in two functions, the angular size distance $y(z) = H_o a_o r(z)$ (eq. [13.29]), and the radial distance $H_o dl/dz$ (eq. [13.41]). The radial function dl/dz is determined by the expansion parameter $a(t)$ alone. As discussed in sections 4 and 5, the form of the differential equation for $a(t)$ follows in a reasonably direct way from the Newtonian limit of the gravity theory (eq. [5.17]), and the resulting function $a(t)$ is a measure of the radius of curvature of space sections only in an indirect way, through the parameter R^{-2} that acts as the constant of integration of the equation for $\ddot{a}$. The angular size distance r depends on the function $a(t)$ and on the spacetime geometry through the trigonometric function in equations (13.5) and (13.29). This section is meant to illustrate the rich variety of ways these relations can be tested.

Evolution of Linear Density Perturbations

This is an appropriate place to look at numerical results for the evolution of mass density fluctuations and the peculiar velocity field in linear perturbation theory, because the equations are related to what appears in the cosmological tests. When nongravitational forces can be neglected, the time variation of the amplitude $\delta = \delta\rho/\rho$ of the growing mode of the mass fluctuation is given by equations (5.87) and (11.68),

$$\delta \propto \frac{\dot{a}}{a} \int_0^t \frac{da}{\dot{a}^3} . \tag{13.77}$$

In the present notation, we can write this as

$$\delta \propto D(z) = E(z)G(z), \qquad G(z) = \frac{5\Omega}{2} \int_z^\infty \frac{1+z}{E(z)^3} \, dz . \qquad (13.78)$$

The normalization for $D(z)$ is chosen so that at high redshift the function is

$$D(z) \rightarrow \frac{1}{1+z} = \frac{a(t)}{a_o} , \qquad (13.79)$$

which is the time variation of the linear density contrast in the Einstein-de Sitter model. Thus the function $D(0)$ evaluated at $z=0$ is the growth factor to the present from high redshift relative to the growth factor from the same initial amplitude and initial redshift in the Einstein-de Sitter model.

The dimensionless factor for the velocity field in terms of the time variation of the amplitude $\delta(t)$ of the mode is (eq. [5.116])

$$f = \frac{\dot{D}}{D} \frac{a}{\dot{a}} . \qquad (13.80)$$

With equation (13.78), this is

$$f(z) = \frac{\ddot{a}a}{\dot{a}^2} - 1 + \frac{5\Omega}{2} \frac{(1+z)^2}{E^3 G} . \qquad (13.81)$$

This expression at $z=0$ with equation (13.7) for the acceleration parameter is

$$f \equiv f(0) = \Omega_\Lambda - \frac{\Omega}{2} - 1 + \frac{5}{2} \frac{\Omega}{G(0)} . \qquad (13.82)$$

If $\Lambda = 0$ the integral in equation (13.78) works out to (LSS, § 11)

$$D(0) = \frac{5}{2} \frac{\Omega}{(1-\Omega)^2} \left[1 + 2\Omega + \frac{3\Omega}{(1-\Omega)^{1/2}} \ln \left[\Omega^{-1/2} - (\Omega^{-1} - 1)^{1/2} \right] \right] . \qquad (13.83)$$

Figures 13.13 and 13.14 show the normalized growth factor $D(0) = G(0)$ in equation (13.78) and the velocity factor f in equation (13.82) for the two standard cases, where space curvature is negligible (solid line) and where the cosmological constant is negligible (dashed line). As it happens, the velocity factor is almost the same in the two sets of models (Peebles 1984c). This means the predicted large-scale velocity field for given Ω and density contrast is almost the same whether one assumes Λ or space curvature can be neglected. The growth factor for density fluctuations is much less sensitive to Ω if space curvature is negligible. This is a result of the more rapid transition to the Einstein-de Sitter limit with increasing redshift, as shown in equation (5.62), which allows the growth of the density fluctuation to continue to lower redshift.

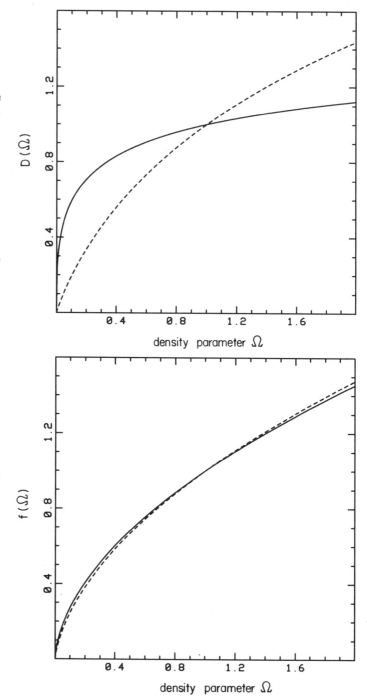

Figure 13.13. Growth factor for density fluctuations in linear perturbation theory. The factor D is evaluated at $z = 0$ and normalized to a fixed amplitude $\delta\rho/\rho$ at a fixed high redshift. The solid line assumes negligible space curvature, the dashed line $\Lambda = 0$.

Figure 13.14. Velocity factor $f = d \log D / d \log a$ (eq. [13.82]). The solid line assumes negligible space curvature, the dashed line $\Lambda = 0$.

14. Cosmology in an Inhomogeneous Universe

The analysis of the cosmological tests in the last section assumes the mass distribution is homogeneous (apart from the local concentration in a lensing object). The clumpy mass distribution in the real world can play quite different roles in different tests. It has little effect on the intersection probability in equation (13.42) applied to the distribution of absorption redshifts in the spectrum of a quasar, or to the fraction of the sky covered by galaxies, because the deflection of a line of sight by the irregular mass distribution does not affect the chance the line intersects an object. The same is true of the count of objects as a function of redshift, for the irregularities in the mass distribution perturb where an object is seen but not the number of objects or the redshift at which they are seen. Zel'dovich (1964), Bertotti (1966), Gunn (1967), Kantowski (1969), and others point out that the irregular mass distribution can produce a systematic error in apparent magnitudes of galaxies, for the mass along the line of sight acts as a lens that determines the rate of change of the convergence of a bundle of light rays, and that fixes the angular size of the image. If the mass distribution is clumpy rather than smooth, it has no effect on the mean value of the solid angle subtended by a sample of objects at a given redshift, because the distortions in angular sizes have to be such as to keep the net area of the sky equal to 4π steradians. However, the observations could be biased to favor objects whose images have been magnified, because they appear brighter (Turner 1980), or the bias could go the other way, for where there is mass there tends to be dust. Thus, if the line of sight to a distant object passes through a large amount of mass, so that gravitational lensing is magnifying the angular size of the object, the image tends to be obscured. Those objects that are readily observable would tend to be on lines of sight that pass through no mass, and their angular sizes therefore would be biased low and their apparent magnitudes biased high relative to the prediction of a homogeneous model with the same mean mass density and space curvature.

There are no firm numbers on the sizes of these effects, but we do have some useful guides that suggest the bias may not be large. There are mass concentrations with surface densities high enough to have a significant effect on the convergence of a light beam, as in observed gravitational lensing events, but we noted in the last section that the observations and what we know of the extragalactic mass distribution both indicate the probability for this to happen along a random line of sight is small. In particular, we will check that the mass in the dusty central parts of spiral galaxies has little effect on the convergence of light rays. The perturbations caused by the mass in the dark halos along a line of sight to a distant object produce shear that adds in quadrature, as in a random walk, while the effect on the convergence of the beam adds linearly. This means that, apart from lensing events, the net shear is expected to be small compared to the expansion. Consistent with this, the images of the luminous arcs produced by the mass concentrations in clusters of galaxies show little evidence of distortion by the tidal shear of mass concentrations along the line of sight (Lynds and Petrosian 1987; Soucail et al. 1988). The same argument suggests the expansion of the light beam along a typical line of sight is likely to be close to the mean value computed in the homogeneous cosmological model. But these are preliminary results that surely will be rediscussed as our understanding of the cosmological tests and the large-scale mass distribution improve.

The Sachs Optical Equations

The Sachs (1961) optical equations describe the evolution of the cross section of a beam of light rays passing through an inhomogeneous mass distribution. One starts from the geodesic deviation equation (9.65), which describes the rate of change of the separation between neighboring lines of sight. The derivation in section 9 assumes particles with nonzero rest mass. For particles such as photons, with vanishing rest mass, the geodesic equation of motion becomes (eq. [9.49])

$$u^i = \frac{dx^i}{d\lambda}, \quad u^i_{;k}u^k = 0, \quad u^i u_i = 0. \tag{14.1}$$

These equations also define the scalar parameter λ, which replaces the proper particle time s for a particle with nonzero rest mass, and which labels positions along the path $x^i(\lambda)$ of a particle. The four-vector u^i is proportional to the energy-momentum vector for a photon in the beam (eq. [9.46]).

The derivation of the equation of geodesic deviation goes through the same way whether we use the particle proper time s or λ. Thus we have for massless particles

$$\frac{D^2\xi^i}{d\lambda^2} = R^i{}_{jkl}u^j u^k \xi^l. \tag{14.2}$$

Here $x^i(\lambda)$ and $x^i(\lambda) + \xi^i(\lambda)$ are neighboring geodesics. The derivative $D/d\lambda$ is the covariant form for what an observer would measure to be the rate of change of a field with respect to position along the path. In particular, if the observer sees that the vector L^i is transported along the path in a parallel way, then in a general coordinate system the vector field satisfies $DL^i/d\lambda = L^i_{;k}u^k = 0$.

Let us set up orthogonal vectors along the path of a light ray as a basis for a coordinate system to describe the evolution of the cross section of a beam of light. At a point on the path we can choose a locally Minkowski coordinate system with the light ray moving toward the positive z axis. In this coordinate system, the local forms for the wanted orthogonal vectors are

$$u^i = (1, 0, 0, 1),$$
$$L_1^i = (0, 1, 0, 0),$$
$$L_2^i = (0, 0, 1, 0),$$
$$w^i = (1, 0, 0, -1).$$

(14.3)

The first four-vector is the direction of propagation of the light ray along the z axis, and the last is an orthogonal null vector. The middle two are orthogonal spacelike vectors. In a general coordinate system, the conditions that the vectors L_α^i, with $\alpha = 1$ and 2, are orthogonal to the light ray and to each other are

$$g_{ik} L_\alpha^i u^k = 0, \qquad g_{ik} L_\alpha^i L_\beta^k = \delta_{\alpha\beta}.$$

(14.4)

We will require that an observer sees that the vectors L_α^i at neighboring points along the path of the light ray are parallel. As noted above, this condition is

$$\frac{DL_\alpha^i}{d\lambda} = L_{\alpha;k}^i u^k = 0.$$

(14.5)

Since the covariant derivative of the metric tensor vanishes, equations (14.1) and (14.5) say the derivatives $D/d\lambda$ of the orthogonality equations (14.4) vanish, that is, the parallel propagation of the vectors L_α^i and u^i along the path of the light ray keeps them orthogonal and normalized.

We are considering neighboring light rays with separation $\xi^i(x)$ at position x. We can choose the ξ^i at a starting point on the path so a freely moving observer sees they are perpendicular to the light ray, $u_i \xi^i = 0$ and $w_i \xi^i = 0$. The geodesic equations for the neighboring paths tell the observer that these vectors are staying orthogonal. Since $u_i \xi^i$ and $w_i \xi^i$ are scalars, they vanish along the path of the light ray.

For practice, let us check this in a more direct way. The four-momentum for the light ray is $u^i = dx^i/d\lambda$ along the path $x(\lambda)$, and $u^i + d\xi^i/d\lambda$ along the path $x + \xi$. The condition that the second vector is null is

$$g_{ij}(x + \xi) \left(u^i + \frac{d\xi^i}{d\lambda} \right) \left(u^j + \frac{d\xi^j}{d\lambda} \right) = 0.$$

(14.6)

To first order in ξ, this is

$$2u_i \frac{d\xi^i}{d\lambda} + g_{ij,k} u^i u^j \xi^k = 0.$$

(14.7)

Since the geodesic equation says $Du_i/d\lambda = 0$, we have

$$\frac{D}{d\lambda} u_i \xi^i = u_i \frac{D\xi^i}{d\lambda} = u_i \left(\frac{d\xi^i}{d\lambda} + \Gamma_{jk}^i \xi^k u^j \right) = 0.$$

(14.8)

The last step indicating that this derivative vanishes follows from equation (14.7) for $d\xi^i/\lambda$ and equation (8.57) for the Christoffel symbol. Since $u_i\xi^i$ is constant along the path, u^i and ξ^i remain orthogonal if that is the initial condition.

The conclusion is that the ξ^i are orthogonal to u^i and w^i, and so are linear combinations of the four-vectors L_1^i and L_2^i,

$$\xi^i(\lambda) = \sum_{\alpha=1,2} d_\alpha(\lambda) L_\alpha^i . \qquad (14.9)$$

The L_α^i are unit vectors normal to each other and to the light ray. An observer who is moving so the L_α^i are seen to have no time component would see that the d_α are the proper orthogonal components of the separation of the two light rays.

On substituting the expansion of ξ^i in this basis into the geodesic deviation equation (14.2), taking the inner product of both sides of the equation with L_β^i, and using the orthogonality (14.4) of the basis vectors L_β^i, we get

$$\frac{d^2 d_\alpha}{d\lambda^2} = \sum_\beta A_{\alpha\beta} d_\beta , \qquad A_{\alpha\beta} = R_{ijkl} L_\alpha^i u^j u^k L_\beta^l . \qquad (14.10)$$

The d_α are scalars, so their covariant derivatives are ordinary derivatives with respect to the parameter λ along the path. The two-by-two tensor $A_{\alpha\beta}$ is symmetric (eq. [8.81]).

The right-hand side of equation (14.10) is reduced by using the decomposition in equation (8.84) of the curvature tensor into the traceless part C_{ijkl} and the Ricci tensor R_{ij}. A short calculation using the fact that vectors u^i and L_α^i are orthogonal and u^i is null gives

$$A_{\alpha\beta} = -\frac{1}{2} R_{jk} u^j u^k \delta_{\alpha\beta} + C_{ijkl} L_\alpha^i u^j u^k L_\beta^l . \qquad (14.11)$$

Let us write the first derivative of the components of proper separation of the two light rays as

$$\frac{d d_\alpha}{d\lambda} = \sum \left(\theta \delta_{\alpha\beta} + w_{\alpha\beta} \right) d_\beta . \qquad (14.12)$$

In the first term θ is the expansion of a beam of light rays. In the second term the shear tensor $w_{\alpha\beta}$ is symmetric and traceless, that is, it is of the form

$$w_{\alpha\beta} = \begin{bmatrix} \rho & \sigma \\ \sigma & -\rho \end{bmatrix} . \qquad (14.13)$$

The expansion and the two shear components are the needed three independent functions corresponding to the three independent components of the symmetric two-by-two tensor $A_{\alpha\beta}$. The shear tensor satisfies

$$\sum w_{\alpha\beta} w_{\beta\gamma} = \begin{bmatrix} \rho & \sigma \\ \sigma & -\rho \end{bmatrix} \begin{bmatrix} \rho & \sigma \\ \sigma & -\rho \end{bmatrix}$$

$$= \begin{bmatrix} \rho^2 + \sigma^2 & 0 \\ 0 & \rho^2 + \sigma^2 \end{bmatrix} \tag{14.14}$$

$$= (\rho^2 + \sigma^2)\delta_{\alpha\beta} = \frac{1}{2} w^2 \delta_{\alpha\beta}.$$

The square of the shear is $\sum w_{\alpha\beta} w_{\beta\alpha} = 2(\rho^2 + \sigma^2) = w^2$.

With the form for the first derivative of d_α in equation (14.12), the second derivative is

$$\frac{d^2 d_\alpha}{d\lambda^2} = \sum (\theta\delta_{\alpha\beta} + w_{\alpha\beta})(\theta\delta_{\beta\gamma} + w_{\beta\gamma})d_\gamma + \left(\frac{d\theta}{d\lambda}\delta_{\alpha\beta} + \frac{dw_{\alpha\beta}}{d\lambda}\right)d_\beta$$

$$= \sum \left[\left(\theta^2 + \frac{w^2}{2} + \frac{d\theta}{d\lambda} \right) \delta_{\alpha\gamma} + \left(\frac{dw_{\alpha\gamma}}{d\lambda} + 2\theta w_{\alpha\gamma} \right) \right] d_\gamma. \tag{14.15}$$

This is the form of the geodesic deviation equation (14.10), with

$$A_{\alpha\beta} = \left(\frac{d\theta}{d\lambda} + \theta^2 + \frac{1}{2} w^2 \right) \delta_{\alpha\beta} + \left(\frac{dw_{\alpha\beta}}{d\lambda} + 2\theta w_{\alpha\beta} \right). \tag{14.16}$$

We can separate the Ricci and Weyl tensors in equations (14.11) and (14.16) with the help of the identity

$$\sum_{\alpha=1,2} C_{ijkl} L_\alpha^i u^j u^k L_\alpha^l = 0. \tag{14.17}$$

To check this expression, note that since it is a scalar we can evaluate it in locally Minkowski coordinates where the u^i and L_α^i are given by equation (14.3) and g_{ij} is the Minkowski form η_{ij}. In this coordinate system the left side of equation (14.17) is

$$\sum_{\alpha=1,2} (C_{\alpha 00\alpha} + 2C_{\alpha 03\alpha} + C_{\alpha 33\alpha}). \tag{14.18}$$

Since the Weyl tensor is traceless and antisymmetric in the first two indices and in the last two indices, we have

$$\sum_{\alpha=1,2} C_{\alpha 00\alpha} = -C_{00i}^i + C_{0000} - C_{3003} = -C_{3003}, \tag{14.19}$$

and similarly,

$$\sum_{\alpha=1,2} C_{\alpha 33\alpha} = C_{3003}, \qquad \sum_{\alpha=1,2} C_{\alpha 03\alpha} = 0. \tag{14.20}$$

This gives equation (14.17).

Equation (14.17) indicates that the second term on the right-hand side of equation (14.11) for $A_{\alpha\beta}$ is traceless, as is the second term in equation (14.16). We have then

$$\frac{d\theta}{d\lambda} + \theta^2 + \frac{1}{2}w^2 = -\frac{1}{2}R_{jk}u^j u^k,$$

$$\frac{dw_{\alpha\beta}}{d\lambda} + 2\theta w_{\alpha\beta} = C_{ijkl}L_\alpha^i u^j u^k L_\beta^l. \tag{14.21}$$

These are the wanted equations for the evolution of the expansion and shear of the beam of light rays (Sachs 1961).

To see how the area of the beam varies along the path, use equation (14.12) to write the components of proper separation of neighboring rays at neighboring positions along the beam as

$$\bar{d}_\alpha \equiv d_\alpha(\lambda + \delta\lambda) = d_\alpha(\lambda) + \left(\theta d_\alpha + \sum w_{\alpha\beta} d_\beta\right)\delta\lambda. \tag{14.22}$$

The proper cross section area A of the beam is an integral over the d_α. The areas at neighboring positions λ and $\lambda + \delta\lambda$ are

$$A(\lambda + \delta\lambda) = \int d\bar{d}_1 d\bar{d}_2 = \int dd_1 dd_2 \frac{\partial(\bar{d})}{\partial(d)} = \frac{\partial(\bar{d})}{\partial(d)}A(\lambda). \tag{14.23}$$

Writing out the Jacobian, and recalling that the shear tensor is traceless, we get

$$\frac{A(\lambda + \delta\lambda)}{A(\lambda)} = \begin{bmatrix} 1 + (\theta + w_{11})\delta\lambda & w_{12}\delta\lambda \\ w_{21}\delta\lambda & 1 + (\theta + w_{22})\delta\lambda \end{bmatrix} = 1 + 2\theta\delta\lambda, \tag{14.24}$$

to first order in $\delta\lambda$. Thus, the proper beam area satisfies

$$\frac{dA}{d\lambda} = 2\theta A. \tag{14.25}$$

With equation (14.21) for $d\theta/d\lambda$, the second derivative is

$$\frac{1}{A}\frac{d^2 A}{d\lambda^2} = 2\frac{d\theta}{d\lambda} + 4\theta^2$$

$$= \frac{1}{2}\left(\frac{1}{A}\frac{dA}{d\lambda}\right)^2 - R_{jk}u^j u^k - w^2. \tag{14.26}$$

This expression is simplified by changing the function to $\sqrt{A}$, which gives

$$\frac{1}{\sqrt{A}}\frac{d^2\sqrt{A}}{d\lambda^2} = -\frac{1}{2}\left(R_{jk}u^j u^k + w^2\right) \tag{14.27}$$

$$= -4\pi G T_{jk}u^j u^k - w^2/2.$$

The second line, which follows from the gravitational field equation with $u^i u_i = 0$ for the light ray, is the form used in the pioneering studies of cosmology in an inhomogeneous universe by Dyer and Roeder (1973). The source term on the right-hand side, which involves the Ricci tensor R_{jk}, is called the Ricci focusing term.

Let us work two examples where it is easy to calculate the evolution of the beam area other ways. The first is the motion of light rays in a quasistatic Newtonian potential $\phi(\mathbf{r})$. This linear approximation deals with the field equations to first order in ϕ, so in the following we drop terms of order ϕ^2.

We will consider an initially parallel beam of light rays moving toward the positive z axis. The gradients of the metric tensor produce relative motions of the rays transverse to the original direction, introducing a convergence or divergence of the beam. We will compute the second derivative of the beam area with respect to distance along the path of the ray, and compare it to what follows from equation (14.26).

In the linear quasistatic approximation, the line element is given by equation (10.84). The time part of the metric tensor is $g_{00} = 1 + 2\phi$, the space part is $g_{\alpha\beta} = -(1 - 2\phi)\delta_{\alpha\beta}$, and the time derivatives of g_{ij} are negligible compared to the space derivatives. It follows that the time part of the geodesic equation (9.47) for a light ray is

$$\frac{d}{dt}(1 + 2\phi)\frac{dt}{d\lambda} = 0. \tag{14.28}$$

This says we can normalize the affine parameter along the path to

$$\frac{dt}{d\lambda} = 1 - 2\phi(\mathbf{x}), \tag{14.29}$$

to order ϕ.

The component of the geodesic equation in the α direction normal to the unperturbed direction of the light ray is

$$\frac{d}{dt}g_{\alpha i}\frac{dx^i}{d\lambda} = \frac{1}{2}g_{jk,\alpha}\frac{dx^j}{d\lambda}\frac{dx^k}{dt}. \tag{14.30}$$

To order ϕ, this is

$$\frac{d^2 x^\alpha}{dt^2} = -2\phi_{,\alpha}. \tag{14.31}$$

Now consider two light rays in the beam that are at coordinate separation δx along the x axis. With the line element (10.84), the proper separation of the rays to order ϕ is

$$d_x = (1 - \phi)\delta x,\tag{14.32}$$

and the time derivative of this separation is

$$\frac{dd_x}{dt} = \delta\frac{dx}{dt} - \frac{\partial\phi}{\partial z}d_x.\tag{14.33}$$

With the geodesic equation (14.31), the second time derivative of the proper separation of the rays is

$$\frac{d^2 d_x}{dt^2} = -\left(2\frac{\partial^2\phi}{\partial x^2} + \frac{\partial^2\phi}{\partial z^2}\right)d_x.\tag{14.34}$$

The proper area defined by beams at separations d_x and d_y along the orthogonal transverse axes is $A = d_x d_y$. The second time derivative of this expression, to first order in ϕ, is

$$\begin{aligned}
\frac{1}{A}\frac{d^2 A}{dt^2} &= \frac{1}{d_x}\frac{d^2 d_x}{dt^2} + \frac{1}{d_y}\frac{d^2 d_y}{dt^2}\\
&= -2\left(\frac{\partial^2\phi}{\partial x^2} + \frac{\partial^2\phi}{\partial y^2} + \frac{\partial^2\phi}{\partial z^2}\right)\\
&= -8\pi G\rho,
\end{aligned}\tag{14.35}$$

where the mass density is $\rho(\mathbf{r})$. This last step follows because ϕ is the Newtonian gravitational potential.

This expression for the beam area can be compared to the Sachs optical equation (14.26). The squared terms are second order, so to the order of this calculation the equation is

$$\frac{1}{A}\frac{d^2 A}{d\lambda^2} = -R_{jk}u^j u^k = -R_{00} - R_{33}.\tag{14.36}$$

The line element we have been using assumes the only appreciable component of the stress-energy tensor is $T_{00} = \rho$, so the gravitational field equation (10.65) is $R_{00} = R_{33} = 4\pi G\rho$, and the Sachs equation (14.36) for the evolution of the beam area with distance along the ray therefore agrees with the calculation in equation (14.35).

For a second example, let us use the optical equation to find the time evolution of the proper area of a beam of light from a distant galaxy converging on an observer in a homogeneous Robertson-Walker model. We will place the observer at the origin in the line element

$$ds^2 = dt^2 - a(t)^2 \left[\frac{dr^2}{1 - r^2/R^2} + r^2 \left(d\theta^2 + \sin^2\theta d\phi^2 \right) \right] . \qquad (14.37)$$

The light rays detected by the observer converge to $r = 0$ along lines of constant θ and ϕ. The space part of the metric tensor scales with time as $g_{\alpha\beta} \propto a^2$, so the time part of the geodesic equation of motion (9.47) for a light ray is

$$\frac{du^0}{dt} = \frac{1}{2} \dot{g}_{\alpha\beta} u^\alpha u^\beta / u^0 = \frac{\dot{a}}{a} g_{\alpha\beta} u^\alpha u^\beta / u^0 . \qquad (14.38)$$

Since u^i is a null vector, this is

$$\frac{1}{u^0} \frac{du^0}{dt} = -\frac{\dot{a}}{a} , \qquad (14.39)$$

with the solution

$$u^0 = \frac{dt}{d\lambda} = \frac{1}{a(t)} . \qquad (14.40)$$

This defines the normalization of the parameter λ. It will be recalled that u^i is the energy-momentum four-vector (eq. [9.46]), so in a locally Minkowski coordinate system u^0 is proportional to the photon energy, $\hbar\omega$. That is, equation (14.40) is just the cosmological redshift.

The radial velocity of the light ray follows from the condition $ds = 0$ in equation (14.37),

$$\frac{dr}{dt} = -\frac{(1 - r^2/R^2)^{1/2}}{a} , \qquad (14.41)$$

so the radial part of the four-vector u^i is

$$u^r = \dot{r} u^0 = \frac{\dot{r}}{a} = -\frac{(1 - r^2/R^2)^{1/2}}{a^2} . \qquad (14.42)$$

As can be checked, this satisfies the radial part of the geodesic equation of motion.

To model the matter we will use an ideal fluid with proper mass density ρ, pressure p, four-velocity $u_f{}^i$, and the stress-energy tensor T^{ij} in equation (10.49). Then the Einstein field equation (10.65) gives

$$R_{ij} u^i u^j = 8\pi G(\rho + p) u_f{}^i u_f{}^j u_i u_j = 8\pi G(\rho + p)/a^2 . \qquad (14.43)$$

The cosmological fluid is at rest in the coordinates of the line element (14.37), so $u_f{}^i u_i = u^0 = 1/a$, from equation (14.40).

Now we can write down the equation for the area as a function of time for a beam of light converging on an observer. The spherical symmetry of the line element eliminates the shear, so if we write the proper area of the beam as $A = d^2$, we have from the first line of equation (14.27)

$$\frac{1}{d}\frac{d^2 d}{d\lambda^2} = -\frac{4\pi G(\rho + p)}{a^2}.$$ (14.44)

With equation (14.40) for $dt/d\lambda$, this is

$$\frac{d^2 d}{dt^2} - \frac{\dot{a}}{a}\frac{dd}{dt} = -4\pi G(\rho + p)d.$$ (14.45)

The cosmological equations (5.15) and (5.18) give the relation

$$\frac{\ddot{a}}{a} - \frac{\dot{a}^2}{a^2} - \frac{1}{a^2 R^2} = -4\pi G(\rho + p),$$ (14.46)

which brings equation (14.45) to

$$\frac{d^2 d}{dt^2} - \frac{\dot{a}}{a}\frac{dd}{dt} = \left(\frac{\ddot{a}}{a} - \frac{\dot{a}^2}{a^2} - \frac{1}{a^2 R^2}\right) d.$$ (14.47)

With equation (14.41) for $\dot{r}$, one can check that the wanted solution to this equation is

$$d = ar\theta,$$ (14.48)

where the two constants of integration are fixed by the angle θ and the condition $r = 0$ at the time t_o of the observation. This is equation (13.46) for the proper length subtended by angle θ at angular size distance r.

Fluxes and Solid Angles

In a detailed analysis of cosmology in an inhomogeneous universe, one would compute $A(\lambda)$ from equations (14.21) and (14.27) for a light ray running from object to observer. The boundary conditions can be applied in two ways: we can imagine a wave front moving from a pointlike object, and compute the energy flux at the observer, or we can follow light rays from a distributed object that converge at the observer, and compute the observed solid angle of the beam that outlines the object.

In the first picture, the wave front initially moves with zero shear radially away from a pointlike source. At small distances from the source gravitational deflections are small, so equation (14.27) is $d^2\sqrt{A}/d\lambda^2 = 0$, with the solution

$$A(t) = K(\delta\lambda)^2 = Ka_s^2(t - t_s)^2 \equiv \Omega_s(t - t_s)^2 .\qquad(14.49)$$

The constant of integration is K. Since the proper time kept by a comoving observer at the source is the world time t, at $t - t_s \ll t_s$ an observer would see that the ray has moved the affine distance $\delta\lambda = a_s(t - t_s)$ given by equation (14.40). The last expression defines the solid angle $\delta\Omega_s$ of a beam of rays as it leaves the source.

Suppose the source is isotropic and radiating $\dot{N}$ photons per unit time. Then in time δt_s the number of photons radiated into the beam is

$$\delta N = \dot{N}\delta t_s\Omega_s/4\pi = n_o\delta t_o A_o .\qquad(14.50)$$

The last expression is the number of photons detected by the observer at the epoch t_o, who sees that the proper beam area $A(\lambda)$ has grown to the area A_o of the observer's detector. The spread in arrival times of the photons is $\delta t_o = (1 + z)\delta t_s$, for the redshift factor $1 + z$ stretches the distance between photons in the same way it stretches the wavelength of each photon. This equation defines the observed number density n_o of photons (number of photons detected per unit time and per unit area normal to the beam). The energy flux f_o is the product of n_o and the energy per photon, $\hbar\omega_o = \hbar\omega_s/(1 + z)$:

$$f_o = \frac{L}{4\pi(1+z)^2}\frac{\Omega_s}{A_o} .\qquad(14.51)$$

The source luminosity is $L = \hbar\omega_e\dot{N}$. In a universe that is close to homogeneous, the beam area at the observer is given by equation (14.48), $A_o = \Omega_s(a_o r_o)^2$, and equation (14.51) then agrees with equation (13.51) for a homogeneous mass distribution.

The density fluctuations in the real world perturb $A(\lambda)$ through the fluctuations in the source term $T_{ij}u^i u^j$ in the area equation (14.27) and the shear term from the tidal field in the second of the optical equations (14.21). The shear causes an initially circular beam to evolve into an ellipse. At a caustic, the ellipse becomes a line with zero area and formally infinite flux per unit area in the approximation of ray optics. The effect is discussed in section 22, in the pancake picture for the first generation of bound systems after decoupling of matter and radiation, and by Blanford and Kochanek (1987) in gravitational lensing events.

In the second picture, one imagines the source is resolved at the observer, and one considers the beam of rays from the source that intersect at the observer and outline the cross section A_s of the source. The observed solid angle of the source is Ω_o, and equation (14.49) applied in the neighborhood of the observer shows that the beam area is

$$A = \Omega_o(t_o - t)^2.$$
(14.52)

As before, we will suppose the luminosity L of the source is isotropic. Then an observer at rest near the source at distance r that is small compared to the Hubble length sees that the source subtends solid angle A_s/r^2, the energy flux is $L/4\pi r^2$, so the mean surface brightness of the source is

$$i_s = L/(4\pi A_s).$$
(14.53)

The surface brightness at the observer is reduced by the redshift factor to $i_o = i_s/(1+z)^4$ (eq. [9.54]), and the observed energy flux is the product of i_o and the observed solid angle of the source in equation (14.52),

$$f_o = \frac{L}{4\pi(1+z)^4}\frac{\Omega_o}{A_s}.$$
(14.54)

The relation between Ω_s/A_o and Ω_o/A_s from equations (14.51) and (14.54) follows because the paths of the photons are described by Hamilton's equations, which give Liouville's theorem for the density in single-particle phase space.

Orders of Magnitude

We can consider separately the rare cases where the expansion θ and shear w are dominated by the effect of a single mass concentration near the line of sight, as in a lensing event, and the sum of the contributions of the numerous smaller perturbations to θ and w by the mass fluctuations along the line of sight (as has been discussed by Refsdal 1970; Vietri 1985; Schneider and Weiss 1988; Futamase and Sasaki 1989; Watanabe and Tomita 1990; and others). We begin by reviewing some useful orders of magnitude characteristic of a lensing event. This is discussed further in section 18, in connection with microlensing by objects with planetary to stellar masses, and in section 20 for lensing by clusters of galaxies.

A compact mass $\mathcal{M}$ halfway to an object at the Hubble distance c/H_o produces multiple images of the object if the impact parameter, r, and the gravitational deflection angle, α, satisfy

$$\alpha \sim \frac{G\mathcal{M}}{rc^2} \sim \frac{H_o r}{c}.$$
(14.55)

Thus, the impact parameter is

$$r \sim (G\mathcal{M}/H_o c)^{1/2}.$$
(14.56)

The volume of space within which we can place the lens to get two similar images of the object is fixed by the product of the cross section $\sim r^2$ and the Hubble length cH_0^{-1}, so if the space number density of lenses is n the lensing probability is

$$P \sim \frac{cr^2n}{H_0} \sim \frac{G\mathcal{M}n}{H_0^2} \sim \frac{\rho_{gl}}{\rho_{crit}} = \Omega_{gl}. \qquad (14.57)$$

Here ρ_{crit} is the Einstein-de Sitter mass density (eq. [5.67]), and Ω_{gl} is the contribution to the density parameter by the mean mass density $\rho_{gl} = \mathcal{M}n$ in the lenses (Press and Gunn 1973). We see from equation (14.56) for the impact parameter r for a lensing event that the characteristic value of the mean mass per unit area at the lens within the path of the deflected ray is

$$\Sigma_{gl} \sim \frac{\mathcal{M}}{r^2} \sim \frac{H_0 c}{G}. \qquad (14.58)$$

This is the critical value of the surface density of a mass concentration capable of lensing objects at the Hubble distance. The mean surface density of the universe to the Hubble length is $\Sigma_u \sim \rho_b c/H_0 \sim \Omega H_0 c/G$. If the density parameter Ω is close to unity, this is the critical lensing density, because the universe is magnifying objects at $z \sim 1$.

If the velocity dispersion in the lensing mass is σ the mass within the lensing radius r is $\mathcal{M} \sim \sigma^2 r/G$, and with equation (14.56) this is $r \sim \sigma^2/cH_0$ and $\mathcal{M} \sim \sigma^4/cH_0 G$. Then the lensing probability (14.57) is $P \sim n\sigma^4/cH_0^3$, as in equation (13.71).

To get more accurate numbers, let us use the limiting isothermal gas sphere model in equation (3.32), which gives a reasonable approximation to the mass distribution in a galaxy or a cluster of galaxies. In this model, the mass density ρ as a function of radius and the surface density Σ as a function of impact parameter are

$$\rho = \frac{\sigma^2}{2\pi Gr^2}, \qquad \Sigma = \frac{\sigma^2}{2Gr}, \qquad (14.59)$$

where the line-of-sight velocity dispersion is σ. Consider a lens at redshift z_l placed in line with an object at redshift z_s. The impact parameter of the line of sight at the lens is given by equation (13.69), and the mass per unit area at the line of sight in the mass model in equation (14.59) is

$$\Sigma_{gl} = \frac{cH_0}{8\pi G} \frac{(1+z_l)y_{os}}{y_{ol}y_{ls}}. \qquad (14.60)$$

In a cosmologically flat model, where the dimensionless angular size distance from observer to object is $y_{os} = y_{ol} + y_{ls}$, this surface density is minimum, and the lensing probability is maximum, when $y_{ol} = y_{os}/2$. In the Einstein-de Sitter model, where

$$y = H_o a_o r(z) = 2 \left[1 - \frac{1}{(1+z)^{1/2}} \right] , \qquad (14.61)$$

an object at redshift $z_s = 3$ is at angular size distance $y_{os} = 1$, the redshift of a lens at distance $y_{ol} = y_{os}/2$ is $z_l = 7/9$, and the critical surface density (14.60) is

$$\Sigma_{gl} = \frac{8}{9\pi} \frac{cH_o}{G} = 0.41 \, h \, \text{g cm}^{-2} . \qquad (14.62)$$

It is a charming but presumed not significant coincidence that this is a laboratory-size quantity.

In the distributed mass model in equation (14.59) the surface density is

$$\Sigma = \frac{0.24}{r_{kpc}} \left(\frac{\sigma}{100 \, \text{km s}^{-1}} \right)^2 \text{g cm}^{-2} . \qquad (14.63)$$

This is equal to the critical density in equation (14.62) at impact parameter

$$r_{gl} = 0.6 h^{-1} \left(\frac{\sigma}{100 \, \text{km s}^{-1}} \right)^2 \text{kpc} , \qquad (14.64)$$

and the mass enclosed at this radius is

$$M_{gl} = 3 \times 10^9 h^{-1} \left(\frac{\sigma}{100 \, \text{km s}^{-1}} \right)^4 M_\odot . \qquad (14.65)$$

For an L_* spiral galaxy, the typical dispersion is $\sigma = 220/2^{1/2}$ km s^{-1} (eq. [3.34]), so the surface density reaches the critical value at $r \sim 1.5 h^{-1}$ kpc, where the enclosed mass is $M_{gl} \sim 10^{10} h^{-1} M_\odot$. The space number density of these galaxies is $n \sim 0.02 h^3$ Mpc^{-3}, so the mean density of the mass active in lensing, $n M_{gl}$, works out to $\Omega_{gl} \sim 0.001$. This can be compared to equation (13.76).

Though the lensing probability, Ω_{gl}, for spiral galaxies is small (and comparable to the probability for lensing by ellipticals), it does happen, as in Huchra's lens (Huchra et al. 1985; Schneider et al. 1988), where a spiral galaxy has produced multiple images of a background quasar. Huchra's lens would have a considerably smaller effect on the energy flux density from a background galaxy, however, because the gravitational lensing is rearranging light rays on the scale $r_{gl} \sim 1 h^{-1}$ kpc, which is smaller than the typical size of the image of a background galaxy. Thus, the conclusion is that most of the mass in and around galaxies contributes to the expansion and shear of a typical light beam through the accumulated perturbations of the many galaxies along the line of sight, rather than in single large events.[5]

In a cluster of galaxies a typical central velocity dispersion is $\sigma = 1000$ km s^{-1}, so equation (14.64) indicates the lensing impact parameter is $r \sim 50h^{-1}$ kpc. This is comparable to or even somewhat smaller than the typical cluster core radius at the inner cutoff for the $\rho \sim r^{-2}$ density run. That is, one expects that not all clusters are capable of lensing, and that in those that are the mass involved is a small fraction of the total in the cluster. Lensing is observed in the giant arcs discussed in section 20, but we can conclude again that lensing by a single mass complex is rare.

Let us consider next the accumulated effect of small contributions to the expansion θ and shear w by the many mass fluctuations along the line of sight. We will estimate the effect on θ and w in two models for a single mass fluctuation, and then find the expected values for the sum along a line of sight.

The first model is based on an order-of-magnitude estimate of the perturbation h_{ij} to the metric tensor by a mass concentration. For this purpose, it is convenient to follow Futamase and Sasaki (1989) in writing the line element as

$$ds^2 = a(t)^2(\eta_{ij} + h_{ij})\, dx^i dx^j , \qquad (14.66)$$

where $a(t)$ is the expansion parameter for the homogeneous model with density equal to the mean value for the lumpy universe, and η_{ij} is the Minkowski metric tensor. The time coordinate $dx^0 \sim dt/a$ in this expression is the conformal time that makes the unperturbed metric tensor proportional to (a conformal mapping of) the Minkowski form. We can ignore the mean curvature of constant time sections, because known mass concentrations are much smaller than the Hubble length. It is reasonable to compute to first order in the h_{ij}, because spacetime gives no sign of large curvature fluctuations (apart from the exceedingly small fraction of space that may be occupied by black holes).

A mass concentration spread over comoving radius x_σ, or physical radius $r_\sigma = ax_\sigma$, and supported by velocity dispersion σ has Newtonian binding energy $\phi \sim \sigma^2$. We see from equation (10.84) that the perturbation to the metric tensor is

$$h_{ij} \sim \phi \sim \sigma^2 . \qquad (14.67)$$

The Christoffel symbol Γ^i_{jk} is a derivative of the metric tensor multiplied by the

---

[5] As discussed in section 18, the dark halo mass may be dominated by low mass stars or by star remnants. If so, there is an appreciable probability for microlensing by the individual dark mass objects, but we see from equation (14.56) that at $\mathcal{M} \sim 1\mathcal{M}_\odot$ the lensing impact parameter is $r \sim 0.01\, h^{-1/2}$ pc, so the rearrangement of rays within the beam of light from a galaxy is negligibly small compared to the beam size. Microlensing can have a significant effect on the much tighter beams from quasars.

reciprocal tensor (eq. [8.57]), so the order of magnitude is

$$\Gamma^i_{jk} \sim \frac{1}{x^0}, \quad \text{or} \quad \frac{h}{x_\sigma}.$$ (14.68)

The terms linear in h in the curvature tensor $R^i{}_{jkl}$ in equation (8.77) are the gradients of the Γ^i_{jk}, so the dominant linear terms in $R^i{}_{jkl}$ from the mass fluctuation are $\sim h/x_\sigma^2$. Lowering the index multiplies this by a^2, so the Weyl tensor (8.84) is

$$C_{ijkl} \sim a^2 h/x_\sigma^2 .$$ (14.69)

The normalization condition (14.4) says the L^i_α are $\sim 1/a$, and the components of the four-momentum $u^i = dx^i/d\lambda$ in equations (14.40) and (14.42) for a light ray are $\sim 1/a^2$ (with $dx^0 \sim dt/a$ in the coordinates in eq. [14.66]). Thus, the source term for the shear in equation (14.21) is

$$C_{ijkl}L^i_\alpha u^j u^k L^l_\beta \sim \frac{\sigma^2}{r_\sigma^2 a^2} .$$ (14.70)

Since $dt/d\lambda \sim 1/a$, the affine distance across the fluctuation is $\delta\lambda \sim ar_\sigma$, so the second line of equation (14.21) says the contribution to the shear by this mass fluctuation is

$$\delta w \sim \frac{\sigma^2}{ar_\sigma} .$$ (14.71)

In a similar way, one sees that the Ricci focusing source term in the area equation (14.27) has two contributions, $R_{jk}u^j u^k \sim 1/(at)^2$ from the mean expansion rate and $R_{jk}u^j u^k \sim h/(ar_\sigma)^2$ from the fluctuating part of the metric tensor. The latter causes a change in the expansion across r_σ that is comparable to δw in equation (14.71).

For a more specific model of $\delta\theta$ and δw, let us use the mass distribution in equation (14.59), where the gravitational bending angle is $\alpha = 4\pi\sigma^2$. Suppose an initially parallel beam of light, with $\theta = w = 0$, is incident at impact parameter r (with $r > 0$ so the beam does not contain the origin). Let d_r be the width of the beam along the projected radius to the center of the mass concentration, and let d_ϕ be the beam width in the orthogonal direction. Since the gravitational deflection angle α is independent of the impact parameter, the initially parallel rays separated by distance d_r along the radial line remain parallel after the deflection, while rays separated by distance d_ϕ at the same impact parameter r end up converging at the rate

$$\frac{d\mathrm{d}_\phi}{dt} = -\frac{\alpha \mathrm{d}_\phi}{r} .$$ (14.72)

Equation (14.12) defines the expansion and shear. We have $\delta\theta + \delta w_{rr} = 0$ along the radial direction, and, from equation (14.72), $\delta\theta + \delta w_{\phi\phi} = -4\pi\sigma^2/ar$ in the transverse direction. Thus, the expansion and shear produced by the mass concentration are

$$\delta\theta = -\delta w_{rr} = \delta w_{\phi\phi} = -\frac{2\pi\sigma^2}{ar}, \tag{14.73}$$

consistent with equation (14.71).

In the integrated effect of many deflections, the shear $w_{\alpha\beta}$ adds as a random walk, because successive contributions are uncorrelated (determined by the position angle of each mass concentration along the line of sight). The contributions $\delta\theta$ to the expansion, computed relative to a line of sight that contains no mass, accumulate monotonically, because $\delta\theta$ in equation (14.73) always is negative. This has two important consequences. First, the integrated effect of many mass fluctuations along the line of sight is much larger in the expansion θ than in the shear w. Second, since θ is the sum of many contributions, the value along a typical line of sight may be expected to be close to the mean computed from the simplified homogeneous mass distribution.

To estimate the relative values of w and θ, suppose all the mass is in randomly placed concentrations, each with velocity dispersion σ and density run $\rho \propto 1/r^2$ truncated at $r \sim R$, so the mass is $\mathcal{M} \sim 2\sigma^2 R/G$ (eq. [14.59]). The number density of concentrations is n, the mean mass density is

$$\rho = \frac{2\sigma^2 Rn}{G} = \frac{3H_o^2\Omega}{8\pi G}, \tag{14.74}$$

so the cutoff is

$$R = \frac{3}{16\pi}\frac{H_o^2\Omega}{\sigma^2 n}. \tag{14.75}$$

The expansion and shear are sums from the concentrations near the line of sight. The ensemble average values of the sums are computed as in equation (7.66). Set $n_i = 1$ if a concentration is found to be centered in the volume element δV_i, making a contribution $\delta\theta_i$ to the expansion. Then $\langle n_i \rangle = \langle n_i^2 \rangle = n\delta V_i$, and for disjoint elements $\langle n_i n_j \rangle = \langle n_i \rangle \langle n_j \rangle$, because we are assuming the concentrations are placed independently, as a Poisson process. The mean value and variance of the expansion are integrals of the form

$$\theta = \langle\theta\rangle = \int 2\pi r \, dr dt \, n\delta\theta(r),$$

$$\Delta\theta^2 = \langle(\theta - \langle\theta\rangle)^2\rangle = \int 2\pi r \, dr dt \, n[\delta\theta(r)]^2. \tag{14.76}$$

For a numerical estimate let us neglect the expansion of the universe and set the path length equal to the Hubble length. Then we have from equation (14.73)

$$\theta \sim 4\pi^2 \sigma^2 n_o H_o^{-1} R/a_o \sim H_o \Omega/a_o, \qquad (14.77)$$

where the last expression uses equation (14.75) for R, and

$$\Delta\theta^2 \sim \langle w^2 \rangle \sim \int 2\pi r \, dr \, dt \, n \frac{4\pi^2 \sigma^4}{a^2 r^2} \sim 8\pi^3 \sigma^4 n_o H_o^{-1} \ln(R/r_c)/a_o^2. \quad (14.78)$$

Here r_c is the critical radius for the rare lensing events. The contribution to the shear by mass concentrations at impact parameter greater than R is small because at large radius $\delta w \propto r^{-2}$.

The mean value of the expansion in equation (14.77) corresponds to what one computes in the homogeneous cosmological model. The ratio of the shear to the expansion is

$$\frac{w^2}{\theta^2} \sim \frac{H_o \ln R/r_c}{\pi n_o R^2} \sim \frac{\sigma^4 n_o}{H_o^3 \Omega^2} \ln R/r_c. \qquad (14.79)$$

As one would expect, this is on the order of the lensing probability (13.71), which we know is small.

The conclusion is that a small fraction of the mass of the universe is in concentrations with surface densities and sizes large enough to have an appreciable effect on the angular size or apparent magnitude of a galaxy at the Hubble distance. The evidence is that the bulk of the mass is in concentrations with lower surface densities, or lower sizes at r_{gl} in equation (14.64), whose integrated effect is considerably larger in the expansion than in the shear of the beam of light from a galaxy at the Hubble distance. This would mean that the value of the expansion along a typical line of sight is close to the average computed from the homogeneous cosmological model.

These arguments are comforting, but it would be well to bear in mind that our picture for the mass distribution is schematic. Perhaps there are serious perturbations from mass concentrated in "dark galaxies," with velocity dispersions and sizes that make r_{gl} comparable to the sizes of seen galaxies, or from mass concentrations on very large scales that approach the lensing criterion in equations (14.63) and (14.65). More detailed computations will be of increasing interest as the cosmological tests and the picture for the mass distribution improve.

III. Topics in Modern Cosmology

The conclusion from the survey of observational developments in parts one and two is that the standard hot evolving cosmological model has survived some impressive tests that seem to have left no credible alternatives among the reasonably well-specified cosmologies. There is ample room for improvement within the standard model, however, and certainly also the chance that will lead us to significant revisions in the world picture. The purpose of this final part is to survey the state of progress in the search for improvements. The observational advances are striking, though as usual they tend to go in unexpected directions. For example, observations of objects at high redshifts have not yet revealed how galaxies formed, but they have given a remarkably detailed picture of the state of the intergalactic medium back to expansion factor $1 + z \sim 6$. The recent directions of evolution of theoretical ideas have been only weakly correlated with the observational developments, but we will see that there are quite promising possibilities for the growth of a tighter interaction between theory and phenomena.

15. Challenges for the Standard Model

This review of conundrums in cosmology, and some of the proposed nostrums (which is based on the earlier survey in Dicke and Peebles 1979), is meant to be a preparation for the remarkable variety of ideas on how the standard model might be made more complete. Most of the conundrums have been recognized for a long time, though not always heavily advertised. A few of the nostrums also are classical, but most were invented in the last fifteen years. The following two sections deal with some of these newer ideas.

Initial Conditions

The standard presumption is that deeper physics yet to be discovered will someday give us an elegant, reasonable, and testable picture for what happened "before the big bang," whatever that means, and "after the big crunch," for the parts of the universe large and small that are fated to collapse back to states that become singular (within general relativity theory). The following comments on initial conditions are meant to emphasize how clear the need is for new physics for a more complete cosmology. The remaining sections in this book are meant to show that progress in cosmology does not await this great discovery, for we have a rich menu of well-posed problems within the standard model and reasonable extensions of it.

Initial conditions play quite different roles in different parts of physics. In quantum theory, initial conditions are the trivial part of the problem, because one is dealing with the ground state or relatively mild excitations. A fully satisfactory physical cosmology would require no initial conditions, for what can it mean to say the universe had an initial condition? Surely whatever happened today flowed out of what things were like yesterday? If this is the correct way to state the goal, we are a long way from it, for as we have noted the standard model applied without adjustment traces back in time to a singularity. The standard and surely correct wisdom is that this paradox ultimately will prove to be a good thing, which will force us to address the fundamental issues of gravitational collapse and lead us to a deeper understanding of the material world. But as a practical matter it does mean that debate in theoretical cosmology tends to center on how models for the physics of the early universe might be tuned, or the initial conditions otherwise postulated, so that the universe ends up with the structure we see around us.

The search for initial conditions for the classical physics of the standard cosmology may seem to be a vacuous exercise in adjusting one set of free functions to fit another, but that need not follow. For example, one way to fit the present mass distribution would assume that at high redshifts the universe was quite inhomogeneous, and that the velocities happened to be directed so as to present us with a distribution that is close to homogeneous now, and moving toward inhomogeneity again. We do not take this seriously because it is contrived, and because it is not consistent with the present galaxy velocity field that is so close to the homogeneous Hubble flow. A more serious challenge will be to decide whether we can find initial conditions for the standard model that lead to the development of mass concentrations comparable to those observed in the great clusters and superclusters of galaxies without overperturbing the spectrum and isotropy of the 3K thermal cosmic background, the isotropy of the X-ray background radiation, and whatever other tests one can devise. If it were shown that this is not possible, it would be an exceedingly important signal that we have missed something important. As far as is now understood (from the analyses of models in sections 21 to 25) this has not yet happened, though the constraints on acceptable scenarios are growing tight.

If it continues to be possible to adjust the standard model to fit the observations, the great challenge will be to find a physical account of the initial conditions for classical cosmology, which have to be quite special. For example, if the rms fluctuations in galaxy counts on the scale of $30h^{-1}$ Mpc in equations (3.24) and (7.73) reflect mass fluctuations, and these fluctuations grew by gravitational instability, then the space curvature fluctuations on this comoving scale at very high redshift are $h = (x/R_i)^2 \sim \delta_h \sim 3 \times 10^{-5}$ (eqs. [5.127] and [10.127]), a small but certainly nontrivial value.

In the search for an explanation of why our universe is close to but not quite homogeneous, one might be tempted by the lesson from nongravitating systems, in which statistical homogeneity is a sign of relaxation: a well-stirred cup of coffee shows no evidence the sugar once was on the bottom. There are two things wrong with this. Misner (1968 and in private communications) emphasized that the standard cosmological model (with pressure that is not negative) has a particle horizon, such that we can see distant objects that cannot have communicated with each other subsequent to the singular start of the expansion (eq. [5.51]). How could different parts of the universe have known to relax to statistically similar states? The second problem is that the Friedmann-Lemaître universe is gravitationally unstable (eqs. [5.87] and [10.123]). That means the universe has to be growing more lumpy (as long as the mass density term in the expansion rate is appreciable, so the universe is self-gravitating), and that the parts of the universe we see had to have been highly ordered at large redshifts in order to have presented us with a state that is close to homogeneous on the scale of the Hubble length. That is, the standard model evolves from order to chaos, not the other way around.

The inflation scenario offers the most elegant way proposed so far to understand why the universe is arranged so remarkably well. As will be discussed in section 17, in this picture we are in a large but finite region outside of which the universe is behaving in quite different ways from what we observe. Dicke (1961) noted that an elementary but essential consistency check in cosmology is that we could only be in a universe that allows galaxies of stars to form and exist for some billions of years, to produce the material out of which we are made and the time to use it. In inflation this could translate into the argument that the universe is chaotic, but that most parts are too hot or dense or dilute to support life that can observe the chaos. Perhaps we could only have lived in a region that happens to be quite close to homogeneous in quite a large volume, as the inflation picture indicates does happen here and there. If so, it may not be surprising that we see none of the chaos. On the other hand, it is difficult to see how our chances for survival would have been lessened, or our ancestors otherwise disturbed, if the mass density fluctuations on the scale of the Hubble length were a good deal larger than they are. An alternative conjecture is that the universe is as close to homogeneous as it can be, for reasons to be found in some deeper theory yet to be discovered, and that structures exist in our universe because the discrete nature of material does not allow exact homogeneity. What might be the effect that breaks homogeneity? As discussed in section 16, a fascinating possibility is that a phase transition in the early universe left structures in some cosmic field, in the forms of strings or global monopoles or textures, which disturbed the baryon distribution and triggered galaxy formation.

In the standard model the predictions for the future seem as unsatisfactory as for our distant past. If the universe eventually stops expanding and collapses, what do we do about the future singularity? If the expansion continues indefinitely we have the curious situation that after the universe commenced classical Friedmann-Lemaître expansion, it passed through a phase lasting a few tens of billions of years, when it was a suitable home for observers as we understand them, and then it will spend the rest of eternity in an asymptotically quiescent state punctuated by the occasional burst of activity as a gravitationally bound system such as a galaxy finally loses enough energy by evaporation of stars or emission of gravitational radiation to collapse into a black hole. This seems wasteful, though perhaps we are not well positioned to judge.

The Dicke Coincidences

Equation (5.18) for the expansion rate is

$$\left(\frac{\dot{a}}{a}\right)^2 = \frac{8}{3}\pi G\rho_b \pm \frac{1}{a^2R^2} + \frac{\Lambda}{3}. \tag{15.1}$$

Each term on the right-hand side varies with time in a different way, so different terms may dominate the expansion rate at different epochs in the history of the universe. We know that in the present epoch the mass density term is no less than about one percent of the total (eq. [5.150] and table 20.1). Since the mass density term varies most rapidly with the expansion parameter $a(t)$, we know that if the expansion of the universe traces back to high redshift, the early universe behaved like a model with negligible space curvature and cosmological constant. (This is the Einstein-de Sitter limit in eq. [5.61]). That means the initial conditions for classical cosmology set two characteristic expansion factors: one for the epoch at which the mass density ceases to dominate the expansion rate, assuming this happens; and one for the epoch at which we have come on the scene. Dicke (1970) emphasized that it would be a curious coincidence if these two exceedingly large expansion factors agreed to an order of magnitude or so.

To get some definite numbers, let us consider the epoch of light element production in the standard model, for this gives a reasonable-looking account of the abundances of the light elements (figure 6.5). The processes relevant to element production start at redshift $z \sim 10^{10}$, when the mass density term is dominated by relativistic matter, in radiation, neutrinos, and thermal electron pairs. The present contribution to the expansion rate by radiation and neutrinos is $\Omega_r \sim 10^{-4}$ (eq. [6.80]). Since the mass density in relativistic matter scales as $(1+z)^4$, at redshift $z \sim 10^{10}$ the mass density term in equation (15.1) is thirty-six orders of magnitude larger than the present value of the sum in equation (15.1). The curvature term on the right-hand side scales as $(1+z)^2$, so it was twenty orders of magnitude larger at $z \sim 10^{10}$. If the curvature term dominates now, then at $z \sim 10^{10}$ the curvature term is down from the mass density term by sixteen orders of magnitude. This means that at the start of light element production the mass density and the expansion rate in equation (15.1) are in balance to an accuracy of at least one part in 10^{16}. If the expansion rate now is dominated by a cosmological constant, the balance of kinetic and gravitational potential energies at $z \sim 10^{10}$ is even tighter. And one presumes that we can apply this calculation at still higher redshifts, to get an even finer balance of kinetic and potential energies. Two questions naturally arise. First, why was there such a remarkable balance between the effective expansion kinetic energy per unit mass, which is proportional to $\dot{a}^2$, and the potential energy, proportional to $G\rho a^2$? Second, is it reasonable to think that we might have come on the scene just as this balance was disappearing? Two lines of thought, both pioneered by Dicke (1961, 1970), seem relevant.

Although the balance of the initial conditions for the kinetic and potential energy of expansion in the very early universe truly is remarkable, consistency with our existence seems to require it. We need a galaxy to contain the debris from a few generation of stars and allow it to collect and form a solar system rich in the heavy elements out of which we and our planet are made. If at the epoch of light element production kinetic and potential energy had not been in very close balance, the universe might have collapsed soon after, leaving no time for any of this. Almost as bad is the alternative, that the kinetic energy is well in excess of the potential energy, so the universe enters a free expansion that suppresses the gravitational instability before radiation drag allows material to collect into galaxies. One might even imagine that the redshift z_{dec} at decoupling of matter and radiation (eq. [6.96]) had to have been about what it is, for as discussed in section 6 it sets a reasonable epoch for the commencement of the sequence of events that led up to galaxy formation. These considerations, which have come to be called the Dicke (1961) anthropic principle, guide our interpretations of what is observed and test our stories for what really happened.

To make more explicit the Dicke coincidences argument for the relative sizes of the terms in the expansion rate equation (15.1), suppose the classical cosmological model first became a useful approximation at epoch a_i. Whatever set the initial conditions at this epoch fixed the epoch a_f at which the mass density ceases to be the dominant term in the expansion rate, and the epoch a_o at which we appear as observers. We noted that the evidence from the successful theory of the production of the light elements is that a_f/a_i and a_o/a_i are very large numbers. It would be a curious coincidence if the initial conditions assigned comparable values to these two numbers. Since consistency with our existence does not allow a_f to be very much smaller than a_o, the likely situation is that $a_f \gg a_o$, that is, that the universe now is well approximated by the Einstein-de Sitter model.

The balance condition on the initial values of the effective kinetic and potential energies of expansion of the universe, and the argument for the Einstein-de Sitter model, were part of the motivation for Guth's invention of the inflation scenario to be discussed in section 17, and the great interest in this scenario in turn made these considerations highly visible. Before inflation, the point was known (and in the group around Dicke at Princeton part of the standard lore for at least a decade before he finally put it into print in 1970), but it was not generally considered compelling. For example, Robertson (1955) noted without explanation that the Einstein-de Sitter case is "of some passing interest." Bondi (1960) indicated that this case "deserves attention" because it is simple, and it has the special property that the ratio of the effective potential and kinetic energies, $\propto G\rho a^2/\dot{a}^2$, is independent of time, but the Einstein-de Sitter case was not the centerpiece for his survey of the relativistic models. Now the power of the argument is fully appreciated, and sometimes even taken as a demonstration that the standard model has to be Einstein-de Sitter. That may turn out to be true, but it is not the situation as now understood. The inflation scenario as an explanation for homogeneity does require that the universe has negligibly small space curvature (eq. [17.23]), consistent with the Einstein-de Sitter model, but despite the elegance of this picture for the early universe there is little objective evidence on which to decide whether the inflation scenario really is valid. On the observational side, as will be discussed beginning in section 18, there are difficulties in accounting for the high mass density predicted in the Einstein-de Sitter model. And one must bear in mind that the argument is based on a coincidence, and that numbers can accidentally coincide. An example may be useful.

The Eddington limit on the luminosity L of a star of mass $\mathcal{M}$ is set by the balance of gravity and radiation pressure in the envelope. At distance r from the center of the star the energy flux is $f = L/(4\pi r^2)$, the momentum flux is f/c, and the rate of transfer of momentum to a free electron is $\sigma_t f/c$, where the Thomson cross section is given in equation (6.120). If the envelope is bound to the star this radiation pressure force has to be balanced by gravity. Since there is about one proton per electron in the fully ionized material in a star, the balance condition is

$$\frac{\sigma_t L}{4\pi r^2 c} < \frac{G\mathcal{M}m_p}{r^2}. \tag{15.2}$$

The bound on the luminosity per unit mass is then

$$\frac{L}{\mathcal{M}} < \frac{4\pi G m_p c}{\sigma_t} = 3 \times 10^4 \frac{L_\odot}{M_\odot} = 6 \times 10^4 \, \text{erg s}^{-1} \text{g}^{-1}. \tag{15.3}$$

My mass is 80 kg, and my luminosity is about 2000 kcal per day. The ratio works out to

$$L/\mathcal{M} = 1 \times 10^4 \, \text{erg s}^{-1} \text{g}^{-1}, \tag{15.4}$$

within a factor of six of the Eddington limit, a remarkable coincidence! But for practical purposes it is only an accident of essentially unrelated numbers.

The moral for cosmology is that although the Einstein-de Sitter model is the most elegant case it is dangerous to rule out the possibility that the density parameter is appreciably different from unity. It would be a double coincidence if at the present epoch evolution had brought both the curvature and Λ terms to values comparable to the mass density term in equation (15.1). Thus it would seem to be rational to consider after the Einstein-de Sitter case the possibility that there is an appreciable contribution to the expansion rate either from space curvature or from a component that acts like a cosmological constant, but not both. This is the basis for the choice of parameters in the numerical examples in section 13.

Phoenix Universe

Before inflation the most popular idea for how one might deal with the conundrum of the very early universe was Lemaître's (1933) phoenix picture, in which our universe is expanding away from a collapse that terminated the end of the previous contracting phase in the last cycle of an oscillating universe. If we could trace the evolution of the universe back through a bounce, it would of course mean the particle horizon has been eliminated. During maximum expansion of a closed oscillating universe, free streaming would smooth the distributions of radiation and relativistic particles, perhaps keeping the universe close to homogeneous deep into the collapse, when new physics outside the standard model might be supposed to cause a bounce. It may be an added bonus that if the bounce preserved magnetic flux the fields from the last generation of galaxies would be present ahead of time, and might even provide nucleation sites for the next generation. As discussed in section 6, it was the entropy production in a phoenix universe that led Dicke to suggest a search for the relict thermal cosmic background radiation. Tolman (1934) showed that if the bounce preserved entropy the entropy production in each cycle would make the next last longer, as follows.

Consider a closed universe with mass dominated by a uniform sea of blackbody radiation. The volume is $2\pi^2(aR)^3$ (eq. [5.11]), where R is the curvature parameter in equation (15.1). The mass density in blackbody radiation is $\rho = a_B T^4$, and the entropy density is $s = 4a_B T^3/3$ (eq. [6.51]), so the total entropy (neglecting other relativistic quanta) is

$$S = \frac{8\pi^2}{3}a_B(aRT)^3.\tag{15.5}$$

Since $T \propto 1/a(t)$, we can write the expansion rate equation (15.1) with $\Lambda = 0$ as

$$\left(\frac{\dot{T}}{T}\right)^2 = \frac{8}{3}\pi G a_B T^4 - \left(\frac{8\pi^2 a_B}{3S}\right)^{2/3} T^2.\tag{15.6}$$

In this model the temperature at maximum expansion, where $\dot{T} = 0$, is $T_m \propto S^{-1/3}$. The time between bounces is $t_m \sim R a_m \propto S^{1/3}/T_m \propto S^{2/3}$, where $a_m R$ is the size of the universe at maximum expansion. The total entropy S increases during each oscillation. If S is conserved in the bounce, we see that each cycle is cooler and lasts longer (Tolman 1934). If there were some provision to create baryons to keep the ratio of baryon to photon number densities constant, one might imagine that the oscillations continue into the indefinite future, although the history would have to trace back to a universe containing just a few quanta.

In the standard Friedmann-Lemaître model an oscillating universe requires that the expansion rate equation (15.1) has two zeros, one of which can be obtained at large expansion parameter by balancing the mass density and space curvature, while the other has to be at a temperature in excess of 10^{10} K if we are to save the standard model for the origin of helium. The latter requires that some new field makes a negative contribution to the total mass density, with magnitude large enough to have a value of the expansion parameter at which the total density vanishes, so $\dot{a} = 0$, and the collapsing universe bounces. That this could happen is indicated by an example given by Bekenstein (1975). For a simpler and more desperate example, suppose we introduce a scalar field with Lagrangian density $\mathcal{L}_\phi = -g_{ij}\phi_{,i}\phi_{,j}$. The field equation (10.4) has the spatially homogeneous solution $\phi \propto a(t)^{-3}$, for which the energy density is $\rho_\phi = -\dot{\phi}^2 \propto a^{-6}$. This varies with the expansion parameter more rapidly than the energy density in radiation, as wanted. It is at best an exceedingly schematic picture, however, because the negative kinetic energy in this Lagrangian means the fluctuations from homogeneity have negative energy, which is not really acceptable.

Planck Limit

In the standard picture for the state of the universe at redshift $z \sim 10^{10}$, which gives what seems to be an observationally successful and believable account of the origin of helium, the expansion of the universe is described by the classical Friedmann-Lemaître model, with the mean mass density dominated by radiation, electron pairs, and neutrinos. This picture has to fail at high enough redshift, for the expansion traces back to a situation in which the mass within the Hubble length is dominated by just a few quanta. When this happens it no longer makes sense to consider a classical gravitational field equation with source term equal to the expectation value of the stress-energy tensor for matter and radiation. Some of the characteristic numbers for the redshift at which this happens are collected here.

The Planck mass is

$$m_{pl} = (\hbar c/G)^{1/2} = 2.18 \times 10^{-5} \, g, \tag{15.7}$$

the Planck energy is

$$m_{pl}c^2 = (\hbar c^5/G)^{1/2} = 1.22 \times 10^{28} \, eV, \tag{15.8}$$

and the Planck time is

$$t_{pl} = (G\hbar/c^5)^{1/2} = 5.38 \times 10^{-44} \, s. \tag{15.9}$$

These expressions are simplified by choosing units with $c = 1$, $\hbar = 1$, and Boltzmann's constant $k = 1$, so mass, energy, and temperature all have units of reciprocal length, and mass density has units of the fourth power of a reciprocal length. In these units the Planck mass is

$$m_{pl} = G^{-1/2}, \tag{15.10}$$

and the characteristic wavelength at the median energy of the blackbody spectrum is (eq. [6.54])

$$\lambda = 0.3/T. \tag{15.11}$$

At high redshifts the energy density in the relativistic components that are in thermal equilibrium at temperature $T(t)$ is

$$u_\gamma = \frac{\pi^2}{30} T^4 g(T), \tag{15.12}$$

where g counts the effective number of degrees of freedom, or fields, contributing to the energy density. Each boson component with mass and chemical potential small compared to T contributes $g = 1$, so the two polarizations of the electromagnetic field give $g = 2$, as in equation (6.14). Each fermion component contributes $g = 7/8$ (eq. [6.70]). Thus in section 6 in the discussion of helium production we had at $T \sim 1$ MeV the sum of two polarizations for the electromagnetic radiation, 7/2 for the four electron-positron spins, and 21/4 for the three families of neutrinos, giving $g = 43/4$. At the end of deuterium burning the situation is a little more complicated, because the neutrinos are no longer in thermal equilibrium with the radiation. At higher redshifts, the number of degrees of freedom increase as the μ and τ leptons are thermally excited and the baryons decompose into almost free quarks and gluons.

Let us compute under the assumption that nothing much is added to g prior to that, so the number of degrees of freedom at very high redshifts is close to constant at $g \sim 100$. When g is constant the temperature scales as $T \propto 1/a(t)$ (eq. [6.3]), so equation (15.1) for the expansion rate is

$$\left(\frac{\dot{a}}{a}\right)^2 = \left(\frac{\dot{T}}{T}\right)^2 = \frac{4\pi^3}{45} \frac{g T^4}{m_{pl}^2}, \tag{15.13}$$

with the solution

$$t = \left(\frac{45}{16\pi^3 g}\right)^{1/2} \frac{m_{pl}}{T^2}. \tag{15.14}$$

The characteristic de Broglie wavelength λ in the thermal sea is $\lambda = (3.5T)^{-1}$ for bosons, and the value is similar for fermions, so the ratio of the wavelength to the expansion timescale (15.14) is

$$\frac{t}{\lambda} = 2\frac{(tm_{\mathrm{pl}})^{1/2}}{g^{1/4}} .\tag{15.15}$$

The typical energy of a field quantum is $\epsilon \sim 2\pi/\lambda$, so its gravitational self-energy ϵ_g is

$$\frac{\epsilon_g}{\epsilon} \sim \frac{G\epsilon}{\lambda} \sim \frac{20}{g^{1/2}m_{\mathrm{pl}}t} .\tag{15.16}$$

In the standard cosmological model, the material world is described by quantum physics operating in the classical geometry of general relativity theory. At modest redshifts this is an excellent approximation, because there is an enormous number of quanta within the Hubble length and the cosmological expansion rate is much slower than the de Broglie frequencies of the fields. Equation (15.15) indicates that the latter fails at the epoch

$$t_1 \sim \frac{g^{1/2}}{m_{\mathrm{pl}}} ,\tag{15.17}$$

and we see from equation (15.16) that the gravitational self-energies of the quanta become significant at

$$t_2 \sim \frac{20}{g^{1/2}m_{\mathrm{pl}}} .\tag{15.18}$$

If $g \sim 100$ both characteristic times are close to the Planck time, $t_{\mathrm{pl}} = 1/m_{\mathrm{pl}}$. If g were very much larger, the temperature at the characteristic time t_1 in equation (15.17) would be smaller, scaling as $T_1 \propto g^{-1/2}$, and this would set the maximum redshift at which one could hope to apply the standard model.

Let us estimate the redshift at the Planck temperature under the assumption that there was negligible entropy production between then and the decoupling of the neutrinos from the radiation at $T \sim 1$ MeV. We can compute the present entropy density in two parts. The present value for the radiation is $4a_BT_o^3/3 = 4\pi^2T_o^3/45$. At $T = 10^{10}$ K this entropy was shared between the radiation and the thermal electron pairs. The present entropy for the three families of neutrinos is down from that of the radiation by the factor $(T_\nu/T)^3 = 4/11$ (eq. [6.76]), down by the factor 7/8 for the integral over momenta (eq. [6.69]), and up by the factor of three families (assumed to have two spin states each, as for photons; and in the computation we are assuming the neutrinos are massless, which introduces no error because T_ν is computed from the entropy density in the neutrinos assuming they are massless). The sum is the entropy density at high redshifts,

$$s = \frac{86\pi^2}{495} T_o^3 (1+z)^3 = \frac{4u}{3T}, \tag{15.19}$$

where the energy density u is given by equation (15.12). If entropy is conserved, the temperature $T(z)$ as a function of redshift in the early universe is

$$\frac{T}{T_o} = \left(\frac{43}{11g}\right)^{1/3} (1+z). \tag{15.20}$$

The present temperature is measured to be $T_o = 2.7$ K. The redshift at $T = m_{pl}$ is

$$z_{pl} = 3 \times 10^{31} g^{1/3}. \tag{15.21}$$

At this redshift, material now separated by the Hubble distance was at separation

$$H_o^{-1} z_{pl}^{-1} = 3 \times 10^{-4} h^{-1} g^{-1/3} \, \text{cm} \sim 1 h^{-1} \, \mu. \tag{15.22}$$

At this epoch the Hubble length is the Planck length (15.9), thirty orders of magnitude smaller than equation (15.22). The fact that these enormous numbers yield a length in a familiar range, the wavelength of light, is not thought to have any particular significance.

In conventional physics, matter and radiation at high redshift were close to thermal equilibrium. It is useful to consider the limiting case where at some starting redshift z_i the material content of the universe was prepared according to the rules for thermal equilibrium in a fixed and given homogeneous background geometry: divide space into small disjoint volume elements, and prepare each by allowing relaxation to equilibrium with a heat reservoir at a universal temperature T_i. That produces thermal fluctuations in the mass distribution, computed as follows.

Suppose a fixed volume with energy levels E_n is in thermal equilibrium with a heat reservoir at temperature T. The probability that the energy is E_n is $P_n \propto \exp -E_n/T$, so the mean energy is

$$U = \langle E \rangle = \frac{\sum E_n e^{-E_n/T}}{\sum e^{-E_n/T}}. \tag{15.23}$$

The derivative of this expression with respect to temperature is

$$T^2 \frac{dU}{dT} = \frac{\sum E_n^2 e^{-E_n/T}}{\sum e^{-E_n/T}} - \left(\frac{\sum E_n e^{-E_n/T}}{\sum e^{-E_n/T}}\right)^2 \tag{15.24}$$

$$= \langle E^2 \rangle - \langle E \rangle^2 = \delta U^2.$$

This is the variance of the energy. Since the volume is fixed, dU/dT is the heat capacity C at fixed volume, and the fractional thermal fluctuation of the energy in the box is

$$\frac{\delta M}{M} = \frac{C^{1/2}T}{U}.$$
(15.25)

If at high redshift g is constant, the energy density is proportional to T^4, so the heat capacity is $C = 4U/T$, the entropy is $S = 4U/3T$, and the thermal noise in equation (15.25) is

$$\delta_i^2 = \left(\frac{\delta M}{M}\right)^2 = \frac{16}{3S}.$$
(15.26)

With equation (15.19) for the present entropy density, the entropy in a sphere that is expanding with the general expansion of the universe and has present proper radius l_o is

$$S = \frac{344\pi^3}{1485}(l_o T_o)^3.$$
(15.27)

This entropy in equation (15.26) gives

$$\delta_i^2 = \frac{990}{43\pi^3}\frac{1}{(l_o T_o)^3}.$$
(15.28)

The numerical value is

$$\delta_i = 3 \times 10^{-39} l_{\text{Mpc}}^{-3/2}.$$
(15.29)

This is the initial thermal fluctuation in the energy in a region that expands with the material to present radius l_o, expressed in megaparsecs. The observations discussed in section 21 indicate that at radius $l_o = 10h^{-1}$ Mpc the present rms mass fluctuation $\delta = \delta M/M$ is close to unity. Under gravitational evolution, δ varies as $a(t)$ back to equality of mass densities in matter and radiation, at $z \sim 1000$, and earlier as $a(t)^2$, so δ_i would grow to the observed value if the universe were thermalized at redshift $z_i \sim 10^{22}$.

Since z_i is much smaller than the Planck redshift (15.21), this model shows that the observed large-scale fluctuations in the mass distribution could have grown out of a universe whose material content was prepared at thermal equilibrium at a suitably chosen epoch within the domain of classical cosmology. The spectrum of mass fluctuations is flat, because we allowed independent thermal fluctuations in disjoint volume elements. With this spectrum, the perturbation to space curvature varies with the comoving length scale as (eqs. [5.126] and [10.127])

$$h \propto \delta_i l_o^2 \propto l_o^{1/2}.$$
(15.30)

This scaling law could not be extrapolated to arbitrarily large values of l_o, for that would produce a diverging perturbation to the mass within the Hubble length relative to the Einstein-de Sitter model, and we can only allow a factor of ten or so. That is, we must introduce a large-scale cutoff for this flat fluctuation spectrum. As discussed in section 21, the large-scale anisotropy of the cosmic background radiation requires that the density fluctuations decrease with increasing wavelength, relative to the flat spectrum model, at scales greater than $l_o \sim 30h^{-1}$ Mpc. The cutoff could have been placed at a scale very much smaller than this by assigning a fixed and given mass to each initial Hubble volume, rather than allowing independent thermal fluctuations. In this case, thermal relaxation would produce density fluctuations that are compensated within the Hubble length, and the result would be very much smaller than the actual fluctuations in the mass distribution on the scale l_o.

The moral is that thermal fluctuations can make the universe as lumpy as it is, but that the simple model considered here offers no useful guide to what the fluctuations ought to be. We turn now to a much more definite and perhaps more promising model for the thermal generation of fluctuations away from homogeneity, in the field structures produced by a cosmic field.

16. Walls, Strings, Monopoles, and Textures

We saw in the last section that the physical picture that gives a successful description of the universe as it is now and back to the epoch of helium production allows the mass distribution to be considerably closer to homogeneous than it is. Popular lines of thought have been that the structure in galaxies and clusters of galaxies might have been recycled from the previous phase of an oscillating universe, or that it grew out of quantum fluctuations in the field that drove an inflation epoch, or that the structure was seeded by the mass distribution in a cosmic field that acquired a definite value as a function of position in the early universe. The advantage of the last two scenarios is that one can see how to develop them into reasonably definite models. As will be discussed here, it is an attractive feature of the cosmic field scenarios that they offer a variety of well-defined and distinctly nontrivial models to compare to the observations.

The cosmic field scenarios follow the concept from condensed matter physics and particle physics that at low energies a field can acquire an almost definite classical value that in turn can fix physical properties such as the mass flux density in a superfluid or the values of particle masses. The field in the ground state can be degenerate, as in a ferromagnet, where the magnetization can point in any direction. If the field cools from high temperature it can settle to different ground states at different positions, and that can produce defects, such as flux tubes in a superconductor or vortices in a superfluid. Zel'dovich, Kobzarev, and Okun (1974) and Kibble (1976, 1980) recognized the possible relevance to the hot big bang cosmology, where the universe expands and cools from exceedingly high temperatures. Since the coherence length of field fluctuations is limited to the Hubble length, it can happen that the cosmic field acquires a value with a spatial gradient that cannot be eliminated without the development of energy concentrations in field fluctuations or defects, as monopoles or strings or the like. These field energy concentrations can in turn gravitationally perturb an initially homogeneous matter distribution, perhaps seeding structure formation. The first detailed explorations of this idea dealt with cosmic strings (Zel'dovich 1980; Vilenkin 1981b; Silk and Vilenkin 1984; Turok 1985), and people since have discussed a fascinating variety of other possible field structures. So far none has yielded a believable picture for structure formation in cosmology, but as will be discussed in section 25 the debate certainly has sharpened our understanding of what is required of a satisfactory model, and has left many ideas still to be explored. The purpose in this section is to present the elements of the physics; more details are in Linde (1990), Kolb and Turner (1990), and Vilenkin and Shellard (1993).

It is well to insert two cautionary reminders. First, the analyses presented here may seem empty because the Lagrangian (16.1) is chosen ad hoc. However, the field behavior it describes is a well-established part of the physics of condensed matter and of particles. The phase transitions predicted by the standard model for particle physics have not been observed, but the successes of the model are so impressive that there is strong reason to accept that something close to the prediction of the standard model really did happen in the early universe, and it is easy to see how that could have left observable effects. What is more, it certainly is a suggestive coincidence that the characteristic value of the field energy that produces cosmologically interesting values for mass density fluctuations is considered an interestng energy in particle physics (eqs. [16.48], [16.69], and [16.71]).

The second, more negative, caution is that there are more candidates for a theory of structure formation in cosmology than there are classes of phenomena, so it is quite evident that most have to be subdominant to the prime source of structure. The remarkably broad variety of possibilities offered by field defects are discussed here in some detail because the effects they could have on the mass distribution coming out of the early universe are important constraints on the particle physics, and because these effects offer fascinating possibilities for solving a very real and pressing problem in cosmology.

Model Lagrangian

We will use a prototype Lagrangian that yields global versions of the cosmic structures expected from symmetry-breaking phase transitions, rather than the local gauge-invariant versions more commonly discussed in particle physics. The names indicate that if the field has more than one component, the Lagrangian is invariant under a phase shift of the components that is constant, or global, in the former case, and a free function of position, or local, in the latter. In addition to simplicity, the global case has the advantage that it offers a somewhat wider range of cases of possible cosmological interest.

Consider a multiplet of N real classical scalar functions $\phi_b(x)$ of position in spacetime $x = \mathbf{x}, t$, with $1 \leq b \leq N$. The Lagrangian density for the fields is taken to be

$$\mathcal{L} = \frac{1}{2} \sum_b g^{ij} \phi_{b,i} \phi_{b,j} - V(\phi),$$

$$V(\phi) = \frac{\lambda}{4} (\phi^2 - \eta^2)^2, \qquad \phi^2 = \sum_b \phi_b^2.$$

(16.1)

We are using units with $\hbar = c = 1$. Since $\mathcal{L}$ has units of energy density, the field ϕ_b and the constant η have units of energy and the constant λ is dimensionless. The expression for the potential energy is arranged so the square of the field tends to relax to a definite constant value,

$$\sum_b \phi_b^2 \rightarrow \eta^2.$$

(16.2)

This ground level or ground state condition can be satisfied by more than one value of ϕ_b, and if the field relaxes to different ground state values in different regions it produces density inhomogeneities that carry energy.

The field energy will be assumed to have a small effect on the geometry, so we can use the Robertson-Walker line element. Length scales of interest for galaxies and clusters of galaxies are small compared to the present Hubble length, which is not smaller than the radius of curvature of space sections, so we can write the line element as

$$ds^2 = dt^2 - a(t)^2 (dx^2 + dy^2 + dz^2).$$

(16.3)

Then the action $S = \int \sqrt{-g} \mathcal{L} \, d^4x$ is

$$S = \int a^3 \, d^4x \sum_b \left[\frac{\dot{\phi}_b^2}{2} - \frac{\nabla \phi_b^2}{2a^2} \right] - V(\phi),$$

(16.4)

and the wave equation (10.5) is

$$\ddot{\phi}_b + 3 \frac{\dot{a}}{a} \dot{\phi}_b - \frac{\nabla^2 \phi_b}{a^2} = -\frac{dV}{d\phi_b} = -\lambda \phi_b (\phi^2 - \eta^2).$$

(16.5)

In locally Minkowski coordinates this simplifies still further to

$$\ddot{\phi}_b - \nabla^2 \phi_b = -\frac{dV}{d\phi_b} = -\lambda \phi_b (\phi^2 - \eta^2).$$

(16.6)

The stress-energy tensor is given by equation (10.37). The energy density is

$$\rho_\phi = \sum_a \left[\frac{\dot{\phi}_a^2}{2} + \frac{(\nabla \phi_a)^2}{2a^2} \right] + \frac{\lambda}{4} \left(\phi^2 - \eta^2 \right)^2, \tag{16.7}$$

and the stress tensor is

$$T_{\alpha\beta} = \phi_{,\alpha} \phi_{,\beta} + a^2 \delta_{\alpha\beta} \left[\frac{1}{2} \phi_{,k} \phi_{,l} g^{kl} - \frac{\lambda^2}{4} \left(\phi^2 - \eta^2 \right)^2 \right]. \tag{16.8}$$

The starting assumption is that at very high redshifts the field has been thermally driven to a fluctuating function of position. The expansion of the universe adiabatically cools the field, leaving it in the ground state $\phi^2 = \eta^2$ except where defects prevent it. This can leave interesting energy distributions, as we now discuss.

Domain Walls

If the field has just one component, $N = 1$, the equilibrium values in equation (16.2) are $\phi = \pm\eta$. Since the field was a random function of position to begin with, the final value may be $+\eta$ in some parts of space, $-\eta$ in others. These regions have to be separated by two-dimensional boundaries, or domain walls, where the field passes through $\phi = 0$ in changing value between $+\eta$ and $-\eta$. If the wall is plane and the field is static, the field equation 16.6 is

$$\frac{d^2\phi}{dl^2} = \frac{dV}{d\phi} = \lambda\phi(\phi^2 - \eta^2), \tag{16.9}$$

where the proper distance l is measured perpendicular to the wall. On multiplying this expression by $d\phi/dl$ we get a total derivative, with the integral

$$\left(\frac{d\phi}{dl} \right)^2 = 2V, \tag{16.10}$$

where the constant of integration follows from the condition that $d\phi/dl = 0$ far from the wall, where ϕ is in the ground state $\phi = \pm\eta$ at $V = 0$.

The components of the stress-energy tensor (16.7) and (16.8) for a static plane wall with normal along the $l = r^3$ axis are

$$\rho_w = T^{00} = -T^{11} = -T^{22} = \frac{1}{2} \left(\frac{d\phi}{dl} \right)^2 + V, \tag{16.11}$$

and

$$T^{33} = \frac{1}{2} \left(\frac{d\phi}{dl} \right)^2 - V. \tag{16.12}$$

With equation (16.10) these are

$$\rho_w = T^{00} = -T^{11} = -T^{22} = 2V , \qquad T^{33} = 0 . \qquad \text{(16.13)}$$

The surface tension is the energy per unit area, σ, as it must be because the work done in increasing the surface area has to be equal to the increase in the surface energy. Bends in the wall thus tend to move at the velocity of light.

Since the third component of the stress tensor vanishes, the active gravitational mass per unit area is the negative of the energy per unit area (eq. [10.68]), so the gravitational acceleration in a neighborhood of the wall (with size small compared to $1/G\sigma$) is

$$g = 2\pi G\sigma , \qquad \text{(16.14)}$$

directed away from the wall (eq. [11.12]).

To estimate the energy per unit area, σ, in terms of the parameters in the potential, note that the value of ϕ changes by the amount $\sim \eta$ in a coherence distance ξ across the wall, so the orders of magnitude of the two sides of equation (16.10) in the neighborhood of the wall are

$$\left(\frac{d\phi}{dl}\right)^2 \sim \frac{\eta^2}{\xi^2} \sim V \sim \lambda\eta^4 , \qquad \text{(16.15)}$$

which tells us the coherence length is

$$\xi \sim \frac{1}{\lambda^{1/2}\eta} , \qquad \text{(16.16)}$$

and the energy per unit area is

$$\sigma \sim V\xi \sim \lambda^{1/2}\eta^3 . \qquad \text{(16.17)}$$

Zel'dovich, Kobzarev, and Okun (1974) pointed out that there is a very significant constraint on the mass per unit area in a network of domain walls left over from the very early universe, from the constraint that the wall must not destroy the observed isotropy of the universe. Adjoining walls with opposite steps in the field can move together and annihilate, the field variation $\phi = \eta \to -\eta \to \eta$ smoothing out to $\phi = \eta$, but since the walls can only move together at the speed of light, this has to leave on the order of one wall stretching across each Hubble length. Since the strongly anisotropic mass distribution in the wall has not appreciably perturbed the isotropy of the diffuse radiation backgrounds, the mass within one wall stretching across the present Hubble length, $\mathcal{M}_w \sim \sigma H_o^{-2}$, must be small compared to the net mass within the Hubble length, $\mathcal{M}_u \sim 1/(GH_o) = m_{\text{pl}}^2/H_o$, where

$m_{\rm pl}$ is the Planck mass in equation (15.7). With equation (16.17) for the mass per unit area, this condition is

$$\delta_h \sim \frac{\mathcal{M}_w}{\mathcal{M}_u} \sim \frac{\lambda^{1/2}\eta^3}{H_o m_{\rm pl}^2} \lesssim 10^{-4}\,, \tag{16.18}$$

from the bound on the anisotropy of the thermal cosmic background radiation. The energy associated with Hubble's constant is

$$\hbar H_o = 3.4 \times 10^{-45} h\,{\rm erg} = 2.1 \times 10^{-33} h\,{\rm eV}\,, \tag{16.19}$$

giving

$$\eta \lesssim 10\lambda^{-1/6}\,{\rm MeV}\,. \tag{16.20}$$

Unless λ is exceedingly small this is well below the range of energies associated with the conventional phase transitions of particle physics. That is, there must be some provision for the elimination of any domain walls that might have been produced in the standard particle physics phase transitions in the very early universe. There is of course the possibility for something new at lower surface energy in domain walls produced in a phase transition at more modest redshifts, though at the time this is written there is no viable and attractive model (Hill, Schramm, and Fry 1989; Press, Ryden, and Spergel 1989).

Cosmic Strings

The next simplest possibility is $N = 2$ components in the Lagrangian (16.1). This produces cosmic strings, the field defects that have been most thoroughly explored for the possible relevance to cosmology.

When the expansion of the universe has relaxed the two components to the potential energy minimum in equation (16.2), the values of the fields satisfy

$$\begin{aligned} \phi_1 &= \eta\cos\theta\,, \\ \phi_2 &= \eta\sin\theta\,, \end{aligned} \tag{16.21}$$

leaving the free angle variable θ. The gradient energy density $(\nabla\phi)^2$ in equation (16.7) encourages the fields to move so as to make θ constant. That is not possible, however, if the value of θ as a function of position is such that in the circuit around a closed path the value of θ increases by 2π. Where this happens, continuity says that in going around a neighboring path the angle variable θ again increases by 2π. But if that continued to be true as we shrank the path to zero area, the field gradient energy would diverge. The field equation solves the problem by producing a region of width ξ within which $\phi^2 = \phi_1{}^2 + \phi_2{}^2$ departs from the value η^2 and passes through a zero, so the components ϕ_a can smoothly switch signs across the region. Since the two equations $\phi_a = 0$ define two surfaces that intersect in a line, the zeros of $\phi_1{}^2 + \phi_2{}^2$ define a line, the core of a cosmic string. (The grownup way to see this uses topological arguments, as in Vilenkin 1985.)

If the field is axially symmetric, translationally symmetric along the axis, and static, we can write the field as

$$\phi_a = v(r)r_a/r ,\tag{16.22}$$

where r_a are orthogonal position coordinates orthogonal to the string, and r is the perpendicular distance. Then the field equation (16.6) works out to

$$\frac{d^2v}{dr^2} + \frac{1}{r}\frac{dv}{dr} - \frac{v}{r^2} = \lambda v(v^2 - \eta^2) .\tag{16.23}$$

The solution $v(r)$ near the origin is $v(r) = \eta r/\xi$, and v approaches η at the coherence length, ξ. The curvature of $v(r)$ at $r \sim \xi$ is $u'' \sim \eta/\xi^2$, so we see from the field equation (16.23) that the coherence length is

$$\xi \sim \frac{1}{\lambda^{1/2}\eta} ,\tag{16.24}$$

as in equation (16.16) for a domain wall.

The energy per unit length within the coherence length is

$$\mu_c \sim \lambda\eta^4\xi^2 \sim \eta^2 .\tag{16.25}$$

At $r \gg \xi$, the gradient of the field in equation (16.22) is $\nabla\phi \sim \eta/r$, so the energy density in the $(\nabla\phi)^2$ term in equation (16.7) is $\rho_\phi \sim \eta^2/r^2$, and the integral of ρ_ϕ is the mass per unit length of string within distance R of the core,

$$\mu \sim \eta^2 \log(R/\xi) .\tag{16.26}$$

This solution is called a global string, because if ϕ_1 and ϕ_2 are considered as the real and imaginary parts of a complex field ϕ, the Lagrangian (16.1) is invariant under the phase transformation $\phi \to \phi \exp i\chi$, where the phase χ is a constant. That is, the Lagrangian is invariant under a global phase transformation. The Yang-Mills generalization of the local phase symmetry of electromagnetism brings the Lagrangian to a form invariant under local phase transformations, where χ is allowed to be a function of position. In the gauge strings that result from this case the field energy well outside the coherence length is suppressed, so the mass per unit length is concentrated within a few coherence lengths of the string, and for a stationary gauge string the mass is approximated by equation (16.25) (e.g., Vilenkin 1985).

Let us consider the main features of the behavior of string defects in an expanding universe under the assumption of local gauge symmetry, where we can drop the logarithmic term in equation (16.26).

The string tension is the work done per unit increase in the length of the string. The string energy per unit length is μ, so the tension has to be $T = \mu$. One recalls from classical mechanics that small transverse waves in a string with tension T and mass per unit length μ move at speed $(T/\mu)^{1/2}$, so it is apparent that waves move along a cosmic string at the velocity of light. A more formal demonstration goes as follows.

Consider first a straight stationary string. Since the time derivatives vanish, the Lagrangian is just the negative of the energy density along the string, so the action is

$$S = \int \mathcal{L} d^4 r = -\mu \int dl dt , \qquad (16.27)$$

where μ is the energy per unit length and l measures the length along the string. If the string is bent, with radius of curvature large compared to the width ξ of the mass distribution within the string, the action for each string segment in the local instantaneous rest frame is $-\mu \, dl dt$. Thus, in this approximation the action for a bent moving string is

$$S = -\mu \int dA , \qquad (16.28)$$

where dA is the element of proper area of the two-dimensional sheet swept out by the motion of the string through spacetime.

To write down the equations of motion for the string, let us set up a coordinate system x^a, $a = 1, 2$, to label events on the world sheet defined by the path of the string. In a four-dimensional locally Minkowski coordinate system the event x^a is at coordinate position $r^i(x^a)$. The Minkowski line element gives the invariant distance between neighboring events on the sheet,

$$ds^2 = g_{ij} dr^i dr^j = g_{ij} \frac{\partial r^i}{\partial x^a} \frac{\partial r^j}{\partial x^b} dx^a dx^b$$

$$\equiv \gamma_{ab} dx^a dx^b . \qquad (16.29)$$

This defines the metric tensor on the sheet,

$$\gamma_{ab} = g_{ij} \frac{\partial r^i}{\partial x^a} \frac{\partial r^j}{\partial x^b} . \qquad (16.30)$$

The coordinate labels x^a can be transformed in the usual way, so that if $x^a = x^a(\bar{x}^c)$ the coordinate intervals are

$$dx^a = \frac{\partial x^a}{\partial \bar{x}^c} d\bar{x}^c , \qquad (16.31)$$

and the metric tensor becomes

$$\bar{\gamma}_{cd} = \gamma_{ab} \frac{\partial x^a}{\partial \bar{x}^c} \frac{\partial x^b}{\partial \bar{x}^d} .$$ (16.32)

The determinant of this equation is

$$\bar{\gamma} = \gamma \left[\frac{\partial(x)}{\partial(\bar{x})} \right]^2 ,$$ (16.33)

for the determinant of a matrix product is the product of the determinants (eq. [8.33]). Under a change of coordinates the two-dimensional area element on the sheet transforms as

$$d^2x = \frac{\partial(x)}{\partial(\bar{x})} d^2\bar{x} .$$ (16.34)

The Jacobian for the coordinate transformation is given by equation (16.33). It follows that the expression

$$(-\gamma)^{1/2} d^2x$$ (16.35)

is a scalar, independent of the coordinate labeling on the sheet, as for the four-dimensional volume element in equation (8.37). Since the proper area is d^2x in the string rest frame with locally proper length and time labels, where $\gamma = -1$, the action (16.28) is

$$S = -\mu \int (-\gamma)^{1/2} d^2x .$$ (16.36)

Let us write out the action for coordinate labels x^a chosen so that $x^0 = t$ is the time variable in the Minkowski coordinate system $t, \mathbf{r}$. With $x^1 \equiv \sigma$, the position of the string is $\mathbf{r}(t, \sigma)$. Since the Minkowski metric tensor is diagonal, equation (16.30) gives

$$\gamma_{00} = 1 - \dot{\mathbf{r}}^2 , \qquad \gamma_{01} = -\dot{\mathbf{r}} \cdot \mathbf{r}' , \qquad \gamma_{11} = -(\mathbf{r}')^2 ,$$ (16.37)

with $\dot{\mathbf{r}} = \partial \mathbf{r}/\partial t$ and $\mathbf{r}' = \partial \mathbf{r}/\partial \sigma$. The determinant gives the Lagrangian density in equation (16.36),

$$\mathcal{L} = -\mu(-\gamma)^{1/2} = -\mu \left[(1 - \dot{\mathbf{r}}^2)(\mathbf{r}')^2 + (\dot{\mathbf{r}} \cdot \mathbf{r}')^2 \right]^{1/2} .$$ (16.38)

Having the Lagrangian, we can write out the equations of motion in the usual way:

$$\frac{\partial}{\partial t} \frac{\dot{\mathbf{r}}(\mathbf{r}')^2 - \mathbf{r}'(\dot{\mathbf{r}} \cdot \mathbf{r}')}{\left[(\mathbf{r}')^2(1 - \dot{\mathbf{r}}^2) + (\dot{\mathbf{r}} \cdot \mathbf{r}')^2 \right]^{1/2}} = \frac{\partial}{\partial \sigma} \frac{\mathbf{r}'(1 - \dot{\mathbf{r}}^2) + \dot{\mathbf{r}}(\dot{\mathbf{r}} \cdot \mathbf{r}')}{\left[(\mathbf{r}')^2(1 - \dot{\mathbf{r}}^2) + (\dot{\mathbf{r}} \cdot \mathbf{r}')^2 \right]^{1/2}} .$$ (16.39)

A solution for $\mathbf{r}(t, \sigma) = x, y, z$ is

$$x = \sigma, \qquad y = f(x \pm t), \qquad z = 0. \tag{16.40}$$

To check, note that in this solution the components of the velocity are $\dot{\mathbf{r}} = 0$, $\pm f'$, 0, and $\mathbf{r}' = 1$, f', 0, from which we see that equation (16.38) is $\mathcal{L} = -\mu$. Since the Lagrangian is independent of the function f, the action is stationary under variations of the path in the plane $z = 0$, and equation (16.40) therefore is a solution to the equation of motion for any f. This represents a wave form $f(x \pm y)$ traveling without dispersion along the x axis at the speed of light.

The equations of motion are simplified by constraining the label σ for the space position along the string. Since $\mathbf{r}'$ points along the string at a fixed time, we can assign σ in neighboring time slices so $\dot{\mathbf{r}}$, the time derivative of the position at fixed σ, is perpendicular to $\mathbf{r}'$,

$$\dot{\mathbf{r}} \cdot \mathbf{r}' = 0. \tag{16.41}$$

At some starting time, the coordinate label σ along the string can be assigned to satisfy

$$\dot{\mathbf{r}}^2 + (\mathbf{r}')^2 = 1. \tag{16.42}$$

Suppose this equation is valid at all times. Then these two equations in the equations of motion (16.39) give

$$\ddot{\mathbf{r}} = \mathbf{r}''. \tag{16.43}$$

The space derivative of the constraint equation (16.41) is

$$\dot{\mathbf{r}}' \cdot \mathbf{r}' = -\dot{\mathbf{r}} \cdot \mathbf{r}'' = -\dot{\mathbf{r}} \cdot \ddot{\mathbf{r}}, \tag{16.44}$$

so the time derivative of the constraint equation (16.42) is

$$\dot{\mathbf{r}} \cdot \ddot{\mathbf{r}} + \dot{\mathbf{r}}' \cdot \mathbf{r}' = 0. \tag{16.45}$$

That is, equation (16.42) agrees with the equation of motion (16.43). Equation (16.40) is a solution to equation (16.43).

In a cosmic string scenario, one imagines that in the early universe a field ϕ_b that initially is a fluctuating function of position settles down to the value $\phi^2 = \eta^2$ everywhere except in the neighborhood of defects, the cosmic strings. The initial mean distance between strings cannot have been much larger than the Hubble length, because the thermal fluctuations that set the local direction of ϕ_a in equation (16.21) must have been independent in separate Hubble volumes. As the universe expands, the increasing Hubble length allows the string to notice that it is a tangled network, and the segments move at relativistic speeds to straighten. The motion overshoots, but the expansion of the universe damps the vibration, just as for electromagnetic waves. Where two strings cross, the fields may reconnect to shorten the string length. This can cut off loops, leaving a still shorter length along the long piece. For interesting values of μ, the vibrating loops shrink away by gravitational radiation.

The conclusion is that the string network relaxes until there is on the order of one long string running across each Hubble length. The mass of string within a Hubble volume H^{-3} is then $\propto \mu H^{-1}$, and the mean mass density in the long string network is

$$\rho_s \sim 50\mu H^2 . \tag{16.46}$$

The scaling argument was given by Kibble (1976); the numerical factor for a radiation-dominated Einstein-de Sitter universe is from the computations of the evolution of a cosmic string network by Bennett and Bouchet (1989), who also conclude that the loops spun off as the string network straightens out are so small that they very quickly disappear by the emission of gravitational radiation.

The ratio of the mean mass density in long strings to the total in an Einstein-de Sitter universe is

$$\frac{\rho_s}{\rho_b} = \frac{8\pi G\rho_s}{3H^2} \sim 300G\mu \sim 300\frac{\eta^2}{m_{\text{pl}}^2} \sim \delta_h , \tag{16.47}$$

where $\mu \sim \eta^2$ is the string mass per unit length (eq. [16.25]). Because the energy is concentrated in a few strings that run across the Hubble length, equation (16.47) gives the fractional perturbation to the mass density on the scale of the Hubble length. It is an attractive feature of this picture that δ_h is independent of epoch, meaning that with an appropriate choice of the energy η in the field potential (16.1) an observer at any epoch would see that a fixed observationally interesting but not catastrophically large mass fluctuation $\delta\mathcal{M}/\mathcal{M} \sim \delta_h$ is appearing in a new generation of long strings on the scale of the Hubble length. This scale invariance appears in some of the other cosmic fields to be discussed below, and in simple versions of the inflation scenario in section 17.

As in equation (16.18), the isotropy of the CBR tells us the universe has expanded in an isotropic way to an accuracy $\delta_h \lesssim 10^{-4}$, so the strongly anisotropic contribution to the mass density by the long strings has to be less than about one part in 10^4 of the total. At this limit, equation (16.47) is

$$300G\mu \sim 300(\eta/m_{\text{pl}})^2 \lesssim 10^{-4} , \qquad \eta \lesssim 10^{16}\,\text{GeV} . \tag{16.48}$$

As we noted at the beginning of this section, it might be counted as an encouraging coincidence that this is the same order of magnitude as the energy that figures in discussions of the unification of the strong and electroweak interactions.

As discussed in section 11, a static string has no active gravitational mass, because the tension along the string just cancels the energy per unit length. When $G\mu \ll 1$ the line element outside the energy concentration in a local string is (eq. [11.25])

$$ds^2 = dt^2 - dr^2 - (1 - 8G\mu)r^2 d\phi^2 - dz^2 . \qquad (16.49)$$

As illustrated in figure 11.1, the circumference of a circle of radius r centered on the string and normal to it is $C = 2\pi(1 - 4G\mu)r$, that is, the angular circumference is smaller than the usual 2π by the constant deficit angle

$$\Phi = 8\pi G\mu . \qquad (16.50)$$

What happens when a long string passes through an object, such as an observer? We can assume that the string is moving at speed v comparable to the velocity of light. The passage of the string causes no immediate rearrangement of material in the observer, because space is flat with no tidal forces outside the very narrow coherence length of the string. However, the deficit angle leaves the two sides of the observer approaching at speed $\sim v\Phi \sim 1$ km s^{-1}, for the value of $G\mu$ in equation (16.48), which would be quite damaging to a person.

Let us estimate the mass per unit area in the wake of a long string passing through matter. We will choose an easy problem, where the string is straight and moving at constant speed v_s through initially homogeneous pressureless material in an Einstein-de Sitter universe. Since the relative velocity induced in the material is small compared to the string speed v_s, we can assume the wake is plane parallel.

Immediately after passage of the string, the material on either side of the wake has closing velocity

$$\Delta v_i = \Phi v_s = 8\pi G\mu v_s . \qquad (16.51)$$

This is a peculiar velocity, relative to the general expansion of the material. We are approximating the wake as plane parallel, and we will let $x(t)$ be the comoving coordinate distance between two planes attached to material on either side of the wake. The planes have initial separation $x_i = x(t_i)$. The string passes at time t_i, and at the later time t the mass per unit area in the wake, in excess of homogeneity, is

$$\Sigma(t) = \rho(t)a(t)[x_i - x(t)] , \qquad (16.52)$$

where the mean mass density is $\rho(t)$. The peculiar closing velocity is

$$\Delta v = -a\dot{x} , \qquad (16.53)$$

and the equation of motion for the peculiar velocity is (eq. [5.109])

$$\frac{d}{dt}\Delta v + \frac{\dot{a}}{a}\Delta v = \Delta g = 4\pi G\Sigma. \tag{16.54}$$

The second step follows from Poisson's equation for the peculiar relative gravitational acceleration, $\nabla \cdot \mathbf{g} = -4\pi G\delta\rho$. Equations (16.52) to (16.54) give

$$\frac{d^2x}{dt^2} + 2\frac{\dot{a}}{a}\frac{dx}{dt} = -4\pi G\rho\,(x_i - x). \tag{16.55}$$

This shows that the coordinate distance $x_i - x$ across the wake satisfies the linear perturbation equation (5.111), until the material reaches the wake. In the Einstein-de Sitter model, the solution is

$$x_i - x = At^{2/3} + B/t. \tag{16.56}$$

The material starts at $x_i - x = 0$ with the velocity in equation (16.51) at redshift $1 + z_i = a_o/a_i = (t_o/t_i)^{2/3}$. These two conditions fix the constants A and B. The present value of the displacement works out to

$$\Delta r = a_o(x_i - x_o) = \frac{16\pi}{5}(1 + z_i)^{1/2}G\mu v_s H_o^{-1}. \tag{16.57}$$

A wake produced at decoupling, at redshift $z_i \sim 1000$, by a string with $G\mu = 10^{-6}$ and $v_s = 1$, amounts to $\Delta r \sim 1h^{-1}$ Mpc, with mass per unit area $\sigma_o \sim 3 \times 10^{11}h^{-1}M_\odot$ Mpc^{-2}. Both quantities are in interesting ranges. Prospects for models for structure formation are considered in section 25.

Hedgehogs and Monopoles

With $N = 3$ components the field has particlelike defects. Polyakov's (1974) hedgehog solution for the three field components has the form

$$\phi_a = u(r)r^a/r, \tag{16.58}$$

where u is a function of the radius r alone and the r^a are the three Cartesian position coordinates. This expression in the field equation (16.6) gives

$$\frac{d^2u}{dr^2} + \frac{2}{r}\frac{du}{dr} - 2\frac{u}{r^2} = \lambda u(u^2 - \eta^2). \tag{16.59}$$

As for equation (16.23) for a global cosmic string, we see that the solution $u(r)$ approaches $u(r) = \eta$ at radius r large compared to the coherence length ξ, and at $r \ll \xi$ the solution varies linearly with radius, $u \sim r\eta/\xi$. At $r \sim \xi$ the second derivative is $u'' \sim \eta/\xi^2 \sim \lambda\eta^3$, from which we see as before that the coherence length is

$$\xi \sim \frac{1}{\lambda^{1/2}\eta} . \tag{16.60}$$

The potential energy density within $r \sim \xi$ is $V \sim \lambda\eta^4$, so the core mass is

$$m_c \sim \lambda\eta^4\xi^3 \sim \eta\lambda^{-1/2} . \tag{16.61}$$

Outside the core, where $u \to \eta$, the square of the field gradient is

$$\sum (\nabla\phi_a)^2 = 2\eta^2/r^2 , \tag{16.62}$$

so the mass within distance R of the core is

$$m(<R) = 8\pi\eta^2 R , \tag{16.63}$$

at $R \gg \xi$. This hedgehog solution is stable, for the only way to remove the zero of the field at $r = 0$ would be to rearrange the topology of the outward-pointing ϕ_a in equation (16.58) at $r \gg \xi$.

The hedgehog solution is also called a global monopole, in analogy to global strings. In a gauge monopole the field is complex, and the Lagrangian is made invariant under local phase transformations by the addition of a Yang-Mills gauge field. This eliminates the field gradient energy well outside the core, so the solution acts like a particle with a definite concentrated mass. The external gauge field has a part that acts like electromagnetism, and with this identification the external field is that of a magnetic monopole ('t Hooft 1974; Polyakov 1974), with mass and coherence length given by expressions similar to m_c and ξ for a global monopole, but modified by the gauge interaction. We need not pause to consider these details, however, for if gauge monopoles were produced in the very early universe, their mean mass density would evolve as $a(t)^{-3}$, which is slower than the evolution of the mass density in the radiation. This leads to a catastrophic addition to the present mass density, as follows (Zel'dovich and Khlopov 1978; Preskill 1979).

In the standard hot cosmological model, one might imagine that at very high redshift a three-component field ϕ_a is thermally excited to a randomly varying function of position, and that as the field expands and adiabatically cools, it settles to the ground state value $\sum \phi_a^2 = \eta^2$, except where the topology of the directions of ϕ_a prevents it and a hedgehog or monopole forms. Let us find a rough estimate of the temperature at which this happens.

We will consider first the thermal fluctuations of a free field, where there is no potential energy term, and then estimate the effect of the potential. We can write the fluctuating part ψ of the ϕ field as a Fourier series with periodic boundary conditions in a box of volume V_u,

$$\psi = \sum \psi_k e^{i\mathbf{k} \cdot \mathbf{r}}. \tag{16.64}$$

The derivative energy (16.7) in the box is

$$E = \frac{1}{2} V_u \langle (\nabla \psi)^2 \rangle = \frac{1}{2} V_u \sum k^2 |\psi_k|^2, \tag{16.65}$$

and a free field has a like amount of energy from the time derivative of the field. At wavenumber $k < T$ the energy per mode in the box is the equipartition value $\sim T$, so the fluctuation spectrum is

$$V_u k^2 |\psi_k|^2 \sim T, \qquad k \lesssim T. \tag{16.66}$$

Then the mean square value of the free field is

$$\langle \psi^2 \rangle = \sum |\psi_k|^2 = V_u \int_{k \lesssim T} |\psi_k|^2 d^3k / (2\pi)^3 \sim T^2. \tag{16.67}$$

The distance between the minima of the potential for a component of ψ is $\delta \psi \sim \eta$, so equation (16.67) indicates that at temperature

$$T_c \sim \eta \tag{16.68}$$

thermal fluctuations are not able to pull the field from one side of the potential to the other, and the field ϕ therefore tends to settle to the ground level, $\phi^2 = \eta^2$.

To judge the effect of the potential $V(\phi)$ on the thermal fluctuations, write the field as $\phi = \bar{\phi} + \psi$, where ψ is the fluctuating part and the mean value is $\bar{\phi} = 0$ if the field is fluctuating around the ensemble average value $\langle \phi \rangle = 0$, or $|\bar{\phi}| = \eta$ if the field is fluctuating around a minimum of the potential. Either way, the leading term in the potential for the field fluctuations is $\sim \lambda \eta^2 \psi^2$, which says the fluctuating field acts as if it had a mass $m \sim \lambda^{1/2} \eta$. If λ is well below unity, the mass is small compared to T_c, so the estimate of $\langle \psi^2 \rangle$ in equation (16.67) properly ignores m.

As the temperature drops to $T \sim T_c \sim \eta$, the field tends to settle to the value $|\phi| = \eta$, with the coherence length $\sim \eta^{-1}$ in equation (16.66). If λ is small this is smaller than the coherence length $\xi = \eta^{-1} \lambda^{-1/2}$ in an equilibrium solution (eq. [16.60]), so the field gradient stress rearranges the field fluctuations until the distance between hedgehogs or monopoles is on the order of ξ. This produces the number density $n \sim \xi^{-3}$ at mass $\sim \eta \lambda^{-1/2}$, giving mass density $\rho_\phi \sim \lambda \eta^4 \sim \lambda T_c^4$ in the monopoles or hedgehogs. The total mass density is $\sim g T_c^4$, if the entropy is spread over g fields, so at T_c a fraction $\sim \lambda g^{-1}$ of the total mass density is placed in the field defects. The consequences for monopoles and hedgehogs are very different.

At $T \ll T_c$ monopole-antimonopole pairs can annihilate, but since they interact only through the relatively weak magnetic field, this is not an efficient way to dispose of them. The mass density in monopoles thus scales as $a(t)^{-3}$, while the density in relativistic fields varies as $a(t)^{-4}$. Since a mass fraction $\sim \lambda g^{-1}$ is placed in monopoles at T_c, after an expansion factor $\sim \lambda^{-1} g$ the mass is dominated by monopoles, and the model predicts an absurdly large present mass density. This is the crisis that led Guth to the idea of an inflation epoch that makes the monopole density negligibly small, as will be discussed in the next section.

Global monopoles, or hedgehogs, behave in a very different way because the field energy outside the core grows linearly with radius (eq. [16.63]). Consider an isolated hedgehog-antihedgehog pair at separation R, where the field ϕ_b is constant well away from the pair, converges on the antihedgehog, and diverges from the hedgehog. The field energy is on the order of $\eta^2 R$, and since this dominates the core mass the pair moves at relativistic speeds, dissipating angular momentum and energy in propagating waves in the direction of ϕ_b, and annihilates in a timescale on the order of R. Thus, the strong interaction between hedgehogs and antihedgehogs leaves only a few per Hubble volume (Barriola and Vilenkin 1989; Linde 1990).

It is interesting to compare the behavior of the mean energy densities in the different classes of fields. There is only on the order of one domain wall across the Hubble length H^{-1}, but the energy is spread over the area $\sim H^{-2}$, with the result that the fraction of the total mass density in the wall grows as $H(t)^{-1}$, as in equation (16.18). This means a domain wall that carries energy is acceptable only if its mass per unit area is considerably smaller than those encountered in particle physics. If a cosmic string network expanded homogeneously, its mean energy density would vary as $a(t)^{-2}$, making the mass problem even worse than for gauge monopoles. That does not happen because the coherence length for the network grows in proportion to the Hubble length, keeping the string mass density a nearly constant fraction of the total. The same happens for hedgehogs, as follows.

Since there would be only a few hedgehogs in a Hubble length H^{-1}, the total energy in hedgehogs within the Hubble length would be $U_m \sim \eta^2 H^{-1}$ (eq. [16.63]). Hence the mean mass density in hedgehogs is $\rho_m \sim U_m H^3 \sim \eta^2 H^2$, and the ratio of the mean mass density in hedgehogs to the total is

$$\delta_h = \frac{\rho_m}{\rho_b} \sim \frac{\eta^2 H^2}{H^2 m_{\text{pl}}^2} = \frac{\eta^2}{m_{\text{pl}}^2} . \tag{16.69}$$

The field energy in the hedgehogs is spread unevenly across the Hubble length, with cores moving toward annihilation with the nearest hedgehog of the opposite sign at close to the speed of light. This means the hedgehogs produce density fluctuations on the scale of the Hubble length, $\delta \mathcal{M}/\mathcal{M} = \delta_h \sim \eta^2/m_{\mathrm{pl}}^2$, independent of the redshift. This is another example of a scale-invariant fluctuation spectrum, as in equation (16.47).

Equation (16.69) indicates that hedgehogs share with cosmic strings the attractive feature that an energy η that is considered interesting in particle physics perturbs the homogeneity of the universe on the scale of the Hubble length by a cosmologically interesting amount, $\delta \mathcal{M}/\mathcal{M} \sim 10^{-4}$. Barriola and Vilenkin (1989) and Bennett and Rhie (1990) have taken the first steps toward the exploration of the possibilities this might offer for structure formation.

Textures

If the field has $N = 4$ components, the field value can be everywhere at the minimum of its potential energy, $\phi^2 = \eta^2$, leaving as initial conditions three free functions of position. Davis (1987) considered the case where the field gradient energy density adds a homogeneous term to the total cosmological mass density. Turok (1989, 1991) pointed out that inhomogeneities in the field gradient energy density can propagate into transient energy concentrations, called texture events, as well as produce smoother density fluctuations on the scale of the Hubble length. Turok notes that this can set up interesting initial conditions for structure formation in an expanding universe.

When the four field components are in the ground state, where $\sum \phi_a^2 = \eta^2$, we are left with three independent components that may be represented by the angular position χ, θ, and ϕ on the surface of a three-sphere, as in equation (12.4). In Turok's spherically symmetric solution for a texture defect, the four fields are

$$
\begin{aligned}
\phi_a &= \eta \sin \chi(r,t) \sin \theta \cos \phi, \\
&= \eta \sin \chi(r,t) \sin \theta \sin \phi, \\
&= \eta \sin \chi(r,t) \cos \theta, \\
&= \eta \cos \chi(r,t).
\end{aligned}
\tag{16.70}
$$

The arguments r, θ, and ϕ are the usual polar position coordinates in our three-dimensional space, and the function $\chi(r,t)$ at fixed time t runs from $\chi = 0$ at $r = 0$ to $\chi \to \pi$ at $r \to \infty$. In this configuration, the field direction on the three-sphere points along the north pole, $\chi = 0$, at small r, and runs to the south pole at great distance from the origin. To understand how this field configuration evolves, suppose the radius at which the field has swung halfway from the north to south poles is $r = \xi$. Then the field gradient energy density is $\sim \eta^2/\xi^2$ within the coherence volume $\sim \xi^3$, so the net gradient energy within this region is $U_t \sim \eta^2 \xi$. The field moves to decrease ξ, because this reduces its potential energy in the gradient. The field thus evolves toward a situation where the components are $\phi_a = (0,0,0,-\eta)$ everywhere except within a shrinking core of size ξ where the field components swing over to $\phi_a = (0,0,0,+\eta)$. When the field gradient energy density $u_t \sim (\eta/\xi)^2$

within this core has grown comparable to the potential energy barrier $V \sim \lambda \eta^4$, it pulls the field in the core through $\phi_a = 0$ to the value outside the core. The energy of the texture radiates away as oscillations in the field direction, the expansion of the universe damps the oscillation, and the field settles down to a homogeneous value.

When $N \leq 3$ the defects in walls, strings, or monopoles come into being at the epoch at which the field can be locked to the minimum of the potential energy, and thereafter evolve through the stress of interaction of different parts of the field. When $N = 4$, we will see that if the ground level degeneracy survives, the mass density is perturbed throughout the course of expansion of the universe by the field gradients appearing across the Hubble length, and by topological texture events that occur wherever the field configuration appearing at the Hubble length can only straighten out by the formation of the structure in equation (16.70). Both effects, which introduce scale-invariant perturbations to the matter distribution, could have interesting consequences (Gooding, Spergel, and Turok 1991; Cen et al. 1991; Chuang et al. 1991).

At any epoch, the field components ϕ_b typically change value by the amount $\delta \phi_b \sim \eta$ across the Hubble length (while keeping $\phi^2 = \eta^2$), so the typical field gradient energy density is $u_t \sim (\eta H)^2$, and the ratio of u_t to the mean mass density $\rho_b \sim (m_{\rm pl} H)^2$ is

$$\delta_h \sim \frac{\eta^2}{m_{\rm pl}^2} \tag{16.71}$$

(Turok 1989). If the field components running across the Hubble length are such that the field can move to straighten out without forming a texture, then it does so, and the expansion of the universe dissipates the field oscillations. Under the initial condition that the mass distribution is homogeneous, so there are no curvature fluctuations, the expansion preserves the homogeneous mass distribution when nongravitational forces can be neglected. This means that where the field gradient energy initially is high it is balanced by a low density of matter and radiation. The field is straightening on the scale of the Hubble length, leaving a fractional perturbation on the order of δ_h in the distribution of matter and radiation averaged across the Hubble length. Since δ_h is independent of redshift, this is another example of scale-invariant initial conditions for structure formation. To estimate the wanted field value, recall that an interesting rms fractional perturbation to the matter distribution on the scale of the Hubble length is $\delta_h \sim 10^{-4}$, which we get if η is about one percent of the Planck mass (eq. [16.71]), or $\eta \sim 10^{17}$ GeV. As before, the number is considered interesting in the particle physics community.

Where a topological texture forms, the core radius ξ (at which the fourth component in the last line in eq. [16.70] has swung halfway through its range) is shrinking at the speed of light. The gradient energy density is $\rho_t \sim \eta^2/\xi^2$, so the mass concentrated within ξ is $\mathcal{M}_t \sim \eta^2 \xi$. At the present epoch, the net mass within the Hubble length is $\mathcal{M}_u \sim 10^{22}\,\mathcal{M}_\odot$, so a texture that originates at low redshift at $\delta_h \sim 10^{-4}$ would have a mass of about $10^{18}\,\mathcal{M}_\odot$ within the Hubble length, and the mass within the core radius would be $\mathcal{M}_\xi \sim 10^{18}(H_o \xi)\mathcal{M}_\odot$. When $\xi \sim 30h^{-1}$ Mpc, the mass concentration in the core would be contracting on a timescale of about 10^8 years, with a core mass about ten times that of the Coma cluster of galaxies. That surely would have an interesting effect on the motions of the nearby galaxies.

With $N = 4$ field components the condition $\phi^2 = \eta^2$ leaves three free functions. As we have discussed, that means we can represent a field value as a position χ, θ, ϕ on the surface of a three-sphere, and we can represent the field as function of position in the three-dimensional space we live in as a map of each point in our space to a point on the three-sphere. If the map is such that our space covers the three-sphere, then there is no way the field can move in a continuous manner so as to assume the same value everywhere, so the field develops a topological texture. If there are $N = 5$ field components, the map of our three-dimensional space is to a four-sphere. The map can only cover a lower dimension part of the four-sphere, so the map can always be slipped along the surface of sphere to bring the field to the same value everywhere, while also keeping $\phi^2 = \eta^2$ everywhere. That is, the field variations that appear at the Hubble length can always straighten out without developing energy concentrations in defects. The field variations appearing at the Hubble length introduce scale-invariant fluctuations in the mass distribution, in the amount $\delta_h \sim (\eta/m_{\mathrm{pl}})^2$, as for strings, hedgehogs, and textures (Turok 1989, 1991). Turok and Spergel (1991) show that the fluctuations in the field gradient energy averaged over the Hubble length are significantly different from Gaussian when $N = 5$, with positive skewness and appreciable excess kurtosis (from the fourth moment). As discussed in section 25, this could help explain why structures are observed in the galaxy space distribution on scales where the galaxy two-point correlation function is small. When $N \gg 5$, the energy distribution is carried in many fields, and the perturbation to the matter distribution accordingly is close to Gaussian. This limiting case would be difficult to distinguish from the quantum field fluctuations in simple models for the inflation scenario to be discussed next.

Lessons

The well-established physics of condensed matter and particles tells us that phase transitions surely happened as our universe expanded and cooled. This offers a way to understand a deep puzzle, the difference between the very clumpy mass distribution observed on relatively small scales and the strikingly smooth distribution observed on the scale of the Hubble length. The examples presented here show how an expanding universe constructed to be as close to homogeneous as it can be might be forced to develop structure that is at least roughly similar to what is observed.

It is a common feature of the cosmic fields as candidates for the source of structure in our universe that the density fluctuations they introduce are scale-invariant, in the sense that the rms mass fluctuation $\delta_h = \delta\mathcal{M}/\mathcal{M}$ that is being introduced on the scale of the Hubble length is independent of redshift. The alternatives are that δ_h is increasing with cosmic time, meaning we are headed toward a manifestly chaotic universe as δ_h approaches unity, or that δ_h is decreasing. The problem with the latter is that we cannot assume δ_h ever was on the order of unity, for that would have promoted the formation of black holes on the scale of the Hubble length, vastly overproducing the dark matter discussed in section 18. The elegance of a scale-invariant spectrum of mass density fluctuations was appreciated before the possible role of cosmic fields was recognized (Harrison 1970a; Peebles and Yu 1970; Zel'dovich 1972), and was particularly studied by Zel'dovich. Cosmic fields quite naturally give us this spectrum, as do simple versions of the inflation scenario to be discussed next. Cosmic fields also offer a remarkable variety of possibilities for the character of the perturbations they introduce, from nearly Gaussian if $N \gg 5$, to distinctly spiky in the case of topological textures, or sheetlike or cylindrical in the wakes of long cosmic strings. This is a fascinating shopping list that seems likely to occupy the community for quite a while to come. At the time this is written there is not a fully worked-out model for the way galaxy formation might be triggered by a cosmic field. Some general considerations on the constraints on such a model are discussed in section 25.

17.   Inflation

Guth hit on inflation as a way to eliminate the gauge monopoles mentioned in the last section. He then saw that inflation offers an elegant way to understand a deeper puzzle: why the universe is so close to homogeneous on the scale of the Hubble length. Linde showed how the inflation concept might lead us to a rational picture for what the universe was like before our region entered the classical Friedmann-Lemaître evolution. These latter two problems are as old as the Friedmann-Lemaître model, for the observed large-scale homogeneity and isotropy of the universe is not required by classical general relativity theory as it is generally understood, but the classical model does say the expansion traces back to a singular state. Because inflation offers such an elegant and richly detailed approach where we previously had no very definite ideas, it quite properly has attracted considerable attention and it has led to much work on specific models to implement the concept.

The great influence of inflation on the directions of research in theoretical cosmology has led people to term it the new paradigm for this subject. As usually understood in physical science, however, a paradigm is a pattern for research that one has reason to believe really is a useful approximation to the physical world, because the pattern has passed nontrivial experimental/observational tests. Since that has not yet happened in inflation, and there is not even a generally accepted and definite inflation model, we will continue to term inflation a scenario. It is notable, though, and perhaps significant, that in the decade since the concept was discovered and the homogeneity puzzle made very visible, nobody has proposed a reasonably definite alternative resolution to the puzzle. Unless or until that happens, or the concept somehow can be shown to be untenable, we must expect that inflation will continue to occupy a central place in the exploration of concepts in theoretical cosmology.

As for many landmark discoveries, one can find precursors. During an inflation epoch the stress-energy tensor would be dominated by a term that acts like a time-variable cosmological constant. Zel'dovich (1968) had pointed out that within quantum mechanics the vacuum certainly could be imagined to have a nonzero energy density, as would be produced by zero-point fluctuations, and that if the vacuum were Lorentz-invariant its stress-energy tensor would have to have the form of a cosmological constant, $T_{ij} = \Lambda g_{ij}/8\pi G$, because this form is not changed by a velocity transformation. Linde (1974) and Bludman and Ruderman (1977) pointed out that the phase transitions that allow nonzero particle masses and fix the character of the interactions of the low energy physics we observe would be accompanied by considerable changes in the mean energy densities in the fields. This means that in the hot early universe, before these phase transitions, the stress-energy tensor is predicted to have contained a term that acts like a cosmological constant that is large and decreases toward its present very small value. Kazanas (1980) and Sato (1981a) noted that if there were a time when this term dominated the stress-energy tensor, it would mean the expansion rate $H = \dot{a}/a$ in the early universe was close to constant, approximating the exponential expansion of the de Sitter model (eq. [5.24]). Starobinsky (1979, 1980) found a similar behavior in a cosmological model that uses the action for gravity with quantum corrections. Kazanas noted that the near exponential expansion could eliminate the particle horizon of the standard model, allowing us to imagine in principle that some causal mechanism produced the observed large-scale homogeneity of our universe. Sato considered the idea that a spontaneous breaking of the symmetry of matter and antimatter in the early universe might have set different signs for the production of matter or antimatter in different regions of space (as discussed, for example, by Brown and Stecker 1979). The distance between the domain walls separating regions of matter and antimatter would be limited by the small Hubble length in the early universe, but if the regions were small the plasma in a great cluster of galaxies would have been drawn together out of patches of matter and antimatter. This would make it very difficult to understand the observed remarkable purity of an intracluster plasma, almost all matter (or all antimatter; eq. [18.114]). Sato's idea was that the decay of the effective cosmological constant could dilute the original ratio of baryon (or antibaryon) number to photon number, bringing the ratio to its present very small value, and considerably enlarge the distance between domain walls, perhaps even making them larger than the present Hubble length. That is, Sato was suppressing or eliminating domain walls in a similar way to what Guth had in mind for monopoles.

Guth (1981) produced the first fully assembled physical picture for inflation, though his version was imperfect because it assumed inflation ends with a first-order phase transition that creates the entropy we observe in the thermal cosmic background radiation (while the CBR in turn produced the baryons). As Sato (1981b) anticipated, this has the problem that the nucleation rate is estimated to be too slow to allow inflation to end, because regions that have completed the phase transition grow in size more slowly than they move apart. This was soon remedied in the pictures developed by Linde (1982, 1983) and Albrecht and Steinhardt (1982), in which the transition from the inflation epoch to the classical Friedmann-Lemaître model is continuous but rapid enough to produce the necessary entropy. Reviews of this idea and the remarkably broad variety of models suggested by the inflation scenario are given by Linde (1990), Olive (1990), and Kolb and Turner (1990).

This short historical survey illustrates the point that there are good physical reasons to suspect that something like inflation happened in the early universe. On the other hand, the physics of the phase transitions are not well enough fixed to give us any assurance that the transitions act as advertised in currently discussed models for inflation, or that an inflation epoch really left observable effects. Thus, the domain walls Sato was considering — and Guth's monopoles — may have been eliminated by inflation, or the physics may be such that they never existed in appreciable abundances.

At the time this is written the inflation scenario offers the only reasonably complete resolution of the puzzle of the large-scale homogeneity of the observable universe. Below we discuss how inflation may also account for the departures from homogeneity. The scenario thus certainly deserves close attention.

Scenario

Let us recall why the standard cosmological model can have a particle horizon, such that galaxies at high redshift are observed as they were when they had not been in causal connection with each other subsequent to the singularity at $a(t) \rightarrow 0$ (as in eq. [5.52]). At high redshift, space curvature is small compared to the mass density, so the cosmological equation (5.18) for the expansion rate is

$$H(t) = \frac{\dot{a}}{a} = \left(\frac{8}{3}\pi G\rho\right)^{1/2}, \qquad (17.1)$$

where the total energy density ρ and pressure p satisfy the local energy conservation equation

$$\dot{\rho} = -3\frac{\dot{a}}{a}(\rho + p). \qquad (17.2)$$

A packet of light moving radially in the Robertson-Walker line element (5.9) moves the distance $aRd\chi = dt$ in the world time interval dt, so the maximum coordinate displacement is

$$R\chi = \int_{a=0}^{a=a_o} \frac{dt}{a(t)} = \int_o^{a_o} \frac{da}{a\dot{a}} \propto \int_o^{a_o} \frac{da}{a^2\rho^{1/2}}. \qquad (17.3)$$

If the pressure is not negative, the energy conservation equation (17.2) shows that as the expansion parameter $a(t)$ approaches zero, the energy density ρ increases at least as rapidly as $a(t)^{-3}$, and the integral in equation (17.3) therefore converges at $a \rightarrow 0$. Galaxies at fixed coordinate separation larger than this limiting value for χ cannot have communicated with each other prior to the present epoch a_o. But looking in opposite directions in the sky we can see galaxies at coordinate distance $\sim \chi$ from us that are separated from each other by the distance $\sim 2\chi$. That is, we can see galaxies that cannot have seen each other. How did they and their environments know to look so similar?

The puzzle is resolved in inflation by the assumption that there was a time when the net pressure was negative,

$$p < -\rho/3. \qquad (17.4)$$

The energy equation (17.2) indicates that under this condition the energy density varies less rapidly than a^{-2}, so the integral in equation (17.3) diverges at $a \rightarrow 0$. If the inflation epoch, when equation (17.4) applies, lasts long enough, the particle horizons among the matter we can see are eliminated. This allows us to imagine, at least in principle, that there is a process that makes the observable universe homogeneous. Identifying a process that would work in practice is not so straightforward, but we will see that there are some reasonable-looking ideas.

The negative pressure in the inflation epoch comes from a single new real scalar field, ϕ, sometimes called an inflaton, with Lagrangian density

$$\mathcal{L} = \frac{1}{2}\phi_{,i}\phi_{,j}g^{ij} - V(\phi). \qquad (17.5)$$

The form is similar to equation (10.2), with the quadratic mass term replaced by the potential energy function $V(\phi)$. The field equation is

$$\frac{1}{\sqrt{-g}} \frac{\partial}{\partial x^i}\left(\sqrt{-g}g^{ij}\frac{\partial\phi}{\partial x^j}\right) + \frac{dV}{d\phi} = 0. \qquad (17.6)$$

The stress-energy tensor belonging to the field is given by equation (10.37) with the mass term replaced by $V(\phi)$. Suppose the field is close to spatially homogeneous, so we can ignore space derivatives compared to time derivatives. If the geometry is cosmologically flat, or close to it, we can write the line element as the Robertson-Walker form,

$$ds^2 = dt^2 - a(t)^2 \left(dr^2 + r^2 d\Omega \right) . \tag{17.7}$$

Then $\sqrt{-g} = a^3$, and the field equation becomes

$$\frac{d^2\phi}{dt^2} + 3\frac{\dot{a}}{a}\frac{d\phi}{dt} = -\frac{dV}{d\phi} . \tag{17.8}$$

In this limit the energy density and effective pressure in the field are (eq. [10.38])

$$\rho_\phi = \dot{\phi}^2/2 + V , \qquad p_\phi = \dot{\phi}^2/2 - V . \tag{17.9}$$

If the potential energy density V is a slowly varying function of the field ϕ, and provided the initial value of the time derivative of ϕ is not too large, the kinetic energy $\dot{\phi}^2/2$ can be small compared to V. If in addition V is large enough to make a significant contribution to the stress-energy tensor, the pressure can satisfy the inflation condition (17.4).

For a simple example of the inflation picture, suppose $\dot{\phi}^2 \ll V$. Then the field energy density and pressure are $\rho = -p = V$. If these are the main contributions to the stress-energy tensor, the universe is dominated by a term that acts like a cosmological constant, as in equation (4.31). If the universe is close to homogeneous within a Hubble length, then after a few Hubble times the solution relaxes to the de Sitter model (because we are assuming the field potential energy density V is evolving relatively slowly). The limiting expanding solution to the expansion rate equation (17.1) is

$$a \propto e^{H_V t} , \qquad H_V{}^2 = \frac{8}{3}\pi G V . \tag{17.10}$$

This is the line element of the steady-state cosmology (eq. [7.4]), in a new situation. As discussed in section 5, in this solution there is no particle horizon, so we are free to search for a causal explanation of the isotropy of the observable universe. Equation (17.10) has an event horizon, but it is eliminated in a more realistic model because the universe has to make the transition to expansion dominated by radiation and then matter well before we come on the scene.

There are other possibilities for the behavior of the expansion rate during an inflation epoch. We get power law inflation by adopting the potential

$$V = Ae^{b\phi} \tag{17.11}$$

(Lucchin and Matarrese 1985), where A and b are constants. Then a solution to equations (17.1) and (17.8) is

$$a \propto t^n, \qquad n = \frac{16\pi}{(bm_{pl})^2}, \qquad (17.12)$$

where the scalar field is

$$e^{b\phi} = \frac{2}{Ab^2t^2} \left[\frac{48\pi}{(bm_{pl})^2} - 1 \right], \qquad (17.13)$$

and the Planck mass is $m_{pl} = G^{-1/2}$ (eq. [15.7]). We can choose the characteristic length b to make $n > 1$. Then the integral in equation (17.3) diverges at $a \rightarrow 0$, which can eliminate the particle horizon among the matter in the galaxies we can observe.

The inflation epoch would end as ϕ approaches a minimum of the potential $V(\phi)$ steep and deep enough for the potential energy of the field to be transferred almost entirely to the kinetic energy term $\dot{\phi}^2/2$, and from there to the excitation of quanta of other fields through their coupling to the rapidly varying ϕ field. This would fill space with a sea of relativistic quanta with positive pressure. It must be assumed that very little energy density is left over in the ϕ field at the minimum of its potential.

The transition from a universe dominated by the energy in the field ϕ to one dominated by entropy would provide the initial conditions for the classical Friedmann-Lemaître model, which expands and cools to the present epoch. If domain walls or gauge monopoles were present in any appreciable abundance before inflation, they would be effectively eliminated by the enormous expansion factor during inflation, and no new ones could be produced if inflation ended at low enough temperatures. The original baryon density also would have become quite negligible during inflation, so baryons would have to form out of the entropy produced during the end of inflation. This last condition is not unique to inflation, of course, for a phoenix universe that produces entropy also has to make baryons.

Inflation might commence when a patch of the primeval chaos happens to expand and the density within the patch drops to the point where the local density of energy (and mass) is dominated by the potential energy V of the ϕ field. The local value of the Hubble length is $H^{-1} \sim (G\rho_\phi)^{-1/2}$, where $\rho_\phi \sim V(\phi)$ is the field energy. This sets the characteristic timescale for the evolution of the spacetime geometry within the patch. Equation (17.6) for field fluctuations $\delta\phi$ is of the form $\delta\ddot{\phi} \sim \delta\phi'' - (d^2V/d\phi^2)\delta\phi$, where the dots and primes mean time and space derivatives. If the potential is sufficiently flat, $|d^2V/d\phi^2| \ll H^2$, the fluctuations in ϕ behave as a massless field. Since the expansion of the universe tends to dissipate fluctuations in the field (through the second term in eq. [17.8]), a natural coherence length for spatial variations in ϕ might be the Hubble length H^{-1}, for this is the distance a field fluctuation can travel in an expansion time. The patch may choose to collapse on the timescale H^{-1}, in which case it is of no use to us, or it may expand on the same timescale. The expansion would make the coherence length of the patch larger than the more slowly changing Hubble length, and therefore larger than the distance gradients in ϕ can propagate in an expansion time, so the field gradients in this patch in effect would be frozen. (This is the inflation condition — that the expansion factor $a(t)$ increases more rapidly than $a/\dot{a}$.) Assuming all this happens, the patch has started to inflate. Its central parts expand in an increasingly better approximation to a section of a Friedmann-Lemaître model, the length scale over which ϕ varies increasing relative to the Hubble length. The process is illustrated by the sketch in figure 5.2, with Guth replacing de Sitter.

To produce the part of the universe we live in, inflation would have had to have lasted long enough to ensure that the central part of our expanding patch, within which the density varies by less than one part in 10^4, is larger than the present Hubble length, and we would have to be in this central region. We still would be in causal contact with the primeval chaos, in the sense that classical null paths would trace from us back to the edges of the patch, but the chaos would be at enormous redshifts. At redshifts where we can hope to receive useful information about the character of the mass distribution, the displacement of a light ray is comparable to the present Hubble length, and for a sufficient expansion factor during inflation this would be a very small fraction of the coherence length of the originally inflating patch. In the inflation scenario this is what makes the universe look homogeneous. It would be expected that the Hubble length in our region eventually will grow to be comparable to the size of the expanding patch, and any observers still on the scene will see that the universe truly is chaotic.

One encounters strong reactions to this scenario from skeptics who point to the special arguments, and from advocates who note that within accepted physics all of this could have happened, and perhaps even did. The proponents have another good argument — that the homogeneity of the observed universe is a very real puzzle for which we have no alternative interpretation spelled out in detail even remotely close to what we have for inflation. It may happen that as the understanding of what the concept has to offer improves, we will arrive at an inflation model that is seamless, demonstrated to be consistent with all the physics and observations we can hope to bring to bear, but tested by no predictions, only by the evidence that went into its construction. Would this be a believable physical science? The worry is not new in cosmology, of course, for this could have happened at any stage in the progress of physics. The impasse has not been encountered so far, and it is sensible to proceed on the assumption that further work will uncover believable evidence on whether inflation (or something else) really is a good approximation to what happened in the very early stages of our expanding universe. There already is a rich lore of ideas on how one might arrive at interesting and maybe even testable models for inflation. The remainder of this section presents a sampler of results.

Constraints

At the present epoch the field ϕ would be homogeneous through our region, at the constant value at a minimum V_o of the potential $V(\phi)$. The residual energy density V_o, if nonzero, would act as a cosmological constant, $\Lambda = 8\pi G V_o$ (eqs. [4.31] and [17.10]). In effect, Einstein's Λ term was large during inflation, dumped most of its energy into entropy at the end of inflation, and then settled down to a value that is very much smaller or perhaps vanishes. (In section 18 we consider a variant picture, in which the field energy density is rolling to the natural value, $\rho_\phi = 0$, but still is appreciable at the present epoch.)

The cosmological tests in sections 13 and 18 indicate that the density parameter in the Λ term is not much larger than unity, so the minimum value of $V(\phi)$ satisfies

$$|V_o| \lesssim \frac{3H_o^2}{8\pi G} = 2 \times 10^{-29} h^2 \text{ g cm}^{-3} = \frac{h^2 (0.003 \text{ eV})^4}{\hbar^3 c^5} . \tag{17.14}$$

The cosmic fields discussed in the last section are allowed to add to $|V_o|$. The small value of the sum is truly striking. The fact that the bound defines an energy $\epsilon_o = 0.003$ eV characteristic of condensed matter physics, rather than particle physics, leads some to conclude that the only natural value for $|V_o|$ is something well below this bound, perhaps identically zero.

If $V(\phi)$ during inflation is close to constant, as in the model in equation (17.10), it defines another characteristic energy, $V_{\text{infl}} = \epsilon^4$ (with units chosen so $\hbar = 1$ and $c = 1$). A popular (but not unique) choice for ϵ has been the energy scale of about 10^{14} GeV at which the strong, electromagnetic, and weak interactions extrapolate to a universal strength, so we will write

$$V_{\text{infl}}^{1/4} = \epsilon = 10^{14} \epsilon_{14} \text{ GeV} = 10^{23} \epsilon_{14} \text{ eV} . \tag{17.15}$$

If $\epsilon_{14} \sim 1$, the characteristic energy belonging to $V(\phi)$ has to drop twenty-five orders of magnitude at the end of inflation, to agree with equation (17.14), and the potential energy has to drop by some one hundred orders of magnitude. The factor is reduced by reducing ϵ, but since inflation has to end well before helium production the required change in the value of $V(\phi)$ is remarkable.

Let us estimate the redshift at the end of inflation, when the bulk of the energy left in ϕ is dumped into the entropy that we observe in the cosmic background radiation. In the approximation that the entropy was produced all at once, we have

$$V_{\text{infl}} = \epsilon^4 = \frac{\pi^2}{30} g_r T_r^4 , \tag{17.16}$$

where T_r is the temperature and g_r is the number of degrees of freedom that share the energy, as in equation (15.12). In the standard particle physics models, g_r is on the order of 100, so to the accuracy we need we can conclude that the temperature at the end of inflation is $T_r \sim \epsilon$. Then the redshift at this epoch is

$$z_r = a_o/a_r \sim \epsilon/T_o \sim 10^{27}\epsilon_{14}. \tag{17.17}$$

We are particularly interested in the material observed now in galaxies at the Hubble distance, at present redshift $z \sim 1$ to 3, as in figures 3.10 and 5.3. Let us assign coordinate radius x_o to the region that contains this material now at the Hubble distance. The present physical distance is

$$x_o a_o = H_o^{-1}, \tag{17.18}$$

where H_o is the present value of Hubble's constant and a_o is the present value of the expansion parameter. There was a time during inflation, at expansion parameter a_p, when the proper radius $a_p x_o$ of this region was equal to the Hubble length at that time,

$$x_o a_p = H_p^{-1}. \tag{17.19}$$

At $a \ll a_p$, the physical size ax_o of the region is small compared to the Hubble length. (This is because we are assuming the expansion rate varies less rapidly than $H \propto a^{-1}$, for we want the integral in eq. [17.3] to diverge going back in time.) An observer at our comoving position at $a \ll a_p$ would see that the edge of this region, at distance ax_o, has redshift $Hax_o \ll 1$. That is, information can bounce back and forth across the region, and a casual process might smooth irregularities in spacetime within the region, to produce the homogeneity we observe in figure 3.10.

An observer present during inflation but at a later time, when $a \gg a_p$, would find that an object at coordinate distance x_o has exceedingly high redshift. The informal way to describe this is to say that objects at separation x_o leave each other's horizons at $a \sim a_p$. A more careful way to put it is that although a message emitted by one object at early enough time can be received by the other during inflation and at $a \gg a_p$, there is no way the message can be returned prior to the end of inflation. After inflation, the Hubble length H^{-1} increases faster than $a(t)$, and the redshift of an element of matter at coordinate distance x_o decreases from an enormous value at the end of inflation to $z \sim 1$ at the present epoch.

The expansion rate during inflation at the expansion parameter $a = a_p$ in equation (17.19) is

$$H_p = \left(\frac{8\pi}{3}\right)^{1/2} \frac{\epsilon_p^2}{m_{\mathrm{pl}}}, \tag{17.20}$$

where the energy density at this epoch has been written as $\rho_\phi = \epsilon_p^4$. The redshift at this epoch is the ratio of equations (17.18) and (17.19),

$$z_p = \frac{H_p}{H_o} \sim 10^{51} \epsilon_{p,14}{}^2 . \tag{17.21}$$

Inflation ends at the redshift z_r in equation (17.17). If ρ_ϕ does not change by a large factor between z_p and z_r, we have

$$\frac{z_p}{z_r} = \frac{a_r}{a_p} \sim 10^{24} \epsilon_{p,14} . \tag{17.22}$$

This is the expansion factor to the end of inflation from the time when the redshift approached unity across the region now observed at the Hubble distance. Since this region is very close to homogeneous, we know it has to be a small part of the original expanding patch. That means the net expansion factor from start to end of inflation has to be much larger than the factor in equation (17.22).

This condition on the net expansion factor during inflation limits the possible size of the space curvature term in the expansion rate equation (5.18),

$$H^2 = \left(\frac{\dot{a}}{a}\right)^2 = \frac{8}{3}\pi G\rho \pm \frac{1}{a^2 R^2} . \tag{17.23}$$

In this equation, the density ρ includes all forms. At the present epoch ρ has an appreciable contribution from ordinary matter, and during inflation ρ is quite dominated by the energy of the ϕ field. We see from equations (17.18) and (17.19) that HaR has the same value now and at the epoch $a = a_p$, when the redshift of the material at comoving distance x_o, which now is unity, passed through unity during inflation. This means space curvature makes the same fractional contribution to the expansion rate in equation (17.23) now and during inflation at $a = a_p$. From the condition that the universe has to have expanded by a large factor from the start of inflation to the epoch a_p, we can conclude that space curvature has to be negligibly small now, as follows (Steinhardt 1990).

Suppose space curvature is negative, so the curvature term in equation (17.23) is positive. If the curvature term had an appreciable effect on the present expansion rate, the same would be true at $a = a_p$. Since the inflation condition is that ρ varies no more rapidly than a^{-2}, it would follow that at a less than a_p the curvature term dominates the expansion rate. Under this condition, equation (17.23) implies that the Hubble length scales as $H^{-1} \propto a(t)$ at $a \lesssim a_p$. This is at the threshold for inflation, but not adequate for the scenario, because the redshift across the region x_o that we can survey now at $z \sim 1$ would be $Hax_o \sim 1$, from $a \sim a_p$ all the way back to the start of inflation. That would mean the original expanding patch is visible at present redshift $z \sim 1$, and the primeval chaos beyond that, which is unacceptable if we want to use inflation to account for the observed homogeneity of the universe. If the curvature were positive and appreciable now, the expansion at $a = a_p$ would trace back to a bounce at a smaller than a_p by a factor on the order of unity. This is not the inflation picture, for prior to that the universe would have been contracting in a homogeneous way. It does resemble the oscillating Phoenix universe discussed in section 15, but with the difference that the energy density in the contracting epoch is dominated by a homogeneous scalar field, rather than the entropy one would have expected to have been produced during the previous oscillation.

The conclusion is that within the inflation scenario the curvature term in the cosmological equation (17.23) has to be negligibly small now, and as long into the future as the universe continues to appear homogeneous on the scale of the Hubble length. For another way to put it, the same expansion during inflation that makes the length scale over which the mass density varies large compared to the present Hubble length $H_o{}^{-1}$ makes the length $a_o R$ large compared to $H_o{}^{-1}$. The immediately interesting observational implication of this prediction is that the mass density in dark matter has to be an order of magnitude larger than the estimates from dynamics on relatively small scales (sections 18 and 20), or else there has to be something that acts like a cosmological constant, to allow the universe we observe to be cosmologically flat.

The last of the general considerations is that the departures from homogeneity introduced during inflation must not exceed what is observed (and may even agree with it). We discuss here the basic elements of the quantum fluctuations that perturb homogeneity. The next subsection presents some elements of a more careful calculation.

A convenient way to start is to write the field that drives inflation as a sum of a classical mean value, $\phi_o(t)$, and a fluctuating part, $\psi(\mathbf{x}, t)$. Substitute the expression $\phi = \phi_o(t) + \psi(\mathbf{x}, t)$ into the action for the Lagrangian density in equation (17.5), and expand the integrand as a series in ψ. The zero order part gives the evolution of the mean field $\phi_o(t)$. The term linear in ψ vanishes, because the action is stationary under linear variations of the field. The first nontrivial terms thus are quadratic, the time part $\psi_{,i}\psi^{,i}/2$ acting as the kinetic energy, and the part $\psi^2 d^2 V/d\phi^2$ acting as the mass term for ψ. We are assuming the potential is flat, so its derivatives are small, so ψ behaves like a massless field.

Suppose the field averaged over the Hubble length H^{-1} during inflation fluctuates by the amount ψ. Then the field gradient in this fluctuation is $\nabla \psi \sim H\psi$, the gradient energy density is $\sim H^2 \psi^2$ (eq. [10.38]), and the energy of this fluctuation integrated over the Hubble volume $\sim H^{-3}$ is $\delta E \sim \psi^2/H$. The timescale for the fluctuation to change is the crossing time, $\sim H^{-1}$. The uncertainty principle tells us the minimum value for δE is on the order of the reciprocal of this time,

$$\delta E \sim \psi^2/H \sim H \, . \tag{17.24}$$

Thus, the rms quantum fluctuation in the field averaged over the Hubble length is

$$\psi \sim H \, . \tag{17.25}$$

In an expansion time, the fluctuation ψ is stretched to a scale larger than the Hubble length. Since the field gradients cannot propagate further than the Hubble length in an expansion time, the fluctuation has become a "frozen" contribution to the space distribution of ϕ for the duration of inflation. A somewhat more elaborate derivation of this result is given in the next subsection.

This argument applies to the fluctuations of any nearly free field, including gravity. If the metric tensor is perturbed from the homogeneous world model by the small fractional amount $\sim h$, then the strain h satisfies a wave equation (eq. [10.80]), so we see from the action (10.50) for the gravitational field that the Lagrangian density belonging to the perturbation is of the schematic form

$$\mathcal{L} \sim m_{\mathrm{pl}}^2 \left[\dot{h}^2 - (\nabla h)^2 / a^2 \right] . \tag{17.26}$$

That is, $m_{\mathrm{pl}} h$ plays the role of ψ in the Lagrangian density in equation (17.5), so equation (17.25) says the rms value of the field averaged over the scale of the Hubble length is

$$h \sim \frac{H}{m_{\mathrm{pl}}} = \left(\frac{8\pi}{3} \right)^{1/2} \frac{\epsilon^2}{m_{\mathrm{pl}}^2} . \tag{17.27}$$

This is the fractional perturbation, or strain, in the spacetime geometry produced during each doubling of the expansion parameter during inflation. The strain at a fixed comoving length scale is frozen until the Hubble length has again grown comparable to the scale of the fluctuation, when h starts to propagate as gravitational radiation. Since ϵ has to be a slowly varying function of the expansion parameter during inflation, the strain appearing on the Hubble length after the end of inflation is nearly independent of redshift, as for the scale-invariant density fluctuations produced by some of the cosmic field defects discussed in the last section (eq. [16.47]).

This gravitational radiation is in principle observable, though not necessarily at the level of sensitivity of present detectors (Grishchuk 1988). To get some numbers, note first that the energy density in gravitational radiation is, from equation (17.26), $\rho_g \sim m_{\mathrm{pl}}^2 h^2 / \lambda^2$, for strain h at proper wavelength λ. When the wavelength is small compared to the Hubble length, the energy density scales as $\rho_g \propto a(t)^{-4}$, and the wavelength scales as $\lambda \propto a(t)$, so the strain has to scale as $h \propto a(t)^{-1}$. The strain starts oscillating when its wavelength is comparable to the Hubble length. For waves that appear at the Hubble length when the universe is radiation dominated, the expansion time scaled from the present density parameter Ω_r in radiation is given by equation (13.18). The waves at comoving wavelength x thus start oscillating, at the strain $h(t_x)$ given by equation (17.27), when the expansion parameter satisfies

$$a(t_x)x \sim t_x = \frac{1}{2H_o\Omega_r^{1/2}} \left[\frac{a(t_x)}{a_o}\right]^2 . \qquad (17.28)$$

Collecting, we have that the present strain on the proper scale $\lambda_o = a_o x$ is

$$h_o \sim \Omega_r^{1/2}\frac{\lambda_o H_o}{c}\frac{\epsilon^2}{m_{pl}^2 c^4} . \qquad (17.29)$$

For a detector operating at wavelength $\lambda_o = 10^3$ km, $H_o\lambda_o/c \sim 10^{-20}$. The density parameter in radiation is $\Omega_r \sim 10^{-4}$ (eq. [6.80]), so $h \sim 10^{-22}(\epsilon/m_{pl}c^2)^2$. If the inflation energy ϵ is well below the Planck mass, the strain is well below current detector limits. For the radiation that appears on the Hubble length at $z > z_{eq}$, the mass density in gravitational radiation per octave of observed wavelength λ_o is independent of λ_o, and the mass fraction per octave is

$$\Omega_g = \frac{\rho_g}{\rho_{crit}} \sim \Omega_r \left(\frac{\epsilon}{m_{pl}}\right)^4 . \qquad (17.30)$$

Stinebring et al. (1990) conclude from pulsar timing that at wavelengths on the order of light years $\Omega_g < 10^{-6}$, which the inflation model satisfies if ϵ is less than about 10% of the Planck mass.

We have another constraint from the effect on the large-scale anisotropy of the thermal cosmic background radiation temperature caused by the strain appearing now on the scale of the present Hubble length. This strain is the fractional difference in the factor by which the universe has expanded in orthogonal directions. As discussed in section 6, a fractional difference h in the expansion factor translates into the same fractional perturbation to the cosmological redshift and to the cosmic background radiation temperature. The large-scale anisotropy of the CBR is

$$\delta T/T \sim 10^{-5} \gtrsim h , \qquad (17.31)$$

as discussed in section 21. We see from equation (17.27) that the energy density ϵ_p^4 at $a = a_p$ (eqs. [17.19] and [17.20]) has to satisfy

$$\epsilon_p \lesssim 10^{-3}m_{pl} \sim 10^{16}\,\text{GeV} \qquad (17.32)$$

(Rubakov, Sazhin, and Veryaskin 1982; Veryaskin, Rubakov, and Sazhin, 1983; Abbott and Wise 1984). In a simple inflation model, in which the expansion is close to exponential because ϵ is close to constant, the temperature at the end of inflation would have to be less than about 10^{16} GeV, which seems easy to imagine.

Now let us estimate the fluctuations in the mass distribution deposited in the Friedmann-Lemaître universe at the end of inflation. This subsection deals with the elements of the physics in some order-of magnitude numerical estimates. A few of the details are considered in the following subsection.

We will suppose inflation ends when the field ϕ reaches a critical value where the potential $V(\phi)$ suddenly becomes very steep and ϕ starts to vary rapidly with time, which dumps much of the energy V into the entropy we observe as the CBR. The fluctuations frozen into ϕ during inflation introduce slight irregularities in the hypersurface on which the entropy is produced. The streaming velocity of the entropy has to be normal to this hypersurface. (An observer moving so the phase transition is seen to happen simultaneously along a local part of the hypersurface must see that the spatial component of the entropy flux density vanishes.) The irregularities in the hypersurface thus produce peculiar velocities that create a growing mode of perturbation to the entropy distribution.

We will assume spacetime is not significantly perturbed prior to the phase transition at this hypersurface; the quantum fluctuations frozen into ϕ just perturb the clocks that tell the field when to convert its energy to entropy. Consider the epoch a_p in equation (17.19) at which the field becomes frozen on the comoving length scale x_o of the material we see at redshift $z \sim 1$. At this epoch the field is rolling toward the minimum at the rate $\dot{\phi}_p$. The field value has been perturbed by the amount $\delta\phi \sim H_p$. That produces a shift,

$$\delta t_p \sim \frac{\delta\phi_p}{\dot{\phi}_p} \sim \frac{H_p}{\dot{\phi}_p}, \tag{17.33}$$

in the time at which the field averaged over the scale x_o finally reaches the critical value at which it dumps its energy into entropy at the end of inflation. Now let us estimate how that affects the mass distribution well after inflation.

The mean value of the expansion parameter at the end of inflation is a_r. The size of the region of the field fluctuation in equation (17.33) that became frozen at $a \sim a_p$ has grown to the physical size $l_p \sim H_p^{-1} a_r/a_p$ at $a \sim a_r$. The time at which the mean value of ϕ averaged over this region reaches the critical value has been shifted by the amount δt_p. As illustrated in figure 17.1, this bends the hypersurface at which inflation ends by the angle $\sim \delta t_p/l_p$. It follows that the perturbation to the streaming velocity field of the entropy is

$$v \sim \delta t_p/l_p, \tag{17.34}$$

on the scale l_p.

The mass conservation equation (5.110) applied at the epoch $a \sim a_r$ at the end of inflation is

$$\dot{\delta} = -\nabla \cdot \mathbf{v}/a \sim v/l_p \sim H_r \delta_r. \tag{17.35}$$

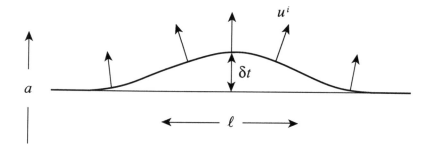

Figure 17.1. Distortion of the hypersurface at the end of inflation.

Here δ_r is the density contrast at $a \sim a_r$ averaged on the proper scale l_p. With equations (17.18) and (17.19), we see that the density contrast on the proper length scale l_p at the end of inflation is

$$\delta_r \sim \frac{v}{H_r l_p} \sim \frac{H_p H_o^2}{H_r \dot{\phi}_p} \frac{a_o^2}{a_r^2} . \tag{17.36}$$

The last step uses $l_p = H_o^{-1}/z_r$, for this is the distance of the material we observe at $z = 1$, redshifted back to the end of inflation.

This perturbation to the mass distribution grows as $\delta \propto t \propto a^2$ while the universe is dominated by radiation, and then as $\delta \propto t^{2/3} \propto a$ while the pressure and the space curvature are negligible (eq. [10.123]). The density perturbation as it appears now on the scale of the Hubble length has thus grown in amplitude by the factor $a_{eq}a_o/a_r^2$, where a_o/a_{eq} is the redshift at equal mass densities in matter and radiation. This brings the rms mass contrast on the scale of the present Hubble length to

$$\delta_h \sim \frac{H_p H_o^2}{H_r \dot{\phi}_p} \frac{a_o^3 a_{eq}}{a_r^4} . \tag{17.37}$$

We can simplify this expression by writing the mass densities at the end of inflation and at the present epoch as

$$\rho_r \sim H_r^2 m_{pl}^2 \sim T_r^4 \sim (T_o a_o/a_r)^4 ,$$
$$\rho_o \sim H_o^2 m_{pl}^2 \sim T_o^4 a_o/a_{eq} . \tag{17.38}$$

The result of collecting these expressions is

$$\delta_h \sim \frac{H_p H_r}{\dot{\phi}_p} . \tag{17.39}$$

Equation (17.39) is the rms mass density contrast observed on the scale of the present Hubble length H_o^{-1}. As we have discussed in part 1, and will consider in more detail in the following sections, a value $\delta_h \sim 10^{-4}$ would be interesting and can be arranged consistent with the other constraints if we are free to adjust the potential $V(\phi)$, as will be shown in the example below.

The freezing epoch during inflation for density fluctuations that appear at the Hubble distance at modest values of the redshift z follows by replacing equation (17.18) with

$$x_z a_z = x_z a_0/(1+z) = H_z^{-1} = H_o^{-1}(1+z)^{-3/2}, \qquad (17.40)$$

for an Einstein-de Sitter universe. We see that the comoving distance x_z of objects observed at the Hubble distance at the epoch at redshift z is $x_z = x_0(1+z)^{-1/2}$. If the expansion rate H is constant during inflation, equation (17.19) indicates that the density fluctuations that appear at the Hubble length at redshift z were frozen in during inflation at epoch

$$a_p(z) \sim a_p(1+z)^{1/2}. \qquad (17.41)$$

The fluctuations that have appeared since decoupling at $z \sim 1000$ thus were produced during an expansion factor ~ 30 during inflation, an exceedingly small part of the total expansion (eq. [17.22]). It might be reasonable therefore to expect that H and $\dot{\phi}$ are nearly constant during the epochs of freezing of the density fluctuations that produced galaxies and their clustering. If so, equation (17.39) for the contrast δ_h appearing at the Hubble length is close to constant, that is, it is scale-invariant.

Since the quantum fluctuations around the mean value of the field that is driving inflation are small, they ought to be accurately described as the Gaussian fluctuations of an almost free field. That is, this model for inflation produces primeval mass density fluctuations that are Gaussian and scale-invariant. The spectrum is quite similar to what follows in some of the cosmic field scenarios discussed in the last section. The character of the fluctuations is similar to what is produced by a cosmic field with more than four components, as discussed in the last section, but quite different from the distinctly non-Gaussian perturbations caused by cosmic strings, global monopoles, or textures. We arrived at Gaussian scale-invariant mass density fluctuations in this discussion because we used a simple model in which inflation is driven by a single field with a potential $V(\phi)$ that varies slowly with the field. Much more elaborate schemes are possible; examples are given in Linde (1990), Olive (1990), and Bond, Salopek, and Bardeen (1988).

Some orders of magnitude for the situation at the end of inflation are amusing and maybe significant. Material observed now at redshift $z = 1$ is at present distance $H_o^{-1} \sim 10^{28}h^{-1}$ cm. At the end of inflation, at the redshift z_r in equation (17.17), this comoving region had a radius of about $10/\epsilon_{14}$ cm. This is a length of human size, probably for no physically interesting reason. The recession velocity of an object at coordinate distance x is $\dot{a}x$. The recession velocity at the end of inflation is

$$\dot{a}x \sim (H_o a_o x) z_r / z_{eq}^{1/2} .$$ (17.42)

For material observed now at redshift $z = 1$, where $H_o a_o x \sim 1$, this velocity at the end of inflation is (eq. [17.17])

$$\dot{a}x_o = H_r l_p \sim 10^{35} \epsilon_{14} \text{ cm s}^{-1} .$$ (17.43)

The density contrast has grown by the factor z_r^2 / z_{eq}, so the present contrast $\delta_h \sim 10^{-4}$ at the Hubble length traces back to the value

$$\delta_r \sim 10^{-54} \epsilon_{14}^{-2}$$ (17.44)

at the end of inflation. The peculiar velocity associated with this density fluctuation is

$$v_r \sim H_r l_p \delta_r \sim 10^{-18} \epsilon_{14}^{-1} \text{ cm s}^{-1} .$$ (17.45)

These remarkable orders of magnitude are not unique to the inflation scenario; they apply to any model that extrapolates back to the redshifts considered here.

Finally, let us work through a somewhat more definite case that illustrates features we might look for in a successful inflation model. There is no general agreement on the best form for the potential energy function for the field that drives inflation, or whether it is reasonable to assume only one field is involved, or on the process by which the field converts its energy to entropy, so we can feel free to adjust and simplify the picture to simplify the calculation. As usual, the evolution at the end of inflation can be analyzed in considerably more detail than is presented here (Ratra 1991).

The starting picture is that inflation commences in an expanding patch when the primeval chaos deposits the ϕ field at a large value of the potential $V(\phi)$, and ends when ϕ rolls to a deep minimum of the potential and dumps its energy into the entropy of other fields. We will model the part of the potential at the field values during inflation and leading up to its end by the function

$$V(\phi) = V_o - \lambda \phi^4 / 4 ,$$ (17.46)

where V_o is a constant energy density and λ is a small dimensionless constant. In a more complete description, one would make $V(\phi)$ round over to a minimum at energy density very close to zero at a value of the field

$$\phi_{min} \sim (V_o / \lambda)^{1/4} .$$ (17.47)

The field equation (17.8) becomes

$$\ddot{\phi} + 3\dot{\phi}\dot{a}/a = \lambda\phi^3 . \tag{17.48}$$

We will check that convenient choices of the parameters allow the use of a "slow roll" approximation, where

$$\ddot{\phi} \ll \dot{\phi}\dot{a}/a , \qquad \dot{\phi}^2 \ll V_o . \tag{17.49}$$

We are assuming that during inflation $V_o \gg \lambda\phi^4$, so the potential is nearly constant and the expansion rate is nearly constant at

$$H = \left(\frac{8\pi}{3}\right)^{1/2} \frac{V_o^{1/2}}{m_{\mathrm{pl}}} . \tag{17.50}$$

That is, the evolution is close to the de Sitter model.

In the slow roll approximation, the field equation for ϕ becomes

$$\frac{\dot{\phi}}{\phi^3} = \frac{\lambda}{3H} , \tag{17.51}$$

and the solution is $\phi \propto (c-t)^{-1/2}$. This solution has to fail when the time t approaches the constant of integration, c, because the slow roll approximation fails. Since our estimates are crude we can assume that c is about equal to the epoch t_r at which ϕ starts to vary really rapidly and inflation ends with the conversion of the last of the ϕ field energy into the entropy we observe in the CBR. That is, we will write the solution as

$$\phi(t) = \left(\frac{3H}{2\lambda}\right)^{1/2} \frac{1}{(t_r - t)^{1/2}} , \tag{17.52}$$

and we will suppose inflation ends when t approaches t_r as this solution fails, the potential bottoms out, and the field settles to the value in equation (17.47) at a very small minimum of the potential.

The first slow roll condition in equation (17.49), $\ddot{\phi}/(H\dot{\phi}) \sim [H(t_r - t)]^{-1} \ll 1$, is consistent with the condition that $H(t_r - t)$ is appreciably greater than unity during inflation. The ratio of the kinetic energy in the ϕ field to the potential energy during inflation is $\dot{\phi}^2/V_o \sim \lambda^{-1}V_o m_{\mathrm{pl}}^{-4}[H(t_r - t)]^{-3}$, and the condition that the potential remains close to V_o until t approaches t_r is $\lambda \gg V_o m_{\mathrm{pl}}^{-4}$. Though we are going to want a small value for λ, we can keep these numbers small by assuming that the potential energy density V_o is sufficiently small compared to the Planck density m_{pl}^4.

The expansion factor during inflation from the time the density fluctuations we can see are frozen to the end of inflation is

$$a_r/a_p \sim e^{H(t_r-t_p)} \sim 10^{24}\epsilon_{14}\,,\qquad (17.53)$$

from equation (17.22). Thus we require

$$H(t_r - t_p) \sim 60\,.\qquad (17.54)$$

The time derivative of the solution in equation (17.52) is then

$$\dot\phi_p = \left(\frac{3H}{8\lambda}\right)^{1/2}\frac{1}{(t_r-t_p)^{3/2}} \sim \left(\frac{3}{8\lambda}\right)^{1/2}\frac{H^2}{(60)^{3/2}}\,.\qquad (17.55)$$

The density contrast appearing now on the horizon is given by equation (17.39):

$$\delta_h \sim H^2/\dot\phi_p \sim (120)^{3/2}3^{-1/2}\lambda^{1/2} \sim 10^{-4}\,.\qquad (17.56)$$

Thus the density fluctuations are roughly in line with what is observed if the parameter λ in the potential energy (eq. [17.46]) is

$$\lambda \sim 10^{-14}\,.\qquad (17.57)$$

The conclusion is that one can construct a model potential that produces the observed level of density fluctuations, although there are tight constraints on the potential. Calculations along these lines were first done by Hawking (1982), Starobinsky (1982), Guth and Pi (1982), and Bardeen, Steinhardt, and Turner (1983). A more complete analysis of the quantum mechanics of the density perturbations is given by Fischler, Ratra, and Susskind (1985).

Let us return to the question at the end of the last subsection: under what conditions might we consider the inflation picture for the early universe believable? The evidence from condensed matter and particle physics is that a phase transition of the kind postulated here is quite reasonable, and perhaps advances in microscopic physics will firm up the details. It may be possible to find some new successful observational prediction of inflationary cosmology that can validate the picture. It could also happen that the classical cosmological tests will show that space curvature makes a significant contribution to the expansion rate, in which case we would have to abandon inflation as an explanation for homogeneity. Failing either eventuality, the theory will have to be judged on its ability to solve the problems it was designed to solve, which evidently is a dangerous operation. It is impressive that inflation seems to be capable of producing a universe as large and close to homogeneous as the one we live in. It would be encouraging for the scenario if it were found that the primeval fluctuations from homogeneity are Gaussian and scale-invariant as in the simplest models for inflation, but the test is not sharp, for the prediction is neither definite nor unique to inflation. The list of tests is short, but the subject is young.

Density Fluctuations

The calculation of the mass density fluctuation amplitude in equation (17.39) may be a little too informal for a full understanding of this very important effect, so a more detailed analysis is presented here.

To highlight the quantum mechanics of the field fluctuations, let us first ignore the expansion of the universe, and derive the equivalent of equation (17.25) for the fluctuations of a free scalar field in Minkowski spacetime. We will write the field that drives inflation as $\phi = \phi_o + \psi$, where the classical part ϕ_o is the mean value, and the field operator ψ, that is real ($\psi^\dagger = \psi$), represents the fluctuations around the mean. The action for the fluctuating field in Minkowski spacetime is

$$S = \int d^3r \, dt \left[\frac{1}{2}\dot{\psi}^2 - \frac{1}{2}(\nabla\psi)^2 \right] . \tag{17.58}$$

This gives the field equation

$$\ddot{\psi} = \nabla^2\psi . \tag{17.59}$$

As in equation (9.5), the canonical momentum $\pi(\mathbf{r}, t)$ conjugate to the field operator $\psi(\mathbf{r}, t)$ is

$$\pi = \frac{\partial \mathcal{L}}{\partial \dot{\psi}} = \dot{\psi} . \tag{17.60}$$

The equal time commutation relations for the field and its momentum are the usual relation between the coordinates and their canonical momenta,

$$[\psi(\mathbf{r}, t), \pi(\mathbf{r}', t)] = i\delta(\mathbf{r} - \mathbf{r}') ,$$
$$[\psi(\mathbf{r}, t), \psi(\mathbf{r}', t)] = 0 = [\pi(\mathbf{r}, t), \pi(\mathbf{r}', t)] . \tag{17.61}$$

The easy way to proceed uses periodic boundary conditions in some large fixed volume, V_u. The plane wave solutions to equation (17.59) are

$$\psi \propto e^{i(\mathbf{k}\cdot\mathbf{r} - kt)} , \tag{17.62}$$

where $k = |\mathbf{k}|$, and the sum over the values of the propagation vector $\mathbf{k}$ allowed by the boundary conditions satisfies the identity

$$\sum_{\mathbf{k}} e^{i\mathbf{k}\cdot(\mathbf{r} - \mathbf{r}')} = V_u \delta(\mathbf{r} - \mathbf{r}') . \tag{17.63}$$

The plane wave expansion of the field operator is

$$\psi(\mathbf{r}, t) = \sum_{\mathbf{k}} \frac{1}{(2kV_u)^{1/2}} \left(a_{\mathbf{k}} e^{i(\mathbf{k} \cdot \mathbf{r} - kt)} + a_{\mathbf{k}}^{\dagger} e^{-i(\mathbf{k} \cdot \mathbf{r} - kt)} \right) , \qquad (17.64)$$

where the annihilation and creation operators $a_{\mathbf{k}}$ and $a_{\mathbf{k}}^{\dagger}$ satisfy the commutation relations

$$[a_{\mathbf{k}}, a_{\mathbf{k}'}^{\dagger}] = \delta_{\mathbf{k}\mathbf{k}'} ,$$
$$[a_{\mathbf{k}}, a_{\mathbf{k}'}] = 0 = [a_{\mathbf{k}}^{\dagger}, a_{\mathbf{k}'}^{\dagger}] . \qquad (17.65)$$

It is a straightforward exercise to check that with these commutation relations the plane wave expansion in equation (17.64) satisfies the canonical commutation relations in equation (17.61).

The normalized vacuum state $|0\rangle$ satisfies $a_{\mathbf{k}}|0\rangle = 0$, because the vacuum does not contain any field excitations to annihilate. Using the plane wave expansion in equation (17.64), we see that the equal time autocorrelation function of the field in the vacuum is

$$\langle 0 | \psi(\mathbf{r}, t) \psi(\mathbf{r}', t) | 0 \rangle = \sum_{\mathbf{k}} \frac{1}{2kV_u} e^{i\mathbf{k} \cdot (\mathbf{r} - \mathbf{r}')} = \frac{1}{16\pi^3} \int \frac{d^3k}{k} e^{i\mathbf{k} \cdot (\mathbf{r} - \mathbf{r}')}$$
$$= \frac{1}{4\pi^2 |\mathbf{r} - \mathbf{r}'|} \int_0^{\infty} dk \, \sin k|\mathbf{r} - \mathbf{r}'| , \qquad (17.66)$$

or finally,

$$\langle 0 | \psi(\mathbf{r}, t) \psi(\mathbf{r}', t) | 0 \rangle = \frac{1}{4\pi^2 |\mathbf{r} - \mathbf{r}'|^2} . \qquad (17.67)$$

The statistical properties of the fluctuations of the field ψ are completely described by its set of correlation functions. The three-point correlation function, $\langle \psi(\mathbf{r}) \psi(\mathbf{r}') \psi(\mathbf{r}'') \rangle$, vanishes because it contains an odd number of creation and annihilation operators; the four-point function reduces to products of the two-point function in equation (17.67), and so on. The result is that the field fluctuations are those of a Gaussian random process.

Let us consider the value of the field averaged over a spherical volume $V = 4\pi r^3/3$,

$$\bar{\psi} = \int_r \psi d^3 r / V . \qquad (17.68)$$

Since $\langle 0|\bar{\psi}|0\rangle = 0$, the variance of $\bar{\psi}$ is, with equation (17.67),

$$(\delta\psi)^2 = \langle 0|\bar{\psi}^2|0\rangle = \frac{1}{4\pi^2 V^2} \int_r \frac{d^3 r_1 d^3 r_2}{(\mathbf{r}_1 - \mathbf{r}_2)^2} = \frac{9}{16\pi^2 r^2} . \tag{17.69}$$

As expected, the rms fluctuation in the field averaged over a region of size r is $\delta\psi \sim 1/r$, as in equation (17.25). (It might be noted that, if the treatment of the integral in the second line of eq. [17.66] seems too crude, one can compute $(\delta\psi)^2$ using the first line and evaluating the position integrals first to get an integral over $\mathbf{k}$ that clearly is well behaved.)

We can do a little better by computing the equal time autocorrelation function in a Friedmann-Lemaître expanding universe. We will simplify the problem by assuming the field energy fluctuations have negligible effect on the spacetime, and we will assume the expansion rate during inflation is close to constant, as in equation (17.50), so we can use the de Sitter line element in equation (17.7) with expansion parameter

$$a = e^{\kappa t} . \tag{17.70}$$

The action (17.58) written in the expanding comoving coordinates of the de Sitter line element is

$$S = \int a^3 d^3 x \, dt \left(\frac{\dot{\psi}^2}{2} - \frac{(\nabla\psi)^2}{2a^2} \right) . \tag{17.71}$$

Then the momentum in equation (17.60) becomes

$$\pi = a^3 \dot{\psi} , \tag{17.72}$$

and the wave equation is

$$\frac{\partial^2 \psi}{\partial t^2} + 3 \frac{\dot{a}}{a} \frac{\partial \psi}{\partial t} = \frac{\nabla^2 \psi}{a^2} . \tag{17.73}$$

A convenient time variable for the plane wave solution is

$$\eta = -e^{-\kappa t} = -a(t)^{-1} \tag{17.74}$$

(Vilenkin and Ford 1982). With the negative prefactor, the value of η increases with increasing world time t. The plane wave solution to the field equation (17.73) is

$$\psi_+ \propto (-k\eta + i\kappa)e^{i(\mathbf{k}\cdot\mathbf{x} - \eta k/\kappa)} , \tag{17.75}$$

as can be checked by substituting into the field equation. In the time interval δt the phase in the exponential changes by the amount $k\delta\eta/\kappa = (k/a)\delta t$. This means that when $k|\eta| \gg \kappa$ the wave is oscillating at proper frequency k/a, as in equation (17.62). This is the appropriate short wavelength limit. The last oscillation of the function is at time t_x such that $-\eta(t_x)k/\kappa = 2\pi$, or

$$\lambda = \frac{2\pi a(t_x)}{k} = \kappa^{-1}, \qquad (17.76)$$

that is, when the proper wavelength is equal to the Hubble length. Thereafter, the solution relaxes to a time-independent function of position. This is the freezing effect discussed in the last subsection.

The generalization of the plane wave expansion of the field operator in equation (17.64) is

$$\psi = \sum_{\mathbf{k}} \frac{-k\eta + i\kappa}{(2V_u)^{1/2}k^{3/2}} e^{i(\mathbf{k}\cdot\mathbf{x} - k\eta/\kappa)} a_{\mathbf{k}} + \text{hc} . \qquad (17.77)$$

The second term is the Hermitian adjoint of the first. The normalization is arranged so that with the commutation relations in equation (17.65) for $a_{\mathbf{k}}$ and $a_{\mathbf{k}}^{\dagger}$ the field and its momentum π in equation (17.72) satisfy the commutation relations (17.61). In the terms in this expansion with wavelength small compared to the Hubble length, where $k|\eta| \gg \kappa$, the numerator in first factor is the proper wavenumber, k/a, and the denominator is the product of the physical volume and the proper wavenumber. These terms thus approach the flat spacetime case in equation (17.64). The interesting new feature in this solution in a de Sitter spacetime is that the modes at long wavelength, $k/a < \kappa$, are frozen with extra power from the extra factor of the wavenumber in the denominator. This is discussed more fully in Ratra (1985).

The equal time autocorrelation function of the field follows as in equation (17.66). In the limit where the separation is large compared to the Hubble length, $\kappa \gg k|\eta|$, the autocorrelation function is

$$\langle 0|\psi(\mathbf{x}, t)\psi(\mathbf{x}', t)|0\rangle = \sum_{\mathbf{k}} \frac{\kappa^2}{2V_u k^3} e^{i\mathbf{k}\cdot(\mathbf{x} - \mathbf{x}')}$$

$$= \frac{\kappa^2}{4\pi^2} \int_{k_{\min}}^{\infty} \frac{dk}{k} \frac{\sin k|\mathbf{x} - \mathbf{x}'|}{k|\mathbf{x} - \mathbf{x}'|} . \qquad (17.78)$$

We have to truncate the integral at small k. If that is not done by the physics, the cutoff is the wavelength $2\pi a/k_{\min}$ of the quantum fluctuations produced at the start of inflation. Whatever might be present at longer wavelengths is of no observational interest to us, assuming the amplitude is not too large, because it is a homogeneous perturbation to the region we are allowed to see. The first line of equation (17.78) shows that, after wavelengths of interest have grown to values much larger than the Hubble length, the magnitudes of the Fourier components of the frozen field as a function of position are

$$|\psi_{\mathbf{k}}| = \frac{\kappa}{(2V_u)^{1/2}k^{3/2}} \, . \tag{17.79}$$

The second line of equation (17.78) shows that the autocorrelation function has equal contribution per octave of wavelength k^{-1} from length scales between the maximum, $k_{\min}^{-1}$, and the lag $|\mathbf{x} - \mathbf{x}'|$. That is, the amplitude of the frozen field fluctuation is independent of length scale, as in equation (17.27).

We are assuming inflation ends when the mean value ϕ_o of the field $\phi = \phi_o + \psi$ reaches a critical value and dumps its energy into the entropy of matter and radiation fields. Also, we are assuming that prior to the end of inflation spacetime is homogeneous, with the perturbation from the field fluctuations entirely in the clocks that tell the field when to make the transition. Since the potential is assumed to be a slowly varying function of ϕ, we can take it that during the range of times when field fluctuations become frozen on the scales observable in the structure of galaxies and their space distribution, the field is varying at the nearly constant rate $\dot\phi_p$. This means that for fluctuations on these scales the hypersurface on which entropy is produced is the surface at time $t = \bar{t}(\mathbf{x})$, where

$$\bar{t} = t_r - \psi/\dot\phi_p \, . \tag{17.80}$$

This is the situation sketched in figure 17.1.

Now let us choose new time-orthogonal coordinates with the new time variable, $\bar{t}$, constant along this hypersurface. Then as shown in equations (9.39) to (9.41), the new coordinates are

$$\bar{t} = t + \zeta \, ,$$
$$\bar{x}^\alpha = x^\alpha + \zeta_{,\alpha} \int_{t_r}^t \frac{dt}{a^2} \, , \tag{17.81}$$

where $\zeta = -\psi/\dot\phi_p$ is the time shift, and the space part of the metric tensor is

$$\bar{g}_{\alpha\beta}(\bar{x}, \bar{t}) = -a(\bar{t})^2 \delta_{\alpha\beta} + 2a\dot a \delta_{\alpha\beta}\zeta + 2a^2\zeta_{,\alpha\beta} \int_{t_r}^t \frac{dt}{a^2} \, . \tag{17.82}$$

The ratio of the last term in this expression to the second one is on the order of $(t_r/\lambda)^2$. Since proper wavelengths λ of interest are much larger than the expansion time t_r, this last term can be ignored. Comparing the perturbation $h_{\alpha\beta}$ to the metric tensor defined in equation (10.98), we have

$$h_{\alpha\beta} = 2\frac{\dot a}{a}\zeta\delta_{\alpha\beta} \, . \tag{17.83}$$

Now choose a term from the Fourier series for ψ, and place the the x^3 axis along the propagation vector. Then a transverse component of the perturbation to the metric tensor is

$$h_{11} = -\frac{2\kappa^2}{(2V_u)^{1/2}\dot{\phi}_p k^{3/2}} e^{ikx^3} = -\frac{20}{9}\left(\frac{a_o}{t_o k}\right)^2 \delta_o(k). \qquad (17.84)$$

The first step uses $\kappa = \dot{a}/a$ and $\zeta = -\psi/\dot{\phi}_p$ with equation (17.79) for $\psi_{\mathbf{k}}$. The last expression uses equation (10.127) for h_{11}. In linear perturbation theory and assuming that the present space curvature, pressure, and the cosmological constant may be neglected, h_{11} is constant, so we can evaluate it at the present epoch, with δ_o the present density contrast belonging to the proper wavelength $2\pi a_o/k$.

With $t_o = 2/(3H_o)$ for the Einstein–de Sitter model assumed in equation (17.84), the present density contrast in linear perturbation theory is

$$\delta_o(\mathbf{x}) = \frac{2}{5}\frac{\kappa^2}{(2V_u)^{1/2}\dot{\phi}_p(H_o a_o)^2}\sum_{\mathbf{k}} k^{1/2} e^{i(\mathbf{k}\cdot\mathbf{x}-\varphi_{\mathbf{k}})}, \qquad (17.85)$$

where the random phases satisfy the reality condition $\varphi_{-\mathbf{k}} = -\varphi_{\mathbf{k}}$. To see what this result means, consider the density contrast smoothed through a Gaussian window,

$$\delta_g = \frac{\int d^3x\, \delta_o(\mathbf{x})e^{-x^2/2x_g^2}}{\int d^3x\, e^{-x^2/2x_g^2}}. \qquad (17.86)$$

We have

$$\frac{\int d^3x\, e^{i\mathbf{k}\cdot\mathbf{x}-x^2/2x_g^2}}{\int d^3x\, e^{-x^2/2x_g^2}} = e^{-k^2 x_g^2/2}, \qquad (17.87)$$

so the smoothed contrast is

$$\delta_g = \frac{2}{5}\frac{\kappa^2}{(2V_u)^{1/2}\dot{\phi}_p(H_o a_o)^2}\sum_{\mathbf{k}} k^{1/2} e^{i(\mathbf{k}\cdot\mathbf{x}-\varphi_{\mathbf{k}})-k^2 x_g^2/2}. \qquad (17.88)$$

The integral for the mean square value is easy to work, giving

$$\langle \delta_g{}^2 \rangle = \frac{1}{50\pi^2}\frac{\kappa^4}{\dot{\phi}_p^2}\frac{1}{(H_o a_o x_g)^4}. \qquad (17.89)$$

If the smoothing length $a_o x_g$ is equal to the Hubble length $H_o{}^{-1}$, the rms density contrast is $\delta_g \sim \kappa^2/\dot{\phi}_p$, which agrees with equation (17.39). This is the scale-invariant fluctuation spectrum discussed in the last section, here in a Gaussian form. Models for galaxy formation inspired by this result are discussed in section 25.

18. Dark Matter

It has been known since the 1930s that the velocities of galaxies in clusters are characteristic of a total cluster mass about an order of magnitude larger than the sum of the luminous masses observed within the galaxies themselves. It has been known since the 1970s that there is a similar situation in the outer parts of spiral galaxies and at least some ellipticals. Before 1980 the usual assumption was that the dark mass is ordinary matter in some not readily detectable form, such as gas, low mass stars, stellar remnants in white dwarfs and neutron stars, or black holes produced by early generations of very massive stars. The 1980s saw the acceptance of another fascinating idea, that the dark mass is in neutrinos or some more exotic nonbaryonic particle, and that a test might be possible by production of these particles in the laboratory or by the laboratory detection of particles from the halo of the Milky Way that pass through the laboratory. Discovering the nature of the dark matter, or explaining why the Newtonian mechanics used to infer its existence has been misapplied, has to be counted as one of the most exciting and immediate opportunities in cosmology today. The problem is well defined, and the material to be interpreted is close at hand on the scale of extragalactic astronomy. It would be hard to overstate the effect on this subject of a convincing demonstration of the nature of the dominant form of matter in our universe.

This section surveys the evolution of opinion on the mass puzzle and the nature of the dark matter that might resolve it. Tests for some of the candidates for dark matter figure here and in the following sections, as we consider the physics and astronomy of the nature and evolution of the distributions of galaxies and mass.

One should bear in mind that the dark mass idea appears in two contexts. The first is the set of dynamical results that indicate most of the mass in galaxies, and in systems of galaxies, is outside the bright central parts where the mass of the luminous stars dominates. As shown in table 20.1, the amount of dark mass estimated from these observations brings the mean density of the universe to about 10% of the critical Einstein-de Sitter value. The second context is the set of arguments in sections 15 and 17 and below that leads one to give very careful consideration to the possibility that there is another factor of ten in dark mass outside systems of galaxies, bringing the total to $\Omega = 1$. If Newtonian mechanics is a useful approximation on the scale of galaxies, the observations unambiguously establish the reality of the first effect, the presence of dark mass in galaxies. And if there is dark mass, it certainly is reasonable to consider the possibility that another factor of ten might be found. However, at the time this is written there is no compelling evidence that this last step follows. Perhaps the extra mass is unrelated to the dark mass in galaxies; perhaps the extra mass is not there. An example of the former is discussed at the end of this section: a term that acts like a cosmological constant could satisfy the condition from inflation that space curvature is negligibly small, but it would have negligible effect on the dynamics of groups and clusters of galaxies.

The evidence for dark mass in galaxies also has a theoretical element: it may be that Newtonian mechanics has been misapplied. The prospects for new physics on the scales of masses and sizes of the galaxies are reviewed by Sanders (1991). This is a possibility to keep in mind, but here we follow the dictum that new physics is to be considered as the main line of approach only when we are driven to it, which certainly is not yet the case for either aspect of the mass puzzle.

The dark mass puzzle also offers a fascinating example of the way ideas evolve in science. It is one of the tasks in a physical science such as cosmology to sift through the clues that might add to our understanding of the physical world. Since the list of possibly helpful empirical evidence is so rich, this seldom is an exercise in pure phenomenology: the way one weighs clues has to be informed by ideas about what is significant. It was known in the cosmology of the 1960s that the mean mass density derived from what is in the readily detectable bright parts of galaxies is at least an order of magnitude below the critical Einstein-de Sitter mass density, and that the dark mass in groups and clusters, if more broadly spread, could bring the total up to the Einstein-de Sitter value. (An example of the state of opinion is to be found in PC.) The Dicke coincidences argument for the Einstein-de Sitter model was known, in at least some parts of the community, as was the point J. A. Wheeler stresses, that Einstein's original argument for a closed universe still seems to be the best way to understand Mach's principle (as discussed in section 2). However, in the 1960s one encountered few heated debates on what the value of the density parameter Ω really ought to be; among those who cared it generally was considered a number to be measured.[1] The 1980s saw a remarkably sharp swing to the widely held belief that the density parameter surely is unity. This was in part due to the growing influence of the Dicke coincidences argument, in larger part a deduction from inflation along with some hopeful thinking from the wealth of dark matter candidates offered by particle physics. That led to a much sharper appreciation of the Dicke coincidences as supporting evidence.

As we have noted, the arguments for the Einstein-de Sitter model are nontrivial but, at the time this is written, they are almost entirely theoretical and certainly not part of a new paradigm. There is appreciable and promising progress in the

---

[1] The very notable exception to the older empirical approach to the values of the cosmological parameters comes from the relation between the mean mass density and the future of the universe within the Friedmann-Lemaître model, whether it is expansion into the indefinite future or collapse back to a "big crunch." The opinion on which is preferable cannot be classified as a debate in science, however, unless we can explain what point is served by the outcome of the debate.

development of observational tests of the model, but not yet to the point where it would be sensible to try to decide whether the weight of the evidence favors the Einstein-de Sitter model or a lower-density case (though that is attempted in section 26). Much of the rest of this book deals with or depends on the outcome of the issue. If the observations are found to support the Einstein-de Sitter model, it will be a success for the standard procedure that looks first to the simplest case. If the evidence favors a low-density model, it will mean the universe is more complicated than many of us thought it had to be. And there always is the very interesting chance that there is something wrong with the choice of models the standard Friedmann-Lemaître cosmology offers us.

We begin with a review of the development of the evidence that the masses of galaxies are dominated by dark matter, and then survey the general constraints on what the dark matter might be, whether baryonic, or neutrinos, or something more exotic. Section 20 deals with the dynamical measures of how much dark matter there might be, and section 21 with the character of its large-scale distribution.

Historical Remarks

The first detailed estimates of the mass of the universe, starting with Hubble (1926b), were based on counts of galaxies and on the galaxy masses derived from the gravitational binding energies needed to contain the motions of stars and gas within galaxies. It is immediately apparent, of course, that this accounting misses the material that is outside the bright parts of the galaxies. The first evidence that a considerable fraction of the mass had been missed came from the measurements of the velocities of galaxies within the Coma and Virgo clusters of galaxies. Zwicky (1933) and Smith (1936) estimated the cluster masses needed to gravitationally bind the galaxies moving at the observed velocity dispersions. The results could be compared to Hubble's estimates of characteristic galaxy masses based on the motion of material within individual galaxies. Zwicky and Smith found that if these clusters are gravitationally bound, the cluster masses are some two orders of magnitude larger than the sum of the masses within the galaxies. It is interesting to read Smith's conclusion:

> The difference between this result and Hubble's value for the average mass of a nebula apparently must remain unexplained until further information becomes available. A statistical study of the relative velocities of close pairs of nebulae may possibly furnish the required data. It is also possible that both values are essentially correct, the difference representing internebular material, either uniformly distributed or in the form of great clouds of low luminosity surrounding the nebula, as suggested by the recent great extension of the boundary of M31. Whatever the correct answer, it cannot be given with certainty at this time.

In the more than five decades since Smith wrote this, the situation has grown somewhat clearer. The old phrase for the effect, the "missing mass," has been abandoned as possibly illogical and replaced with a new one, the "dark matter."[2] We have considerably more information on how the dark mass is distributed, but still no reasonably convincing picture for its nature or whether the amount agrees with the critical Einstein-de Sitter value.

If Newtonian mechanics gives a useful approximation to the dynamics of galaxies and clusters of galaxies, their masses certainly are dominated by dark matter, that is, by material with a mass-to-light ratio considerably higher than that of the matter in and around the stars seen in the central parts of galaxies. In the examples shown in figure 3.12, the solid lines show the rotation curves expected under the assumption that the distribution of the mass follows that of the light, with the constant of proportionality adjusted to fit the inner part of the observed rotation curves. The fit yields a mass-to-light ratio consistent with that of a standard mix of luminous star masses. That is, the evidence is that the masses of the inner parts of the galaxies are dominated by visible stars. Massive halos with considerably higher mass-to-light ratios are needed to account for the nearly flat outer parts of rotation curves of isolated spiral galaxies.

Smith (1936) noted that to detect the total mass in a cloud of matter surrounding a galaxy — now called dark halo — one needs to measure its gravitational effect on motions in or beyond the halo, for under the inverse square law the gravitational acceleration in the central part of the galaxy is unaffected by a spherical external halo, and little affected by a flattened one. By the middle 1970s the measurements of rotation velocities in the disks of nearby spiral galaxies, from the Doppler shifts of the atomic hydrogen 21-cm line and optical lines from gas in the disk, had given reasonably convincing evidence that under Newtonian mechanics there are spiral galaxies in which the mass is more broadly spread than the starlight (Rubin and Ford 1970; Rogstadt 1971; Rogstadt and Shostak 1972; Roberts and Rots 1973; Roberts 1976), as Smith's comment anticipated. Another indication that the mass of a galaxy need not be where the starlight is came from the demonstration that a flat, cold, self-gravitating disk, with the nearly flat rotation curve of a spiral galaxy, is dynamically unstable to barlike collapse. Since spiral galaxies are common, the indication is that most of the mass is in a component with a more stable distribution, with a larger random velocity dispersion than in the seen material in the disk (Ostriker and Peebles 1973). This has no direct bearing on extended halo masses, for the mass needed to produce the gravitational acceleration v_c^2/r in a given rotation curve $v_c(r)$ is nearly the same whether the mass is distributed in a thin disk or in a more nearly spherical halo. But if, as this instability suggests, much of the mass in the inner parts of a spiral galaxy is

---

[2] Since historical names should not be abandoned lightly, it is good to note that the old term, "missing mass," can be recycled to describe the problem of reconciling many of the observations with the large mass density required by the Einstein-de Sitter model. This is discussed in sections 20, 25, and 26.

in a dark halo, it does invite the speculation that the run of mass with radius could continue to an extended massive halo. Early discussions of the possible benefits of dark halos as an aid to understanding the internal dynamics of galaxies, and the motions of galaxies in systems such as the Local Group, were given by Freeman (1970), Peebles (1971a), Einasto, Kaasik, and Saar (1974), and Ostriker, Peebles, and Yahil (1974).

Let us construct a simple model for the contribution to the mean mass density by massive halos. Galaxies come in a large range of luminosities and velocity dispersions, and most are dwarfs, as we see in figure 3.3. But the standard estimates of the luminosity function indicate that most of the luminosity of the universe comes from the galaxies near the knee of the luminosity function, at $L \sim L_*$ (eq. [5.131]), and since mass-to-light ratios do not scale rapidly with luminosity, the same is likely true of the mass. This means we get a reasonable model by imagining the mass is in $L \sim L_*$ galaxies at number density $n_* \sim 0.01 \, h^3 \, \mathrm{Mpc}^{-3}$ (eq. [5.144]).

Let us suppose each of these L_* galaxy has a flat rotation curve, as in figure 3.12, with the mass density run $\rho \propto r^{-2}$ cut off at radius r_h. Then the mass per galaxy is (eq. [3.29])

$$\mathcal{M}(<r_h) = \frac{v_c^2 r_h}{G} = 1.1 \times 10^{10} r_h(\mathrm{kpc}) \, \mathcal{M}_\odot . \tag{18.1}$$

The cutoff radius is expressed in kiloparsecs, and the numerical value assumes the rotation velocity $v_c = 220 \text{ km s}^{-1}$ characteristic of an L_* galaxy (eqs. [3.34] and [3.35]). The mean mass density contributed by these halos is the product of the halo mass $\mathcal{M}(<r_h)$ with the number density n_*. The resulting density parameter (5.55) is

$$\Omega_{\mathrm{halo}} = \frac{\rho_{\mathrm{halo}}}{\rho_{\mathrm{crit}}} \sim \frac{n_* v_c^2 r_h}{G \rho_{\mathrm{crit}}} = 0.4 \, h \, r_h(\mathrm{Mpc}) . \tag{18.2}$$

This says the dark halos would provide the critical Einstein-de Sitter mass density if the halos typically extended to $2.5h^{-1}$ Mpc radius. It is a suggestive coincidence that this cutoff radius is comparable to the mean distance between L_* galaxies (eq. [5.145]). That is, if the $\rho \propto r^{-2}$ halo of each galaxy ended about where the next one begins, the density parameter would be close to the theoretically preferred value, $\Omega = 1$.

As discussed in section 20, this could not quite be what happens, for the mass per galaxy obtained this way exceeds what is observed in groups and clusters; the observed effective halo cutoff is $r_h \sim 200h^{-1}$ kpc (eq. [20.31]). It is possible that much of the mass is in a more broadly distributed component; the test is to estimate masses from dynamics on larger scales. The review in section 20 indicates the present observational results would allow the density parameter to be as low as $\Omega \sim 0.05$, or as high $\Omega = 1$, though in the latter case most of the mass of the universe would have to be outside the groups and clusters of galaxies.

Let us turn now to two questions: What is the nature of the dark mass, and could there be enough to agree with the Einstein-de Sitter model? It is natural to consider first the possibility that the dark mass is something baryonic, such as low mass stars or star remnants. However, Gott et al. (1974) emphasized that, in the standard model for the origin of the light elements (section 6), the residual deuterium abundance is quite low if the density parameter Ω_B in baryons is near unity, because a high density at the epoch of nucleosynthesis causes almost all the deuterium to burn to helium (as is illustrated in figure 6.5). Gott et al. estimated that to produce the observed deuterium we would need $\Omega_B h^2 \sim 0.02$ (which is quite close to the recent estimate in eq. [6.27]). One way around this constraint is to assume most of the mass is in a form that cannot participate in the element-building reactions, as baryons locked in black holes, or exotic nonbaryonic matter (as discussed, for example, by Gott et al. 1974; and Chapline 1975).

The first widely discussed candidate for nonbaryonic dark matter was a family of neutrinos with a rest mass of a few tens of electron volts. Neutrino pairs are produced in thermal equilibrium with the cosmic background radiation in the early universe, as discussed in section 6, and the present density of these relict neutrinos is fixed by the observed CBR temperature (eqs. [6.71], [6.76], and [6.78]). The present mean number density of neutrinos plus their partners in one family is

$$n_\nu = 113 \text{ neutrinos cm}^{-3}. \tag{18.3}$$

If the neutrino rest mass in this family is m_ν, the mean mass density is $n_\nu m_\nu$. In terms of the density parameter Ω_ν belonging to $n_\nu m_\nu$, the neutrino mass is

$$m_\nu = 1.88 \times 10^{-29} \Omega_\nu h^2 \text{ g cm}^{-3}/n_\nu$$
$$= 93 \Omega_\nu h^2 \text{ eV}. \tag{18.4}$$

Gershtein and Zel'dovich (1966) noted that relict neutrinos could make an appreciable contribution to the present cosmic mean mass density, and Marx and Szalay (1972) introduced the detailed calculation of the bound on the neutrino rest mass from the condition that the neutrinos not overcontribute dark matter. Cowsik and McClelland (1973) found that neutrinos would be appropriate dark matter for rich clusters of galaxies. As will be described, the Cowsik-McClelland approach shows that massive relict neutrinos would give a very natural fit to the mass distributions in the dark halos of large galaxies, but that the dark mass in some dwarf galaxies does not fit this picture. Szalay and Marx (1976) pioneered the study of the gravitational growth of mass concentrations in a cosmology in which at redshifts less than z_{eq} (the redshift at equality of the mass densities in matter and radiation) the mass is dominated by neutrinos. In the early 1980s there was an intense interest in this subject, because laboratory evidence suggested the dominant neutrino family in nuclear beta decay might have a mass of a few tens of electron volts (Lubimov et al. 1980), comparable to the number in equation (18.4).

Interest soon shifted away from massive neutrinos, under the influence of three developments. The first was that the experimental bound on the mass for the electron-dominated family was reduced below the number in equation (18.4) for interesting values of Hubble's constant, $h \gtrsim 0.5$. (The present upper bound is about 9 eV, and the lower bound still includes zero.) Second, it was recognized that if the primeval fluctuations in the mass distribution are Gaussian, scale-invariant, and adiabatic (meaning the dominant mass components, in neutrinos, photons, and baryons, all have the same primeval space distribution), the picture for galaxy formation is problematic, as will be discussed in section 25. Neither of these points really is evidence against massive neutrinos as the dark matter, for the mass bound eliminates only one family, leaving two other possibilities for the mass in equation (18.4), and the more manifest problems with galaxy formation are eliminated by going to a nonadiabatic scheme, as in the cosmic field scenarios in section 16. The most immediately influential development was the third — the recognition that there are many other nonbaryonic dark mass candidates, with acceptable particle physics pedigrees, that offer possibilities for galaxy-formation models that arguably are as elegant as the massive neutrinos picture, and can accommodate the adiabatic initial conditions that naturally come out of the simplest models of the inflation scenario discussed in the last section.

Following R. Bond and A. Szalay, neutrinos have come to be called hot dark matter, because at redshift $z \sim z_{eq}$ the neutrino velocities would be close to relativistic, preventing early formation of bound systems with the binding energy typical of galaxies. In the opposite extreme is nonbaryonic cold dark matter with negligible pressure (until it is mixed by orbit crossings).

The growth of influence of these ideas can be traced in the proceedings of the Texas Symposia on Relativistic Astrophysics. At the ninth conference (Ehlers, Perry, and Walker 1980), I argued for a density parameter $\Omega \sim 1$, partly because of the Dicke coincidences argument, partly because it seemed difficult to understand the frequent occurrence of galaxy concentrations with mean density contrast $\delta\rho/\rho$ on the order of unity in a low density universe, where gravitational instability is suppressed at low redshift. I still was optimistic that the galaxy relative velocity dispersions discussed in section 20 would prove to be high enough to be consistent with a high-density universe.[3] Most who had opinions tended to

---

[3] Shortly after this it became clear from the data from the CfA redshift survey that this is not so. The measure is illustrated in figure 20.1. This development was influential in leading N. Kaiser and J. Bardeen to the invention of the biasing concept discussed in section 25. As indicated in figure 13.13, the suppression of gravitational instability in a low-density universe is much less significant in a cosmologically flat model with a nonzero cosmological constant than in an open universe. That, plus the argument from inflation, is what leads to the interest in low-density, cosmologically flat models with a positive cosmological constant.

favor a low value for Ω, however, on the sensible ground that this was (and is) the straightforward reading of most of the dynamical estimates. The tenth conference (Ramaty and Jones 1981) introduced massive neutrinos, the eleventh (Evans 1984) inflation and the zoo of dark-matter candidates from particle physics, and the twelveth (Livio and Shaviv 1986) cosmic strings. By this time a commonly held opinion was that the density parameter really ought to be unity, and that the mass is likely dominated by some form of nonbaryonic matter. A measure is to be found in the proceedings of the symposium, *Dark Matter in the Universe*, held in June of 1985 (Kormendy and Knapp 1987), where the concluding discussion included a poll of opinion on the value of Ω. Of the 132 voting, 71 registered as undecided, two voted for $\Omega \leq 0.05$, 31 voted for the interval $0.05 < \Omega \leq 0.999$, 28 for $0.999 < \Omega \leq 1.001$, and there were two votes for $\Omega > 1.001$. As one might expect, theorists tended to favor $\Omega = 1$, though there certainly were exceptions, and it was natural that observers tended to register as undecided, though again thoughtful and experienced observers registered for $\Omega = 1$ to an accuracy of one part in 10^3. (And it is difficult to decide what fraction of those voting were taking the proceedings seriously.) At the time this is written, six years after the poll, the situation still is quite unclear (as is illustrated in table 20.1 below).

It may be reasonable to conclude that there was a paradigm shift in cosmology in the decade from 1975 to 1985, for there was a distinct shift in the general opinion of what are reasonable cosmologies to study. This was not driven by the phenomena, however, and indeed the major advance in the interaction of theory and observation was the improving evidence that the mass density estimated from dynamics on relatively small scales is significantly less than the Einstein-de Sitter value.

Perhaps this is not an entirely edifying example of the way to do science, but one must bear in mind that the high-density Einstein-de Sitter cosmology dominated by nonbaryonic matter may very well turn ought to be right, and if otherwise, the debate on why not has been an invaluable stimulus to the development of the present state of understanding of the issues. And there remains the very real problem of the dark mass in galaxies and systems of galaxies. We turn now to candidates for the dark matter.

Baryonic Matter in Our Neighborhood

Baryons as we know them can be in dilute gaseous clouds of plasma or neutral atoms and molecules, or condensed in snowballs or more massive icy bodies similar to comets, or in stars and planets, or locked up in stellar remnants, including white dwarfs, neutron stars, and presumably also black holes. The purpose of this subsection is to survey what is known of the baryonic mass in our neighborhood of the Milky Way, and what this might teach us about baryonic mass in dark halos and in clusters of galaxies. Section 5 dealt with the constraint on the abundance of intergalactic stars from the upper bound on their contribution to the surface brightness of the sky (eqs. [5.167] and [5.168]). Measures of neutral and ionized intergalactic gas are reviewed in section 23. More details of the possibilities for baryonic matter are discussed by Silk (1991).

Apart from stellar remnants and terrestrial-type planets, known baryonic matter has a close-to-standard cosmic composition of about 70% hydrogen by mass, usually less than about 3% in the elements heavier than helium, and the rest helium. It seems reasonable to look for a similar composition in any dark baryonic matter outside stellar remnants.[4] The remarkable observations reviewed in section 23 detect an amount of gaseous hydrogen at redshift $z \sim 3$ that is comparable to what is known to be present at low redshifts in the stars and gas in galaxies. The X-ray luminosities show that rich clusters of galaxies contain ionized hydrogen, with close to cosmic abundances of the heavier elements (section 20). The plasma is not dense enough to be the dominant form of dark matter within clusters, and its presence eliminates the possibility that there is an appreciable amount of diffuse neutral atomic or molecular hydrogen. Neutral and ionized hydrogen in the interstellar medium in the disk of the Milky Way contribute about 20% of the net mean mass per unit area; the dominant part is in the stars listed in table 18.1 below.

On the face of it, stellar remnants do not seem to be good places to hide an appreciable fraction of the baryons, because a star tends to shed most of its material before collapsing to a white dwarf or neutron star. Thus, many generations of stars would be needed to sequester most of the baryons in white dwarfs or neutron stars, and it seems unlikely that the process could be efficient enough to produce the large mass-to-light ratios seen in dark halos without overproducing heavy elements. An exception may be stars with mass $\gtrsim 10^4 \mathcal{M}_\odot$, which computations indicate can collapse to black holes without shedding much mass. As will be discussed in this subsection, the evidence is that there is not a large mass fraction in stellar remnants in our neighborhood of the Milky Way.

The Solar System is surrounded by a remarkably fragile system of comets, at about 10^4 times our distance from the Sun. Since a large part of a comet is volatile materials such as water, ammonia, and methane, it is natural to ask whether an appreciable fraction of the baryonic mass in our galaxy could be similarly sequestered in cometlike snowballs. Real comets presumably formed by chemical condensation. That is not possible for hydrogen, which we are assuming is the dominant form of the baryons, because the vapor pressure of condensed molecular hydrogen at the cosmic background radiation temperature $T = 2.7$ K (eq. [6.1]) is too high (PC, p. 107; Hegyi and Olive 1986). Thus within the standard model, in which the universe is expanding and cooling, the snowballs

---

[4] Exceptions are the inner planets in the Solar System, which are thought to have formed out of dust segregated from the hydrogen by gravitational settling to the plane of the rotating gas cloud out of which the Solar System formed. In white dwarf stars virtually all the hydrogen has been burned at least as far as helium.

would have had to have formed by gravity, in some variant of the way stars form. Since star formation is not understood we cannot be very specific about the possible properties of gravitationally bound cometlike bodies, but there are a few useful general considerations.

The condition for gravitational binding of molecular hydrogen (with mass $2m_p$) at surface temperature T in a snowball of radius r is

$$\frac{G\mathcal{M}}{r} = \frac{4}{3}\pi G\rho r^2 \gtrsim \frac{kT}{2m_p} . \tag{18.5}$$

If the mass, $\mathcal{M}$, is below that of Jupiter, the density of the snowball is that of condensed molecular hydrogen, $\rho = 0.07$ g cm^{-3}. Since the surface could not be cooler than the cosmic background radiation, equation (18.5) gives $r \gtrsim 10^8$ cm and $\mathcal{M} \gtrsim 10^{24}$ g. This is some six orders of magnitude greater than the mass of Halley's comet. Comet masses can be small because they are held together by chemistry, not gravity.

The smallest hydrogen bodies are likely to be considerably larger than the bound from this argument, because the snowball has to form by contraction, and at a larger radius the gravitational binding energy is lower. One could imagine dense condensed cores form by evaporation of an envelope, as perhaps was the case for Jupiter, but that does not seem to be an efficient way to sequester an appreciable fraction of the hydrogen. The more likely route follows the picture for star formation by fragmentation up to the development of gas clouds that can quasistatically contract as they cool (Hoyle 1953). The orders of magnitude go as follows.

Let us characterize a gravitationally bound gas cloud by a mean temperature T and density ρ. Then the mass and radius are

$$r \sim \left(\frac{kT}{G\rho m_p}\right)^{1/2} , \qquad \mathcal{M} \sim \rho r^3 ,$$

$$T \sim Gm_p k^{-1}\mathcal{M}^{2/3}\rho^{1/3} . \tag{18.6}$$

The first relation is equation (18.5). The second line is the result of eliminating the radius. It would be difficult to get the matter temperature below that of the CBR. Thus the hydrogen temperature T is not less than about 10 K, and we have from the second line

$$\frac{\mathcal{M}}{\mathcal{M}_\odot} \gtrsim \frac{10^{-9}}{\rho^{1/2}} , \tag{18.7}$$

where $\mathcal{M}_\odot$ is the mass of the Sun, and the density ρ is expressed in grams per cubic centimeter. At $\rho \sim 0.1$ this is the limit obtained above. At lower densities the minimum allowed mass is larger.

Next, let us compare the characteristic times for cooling and free collapse of the cloud. The latter is $t_{\text{collapse}} \sim (G\rho)^{-1/2}$. The characteristic cooling time, t_{cooling}, is the ratio of the internal energy, $\sim \mathcal{M}kT/m_p$, to the rate L of loss of energy by radiation. We get the minimum possible cooling time by assuming the cloud is opaque, black, and radiating at the characteristic temperature T in equation (18.6), $L \sim \sigma T^4 r^2$, where $\sigma = ca/4$ is the Stefan-Boltzmann constant. Since the temperature at the surface, where the optical depth is unity, is no greater than T, and eventually is considerably less, we have not underestimated the cooling rate. The same is true of the loss of luminosity when the density is so low that the cloud is optically thin. The conclusion is that the ratio of the cooling and collapse times is

$$\frac{t_{\text{cooling}}}{t_{\text{collapse}}} \gtrsim \frac{k^4}{\sigma m_p{}^4 G^{5/2}} \frac{\rho^{1/6}}{\mathcal{M}^{5/3}}$$

$$\sim \left(\frac{0.1 \mathcal{M}_\odot \rho^{1/10}}{\mathcal{M}} \right)^{5/3}, \tag{18.8}$$

in the units of equation (18.7).

Following Hoyle (1953), we note that if the cooling time is shorter than the collapse time, the cloud contracts more or less freely and irregularities in the matter distribution are free to grow, which may be expected to cause the cloud to fragment into less massive pieces. Equation (18.8) indicates that a gas cloud contracts quasistatically as it radiates when

$$\mathcal{M} \lesssim 0.1 \rho^{1/10} \mathcal{M}_\odot. \tag{18.9}$$

As we have noted, this is a conservative bound (because we have overestimated the surface temperature). Under quasistatic contraction, the cloud fragments only when it spins up and fissions. This is not likely to happen a large number of times in the course of contraction, so equation (18.9) is the wanted lower bound on the cloud mass.

The bounds in equations (18.7) and (18.9) agree at $\rho \sim 10^{-14}$ g cm^{-3} and $\mathcal{M} \sim 0.01 \mathcal{M}_\odot$. The surface density at this point is $\Sigma \sim \rho^{2/3} \mathcal{M}^{1/3} \sim 10$ g cm^{-2}, thick enough so that a dirty cloud might be expected to radiate like a black body.

The mass bound is not very sensitive to the density, and at $\rho \sim 10^{-24}$ g cm^{-3} amounts to about 10^{-3} solar masses, roughly the mass of Jupiter. That does leave an interesting and not very broad window below the minimum at $0.08 \mathcal{M}_\odot$ for the hydrogen burning that makes ordinary stars readily visible. Let us turn now to the distribution of star masses, with particular attention to what is known about the low mass end.

Table 18.1 shows the distribution of star masses in our neighborhood in the Milky Way. The numbers are adapted from Scalo's (1986) survey of what is

Table 18.1
Local Distribution of Star Masses and Luminosities

$\log m$	$\mathcal{M}/L$	$dn/d\log m$	$m\,dn/d\log m$	$L\,dn/d\log m$
-1.0	4000	0.05	0.005	—
-0.8	1500	0.20	0.031	—
-0.6	370	0.29	0.073	0.0002
-0.4	100	0.25	0.10	0.0011
-0.2	14	0.13	0.084	0.0059
0.0	1	0.065	0.065	0.065
0.2	0.17	0.006	0.010	0.060
0.4	0.073	0.001	0.002	0.024
0.6	0.038	—	0.0003	0.0083
0.8	0.021	—	—	0.0041
1.0	0.011	—	—	0.0027
1.2	0.006	—	—	0.0016
1.4	0.004	—	—	0.0010
1.6	0.003	—	—	0.0010
—	—	1.0	0.37	0.17

known about the stars in the disk at distances less than a few hundred parsecs from us. The table shows the present-day mass function (so called because the mass function has to change as new stars are born and the massive ones exhaust the supply of hydrogen in their cores and evolve off the main sequence to pass through luminous, relatively short-lived phases on their way to dark remnants). The table gives the number of disk stars per unit area, projected onto the plane of the disk, and counting only the stars still burning hydrogen on the main sequence. The units are solar masses and solar luminosities.

The second column is the mass-to-light ratio at the central star mass listed in the first column.[5] The third column is the relative number of stars per unit area of the disk, with mass in a fixed range $\Delta \log m$ around the central value of $\log m$ for the bin, and normalized to unit sum over the mass range in the table. The last two columns are the products of this relative number of stars with the central values of the star mass and luminosity for the mass bin.

The sums in the bottom rows of the last two columns in the table are the mean mass per star and the mean luminosity per star for the distribution and mass range represented in the table. The ratio is

---

[5] The luminosities in this section are measured in the V band centered at ~ 5500 Å. As indicated in equation (3.49), this is somewhat longer than the B band used in most of this book.

$$\frac{\mathcal{M}}{L} = 2.2 \frac{\mathcal{M}_\odot}{L_\odot}, \tag{18.10}$$

in units of solar masses per solar luminosity. This is the mass-to-light ratio in the V band for the mix of stars in our neighborhood. The median star mass (where $dn/d\log m$ sums to 0.5) is $m = 0.3$ solar masses. Half the mass is contributed by stars at $m \lesssim 0.5$, where the star mass-to-light ratio is $m/L(m) \gtrsim 30$ solar units. Half the light is contributed by stars at $m \gtrsim 1.5$, where $m/L(m) \lesssim 0.3$. For our purpose, the point of these numbers is that $\mathcal{M}/L$ in equation (18.10) for the population in our neighborhood is close to the solar value by an accident: most of the mass comes from low mass stars with large $m/L(m)$, most of the light comes from the more massive stars with low $m/L(m)$, and the average across the distribution happens to be close to the solar value. That means a relatively modest change in the form of the mass function could have a considerable effect on the mass-to-light ratio for the population.

In an old system of stars, where star formation ceased a Hubble time ago, the stars more massive than the Sun would have exhausted their fuel and died. We can estimate the mass-to-light ratio of the remaining low-mass stars, ignoring the light from the dying ones, by summing the last two columns over the range $m < 1$. The ratio of these sums is

$$\frac{\mathcal{M}}{L} \sim 10 \frac{\mathcal{M}_\odot}{L_\odot}, \quad \text{for } m \lesssim 1 \mathcal{M}_\odot, \tag{18.11}$$

for an old population with the local mass function. Within the uncertainties, this agrees with the mass-to-light ratios found in the central parts of other galaxies, as discussed in the next subsection, and with the value in equation (5.148) that we used to find the density parameter contributed by the bright parts of galaxies (eq. [5.150]).

The table includes only stars on the hydrogen-burning main sequence. Those that have exhausted the hydrogen in their cores and are still evolving through luminous phases away from the main sequence add very little to the mass — because they remain luminous for a relatively short time — and considerably more to the luminosity, but not enough to make a substantial change in the mass-to-light ratio. The mass in white dwarf star remnants is estimated from counts of white dwarfs as a function of luminosity, together with a check for the dwarfs that may be too cold to be detectable, based on the theory for how these stars cool. The results suggest white dwarfs make a relatively small contribution to the local mass density. We get a feeling for what is involved from Mestel's (1952) elegant argument for the cooling law, as follows.

Apart from a relatively thin envelope, white dwarfs are supported by the degeneracy energy of the electrons. Degenerate electrons at number density n act as if each were confined to a box of width $r \sim n^{-1/3}$, so they can avoid each other, and the uncertainty principle therefore says the electrons have momenta

$p_e \sim \hbar/r = \hbar n^{1/3}$. Since the electrons are nonrelativistic in a stable star, the degenerate electron pressure is

$$P_e = m_e n v_e^2 \propto \rho^{5/3} , \qquad (18.12)$$

where $\rho \propto n$ is the mass density.

The degenerate core of a white dwarf is hot, so radiation and the tail of the distribution of thermally excited electrons both act as efficient thermal conductors that keep the core at a nearly uniform temperature. In the nondegenerate outer envelope, the temperature drops from the core value to the much lower value at the surface. The radiation energy density as a function of radius r is $u(r) = aT(r)^4$, where $T(r)$ is the local temperature. The radiation energy flux density is $f = -\kappa^{-1} c \, du/dr$, where κ is the opacity (the product of the scattering cross section and the number density of scatterers, so κ is the fraction of energy absorbed per unit length along the path of a light ray). At high temperatures, κ is dominated by the absorption and emission of photons by electrons accelerated in the electric fields of the ions. In section 24 it is shown that the opacity for this process is (eq. [24.15] with $\lambda \propto T^{-1}$)

$$\kappa \propto \rho^2 / T^{7/2} . \qquad (18.13)$$

It follows that the luminosity of the star (the net energy flux through the envelope) is

$$L = 4\pi r^2 f \propto -\frac{du}{dr}\frac{r^2}{\kappa} \propto -T^3 \frac{dT}{dr}\frac{T^{7/2}}{\rho^2} r^2 . \qquad (18.14)$$

In the nondegenerate envelope the pressure is $P = nkT$, so the equation of hydrostatic support is

$$\frac{d(\rho T)}{dr} \propto -\frac{G\mathcal{M}}{r^2}\rho . \qquad (18.15)$$

The mass $\mathcal{M}$ within radius r is nearly constant in the envelope, because most of the mass is in the degenerate core, so the result of dividing $d(\rho T)/dr$ by dT/dr from equation (18.14) is

$$\frac{d(\rho T)}{dT} \propto \frac{\rho}{r^2}\frac{T^{13/2} r^2}{\rho^2 L} . \qquad (18.16)$$

The radius drops out, and L is independent of the radius, because most of the heat comes from the cooling core. The result of multiplying this expression by ρT and integrating is

$$\rho \propto T^{13/4} / L^{1/2} . \qquad (18.17)$$

This is the relation between the luminosity and the temperature and density at the inner edge of the envelope. The pressure in the degenerate core (eq. [18.12]) has to match the pressure at the bottom of the envelope, $P_e \sim \rho^{5/3} \sim \rho T$. The result of using this relation to eliminate ρ from equation (18.17) is

$$L \propto T^{7/2}. \tag{18.18}$$

The dominant heat capacity in the star is in the thermal vibrations of the ions in the core. At the temperatures of interest the ions act as classical particles, with equipartition energy $3kT/2$ per ion. This means we can approximate the rate of loss of energy as $L \propto -dT/dt$. This with equation (18.18) gives

$$T \propto t^{-2/5}. \tag{18.19}$$

On putting this back into equation (18.18) we arrive at Mestel's cooling law,

$$L \propto t^{-7/5}. \tag{18.20}$$

Liebert (1980) gives a guide to more detailed computations; the results are not far from this power law.

To get an estimate of the expected white dwarf luminosity function in our neighborhood, let us suppose stars are dying and producing white dwarfs at a roughly constant rate. Then the number of white dwarfs with ages in the range t to $t + dt$ is $dN \propto dt$, and equation (18.20) says the luminosity function is

$$\frac{dN}{dL} \propto \frac{dt}{dL} \propto L^{-12/7}. \tag{18.21}$$

The local white dwarf luminosity function is not far from this power law at $L \gtrsim 3 \times 10^{-5} L_\odot$, with a rather abrupt cutoff at fainter luminosities. The cooling age at this cutoff is about 10^{10} y, indicating stars started forming in our region of the disk about this long ago (Winget et al. 1987; Noh and Scalo 1990). The similarity of the observed luminosity function to the model in equation (18.21) suggests stars have been forming and dying at a mean rate that has not greatly changed with time. (The models of Noh and Scalo indicate that the rate smoothed through a window of 10^9 y has not changed by more than an order of magnitude in 10^{10} y.) The local mass density in these seen white dwarfs is about 10% of the density from the main sequence stars listed in table 18.1.

It would have been easy to imagine that when our galaxy was young, and the supply of gas plentiful, star formation was quite rapid and a good deal of mass was sequestered either in very low mass stars or in the remnants of massive short-lived stars. The evidence is that neither happened in our neighborhood. As we have noted, the white dwarf luminosity function suggests the mean star formation rate has not changed by a large factor in the past ten billion years, and has left only a modest density of remnants in white dwarfs. The fourth column of table 18.1 gives the relative contribution to the mass per unit area of the disk by stars in equally spaced bins of $\log m$. We see that most of the mass is coming from stars of about 0.4 solar masses, and that an extrapolation to lower masses would add very little from objects below the last bin of the table. The last bin has a lower bound at $m = 0.08 \mathcal{M}_\odot$, the limit for hydrogen burning.

One could suppose that in the massive dark halos of galaxies, star formation produced a second, low mass, peak in $m\, dn/d \log m$, or that star formation shifted the single peak to favor lower mass. Stars with masses below the limit $m \sim 0.08 \mathcal{M}_\odot$ for nuclear burning and above the formation bound in equation (18.9) have been called black dwarfs (Kumar 1963), jupiters (because the composition of the planet Jupiter is that of a star), and brown dwarfs (Tarter 1975). As for white dwarfs, the cooling radiation of brown dwarfs is detectable if they are young and massive enough. Unlike white dwarfs, the heat transfer to the surface in the cooler brown dwarfs is by convection rather than radiative diffusion, as analyzed by Kumar (1963). The status of the cooling theory is reviewed by Stevenson (1991), and Stevenson (1991) and Daly and McLaughlin (1992) consider the prospects for detection of the cooling radiation from these low mass stars, as a function of their age and mass. Below we consider the possibility of detecting brown dwarfs by their gravitational lensing.

We have a check on the sums for the local mass from its gravitational effect on the orbits of stars in and near the disk (Oort 1932). To see how this goes, note that the mass distribution in the seen disk stars is quite thin compared to our distance from the center of the galaxy. Thus it is reasonable to neglect rotation and model the disk as a plane symmetric sheet with mass per unit area Σ. Then Poisson's equation for the gravitational potential as a function of distance z from the midplane is

$$\frac{d^2\phi}{dz^2} = 4\pi G\rho(r).$$ (18.22)

The solution is

$$g = \frac{d\phi}{dz} = 2\pi G\Sigma(z),$$ (18.23)

where $\Sigma(z)$ is the mass per unit area between the sheets at $\pm z$ from the midplane in the assumed symmetric mass distribution. We will simplify the calculation by assuming the mass sheet is very thin, so at interesting distances z from the plane Σ is constant and the gravitational acceleration, g, is independent of z.

If a test population of stars has a Gaussian distribution in the component v of velocity normal to the disk, with mean square value $\langle v^2 \rangle$, the equilibrium distribution of distances is

$$\frac{dn}{dz} \propto e^{-2\pi G\Sigma z/\langle v^2\rangle} \, . \tag{18.24}$$

The easy way to see this is to recall that the Boltzmann distribution is $\exp -E/kT$, where the energy is $E = m\phi$ for a star of mass m, and equipartition says $kT/2 = m\langle v^2\rangle/2$ for the kinetic energy normal to the disk. If a class of stars has measured distributions in v and z, one can estimate Σ from this equation (or its generalizations to allow for the thickness of the disk, or even for a non-Gaussian velocity distribution).

Bahcall, Flynn, and Gould (1992) have applied this method to a star sample selected to be as close as feasible to an unbiased representation of a single class of stars moving in a well-mixed way through the central plane of the disk. Let us consider the orders of magnitude. The velocity dispersion in the sample is

$$\langle v^2\rangle^{1/2} = 18.8 \pm 1.3 \, \text{km s}^{-1} \, , \tag{18.25}$$

the distribution in z drops to half the value at $z = 0$ at

$$z_{1/2} = 170 \pm 20 \, \text{pc} \, , \tag{18.26}$$

and these two numbers in equation (18.24) give

$$\Sigma = 55 \pm 10 \, \mathcal{M}_\odot \, \text{pc}^{-2} \, . \tag{18.27}$$

The estimates from star counts for the mean mass per unit area in hydrogen-burning stars are

$$\Sigma_* = 30 \pm 5 \, \mathcal{M}_\odot \, \text{pc}^{-2} \tag{18.28}$$

(Scalo 1986; Rana 1991). The mass per unit area in neutral and ionized gas is

$$\Sigma_{\text{gas}} = 12 \pm 3 \, \mathcal{M}_\odot \, \text{pc}^{-2} \tag{18.29}$$

(Bahcall et al. 1992; Rana 1991), and the white dwarf contribution is

$$\Sigma_{\text{wd}} = 4 \, \mathcal{M}_\odot \, \text{pc}^{-2} \, . \tag{18.30}$$

The sum is $46 \pm 6 \, \mathcal{M}_\odot \, \text{pc}^{-2}$, less than one standard deviation from the dynamical mass in equation (18.27). This is in line with the much more careful analyses by Kuijken and Gilmore (1991) and Bahcall et al. (1992) in indicating that if dark matter is present in the local disk of the Milky Way it is subdominant; the main ingredient of our neighborhood is baryons.

Cores of Galaxies

The mass-to-light ratios from the star mass function in our neighborhood are comparable to what is derived from the velocity dispersions and luminosities in the bright parts of other galaxies. The example presented here uses the core-fitting technique pioneered by Ivan King; the history of the method is traced by Richstone and Tremaine (1986). It will be useful also for the discussion below of massive neutrinos.

The star system is taken to be spherically symmetric, with mean mass-to-light ratio $\mathcal{M}/L$ that is independent of position. Then the surface brightness $i(h)$ (energy flux per unit area and solid angle) as a function of projected distance h from the center is given by the integral

$$i(h) = \int_{-\infty}^{\infty} \frac{j}{4\pi} \, dy, \qquad j = \rho((h^2 + y^2)^{1/2}) \frac{L}{\mathcal{M}}, \qquad (18.31)$$

where $j(r)$ is the luminosity per unit volume as a function of radius r. A useful model for the mass density $\rho(r)$ is

$$\rho(r) = \frac{\rho_c}{(1 + r^2/r_c^2)^{3/2}}, \qquad (18.32)$$

where r_c is a constant. This expression in the integral in equation (18.31) gives

$$i(h) = \frac{L}{\mathcal{M}} \frac{\rho_c r_c}{2\pi} \frac{1}{1 + h^2/r_c^2}. \qquad (18.33)$$

In this model, the surface brightness falls to half its central value at projected radius h equal to the core radius r_c.

Next, let us write down the equation for gravitational equilibrium in the neighborhood of the core. At $r \ll r_c$ the density run in equation (18.32) is

$$\rho = \rho_c \left(1 - \frac{3}{2} \frac{r^2}{r_c^2} + \cdots \right). \qquad (18.34)$$

We will assume the velocity distribution is isotropic and independent of position. The velocity dispersion is $\sigma = \langle v^2 \rangle^{1/2}$ in one dimension, and the pressure is $p = \rho\sigma^2$. The pressure force per unit volume is $-\partial p/\partial r$, and the condition for gravitational equilibrium is that this is balanced by $g\rho$, where g is the gravitational acceleration. That gives

$$-\frac{\partial p}{\partial r} = \frac{3\rho_c \sigma^2 r}{r_c^2} = \frac{G\mathcal{M}(r)\rho}{r^2} = \frac{4}{3}\pi G\rho_c^2 r. \qquad (18.35)$$

The second expression uses the leading correction in equation (18.34). To the same order, the mass within radius r in the third expression is $4\pi\rho_c r^3/3$. The result is that central mass density is

$$\rho_c = \frac{9\sigma^2}{4\pi G r_c^2}.$$ (18.36)

With equation (18.33) for the central surface brightness, i_o, the mass-to-light ratio is

$$\frac{M}{L} = \frac{9\sigma^2}{8\pi^2 G r_c i_o}.$$ (18.37)

Equation (18.36) for the central mass density ρ_c depends on the model, but Richstone and Tremaine (1986) have shown that the result is insensitive to the details if the velocity dispersion is isotropic, as seems reasonable to expect in the central parts of a galaxy.

Lauer (1985) gives a useful set of measurements of core radii for giant elliptical galaxies. His highest-quality measurements favor galaxies with larger core radii, because the correction for atmospheric seeing is relatively smaller, and we will simplify the discussion by limiting our attention to this set. Median values for the radius r_c at half the maximum surface brightness and the line-of-sight velocity dispersion σ in the Lauer galaxies with highest quality measurements are

$$\sigma = 275 \text{ km s}^{-1}, \qquad r_c = 225h^{-1} \text{ pc}.$$ (18.38)

This works out to a typical central density

$$\rho_c \sim 250h^2 M_\odot \text{ pc}^{-3}.$$ (18.39)

Lauer's much more careful analysis yields a median mass-to-light ratio

$$M/L = 20h M_\odot/L_\odot.$$ (18.40)

Still more detailed analyses that take account of the run of velocity dispersion with radius and the constraints on the shapes of the star orbits from the surface brightness distribution give similar values for the central mass-to-light ratio.

Since at the present epoch star formation in giant elliptical galaxies is not rapid, it is reasonable to compare this result to equation (18.11) for the mass-to-light ratio in the stars in our neighborhood that live longer than 10^{10} years, $M/L \sim 10$. Within the uncertainty in the Hubble parameter, there is no more than a factor of two difference. The similarity is quite remarkable, for the physical conditions differ in at least four ways, each of which one would imagine could have had a substantial effect on the stellar initial mass function. First, the local mass density at the midplane of the disk in the Milky Way is $\rho \sim 0.1 M_\odot \text{ pc}^{-3}$, some three orders of magnitude less than the value in equation (18.39) for giant elliptical galaxies. Second, the local mass distribution is supported by rotation in a thin disk, rather than by the much higher random velocity dispersion within a nearly spherical system in an elliptical galaxy. Third, the star-formation history is thought to be quite different: the stars in giant ellipticals look old, while stars have been forming at a more nearly uniform rate in the disk of our galaxy. And fourth, the heavy-element abundances in the stars in the cores of elliptical galaxies are larger than in our neighborhood. Since heavy elements play an important role in helping a protostellar gas cloud radiate away its binding energy, it is easy to imagine that the larger heavy-element abundance changes the mix of star masses. These differences in conditions surely play a role in star formation, but the straightforward interpretation is that the effects are not large, that the mass-to-light ratios are similar because star formation is not much affected by the present ambient conditions. And the reasonable presumption therefore is that the central parts of giant elliptical galaxies are dominated by baryons in stars, as for the solar neighborhood.

Microlensing

On the usual basis of Newtonian mechanics, the mass-to-light ratio is considerably higher in the matter in dark halos at radii $r \gtrsim 10h^{-1}$ kpc. Could this be because halos are made of baryons in stars with a mass function shifted so as to put the peak of $m \, dn/d \log m$ below $0.1\mathcal{M}_{\odot}$ in the brown dwarf sector? The stability of $\mathcal{M}/L$ in the bright parts of galaxies could be taken to argue against the idea. On the other hand, some massive halos have detectable optical luminosities, and the reasonable presumption is that this is starlight, because it continuously joins the light from the central regions where the spectrum is known to have the Fraunhofer absorption lines characteristic of stars. Other aspects of this continuity are shown in figure 3.12. If stars really are present in massive halos, perhaps it is not an extreme extrapolation to imagine that the halos consist only of stars and the remnants of stars. It is remarkable that there is a feasible way to check: by the gravitational lensing effect the lumpy mass distribution in stars would have on the flux density from background pointlike objects.

The lensing effect of objects (or small numbers of objects) with masses characteristic of a star or planet is called microlensing, to distinguish it from the effect of the mass concentration in a galaxy or a cluster of galaxies. The effect was discussed by Liebes (1964) and Refsdal (1964); Gott (1981) and Young (1981) showed that the microlensing rate is expected to be particularly large for a light ray from a high redshift quasar that passes through a galaxy about halfway along the path, and Paczyński (1986) showed that it is a feasible though large project to look for microlensing events in the massive halo of the Milky Way, if it is made of stars or brown dwarfs. Here we consider the simple limiting case where the lensing probability along a line of sight is small, so we only need to deal with the effect of a single pointlike gravitational lens. In a line of sight through the center of a cluster of galaxies, the lensing optical depth may approach unity (eq. [18.58]), and the magnification becomes a more complicated cooperative effect in the mass distribution.

In a single microlensing event, the flux density from a compact source such as a quasar temporarily increases when a star passes close enough to the line of sight to act as a gravitational lens. The geometry is illustrated in figures 13.11 and 18.1. We are considering lensing by a pointlike mass m, where the gravitational bending angle α is given by the Einstein equation (10.94). Let Θ be the angular distance between the lensing mass and the position of the source in the sky in the absence of the lensing effect, and let θ be the angular distance between the lens and the image, as indicated in figure 18.1. Then the angle between the image and the unperturbed line to the source is $\theta - \Theta$, and equation (13.68) changes to $\alpha r_{ls} = (\theta - \Theta) r_{os}$, where r_{os} and r_{ls} are the coordinate angular size distances from observer and lens to the source. The proper impact parameter of the line of sight at the lens is $d = a_o r_{ol} \theta / (1 + z_l)$, so the bending angle of the line of sight at the lens satisfies the equation

$$\alpha = \frac{4Gm}{r_{ol}\theta} \frac{1+z_l}{a_o} = (\theta - \Theta)\frac{r_{os}}{r_{ls}} . \tag{18.41}$$

This is

$$\theta^2 - \Theta\theta - \frac{4Gm}{R} = 0, \qquad R \equiv \frac{r_{ol} r_{os}}{r_{ls}} \frac{a_o}{1+z_l} . \tag{18.42}$$

If the lens is much closer than the source, so $r_{os} \to r_{ls}$, and the redshift z_l of the lens is small, the length R is the proper distance to the lens.

The solutions to equation (18.42) for the angle θ between the lens and image are

$$\theta_{\pm} = \frac{\Theta}{2} \pm \left[\frac{\Theta^2}{4} + \frac{4Gm}{R} \right]^{1/2} . \tag{18.43}$$

The formal solutions for the positions of two images, one on each side of the pointlike lens, exist whatever the impact parameter, but as indicated below the flux density in one image is negligibly small when Θ is large. If the lensing mass were transparent, there would be a third solution, with a line of sight that runs through the central region, but that is blocked by the matter in an isolated microlens. If the optical depth for microlensing is large, there can be images from two or more lenses, producing more complex time histories of the observed flux density in a lensing event.

A convenient way to compute the flux densities in the two images follows the geometry of the images in the right-hand part of figure 18.1. A part of the resolved

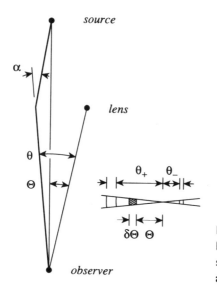

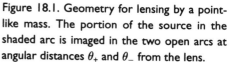

Figure 18.1. Geometry for lensing by a point-like mass. The portion of the source in the shaded arc is imaged in the two open arcs at angular distances θ_+ and θ_- from the lens.

image of the source in the absence of the lens is shown as the shaded arc segment at angular radius Θ to $\Theta + \delta\Theta$. Since the lens is axially symmetric, it moves the images along the radial lines to the two open segments at angles $\theta_\pm$. The widths of the segments follow by differentiating equation (18.42):

$$\frac{d\theta_\pm}{d\Theta} = \frac{\theta_\pm}{2\theta_\pm - \Theta} = \frac{1}{2} \pm \frac{\Theta}{4} \left[\frac{\Theta^2}{4} + \frac{4Gm}{R}\right]^{-1/2} . \tag{18.44}$$

Since microlenses in a dark halo would be moving at speeds well below the velocity of light, we can ignore the frequency shift of the light caused by the changing path lengths along the two lines of sight. That means the surface brightness is the same in the two images (eq. [9.53]), so the flux density is proportional to the solid angle of the image. Thus, the ratio of the flux density $f_\pm$ in an image to the flux density f_o in the absence of the lens is

$$\frac{f_\pm}{f_o} = \left|\frac{\theta_\pm}{\Theta}\frac{d\theta_\pm}{d\Theta}\right| . \tag{18.45}$$

The result of combining these relations is

$$\frac{f_\pm}{f_o} = \frac{1}{4}\frac{(1 \pm F)^2}{F} , \qquad F = \left(1 + \frac{16Gm}{R\Theta^2}\right)^{1/2} . \tag{18.46}$$

The magnification factor for the net flux density from the two images is

$$A = \frac{f_+ + f_-}{f_o} = \left(1 + \frac{8Gm}{R\Theta^2}\right)\left(1 + \frac{16Gm}{R\Theta^2}\right)^{-1/2}. \tag{18.47}$$

If lens and source are in line, $\Theta = 0$, the source is imaged as an Einstein ring with angular radius in the sky (eq. [18.43])

$$\theta_e = (4Gm/R)^{1/2}. \tag{18.48}$$

In terms of this critical Einstein radius, equation (18.47) for the magnification of the flux density received from the source is

$$A = \frac{1 + 2\theta_e^2/\Theta^2}{(1 + 4\theta_e^2/\Theta^2)^{1/2}}. \tag{18.49}$$

The magnification factor approaches unity when the angular distance Θ from the lens to the unperturbed line to the source is much larger than θ_e. At $\Theta = \theta_e$ the magnification is $A = 3/5^{1/2} = 1.34$, large enough for a reasonable chance of detection when the duration of the magnification event is favorable. At $\Theta = \theta_e$ the ratio of the flux densities in the two images is $f_+/f_- = 7$, close enough to unity for there to be a reasonable chance of detection of both images if the angular separation allows resolution of the images (as in the lensing events caused by the mass concentrations in galaxies and clusters of galaxies).

Let us pause to note some characteristic numbers useful for the analysis of microlensing and lensing by larger mass concentrations. If the mean proper number density of lensing objects of mass m along a line of sight is $n(r)$, the probability for a lensing event at magnification greater than a value fixed by the choice of Θ/θ_e is

$$P = \int n(r) \frac{a_o}{1+z} \frac{dr}{(1+r^2/R^2)^{1/2}} \frac{\pi a_o^2 r^2 \Theta^2}{(1+z)^2} \cdot \frac{4Gm(1+z)r_{ls}}{\theta_e^2 a_o r_{os} r}$$

$$= 4\pi Gm(\Theta/\theta_e)^2 a_o^2 \int \frac{dr}{(1+r^2/R^2)^{1/2}} \frac{r_{ls} r}{r_{os}} \frac{n(r)}{(1+z)^2}. \tag{18.50}$$

The first factors in the first line are the proper number density and the proper volume element at angular distance Θ from the line to the source (and in this expression R is the coordinate radius of curvature of constant time sections in the cosmological model). The last factor is unity, by the definition of the Einstein angle θ_e in equation (18.48). The ratio Θ/θ_e is constant at a fixed choice for the magnification. The abundance of lenses enters as the product $nm = \rho_l(r)$, which is the mean mass density in the lensing objects.

If the source is at the Hubble distance $r \sim H_o^{-1}$, the order of magnitude of this lensing probability in a cosmologically flat model is

$$P \sim 4\pi GmnH_o^{-2}(\Theta/\theta_e)^2 \sim \Omega_l(\Theta/\theta_e)^2 . \tag{18.51}$$

That is, the probability that the impact parameter is small enough for a prominent lensing event, at $\Theta \lesssim \theta_e$ that might be observable by the doubling of the image or by the time variation of the net flux density from the source is on the order of the density parameter Ω_l in the lensing objects. Other aspects of this result, which was discovered by Press and Gunn (1973), are discussed in sections 13 (eq. [13.72]) and 14 (eq. [14.57]).

A lensing event requires that the lens mass and the source be compact. To see the orders of magnitude, let us suppose the lens is much closer than the source and at low redshift, so the length R in equations (18.42) and (18.48) is the proper distance to the lens. The characteristic impact parameter of the line of sight at the lens in a lensing event is

$$p = R\theta_e = (4GmR)^{1/2}/c . \tag{18.52}$$

For a brown dwarf with the mass of Jupiter, $m \sim 10^{31}$ g, at distance $R = 10$ kpc in our halo, the impact parameter is $p = 10^{13}$ cm, three orders of magnitude larger than the radius of Jupiter. That is, even at this low distance and mass a starlike object is compact enough to lens a sufficiently compact source. If the source is a background quasar, the lensing effect is erased when the unperturbed quasar image subtends an angular size larger than the Einstein angle (18.48), for in this case the lens rearranges lines of sight within the image without greatly altering the observed angular size or flux density. For a quasar at the Hubble distance, the Einstein angle subtends the length $\sim \theta_e c/H_o$. The characteristic light travel time across this length is

$$t_s = \frac{\theta_e}{H_o} = \left(\frac{4Gm}{H_o^2 Rc^2} \right)^{1/2} \sim \frac{10^6}{(hz)^{1/2}} \left(\frac{m}{m_\odot} \right)^{1/2} \text{s}, \tag{18.53}$$

for a lens at redshift $z \ll 1$ and distance $a_o r = cz/H_o$. If the microlens mass m is on the order of one solar mass, this is comparable to observed timescales for fluctuations in the luminosities of quasars, indicating that star remnants in distant clusters may cause observable microlensing events in the flux densities from background quasars.

Finally, let us consider some characteristic numbers for the probability of microlensing in our galaxy and in distant clusters of galaxies. A line of sight at 45° from the direction to the center of the Milky Way to a star at distance $R \sim 10$ kpc passes through matter at mean density

$$\rho \sim \frac{v_c^2 R}{G} \frac{3}{4\pi R^3} \sim 10^{-24} \text{ g cm}^{-3}, \tag{18.54}$$

where $v_c \sim 200$ km s^{-1} is the circular velocity. If this halo mass is in brown dwarfs, the lensing probability (18.50) at $\Theta = \theta_e$ amounts to

$$P = 2\pi G\rho R^2/c^2 \sim 10^{-6}. \tag{18.55}$$

The halo stars move at velocities $v \sim 300$ km s^{-1}, so the duration of a lensing event is

$$t_l \sim p/v \sim 30(m/m_\odot)^{1/2} \text{ days}. \tag{18.56}$$

One looks for microlensing by brown dwarfs in the massive halo of our galaxy through their characteristic effect on the apparent magnitudes of seen stars at distances greater than about 10 kpc. The lensing duration is observationally convenient for a considerable range of brown dwarf masses. The lensing probability is small, but there are a lot of stars to monitor (and the searches for microlensing will at the very least yield a wealth of information on variable stars).

The microlensing probability (18.50) in a line of sight through a cluster of galaxies, at redshift $z \ll 1$ and distance $R = cz/H_o$, to a considerably more distant quasar is

$$P = 4\pi GR(\Theta/\theta_e)^2 \int dl\rho/c^2. \tag{18.57}$$

The integral is the projected mass per unit area, $\Sigma(r)$, in the cluster. In the singular isothermal gas sphere model (eqs. [3.32], [14.59]), the surface density is $\Sigma = \sigma^2/2Gr$, and the lensing probability at $\Theta = \theta_e$ is

$$P = \frac{2\pi z\sigma^2}{H_o cr} = \frac{0.2z\sigma_3^2}{hr(\text{Mpc})}. \tag{18.58}$$

Here σ_3 is the line-of-sight velocity dispersion in units of 10^3 km s^{-1}, and the projected distance r of the line of sight from the cluster center is measured in megaparsecs. The considerably larger probability is mainly the result of the smaller gravitational bending angle at the much greater distance. Since one must look for microlensing in quasars rather than stars, the constraint (18.53) on the lens mass for an observable microlensing event is tighter, but still interesting. Since stars in clusters are moving at speeds $v \sim 1000$ km s^{-1}, the duration of a microlensing event is $p/v \sim 20(z\, m/m_\odot)^{1/2}$ years, which again can be in an observationally interesting range.

More serious computations are quite feasible and have been done. The lesson from the order-of-magnitude estimates presented here is that one can use microlensing to test the idea that the dark matter in the halo of our galaxy and in clusters of galaxies is in compact objects such as stellar remnants or brown dwarfs with interesting ranges of masses.

Hot Dark Matter

There is a good case for considering massive neutrinos as the first non-baryonic candidate for dark matter, because neutrinos really are known to exist. We consider here the constraints on this idea under the assumptions that the main contributions to the mass of the universe now and back to the epoch of light element production are baryons, the cosmic background radiation, and the known three families of neutrinos, each with two spin states, and each having chemical potentials small compared to the temperature (so the numbers of neutrinos and antineutrinos are very nearly the same). One family is assumed to have the rest mass m_ν in equation (18.4) with density parameter Ω_ν close to unity, so the net density parameter is $\Omega = 1$, and the present value is dominated by these neutrinos. This rest mass is greater than the upper bound for the main neutrino component in nuclear beta decay, but well within the laboratory upper bounds for the neutrinos of the other two known families, 250 keV and 35 MeV.

The number of neutrinos in this massive family at a given position and momentum in the range d^3r and d^3p in single-particle phase space may be written as

$$dN = 2\mathcal{N}(\mathbf{r}, \mathbf{p}, t)\, d^3r\, d^3p/(2\pi\hbar)^3 . \tag{18.59}$$

The mean number of neutrinos per mode is $\mathcal{N}$, and the factor of two takes account of the two spin states. At redshifts less than about $z \sim 10^{10}$ the neutrinos move as free particles, and Liouville's theorem (9.50) tells us the density $\mathcal{N}$ is constant along orbits in phase space, with the initial value

$$\mathcal{N} = \frac{1}{e^{p/kT_\nu} + 1} , \tag{18.60}$$

because the neutrinos were relativistic when they last were in thermal equilibrium with the radiation. The momentum p in this expression is evaluated at a redshift z chosen to be high enough so that the neutrinos have not been significantly perturbed from their original thermal distribution by density inhomogeneities. The parameter is

$$T_\nu = \left(\frac{4}{11}\right)^{1/3} T_0(1 + z), \tag{18.61}$$

where the present CBR temperature is $T_0 = 2.736$ K, and the numerical prefactor comes from the entropy deposited in the CBR by the thermal electron pairs after the neutrinos have decoupled (eq. [6.76]).

If we ignore the gravitational perturbation to the neutrino motions by the irregularities in the mass distribution we can apply equation (18.60) at low redshift, where the neutrinos are nonrelativistic, and the distribution in the neutrino peculiar velocity $v = p/m_\nu$ is

$$\frac{dn}{dv} \propto \frac{v^2}{e^{m_\nu v / kT_\nu} + 1} . \tag{18.62}$$

The median velocity, where the integral over v in this equation reaches half its total value, is

$$v_m = 2.84 \frac{kT_\nu}{m_\nu c} = 1.5 h^{-2}(1 + z) \, \text{km s}^{-1} . \tag{18.63}$$

At the redshift z_{eq} of equality of mass densities in matter and radiation (eq. [6.81]), $v_m/c = 0.12$. The name "hot dark matter" comes from the fact that at z_{eq} the neutrino velocities are close to relativistic. The present value of the characteristic velocity v_m in equation (18.63) is small compared the escape velocity from a large galaxy, so the velocities of the neutrinos in a galaxy would come from the gravitational field of the mass distribution.

A useful constraint on the density of the neutrinos in the massive halo of a galaxy follows from the mean of the distribution in single-particle phase space (Tremaine and Gunn 1979). Gravity conserves the value of the distribution function $\mathcal{N}$ along a neutrino orbit, meaning that if the neutrinos at present position $\mathbf{r}_f$ and momentum $\mathbf{p}_f$ trace back to initial momentum p, the present local density in phase space is $\mathcal{N}_f(\mathbf{r}_f, \mathbf{p}_f) = \mathcal{N}(p)$, with $\mathcal{N}(p)$ the thermal distribution in equation (18.60). A neighboring value for the present position and momentum may trace back to a very different value of p, in which case the present local value of $\mathcal{N}$ is quite different at these neighboring points in phase space. We are interested in the mean distribution smoothed over this fine-structure. The Tremaine-Gunn constraint follows from the fact that the mean cannot exceed the maximum value of the fine-grain distribution,

$$\mathcal{N}_f = \langle \mathcal{N}_f(\mathbf{r}_f, \mathbf{p}_f) \rangle = \langle [\exp(pc/kT_\nu) + 1]^{-1} \rangle \leq 0.5 . \tag{18.64}$$

A convenient model for the present mean distribution of neutrinos in a dark halo is the spherically symmetric isothermal form

$$\mathcal{N}_f = \mathcal{N}_o \exp - \frac{1}{\sigma^2} \left(\frac{v^2}{2} + \phi(r) \right) . \tag{18.65}$$

The one-dimensional (line-of-sight) velocity dispersion is σ. The gravitational potential energy per unit mass at radius r is $\phi(r)$, with $\phi(0) = 0$ at the minimum of the potential well, so the prefactor has to satisfy $\mathcal{N}_o \leq 0.5$. The neutrino number density at radius r is the integral of $\mathcal{N}_f$ over the momentum $p = m_\nu v$, with a factor of two to count neutrinos and their partners. For this distribution, the mass density is

$$\rho_\nu(r) = \frac{2m_\nu}{(2\pi\hbar)^3} \int \mathcal{N} \, d^3p = \frac{\mathcal{N}_o m_\nu^4 \sigma^3}{2^{1/2}\pi^{3/2}\hbar^3} e^{-\phi(r)/\sigma^2}, \tag{18.66}$$

with $\phi(0) = 0$.

Suppose these neutrinos dominate the mass of the halo of the galaxy. Then Poisson's equation for the gravitational potential ϕ is

$$\nabla^2\phi = \frac{d^2\phi}{dr^2} + \frac{2}{r}\frac{d\phi}{dr} = 4\pi G\rho_\nu(r). \tag{18.67}$$

With the change of variables

$$\phi(r) = -u(r)\sigma^2, \qquad r = \alpha z, \tag{18.68}$$

where the length scale factor α satisfies

$$\alpha^2 = \frac{\pi^{1/2}}{2^{3/2}} \frac{\hbar^3}{G\mathcal{N}_o m_\nu^4 \sigma}, \tag{18.69}$$

Poisson's equation (18.67), with equation (18.66) for the mass density, becomes

$$\frac{d^2u}{dz^2} + \frac{2}{z}\frac{du}{dz} + e^u = 0. \tag{18.70}$$

This is the standard form for an isothermal gas sphere (Emden 1907). The boundary conditions at $z=0$ are $u=0$ and $du/dz=0$. At $z \gg 1$ the solution approaches the limiting form $e^u = 2/z^2$. With equations (18.66), (18.68), and (18.69), this limit is

$$\rho_\nu(r) = \frac{\sigma^2}{2\pi Gr^2}, \tag{18.71}$$

at $r \gg \alpha$. This is the limiting isothermal gas sphere in equation (3.32).

Since $\rho \propto \exp{-\phi/\sigma^2} = e^u$, the mass within radius r varies with the radius as

$$\mathcal{M}(z) \propto \int_0^z e^u z^2 \, dz = -z^2 \, du/dz \to 2z \quad \text{at } z \gg 1. \tag{18.72}$$

The third expression follows from equation (18.70), the last from the limiting solution $e^u \to 2/z^2$. The rotation curve for a disk of material with negligible mass moving in the potential well of the neutrinos is $v_c^2 \propto \mathcal{M}(z)/z$. Emden's tables show that the rotation curve reaches half its large-z asymptotic flat value $v_c = 2^{1/2}\sigma$ at $z_{1/2} = 1.3$. Thus a measure of the core radius for the neutrino distribution is $r_{1/2} = \alpha z_{1/2}$. This expression in equation (18.69) for α gives the neutrino mass in terms of the core radius and the circular velocity in the flat part of the rotation curve for this model:

$$ m_\nu = \frac{\pi^{1/8}\hbar^{3/4}}{(2G\mathcal{N}_o v_c)^{1/4}} \left(\frac{z_{1/2}}{r_{1/2}}\right)^{1/2}. \tag{18.73} $$

At the bound $\mathcal{N}_o = 0.5$ in equation (18.64), the numerical value is

$$ m_\nu = \frac{70\,\text{eV}}{[r_{1/2}\,(\text{kpc})]^{1/2}} \left(\frac{200\,\text{km s}^{-1}}{v_c}\right)^{1/4}. \tag{18.74} $$

This bound traces back to the exclusion principle that fixes the thermal expression for the initial occupation number in equation (18.60). Because the neutrino de Broglie wavelengths are small,

$$ \lambda = \frac{h}{m_\nu v} \sim 0.001\,\text{cm} \tag{18.75} $$

at $v = 300$ km s^{-1}, classical physics gives us an excellent approximation to the motions of the neutrinos, and Liouville's theorem assures us the classical orbits will not violate the exclusion principle, because the occupation number $\mathcal{N}$ is conserved.

The conclusion from equation (18.74) is that typical dark halos of giant galaxies, with $v_c \sim 200$ km s^{-1} and core radii $r_{1/2}$ of a few kiloparsecs, require a neutrino mass similar to the mass at which neutrinos close the universe at an acceptable value of Hubble's constant (eq. [18.4]).

The beautiful coincidence of neutrino masses derived from cosmology and galaxy structure unfortunately does not apply in a straightforward way to some of the dwarf spheroidal galaxies in the halo of the Milky Way. These seven dwarfs have no observed interstellar gas or dust, low luminosities, and surface brightnesses so low they resolve into individual stars. Pryor (1992) surveys what is known of the mass-to-light ratios of these systems; the evidence is that there is a considerable range in central values of $\mathcal{M}/L$. We will consider numbers for the two best studied cases with large apparent $\mathcal{M}/L$, Draco and Ursa Minor. Both are at distances ~ 70 kpc. Pryor's survey indicates they have similar core radii (the radius at which the surface brightness is half the central value) and line-of-sight velocity dispersions,

$$ r_c = 125\,\text{pc}, \qquad \sigma_* = 10.5\,\text{km s}^{-1}. \tag{18.76} $$

The velocity dispersions are based on precision measurements of the redshifts of the individual most luminous stars. Aaronson (1983) showed this could be done, and obtained the first estimate of the velocity dispersion. Aaronson and Olszewski (1987) and others have added observations of more stars at different times (to eliminate variable velocities, as in binary stars). If the mass has the same radial distribution as the starlight, the central mass density is given by the King equation (18.36),

$$\rho_c = 1.2 \, \mathcal{M}_\odot \, \text{pc}^{-3} . \tag{18.77}$$

This is two orders of magnitude down from the central density in a giant elliptical galaxy (eq. [18.39]), and ten times the local density in the disk of the Milky Way.

The central surface brightnesses of these galaxies are about $\mu = 16.4$ mag per square arc minute. This is difficult to measure because the surface brightnesses are so low; Pryor allows a factor of about two uncertainty. The conversion to surface brightness in units of energy flux density per steradian is (eq. [3.52])

$$i_o = 10^{0.4(4.83-\mu)} \frac{L_\odot}{4\pi(10\,\text{pc})^2} \left(\frac{60 \times 180}{\pi} \right)^2 \text{ster}^{-1} . \tag{18.78}$$

The energy flux density received in a one square arc minute element of solid angle is that of a star of absolute magnitude $M = \mu$ at 10 parsecs distance. The absolute magnitude of the sun is $M_V = 4.83$. The first two factors give the energy flux density, and the last factor converts the solid angle to one steradian.

With these numbers, the central mass-to-light ratio is (eq. [18.37])

$$\frac{\mathcal{M}}{L} = 100 \frac{\mathcal{M}_\odot}{L_\odot} , \tag{18.79}$$

an order of magnitude above the values for the old stars in the Milky Way and for the central parts of large elliptical galaxies (eqs. [18.11] and [18.40]), and comparable to the mass-to-light ratios in clusters of galaxies (section 20). If this is because the masses of the central parts of Draco and Ursa Minor are dominated by neutrinos, and if the neutrino orbits are similar to those of the stars, the parameters in equation (18.76) in equation (18.74) yield a neutrino mass

$$m_\nu \sim 400\,\text{eV} , \tag{18.80}$$

an order of magnitude above what is allowed by the mean mass density (eq. [18.4]). Since the discrepancy is thought to be well outside the uncertainties, the alternatives are that the core radii of the neutrino distributions in these galaxies are considerably broader than for the visible stars, or else that these galaxies are not made of neutrinos. As we will now discuss, the first idea is not reasonable.

To streamline the calculation, let us neglect the star masses and suppose the stars see the gravitational potential of the broad flat central core of a massive halo,

$$\phi(r) = \frac{2}{3}\pi G\rho_c r^2 , \tag{18.81}$$

where ρ_c is the central mass density. If the star orbits are isotropic, with a line-of-sight velocity dispersion σ_* that is independent of radius, the pressure gradient force per unit volume in the gas of stars, $-dp/dr$, is balanced by the gravitational force, $-\sigma_*^2 dn_*/dr = n_* d\phi/dr$. The solution for the star number density as a function of radius is the Maxwell-Boltzmann form

$$n_* \propto \exp -2\pi G\rho_c r^2/3\sigma_*^2 . \tag{18.82}$$

The core radius r_c for the stars is approximately the radius at which n_* is down from the central value by a factor of e. With equation (18.76) for σ_* and r_c, the central mass density in Draco and Ursa Minor in this model is

$$\rho_c = \frac{3}{2\pi} \frac{\sigma_*^2}{Gr_c^2} = 0.8 \mathcal{M}_\odot \, pc^{-3} , \tag{18.83}$$

quite similar to the previous model (eq. [18.77]).

We can use this number for the central mass density with the cosmologically interesting neutrino mass in equation (18.4) to derive the required neutrino core radius and velocity dispersion. Equations (18.66) for the central mass density and (18.69) for the characteristic radius α for the core of the neutrino distribution give

$$\alpha = \frac{\hbar}{2^{5/6}G^{1/2}\mathcal{N}_o^{1/3}m_\nu^{4/3}\rho_c^{1/6}} = 500h^{-8/3} \, pc . \tag{18.84}$$

The numerical value assumes the maximum allowed mean occupation number from the initially thermal distribution, $\mathcal{N}_o = 0.5$. The result is well above the core radius for the stars, so the model in equation (18.81) for the potential seen by the stars is consistent. The neutrino velocity dispersion from equations (18.69) and (18.84) is

$$\sigma = \frac{2^{1/6}\pi^{1/2}\hbar\rho_c^{1/3}}{\mathcal{N}_o^{1/3}m_\nu^{4/3}} = 110\,h^{-8/3} \, km \, s^{-1} . \tag{18.85}$$

We have assumed $\Omega = 1$, so for a reasonable expansion timescale in this Einstein-de Sitter universe the Hubble parameter would have to be close to the lower end of the range of estimates, $h = 0.5$. This would mean $\sigma \sim 700$ km s^{-1}, considerably larger than the velocity dispersion in the Milky Way, and the mass within the radius α would be comparable to the mass of the Milky Way within our position. These numbers seem quite unreasonable for such inconspicuous galaxies (Gerhard and Spergel 1992).

This problem with the dark matter in dwarf spheroidal galaxies is one reason to consider the possibility that the dark mass is a colder still more exotic class of particles.

Cold Dark Matter

Cold dark matter is hypothetical nonbaryonic particles that have primeval velocities (before structures form) small enough for fluctuations in the mass distribution on the length scales of cosmologically interesting objects to start to grow once the dark matter becomes self-gravitating, at redshift $z \sim z_{\text{eq}}$. The primeval rms particle velocity may be low because the particles were thermally produced, as for hot dark matter, but with a considerably higher rest mass, or it may be low because the particles were created with momenta well below kT, as for axions. The low primeval pressure eliminates the phase space problem in dwarf spheroidal galaxies. As described in section 25, the standard cold dark matter model for galaxy formation has some attractive features, as well as a complement of problems. The recognition of the former was the main reason for the general swing of attention in the mid-1980s from hot to cold dark matter. It should be noted, however, that cold dark matter as a candidate for the mass in the dark halos of galaxies is not the same as the cold dark matter model for galaxy formation, because the model uses cold dark matter as an ingredient to which it adds a definite and specific list of assumptions on how all the significant forms of mass are distributed at high redshift.

Considerable beautiful physics underlies the theory of the cold dark matter particle candidates and the laboratory searches for them; Kolb and Turner (1990) and Primack, Seckel, and Sadoulet (1988) offer useful guides. Here we consider only an elegant set of order-of-magnitude relations among the particle rest mass and the annihilation and scattering cross sections for thermally produced massive particles (Lee and Weinberg 1977; for other references see Kolb and Turner 1990).

The starting assumptions are that at high redshift the dark matter particles are in thermal equilibrium at the cosmic background radiation temperature T and at chemical potential small compared to kT, so the occupation number at energy ϵ is

$$\mathcal{N} = \frac{1}{e^{\epsilon/kT} \pm 1}. \tag{18.86}$$

It is assumed that the annihilation cross section is such that thermal equilibrium is broken when the particles are nonrelativistic. The condition that the particle mass and the annihilation cross section are such as to leave the relict density at the Einstein-de Sitter value yields some interesting and perhaps significant coincidences.

For definiteness, we will suppose the dark matter acts as a Dirac particle with four states at given momentum, and we will assume the particles annihilate from zero orbital angular momentum to much less massive particles, so the annihilation rate coefficient $\langle \sigma v \rangle$ is independent of energy (because the final distribution in phase space is almost entirely fixed by the rest mass energy $2m$ released in the annihilation). The particle abundance is frozen at world time t_f, when the thermal particle number density is n_f, and

$$\langle \sigma v \rangle n_f t_f \sim 1 . \tag{18.87}$$

The product $tn(t)$ increases with increasing redshift, so at $t \ll t_f$ there are many creation and annihilation events per dark matter particle in an expansion time, and that keeps the distribution close to thermal. At $t \gg t_f$, annihilation is essentially eliminated. That is, the residual number of particles is the thermal equilibrium number at $t \sim t_f$, when the CBR temperature is T_f.

We are assuming the particles are nonrelativistic at the freezing temperature T_f, so the energy is $\epsilon = mc^2 + p^2/2m$, and the particle number density at the freezing temperature is

$$n_f = \int \frac{4 \, d^3 p}{(2\pi \hbar)^3} e^{-p^2/2mkT_f} e^{-mc^2/kT_f}$$
$$= 4 \left(\frac{mkT_f}{2\pi \hbar^2} \right)^{3/2} e^{-mc^2/kT_f} . \tag{18.88}$$

The present mean number density of relict particles is down from n_f by the redshift factor $(1 + z_f)^3$, where $1 + z_f \sim T_f/T_o$. To get a more accurate expression for the redshift, note that the entropy density at T_f is

$$s_f = \frac{4}{3} a T_f^3 \frac{g_f}{2} = (1 + z_f)^3 \frac{4}{3} a T_0^3 \left(1 + 3 \frac{7}{8} \frac{4}{11} \right) . \tag{18.89}$$

As in equation (15.12), the factor g_f counts the number of fields that share the entropy at T_f, with $g = 1$ for each relativistic boson component and $g = 7/8$ for each fermion component. The present entropy density on the right-hand side is computed as in equation (15.19), taking account of the CBR and three neutrino families at $T_\nu = (4/11)^{1/3} T_o$. This gives

$$\frac{T_f}{T_o} = \left(\frac{43}{11 g_f} \right)^{1/3} (1 + z_f) . \tag{18.90}$$

Table 18.2
Lee-Weinberg Criterion

m (GeV)	T_f (GeV)	$\langle \sigma v \rangle (\mathrm{cm^3\,s^{-1}})$
2	0.09	5.9×10^{-26}
4	0.17	6.0×10^{-26}
10	0.40	6.3×10^{-26}

Then the present mass density in the relict particles is

$$\rho_d = \frac{n_f m}{(1+z_f)^3} = \frac{3 H_o^2 \Omega_d}{8 \pi G} , \tag{18.91}$$

where Ω_d is the density parameter in this component, and equation (18.88) gives n_f as a function of T_f. The result of collecting these relations is

$$\frac{m}{T_f} = 18.9 + \ln \left[\frac{m^{5/2}}{T_f^{3/2} \Omega_d h^2 g_f} \right] , \tag{18.92}$$

with T_f and m expressed in GeV. If these two quantities are on the order of one GeV, and the density parameter in this component is on the order of unity, then this expression is insensitive to the quantities in the logarithm. Table 18.2 shows the freezing temperature T_f as a function of the assumed particle mass m for $\Omega_d = 1$, $g_f = 10$, and $h = 0.5$.

Now we can get the annihilation cross section. Since the universe is radiation dominated at T_f, the expansion time is

$$\frac{1}{t_f^2} = \frac{32 \pi G}{3} \frac{g_f}{2} \frac{a T_f^4}{c^2} , \qquad t_f = \frac{2.4 \times 10^{-6}}{g_f^{1/2} T_f^2} \, \mathrm{s} , \tag{18.93}$$

with the temperature in GeV. The annihilation rate coefficient is then

$$\langle \sigma v \rangle = \frac{1}{n_f t_f} \sim \frac{2 \times 10^{-27}}{\Omega_d h^2} \frac{m(\mathrm{GeV})}{g_f^{1/2} T_f(\mathrm{GeV})} \, \mathrm{cm^3\,s^{-1}} . \tag{18.94}$$

The numbers in the table for the annihilation rate coefficient $\langle \sigma v \rangle$ given by this equation are insensitive to the particle mass if it is on the order of 1 GeV, because equation (18.92) shows that m/T_f is insensitive to m.

The annihilation cross section can be compared to a characteristic value from the weak interactions. If the matrix element between initial and final states is independent of energy, the rate coefficient varies as the square of the energy in the reaction, here $2m$ (as in eq. [18.97] below). If we write

$$\langle \sigma v \rangle = G_d^2 m^2 \hbar^2 c^7 , \tag{18.95}$$

the rate coefficient in the table gives

$$G_d \sim 10^{-4} m^{-1} \, \text{GeV}^{-2} , \tag{18.96}$$

for m in GeV. If the mass is a few GeV, the coupling constant G_d is that of the weak interactions, which is an interesting and maybe significant coincidence.

Finally, we can estimate the cross section for scattering of these relict particles in ordinary matter. If they annihilate to much less massive particles, the rate is

$$\langle \sigma v \rangle \propto \sum |\langle V \rangle|^2 \int d^3 p \, \delta(\epsilon - 2m) \propto m^2 N , \tag{18.97}$$

where $\langle V \rangle$ is the matrix element for the transition. The sum is over the number N of kinds of annihilation products, the annihilation energy for the nonrelativistic dark matter particles is $2m$, the integral is over the phase space for the relative momentum of the decay products, and we have assumed the decay products are relativistic, $\epsilon = p$. If the particles annihilate into components of ordinary matter, such as quarks, the scattering cross section for a dark matter particle incident at velocity v on the material of a laboratory detector is

$$\sigma_s v \propto |\langle V \rangle|^2 \int d^3 p_1 \delta(p^2/2m - p_1^2/2m) \propto m^2 v . \tag{18.98}$$

The ratio gives the scattering cross section,

$$\sigma_s \sim \frac{\langle \sigma v \rangle}{Nc} \sim 10^{-37} \, \text{cm}^2 , \tag{18.99}$$

for $N \sim 10$ and the numbers in table 18.1.

If these particles dominate the mass of the Milky Way at our position ~ 8 kpc from the center, at circular velocity 220 km s^{-1}, their local number density is

$$n \sim \frac{v_c^2}{4\pi G r^2 m} \sim \frac{1}{m \, (\text{GeV})} \text{ particles cm}^{-3} . \tag{18.100}$$

The rms particle velocity is $v \sim (3/2)^{1/2} v_c \sim 300$ km s^{-1}. In a one kilogram detector there are $N_n = 6 \times 10^{26}$ nucleons; if the dark matter particles with mass on the order of 1 GeV scattered incoherently off the nucleons, the scattering rate $\sigma_s v n N_n$ would be about 100 per day. The recoil energy deposited in the material and the scattering rate both are small, but considered within experimental reach.

As usual, one certainly can do better than these order-of-magnitude considerations (for example, Primack, Seckel, and Sadoulet 1988). For our purpose, the point of the exercise is that one can find sensible-looking parameters for exotic dark matter candidates, and that one can imagine significant laboratory searches for the presence of these particles in the massive halo of our galaxy. A positive outcome would have a marked effect on physical cosmology.

Cosmological Constant

A cosmological constant, Λ, or a field energy that acts like it, is one way to make a low-density universe consistent with the condition from inflation that space sections have negligible curvature. This would be a two-step solution to the mass problems, for the inflation condition would be met by Λ and we would still need movable dark matter to account for dynamics in galaxies and systems of galaxies. (A positive Λ tends to pull a cluster apart, but for interesting values of Λ the effect is negligibly small.) A single family of particles that satisfies both conditions would be more elegant, but it is dangerous to ignore the possibility that Nature has found another way to elegance.

Trends of opinion on Λ have been noted in sections 4, 13, and 15. Einstein introduced this term as a way to get a static universe, and once he became convinced that the universe is expanding he saw no further logical need for it (Einstein 1945). The paper by Einstein and de Sitter (1932) offered a somewhat more pragmatic point of view: until the observational situation improves to the point that there is some chance of measuring space curvature and the cosmological constant, it makes sense to consider the simplest acceptable case, what is now called the Einstein-de Sitter universe, where both Λ and space curvature are negligible. Others have taken note of the possible merits of Λ as an aid to fitting the cosmological models to the observations. Lemaître (1933) liked a cosmological constant adjusted to produce a quasistatic phase in the expansion, at least in part because it allowed him to reconcile evolution from a primeval atom (or big bang) with the short timescale H_o^{-1} indicated by the early estimate of Hubble's constant. It is now known that H_o was considerably overestimated, but the timescale is still under discussion. This was one factor in the survey of the cosmological tests that led Gunn and Tinsley (1975) to consider a cosmological constant. At the time this is written it still seems quite possible that Hubble's constant and the age of the universe measured from high redshift will be found to satisfy the bound $H_o t_o > 1$. If so, it would require a term in the expansion rate equation that acts like a cosmological constant (figure 13.1).

As we noted in the last section, if zero-point fluctuations produced a vacuum energy density that is Lorentz-invariant, its stress-energy tensor would have to be proportional to the metric tensor, because that is the only form invariant under a Lorentz transformation. That means the vacuum energy density would act as a cosmological constant. The problem, as indicated in equation (17.14), is that the characteristic energy, $\epsilon \sim 3$ meV, associated with an interesting value of Λ does not suggest anything interesting in particle physics (Carroll, Press, and Turner 1992). Inflation indirectly led to yet another revival of interest in a cosmological constant, for it provides an example of a term in the stress-energy tensor that acts like a dynamic Λ and a reason why we might want one. If inflation is to account for the observed homogeneity of the universe, space curvature has to be negligibly small (eq. [17.23]). The evidence in section 20 shows that the mass this would require is not clustered with the galaxies on scales less than about 10 Mpc. Possibilities this leaves are that most of the mass is in movable matter that has avoided clustering with the galaxies, or in a form that cannot move. The first is discussed in section 25. One way to the latter is a term that acts like a cosmological constant.

One might imagine that the evolving cosmological constant of inflation, which initially is large and ends up at $\Lambda = 0$, is present today in a remnant evolving toward its natural value, slowly enough that Λ still is appreciable, yet quite small by particle physics standards because Λ has been rolling toward zero for a long time. (This philosophy follows the Brans-Dicke 1961 picture, that the Planck mass is large because the strength of the gravitational interaction has been decreasing for a long time.) Pictures for a rolling $\Lambda(t)$ assume the cosmological "constant" is decaying into particles (Özer and Taha 1986; Freese et al. 1987), or else it is rolling to zero without dissipation (Ratra and Peebles 1988).

The purpose of this subsection is to present a simple model for a dissipationless rolling Λ, but it is useful to pause first to note that a cosmological constant can be described as an ideal fluid, with energy density and pressure, but satisfying very specific constraints. The stress-energy tensor for an ideal fluid with energy density ρ, pressure p, and velocity u^i is (eq. [10.49])

$$T^{ij} = (\rho + p)u^i u^j - g^{ij}p . \qquad (18.101)$$

If the fluid couples only to the gravitational field, the conservation law $T_i{}^j{}_{;j} = 0$ is (eq. [10.107])

$$\frac{1}{\sqrt{-g}} \frac{\partial}{\partial x^j} [\sqrt{-g}(\rho + p)u_i u^j] = \frac{\partial p}{\partial x^i} + \frac{1}{2} g_{jk,i}(\rho + p)u^j u^k . \qquad (18.102)$$

We can interpret this in the terms of an ordinary momentum equation, in which the first expression on the right-hand side is the pressure gradient force per unit volume, and the second acts as a gravitational force per unit volume. The passive gravitational mass density in this latter term, which determines the gravitational force on a static fluid in a given gravitational field, is $\rho + p$. Consistent with the equivalence principle, the inertial mass density that appears on the left side is also $\rho + p$. The active gravitational mass density is $\rho + 3p$ (eq. [10.70]), and the energy density is ρ.

If gravity can be neglected, we can set g_{ij} equal to the Minkowski form η_{ij}. If in addition the fluid velocity is small and ρ and p are close to homogeneous, the leading terms in the time and space parts of equation (18.102) are

$$\frac{\partial \rho}{\partial t} + (\rho + p)\nabla \cdot \mathbf{v} = 0, \qquad (\rho + p)\frac{\partial v^\alpha}{\partial t} = -p_{,\alpha}. \tag{18.103}$$

These combine to the equation

$$\frac{\partial^2 \rho}{\partial t^2} = \nabla^2 p. \tag{18.104}$$

If $dp/d\rho$ is positive, this is a wave equation. The stress-energy tensor (18.101) looks like the cosmological constant term if we set $p = -\rho$, but in this case (18.104) indicates that irregularities in the energy distribution are unstable, with exponential growth time equal to the light travel time across the fluctuation, which is quite unacceptable. We can prescribe this away by requiring as an initial condition that the density ρ in this fluid is homogeneous. In general, this prescription is not permanent, because the fluid acts on the gravitational field, and the action back on the fluid by the gravitational field of the lumpy mass distribution around us in general perturbs the mass distribution in the fluid. We see from equation (18.102) that this does not happen in the special case $p = -\rho$, for the passive gravitational mass density vanishes, so the fluid cannot be perturbed. That leaves the active gravitational mass -2ρ of the cosmological constant.

A simple model for a rolling $\Lambda(t)$ follows inflation, and before that the Brans-Dicke (1961) theory, in adding to the material content of the universe a scalar field with the action

$$S_\phi = \int \sqrt{-g} d^4x \left[\frac{1}{2} m_{\text{pl}}^2 g^{ij}\phi_{,i}\phi_j - \frac{\kappa m_{\text{pl}}^4}{\phi^\alpha} \right] \tag{18.105}$$

(Ratra and Peebles 1988). The constants α and κ are dimensionless and positive. The field $\phi(\mathbf{x}, t)$ is dimensionless; the Planck mass gives the action the right units. If the line element is the cosmologically flat Friedmann-Lemaître form with expansion parameter $a(t)$, and ϕ is a function of world time t only, the field equation (17.8) is

$$\ddot{\phi} + 3\frac{\dot{a}}{a}\dot{\phi} = \frac{\alpha\kappa m_{\text{pl}}^2}{\phi^{1+\alpha}}. \tag{18.106}$$

The energy density in this field is the sum of the kinetic and potential parts, as in equation (10.38),

$$\rho_\phi = \frac{1}{2} m_{\text{pl}}^2 \dot{\phi}^2 + \frac{\kappa m_{\text{pl}}^4}{\phi^\alpha}. \tag{18.107}$$

Suppose at redshift $z \gg 1$ the mass density ρ in ordinary matter and radiation is large compared to the field contribution ρ_ϕ, and suppose the expansion parameter is varying as a power of time,

$$a \propto t^n, \qquad \left(\frac{\dot{a}}{a}\right)^2 = \frac{n^2}{t^2} = \frac{8\pi\rho}{3m_{\rm pl}^2}. \qquad (18.108)$$

Then the field equation (18.106) is

$$\ddot{\phi} + \frac{3n}{t}\dot{\phi} = \frac{\alpha\kappa m_{\rm pl}^2}{\phi^{1+\alpha}}. \qquad (18.109)$$

This has a power law solution, $\phi \propto t^p$, with

$$p = \frac{2}{2+\alpha}. \qquad (18.110)$$

In this solution both terms in the energy density scale as $\rho_\phi \propto t^{2p-2}$. Since the density in ordinary material varies as $\rho \propto t^{-2}$, as in equation (18.108), the ratio of the mass density in the field to the mass density in ordinary material scales as

$$\frac{\rho_\phi}{\rho} \propto t^{4/(2+\alpha)}. \qquad (18.111)$$

The mass density in the scalar field in this model is rolling to zero, consistent with the opinion that this is its natural value, but the rate is less rapid than for ordinary material, so there comes a time when the scalar field dominates. The field is allowed to have spatial gradients, but the gradients with wavelengths much less than the Hubble length behave as massless waves, so they would not appear as significant dark mass concentrations. If one imagined that this field is a relict of inflation, one would of course have to devise a considerably more complicated scenario for its behavior at high redshift.

Perhaps the central moral from this model is that there are many ways to approach the dark mass puzzles.

Antimatter

This is an appropriate place to take note of the one form of exotic matter whose properties are experimentally very well characterized. Antibaryonic matter is no darker than ordinary baryonic matter and can be considerably easier to detect when mixed with the opposite sign, but otherwise it is difficult to distinguish from the matter we are made of. It is natural to ask whether distant objects may be made of antimatter (Stecker 1989); one might even hope the answer will teach us something about the way baryons and the other forms of matter were created.

Steigman (1976) gave the classical survey of the observational tests for the presence of antimatter concentrations in the observable universe. Here we consider one of the tests — the bound from annihilation radiation in the plasma in clusters of galaxies. Intracluster plasma is discussed in sections 20 and 24. It is a particularly useful probe because one would expect that the galaxies keep the plasma in the center of a cluster well stirred, making it easy to get an unambiguous estimate of the annihilation rate in a cluster made out of a mixture of matter and antimatter.

Steigman finds that at the plasma temperatures in clusters the rate coefficient for $p\bar{p}$ annihilation is usefully approximated as

$$\alpha_a = 10^{-10}T^{-1/2}\,\mathrm{cm}^3\,\mathrm{s}^{-1}\,, \tag{18.112}$$

with the temperature measured in degrees Kelvin. The annihilation rate per unit volume in a plasma with proton number density n_p and antiproton density $n_{\bar{p}}$ is $\alpha_a n_p n_{\bar{p}}$. The X-ray luminosity density from thermal bremsstrahlung in the plasma is proportional to the product of the ion and electron number densities. Since an annihilation produces a few gamma rays at energies ~ 100 MeV, the ratio of the gamma ray flux density to the X-ray flux density from a well-mixed intracluster plasma is

$$\frac{F_\gamma}{L_x} \sim \frac{\alpha_a \int n_p n_{\bar{p}}\,dV}{10^{-27}T^{1/2}\int n_p^{\,2}\,dV}\,. \tag{18.113}$$

In the denominator, the electron density is close to the density of protons (or antiprotons). The numerical factor for the X-ray luminosity density is worked out in section 24 (eq. [24.16]). The temperature again is expressed in degrees Kelvin, the density and volume element in cubic centimeters. The ratio of the integrals is close to $n_p/n_{\bar{p}}$ or $n_{\bar{p}}/n_p$, whichever is smaller, and independent of the cluster distance. Edge's (1989) survey indicates typical cluster luminosities are $L_x \sim 3 \times 10^{44}$ erg s^{-1} at $T \sim 5$ keV. Houston, Wolfendale, and Young (1984) find that the mean cluster luminosity at gamma ray energies greater than 35 MeV is less than about 10^{47} photons s^{-1}. These numbers give

$$n_{\bar{p}}/n_p \lesssim 10^{-6}\,. \tag{18.114}$$

This samples the cluster mass that is drawn from a region at least $15h^{-1}$ Mpc across (eq. [20.52]).

The ratio in equation (18.114) could of course be the other way around; perhaps some clusters are made of antimatter. We can conclude that if baryons were created in domains of matter and antimatter, the domains would have to have been considerably larger than $15h^{-1}$ Mpc, or they certainly would have been mixed and revealed themselves where the great clusters formed.

19. Measures of the Galaxy Distribution

Tests of models for the origin of the structures observed in our universe ultimately depend on a comparison of model predictions with numerical measures of the structures. The statistical measures to be discussed here are the low order N-point correlation functions. Others that have proved useful include group-finding algorithms that connect neighbors of neighbors (Turner and Gott 1976; Rood 1976; Kuhn and Uson 1982; Einasto et al. 1982; Barrow, Bhavsar, and Sonoda 1985), the frequency distribution of counts in cells (Saslaw 1985), and the topology of surfaces of constant mean number density (Gott, Melott, and Dickinson 1986). There is no unambiguous guide to which measure will be most revealing and useful; that may well depend on the character of the model and the structures to be modeled. One can identify three attractive features of the low-order correlation functions. First, we noted in section 7 that until there are considerably better measures of galaxy distances, studies of structure must deal with the distortion in redshift distance caused by peculiar motions (LSS, § 76; Kaiser 1987). The correlation functions are designed for a straightforward relation between the wanted spatial statistics and the statistics of the two-dimensional distribution projected to eliminate redshift smearing. Second, as will be discussed in the next sections, the correlation functions conveniently lend themselves to the measurement of the mass distribution and the analysis of its dynamical evolution. The third feature, which is more subjective, is that the two-point function is the analog of the second moment, or variance, of a distribution, the three-point function is the analog of the third moment, or skewness, and so on. It would be difficult to count the cases in quantitative sciences where the first defining measure is the standard deviation or second central moment, and the same measure surely is worth trying here.

Correlation Functions

The two-point correlation function was used in the analysis in section 7 of a fractal picture for the large-scale distribution of luminous matter. Here we consider the extension to higher-order correlation functions, and a few of the numerical results.

The evidence reviewed in sections 3 and 7 rules out a fractal model for the large-scale distribution of luminous matter within the Hubble distance (unless the fractal dimension is very close to $d = 3$), and as far as is known it agrees with the standard assumption of large-scale homogeneity. Consistent with this, we will model the space distributions of galaxies and mass as realizations of statistically homogeneous and isotropic (stationary) random point processes.

The number density, n, for a stationary point process is defined by the probability that a galaxy is found centered in the volume element dV (eq. [7.30]),

$$dP = n\, dV .\tag{19.1}$$

The two-point correlation function, $\xi(r)$, is defined by the joint probability that galaxies are found in the two volume elements dV_1 and dV_2 placed at separation r (eq. [7.31]),

$$dP = n^2 \, dV_1 dV_2 [1 + \xi(r)] . \qquad (19.2)$$

This is equivalent to the conditional probability of finding a galaxy in the element dV at distance r from a galaxy,

$$dP = n[1 + \xi(r)] \, dV . \qquad (19.3)$$

These two statistics define the process only in the special case of the Gaussian model to be presented below. One way to get more detail on the character of a more general process is to go to higher-order moments. The three-point correlation function, ζ, is defined by the joint probability of finding galaxies centered in each of the three volume elements dV_1, dV_2, and dV_3:

$$dP = n^3 \, dV_1 dV_2 dV_3 [1 + \xi_{12} + \xi_{23} + \xi_{31} + \zeta] . \qquad (19.4)$$

Since the process is homogeneous and isotropic, ζ can only be a function of the lengths of the three sides of the triangle defined by the positions of the volume elements, $\zeta(r_{12}, r_{23}, r_{31})$. With the added terms in this expression, ζ vanishes when two points are close together and the third is far away, and the expression reduces to a product of the one- and two-point probabilities in equations (19.1) and (19.2). That is, the reduced function ζ vanishes when the points are statistically independent. The extra terms also play a useful role in the relation to the distribution projected along the line of sight, as we now discuss.

In the projected angular distribution, the three-point function, $z(\theta_{12}, \theta_{23}, \theta_{31})$, similarly is defined by the equation

$$dP = \mathcal{N}^3 d\Omega_1 d\Omega_2 d\Omega_3 [1 + w(\theta_{12}) + w(\theta_{23}) + w(\theta_{31}) + z] , \qquad (19.5)$$

where $\mathcal{N}$ is the mean number per unit solid angle and w is the angular two-point correlation function (eq. [7.44]). All these angular statistics can be considered as applying to ensembles of angular distributions, but they are only useful as precision descriptive statistics when the sample is deep enough so that what is seen in different parts of the sky gives a useful approximation to what would be found in the theoretical ensemble of skies. (And the same of course applies to the estimates of the spatial functions from the point distribution in a limited volume of space.)

The angular three-point probability in equation (19.5) includes triplets of points accidentally seen close together in projection in the sky but at quite different radial distances. These are accounted for in the first term in brackets in equation (19.5). Other triplets consist of a physical pair close in space with a third point accidentally seen close in projection. These are described by the next three terms. That leaves the last part, z, as an integral over the spatial function, ζ. The derivation of this integral is an easy exercise that follows Limber's equation (7.47) for the relation between the spatial and angular two-point functions $\xi(r)$ and $w(\theta)$ (Peebles and Groth 1975). This result — that the correlation functions conveniently sort out projection effects — is particularly useful in the application of these statistics in extragalactic astronomy.

The defining equation for the four-point correlation function is the joint probability of finding galaxies centered in four volume elements,

$$dP = n^4[1 + \xi_{12} + \xi_{13} + \xi_{14} + \xi_{23} + \xi_{24} + \xi_{34}$$
$$+ \xi_{12}\xi_{34} + \xi_{13}\xi_{24} + \xi_{14}\xi_{23} \qquad \textbf{(19.6)}$$
$$+ \zeta_{123} + \zeta_{124} + \zeta_{134} + \zeta_{234} + \eta] \, dV_1 dV_2 dV_3 dV_4$$

(Fry and Peebles 1978). The reduced four-point function, η, vanishes when two close pairs of points are well separated from each other, and this expression reduces to the product of two two-point probabilities, and η vanishes when one point is well away from the other three and the expression becomes the product of a three-point probability and a one-point probability. The projected function is defined by a similar form. As for the three-point function, one finds that all the added terms take account of projection effects, leaving the four-point angular function as an integral over η.

A more detailed description of the galaxy space distribution would label the galaxies by parameters such as the optical and radio luminosity, the morphological type, and the peculiar velocity, and the correlation functions would contain these arguments as well as the spatial separation. An example discussed in section 3 is that early-type galaxies tend to be found in denser environments, meaning they are more strongly clustered than late-type spirals and irregulars. The difference of the two-point functions for different morphological types was demonstrated by Davis and Geller (1976). Since the gas-rich late-type galaxies that prefer less dense environments tend to have appreciable luminosities from bright young stars, one would expect that the correlation functions depend on luminosity. The effect is quite weak, however, for we saw in section 3 that giant and dwarf galaxies have quite similar-looking space distributions. The marked exceptions are the cD galaxies found in the centers of clusters of galaxies, but they are rare. The result is that it has proved to be useful to consider the galaxy position correlation functions without discriminating by luminosity. Some aspects of the velocity correlation functions are discussed in the next section.

The following model for a point process helps illustrate the meaning of the N-point correlation functions, and can be useful in computations. Let $f(\mathbf{r})$ be a given positive function of position, and suppose a galaxy is placed in the volume element dV at position $\mathbf{r}$ with probability

$$dP = f(\mathbf{r})\,dV \,. \tag{19.7}$$

This is a Poisson process, because the equation means the probability is independent of what happened in neighboring elements. The point process is not stationary; the mean density is the function $f(\mathbf{r})$. We get a stationary point process by replacing the given $f(\mathbf{r})$ with a stationary random function. This is a model, rather than a descriptive statistic, because a Poisson distribution with continuous $f(\mathbf{r})$ does not exist for all stationary random point processes. (An example is a gas of hard spheres, where particles avoid each other, so $\xi(0) < 0$; the trivial exception to the example is a set of realizations of $f(\mathbf{r})$ that are sums of delta functions.)

In this Poisson model, the mean number density is

$$n = \langle f \rangle \,, \tag{19.8}$$

and the joint probability that galaxies are placed in the elements dV_1 and dV_2 is

$$dP = \langle f(\mathbf{r}_1)f(\mathbf{r}_2) \rangle \, dV_1 dV_2 \,, \tag{19.9}$$

so we see from equation (19.2) that

$$n^2 \xi(r_{12}) = \langle f_1 f_2 \rangle - n^2 = \langle (f_1 - n)(f_2 - n) \rangle \,. \tag{19.10}$$

Here and in what follows the subscript is the argument: $f_1 = f(r_1)$ and $r_{12} = |\mathbf{r}_1 - \mathbf{r}_2|$. Equation (19.10) says that, when the random function f exists, the reduced two-point function $\xi(r)$ is the dimensionless autocorrelation function of $f(\mathbf{r})$, and the value at zero lag is the second central moment or variance of f, $n^2 \xi(0) = \langle (f - n)^2 \rangle$. The reduced three-point function defined in equation (19.4) similarly works out to

$$n^3 \zeta(r_{12}, r_{23}, r_{31}) = \langle (f_1 - n)(f_2 - n)(f_3 - n) \rangle \,. \tag{19.11}$$

That is, ζ is the third lagged moment of $f(\mathbf{r})$, and at zero lag it is the third central moment, or skewness. The reduced four-point function defined in equation (19.6) is

$$\begin{aligned}
n^4 \eta = & \langle (f_1 - n)(f_2 - n)(f_3 - n)(f_4 - n) \rangle \\
& - \langle (f_1 - n)(f_2 - n) \rangle \langle (f_3 - n)(f_4 - n) \rangle \\
& - \langle (f_1 - n)(f_3 - n) \rangle \langle (f_2 - n)(f_4 - n) \rangle \\
& - \langle (f_1 - n)(f_4 - n) \rangle \langle (f_2 - n)(f_3 - n) \rangle \,.
\end{aligned} \tag{19.12}$$

At zero lag this is the reduced fourth moment, or excess kurtosis, $n^4 \eta = \langle (f - n)^4 \rangle - 3\langle (f - n)^2 \rangle^2$. If $f(\mathbf{r})$ has a Gaussian probability distribution, these two terms cancel and the excess kurtosis η vanishes.

If $f(\mathbf{r}) - n$ is a Gaussian random process, as for the density fluctuations produced in the simple model for inflation discussed in section 17, we can think of it as a Fourier series (with periodic boundary conditions in some large volume),

$$f(\mathbf{r}) - n = \sum |\delta_k| e^{i[\mathbf{k} \cdot \mathbf{r} - \phi(\mathbf{k})]}, \qquad (19.13)$$

where the phases ϕ are random (with the reality condition $\phi(-\mathbf{k}) = -\phi(\mathbf{k})$). This makes the three-point function ζ vanish, because it contains an odd number of phase factors, and it reduces the fourth product to the last three terms on the right-hand side of equation (19.12), meaning η vanishes. All the extra terms added to the reduced functions in equations (19.2), (19.4), and (19.6) thus serve both to eliminate projection effects in the angular function and to make the reduced functions vanish for a Gaussian process.

A related statistic is the power spectrum $|\delta_k|^2$. It is an easy exercise to check that the Fourier transform of the power spectrum is the autocorrelation function $\xi(r)$ (eq. [21.40]). Since the stationary random Gaussian process in equation (19.13) is completely specified by its power spectrum it is completely specified by its two-point correlation function $\xi(r)$.

Another related statistic, which may be easier to visualize, is the moments of the frequency distribution in the count N of galaxies found within a randomly placed cell of fixed size and shape. The relation between the variance of the counts and the two-point correlation function was obtained in section 7 (eqs. [7.64] to [7.66]) and is easily generalized to higher moments. The mean value of the count in the cell is

$$\bar{N} = nV, \qquad (19.14)$$

where V is the cell volume. The second central moment in equation (7.71) is

$$\langle (N - \bar{N})^2 \rangle = n^2 \int d^2 V \xi_{12} + \bar{N}, \qquad (19.15)$$

where the integral is over the cell volume. The third moment similarly works out to (LSS, eq. [36.13])

$$\langle (N - \bar{N})^3 \rangle = n^3 \int d^3 V \zeta + 3 \langle (N - \bar{N})^2 \rangle - 2\bar{N}. \qquad (19.16)$$

When the shot noise terms on the right-hand side can be neglected, the variance and skewness of the count distribution are integrals of the two- and three-point correlation functions. The fourth moment is (LSS, eq. [36.20])

$$\langle (N - \bar{N})^4 \rangle - 3 \langle (N - \bar{N})^2 \rangle^2 = n^4 \int d^4 V \eta$$
$$+ 6 \langle (N - \bar{N})^3 \rangle - 11 \langle (N - \bar{N})^2 \rangle + 6\bar{N}. \qquad (19.17)$$

The left-hand side is the excess curtosis of the distribution. Apart from the shot noise terms, it is the integral of the reduced four-point function.

Galaxy Correlation Functions

The purpose of this subsection is to list some numerical results for the low-order galaxy correlation functions. Estimates of the galaxy two-point spatial correlation function $\xi(r)$ are discussed in section 7; they show that ξ is usefully approximated as

$$\xi(r) = \left(\frac{r_o}{r}\right)^{\gamma}, \quad r_0 = 5.4 \pm 1\, h^{-1}\, \text{Mpc}, \quad \gamma = 1.77 \pm 0.04, \quad \textbf{(19.18)}$$

at the range of separations

$$10\, \text{kpc} \lesssim hr \lesssim 10\, \text{Mpc}. \quad \textbf{(19.19)}$$

This power law model is a remarkably good approximation down to separations where the visible parts of the galaxies start to overlap (Gott and Turner 1979). At the large-scale end, $\xi(r)$ shows a rise above the power law at $hr \sim 10$ Mpc, and then a rather distinct break below the power law at $hr \sim 20$ to 30 Mpc (Groth and Peebles 1977, 1986; Soneira and Peebles 1978; Maddox et al. 1990). More detailed characterizations of the large-scale end of the correlation function are still under discussion.

In the range of useful estimates, the galaxy three-point function is well approximated by the form

$$\zeta = Q[\xi(r_{12})\xi(r_{23}) + \xi(r_{23})\xi(r_{31}) + \xi(r_{31})\xi(r_{12})] \quad \textbf{(19.20)}$$

at $100\, \text{kpc} \lesssim hr \lesssim 3$ Mpc (Groth and Peebles 1977). The surveys by Fry (1984), Sharp, Bonometto, and Lucchin (1984), Szapudi, Szalay, and Boschán (1992), and Meiksin, Szapudi, and Szalay (1992) give

$$Q = 1.0 \pm 0.2. \quad \textbf{(19.21)}$$

This applies to optically selected galaxies. The indications are that Q is smaller for the IRAS galaxies plotted in figures 3.7 and 3.8, consistent with the observation that IRAS galaxies avoid the dense central parts of clusters.

The model in equation (19.20) was introduced as one of the two simplest forms for ζ one can construct out of the two-point function consistent with the conditions that ζ has to be a symmetric function of its three arguments, and it has to approach zero when one point is moved well away from the other two. The other possibility, $\xi(r_{12})\xi(r_{23})\xi(r_{31})$, is quite inconsistent with the data.

The estimates of the galaxy four-point function, η, are still quite crude. At separations in the range $0.5 \lesssim hr \lesssim 3$ Mpc they agree with the simplest model one can construct out of triple products of the two-point function,

$$\eta = R_a \left[\xi_{12}\xi_{23}\xi_{34} + \text{sym. (12 terms)}\right]$$
$$+ R_b \left[\xi_{12}\xi_{13}\xi_{14} + \text{sym. (4 terms)}\right] . \qquad (19.22)$$

The twelve terms in the first line are the ways of joining all four points by an unbroken line (for there are $4 \times 3 \times 2/2$ distinct ways to assign any point to an end, any one of the remaining three points to the next step, and any of the remaining two to the last step). The four terms in the second line join one point to the other three. The surveys by Fry (1983, 1984), Szapudi, Szalay, and Boschán (1992), and Meiksin, Szapudi, and Szalay (1992) indicate

$$\frac{3R_a + R_b}{4} = 2.5 \pm 0.5 . \qquad (19.23)$$

The estimates of the two coefficients are tightly correlated and individually only poorly constrained.

The simple forms of the low-order galaxy correlation functions might mean one can derive them from a scaling pattern in the underlying dynamics of mass clustering in an expanding universe. The search for the pattern is outside the bounds of this book; samplers of the state of the art are in Schaeffer (1984) and Hamilton (1988a).

The practical art of estimating the correlation functions of extragalactic objects will not be surveyed here, but some general points are worth listing. The galaxy correlation functions have proved to be useful, and the measurements believable, because we have reproduceable estimates from independent samples. This is illustrated in figures 7.2 and 7.3. At larger separations, where the correlation functions are small, the estimates are uncertain and the interpretation is difficult because statistical fluctuations in the estimates are correlated. For example, in a deep narrow sample of galaxy positions the autocorrelation function at large separations along the line of sight has prominent bumps caused by the strong clustering of galaxies on small scales: where the sample intersects one galaxy it is likely to intersect many neighbors that make many pairs with all the galaxies in other unrelated clumps along the line of sight (Broadhurst et al. 1990). Distinguishing these statistical bumps from true large-scale structure is not straightforward. Blackman and Tukey (1958) advise that in this case one may do better to use the power spectrum, in which statistical fluctuations are uncorrelated. (In a Gaussian process the components of the power spectrum are statistically independent. In general, the components are uncorrelated in the sense that the second moments vanish, as in eq. [21.83]. If there are gaps in the sample, the expansion functions are not orthogonal, which introduces correlations, but this can be controlled by a sensible choice of the basis functions.) When the correlation function is small, the estimates of it and the power spectrum are readily compromised by systematic errors in the mean density as a function of position that may result from variable obscuration across the sky or variable efficiency of detection as a function of red-shift. If this causes the apparent mean density to vary across the sample by the rms amount Δ, it adds Δ^2 to the apparent correlation function on small scales, and it subtracts a term on the order of Δ^2 from the apparent correlation function at separations comparable to the sample depth.

Bounded Galaxy Clustering Hierarchy

The power law models for ξ, ζ, and η have the forms one would expect if the small-scale galaxy distribution were a segment of a scale-invariant cluster-ing hierarchy. One way to visualize this is to imagine that the galaxies are placed in clumps, or cluster balls, and the cluster balls are placed in space independently and uniformly at random. If this distribution is averaged through a large smooth-ing length, it looks smooth, as in the maps in figures 3.8 to 3.11. When the resolu-tion is improved to a scale comparable to the cluster ball size, one sees structure. In a bounded clustering hierarchy, the galaxy distribution viewed with resolution $r \lesssim 10h^{-1}$ Mpc shows that the galaxies are concentrated in clumps of size r, and when the resolution is improved to $r' < r$ these clumps are seen to break up into clumps of size r'. We will see that a model galaxy distribution constructed ac-cording to this plan, and designed to reproduce the low-order galaxy correlation functions, successfully predicts some striking features of the real galaxy distribu-tion, and quite misses others.

In a scale-invariant range of this clustering hierarchy, the typical density $n(r)$ within the clumps seen with resolution r scales with r as a power law,

$$n(r) \sim (r_o/r)^\gamma . \tag{19.24}$$

It is easy to see that this reproduces the general features of the low-order galaxy correlation functions. The chance of finding a neighbor at distance $r \lesssim 10h^{-1}$ Mpc from a galaxy is the conditional probability in equation (19.3),

$$dP = n[1 + \xi(r)] \, dV \sim n(r) \, dV , \tag{19.25}$$

because the two galaxies tend to be in a cluster of size $\sim r$, density $\sim n(r)$. At $r \lesssim r_o \sim 5h^{-1}$ Mpc, where $\xi > 1$, the power law behavior of the hierarchy in equation (19.24) reproduces the observed power law shape of $\xi(r)$ in equation (19.18). The chance of finding that a galaxy has neighbors in the volume elements dV_1 and dV_2 at distances r_1 and r_2 is

$$dP \sim n(r_1)n(r_2)\,dV_1 dV_2\,, \qquad\qquad (19.26)$$

because the lengths r_1 and r_2 typically are in levels of the hierarchy with densities $n(r_1)$ and $n(r_2)$. At $r \lesssim r_o$ the reduced part ζ dominates the probability in equation (19.4), and we see that in this model ζ scales as ξ^2, as observed (eq. [19.20]). The chance of finding three neighbors similarly is seen to scale as ξ^3, in agreement with equation (19.22) for the four-point function.

One way to test a model for the galaxy distribution is to use it as a prescription for placing a space distribution of points, assign the points luminosities from the observed distribution for galaxies, select points according to the conditions of a galaxy catalog, and then make a map that can be compared to that of the real galaxies. This was first done by Scott, Shane, and Swanson (1954). The examples in figures 19.1 and 19.2 follow the bounded clustering hierarchy prescription in equations (7.25) and (7.26), as follows (Soneira and Peebles 1978).

In the model construction, galaxies are placed in bounded clumps, or cluster balls. The construction of a cluster ball commences with a stick of length $d_1 = 11.35h^{-1}$ Mpc placed in space at random (as a stationary random Poisson process), at random orientation, and inside a region large enough so that the boundaries cannot affect what is mapped. At each end of the stick is centered a stick of length $d_2 = d_1/1.1$, randomly oriented. On each of these ends are centered sticks of length d_2/λ, on these ends sticks of length d_2/λ^2, and so on until there are enough ends to place the wanted number of galaxies. The construction then repeats at a new independently chosen center, and continues until it has produced the wanted mean number density of galaxies. The number λ is chosen to get the observed scaling of density with length within a cluster ball (eq. [7.26]). The smaller value of the length ratio d_1/d_2 on the first level of the hierarchy of sticks shapes $\xi(r)$ so the function rises somewhat above the power law at $r \sim 10h^{-1}$ Mpc. All the reduced correlation functions vanish at $r \gtrsim 20h^{-1}$ Mpc, the maximum diameter of a cluster ball, because the positions of the cluster balls are statistically independent. Finally, the number of levels in a cluster ball is a random variable, with distribution shaped to reproduce the amplitudes of ξ, ζ, and η.

In figure 19.1, panel A is a segment of the Lick map in figure 3.9, and panel B shows the result of the model construction projected at the depth of the Lick map. People tend to agree that the model lacks a certain crispness of detail, but that it might pass as a reasonable first approximation to the real thing. Note that in the model map the eye can pick out accidental linear features in a distribution that has no preference for them, no true walls.

Figure 19.2 is a redshift map based on this model construction (Soneira 1978). The redshift includes motions within the clustering hierarchy, computed as follows (Peebles 1978). At the level l in a hierarchy, counting from the largest scale, there are 2^{l-1} clusters, each consisting of two subclusters with mass and separation

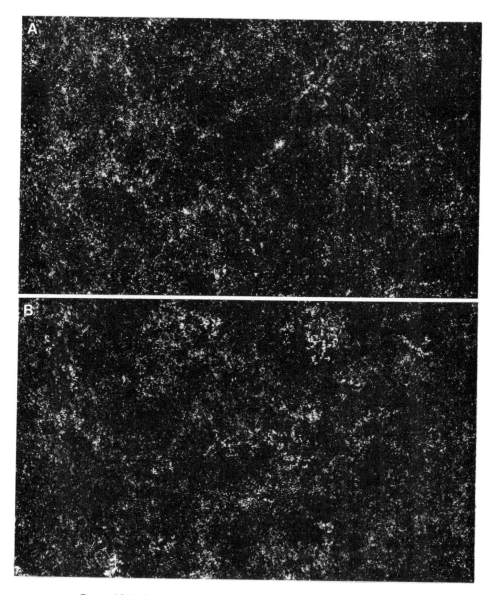

Figure 19.1. A model for the galaxy distribution. Panel A is a section from the Lick map in figure 3.9. Panel B is a map at the same depth and scale from a numerical model designed to reproduce the observed correlation functions through the four-point function (Soneira and Peebles 1978).

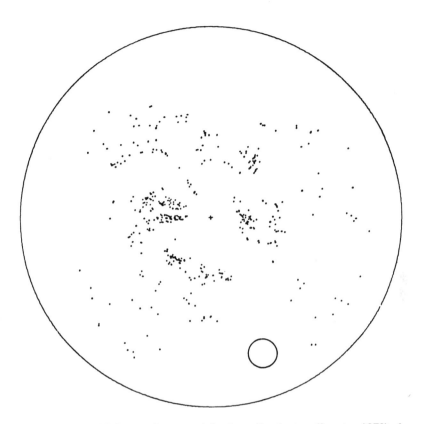

Figure 19.2. Redshift map for a model galaxy distribution (Soneira 1978). A deeper angular map based on the same construction is shown in figure 19.1. The small circle has a diameter of $10h^{-1}$ Mpc, the radius of the outer circle is 7000 km s^{-1}, and the wedge runs from 50° to 60° declination.

$$\mathcal{M}_l \propto 2^{-l}, \qquad r_l \propto \lambda^{-l}, \tag{19.27}$$

where $\lambda = 2^{1/(3-\gamma)}$ is the ratio of separations on successive levels of the hierarchy (eq. [7.26]). The virial theorem says the mean square relative velocity of the two subclusters is

$$v_l \propto (\mathcal{M}_l/r_l)^{1/2} \propto r_l^{1-\gamma/2}. \tag{19.28}$$

In the model construction, the two subclusters at each level are assigned equal and opposite velocities, with random orientation and magnitude v_l. This means the velocity structure in this model is constructed on the same plan as the positions:

at each step in the level of the hierarchy, the separations decrease by the factor λ, and the velocities decrease by the factor $\lambda^{(1-\gamma/2)}$.

The velocity normalization, based on the very limited information available when figure 19.2 was constructed, made the small-scale rms relative velocities about twice the observed value. The construction neglects the gravitational attraction of neighboring cluster balls. Thus the redshift map in figure 19.2 underestimates the large-scale distortion due to relative motions of cluster balls, and it overstates the small-scale distortion.

The map was made before the first substantial redshift surveys, and it illustrates what the low-order correlation functions did and did not lead us to expect in advance of the three-dimensional redshift maps that give a more direct picture of the large-scale galaxy and mass distributions.

The velocity scaling law in equation (19.28), with $\gamma = 1.8$, agrees with the observed galaxy rms relative velocity dispersion as a function of separation, shown in figure 20.1. In the model construction the distribution of relative line-of-sight peculiar velocities is found to be close to exponential, which again is reasonably close to the observations. As discussed in the next section, this checks both the interpretation of the galaxy space distribution as an approximation to a clustering hierarchy and the assumption that the clustered mass distribution is traced by the galaxies.

The limiting magnitude in the model in figure 19.2 is about 1.5 magnitudes shallower than the real redshift maps in figure 3.6, and the sampling is correspondingly more sparse, but one can make some sense of a comparison of the distributions. Prominent in the model redshift map are what Soneira called "empty regions," and now are called voids. They are the result of the gathering of points to make the small-scale cluster balls: there had to be large accidental empty regions because the cluster balls had to be placed at random to make the correlation functions vanish at large separations. The diameter of the small circle marked on the map is $10h^{-1}$ Mpc. It is easy to find empty regions $30h^{-1}$ Mpc across and comparable in size to the voids seen in the galaxy maps in figure 3.6. The empty regions would not shrink much if they were mapped to a fainter limiting magnitude, because we know that in the real world faint galaxies cluster with the bright ones, and the same was assumed in the model.

A second striking feature of the real galaxy maps is the tendency for galaxies to appear in large-scale sheetlike arrangements. The willing believer can see linear structures in the real and model angular maps in figure 19.1. We know the latter are the accidents of a construction that has no preference for linear structures. As discussed in section 3, perceptive observers did see evidence of a significant difference, but the effect became manifest only with the large-scale redshift surveys. In the galaxy maps in figure 3.6 the voids tend to have rather smooth edges, while in the model the edges are highly irregular. The origin of this remarkable behavior is still under discussion; some ideas are surveyed in sections 22 and 25.

Walls, Clusters, and Scaling

The simple bounded clustering hierarchy in figures 19.1 and 19.2 misses two distinctive features of the large-scale galaxy distribution: the great clusters, and the tendency of galaxies to lie on two-dimensional structures. We consider here a scaling relation for the former, and the contribution of the latter to the galaxy two-point correlation function.

The Great Wall in figure 3.6 has a breadth of at least $100h^{-1}$ Mpc. Let us see why the existence of such structures does not conflict with the observation that the galaxy clustering length defined by the separation at which the two-point correlation function falls to unity is only $r_o \sim 5h^{-1}$ Mpc. As a simple model for walls, suppose a fraction f of all galaxies are placed on very large flat sheets, and that the sheets are placed independently at random, with statistically homogeneous and isotropic space positions and orientations. In keeping with the simplicity of the model, we will suppose that the galaxies on the sheets are distributed in a statistically homogeneous way, with mean surface density σ, and that the remaining fraction $1 - f$ are field galaxies that are placed in space uniformly at random.

We can define a characteristic distance L between sheets by the expression for the probability that an observer at a randomly chosen position in space sees that there is a sheet at perpendicular distance h in the range dh,

$$dP = dh/L. \tag{19.29}$$

Now let us get the relation between L, σ, the mean number density n of galaxies, and the fraction f of galaxies on sheets. If a sphere with radius r is centered at distance $h < r$ from a sheet, the mean number of sheet galaxies inside the sphere is $\sigma\pi(r^2 - h^2)$. If the sphere radius r is small compared to L, there is negligible chance the sphere contains more than one sheet, so we see from equation (19.29) that the mean number of sheet galaxies found in a randomly placed sphere of radius $r \ll L$ is

$$\langle N \rangle = \sigma\pi \int_0^r (r^2 - h^2)\frac{dh}{L} = \frac{2\pi\sigma r^3}{3L} = \frac{4}{3}\pi r^3 fn. \tag{19.30}$$

In the last expression, n is the mean galaxy number density and fn is the mean space number density of sheet members. This shows that the mean surface density of galaxies on sheets is

$$\sigma = 2Lfn. \tag{19.31}$$

To get the two-point correlation function $\xi(r)$ for this model distribution, consider the probability that a randomly chosen galaxy has a neighbor at distance r in the range dr:

$$dP = 4\pi(1 - f)nr^2dr + f[4\pi nr^2dr + 2\pi rdr \cdot 2Lfn]$$
$$= 4\pi nr^2dr[1 + \xi(r)]. \tag{19.32}$$

The probability of choosing a field galaxy is $1 - f$. Since the field galaxies are placed uniformly at random, a field galaxy on average sees the mean number density n of neighbors, so the probability that it has a neighbor is n multiplied by the volume of the shell at distance r to $r + dr$. As indicated in the second term, a sheet galaxy sees mean number density n from the field galaxies and all the other independently placed sheets, and it sees the other members of its own sheet. The second line defines the two-point correlation function. The result is

$$\xi(r) = f^2L/r. \tag{19.33}$$

The Great Wall pictured in figure 3.6 is at distance $\sim 70h^{-1}$ Mpc. To get a definite number, let us suppose the characteristic distance between walls (as defined in eq. [19.29]) is $L = 100h^{-1}$ Mpc. Most galaxies are not on prominent walls, so let us take the wall fraction to be $f \sim 0.1$. Then at $r = 10h^{-1}$ Mpc the correlation function (19.33) is $\xi \sim 0.1$, and at $r = 30h^{-1}$ Mpc it is $\xi \sim 0.03$, both below presently available bounds. More elaborate models could further suppress $\xi(r)$ at large r by introducing an anticorrelation of field and wall galaxy positions, or increase ξ by correlating wall positions; the possible sizes of both effects are still under discussion. The point of the simple model presented here is that there can be structures in the galaxy distribution that have large scales and are quite prominent in the three-dimensional maps, but have little effect on the rms fluctuation in the numbers of galaxies contained in a randomly placed large sphere, as measured by $\xi(r)$. The latter is what is relevant for the fluctuations in the gravitational field, as measured by large-scale peculiar velocities and gravitationally induced anisotropies in the cosmic background radiation temperature. This is discussed in section 21.

Let us consider next clusters of galaxies. Figure 3.5 shows an example, the Coma cluster. One typically sees large irregularities in the galaxy distribution in the outer regions of a cluster, and this can extend to substructures in the central parts. An example is the condensations around the two giant elliptical galaxies in the Coma cluster (Fitchett and Webster 1987; Geller 1990). This substructure could be interpreted to mean that gravitational interactions are causing the richest of the cluster balls to relax toward equilibrium gas spheres, and that the relaxation is close to complete only in some of the densest central regions.

The mean concentration of galaxies around a cluster is measured by the cluster-galaxy cross correlation function, $\xi_{cg}(r)$, defined by the joint probability that a cluster center is in the volume element dV_c and a galaxy is found in the volume element dV_g at distance r,

$$dP = n_c n \, dV_c dV_g [1 + \xi_{cg}(r)], \qquad (19.34)$$

where n and n_c are the mean number densities of galaxies and clusters. Another way to put it is that $n\xi_{cg}(r)$ is the mean galaxy density run found by stacking galaxy counts as a function of radius around a fair sample of clusters, and averaging across the collection.

The observed cross correlation function on scales of a few megaparsecs is usefully approximated as a power law,

$$\xi_{cg} = \left(r_{cg}/r \right)^2, \qquad (19.35)$$

with clustering length

$$r_{cg} = 15 \pm 3h^{-1} \, \text{Mpc}, \qquad (19.36)$$

for Abell clusters (in richness classes $R \geq 1$). The shape of $\xi_{cg}(r)$, from the cross correlation of the Lick galaxy catalog and the Abell cluster catalog (Seldner and Peebles 1977; Lilje and Efstathiou 1988), is reasonably well determined and is slightly steeper than the galaxy-galaxy correlation function. The amplitude depends on the still somewhat uncertain distribution of Lick galaxy distances; the number here is based on the contrast of the typical cluster luminosity relative to the mean luminosity density. It is not surprising that ξ_{cg} is larger than the galaxy correlation function ξ_{gg} (eq. [19.18]), for clusters are chosen because they are dense concentrations.

The dense regions preferred by clusters tend to be wall-like structures; examples are the Coma cluster on the Great Wall and the Virgo cluster on the local sheet of galaxies (figure 3.3). The lowest-order statistical measure of this effect is the cluster-galaxy-galaxy three-point cross correlation function, defined by the joint probability that a cluster is centered in the volume element dV_c and galaxies are centered in the elements dV_a and dV_b at distance r_{ab} from each other and distances r_a and r_b from dV_c:

$$dP = n^2 n_c [1 + \xi_{cg}(r_a) + \xi_{cg}(r_b) + \xi(r_{ab}) + \zeta_{cgg}(r_a, r_b, r_{ab})] \, dV_c dV_a dV_b. \quad (19.37)$$

The added terms in the brackets assure that the reduced function ζ_{cgg} vanishes when either or both of the galaxy volume elements are well removed from the cluster. The function ζ_{cgg} measures the variance of the number of galaxies around a cluster. As one might expect, the scatter is considerable, $\delta N/N \sim 0.6$ for Abell clusters (Fry and Peebles 1980). The tendency for clusters to lie on sheets ought to be seen as a peak in the value of ζ_{cgg} as a function of the angle θ subtended by the galaxies at the cluster, at $\theta = 180°$. The effect has not yet been detected in this statistic; it will be interesting to see what the newer catalogs yield.

The cluster-cluster correlation function $\xi_{cc}(r)$ is defined by equation (19.2), using the mean cluster number density n_c. We noted that ξ_{cg} is greater than the galaxy-galaxy correlation function ξ at the same radius, because clusters are identified as regions of unusually high densities, and the same argument might suggest that ξ_{cc} is larger still, as was found by Hauser and Peebles (1973). Bahcall and Soneira (1983) and Klypin and Kopylov (1983) showed that ξ_{cc} has about the same power law shape as the galaxy function ξ, and that the amplitude of ξ_{cc} increases with increasing cluster richness (here defined by the count of bright galaxies within the Abell radius $1.5h^{-1}$ Mpc). Since there are more poor clusters than rich ones, this means ξ_{cc} tends to increase with decreasing mean number density of the class of objects. Szalay and Schramm (1985) noted that this is in the direction one might expect if structure formed by a truly scale-invariant process, such that the clustering length r_i for a class of clusters scales as the characteristic length defined by the mean number density n_i in this class,

$$r_i \propto d_i \equiv n_i^{-1/3} . \tag{19.38}$$

The surveys (Bahcall 1988; Bahcall and West 1992; and Luo and Schramm 1992) indicate that this *ansatz* is remarkably successful. Figure 19.3 shows measures of the correlation functions for clusters of different mean number densities. The vertical axis is the clustering amplitude, defined as the two-point correlation function fitted to $\xi \propto r^{-1.8}$, and scaled to the fiducial separation $r = 1h^{-1}$ Mpc. The horizontal axis is the mean distance between objects in the sample. The solid line is the scaling relation

$$\xi_i(r) = \left(\frac{0.4}{n_i^{1/3}r} \right)^{1.8} . \tag{19.39}$$

Some scatter surely has been introduced by the considerable differences in procedures by which different classes of clusters have been identified, but the results do reasonably well fit this universal dimensionless correlation function. At the cluster space number density in equation (20.33), $n_{cl} \sim 1 \times 10^{-5}h^3$ Mpc^{-3}, the length at which ξ_i in equation (19.39) extrapolates to unity is the cluster-cluster clustering length $r_{cc} \sim 18h^{-1}$ Mpc.

Galaxies do not fit this scaling relation in a natural way, as is indicated by the open triangle in the lower left part of the figure. It is placed at the mean distance between L_* galaxies, but one could move the triangle for galaxies a good deal farther to the left, for as we saw in section 3 there are many faint galaxies for every L_* one, all with quite similar space distributions, that is, similar correlation functions $\xi(r)$ as a function of the physical separation r.

Equation (19.39) also indicates that the clustering amplitude increases with

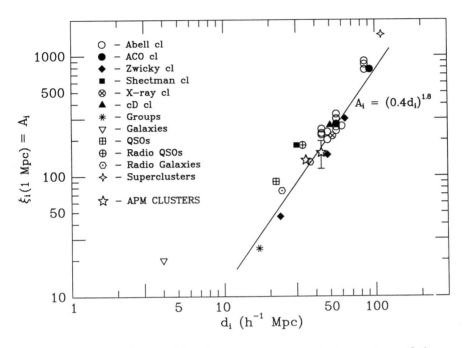

Figure 19.3. Scaling of the clustering amplitude with the density n_i of objects (Bahcall and West 1992). The vertical axis is the amplitude of the two-point correlation function scaled to $r = 1h^{-1}$ Mpc. The horizontal axis is the mean distance $n_i^{-1/3}$ between clusters. The straight line is the scaling relation in equation (19.39).

increasing richness of the system, and a version of this trend is seen in the galaxy distribution, in the sense that close pairs of galaxies cluster more strongly than do individual ones. To get a measure of this effect, let us define a characteristic number density n_2 of close pairs of galaxies. The probability δP_2 that two or more galaxies are found to be in a randomly placed sphere of radius $r \ll r_o$, volume $\delta V_2 = 4\pi r^3/3$, is the integral of equation (19.2) over the radius of the sphere,

$$\delta P_2 = \frac{1}{2} n^2 \int^r d^3 r_a \, d^3 r_b [1 + \xi(r_{ab})] \equiv n_2 \delta V_2. \qquad \textbf{(19.40)}$$

With the factor of one-half we are counting each pair once, and we will take r to be small enough for there to be a negligible chance that the sphere contains more than two galaxies. The volume of the sphere is δV_2, and the factor multiplying it is the characteristic number density of close pairs of galaxies.

Consider next the cross correlation between the positions of close pairs of galaxies and the positions of the full set of galaxies. The joint probability that the sphere δV_2 contains a galaxy pair and the volume element dV' at distance $R \gg r$ from the sphere contains a galaxy is

$$dP_{12} = \frac{1}{2} dV' n^3 \int^r d^3 r_a\, d^3 r_b [1 + \xi(r_{ab}) + \xi(r_a) + \xi(r_b) + \zeta]$$

$$\equiv n_2 n \delta V_2\, dV' [1 + \xi_{12}(R)].$$

(19.41)

The condition $R \gg r$ means the distances r_a and r_b to dV' from $d^3 r_a$ and $d^3 r_b$ are nearly equal to R. The dominant terms in the integral are the second, $\xi(r_{ab})$, and parts of the three-point function in equation (19.20), $\zeta = 2Q\xi(r_{ab})\xi(R)$. This gives

$$\xi_{12}(R) = 2Q\xi(R).$$

(19.42)

That is, the chance that a close pair of galaxies is found to have a neighboring galaxy is $2Q \sim 2$ times the probability for a randomly chosen galaxy.

The two-point correlation function for close pairs of galaxies is defined by the probability of finding a pair of galaxies in each of the two spheres of radius r at separation $R \gg r$,

$$\delta P_{22} = n^4 \int^r \frac{d^3 r_a\, d^3 r_b}{2} \int^r \frac{d^3 r_c\, d^3 r_d}{2} [1 + \cdots + \xi_{ab}\xi_{cd} + \cdots + \eta]$$

$$\equiv n_2{}^2 \delta V_2 \delta V_2' [1 + \xi_{22}(R)].$$

(19.43)

The dominant terms in the integral at small sphere radius r are $\xi_{ab}\xi_{cd}$, and the part of the four-point function (19.22) that contains only one factor of the two-point galaxy function at the sphere separation, $\eta = 4R_a\xi_{ab}\xi_{cd}\xi(R)$. In this limit, the two-point function for the clustering of pairs of galaxies is

$$\xi_{22}(R) = 4R_a\xi(R).$$

(19.44)

The coefficient in equation (19.23) is $R_a \sim 2.5$, indicating that tight pairs of galaxies have clustering amplitude roughly ten times that of individual galaxies. This is in the direction observed for richer collections of galaxies, where Bahcall (1988) finds that the clustering amplitude scales roughly as the number of galaxies in the system.

The trend of the clustering amplitude with richness could be a gravitational effect, for it would be reasonable to expect that a richer and more massive system is better able to attract neighbors. It could also be an effect of initial conditions, as in the following example given by Kaiser (1984a). Suppose structures grew by gravity out of small initial fluctuations in the mass distribution. Then rare rich clusters would be expected to form in regions where the initial mass fluctuations happened to be positive and unusually large. If the initial density fluctuations are a random Gaussian process, with a positive large-scale autocorrelation function, then the rare extreme peaks in the distribution are more strongly clustered than is the mass (as shown in eq. [25.13]). If gravity has not much rearranged the distribution of the clumps, we would expect to see that rarer, more massive ones appear in stronger concentrations, which certainly goes in the right direction: the richer and rarer the concentration, the larger the observed clustering amplitude. There is the problem that the same picture might lead one to expect that giant galaxies cluster more strongly than dwarfs, which, as we have seen in section 3, is not observed. One might imagine this is because gravity caused the development of the hierarchy of galaxy clustering after protogalaxies had formed as distinct entities dense enough to resist merging as the galaxy clustering pattern develops. That would empty the voids by drawing together dwarfs and giants alike, while the Kaiser effect might operate on larger scales. It remains to be seen whether this can agree with the remarkable velocity scaling relation that connects the clustering of galaxies and their internal structures, as is illustrated in figure 20.1 and discussed in the following sections.

20. Dynamical Mass Measures

Mass measures deal with the related questions, how much mass is there, and how is it distributed through the observable universe? The dynamical estimates discussed in this section use the gravitational effect of departures from a strictly homogeneous distribution on the motions of objects such as stars and galaxies considered as test particles. Estimates of the fluctuations in the mass distribution on larger scales are considered in the next section.

A goal in this section is to discover whether there is a consistent and believable case for some definite set of values for the parameters in the standard cosmological model, based on the dynamical measures and the cosmological tests surveyed in sections 13 and 14. In the past decade there has been considerable progress toward believable answers, but it may be well to bear in mind that people have been searching for ways to determine the cosmological parameters since the late 1920s. The history suggests progress will not always be as rapid as one might hope, nor always go in the expected directions.

We begin with a summary of what has been learned about the density parameter from the dynamical measures, and in the following subsections review some of the details.

Estimates of the Density Parameter

The first line in table 20.1 is the density parameter in baryons in the observationally successful model for the origin of the light elements at high redshift

Table 20.1
The Density Parameter

Baryons, based on BBNS	$\Omega_B = (0.013 \pm 0.005)\, h^{-2}$
Stars in galaxies	$\Omega_* = 0.004$
Intergalactic stars	$\Omega_{\text{luminous}} \lesssim 0.04$
Rich Clusters	$\Omega_{\text{clusters}} = 0.01$
Dynamics at $r \lesssim 10h^{-1}\,\text{Mpc}$	$\Omega_d \sim 0.05$ to 0.2
Dynamics at $r \gtrsim 30h^{-1}\,\text{Mpc}$	$\Omega_l \sim 0.05$ to 1

(section 6). The density parameter in this computation scales with Hubble's constant as $\Omega_B \propto h^{-2}$ (because the critical Einstein-de Sitter density is proportional to h^2, and since the present CBR temperature is known the theory predicts the present baryon number density). The other entries are based on dynamical estimates in which Hubble's constant scales out of the density parameter.

The entry for Ω_B assumes a homogeneous baryon distribution with no neutrino degeneracy. It is thought that if the primeval baryon distribution were inhomogeneous the value of $\Omega_B h^2$ might be increased by a factor of two or three, which could bring it into consistency with the relatively well-established number for the mass clustered with galaxies on scales less than about one megaparsec, $0.05 \lesssim \Omega_d \lesssim 0.2$. It is considered very difficult to see how the theory and observations of the light elements could be reconciled with a value for the baryon density parameter Ω_B near unity (Kurki-Suonio et al. 1990).

The second line in the table is the density parameter Ω_* in the material in the luminous central parts of galaxies (eq. [5.150]). It is plausible to assume this material is baryonic — stars and gas (section 18). The fact that Ω_* is well below the baryon density Ω_B suggests most of the baryons are outside the bright central parts of galaxies, perhaps in brown dwarfs in the massive halos of galaxies.

The upper bound on the integrated extragalactic background light allows a total luminosity density about an order of magnitude larger than the contribution from the bright parts of galaxies (eq. [5.168]). The entry in the table for intergalactic luminous matter assumes it has the mass-to-light ratio characteristic of the central parts of galaxies. If the density parameter is unity the dominant mass has to be dark, with mass-to-light ratio greater than about $100h$ in solar units.

Rich clusters of galaxies within the Abell radius $r_a = 1.5h^{-1}$ Mpc occupy about one part in 10^4 of space, and contribute one percent of the Einstein-de Sitter mass density. The mean cluster mass within the Abell radius is reliably measured, for as discussed below there are consistent results from the independent measures from galaxy velocity dispersions, the pressure of the X-ray emitting plasma gravitationally confined in clusters, and in a few cases gravitational lensing of the images of background objects. Thus the density parameter is reliably bounded at $\Omega > 0.01$.

The estimate of the mass density from the dark halos of galaxies is reasonably well fixed, because as discussed in this section we have independent and consistent results from the rms relative velocities of galaxies as a function of separation, the motions of galaxies in and near the Local Group, the motions of the galaxies gravitationally bound in rich clusters, and the rate of contraction of the Local Supercluster. The consistency of the measures from quite different kinds of systems, ranging from loose groups to rich clusters, and length scales ranging from $10h^{-1}$ kpc to $10h^{-1}$ Mpc, argues that the result is believable: the mass clustered with galaxies on comoving scales less than about $10h^{-1}$ Mpc is within a factor of two of $\Omega_d = 0.1$.

The entry in the last line refers to dynamical measures of the mass clustered with galaxies on scales $\sim 30h^{-1}$ Mpc, from the gravitational acceleration that would drive the large-scale velocity field. The current flow of preprints favors Ω close to unity, though still with some scatter (for example, Kaiser 1991; Shaya, Tully, and Pierce 1992). If the density of the matter clustered with galaxies on large scales is close to the Einstein-de Sitter value, then the dominant mass component has to have a coherence length broader than about $10h^{-1}$ Mpc, because it is not detected in the dynamical tests on smaller scales. Some ideas on how this could happen are explored in sections 25 and 26. And at the time this is written it seems prudent to bear in mind the option that what is securely detected on smaller scales, $\Omega \sim 0.1$, is all there is.

Galaxy Relative Velocity Dispersion

The relative velocity of a gravitationally bound pair of galaxies is a measure of the masses of the galaxies. It is not easy to estimate relative velocities in individual cases, however, because the cosmological redshift differences of accidental pairs can bias the velocity high, while overpruning of accidentals can bias it low. The following statistical approach uses the distribution of galaxies in redshift space (Geller and Peebles 1973; Peebles 1976, 1979; Davis and Peebles 1983).

Redshift space uses galaxy angular positions with redshifts as the radial coordinate, as in the maps in section 3. Galaxy peculiar motions distort the true spatial distribution: motions within bound systems tend to make structures appear elongated along the line of sight, while the systematic motion of collapse of newly forming structures makes them appear flattened along the line of sight (LSS § 76; Kaiser 1987). This distortion is a problem if one wishes to study spatial structures in the galaxy distribution, but it is a useful probe of the dynamics of the clustering.

A measure of the velocity distortion effect is the galaxy two-point correlation function in redshift space. It is defined by the probability of finding a pair of galaxies at separations r_π and r_σ along and perpendicular to the line of sight,

$$dP = n^2[1 + \xi(r_\sigma, r_\pi)]2\pi r_\sigma \, dr_\sigma \, dr_\pi \, . \tag{20.1}$$

The two-point correlation function $\xi(r_\sigma, r_\pi)$ is anisotropic, a function of separation parallel and perpendicular to the line of sight. It is a convolution of the true isotropic spatial function $\xi(r)$ with the distribution of relative line-of-sight velocities, which we will model as follows.

Consider the statistical ensemble of pairs of galaxies at projected separation r_σ, true spatial separation r, and true separation $y = (r^2 - r_\sigma{}^2)^{1/2}$ along the line of sight. The separation r_π along the line of sight in redshift space is a random variable determined by the distribution of relative peculiar velocities of the pairs of galaxies. We will write the mean value of the relative peculiar velocity as

$$\mathbf{v} = -H_o g(r)\mathbf{r} \, . \tag{20.2}$$

On small scales, where the clustering pattern is close to stable, $g(r)$ is close to unity because the mean peculiar velocity cancels the Hubble flow. At large r, where the galaxy distribution is expanding with the general expansion of the universe, $g(r) = 0$. The distribution around the mean will be modeled as isotropic, with the distribution in one dimension

$$dP = F(V) \, dV \, , \tag{20.3}$$

where $F(V)$ has zero mean value. Then at true separation y along the line of sight the redshift separation is

$$r_\pi = y[1 - g(r)] + H_o{}^{-1}V_\pi \, , \tag{20.4}$$

where

$$r = (r_\sigma^2 + y^2)^{1/2} \, , \tag{20.5}$$

and V_π is drawn from the distribution (20.3).

The probability of finding that a galaxy has a neighbor at distance r_σ, r_π in the range $r_\sigma + dr_\sigma$ and $r_\pi + dr_\pi$ in redshift space is

$$dP = 2\pi r_\sigma dr_\sigma \, dy \cdot n[1 + \xi(r)] \cdot F(V_\pi) \, dV_\pi \, \cdot \delta(r_\pi - y + g(r)y - H_o{}^{-1}V_\pi) dr_\pi \, . \tag{20.6}$$

The first factor is the true space volume element. That multiplied by the second factor is the ordinary two-point probability in equation (19.3). The next factor is the probability for the random part of the relative velocity. The delta function in the last factor selects cases with given radial separation r_π to $r_\pi + dr_\pi$ in redshift space (eq. [20.4]). Integration over the unobserved variables V_π and y eliminates the delta function. Comparing equations (20.1) and (20.6), we see that the two-point function in redshift space is

$$\xi(r_\sigma, r_\pi) = H_o \int_{-\infty}^{\infty} dy \xi(r) F(H_o(r_\pi - y + g(r)y)), \qquad (20.7)$$

where $r = (r_\sigma^2 + y^2)^{1/2}$.

Equation (20.7) is a convolution of the true space correlation function $\xi(r)$ with the distribution of relative peculiar velocities. The random part causes the small-scale galaxy distribution to appear elongated along the line of sight, the contours of $\xi(r_\sigma, r_\pi)$ elongated along the r_π direction. At larger separations the dominant distortion is the streaming motion $g(r)$, which causes the distribution in redshift space to appear flattened along the line of sight.

One estimates the rms galaxy relative peculiar velocity dispersion $\sigma = \langle V^2 \rangle^{1/2}$ by fitting the functions $F(V)$ and $H_o g(r)r$ to estimates of the two-point correlation function in redshift space. For the data in figure 20.1, at $r_\sigma \lesssim 1h^{-1}$ Mpc, it is reasonable to take $g(r) = 1$, because the clustering pattern looks stable (and the estimates are not sensitive to this *ansatz*, because the velocity dispersion is larger than a reasonable streaming component). A model for $F(V)$ that gives a reasonable-looking fit to the correlation function $\xi(r_\sigma, r_\pi)$ may be expected to give a reasonable measure of σ, except if the true $F(V)$ has a small tail extending to large V. That would be easy to miss in $\xi(r_\sigma, r_\pi)$, and could make a consider-able addition to the standard deviation σ. However, galaxy peculiar velocities are not observed to scatter by large amounts, so a large error from this effect seems unlikely.

The estimates of $\xi(r_\sigma, r_\pi)$ at small r_σ are consistent with an exponential model,

$$F(V) \propto \exp{-2^{1/2} |V| / \sigma(r)} \qquad (20.8)$$

(Davis and Peebles 1983). At larger separations a more rounded distribution is indicated (Bean et al. 1983). The numerical results for the galaxy relative velocity dispersion shown in figure 20.1 are from the summary in Peebles (1984a). The velocity dispersion estimates from the Southern Sky Redshift Survey are in close agreement with figure 20.1 (da Costa et al. 1991; Davis 1988).

Now let us consider the relation between the rms relative velocity, $\sigma(r)$, and the mass distribution. We can think of the observed motion of the bright central part of a galaxy as the motion of a test particle, because the mass in the bright part is a small fraction of the total concentrated around the galaxy. We will assume that the sample of galaxy pairs at separation r is in statistical equilibrium, so the mean relative acceleration of the pairs is $g \sim \sigma^2/r$, where σ is the rms relative velocity. This need not be true of the galaxy motions at the time of formation, but it seems

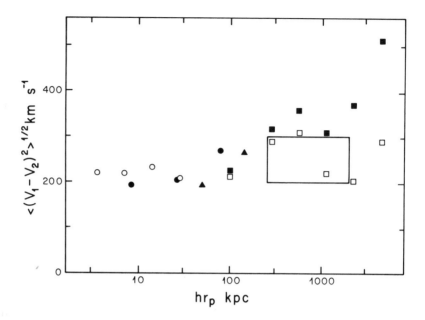

Figure 20.1. Relative velocity dispersion of galaxy pairs as a function of projected separation r_σ.

safe to assume that if the crossing time is less than the Hubble time the galaxies will have forgotten their initial relative velocities. Carlberg (1991) notes that when dynamical drag is important, σ^2/r is significantly smaller than the mean relative gravitational acceleration $\sim G\mathcal{M}/r^2$, where $\mathcal{M}$ is the mass concentrated around the pair within distance $\sim r$, because the drag would lower the velocity. Carlberg observes this dynamical biasing effect in numerical N-body computations. If the effect were important in groups of galaxies, it would mean the groups are coalescing in a timescale comparable to the crossing time r/σ. This seems unlikely, because it would require that there be a substantial population of remnants of already merged groups. (A galaxy has on average one neighboring L_* galaxy within distance $0.3h^{-1}$ Mpc, which would have to be replaced in a crossing time $\sim 0.1H_o^{-1}$ if dynamical drag dominated the relative motion.) It seems reasonable therefore to assume that the mean relative gravitational acceleration of the galaxy pairs can be estimated from the rms relative velocity.

A detailed computation of the relative gravitational acceleration uses the galaxy-galaxy-mass cross correlation function, defined as in equation (19.37), because that determines the mean distribution of mass averaged across a fair sample of pairs of galaxies at a given separation r. The theory is worked out in LSS, § 74. Here we give a streamlined approximation that is reasonably close to the right answer.

At galaxy separation $r \lesssim 1h^{-1}$ Mpc the crossing time $r/\sigma(r)$ is smaller than the Hubble time, and we are assuming the distribution of galaxy-pair separations is in statistical equilibrium. The velocity dispersion acts as a pressure, $p = \rho\sigma^2$, assuming the relative velocities are roughly isotropic. Since figure 20.1 indicates that σ is nearly independent of separation, the pressure scales as $p \sim r^{-2}$. On small scales, where we are assuming the clustering pattern is in statistical equilibrium, the pressure gradient force per unit mass is balanced by the mean relative gravitational acceleration g of galaxies at separation r,

$$g = -\frac{1}{\rho}\frac{dp}{dr} = \frac{2\sigma^2}{r} \sim \frac{G\rho_b}{r^2} \cdot 2\int_0^r \xi_{g\rho}(r)d^3r = \frac{8\pi}{3-\gamma}G\rho_b r_o{}^\gamma r^{1-\gamma}. \quad (20.9)$$

Here ρ_b is the mean mass density, and $\xi_{g\rho}(r)$ is the galaxy-mass cross correlation function defined by the mean value of the mass density at distance r from a galaxy,

$$\langle \rho \rangle = [1 + \xi_{g\rho}(r)]\rho_b. \quad (20.10)$$

There is a factor of two in front of the integral because the mass around each galaxy is pulling on the other. The last step assumes galaxies trace mass, that is, $\xi_{g\rho}$ is usefully approximated by the galaxy correlation function given by the power law model in equation (19.18). Equation (20.9) for ρ_b divided by the critical Einstein-de Sitter density is

$$\Omega_d \sim \frac{2(3-\gamma)}{3}\frac{\sigma(r)^2}{H_o{}^2 r_o{}^\gamma r^{2-\gamma}}. \quad (20.11)$$

A more careful calculation uses the galaxy-galaxy-mass three-point correlation function (LSS, § 74). If this statistic can be approximated by the galaxy three-point correlation function (19.20), with $Q \sim 1$, the "cosmic virial theorem" relating the relative velocity dispersion and the mean mass density is usefully approximated by equation (20.11).

In this model, the velocity dispersion scales with separation as

$$\sigma \propto r^{1-\gamma/2} = r^{0.12}. \quad (20.12)$$

This agrees with figure 20.1, which shows a slow increasing trend of σ with increasing separation r. It also agrees with the velocity scaling relation in the bounded clustering hierarchy model in equation (19.28).

A reasonable approximation to the estimates of the galaxy relative velocity dispersion is

$$\sigma(hr = 1\,\text{Mpc}) = 300\,\text{km s}^{-1} \quad (20.13)$$

(Peebles 1984a; Hale-Sutton et al. 1989). With the parameters in equation (19.18), equation (20.11) gives $\Omega_d \sim 0.2$. The more detailed analysis (Davis and Peebles 1983), with the dispersion in equation (20.13), yields

$$\Omega_d = 0.15e^{\pm 0.4}, \qquad \mathcal{M}_d/L \sim 300h\mathcal{M}_\odot/L_\odot. \qquad (20.14)$$

The mass-to-light ratio uses the mean luminosity density in equations (5.143) and (5.151).

Another approach to the measurement of the mass concentrated in systems with sizes on the order of one megaparsec identifies groups of galaxies close enough in angular position and redshift that it is reasonable to identify them as gravitationally bound systems. The typical velocity dispersions within groups are quite similar to equation (20.13), and the masses derived from the condition that the groups are gravitationally bound yield similar values for the mass-to-light ratio (Tully 1987a). This number is an order of magnitude larger than what is found in the central parts of galaxies (eq. [18.40]), consistent with the idea that galaxy masses are dominated by the dark halos, but it is an order of magnitude below what is needed to close the universe.

The results in figure 20.1 for the galaxy relative velocity dispersion as a function of separation have interesting consequences for two issues: the extent of the dark halos of galaxies, and the use of galaxies as mass tracers.

At the small separation limit in figure 20.1, $r_\sigma \sim 10h^{-1}$ kpc, the galaxy pairs are nearly touching, and their relative velocities therefore can be compared to the rotation velocities v_c of gas in the disks of spiral galaxies. For isotropic orbits in a $\rho \propto r^{-2}$ halo the line-of-sight dispersion is $v_c/2^{1/2}$ (eq. [3.33]), and for relative galaxy motions that should be multiplied by $2^{1/2}$ to take account of the two masses, giving $\sigma \sim 220$ km s^{-1} (eq. [3.34]). This agrees with the measurements of galaxy relative motions in figure 20.1. The nearly flat rotation velocity curve in a typical L_* galaxy, as in figure 3.12, means the mass density scales with radius as $\rho \sim r^{-2}$. The slow variation of $\sigma(r)$ with separation r in figure 20.1 means that at larger radii the average value of the mass density as a function of distance r from an L_* galaxy continues to follow this power law, in a smooth continuation of the massive halo revealed by the rotation curves at smaller radii. A picture of the mass distribution would show that at distances $r \lesssim 10h^{-1}$ kpc from the center of a typical L_* galaxy the density is dropping about as $\rho(r) \sim r^{-2}$, in a distribution that has to be smooth because the rotation curves usually are quite symmetric. At larger radii, the mean density run continues to scale with the same power law, but the distribution breaks up into the concentrations in the halos around neighboring galaxies. This is a statistical continuity between the monolithic mass structures of individual galaxies and the fractal distribution on larger scales.

The other issue is the relation between the space distributions of galaxies and mass. In a large spiral galaxy the starlight does not trace the mass: under Newtonian mechanics the latter is in the outskirts in a dark halo. If an analogous effect applied at a larger scale, the mass per galaxy derived from motions within groups and clusters could underestimate the net mass per galaxy, because most of the mass would be in more broadly spread superhalos, perhaps explaining why the dynamical estimates from figure 20.1 point to a mass density that is about 10% of the critical Einstein-de Sitter value. This biasing picture became popular in the mid-1980s, as people came to appreciate the power of the arguments for an Einstein-de Sitter universe. As it happened, at about the same time the improving observational situation was offering evidence that galaxies do give useful measures of the clustered mass distribution. The example in figure 20.1 is the scaling of the velocity dispersion with separation, which follows the trend expected if galaxies trace mass (eqs. [19.28], [20.12], and [20.13]). Some other examples are reviewed in the following subsections.

The conclusion is that under Newtonian mechanics there is a dark mass component, with a distribution on scales 10 kpc $\lesssim hr \lesssim$ 1 Mpc that follows that of the galaxies, and an amount that is about 10% of the Einstein-de Sitter density. Still under discussion is the possibility that there is another more smoothly distributed component of dark matter, perhaps with ten times the mass.

The Local Group

The Local Group offers a special opportunity to study galaxy masses, because the system is close enough to be measured and modeled in some detail. The dynamical models indicate that the mass-to-light ratio in the Local Group is similar to the above result from the galaxy relative velocity dispersion.

As discussed in section 3, the dominant members of the group are the Milky Way and the Andromeda Nebula, M31. Their separation is

$$l_o = 770 \pm 30 \, \text{kpc} , \tag{20.15}$$

and the rate of change of the separation (measured relative to the centers of the galaxies) is

$$\frac{dl_o}{dt} = -123 \pm 20 \, \text{km s}^{-1} . \tag{20.16}$$

The relative peculiar velocity of approach is the sum of dl_o/dt and the cosmological redshift. It is toward the low side of the rms value shown in figure 20.1, which seems reasonable because the Local Group is not very dense or rich.

Kahn and Woltjer (1959) pointed out that the likely explanation for the motion of approach of M31 is the mutual gravitational attraction of the masses of the two galaxies. The alternative would be that the Milky Way and M31 are accidentally passing by, but if the galaxies had moved at constant speed for a Hubble time, their separation would have changed only by $\sim \dot{l}_o/H_o \sim 1$ Mpc, less than the distance to the next large galaxy, so this alternative does not seem promising. In the Kahn-Woltjer dynamical picture we can use the relative motion to estimate the galaxy masses. The simplest model within the Friedmann-Lemaître cosmology goes as follows (PC, § IV).

In the standard cosmological model, the mass distribution at high redshift had to have been very close to homogeneous, for otherwise gravitational instability would have made the present distribution even more clumpy than it is. In the gravitational instability picture, gravity gathered mass into protogalaxies, by this instability, and these mass concentrations pulled on each other to generate the galaxy peculiar velocities. The simplest model of the Local Group is an isolated two-body system, the Milky Way and M31. In this model, the relative angular momentum is negligible because the system is isolated, and at high redshift the two bodies are close together and the rate of separation is that of the background cosmological model. The equation of motion for the separation $l(t)$ of the centers of mass is

$$\frac{d^2 l}{dt^2} = -\frac{G\mathcal{M}}{l^2} , \tag{20.17}$$

where $\mathcal{M}$ is the sum of the masses of the two galaxies. This is a reasonable approximation at high redshifts as well as low, because the equation has the same form as the cosmological acceleration equation (5.14) with $\rho \propto 1/a(t)^3$. The solution is (eq. [13.10])

$$l = A(1 - \cos \eta) , \qquad t = B(\eta - \sin \eta) , \tag{20.18}$$

where the constants A and B satisfy

$$A^3 = G\mathcal{M}B^2 . \tag{20.19}$$

One constant of integration has been used to make l approach zero at $t \to 0$, consistent with the picture that the material at high redshift is expanding with the general expansion. The other is fixed by the dimensionless constant

$$\alpha = \frac{t_o}{l_o} \frac{dl_o}{dt} = \frac{\sin \eta_o (\eta_o - \sin \eta_o)}{(1 - \cos \eta_o)^2} , \tag{20.20}$$

where η_o is the present value of the parameter in equation (20.18).

A reasonable value for the present time is $t_o = 15$ Gy. Then the above numbers for l_o and dl_o/dt give $\alpha = -2.4$; we have from equation (20.20) $\eta_o = 4.3$; and from equations (20.18) and (20.19) the mass of the Local Group is

$$\mathcal{M} = \frac{l_o^3}{Gt_o^2} \frac{(\eta_o - \sin \eta_o)^2}{(1 - \cos \eta_o)^3} = 4 \times 10^{12} \mathcal{M}_\odot . \tag{20.21}$$

The sum of the luminosities of the two galaxies gives

$$L = 5 \times 10^{10} L_\odot, \qquad \mathcal{M}/L = 100 \mathcal{M}_\odot/L_\odot. \qquad (20.22)$$

A more detailed dynamical model takes account of the effect of the mass concentrations in neighboring groups and uses the observed motions of the dwarf members of the Local Group as additional tests of the model. The following calculation assumes the picture of hierarchical gravitational growth of clustering, in which small-amplitude fluctuations in the initial distribution cause mass to collect into protogalaxies, and fluctuations on larger scales then cause the protogalaxies to move together to form systems such as the Local Group. The process can be followed in detail in numerical N-body computations of the evolution of the mass distribution (Peebles et al. 1989), but that is labor-intensive, so we will use the numerical results as a guide to a model that is simple enough for easy computation.

We will assume the mass within the Local Group is now concentrated around the galaxies, and the mass now belonging to a galaxy was in a contiguous patch in the initial nearly homogeneous mass distribution. Then we get a useful approximation to the dynamics of the system at early and late times by assigning to the mass belonging to galaxy i a central position $\mathbf{x}_i(t)$. The gravitational effect of this mass on the motions of the central positions $\mathbf{x}_j(t)$ representing neighboring patches of material can be taken to be the same as if each mass were concentrated at the central position. That is, we are replacing the initial nearly homogeneous mass distribution with a set of particles of masses m_i with positions $\mathbf{x}_i(t)$ at high redshifts that are such that the masses are moving under their mutual gravitational interaction with the general expansion of the universe. The small initial departures from homogeneous expansion have to be such that the central positions $\mathbf{x}_i(t)$ end up now where the galaxies are. A solution gives a sensible approximation to the initial conditions, when the system is behaving like a homogeneous section of the expanding universe, and to recent epochs, when the mass is concentrated around the individual galaxies.

This model replaces the more usual initial conditions — the six components of the position and velocity of each galaxy — with the three components of the present position and the three components of the condition that the peculiar velocity vanishes at high redshift. The model predicts the present galaxy peculiar velocities, which can be compared to what is observed.

Within this model the numerical computation of the orbits $\mathbf{x}_i(t)$ is easy; the difficult part is to find the initial conditions at high redshift. The trick uses the action principle of mechanics (Peebles 1990). To see how this works, neglect for the moment the expansion of the universe, so the action for a set of particles interacting by the potential V is

$$S = \int_0^{t_o} dt \left[\sum \frac{m_i}{2} \left(\frac{d\mathbf{x}_i}{dt} \right)^2 - V \right].$$ (20.23)

Under the infinitesimal change of orbits $\mathbf{x}_i \to \mathbf{x}_i + \delta\mathbf{x}_i(t)$, the action changes by the amount

$$\delta S = \int_o^{t_o} dt\, \delta\mathbf{x}_i \cdot \left[-m_i \frac{d^2\mathbf{x}_i}{dt^2} - \frac{\partial V}{\partial \mathbf{x}_i} \right] + \left[m_i \delta\mathbf{x}_i \cdot \frac{d\mathbf{x}_i}{dt} \right]_o^{t_o}.$$ (20.24)

One usually fixes the orbit at the end points, so $\delta\mathbf{x}_i = 0$ at $t = 0$ and $t = t_o$. Then the last term resulting from the integration by parts vanishes, and $\delta S = 0$ for orbits that satisfy the equation of motion. This is the action principle. But note that the action principle also applies if the initial position is free at $t = 0$, so $\delta\mathbf{x}_i \neq 0$, and the initial velocity vanishes, $d\mathbf{x}_i/dt = 0$ at $t = 0$. In the cosmological problem, this is equivalent to the condition that structures grow by gravity out of small initial departures from homogeneous expansion.

The action for orbits expressed in comoving coordinates in an expanding universe with mean mass density $\rho(t)$ is

$$S = \int_0^{t_o} dt \left[\sum \frac{m_i a^2}{2} \left(\frac{d\mathbf{x}_i}{dt} \right)^2 + \frac{G}{a} \sum_{i \neq j} \frac{m_i m_j}{|\mathbf{x}_i - \mathbf{x}_j|} + \frac{2}{3} \pi G \rho a^2 \sum m_i \mathbf{x}_i^2 \right].$$ (20.25)

The change in the action under an infinitesimal change $\delta\mathbf{x}_i(t)$ of an orbit is

$$\delta S = \int_0^{t_o} dt\, \delta\mathbf{x}_i \cdot \left[-\frac{d}{dt} m_i a^2 \frac{d\mathbf{x}_i}{dt} + m_i a \mathbf{g}_i \right] + \left[m_i a^2 \delta\mathbf{x}_i \cdot \frac{d\mathbf{x}_i}{dt} \right]_0^{t_o},$$ (20.26)

where

$$\mathbf{g}_i = \frac{G}{a^2} \sum_j m_j \frac{\mathbf{x}_j - \mathbf{x}_i}{|\mathbf{x}_j - \mathbf{x}_i|^3} + \frac{4}{3} \pi G \rho_b a\, \mathbf{x}_i.$$ (20.27)

This is the peculiar gravitational acceleration measured by a comoving observer at rest relative to distant matter. The second term in the expression for $\mathbf{g}_i$ cancels the first in the limit of a homogeneous mass distribution.

As in equation (20.24), the term in equation (20.26) that comes from the integration by parts is eliminated by the boundary conditions

$$\delta\mathbf{x}_i = 0 \quad \text{at } t = t_o, \qquad a^2 d\mathbf{x}_i/dt \to 0 \quad \text{at } a \to 0.$$ (20.28)

The first condition means the present positions are fixed at the known positions of the galaxies. In linear perturbation theory the peculiar velocity in the growing density mode in equation (5.105) scales with time as

$$v = a \, dx_i/dt \propto a^{1/2} \quad \text{at} \quad a \to 0, \tag{20.29}$$

consistent the second condition. The decaying mode, with $v \propto a(t)^{-2}$, is not allowed or wanted.

In a numerical application of the action principle, the orbits are modeled as functions of time with free parameters, $x_i(t, \alpha)$. Then the action is numerically evaluated as a function of the parameters α and the computer is asked to seek a stationary point of S, either by walking down the gradient $\partial S/\partial \alpha$ or by inverting the matrix of second derivatives of S. When the trial orbits have been adjusted to make the action stationary under the boundary conditions (20.29), the equation of motion is satisfied in the sense of a time average with weight function $\delta x_i = \partial x_i/\partial \alpha$ in the integral in equation (20.26).

The velocities in the trial orbits at $t \to 0$ have to satisfy equation (20.28). That means the free initial positions have to adjust themselves to eliminate the initial peculiar gravitational acceleration. That is, the significance of this somewhat awkward numerical solution of the N-body problem is that it yields orbits that satisfy the cosmological boundary condition.

Two solutions are shown in figure 20.2.[6] The galaxies in this model are the Milky Way, M31, six outlying members of the Local Group, and the nearby Maffei and Sculptor groups, each modeled as a point particle with the mass of the Local Group. The mass ratios of Local Group members are fixed ahead of time, based on the luminosity ratios, and the mass scale is adjusted to give the observed velocity of approach of the Milky Way and M31.

The action solution is not unique: there can be many parameter choices that make the action an extremum. Trials suggest only NGC 6822 has solutions with seriously different orbits, in one of which it is falling toward us (in expanding coordinates) for the first time, while in the other it has passed and at the moment is nearly at rest. In the former solution the velocity of NGC 6822 is much too large, while the latter is not far from the observations. The orbit shown in panel (a) of figure 20.2 is the observationally acceptable passing solution.

Both solutions in figure 20.2 assume the universe is cosmologically flat. In panel (a), the density parameter is $\Omega = 0.1$ (with the rest in a cosmological constant), and the Hubble parameter is $h = 0.75$, making the age of the universe $t_0 = 17$ Gy (eq. 13.20). In panel (b), $\Omega = 1$, $h = 0.5$, and the age t_o is nearly the same as in the low density model. The circle is the volume of the sphere that contains the mass now in the galaxies in the three groups. In these solutions the mass

---

[6] This solution assumes the distance to M31 is 725 kpc. The larger distance in equation (20.15) would give slightly larger masses, but the correction is small compared to the main uncertainty, in the effect of the external masses.

Figure 20.2. Orbits for nearby galaxies. In the solution in panel (a) the density parameter is $\Omega = 0.1$, and in panel (b) $\Omega = 1$. The orbits in expanding coordinates commence at high redshift at the names of the galaxies, and end at present positions at the boxes.

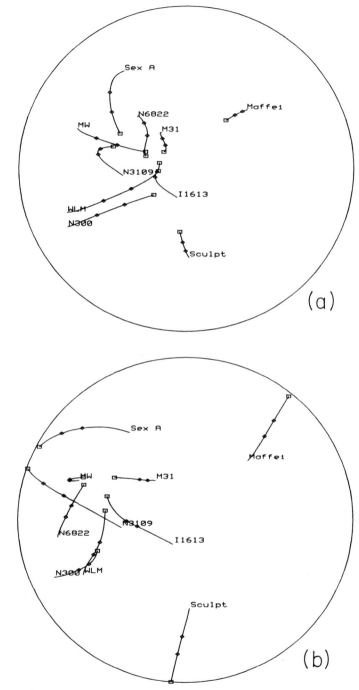

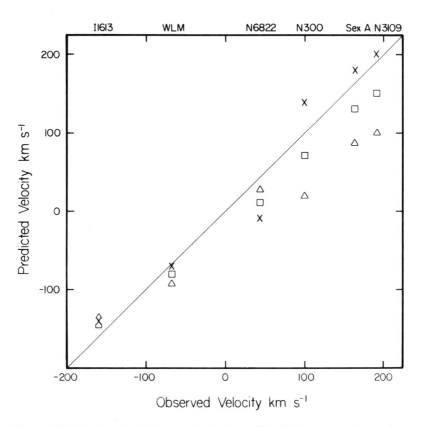

Figure 20.3. Predicted and observed velocities of Local Group members relative to the center of the Milky Way, for $\Omega = 0.1$ and Hubble constant $H = 50$
(triangles), 75 (squares), and 100 km s^{-1} Mpc^{-1} (crosses).

of the Local Group and the sphere diameter that contains the mass of the three
groups are

$$\mathcal{M}_{LG} = 6.4 \times 10^{12} \mathcal{M}_{\odot}, \quad H_o D = 1000\,\text{km s}^{-1}, \quad \text{for} \ \ \Omega = 0.1 \ , \tag{20.30}$$
$$\mathcal{M}_{LG} = 5.7 \times 10^{12} \mathcal{M}_{\odot}, \quad H_o D = 390\,\text{km s}^{-1}, \quad \text{for} \ \ \Omega = 1 \ .$$

The mass-to-light ratio of the Local Group in this model is about the same as in
equation (20.22).

Figure 20.3 shows observed and predicted velocities of the dwarf members
of the Local Group, with the passing orbit solution for NGC 6822, for the low-
density cosmological model and three choices of Hubble's constant. The consistency of theory and observations for the velocities is encouraging.

The model masses and velocities within the Local Group are insensitive to the density parameter, because for a given age t_o the group acts almost like a closed system, cosmology entering mainly through the perturbations from the neighboring groups. The cosmology does have a marked effect on the relative motions of the three groups, for the solution fixes the present mean mass density within the neighborhood. If $\Omega = 0.1$, this local density is higher than the background, and the solution finds that the three groups are moving together, as one sees in panel (a) of figure 20.2. This is reflected in the relatively large initial comoving diameter occupied by the matter, shown in the first line of equation (20.30). If $\Omega = 1$ the background density is so high that the three groups are in a local density minimum, and the solution accordingly finds that the groups are moving away from this local hole in the mass distribution. This seems unreasonable, for we see from figure 3.3 that we are in a very distinct local concentration of galaxies.

The velocity test in figure 20.3 indicates that we have a reasonable-looking dynamical picture for the formation of the Local Group. The mass in equation (20.30) is a little greater than what is obtained from the isolated two-body solution in equation (20.21), mainly because the relative motion of the Milky Way and M31 is slowed by the angular momentum they get from the two other groups.

The conclusion from this picture is that the mean value of the masses of the Milky Way and M31 is $\mathcal{M} = 3 \times 10^{12} \mathcal{M}_\odot$. If the rotation curve in each galaxy were flat out to a sharp cutoff in the density run at radius r_h, the cutoff radius would be

$$r_h = G\mathcal{M}/v_c^2 = 200\,\text{kpc} . \tag{20.31}$$

We noted in section 18 that if r_h were about $2h^{-1}$ Mpc, the halo masses would close the universe (eq. [18.2]). The Local Group galaxies give about 10% of this value, consistent with the galaxy relative velocity dispersions that indicate the mass clustered with galaxies is about 10% of the critical Einstein-de Sitter value (eq. [20.14]).

Rich Clusters

Rich clusters of galaxies are of particular interest for the purpose of this section because there are believable and consistent measures of the central masses from the galaxy motions and distributions, the intracluster plasma, and gravitational lensing. The net mass of a cluster is not known, and perhaps not even well defined, for the mass concentration around a typical cluster may be present halfway to the next one. What can be done is to measure the mass within one or two megaparsecs of the cluster center. The results indicate that the dark mass per bright galaxy is about the same as for the less dense systems discussed above.

Abell (1958) cataloged the richest of the clusters of galaxies found in the first examination of the Palomar Observatory Sky Survey plates. His statistical sample of clusters, rich enough for him to feel that the identification is reasonably complete (to what turned out to be a depth of about $600h^{-1}$ Mpc), contains some 1700 clusters. Abell's cluster richness criterion is based on the galaxy count within the Abell radius,

$$r_a = 1.5h^{-1} \text{ Mpc} . \tag{20.32}$$

Bahcall (1988) estimates the space number density in Abell's statistical sample of clusters in richness classes $R \geq 1$ is

$$n_{cl} \simeq 1 \times 10^{-5} h^3 \text{ Mpc}^{-3} . \tag{20.33}$$

Many of these clusters were later found to contain plasma dense and hot enough to be detectable by its thermal X-ray bremsstrahlung emission. The X-ray detections have shown that Abell was remarkably good at identifying significant concentrations of mass, for most of the nearer Abell clusters are detectable as X-ray sources, and most galaxy concentrations that are dense, hot, and near enough for detection of the X-ray thermal emission from the intracluster plasma are in Abell's catalog. This is particularly humbling to any theorist who has tried to spot Abell's clusters on the sky survey plates.

The rms value of the line-of-sight galaxy velocity dispersion estimates (based on ten or more galaxy redshifts) in the $R \geq 1$ clusters in the compilation of Struble and Rood (1991) is

$$\bar{\sigma} = \langle \sigma^2 \rangle^{1/2} = 870 \text{ km s}^{-1} . \tag{20.34}$$

One must be cautious in using this quantity, because the measurements of cluster velocity dispersions can be biased high by the cosmological redshift differences of clusters seen close in projection and counted as single systems, and biased low if parts of a single gravitationally bound system are counted as independent. Also, the conversion of a velocity dispersion to a mass depends on a model for the galaxy orbits. In the extreme, the mass can be negligibly small for given σ if the galaxies move at nearly constant velocities on radial orbits, producing a density run $n(r) \sim r^{-2}$, not far from what is observed (eq. [22.40]). Fortunately, we have two checks of consistency of the measurement of $\bar{\sigma}$ and the simplest sensible model, isotropic orbits, from the gravitational binding of the intracluster plasma and gravitational lensing. Let us consider some implications of the simplest model and then review the tests.

A typical value for the cluster core radius at which the mean density drops to half the central value is

$$r_c = 0.2h^{-1} \text{ Mpc} \tag{20.35}$$

(Bahcall 1975; Dressler 1978; Jones and Forman 1984). If the velocities are isotropic the central mass density is given by equation (18.36), and we have from equations (20.34) and (20.35) for σ and r_c the mean value

$$\rho_c = 2 \times 10^{-25} h^2 \text{ g cm}^{-3} \sim 0.003 h^2 \mathcal{M}_\odot \text{ pc}^{-3}$$
$$\sim 1 \times 10^4 \rho_{\text{crit}}.$$

(20.36)

This is some five orders of magnitude below the core density in a giant elliptical galaxy (eq. [18.39]), and four orders of magnitude above the critical Einstein-de Sitter density.

Estimates of the central mass-to-light ratios in clusters are

$$\mathcal{M}/L = 300 \pm 100 \, h\mathcal{M}_\odot/L_\odot.$$

(20.37)

Since the galaxies in the central parts of clusters are gas-poor, one might expect they are less luminous for a given baryon mass than the field spirals that still are producing massive luminous stars. (The exception may be in cooling flows, where the intracluster plasma is dense enough to radiatively cool in a Hubble time, meaning the plasma needs an energy source or else it has to be settling to something dissipationless, such as stars; Fabian, Nulsen, and Canizares 1991.) Consistent with this, the cluster central mass-to-light ratio is similar to but somewhat higher than the estimates found above for field galaxies. Thus the evidence is that here too the mass is not in stars with a standard mass function, and that the mean value of the mass per galaxy is about 10% of the global value in the Einstein-de Sitter cosmology.

Clusters contain plasma, with space distribution similar to that of the galaxies, and typical core density derived from the X-ray luminosity

$$n_c = 3 \times 10^{-3} h^{1/2} \text{ protons cm}^{-3}, \qquad m_p n_c/\rho_c = 0.03 h^{-3/2}$$

(20.38)

(Jones and Forman 1984; eqs. [24.50] to [24.55] below). For acceptable values of Hubble's constant the plasma density is no more than about 10% of the total core density in equation (20.36), and similar to the contribution from the seen stars. That is, under Newtonian mechanics the bulk of the mass in a cluster is dark matter whose nature is yet to be identified.

The galaxy line-of-sight velocity dispersion within a cluster typically decreases with increasing projected distance from the cluster center (Regös and Geller 1989); a reasonable rms value at the Abell radius for Abell clusters is $\sigma = 700 \text{ km s}^{-1}$. In the limiting isothermal model in equation (3.32), this translates to mean cluster mass

$$\mathcal{M}_{\text{cl}}(<r_a) = 2\sigma^2 r_a/G = 3 \times 10^{14} h^{-1} \mathcal{M}_\odot.$$

(20.39)

With equation (20.33) for the number density n_{cl} of rich clusters, the density parameter contributed by the mass within the Abell radii of rich clusters is

$$\Omega(r < r_a) = n_{cl}\mathcal{M}_{cl}(< r_a)/\rho_{crit} = 0.01 . \tag{20.40}$$

Thus we can conclude that the total mass density parameter certainly cannot be less than about one percent.

We have two checks on these numbers, one from the intracluster plasma and the other from gravitational lensing. The former is detected by its X-ray luminosity (Gursky et al. 1971); the identification as thermal bremsstrahlung is confirmed by the observation of emission lines from highly ionized iron, with abundance on the order of the present cosmic value (Mitchell et al. 1976).

Cavaliere and Fusco-Femiano (1976) introduced the standard way to analyze the distribution of the intracluster plasma in a way that is simple but still allows for some of the complexity of a real cluster. The model assumes the plasma has a uniform temperature T, the galaxies have an isotropic Gaussian velocity distribution with line-of-sight dispersion σ, and both components are in dynamical equilibrium in the gravitational potential $\phi(\mathbf{r})$. Then the Boltzmann distributions for gas and galaxies are

$$\rho_{gas} \propto e^{-\overline{m}\phi/kT} , \qquad \rho_{galaxies} \propto e^{-\phi/\sigma^2} , \tag{20.41}$$

where $\overline{m}$ is the mean mass per plasma particle (slightly larger than half the mass of a proton, because of the heavier elements). The result of eliminating the potential is

$$\rho_{gas} \propto \rho_{galaxies}^{\beta} , \tag{20.42}$$

with

$$\beta = \frac{\overline{m}\sigma^2}{kT} . \tag{20.43}$$

If $\beta = 1$ the galaxies and plasma particles have the same mean velocity dispersion and the same space distribution. This would be expected if the galaxies and plasma both fell into the cluster potential from similar distances with little energy dissipation, or if the plasma were shed by the cluster members at low relative velocity.

Estimates of β from the relative space distributions of galaxies and X-ray emitting plasma give

$$\langle \beta \rangle = 0.7 \tag{20.44}$$

(Jones and Forman 1984), meaning the space distributions are reasonably similar. There is a good correlation between the cluster plasma temperature derived from the X-ray spectrum and the galaxy velocity dispersion; Edge (1989) finds that the mean of the ratio is

$$\langle \beta \rangle = \langle \overline{m}\sigma^2/kT \rangle = 0.9 \,. \tag{20.45}$$

The correlation of the plasma temperature with the galaxy velocity dispersion and the similarity of the space distributions of plasma and galaxies both indicate that we have a believable picture for the mass distribution within a typical rich cluster of galaxies.

Gravitational lensing offers a remarkable check of the estimates of the mass distribution in clusters (Soucail et al. 1988; Tyson, Valdes, and Wenk 1990). Here we work through the elements of the test of the cluster mass based on the the giant luminous arcs of Lynds and Petrosian (1987).

Paczyński (1987) suggested a luminous arc is the image of a background object that happens to be near the line of sight through the cluster center and is lensed by the cluster mass. The test is that the arc ought to have the spectrum of an extragalactic object at a redshift well above that of the cluster. Soucail et al. (1988) showed that the luminous arc in the Abell cluster A370 does have the emission and stellar absorption lines characteristic of a spiral galaxy. The redshifts of the cluster and the background galaxy are

$$z_l = 0.375 \,, \qquad z_s = 0.724 \,, \tag{20.46}$$

and the radius of the arc of the galaxy image caused by lensing in the cluster is

$$\theta = 25 \text{ arc sec} \,. \tag{20.47}$$

To estimate the cluster mass in this picture, let us model the mass distribution as a limiting isothermal gas sphere, and seek the cluster velocity dispersion σ that gives the observed arc radius. In this model the gravitational bending angle is $\alpha = 4\pi(\sigma/c)^2$ (eq. [10.97]). As illustrated in figure 13.11, a source on the line of sight through the center of an axially symmetric mass distribution is seen as an Einstein ring of angle θ, with $\theta r_{os} = \alpha r_{ls}$, where the angular size distance (eq. [13.29]) from the observer to the lensed source galaxy is r_{os} and the angular size distance from the lensing cluster to the source is r_{ls}. The observed images are arcs rather than circles; this is interpreted to be the perturbing effects of an eccentric cluster mass distribution and of the mass concentrations in the individual galaxies.

In our simplified model the line-of-sight velocity dispersion in the cluster is

$$\frac{\sigma}{c} = \left(\frac{\theta}{4\pi} \frac{y_{os}}{y_{ls}} \right)^{1/2} \,, \tag{20.48}$$

with $y = H_o a_o r$. In the Einstein-de Sitter model, $y = 2 - 2(1+z)^{-1/2}$; in a low-density model with $\Lambda = 0$ the angular size distances are given by equations (13.36) and (13.70). For the arc in A370 we get from the redshifts in equation (20.46) $y_{os}/y_{ls} = 2.61$ if $\Omega = 1$, and 2.51 if $\Omega = 0$ with $\Lambda = 0$, indicating the test is not sensitive to the cosmological parameters within the standard model. The numerical result for the arc radius in equation (20.47) is

$$\sigma_{\text{arc}} = 1500 \,\text{km s}^{-1}. \tag{20.49}$$

Struble and Rood (1991) estimate that the galaxy velocity dispersion from the measured redshifts in this cluster is

$$\sigma_{\text{galaxies}} = 1360 \,\text{km s}^{-1}. \tag{20.50}$$

This number is not easy to measure and is still under discussion, but it does seem to be in reasonable agreement with what is needed to make the arc. The impact parameter of the galaxy image at the cluster is

$$d = \theta r_{ol} a_l = \frac{c}{H_o} \frac{\theta y_{ol}}{1 + z_l} = 80h^{-1} \,\text{kpc}. \tag{20.51}$$

The main part of the bending angle would be produced at radii comparable to this.

The lensing effect requires a relatively large velocity dispersion and small core radius, so it is not surprising that the velocity dispersion in this cluster is well above the rms value in equation (20.34). As for the X-ray measures, the consistency of the velocity dispersions derived from the galaxy redshifts and the arc radius indicates we have a useful approximation to the mass distribution.

Considerably more detailed models for the mass structures of clusters take account of the run of velocity dispersion with projected distance from the cluster center, the plasma distribution, the bounds on the radial variation of the plasma temperature, and the constraints from lensing arcs. The state of the art is reviewed by Fitchett (1990) and Mushotzky (1991).

For our purpose, there are three main lessons to be drawn from the studies of the masses of the great clusters of galaxies. The first is that the lensing test has used general relativity theory and the redshift-distance relation on scales where they have not previously been tested in such a direct way, with a result that agrees with the observations. Such events do not occur so often in cosmology that we should allow them to pass without comment. This test is discussed further by Soucail and Fort (1991) and Dar (1992).

The second lesson is that we have a believable lower bound on the cosmological density parameter, $\Omega > 0.01$, from the contribution by the mass within the Abell radius of rich clusters of galaxies.

The third lesson is that the mean value of the mass-to-light ratio of the material in rich clusters, after allowance for a little fading in the gas-poor cluster members, is similar to the estimates from field galaxies based on the relative velocity dispersion and on models for groups of galaxies, as in the calculations presented here for the Local Group (eqs. [20.14] and [20.22]). This is what one would expect if L_* protogalaxies were born with a universal value for the net dark mass they bring to their group or cluster.

It is worth pausing to consider the size of the region from which the mass in a rich cluster is drawn. In a homogeneous mass distribution the present radius of the sphere that contains the cluster mass (20.39) inside the Abell radius is

$$R_{cl} = \left[\frac{3\mathcal{M}_{cl}(<r_a)}{4\pi\Omega\rho_{crit}} \right]^{1/3} = \left(\frac{4\sigma^2 r_a}{H_o^2\Omega} \right)^{1/3}$$

$$= 7h^{-1}\Omega^{-1/3} \, \text{Mpc} \, . \tag{20.52}$$

Suppose we have been underestimating the density parameter because protogalaxies are born with superhalos that initially extend from 1 to 5 Mpc. These halos would not be detected in the relative velocity dispersion in figure 20.1 (and we would have to assume they are not detected in the Local Group because these galaxies are unusual). But the superhalos would have been incorporated in the clusters, which would have made the great clusters unacceptably massive.

We consider next another measure for a mass component with a broad coherence length.

Large-Scale Velocity Fields

In linear perturbation theory, the peculiar velocity field associated with the growing mass density distribution $\delta\rho/\rho = \delta$ is (eq. [5.117])

$$\mathbf{v}(\mathbf{r}) = \frac{2}{3} \frac{f}{\Omega H_o} \mathbf{g}$$

$$= \frac{H_o f}{4\pi} \int \frac{\mathbf{y} - \mathbf{r}}{|\mathbf{y} - \mathbf{r}|^3} \delta(\mathbf{y}) d^3 y \, . \tag{20.53}$$

This assumes the density contrast is scaling with time according to linear perturbation theory, $\delta\rho/\rho = \delta(\mathbf{r}, t) \propto D(t)$. The peculiar gravitational acceleration is $\mathbf{g}$, and the dimensionless factor is

$$f = \frac{a}{D} \frac{dD}{da} \sim \Omega^{0.6} \, , \tag{20.54}$$

if $\Lambda = 0$ or if the universe is cosmologically flat (fig. 13.14). For a spherical mass distribution, the velocity is

$$v(r) = \frac{f H_o}{r^2} \int_0^r y^2 \, dy \delta(y) = \frac{1}{3} f H_o r \bar{\delta} \, , \tag{20.55}$$

where $\bar{\delta}(r)$ is the mass density contrast averaged over the volume within radius r.

Let us use this result to estimate the mass concentration in and around the Virgo cluster of galaxies (the de Vaucouleurs Local Supercluster discussed in section 3). The contrast in IRAS galaxy counts in a sphere centered on the Virgo cluster and with us on the edge is

$$\bar{\delta}_v = \frac{\delta N}{N} = 1.4 \tag{20.56}$$

(Strauss et al. 1992a). If mass follows IRAS galaxies this is a mildly nonlinear mass fluctuation, but the arguments in section 22 suggest the linear expression (20.55) still is a useful approximation to the velocity.

The virgocentric peculiar velocity of the Local Group produced by this mass concentration is estimated to be

$$v_v = 170 \pm 50 \, \text{km s}^{-1} \tag{20.57}$$

(Sandage and Tammann 1990). If our velocity relative to the cluster is not strongly disturbed by the tidal fields of more distant mass concentrations, then we can use equation (20.55) to estimate the cluster mass in excess of homogeneity. The result is that within our distance $H_o r_v = 1200$ km s^{-1} from the center of the cluster the mass excess is

$$\mathcal{M} = \frac{3\Omega H_o r_v^2 v_v}{2f(\Omega)G} = 8 \times 10^{14} h^{-1} \Omega^{0.4} \mathcal{M}_\odot . \tag{20.58}$$

This is comparable to the mean value of the mass within the Abell radius of a richness class 1 cluster (eq. [20.39]), a reasonable result because the Virgo cluster is less rich than the typical $R = 1$ cluster, but here we are counting the mass to about ten Abell radii. Under the assumption that v_v in equation (20.57) is unbiased, the contrast in equation (20.56) in equations (20.54) and (20.55) gives

$$\Omega_v = 0.14 \, e^{\pm 0.5} . \tag{20.59}$$

The estimate of v_v we have used in equation (20.59) would be biased low if our velocity happened to be reduced by the tidal field of more distant and massive concentrations such as the Great Attractor (figs. 3.7 and 5.4). If Ω were unity, the above numbers say the mean virgocentric radial velocity averaged over a sphere centered on the cluster at our radius would be $v_v \sim 600$ km s^{-1}. If a tidal field had reduced our velocity to 200 km s^{-1}, it would mean other regions at our distance from the cluster are falling in at 1000 km s^{-1}, and others have comparable transverse velocities, with an rms dispersion of about 400 km s^{-1} at our distance from the Virgo cluster. That certainly is not suggested by the quiet local galaxy flow illustrated in figure 5.5. And we note that equation (20.59) is the result familiar from all the other measures we have discussed.

The mass fluctuations on still larger scales are estimated from the large-scale velocity field. It is presumed that the motion of the Local Group relative to the preferred frame defined by the cosmic background radiation (table 6.1) is the result of the peculiar gravitational acceleration of the large-scale irregularities in the mass distribution. The effect is conveniently expressed in terms of the dipole anisotropy of the sky surface brightness, as follows.

Let us write proper vector positions relative to the Local Group as

$$y^\alpha = y\gamma^\alpha , \tag{20.60}$$

where y is the length of the vector and the direction cosines are $\gamma^x = \sin\theta\cos\phi$ and so on. Then equation (20.53) is

$$v^\alpha = \frac{H_o f}{4\pi} \oint \gamma^\alpha d\Omega \int_0^\infty dy\, \delta(y,\theta,\phi) . \tag{20.61}$$

If the large-scale distribution of light traces the mass, we can write the density contrast as

$$1 + \delta = j/j_b , \tag{20.62}$$

where $j(\mathbf{y})$ is the luminosity density as a function of position and j_b is the mean value. Then the line-of-sight integral is the surface brightness as a function of position on the sky,

$$i(\theta,\phi) = \int dy\, j/4\pi . \tag{20.63}$$

(This expression does not have the redshift factors of eq. [5.159], but this does not matter if the dipole moment is generated by density fluctuations on scales small compared to the Hubble length.) In terms of the surface brightness, the velocity is

$$v^\alpha = \frac{H_o f}{j_b} \oint d\Omega\, i(\theta,\phi)\gamma^\alpha . \tag{20.64}$$

The conclusion is that if light traces mass, the velocity is fixed by the dipole moment of the sky surface brightness. The distances of the gravitating masses do not matter, as long as they are well within the Hubble distance, because the gravitational effect of a mass and the contribution of its light to the received energy flux density both scale as the inverse square of the distance.

One can estimate the gravitational acceleration of the Local Group by a sum over a catalog of objects, rather than from the dipole moment of their mean surface brightness. Preliminary applications indicate that a substantial part of our peculiar velocity originates at distances $H_o y \gtrsim 2000$ km s^{-1}, consistent with the idea that the peculiar velocity field has a broad coherence length, as was discussed in section 5 and will be considered further in the next section. This suggests that if the Einstein-de Sitter mass really is present in movable matter then the large-scale growing density fluctuations will have caused the mass to cluster with the light, giving a fair indication of the total in this dynamical measure. The history of this method may be traced in Rubin and Coyne (1988), Vittorio (1988), Juszkiewicz, Vittorio, and Wyse (1990), and Strauss et al. (1992b).

Another approach compares the relative galaxy velocity field averaged over large scales to what would be expected from the relative gravitational acceleration produced by the fluctutations in the galaxy distribution. Bertschinger and Dekel (1989) introduced an elegant way to estimate the galaxy three-dimensional relative peculiar velocity field from the observed line-of-sight galaxy velocities. Since the peculiar velocity $\mathbf{v}(\mathbf{r})$ in the linear perturbation equation (20.53) is proportional to the gravitational acceleration, it can be derived from a velocity potential, $\chi(\mathbf{r})$:

$$\mathbf{v} = -\nabla\chi. \tag{20.65}$$

The line-of-sight component is

$$v_r = -\frac{\partial\chi}{\partial r} = c\frac{\delta\lambda}{\lambda} - H_o r. \tag{20.66}$$

This is the difference between the observed galaxy radial velocity, which is written in terms of the observed first-order Doppler shift $\delta\lambda/\lambda$, and the cosmological recession velocity $H_o r$, where r is an estimate of the true distance to the galaxy. An integral of the estimates of v_r along the line of sight gives an estimate of $\chi(\mathbf{r})$, and the gradient of this function in the transverse directions gives an estimate of the full three-dimensional peculiar velocity field, which can be compared to what is expected from the galaxy distribution. The peculiar velocity field is irrotational only until orbits cross, so one has to smooth the radial field before computing $\chi(\mathbf{r})$. One has a check on the results from the condition that the derived transverse components of the gradient of χ have to be statistically equivalent to the radial peculiar velocity.

As discussed in Rubin and Coyne (1988), Lahav (1991), Strauss et al. (1992b), and Shaya, Tully, and Pierce (1992), the observed large-scale peculiar velocity field does correlate reasonably well with the gravitational field computed from the galaxy distribution, and the direction of our motion relative to the CBR agrees with the acceleration derived from equation (20.64). These important positive results suggest the large-scale galaxy distribution does usefully trace the large-scale fluctuations in the mass distribution, and that these methods therefore are capable of giving a believable measure of the density parameter. At the time this is written, it is not quite so clear whether the final answer will agree with the Einstein-de Sitter model or with the lower values of the density parameter found from the dynamical measures on smaller scales. Discussion of this issue continues in section 26.

21. The Large-Scale Mass Distribution

The central issue for this section is whether there is within the standard cosmology a world picture that reconciles the dramatic structures observed on scales of tens of megaparsecs, and illustrated in figures 3.3 to 3.6, with the remarkable large-scale uniformity seen in figures 3.8 to 3.10 and in the high degree of isotropy of the cosmic radiation backgrounds. If this proved to be impossible it would show there is something very wrong with the standard model. As will be described, there is no such crisis in theory and observation as now understood: it is easy to find prescriptions for the mass distribution to fit the observational constraints and the theory for how the distributions of mass and radiation evolve in an expanding universe. The constraints include the large-scale anisotropy of the thermal cosmic background radiation, at $\delta T/T \sim 1 \times 10^{-5}$, the velocity field on scales $\sim 30h^{-1}$ Mpc, and the clustering of galaxies and mass on smaller scales.

Assuming this phenomenological concordance continues to hold, we are led to a second issue: is there a believable physical theory for the early universe that reproduces the prescription for an observationally acceptable world picture? We have a considerable variety of interesting ideas on how this might be done, though it will be argued in sections 25 and 26 that none so far has been shown to be compelling. One of the hopes for narrowing the range of the search for an acceptable and believable theory is the continued tightening of the observational constraints on what the mass distribution is like now and at modest redshifts.

We begin with some general relations among measures of the large-scale mass distribution, and then consider the orders of magnitude for what is allowed.

Sachs-Wolfe Relation

Sachs and Wolfe (1967) showed that in the standard cosmological model the large-scale fluctuations in the cosmic background radiation temperature as a function of position across the sky provide a sensitive measure of the large-scale fluctuations in the mass distribution. There are two ways to think of this Sachs-Wolfe effect. We noted in section 6 that the CBR temperature scales inversely with the redshift along the line of sight to last scattering, and the redshift is proportional to the expansion factor along the line of sight. If large-scale departures from homogeneity caused the expansion of the universe across the Hubble length to differ by the fractional amount δ_h, measured in orthogonal directions, it would produce a quadrupole CBR anisotropy $\delta T/T = \delta_h$. Since the CBR is isotropic to better than one part in 10^4, the large-scale expansion of the universe had to have been remarkably close to isotropic.

Another way to think of the effect is in terms of the Newtonian gravitational potential of the fluctuations from a homogeneous mass distribution. Suppose a line of sight traces back to an origin at high redshift at a position where the mass density contrast averaged over the comoving scale x is $\delta\rho/\rho = \delta_x$. The Newtonian gravitational potential energy produced by this fluctuation is

$$\phi \sim \frac{G\delta\mathcal{M}}{ax} = \frac{G}{ax}\frac{\delta\mathcal{M}}{\mathcal{M}}\frac{4}{3}\pi\rho_b(ax)^3$$

$$= \frac{1}{2}\Omega H^2(ax)^2\delta_x. \tag{21.1}$$

In the Einstein-de Sitter limit, the expansion rate scales as $H \propto a^{-3/2}$, and the density contrast grows as $\delta\mathcal{M}/\mathcal{M} = \delta_x \propto a(t)$ (eq. [5.113]), so the gravitational potential ϕ belonging to the mass fluctuation is independent of time. The radiation we receive had to move out of the potential. That causes a gravitational redshift, $\delta\nu/\nu \sim \phi$, and a like perturbation to the CBR temperature. That is, the CBR anisotropy caused by the mass density fluctuation δ_x on the comoving scale x at the Hubble distance is

$$\delta T/T \sim (H_o a_o x)^2 \delta_x(0). \tag{21.2}$$

We have noted that if the cosmological model is Einstein-de Sitter or close to it, the potential ϕ is constant, so we can evaluate this expression at the present epoch, where the density contrast extrapolated to the present in linear perturbation theory is $\delta_x(0)$. With an averaging length $a_o x$ equal to the present Hubble length H_o^{-1}, this is $\delta T/T \sim \delta_x(0)$, as indicated in the first argument. Since the bounds on $\delta T/T$ are roughly similar on all angular scales down to a few tens of seconds of arc, we see that the constraint on δ_x from the Sachs-Wolfe relation decreases with decreasing scale. At small enough scales the anisotropy may be masked by foreground sources or erased by scattering; the effects are discussed in section 24.

This subsection presents a more detailed computation of the relation between the mass distribution and the perturbation $\delta T/T$ to the CBR anisotropy, under some simplifying approximations. We shall assume that at relatively low redshifts, during which the displacement of the radiation has been appreciable, the inhomogeneously distributed mass component has been dominated by material whose pressure may be neglected. Since $\delta T/T$ is small, it is reasonable to use linear perturbation theory. This means we can select for discussion the effect of the growing density perturbation. The computation is greatly simplified by adopting a cosmologically flat model. And finally, we can work the integrals analytically in the special case of an Einstein-de Sitter model.

Under these assumptions we can assign time-orthogonal coordinates fixed to the matter, [10.98], so the matter coordinate velocity vanishes, and the line element is (eq. [9.37])

$$ds^2 = dt^2 - a(t)^2(\delta_{\alpha\beta} - h_{\alpha\beta}) \, dx^\alpha dx^\beta . \tag{21.3}$$

The coordinates have been assigned so $g_{00} = 1$ and the time-space part of the metric tensor vanishes. The perturbation to the space part of the metric tensor from the cosmologically flat background is $h_{\alpha\beta}$.

In the unperturbed background, the path of a light ray that reaches our position at the present epoch, t_o, is

$$x^\alpha(t) = \gamma^\alpha x(t), \qquad x(t) = \int_t^{t_o} \frac{dt'}{a(t')}, \tag{21.4}$$

where the constant direction cosines satisfy the normalization $\delta_{\alpha\beta}\gamma^\alpha\gamma^\beta = 1$.

The first step is to get the perturbation to the redshift in terms of the $h_{\alpha\beta}$. Suppose the packet of radiation we receive passes a comoving observer at time t and a second observer at time $t + \delta t$. The coordinate separation of these two observers is

$$\delta x^\alpha = \gamma^\alpha \delta x . \tag{21.5}$$

The proper distance between the observers at fixed time t is given by the line element in equation (21.3),

$$\delta l = (-g_{\alpha\beta}\delta x^\alpha \delta x^\beta)^{1/2} = a(1 - h_{\alpha\beta}\gamma^\alpha\gamma^\beta/2)\delta x , \tag{21.6}$$

to first order. Since the two observers are at rest in the coordinate system, the coordinate interval δx^α connecting them is constant, and equation (21.6) gives the time dependence of the proper distance between the observers. Thus their relative velocity, to first order in $h_{\alpha\beta}$, is

$$\frac{1}{\delta l}\frac{d\delta l}{dt} = \frac{\dot{a}}{a} - \frac{1}{2}\frac{\partial h_{\alpha\beta}}{\partial t}\gamma^\alpha\gamma^\beta . \tag{21.7}$$

The brightness theorem in equation (9.53) says the fractional difference of the CBR temperatures measured by neighboring observers moving apart at speed $\delta v \ll 1$, and measured along the line connecting them, is $\delta T / T = -\delta v$. The proper separation of the observers in equation (21.7) at time t is $\delta l = \delta t$, so the fractional difference in the CBR temperatures they measure is

$$\frac{\delta T}{T} = \left(-\frac{\dot{a}}{a} + \frac{1}{2}\frac{\partial h_{\alpha\beta}}{\partial t}\gamma^\alpha\gamma^\beta \right) \delta t . \tag{21.8}$$

The first term is the usual cooling law in a homogeneous universe. The second is the wanted effect of the perturbation to the line element, which in turn is fixed by the irregular mass distribution.

The observed fractional perturbation to the CBR temperature is the result of summing the perturbation (21.8) over a sequence of observers along the path of the wave packet from high redshift to the present,

$$\frac{\delta T}{T} = \frac{1}{2}\gamma^{\alpha}\gamma^{\beta} \int_0^{t_o} dt \frac{\partial}{\partial t} h_{\alpha\beta}(t, \vec{\gamma}x(t)). \qquad (21.9)$$

The spatial argument of $h_{\alpha\beta}$ is constructed in equation (21.4).

The easiest way to find the perturbation to the metric tensor starts from the peculiar velocity field. Equations (20.53) and (20.54) are

$$v^{\alpha} = -\frac{1}{4\pi G\rho_b} \frac{\dot{D}}{D} \frac{\phi_{,\alpha}}{a}, \qquad (21.10)$$

where the peculiar gravitational acceleration is $\mathbf{g} = -\nabla\phi/a$, and Poisson's equation for the potential is

$$\nabla^2\phi = 4\pi G\rho_b a^2 \delta. \qquad (21.11)$$

Equation (21.10) says that, relative to the general expansion, the rate of change of the proper separation of the material elements at the fixed coordinate separation $\delta x^{\beta} = \gamma^{\beta}\delta x$ is

$$v = \gamma^{\alpha}\delta v^{\alpha} = -\frac{1}{4\pi G\rho_b} \frac{\dot{D}}{D} \frac{\phi_{,\alpha\beta}}{a}\gamma^{\alpha}\gamma^{\beta}\delta x. \qquad (21.12)$$

This is the same as the second term on the right-hand side of equation (21.7). Since the tensors multiplying the direction cosines in these two expressions are symmetric, the tensors have to be the same. That gives

$$\frac{\partial h_{\alpha\beta}}{\partial t} = \frac{1}{2\pi G\rho_b a^2} \frac{\dot{D}}{D} \frac{\partial^2\phi}{\partial x^{\alpha}\partial x^{\beta}} \equiv \frac{dD}{dt} \frac{\partial^2 k(\mathbf{x})}{\partial x^{\alpha}\partial x^{\beta}}, \qquad (21.13)$$

where $dt = a\,dx$ (eq. [21.4]). The last step defines the function $k(\mathbf{x})$,

$$k(\mathbf{x}) = \frac{\phi}{2\pi G\rho_b a^2 D} = -\frac{1}{2\pi} \int \frac{\delta(\mathbf{x}',t)}{D(t)} \frac{d^3x'}{|\mathbf{x}-\mathbf{x}'|}. \qquad (21.14)$$

This is proportional to the gravitational potential associated with the departure from homogeneity. It is a function of position $\mathbf{x}$ alone, because we are assuming the density perturbation is the pure mode, $\delta(\mathbf{x},t) \propto D(t)$.

With equation (21.13), the perturbation (21.9) to the CBR temperature as a function of direction γ^α is

$$\frac{\delta T}{T} = \frac{1}{2}\gamma^\alpha\gamma^\beta \int_0^{t_o} dt\, \frac{dD}{dt}\, \frac{\partial^2 k}{\partial x^\alpha \partial x^\beta}\,. \tag{21.15}$$

The integral is evaluated along the path $x^\alpha(t) = x(t)\gamma^\alpha$ in equation (21.4), so we can consider k to be a function of x, with

$$\frac{dk}{dx} = \frac{\partial k}{\partial x^\alpha}\frac{dx^\alpha}{dx} = \gamma^\alpha \frac{\partial k}{\partial x^\alpha}\,, \tag{21.16}$$

and similarly for the second derivative. This brings equation (21.15) to

$$\frac{\delta T}{T} = \frac{1}{2}\int_0^{t_o} dt\, \frac{dD}{dt}\frac{d^2 k}{dx^2}\,, \qquad dx = -dt/a\,. \tag{21.17}$$

The result of integrating by parts twice is

$$\frac{\delta T}{T} = -\frac{a_o}{2}\frac{dD}{dt_o}\frac{\partial k}{\partial x^\alpha}\gamma^\alpha - \left[\frac{a}{2}\frac{d}{dt}\left(a\frac{dD}{dt}\right)k\right]_0^{t_o}$$
$$+ \frac{1}{2}\int_0^{t_o} dt\, k\frac{d}{dt}a\frac{d}{dt}a\frac{dD}{dt}\,. \tag{21.18}$$

Sachs and Wolfe (1967) discovered this relation between the fractional perturbation to the CBR temperature and the mass density perturbation that appears in the function k (eq. [21.14]).

The first term in equation (21.18) with equation (21.14) for k is

$$\frac{\delta T}{T} = \frac{a_o}{4\pi}\frac{\dot{D}_o}{D_o}\gamma^\alpha \int \frac{(\mathbf{x}' - \mathbf{x})^\alpha}{|\mathbf{x}' - \mathbf{x}|^3}d^3 x'\delta(\mathbf{x}', t_o)\,. \tag{21.19}$$

This is the dipole anisotropy caused by our peculiar motion, as discussed in the last section (eqs. [20.53] and [20.54]).

The last term in equation (21.18) is the effect of the time-variable gravitational potential along the path of the radiation. The contribution to the integral by a region along the line of sight within which the potential is $\sim \phi$, when averaged over a proper distance $\delta t = l$, is

$$\frac{\delta T}{T} \sim l \cdot \frac{\phi}{G\rho_b a^2 D} \cdot \frac{a^2 D}{t^3} \sim \phi\frac{l}{t}\,. \tag{21.20}$$

The first factor in this expression is the path length for the integral through the patch. The second is the expression for k in equation (21.14). The last factor is the order of magnitude of the time derivatives, because the characteristic timescales for the variation of D and a are the expansion time t. We see that this expression is on the order of the change in the potential during the time taken for the packet of radiation to move across the region. That is, if the wave packet falls into a potential well that differs from the one it moves out of as it leaves the region, the result is a gravitational redshift or blueshift of the photon energy, and a perturbation to the CBR temperature (Rees and Sciama 1968). At low redshifts, this effect is down from the Sachs-Wolfe effect in equation (21.1) by the factor l/t. In the Einstein-de Sitter model, the growing density perturbation mode is $D \propto t^{2/3} \propto a(t)$, and the last term in equation (21.18) vanishes. This special case follows because in the Einstein-de Sitter model in linear perturbation theory the gravitational potential expressed in expanding coordinates is independent of time. In other cosmologies this term can make a significant contribution to the CBR anisotropy (Kofman and Starobinsky 1985; Gouda, Sugiyama, and Sasaki 1991).

The second term in the right-hand side of equation (21.18) has a part evaluated at the present epoch that makes no contribution to the CBR anisotropy. The CBR was last scattered at relatively high redshift, so we can use the Einstein-de Sitter limit in the other part of the term, where $D \propto a \propto t^{2/3}$. This brings the term to

$$\frac{\delta T}{T} = -\frac{1}{18\pi} \frac{D}{D_o} \frac{a^2}{t^2} \int \frac{d^3x'}{|\mathbf{x} - \mathbf{x}'|} \delta(\mathbf{x}', t_o). \tag{21.21}$$

The density contrast has been scaled to the present epoch, where the present value of the function $D(t)$ is D_o. With the normalization in equation (13.79) and in figure 13.13, at high redshift $D = a/a_o$. In the Einstein-de Sitter limit, the time is (eq. [5.61])

$$t = \frac{2}{3H_o \Omega^{1/2}} \left(\frac{a}{a_o}\right)^{3/2}. \tag{21.22}$$

This brings equation (21.21) to

$$\frac{\delta T}{T} = \frac{\phi_o}{3D_o}, \qquad \phi_o(x\gamma^\alpha) = -G\rho_o \int \frac{d^3r'}{|\mathbf{r}' - r\vec{\gamma}|} \delta(\mathbf{r}, t_o). \tag{21.23}$$

The gravitational potential in this expression is evaluated for the mass density fluctuation $\delta\rho = \rho_o \delta_o(\mathbf{r})$ extrapolated to the present epoch in linear perturbation theory (with ρ_o the present mean density of movable matter, not counting the smoothly distributed component if Ω belonging to movable mass is not equal to unity). Apart from the factor $3D_o$, this is just the gravitational redshift one would write down in a static spacetime for radiation that originated at position $\vec{\gamma} r$ relative to us. Recall that we have assumed a cosmologically flat model, where the coordinate angular size distance x to the point of origin of the radiation at high redshift is given by equation (21.4), and the proper distance in equation (21.23) is $r = a_o x$.

We see that a mass density fluctuation at the Hubble distance, with present contrast $\delta\rho/\rho = \delta_o$, that extends over a region of present proper size l gravitationally perturbs the CBR by an amount comparable to the present gravitational potential energy of the fluctuation, $\delta T/T \sim \phi \sim G\rho_o l^2 \delta$, as in equations (21.1) and (21.2). More formal relations between the spectrum of mass fluctuations and the anisotropy of the CBR are considered below, after we deal with some aspects of the large-scale peculiar velocity field.

Layzer-Irvine Equation

Irvine (1961) found an elegant relation between the mass autocorrelation function and the rms peculiar velocity when nongravitational forces can be ignored. The relation was further discussed by Layzer (1963) and Dmitriev and Zel'dovich (1963); Fall (1975) introduced its use as a constraint on the mean cosmological mass density.

The relation neglects nongravitational interactions, so we can model the mass distribution as a collection of particles of masses m_i at coordinate positions $\mathbf{x}_i$. It assumes the fluctuations from homogeneity are appreciable only on scales small compared to the Hubble length, and the peculiar motions are nonrelativistic. In this limit, we can write the Lagrangian for the system of particles as

$$\mathcal{L} = \frac{1}{2} \sum m_i a^2 \dot{\mathbf{x}}_i^2 - MW \,. \qquad (21.24)$$

The sum is over the kinetic energies of the particles in some large comoving volume. The mass in this volume is $M = \sum m_i$. The gravitational potential energy per unit mass is

$$W = -\frac{1}{2} \frac{Ga^5}{M} \int d^3x_1 \, d^3x_2 [(\rho(\mathbf{x}_1) - \rho_b)(\rho(\mathbf{x}_2) - \rho_b)]/x_{12} \,. \qquad (21.25)$$

The gravitational energy is determined by the difference between the local mass density $\rho(\mathbf{x})$ and the mean value, ρ_b, as in equation (20.53).

The momentum belonging to the ith particle in the Lagrangian (21.24) is

$$\mathbf{p}_i = \frac{\partial \mathcal{L}}{\partial \dot{\mathbf{x}}_i} = m_i a^2 \dot{\mathbf{x}}_i \,, \qquad (21.26)$$

and the Hamiltonian is

$$H = \sum \frac{p_i^2}{2m_i a^2} + MW \equiv M(K + W) \,. \qquad (21.27)$$

The kinetic energy per unit mass is

$$K = \frac{1}{2} \frac{\sum m_i (a \dot{x}_i)^2}{\sum m_i} = \frac{1}{2} v_a^2 . \tag{21.28}$$

That is, $2K$ is the mass-weighted mean square peculiar velocity.

The dimensionless mass autocorrelation function is

$$\xi(|\mathbf{r}_1 - \mathbf{r}_2|) = \frac{\langle (\rho(\mathbf{x}_1) - \rho_b)(\rho(\mathbf{x}_2) - \rho_b) \rangle}{\rho_b^2} , \tag{21.29}$$

as in equation (19.10). With the change of variables $\mathbf{x} = \mathbf{x}_1 - \mathbf{x}_2$, and the mass $M = a^3 \int \rho_b d^3 x$, equation (21.25) is

$$W = -\frac{1}{2} G \rho_b a^2 \int \frac{d^3 x}{x} \xi(x) = -2\pi G \rho_b J_2 , \tag{21.30}$$

where

$$J_2 = \int_o^\infty \xi(r) r \, dr . \tag{21.31}$$

The energy equation is

$$\frac{dH}{dt} = \frac{\partial H}{\partial t} , \tag{21.32}$$

where the time derivative on the right-hand side is computed at fixed particle coordinates and momenta. At fixed $\mathbf{x}_i$ and $\mathbf{p}_i$ we have $K \propto a^{-2}$ (eq. [21.27]), and $W \propto \rho_b a^2 \propto a^{-1}$ (eq. [21.30]), so the energy equation is

$$\frac{d}{dt}(K + W) + \frac{\dot{a}}{a}(2K + W) = 0 . \tag{21.33}$$

This is the Layzer-Irvine equation relating the rms peculiar velocity and the gravitational potential energy, which is an integral over the mass autocorrelation function. The relation assumes the clustering is nonrelativistic, and it ignores nongravitational forces, but it is fully nonlinear in the gravitational interaction.

We can rewrite the equation as

$$\frac{d}{dt} a(K + W) = -K \dot{a} . \tag{21.34}$$

Since $K > 0$ and $\dot{a} > 0$ in an expanding universe, the right-hand side is negative. Thus, if K and W grow from small initial values, the sum is negative:

$$K + W < 0 . \tag{21.35}$$

This means W has to be negative (which is not a trivial constraint, for the mass density in W is $\rho(\mathbf{x}) - \rho_b$, which is negative as well as positive). If K and W are dominated by the kinetic and potential energies within clusters that have relaxed to dynamical equilibrium, then K and W are nearly independent of time, and the Layzer-Irvine equation becomes

$$K + W/2 = 0, \tag{21.36}$$

which is the Newtonian virial theorem relating the velocity dispersion and the mean gravitational potential energy per unit mass. We can generalize this by writing the Layzer-Irvine equation as

$$\frac{d}{dt}a^2(K + W/2) = -a^2\dot{W}/2. \tag{21.37}$$

If the universe is dense enough for $\int x\,dx\xi$ to grow at least as fast as $a(t)$, then $\dot{W}$ is negative, and $K + W/2 > 0$. These two conditions are

$$-W/2 < K < -W. \tag{21.38}$$

More details, and a scaling relation between K and W, are in LSS, §§24 and 74.

The bounds in equation (21.38) give a relation between the mean square peculiar velocity $v_a^2 = \langle v^2 \rangle$ in K (eq. [21.28]) and the mass autocorrelation function $\xi(r)$ in W (eq. [21.30]).

Second Moments

We noted at the beginning of section 19 that it is natural to turn to the second moment, or autocorrelation function $\xi(r)$, as a first measure of the mass distribution, though it is prudent to bear in mind that this statistic can miss features such as walls of galaxies. A more specific reason why $\xi(r)$ is well suited to the analysis of large-scale fluctuations in the mass distribution is that $\xi(r)$ averages over features such as walls in much the same way as does the long-range effect of gravity. This subsection reviews the use of $\xi(r)$ and its Fourier transform, the power spectrum, as a way to explore the relation between the fluctuations in galaxy counts, the galaxy peculiar velocity field, and the Sachs-Wolfe effect.

Let us begin with the relation between $\xi(r)$ and the power spectrum. Suppose we represent the present mass density as a continuous function $\rho(\mathbf{r})$, with mean value ρ_o. To keep the equations simple, we will use periodic boundary conditions in some large volume V_u (which always cancels out of the final statistics).[7] Then

---

[7] It may be worth emphasizing that the periodicity is not physical, only a convenient — and for some simplifying — mathematical device. Those who do not agree will mentally change the Fourier sums to integrals.

the Fourier expansion of the mass density contrast is

$$\delta(\mathbf{r}) = \frac{\rho}{\rho_0} - 1 = \frac{(2\pi)^{3/2}}{V_u^{1/2}} \sum \delta_{\mathbf{k}} e^{i\mathbf{k}\cdot\mathbf{r}}. \tag{21.39}$$

The reality condition is $\delta_{\mathbf{k}}^* = \delta_{-\mathbf{k}}$.

The autocorrelation function $\xi(r)$ for the density contrast is the mean value of the lagged product (eq. [21.29]),

$$\xi(r) = \langle \delta(\mathbf{r}_1)\delta(\mathbf{r}_1 + \mathbf{r}) \rangle = \frac{(2\pi)^3}{V_u} \sum \langle |\delta_{\mathbf{k}}|^2 \rangle e^{i\mathbf{k}\cdot\mathbf{r}}$$

$$= \int d^3k \langle |\delta_{\mathbf{k}}|^2 \rangle e^{i\mathbf{k}\cdot\mathbf{r}} = 4\pi \int_0^\infty k^2 \, dk \, \mathcal{P}(k) \frac{\sin kr}{kr}. \tag{21.40}$$

The average in the first line eliminates the cross terms in the Fourier series, leaving ξ a function of the lag (the distance r between the two points in the autocorrelation function). The sum is changed to an integral in the usual way (eq. [6.8]). The last step computes the angular integral under the standard assumption that the mass fluctuations are statistically independent of direction.

Equation (21.40) says the autocorrelation function is the Fourier transform of the power spectrum,

$$\mathcal{P}(k) = \langle |\delta_{\mathbf{k}}|^2 \rangle, \tag{21.41}$$

where the Fourier amplitude in equation (21.39) is

$$\delta_{\mathbf{k}} = \frac{1}{(2\pi)^{3/2} V_u^{1/2}} \int \delta(\mathbf{r}) e^{-i\mathbf{k}\cdot\mathbf{r}} d^3r. \tag{21.42}$$

For a point process, such as the distribution of the central positions of galaxies, we can define the Fourier amplitude by equation (21.42) with $\delta(\mathbf{r})$ a sum of delta functions. That makes the amplitude a sum over the point positions,

$$\delta_{\mathbf{k}} = \frac{1}{(2\pi)^{3/2} V_u^{1/2} n \langle m \rangle} \sum m_i e^{-i\mathbf{k}\cdot\mathbf{r}_i}. \tag{21.43}$$

The galaxy at position $\mathbf{r}_i$ is assigned weight m_i, the mean number density of galaxies is n, and the mean weight of a galaxy is $\langle m \rangle = \sum m_i / n V_u$. The power spectrum is the average of $|\delta_{\mathbf{k}}|^2$ across an ensemble, or over a fair sample of one realization:

$$\mathcal{P}(k) = \langle |\delta_{\mathbf{k}}|^2 \rangle = \frac{\sum m_i^2}{(2\pi)^3 V_u n^2 \langle m \rangle^2}$$

$$+ \frac{\langle m \rangle^2}{(2\pi)^3 V_u n^2 \langle m \rangle^2} \int n^2 d^3r_1 \, d^3r_2 e^{-i\mathbf{k}\cdot(\mathbf{r}_1 - \mathbf{r}_2)} (1 + \xi_{12}). \tag{21.44}$$

The first term from the sum of the squared terms in $|\delta_{\mathbf{k}}|^2$ is proportional to the mean square value of the weight. The expression for the cross terms assumes the weights are statistically independent of the positions, so $\langle m_i m_j \rangle = \langle m \rangle^2$ for $i \neq j$. The mean of the sum over cross terms is the integral over the distribution of point pairs in equation (19.2) for the two-point correlation function. The constant term in parentheses in the integral vanishes when $\mathbf{k} \neq 0$. Since ξ is a function of relative position, we eliminate one volume integral by a change of variables to $\mathbf{r} = \mathbf{r}_1 - \mathbf{r}_2$. That brings the power spectrum (21.44) to

$$\mathcal{P}(k) = \frac{\langle m^2 \rangle}{(2\pi)^3 n \langle m \rangle^2} + \frac{1}{(2\pi)^3} \int d^3 r e^{-i\mathbf{k}\cdot\mathbf{r}} \xi(r) . \qquad (21.45)$$

With the Fourier inverse relation

$$\int e^{i\mathbf{k}\cdot\mathbf{r}} d^3 r = (2\pi)^3 \delta^3(\mathbf{r}) \qquad (21.46)$$

(where the last factor is a Dirac delta function, not the density contrast), we see that the transform of equation (21.45) is

$$\int d^3 k \, \mathcal{P}(k) e^{i\mathbf{k}\cdot\mathbf{r}} = \xi(r) + \frac{\langle m^2 \rangle}{\langle m \rangle^2} \frac{\delta^3(\mathbf{r})}{n} . \qquad (21.47)$$

This is just equation (21.40) with the delta function at zero lag representing the second moment of the weight within points. The constant term in the spectrum $\mathcal{P}(k)$ in equation (21.45) is the white shot noise from the pointlike concentrations. In most applications to be discussed here, this shot noise term is subdominant and will be dropped. (And where the shot noise term is appreciable it is easily calculated and subtracted from the spectrum.)

The power spectrum is represented in a statistic that is more readily visualized by considering the mass distribution smoothed through a window by the equation

$$\bar{\delta}(\mathbf{r}) = \int d^3 r' \, W(|\mathbf{r}' - \mathbf{r}|) \delta(\mathbf{r}') = \frac{(2\pi)^{3/2}}{V_u^{1/2}} \sum \tilde{W}(k) \delta_{\mathbf{k}} e^{i\mathbf{k}\cdot\mathbf{r}} . \qquad (21.48)$$

The last step follows from the series expansion in equation (21.39). The Fourier transform of the window function is

$$\tilde{W}(k) = \int d^3 r \, W(r) e^{i\mathbf{k}\cdot\mathbf{r}} , \qquad (21.49)$$

and the normalization condition is

$$\tilde{W}(0) = \int d^3 r \, W(r) = 1 . \qquad (21.50)$$

A Gaussian smoothing window, with characteristic smoothing length r_g, is

$$W_g(r) \propto e^{-r^2/2r_g^2}, \qquad \widetilde{W}_g(k) = e^{-k^2r_g^2/2}. \qquad (21.51)$$

A square window, W_s, is constant at $W_s = 1/V_s$ inside a volume V_s and vanishes outside this volume. A window that is square, in the sense that the value of W is either zero or V_s^{-1}, and spherical, meaning the window shape is a sphere of radius r_s, often is called a top hat. The transform of a spherical square, or top hat, window is

$$\widetilde{W}_s(k) = 3 \left[\frac{\sin kr_s}{(kr_s)^3} - \frac{\cos kr_s}{(kr_s)^2} \right] = \frac{3}{kr_s} j_1(kr_s), \qquad (21.52)$$

where j_1 is the spherical Bessel function.

The variance (mean square value) of the smoothed mass contrast in equation (21.48) follows as in equation (21.40):

$$\bar{\delta}^2 = \langle \bar{\delta}(\mathbf{r})^2 \rangle = \frac{(2\pi)^3}{V_u} \sum \widetilde{W}(k)^2 |\delta_{\mathbf{k}}|^2$$
$$= \int d^3k \, \widetilde{W}(k)^2 \mathcal{P}(k). \qquad (21.53)$$

Since $\widetilde{W}(0) = 1$ and $\widetilde{W}(k)$ is cut off at about the reciprocal of the width of the averaging window, we see that the power spectrum is the variance of the unsmoothed mass distribution per unit volume in $\mathbf{k}$ space. The equivalent relation in terms of the mass autocorrelation function is (eqs. [21.40] and [21.48])

$$\bar{\delta}^2 = \int d^3r_1 \, d^3r_2 \, W(\mathbf{r}_1) W(\mathbf{r}_2) \xi(\mathbf{r}_1 - \mathbf{r}_2). \qquad (21.54)$$

For the point process in equation (21.45), the variance $\bar{\delta}^2$ is the sum of this integral and the shot noise term (as displayed in eq. [7.71]).

For a square spherical window with radius r_s, and the power law model $\xi = (r_0/r)^\gamma$ for the correlation function, the variance $\bar{\delta}^2$ in equation (21.54) is

$$\bar{\delta}^2 = \left(\frac{\delta N}{N} \right)^2 = \int_0^{r_s} \frac{dV_1 dV_2}{V_s^2} \xi_{12} = C \left(\frac{r_0}{r_s} \right)^\gamma. \qquad (21.55)$$

The dimensionless integral C is given in equation (7.72). For $\gamma = 1.8$, the observed value for galaxies, C is on the order of unity, and $\bar{\delta}^2$ is comparable to the correlation function evaluated at the smoothing radius.

Because the autocorrelation function $\xi(r)$ is the Fourier transform of the spectrum $\mathcal{P}(k)$ (eq. [21.40]), the integral of $\xi(r)$ over all separations is (with eq. [21.46])

$$4\pi J_3 \equiv \int d^3r\, \xi(r) = (2\pi)^3 \mathcal{P}(0).\qquad (21.56)$$

If the power spectrum $\mathcal{P}(k)$ approaches zero at $k \to 0$, the integral of ξ vanishes. This means that if $\xi(r) > 0$ at small r, as is the case for the galaxy distribution, it has to be negative at larger r to make the integral of ξ vanish. In this case, the standard deviation $\bar{\delta}$ vanishes with increasing averaging volume more rapidly than for a random Poisson process, because the correlated fluctuations on small scales are balanced by anticorrelated fluctuations at larger separations.

It may happen that the density fluctuations are uncorrelated at large separations, so $\xi(r) = 0$ at $r > R$. Then the Fourier transform is flat (white) at $k \ll R^{-1}$, approaching the value $\mathcal{P}(k) \to P_o$ at $k \to 0$. If the width of the smoothing window is large compared to the correlation length R, equations (21.54) and (21.56) for the variance of the smoothed mass distribution are

$$\bar{\delta}^2 = \int d^3r\, \xi(r) \int d^3r\, W(r)^2 = (2\pi)^3 P_o \int d^3r\, W(r)^2$$
$$= 4\pi J_3/V_s.\qquad (21.57)$$

The second line assumes a square window, where $W = 1/V_s$ inside the window volume V_s.

Equation (21.57) can be compared to the fractional fluctuation in the mass found within the square window V_s when the mass is in particles with masses m_i placed uniformly at random (in a stationary Poisson process) at number density n_{eff},

$$\bar{\delta}(r_s)^2 = \frac{\langle m^2 \rangle}{V_s n_{\text{eff}} \langle m \rangle^2}.\qquad (21.58)$$

This follows as in equation (21.47). We see that, when the spectrum is flat at large wavelengths, the rms fluctuation in the large-scale mass distribution is the same as that produced by placing particles at random at mean number density

$$n_{\text{eff}} = \frac{1}{4\pi J_3} \frac{\langle m^2 \rangle}{\langle m \rangle^2},\qquad (21.59)$$

where J_3 is the integral in equation (21.56).[8] The mean mass density is $\rho_o =$

----

[8] The normalizations for J_2 and J_3 in equations (21.31) and (21.56) are a historical accident from early applications (Clutton-Brock and Peebles 1981).

$\langle m \rangle n_{\text{eff}}$, so the effective mass per independent particle in this model is

$$\mathcal{M}_{\text{eff}} = \frac{\langle m^2 \rangle}{\langle m \rangle} = \rho_0 \int d^3r \xi(r). \tag{21.60}$$

We see from this expression that we can interpret the statistic J_3 in two ways. Since $\rho_0[1 + \xi(r)]$ is the mean value of the mass density at distance r from a mass element, the net mass in excess of homogeneity around a mass element is $\mathcal{M}_{\text{eff}} = 4\pi J_3 \rho_0$. Consistent with this, the large-scale fluctuations in the mass distribution in a universe with this value of J_3 have the rms value that would be produced if cluster balls with mean effective mass $\mathcal{M}_{\text{eff}}$ were placed uniformly and independently. This is the construction in figures 19.1 and 19.2.[9]

The galaxy two-point correlation function is known to be positive at $r \lesssim 20h^{-1}$ Mpc; at larger separations we have only bounds. The integral over the known or conjectured part of the two-point function is

$$J_3(r) \equiv \int_0^r r^2 dr\, \xi(r). \tag{21.61}$$

This form is convenient because it is simpler than the double integral in equation (21.54), and nearly equivalent when $\xi(r)$ is a slowly varying function of r. It defines an effective clustering mass that is a function of scale, $\mathcal{M}_{\text{eff}}(r) = 4\pi \rho_0 J_3(r)$. This is the mean value of the mass in excess of homogeneity within distance r of a mass element.

The large-scale part of the mass fluctuation spectrum $\mathcal{P}(k)$ is constrained by the measures or bounds on the velocity field and the CBR anisotropy, for in linear perturbation theory the second moments of these variables are linear integrals over $\mathcal{P}(k)$ (or the mass autocorrelation function). We complete this subsection by writing down some of these integrals. To begin, equation (21.40) for the autocorrelation function as an integral over $\mathcal{P}$ gives

$$J_3(r) = \frac{4\pi r^3}{3} \int_0^\infty k^2 dk\, \mathcal{P}(k) \widetilde{W}_s(kr), \tag{21.62}$$

where $\widetilde{W}_s$ is the transform of a spherical square window function (eq. [21.52]).

---

[9] The meaning of the statistics J_3 and $\mathcal{M}_{\text{eff}}$ estimated from a catalog of positions may be confusing, for in a sample of N particles each has just $N - 1$ neighbors. That means that if the mean density is estimated from the same sample there is a bias in the estimated correlation function, such that the integral of the estimate of $\xi(r)$ has to vanish. But in the background random process of which the catalog is a realization, the net excess mean number of neighbors can be nonzero, as one sees in the construction in the model in figure 19.2.

In the Layzer-Irvine equation (21.33) for the rms matter velocity, the integral (21.31) is

$$J_2 = \int_0^\infty r\, dr\, \xi(r)$$

$$= 4\pi \int_0^\infty r\, dr \int_0^\infty k^2 dk \mathcal{P}(k) \frac{\sin kr}{kr} \qquad \textbf{(21.63)}$$

$$= 4\pi \int_0^\infty \sin y\, dy \int_0^\infty dk \mathcal{P}(k).$$

The result of changing the order of integration is an integral over $\sin y$ that is unity (if we put in a little smoothing to damp the oscillations at $y \to \infty$), so

$$J_2 = 4\pi \int_0^\infty dk \mathcal{P}(k). \qquad \textbf{(21.64)}$$

In linear perturbation theory, and assuming the density fluctuations have been growing long enough so that the growing mode dominates, the velocity field belonging to the mass density contrast $\delta(\mathbf{r})$ is given by equation (20.53). With the representation of $\delta(\mathbf{r})$ in equation (21.39), this is

$$v^\alpha = iH_o f \frac{(2\pi)^{3/2}}{V_u^{1/2}} \sum \frac{k^\alpha}{k^2} \delta_{\mathbf{k}} e^{i\mathbf{k}\cdot\mathbf{r}}. \qquad \textbf{(21.65)}$$

The easy way to check this expression is to work out its divergence:

$$\nabla \cdot \mathbf{v} = -H_o f \delta = -\delta \dot{D}/D = -\partial \delta / \partial t, \qquad \textbf{(21.66)}$$

where $f = d\log D / d\log a$ (eq. [20.54]), and we are assuming the density contrast is scaling as a single density perturbation mode, $\delta \propto D(t)$. The result is the mass conservation equation in linear perturbation theory. Recall that if space curvature or the cosmological constant may be neglected, $f \sim \Omega^{0.6}$ (fig. 13.14).

The velocity field smoothed through a window is

$$\bar{\mathbf{v}}(\mathbf{r}) = \int d^3 r'\, W(\mathbf{r}' - \mathbf{r}) \mathbf{v}(\mathbf{r}')$$

$$= iH_o f \frac{(2\pi)^{3/2}}{V_u^{1/2}} \sum \frac{\mathbf{k}}{k^2} \delta_{\mathbf{k}} \widetilde{W}(k) e^{i\mathbf{k}\cdot\mathbf{r}}. \qquad \textbf{(21.67)}$$

Thus, the mean square value of the smoothed velocity is

$$\bar{v}^2 = \langle \bar{\mathbf{v}}^2 \rangle = 4\pi (H_o f)^2 \int_0^\infty dk \mathcal{P}(k) \widetilde{W}(k)^2. \qquad \textbf{(21.68)}$$

Equation (21.57) for the rms fluctuation in the mass in a square window assumes the mass autocorrelation function is negligibly small at $r > R$ and the width of the window is large compared to R. In this same limit, equation (21.68) is

$$\bar{v}^2 = 4\pi(H_of)^2 \mathcal{P}_o \int_0^\infty dk \widetilde{W}(k)^2$$

$$= H_o^2 f^2 J_3 \int d^2V \, W_1 W_2/r_{12}. \tag{21.69}$$

The first line assumes the spectrum is flat at large scales, and the second follows by a short calculation with $\widetilde{W}(k)$ as the Fourier transform of $W(\mathbf{r})$. For a square spherical window the integral works out to (eq. [7.72])

$$\bar{v}^2 = \frac{6}{5} \frac{H_o^2 f^2 J_3}{r_s}. \tag{21.70}$$

This is the mean square value of the peculiar velocity averaged over a square window of radius r_s that is large compared to the correlation length of the mass distribution.

The autocorrelation function of the peculiar velocity field is

$$\langle v^\alpha(\mathbf{r}_1)v^\beta(\mathbf{r}_2)\rangle = \xi_v^{\alpha\beta}(\mathbf{r}_2 - \mathbf{r}_1). \tag{21.71}$$

For a homogeneous and isotropic random process the general form of this function is

$$\xi_v^{\alpha\beta}(\mathbf{r}) = \Pi(r)\hat{r}^\alpha\hat{r}^\beta + \Sigma(r)(\delta^{\alpha\beta} - \hat{r}^\alpha\hat{r}^\beta) \tag{21.72}$$

(Davis and Peebles 1977), where $\hat{r}^\alpha = r^\alpha/r$ is the unit vector along the line joining the particles, and $\Pi(r)$ and $\Sigma(r)$ are the one-dimensional velocity dispersions in directions parallel and perpendicular to the line joining the particles. With equation (21.67) for the velocity, the autocorrelation function (21.71) is

$$\xi_v^{\alpha\beta} = 4\pi H_o^2 f^2 \int_0^\infty dk \mathcal{P}(k) \left[\delta^{\alpha\beta}\frac{j_1(kr)}{kr} - \hat{r}^\alpha\hat{r}^\beta j_2(kr)\right] \tag{21.73}$$

(Górski 1988; Groth, Juszkiewicz, and Ostriker 1989), where the j_l are spherical Bessel functions. In terms of the mass autocorrelation function, this is

$$\xi_v^{\alpha\beta} = \frac{H_o^2 f^2}{2} \left[\left(\frac{J_3(r)}{r} - \frac{J_5(r)}{3r^3} + \frac{2K_2(r)}{3}\right)\delta^{\alpha\beta} + \left(\frac{J_5(r)}{r^3} - \frac{J_3(r)}{r}\right)\hat{r}^\alpha\hat{r}^\beta\right], \tag{21.74}$$

where the integrals are

$$J_n(r) = \int_0^r \xi r^{n-1} dr , \qquad K_n(r) = \int_r^\infty \xi r^{n-1} dr . \tag{21.75}$$

The velocity autocorrelation function as a probe for the large-scale structure of the mass distribution was first considered in the scalar version

$$\xi_v(r) = \langle \mathbf{v}(\mathbf{r}_1) \cdot \mathbf{v}(\mathbf{r}_1 + \mathbf{r}) \rangle = 4\pi (H_o f)^2 \int_0^\infty j_0(kr) \mathcal{P}(k) dk$$

$$= (H_o f)^2 \left[\frac{1}{r} \int_0^r r^2 dr \, \xi(r) + \int_r^\infty r dr \, \xi(r) \right] \tag{21.76}$$

(Clutton-Brock and Peebles 1981). The third expression follows from equation (21.73) because $3j_1/z - j_2 = j_0$. The second line is another way to equation (21.70).

In linear perturbation theory in a cosmologically flat universe, the Sachs-Wolfe relation (21.23) between the mass density contrast $\delta(\mathbf{r})$ evaluated at the present epoch and the CBR temperature as a function of position in the sky is

$$\tau = \frac{\delta T}{T} = -\frac{1}{3} \frac{G\rho_o}{D_o} \int \frac{\delta(\mathbf{r}')}{|\mathbf{r} - \mathbf{r}'|} d^3 r' . \tag{21.77}$$

The vector $\mathbf{r}$ points in the direction of measurement of the CBR temperature, and the length of $\mathbf{r}$ is the proper distance at the present epoch from the point of origin of the detected radiation (eq. [21.4]).

The fluctuations in the CBR temperature as a function of position across the sky yield a particularly sensitive measure of the large-scale mass distribution, because the integral in equation (21.77) scales as the square of the linear extent of the density fluctuation (eq. [21.2]). The large-scale CBR anisotropy is best represented by its angular power spectrum, which is the analog on a sphere of the power spectrum we have been using for an expansion in plane waves in flat space. Let us pause to note the main features of the angular fluctuation spectrum.

The spherical harmonic expansion of the sky temperature as a function of angular position is

$$\tau(\theta, \phi) = \sum_{l,m} a_l^m Y_l^m(\theta, \phi), \qquad a_l^m = \int d\Omega \, Y_l^{-m} \tau(\theta, \phi) . \tag{21.78}$$

The spherical harmonics satisfy the usual normalization,

$$\int d\Omega \, Y_l^m Y_{l'}^{-m'} = \delta_{ll'} \delta_{mm'} , \tag{21.79}$$

and the addition theorem

$$\sum_m Y_l^m(1)Y_l^{-m}(2) = (2l+1)P_l(\cos\theta_{12})/(4\pi).$$ (21.80)

The Legendre polynomials are P_l, and θ_{12} is the angular distance between the directions $\vec{\theta}_1$ and $\vec{\theta}_2$ in the arguments of the spherical harmonics.

The second moment of the expansion coefficients in equation (21.78) is

$$\langle a_l^m a_{l'}^{-m'}\rangle = \int d\Omega_1 d\Omega_2 Y_l^{-m} Y_{l'}^{m'} w(\theta_{12}),$$ (21.81)

where the angular two-point correlation function is

$$w(\theta_{12}) = \langle\tau(1)\tau(2)\rangle.$$ (21.82)

If the fluctuations are a stationary process, w is a function of the angular distance θ_{12} between the two directions in the integral, so it can be expanded as a series in Legendre polynomials, and since the addition theorem (21.80) represents these polynomials as sums over the orthogonal spherical harmonics, we see that the second moment is

$$\langle a_l^m a_{l'}^{-m'}\rangle = \delta_{ll'}\delta_{mm'}a_l^2,$$

$$a_l^2 = 2\pi\int d\cos\theta\, P_l(\cos\theta)w(\theta),$$ (21.83)

$$w(\theta) = \sum \frac{2l+1}{4\pi}a_l^2 P_l(\cos\theta).$$

This is the analog of the last part of equation (21.40), which relates the space autocorrelation function to the power spectrum.

The angular spectrum is independent of m,

$$(a_l)^2 = \langle|a_l^m|^2\rangle,$$ (21.84)

so from an all-sky map of $\delta T/T$ one has $2l+1$ estimates of a_l. Since the fluctuation spectrum a_l^2 is the transform of the correlation function $w(\theta)$ in equation (21.83), the functions a_l and $w(\theta)$ contain the same information, but the interpretation can be easier with one of the functions, depending on the form of the fluctuation spectrum. Of particular importance for measurements on large angular scales is the relatively large dipole moment a_1, which very likely is the effect our peculiar motion. If the dipole moment is left in $\delta T/T$, it dominates the correlation function $w(\theta)$. If the a_1^m are removed, it removes the dipole moment of the sky fluctuations, which biases $w(\theta)$ away from zero. The easy way out of this problem is to consider the angular fluctuation spectrum a_l. We noted in section

19 another advantage of the fluctuation spectrum, that the estimates a_l^m from a full sky map are uncorrelated in the sense that the correlation coefficients vanish unless $l = l'$ and $m = m'$. Also, the central limit theorem tells us the $|a_l^m|^2$ have an exponential distribution, with mean a_l^2 (Yu and Peebles 1969).[10]

Finally, it is useful to have the relation between the fluctuation spectrum a_l and the rms value of τ averaged through a window of size θ_c, as in equation (21.53) for a spatial distribution. We will consider a Gaussian window, where $\tau = \delta T / T$ smoothed by the window is

$$\bar{\tau}(\vec{\theta}) = \int d\Omega_1 \tau(\vec{\theta}_1) e^{-(\vec{\theta} - \vec{\theta}_1)^2 / 2\theta_c^2} / N , \tag{21.85}$$

and N is the usual normalization. For the mean square value of $\bar{\tau}$ we need the integral

$$\int d\Omega_1 e^{-[\vec{\theta}_1{}^2 + (\vec{\theta}_1 + \vec{\theta})^2]/2\theta_c^2} = e^{-\theta^2/4\theta_c^2} N / 2 \tag{21.86}$$

for $\theta_c \ll 1$. With equations (21.82) and (21.83), we have

$$\langle \bar{\tau}^2 \rangle = \frac{1}{2N} \int d\Omega \, w(\theta) e^{-\theta^2/4\theta_c^2}$$
$$= \frac{1}{2N} \sum \frac{2l+1}{4\pi} a_l^2 I_l , \tag{21.87}$$

with

$$I_l = \int d\Omega \, P_l(\cos\theta) e^{-\theta^2/4\theta_c^2} . \tag{21.88}$$

At $\theta \ll 1$, the Legendre polynomials are

$$P_l(\cos\theta) = \frac{1}{2\pi} \oint d\psi \, e^{i[(l+1/2)\theta \cos\psi]} = J_0((l+1/2)\theta) \tag{21.89}$$

(Jahnke and Emde 1945, p. 116), with J_0 the Bessel function of zero order. Now it is an easy exercise to check that at small smoothing window θ_c the sum in equation (21.87) for the mean square temperature fluctuation is

$$\langle \bar{\tau}^2 \rangle = \sum_l \frac{2l+1}{4\pi} a_l^2 e^{-\theta_c^2 l^2} . \tag{21.90}$$

---

[10] Note that the lack of correlation of the a_l^m is not the same as statistical independence: the second moments vanish by the assumption that the fluctuations are a stationary (statistically isotropic) process, just as stationarity implies that the Fourier coefficients in flat space are uncorrelated, $\langle \delta_{\mathbf{k}} \delta_{-\mathbf{k}'} \rangle = \mathcal{P} \delta_{\mathbf{k}\mathbf{k}'}$. Also, if the full sky is not sampled, the a_l^m are correlated, but the effect is small if the sky fraction sampled is reasonably large. This is discussed in Hauser and Peebles (1973).

Returning to the Sachs-Wolfe effect, the series expansion of the mass density contrast $\delta(\mathbf{r}')$ in equation (21.39) brings equation (21.77) to

$$\tau = -\frac{H_o^2}{2} \frac{\Omega}{D_o} \frac{(2\pi)^{3/2}}{V_u^{1/2}} \sum \frac{\delta_\mathbf{k}}{k^2} e^{i\mathbf{k}\cdot\mathbf{r}}, \tag{21.91}$$

with $\Omega H_o^2 = 8\pi G\rho_0/3$. Since this is a sum over plane waves, the expansion coefficients a_l^m follow from the plane wave expansion

$$e^{i\mathbf{k}\cdot\mathbf{r}} = 4\pi \sum i^l j_l(kr) Y_l^m(\Omega_\mathbf{r}) Y_l^{-m}(\Omega_\mathbf{k}), \tag{21.92}$$

where the directions $\Omega_\mathbf{r}$ and $\Omega_\mathbf{k}$ point along $\mathbf{r}$ and $\mathbf{k}$, and the spherical Bessel function is

$$j_l(x) = \sqrt{\frac{\pi}{2x}} J_{l+1/2}(x). \tag{21.93}$$

This yields

$$a_l^m = -2\pi H_o^2 \frac{\Omega}{D_o} \frac{(2\pi)^{3/2}}{V_u^{1/2}} \sum_\mathbf{k} i^l j_l(kr) Y_l^{-m}(\Omega_\mathbf{k}) \delta_\mathbf{k}/k^2. \tag{21.94}$$

The mean square value is

$$(a_l)^2 = \langle |a_l^m|^2 \rangle = 4\pi^2 H_o^4 \frac{\Omega^2}{D_o^2} \int_0^\infty \frac{dk}{k^2} \mathcal{P}(k) j_l(kr)^2, \tag{21.95}$$

where the angular size distance r in the argument of the spherical Bessel function is given by equation (21.4).

If the power spectrum can modeled as a power law,

$$\mathcal{P} = Ak^n, \tag{21.96}$$

and the integral for a_l converges, it is given by the expression

$$\int_0^\infty \frac{dz}{z^m} j_l(z)^2 = \frac{\pi}{2^{m+2}} \frac{m!}{(m/2)!^2} \frac{(l-m/2-1/2)!}{(l+m/2+1/2)!} \tag{21.97}$$

(Gradshteyn and Ryzhik 1965). This gives the analytic solution for the spectrum a_l of the fluctuations of the CBR temperature caused by the gravitational potential of the mass fluctuation spectrum (21.96) (Peebles 1981a and 1982a; Bond and Efstathiou 1987).

It might be noted finally that useful expressions for the integral of the mass autocorrelation function for the power law power spectrum (21.96), with different ranges of convergence, are

$$J_3(r) = \int_0^r r^2 dr \xi(r) = 4\pi \frac{(n+1)!}{n} \sin\left(\frac{\pi n}{2}\right) \frac{A}{r^n}$$

$$= 2^{n+2} \pi^{3/2} \frac{(n/2 + 1/2)!}{(-n/2)!} \frac{A}{r^n} .$$

(21.98)

Numerical Estimates

There are many open questions about the nature of the mass distribution in the observable universe, and doubtless many more yet to be framed, but we do have a useful if schematic outline of the picture from the fluctuations in galaxy counts, the galaxy peculiar velocity field, and the large-scale anisotropy of the CBR. Since all these measures are improving at an impressive rate, detailed models may be of transient interest. The strategy used here is to compare orders of magnitude for three simple models for the spectrum of the mass fluctuations on large scales. Other aspects of this approach are discussed in Juszkiewicz, Górski, and Silk (1987) and Suto et al. (1988). Models that are more elaborate and perhaps more realistic are considered in section 25.

The first model assumes the mass autocorrelation function is usefully approximated by the galaxy two-point function on scales of about $10h^{-1}$ Mpc, and that at $r > 30h^{-1}$ Mpc, where the galaxy function is known only to be quite small, the mass function vanishes. This last assumption is not immediately ruled out by the presence of structures that are more than $30h^{-1}$ Mpc wide. Even in a model with no large-scale structure in the statistical sense that the clustering pattern has been built by placing cluster balls at random, the statistical accidents of where the cluster balls are placed builds connected entities much larger than a cluster ball. One sees examples in figures 19.1 and 19.2. A sheet of galaxies is not a statistical accident, but it could be shaped out of one without appreciably changing the mass autocorrelation function or the large-scale velocity field that the mass distribution produces.

We will see that this model, with zero correlation of the mass fluctuations at large separations, gives a not unreasonable picture for the large-scale velocity field, but overproduces density fluctuations on the scale of the Hubble length. A simple way out assumes the mass fluctuations are anticorrelated at large separations, as in the second model discussed here, and in the isocurvature models discussed in section 25.

Here we are approximating the mass autocorrelation function as

$$\xi(r) = \begin{cases} (r_o/r)^\gamma & \text{at } r < r_x, \\ 0 & \text{at } r > r_x = 30h^{-1} \text{ Mpc}. \end{cases}$$

(21.99)

The parameters are taken from the galaxy function in equation (19.18), and the choice for the cutoff is about at the end of the break in the galaxy function in figures 7.2 and 7.3. In this model the integrals of the two-point function are

$$J_3 = \frac{r_0^\gamma r_x^{3-\gamma}}{3-\gamma} = 1000h^{-3}\,\mathrm{Mpc}^3,$$

$$J_2 = \frac{r_0^\gamma r_x^{2-\gamma}}{2-\gamma} = 200h^{-2}\,\mathrm{Mpc}^2. \tag{21.100}$$

These numbers are in line with the estimates from the Lick (Clutton-Brock and Peebles 1981) and CfA (Davis and Peebles 1983) catalogs. The variance of the mass within a cube with side l large compared to r_x is (eq. [21.57])

$$\bar{\delta} = \left(\frac{4\pi J_3}{l^3}\right)^{1/2} = 0.3 \quad \text{for } l = 50h^{-1}\,\mathrm{Mpc}. \tag{21.101}$$

This is reasonably close to the variance of counts in cells in the IRAS catalog mapped in figure 3.7, as indicated in equation (3.24) (Efstathiou 1991). The rms fluctuation in the mass in a sphere with radius equal to the Hubble length is

$$\delta_h = [3(H_o/c)^3 J_3]^{1/2} = 3 \times 10^{-4}. \tag{21.102}$$

This sets the scale for the large angular scale fluctuations in the X-ray and thermal cosmic backgrounds. The bounds on the former are consistent with this number. The latter requires a more careful analysis, as will be discussed below.

In this model, the effective cluster ball mass and number density (eqs. [21.59] and [21.60]) are

$$\mathcal{M}_{\mathrm{eff}} = 3 \times 10^{15}\Omega h^{-1}\mathcal{M}_\odot, \qquad n_{\mathrm{eff}} = 8 \times 10^{-5}h^3\,\mathrm{Mpc}^{-3}. \tag{21.103}$$

Depending on the density parameter, this is between one and ten times the mass within the Abell radius of an Abell cluster (eq. [20.39]) and comparable to the mass of the Local Supercluster in a sphere centered on the Virgo cluster and at our radius (eq. [20.58]). The number density n_{eff} is an order of magnitude larger than the number density of Abell clusters (eq. [20.33]).

With equation (21.100) for J_2, the Layzer-Irvine relation gives the rms peculiar velocity (eqs. [21.28], [21.30], and [21.38])

$$v_a \sim (3\pi G\rho_b J_2)^{1/2} = 1500\,\Omega^{1/2}\,\mathrm{km\ s}^{-1}. \tag{21.104}$$

For any reasonable choice of the density parameter, this net rms peculiar velocity is considerably larger than the relative velocity dispersion in figure 20.1, meaning the peculiar velocities are dominated by the large-scale component. This is considered realistic. For example, our motion relative to the CBR is 600 km s^{-1}, considerably larger than the velocity dispersion in our neighborhood or the small-scale relative velocity dispersion in figure 20.1.

For a measure of the large-scale velocity field in this model, let us use equation (21.70) for the mean square value of the velocity averaged over a sphere of radius r_s. With equation (21.100) for J_3, the predicted rms velocity is

$$\bar{v} = 500\Omega^{0.6} \text{ km s}^{-1} \quad \text{at } r_s = 50h^{-1} \text{ Mpc}. \tag{21.105}$$

This is appreciably smaller than the single-particle rms velocity, v_a, because the contribution to the velocity field peaks at r_x, half the radius of the spherical averaging volume in equation (21.105).

The rms smoothed velocity, $\bar{v}$, applies to a randomly placed window. In a fluctuating velocity field it would be easy to find a place to center the sphere, with us near the edge, and placed so the mean velocity averaged within the sphere is two standard deviations, or about $1000 \Omega^{0.6}$ km s^{-1}. With the density parameter in the observationally acceptable range discussed in the last section, $0.1 \lesssim \Omega \lesssim 1$, this is comparable to what is observed in velocity fields such as the Great Attractor, as discussed in Rubin and Coyne (1988). If $\Omega = 1$ the predicted large-scale velocities may be somewhat high, but that could be remedied by the biasing idea to be discussed in section 25, which assumes the mass autocorrelation function on scales of the order of $10h^{-1}$ Mpc is smaller than the galaxy two-point function, rather than similar in value, as assumed in equations (21.104) and (21.105).

In this model the mass fluctuation spectrum at long wavelengths is flat, so the multipole moments of the CBR anisotropy from the Sachs-Wolfe effect are given by equations (21.95) and (21.97) with $m = 2$:

$$(a_l)^2 = \frac{2\pi\Omega^2 D_o^{-2} H_o^4 r_h J_3}{(2l+3)(2l+1)(2l-1)}. \tag{21.106}$$

The proper distance to the point of last scattering at high redshift is r_h, and the Einstein-de Sitter value is $r_h = r_e = 2H_o^{-1}$. With the normalization in equation (21.100), the quadrupole moment in the Einstein-de Sitter model is

$$a_2 = \left(\frac{4\pi J_3 H_o^3/c^3}{105} \right)^{1/2} = 7 \times 10^{-5} \tag{21.107}$$

(Peebles 1981a). This scales with the density parameter as $a_2 \propto \Omega D_o^{-1}(r_h/r_e)^{1/2}$. The first factor comes from the mass density, which fixes the gravitational potential for a given present density contrast, and the factor $D_o(\Omega)$ accounts for the suppression of the growth factor of the contrast relative to the Einstein-de Sitter model. If $\Omega \neq 1$ there is a contribution of comparable size from the last term in equation (21.18). Since D_o is not very sensitive to Ω in a cosmologically flat model, the predicted quadrupole moment is not very sensitive to the density parameter.

The angular fluctuation spectrum (21.106) at large l scales as $a_l^2 \sim l^{-3}$, so the sum in equation (21.90) for the rms value of $\delta T/T$ smoothed through a window of radius θ_c is dominated by the smallest values of l, and thus is almost independent of θ_c. That is, the anisotropy caused by the white noise mass fluctuations is dominated by the largest observable scales.

There have been important contributions to the measurement of the quadrupole anisotropy from ground-based and balloon observations and the RELICT and COBE satellites (Strukov 1990; Smoot et al. 1991). At the time this is written the best number is

$$a_2 = 7.5 \pm 2.5 \times 10^{-6} \qquad (21.108)$$

(Smoot et al. 1992). This result is outside the uncertainties in the parameters in the model in equation (21.107), so the model fails the test by a convincing factor.

The quadrupole moment is relatively large in this model because we assumed the mass density fluctuations are uncorrelated at separations greater than $a_x = 30h^{-1}$ Mpc, and we normalized to the observed fluctuations in galaxy counts on this scale. The latter assumption is relaxed in the biasing picture to be discussed in section 25, but the biasing factors that are considered reasonable are much closer to unity than the factor ~ 10 discrepancy here. Thus the assumption of uncorrelated mass fluctuations at large separations has to be wrong; on large scales the fluctuations have to be anticorrelated to suppress the rms mass contrast on the scale of the Hubble length.

Let us check that a very modest anticorrelation can considerably decrease $J_3(r)$, and with it the quadrupole moment a_2. This anticorrelation is present in the cold dark matter and baryonic isocurvature models discussed in section 25. We can introduce the effect in the present calculation by adjusting the assumed primeval form of the mass fluctuation spectrum.

Let us suppose the power spectrum is

$$\mathcal{P}(k) = Ake^{-ks}. \qquad (21.109)$$

At large wavelengths (small wavenumber k), this is the scale-invariant form produced in many cosmic field models and simple models for inflation (sections 16 and 17). The cutoff represents the suppression of the growth of clustering on small scales, where virialization has slowed the growth of the clustering pattern. As indicated in equation (21.56), since this spectrum approaches zero at large wavelength the mass autocorrelation function is negative at large separation, meaning the fluctuations are anticorrelated. That reduces the large-scale mass fluctuations and the Sachs-Wolfe CBR anisotropy.

To evaluate the Fourier transform of $\mathcal{P}(k)$ in equation (21.109), write $\sin kr$

as the imaginary part of an exponential. The result for the mass autocorrelation function (eq. [21.40]) is

$$\xi(r) = 4\pi A \int_0^\infty k^3 dk \, e^{-ks} \sin kr / (kr)$$

$$= 8\pi A \frac{3s^2 - r^2}{(s^2 + r^2)^3} \, . \tag{21.110}$$

The correlation function is $\xi = -8\pi A / r^4$ at $r \gg s$, and at $r \ll s$ the correlation function is positive and constant at $\xi = 24\pi A / s^4$. A more realistic model would bend the spectrum to $\mathcal{P} \propto k^{-1.2}$ at large k, to get the power law part of the correlation function, but the simpler model in equation (21.109) will do to fix the large-scale measures.

A realistic choice for the cutoff length s in the spectrum (21.109) is small compared to the Hubble length, so the low-order moments of the CBR angular fluctuation spectrum are insensitive to the cutoff. We have from equations (21.95) and (21.97) for the Einstein-de Sitter model, where $r = 2H_o^{-1}$,

$$a_l^2 = \frac{2\pi^2 A H_o^4}{l(l+1)} \, . \tag{21.111}$$

A reasonable number for the cutoff is $s = r_x / 3^{1/2}$, so the mass autocorrelation function (21.110) vanishes at $r = r_x = 30h^{-1}$ Mpc. Then, if the fluctuations in the galaxy counts trace the mass, we can normalize the mass fluctuations to $\xi = 1$ at r equal to the galaxy clustering length, r_o. That makes the quadrupole moment

$$a_2 = 8 \times 10^{-6} \tag{21.112}$$

(Peebles 1982a). This is an order of magnitude below the flat spectrum case, and quite close to the measured value (eq. [21.108]).

At large l the spectrum (21.111) is

$$a_l^2 = \frac{6}{l^2} a_2^2 \, . \tag{21.113}$$

At small smoothing radius θ_c the sum in equation (21.90) may be approximated as an integral:

$$\langle \bar{\tau}^2 \rangle = \frac{3a_2^2}{\pi} \int \frac{dl}{l} e^{-\theta_c^2 l^2}$$

$$\sim \frac{3}{\pi} a_2^2 \ln \theta_c^{-1} \, . \tag{21.114}$$

The measured rms CBR temperature anisotropy is

$$\bar{\tau}_{\text{COBE}} = 1.1 \pm 0.2 \times 10^{-5} \quad \text{at} \quad \theta_c \sim 4° \tag{21.115}$$

(Smoot et al. 1992). Equation (21.114) with equation (21.112) for a_2 is $\bar{\tau} \sim 1 \times 10^{-5}$ at this smoothing radius, in reasonable agreement with the measurement in equation (21.115).

With $\mathcal{P} = Ak$, the large-scale peculiar velocity field averaged within a square spherical window of radius r_s works out to (eqs. [21.52], [21.68], and [21.97])

$$\bar{v}^2 = 9\pi H_o^2 f^2 A / r_s^2. \tag{21.116}$$

The more rapid decrease with increasing averaging radius r_s compared to the flat spectrum case in equation (21.70) is a result of the anticorrelation at large separations that suppresses the mass fluctuations. With the normalization used in equation (21.112), the rms velocity at window radius $r_s = 50h^{-1}\text{Mpc}$ is $\bar{v} = 400\Omega^{0.6}$ km s^{-1}, almost the same as for the flat spectrum case in equation (21.105).

The conclusion is that we can adjust the primeval mass fluctuation spectrum to get reasonable-looking orders of magnitude for the statistics of the large-scale velocity field and the mass fluctuations that produce them, and, by suppressing the spectrum at large scales (small k), reasonable agreement with the large-scale anisotropy of the CBR. It will be interesting to see whether it continues to be possible to tune the shape of the mass fluctuation spectrum to fit the improving measurements of the low-order multipole moments of the CBR and the peculiar velocity field.

In the absence of the constraint from the CBR, it would have been easy to imagine that the flat spectrum model seriously underestimates the large-scale fluctuations in the mass distribution, rather than overestimates them. The mass fluctuations are measured by the integral $J_3(r) = \int \xi r^2 dr$. Since the weight multiplying $\xi(r)$ in the integral increases as r^3, only a small positive value of $\xi(r)$ at large r would be needed to make a considerable addition to J_3. To see the size of this effect, consider how the statistical measures scale with the power law cutoff r_x in the flat spectrum model in equation (21.99) (at fixed r_o). The rms velocity scales as $v_a \propto J_2^{1/2} \propto r_x^{1-\gamma/2}$ (eq. [21.104]). With the power law observed for the galaxy distribution, $\gamma = 1.8$, this is not sensitive to the cutoff. But the rms mass fluctuation at the Hubble length scales as $\delta_h \propto J_3^{1/2} \propto r_x^{(3-\gamma)/2}$, so increasing r_x considerably increases δ_h, and with it the large-scale anisotropy of the cosmic background radiation.

For a more direct model, suppose the spectrum at long wavelengths is

$$\mathcal{P}(k) = B/k, \tag{21.117}$$

where B is a constant. Equation (21.76) for the autocorrelation function of the peculiar velocity field is

$$\langle \mathbf{v}(\mathbf{r}_1) \cdot \mathbf{v}(\mathbf{r}_1 + \mathbf{r}) \rangle = 4\pi H_o^2 f^2 B \int \frac{dk}{k} \frac{\sin kr}{kr} . \tag{21.118}$$

The integral diverges, but only as the logarithm of the ratio of the Hubble length to the long wavelength cutoff of the power law model, so this picture for the large-scale velocity field is not unreasonable. The Fourier transform of equation (21.117) is the mass autocorrelation function, which gives the rms fluctuation $\bar{\delta}(r)$ in the mass contained within a sphere of radius r_s,

$$\xi(r) = \frac{4\pi B}{r^2} , \qquad \bar{\delta}^2 = \frac{9\pi B}{r_s^2} . \tag{21.119}$$

We can make this an extrapolation of the galaxy two-point correlation function, $\xi_{gg} = (r_o/r)^\gamma$, by normalizing the spectrum to $4\pi B = r_o^2$. In this model, equations (21.95) and (21.97) for the CBR quadrupole moment are

$$a_2 = \left(\frac{\pi}{18} \right)^{1/2} \frac{H_o r_o}{c} = 8 \times 10^{-4} , \tag{21.120}$$

well above the measurement in equation (21.108).

  The conclusion from the model in equation (21.117) is that the power law form for the galaxy two-point function observed at $r \lesssim 10h^{-1}$ Mpc, extrapolated to the Hubble distance, makes $\bar{\delta} = \delta \mathcal{M}/\mathcal{M}$ well below unity on large scales, but the resulting potential energy fluctuations are two orders of magnitude too large.[11] It is encouraging, and perhaps significant, that the measured galaxy two-point function agrees with this conclusion: it is a power law at small separations, but shows a distinct break down from the power law about at the cutoff in equation (21.99), as is illustrated in figures 7.2 and 7.3. It still may turn out that ξ is positive at very large separations, as might be suggested by the sheet model in equation (19.33). If this proved to be so, and if there were not a very special arrangement to cancel the mass fluctuations, it would be a crisis for the standard model.

---

[11] Another illustration of this effect was encountered in the construction of the model galaxy distribution mapped in figure 19.1. In the first attempts, the two-point correlation function in the model had a large-scale tail that was presumed not to matter, because the value of ξ in the tail was quite small. But the tail made the integral J_3 large, and with it made very visible and unrealistic fluctuations in the surface density of galaxies in the model map at the Lick depth. The improvement afforded by a cutoff of $\xi(r)$ was first checked by cutting the model map into pieces that were reassembled at random.

Lessons

Two main general lessons emerge from these considerations. First, equations (21.105) and (21.116) show that large-scale velocity fields comparable in magnitude to what is observed are produced by the known fluctuations in the galaxy distribution, if mass follows galaxies (Clutton-Brock and Peebles 1981). The coherence length of the peculiar velocity field is considerably broader than that of the mass, because the gravitational acceleration is a long-range field. Thus, when the large-scale fluctuations in the mass distribution are uncorrelated, the rms velocity averaged through a square spherical window scales with the window radius as $\bar{v} \propto r_s^{-1/2}$ (eq. [21.70]). This broadening of the velocity coherence length relative to that of the galaxies is similar to what is suggested by the measurements of the galaxy peculiar velocity field (as reviewed in Rubin and Coyne 1988).

The second lesson is that the measurements of the large-scale anisotropy of the CBR (eqs [21.108] and [21.115]) require that the large-scale fluctuations in the mass distribution are anticorrelated. This constraint comes about because the gravitational potential has an even broader range than that of the gravitational acceleration, and indeed uncorrelated mass density fluctuations drive rms fluctuations in the potential that diverge with increasing length scale. The scale-invariant model in equation (21.109) gives a reasonable fit to the presently known constraints on large-scale structure, and it is notable that this spectrum naturally appears in cosmic field models and simple models for inflation. The cold dark matter model assumes this spectrum; some other aspects of this model are discussed in section 25. A second way to anticorrelated large-scale mass fluctuations is the isocurvature baryonic dark matter model also discussed in section 25.

It is of course feasible to make much more detailed analyses, analytic and numerical, of the predicted observational effects of a postulated primeval mass fluctuation spectrum, and much closer comparisons to the growing base of observations of the large-scale velocity field and the large-scale distributions of mass and galaxies (Frenk 1991 and references therein). Out of this could come a demonstration that the available parameters do not allow us to reconcile theory and observation within the standard model, which would be an exceedingly interesting result. The alternative would be that we are led to a reasonably definite prescription for the primeval mass fluctuations, which surely would be a helpful guide to the search for a fundamental theory of where they came from.

22. Gravitational Evolution

The mass concentrations in galaxies and clusters of galaxies could have been present in the early universe only as exceedingly small departures of the mass distribution from the mean, because the expanding universe is gravitationally unstable. It is generally believed that the mass structures observed on scales greater than a few megaparsecs grew by this gravitational instability out of much smaller density fluctuations present at the decoupling of matter and radiation at redshift $z_{dec} \sim 1000$, because no one has been able to think of any other reasonably effective force on such large scales. Still under discussion is the mix of stresses responsible for the assembly of galaxies, and for the origin of the small departures from homogeneity at very high redshift out of which the present large-scale structure is thought to have grown.

The study of the evolution of the mass distribution in an expanding universe has become a rich topic for analytic methods, with emphasis on scaling solutions, and for numerical experiments, such as N-body computations in which particles serve as coarse-grain tracers for the evolution of the fine-grain mass distribution. The analysis of the evolution in linear perturbation theory is presented in sections 5 and 10. This section continues the discussion to a few results beyond linear theory that are simple and in general use. Some of these results are well established, some still conjectural. Still other more model-dependent and phenomenological issues are addressed in the following sections.

Evolution of the Power Spectrum

In analytic and numerical studies of the evolution of the mass distribution, it can happen that one arrives at an estimate of the power spectrum $\mathcal{P}(k)$ that serves as a reasonably close approximation to the true fluctuation spectrum within the model, but that has a Fourier transform $\xi(r)$ with unfortunate properties, perhaps oscillating through zero where one is sure the true mass autocorrelation function is positive, or even violating the constraint $\xi \geq -1$. The problem with the instability of the Fourier transform is avoided by working with the power spectrum itself, or with nearly equivalent statistics, such as the integral $J_3(r)$ and the standard deviation $\bar{\delta}(r)$ of the density contrast smoothed through a window of radius r, which are integrals over $\mathcal{P}(k)$ cut off at $k \lesssim 1/r$ (eqs. [21.53] and [21.62]). A particular advantage of this approach is that in the limit of large r, and neglecting nongravitational interactions, there is a simple and useful theory for the time evolution of the spectrum that is valid even when the mass clustering is strongly nonlinear on smaller scales.

The starting assumption is that the departures from a homogeneous expanding universe are large only on scales small compared to the present Hubble length, and grow by gravity alone out of small initial fluctuations. At comoving wavelength $2\pi/k$ large compared to the mass clustering length, the Fourier amplitudes are in the linear regime, and so vary with time as $\delta_{\mathbf{k}}(t) \propto D(t)$, where $D(t)$ is the growing solution to the linear perturbation equation (5.111). Since J_3 and $\bar{\delta}^2$ are integrals of $\mathcal{P}(k) = |\delta_{\mathbf{k}}(t)|^2$, it follows that when x is large compared to the clustering length — so that there is a small value of the rms density contrast $\bar{\delta}(x, t)$ smoothed over a sphere of radius x — these statistics scale with time at fixed comoving x as

$$J_3(t,x) \propto \bar{\delta}(t,x)^2 \propto D(t)^2 . \tag{22.1}$$

To see why this result from linear perturbation theory applies even when the clustering is strongly nonlinear on scales small compared to x, suppose the mass distribution is represented as a collection of particles with masses m_j at positions x_j. Then the Fourier amplitudes for the mass distribution are (eq. [21.43])

$$\delta_{\mathbf{k}} \propto \sum_j m_j e^{i\mathbf{k}\cdot\mathbf{x}_j} . \tag{22.2}$$

Consider a Fourier component $\delta_{\mathbf{k}}$ with wavelength large compared to the clustering length x_o at which the structure is nonlinear. Since the nonlinear interactions displace particles by distances $\Delta x_j \lesssim x_o$, we can expand the effect on $\delta_{\mathbf{k}}$ of the motions associated with the nonlinear interactions as

$$\Delta\delta_{\mathbf{k}} \propto \sum_j m_j [e^{i\mathbf{k}\cdot(\mathbf{x}_j+\Delta\mathbf{x}_j)} - e^{i\mathbf{k}\cdot\mathbf{x}_j}]$$
$$= \sum_j m_j [i\mathbf{k}\cdot\Delta\mathbf{x}_j - (\mathbf{k}\cdot\Delta\mathbf{x}_j)^2/2 + \cdots]e^{i\mathbf{k}\cdot\mathbf{x}_j} . \tag{22.3}$$

The interactions conserve momentum, so the center of mass shift $\propto m_j\Delta x_j$ from particle j has to be balanced by net opposite shifts of nearby particles. This cancels the leading term in the expansion to order $\mathbf{k}\cdot\Delta\mathbf{x}$. Thus, the perturbation to the power spectrum $\mathcal{P}(k) = |\delta_{\mathbf{k}}|^2$ at $k^{-1} \gg x_o$ caused by the nonlinear interactions that redistribute matter on scales smaller than the clustering length x_o is on the order of $(kx_o)^4$. If the power spectrum at long wavelengths approaches zero less rapidly than $\mathcal{P} \propto k^4$, the linear terms dominate the evolution of $\delta_{\mathbf{k}}(t)$, and the time evolution in equation (22.1) applies.

Zel'dovich (1965) showed that nonlinear interactions build the tail $\mathcal{P} \propto k^4$ in the power spectrum, if a larger one is not already present. The argument for the evolution in equation (22.1) was given by Peebles (1974a) and Peebles and Groth (1976). The last reference also shows that in a numerical N-body model with a flat initial power spectrum the relation is quite a good approximation up to $\bar{\delta} = 1$.

In the Einstein-de Sitter limit, the linear growth factor is $D \propto a(t) \propto (1+z)^{-1}$. This means the effective mass in equation (21.60) at fixed comoving length x varies with time as

$$\mathcal{M}_{\text{eff}} \propto (1+z)^{-2} . \tag{22.4}$$

With the present value in equation (21.103), the value at decoupling of baryons from the cosmic background radiation, at $z \sim 1400$, is

$$\mathcal{M}_{\text{eff}} \sim 10^9 \Omega h^{-1} \mathcal{M}_\odot , \tag{22.5}$$

a mass characteristic of dwarf galaxies. This does not mean dwarf galaxies were present at decoupling, only that the rms value of the mass fluctuations present then on the comoving length scale $a_o x_x \sim 30h^{-1}$ Mpc at which J_3 is evaluated are of the size that would result by placing at random clumps with the mass in equation (22.5).

The First Generation: Zel'dovich Pancakes

What was the nature of the first generation of gravitationally bound objects? The answer depends on the picture for what happened at very high redshifts. In the isocurvature scenarios to be discussed in section 25, the first generation may have been present before matter had decoupled from the radiation drag at redshift $z_{\text{dec}} \sim 1400$ (eq. [6.96]), or may have formed relatively soon after at the matter Jeans mass (eq. [6.146]). In other scenarios the coherence length of the mass distribution at decoupling is much larger than the matter Jeans length, so the first generation collapses as a nearly pressureless gas. Zel'dovich (1970) showed that in the second case the first nonlinear bound systems form as sheetlike mass concentrations, or pancakes.

The starting assumptions for this result are that before orbits cross in the first collapse, nongravitational forces may be neglected, and that the initial mass distribution is smooth enough so that the initial peculiar velocity field is a twice-differentiable function of position.

Let $\mathbf{x}_i$ be the coordinate position of a mass element at high redshift, when the mass distribution is close to homogeneous, and let the position of this element at a time t when the first regions are collapsing be

$$x^\alpha = x^\alpha(\mathbf{x}_i, t). \tag{22.6}$$

The mass elements initially at coordinate separation $\delta x_i^\alpha = s_i^\alpha$ are at time t at separation

$$\delta x^\alpha = s^\alpha = A_{\alpha\beta} s_i^\beta , \qquad A_{\alpha\beta} = \frac{\partial x^\alpha}{\partial x_i^\beta} . \tag{22.7}$$

The set of points with separations initially on a small sphere of radius s_i are at time t on the ellipsoid of relative positions s^α defined by the equation

$$s_i^2 = \sum (s_i^\alpha)^2 = \sum A_{\alpha\gamma}^{-1} A_{\alpha\delta}^{-1} s^\gamma s^\delta . \tag{22.8}$$

Where matter is contracting to the first crossings of neighboring orbits, this ellipsoid generally contracts to a sheet, a Zel'dovich pancake. (A collapse to a line is a special case that requires the flow to be axially symmetric as it collapses. A collapse to a line or point is equivalent to the condition that eq. [22.11] has more than one linearly independent solution at the same instant of time.)

To follow the mass density in the neighborhood of the pancake, note that the mass in a volume element is

$$dm = \rho\, d^3x = \rho \left|\frac{\partial x}{\partial x_i}\right| d^3x_i = \rho_b(t)\, d^3x_i\,. \tag{22.9}$$

The first step defines the local mass density ρ at time t, the second is the coordinate transformation to the initial comoving positions, and the third step uses the fact that the initial positions are referred to the homogeneous mean mass density, ρ_b. We see that the local mass density is

$$\rho(\mathbf{x}, t) = \rho_b(t)/A\,, \tag{22.10}$$

where the Jacobian A of the transformation from initial positions to positions at time t is the determinant of the matrix $A_{\alpha\beta}$ in equation (22.7). Where neighboring orbits cross, the determinant A vanishes and the local mass density rises to a cusp. Since the determinant vanishes, the linear equation

$$A_{\alpha\beta}d^\beta = 0 \tag{22.11}$$

has the nontrivial solution d^β. It follows from equation (22.7) that mass elements initially along the line defined by $s_i^\alpha \propto d^\alpha$ have contracted to the point where the density diverges. The separation between nearby mass elements not on this line has two components normal to d^α that at the cusp define the plane of the flattened ellipsoid or pancake.

To find the character of the mass distribution in the neighborhood of a cusp at $\mathbf{x} = \mathbf{x}_o$, continue the expansion in equation (22.7) to the next order in relative position,

$$x^\alpha = x_o^\alpha + A_{\alpha\beta}(x_i^\beta - x_{o,i}^\beta) + \frac{1}{2}\frac{\partial A_{\alpha\beta}}{\partial x_i^\gamma}(x_i^\beta - x_{o,i}^\beta)(x_i^\gamma - x_{o,i}^\gamma)\,, \tag{22.12}$$

where $A_{\alpha\beta}$ and its derivatives are evaluated at $\mathbf{x}_{o,i}$. The first terms in the power series expansion of the determinant of $A_{\alpha\beta}$ are

$$A = A_o + \frac{\partial A_{\alpha\beta}}{\partial x_i^\gamma} B^{\alpha\beta}(x_i^\gamma - x_{o,i}^\gamma)\,, \tag{22.13}$$

where B_β^α is the matrix of the subdeterminants obtained by deleting row and column α and β from the matrix A (the cofactors). Suppose the determinant vanishes at the point $\mathbf{x}_i = \mathbf{x}_{o,i}$, and choose points along the initial line of collapse,

$$x_i^\alpha(s) = x_{o,i}^\alpha + sd^\alpha\,, \tag{22.14}$$

where $A_{\alpha\beta}d^\alpha = 0$. Then the separation in equation (22.12) evolves forward to

$$x^\alpha = x_o^\alpha + \frac{1}{2}\frac{\partial A_{\alpha\beta}}{\partial x_i^\gamma}d^\beta d^\gamma s^2, \qquad (22.15)$$

and equation (22.13) with $A_o = 0$ is

$$A = \frac{\partial A_{\alpha\beta}}{\partial x_i^\gamma}B^{\alpha\beta}d^\gamma s. \qquad (22.16)$$

Equation (22.15) says the distance $x = |\mathbf{x} - \mathbf{x}_o|$ away from the zero of A varies as $x \propto s^2$, so A in equation (22.16) varies as $s \propto x^{1/2}$, and the density in equation (22.10) varies as

$$\rho \propto \frac{1}{x^{1/2}}. \qquad (22.17)$$

The mass per unit area integrated across the cusp converges, so the gravitational acceleration is finite.

This discussion has not used dynamics; it simply follows the consequences of the continuity of the velocity field. In what usually is called the Zel'dovich approximation one writes equation (22.6) for the position of the mass element x_i^α at time t as $x^\alpha(\mathbf{x}_i, t) = x_i^\alpha + S^\alpha(\mathbf{x}_i)D(t)$, where $D(t)$ is the solution to the linear perturbation equation (5.111). This is the exact solution in a plane parallel mass distribution (eq. [16.56]), and it generally is a good approximation for the motion leading up to orbit crossing at a pancake.

The pancake effect was first demonstrated in numerical models for the evolution of the three-dimensional mass distribution by Doroshkevich et al. (1980), Klypin and Shandarin (1983), and Centrella and Melott (1983). Melott and Shandarin (1990) give an elegant demonstration of the effect by using two-dimensional computations that afford considerably better resolution for given total particle number. Shandarin (1988) and Kofman, Pogosyan, and Shandarin (1990) present a powerful semianalytic method for predicting the positions of the pancakes from the initial conditions.

The numerical example in figure 22.1 shows the distribution of particles in a slice 3/64 of the box width, with periodic boundary conditions, in a computation with 64^3 particles in the full three-dimensional box, and the gravitational interactions computed in a three-dimensional 64^3 lattice in wave number space. The positions are shown when the standard deviation in the mass density, scaled from the initial value in the approximation of linear perturbation theory, is

$$\langle(\delta\rho/\rho)^2\rangle^{1/2} = 0.5 \quad \text{and} \quad 1, \qquad (22.18)$$

in panels (a) and (b), respectively. The overall density fluctuations thus are

Figure 22.1. Pancake formation (Melott 1992). The 64^3 mass tracers initially are on a square lattice distorted to correspond to Gaussian density fluctuations with spectrum $\mathcal{P} \propto k$ on large scales, and a short wavelength cutoff of the form expected in the adiabatic hot dark matter model discussed in section 25. The two panels show the same slice of space, three initial cell widths deep, at the times defined in equation (22.18).

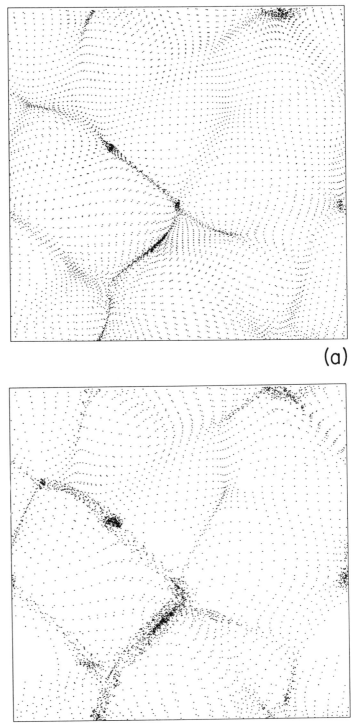

(a)

(b)

fairly small, but the peaks at three standard deviations have collapsed to the first strongly nonlinear structures, the Zel'dovich pancakes.

The pancakes look a good deal like the local plane of galaxies shown in figure 3.3. Perhaps such sheetlike features in the present galaxy distribution are Zel'dovich pancakes, but this could not be the whole story, for the pancakes shown in in figure 22.1 are a transient effect: with increasing time the mass in pancakes drains into clumps that are concentrated in all three dimensions. One sees examples of the effect by comparing the neighboring time slices in the figure. This means that if the local sheet of galaxies were a pancake it must have formed fairly recently. This is consistent with the model in figures 20.2 and 20.3 that indicates the Local Group is forming now, but quite inconsistent with the fact that the major galaxies in the Local Group are old, not now in the process of forming.

This argument, which is discussed further in section 25, leads to this question: Could there be a second generation of pancakes that form by the collective collapse of groups of the clumps that formed out of the first generation? This does not follow by the analysis given above, for it depends on the continuity of the velocity field that allows us to write down a series expansion for the evolution of relative positions. After the formation of the first generation of clumps, which might be the galaxies or their progenitors, the velocity field in general does not have a coherence length, and the analysis from continuity does not apply.

To see how the growth of clustering of the first-generation clumps might be expected to go, consider the smoothed velocity field $v(x)$ defined by the newly formed clumps. The characteristic time for the formation of clusters of clumps on the scale x is

$$t_x \sim \frac{ax}{v_{,x}x} \propto \frac{1}{\bar{\delta}_x} . \tag{22.19}$$

The denominator in the second expression is the typical relative velocity of clumps at separation x. This, divided into the relative separation, fixes the characteristic closing time for the formation of mass concentrations on this scale. The continuity equation indicates that this closing time is inversely proportional to the local density contrast $\bar{\delta}_x$ on the scale x. In general, the contrast decreases with increasing x, meaning the closing time increases with increasing x, meaning small collections of clumps tend to collapse before larger ones. That is, structure tends to grow in a clustering hierarchy rather than by the formation of pancakes. This does not apply to the classical Zel'dovich pancake theory, for it assumes the mass distribution has a coherence length within which $\bar{\delta}_x$ is nearly independent of x, implying near simultaneous collapse within a coherence length. A noncontroversial condition for the formation of a second generation of pancakes is that in the initial conditions there is a range of scales in which the power spectrum varies as $\mathcal{P}(k) \sim k^n$, with $n \lesssim -3$, so that in this range $\bar{\delta}_x$ does not increase appreciably with increasing x. In this case the relative velocity field allows the collapse of collections of clumps by a smooth flow to a new generation of pancakes at the scale set by the long wavelength cutoff of the $\mathcal{P}(k) \sim k^n$ part of the spectrum.

Accretion Models

The mass concentration in a galaxy or a cluster of galaxies, once formed, can continue to grow by gravitationally attracting more matter. The spherical accretion model, which was first discussed by Gunn and Gott (1972), gives a useful guide to how this growth might go.

A particularly simple and convenient version of the spherical accretion model assumes an Einstein-de Sitter universe in which pressureless noninteracting dust is accreting onto an isolated mass concentration outside of which the initial mass distribution is homogeneous, with no initial peculiar motion. This means the initial growing density contrast averaged over a sphere of initial comoving radius x at the starting time t_i is of the form

$$\delta_i = \frac{3m_i}{4\pi \rho_i a_i{}^3 x^3}. \tag{22.20}$$

The initial values of the density and expansion parameter are ρ_i and a_i, and the initial value of the isolated mass concentration belonging to the growing perturbation mode is m_i.

Now consider the density run that has built up around this seed well after the starting time, at expansion parameter $a \gg a_i$. We will use a scaling solution for the density run at small radii, where shells have crossed many times. At large radii, where shells have not yet crossed, we have the analytic solution discussed in sections 13 and 20 for the Local Group and the background cosmological model,

$$r = A(1 - \cos \eta), \qquad t = B(\eta - \sin \eta), \qquad A^3 = G\mathcal{M}B^2, \tag{22.21}$$

where $r(t)$ is the physical radius as a function of proper world time t for the shell that contains mass $\mathcal{M}$.

To apply the analytic solution, we need to find how the constants of integration A and B scale with the initial comoving radius x of the shell that contains mass $\mathcal{M} \propto x^3$. The Newtonian energy belonging to the mass shell with initial comoving radius x is

$$-E = \frac{G\mathcal{M}(x)}{2A} \propto \frac{1}{x}. \tag{22.22}$$

The second expression is the potential energy per unit mass at maximum expansion, when the kinetic energy momentarily vanishes and the radius is $r = 2A$. The last expression is the condition that the energy is perturbed from the Einstein-de Sitter model, where E vanishes, by the gravitational potential energy of the initial isolated mass concentration at distance $r_i = a_i x$ from the shell. Since the mass within the shell x varies as $\mathcal{M} \propto x^3$, equation (22.22) shows that the constants in the solution (22.21) vary with the initial shell radius as

$$A \propto x^4, \qquad B = (A^3/G\mathcal{M})^{1/2} \propto x^{9/2}. \tag{22.23}$$

Following Bertschinger (1985), it is convenient to scale the lengths at a chosen time t to the physical radius r_{ta} of the mass shell at $\eta = \pi$ that has just stopped expanding at this time. The initial comoving length labeling the shell at the turnaround radius at time t is x_{ta}. The scaling relations in equation (22.23) say that at the chosen time t the physical radius r of the shell labeled with parameter η and initial comoving radius x is

$$\frac{r}{r_{\mathrm{ta}}} = \left(\frac{x}{x_{\mathrm{ta}}}\right)^4 \frac{1 - \cos\eta}{2}, \qquad \left(\frac{x}{x_{\mathrm{ta}}}\right)^{9/2} = \frac{B(x)}{B(x_{\mathrm{ta}})} = \frac{\pi}{\eta - \sin\eta}. \tag{22.24}$$

This gives

$$\frac{r}{r_{\mathrm{ta}}} = \frac{1 - \cos\eta}{2} \left(\frac{\pi}{\eta - \sin\eta}\right)^{8/9}. \tag{22.25}$$

Now we can compute the mass density run at time t, from $d\mathcal{M}/dr = 4\pi\rho(r)r^2$, for we see that the mass within comoving radius x is $\mathcal{M}(x) \propto x^3$, the second part of equation (22.24) fixes x as a function of η, and equation (22.25) fixes η as a function of the physical radius r. The mass density $\rho(r)$ as a function of radius r, relative to the background density $\rho_b(t) = 1/(6\pi G t^2)$, works out to

$$\frac{\rho(r, t)}{\rho_b(t)} = \frac{9}{2} \frac{(\eta - \sin\eta)^2}{(1 - \cos\eta)^3} \left[4 - \frac{9}{2} \frac{\sin\eta(\eta - \sin\eta)}{(1 - \cos\eta)^2}\right]^{-1}. \tag{22.26}$$

The solution in equations (22.25) and (22.26) for the mass density as a function of proper radius is plotted as the curved line in figure 22.2.

At small radius, where mass shells have crossed many times, the density run in this model relaxes to a scaling solution obtained as follows. The density contrast $\bar{\delta}$ averaged within the shell x grows as the linear perturbation solution $\bar{\delta} \propto a(t)$ until $\bar{\delta}$ approaches unity, at expansion factor a_x, with

$$\bar{\delta}_i a_x / a_i \sim 1. \tag{22.27}$$

Then this mass shell collapses and oscillates around a mean physical radius $r \sim x a_x$, with mean density comparable to the background value at the epoch of collapse,

$$\rho(r) \sim \rho_i (a_i/a_x)^3. \tag{22.28}$$

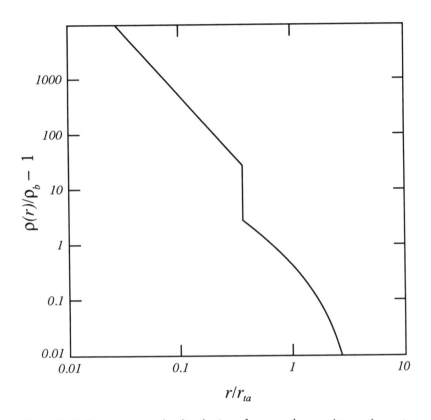

Figure 22.2. Density run in the distribution of pressureless noninteracting matter that accretes around an isolated mass concentration in an Einstein-de Sitter universe. The curve at low density contrast is the analytic solution in equations (22.25) and (22.26), valid where mass shells have not yet crossed. The straight line at high density is the scaling solution in equation (22.29). The two are joined at the maximum radius of expansion of the mass shell after first collapse and expansion.

The result of eliminating x and a_x from these equations, with equation (22.20) for δ_i, is

$$\rho(r) \sim \rho_i^{1/4} m_i^{3/4} r^{-9/4} . \tag{22.29}$$

The velocity dispersion scales as

$$\sigma \sim [G\rho(r)r^2]^{1/2} \propto r^{-1/8} . \tag{22.30}$$

This scaling solution was found by Gott (1975).

To visualize the character of the exact solution for $\rho(r)$ in this pressureless spherical accretion model, consider the single-particle phase space in radius and radial momentum. Since the material is cold, it occupies a line in phase space, which circles the origin once for every time the material has collapsed and re-expanded. The mass density $r^2\rho(r)$ is the integral of the density in phase space over the radial momentum. This integral is the projection of the line onto the radius axis. The result is that $\rho(r)$ has a series of cusps where the mass shells are instantaneously at rest after some number of radial oscillations, and the tangent to the line in phase space is parallel to the momentum axis. The vertical line in figure 22.2 is the radius at the first cusp, where the matter is at rest after having collapsed, expanded, and momentarily come to rest before collapsing for a second time. Bertschinger's (1985) numerical solution for the density as a function of radius fixes the normalization of the scaling law (22.29):

$$\rho(r)/\rho_b = 2.79 (r_{\mathrm{ta}}/r)^{9/4}. \tag{22.31}$$

In this model, the density rises by an order of magnitude at a shoulder where material falling in for the first time encounters material that has bounced back. This is because material that has not yet crossed expanding mass shells sees an isolated mass concentration that produces a tidal field that has little effect on its mass density: the material collapses with large shear but little increase in density. The density run has a prominent rise where the material has collapsed and reexpanded to join the inner part of the distribution that has approached dynamical equilibrium.

The scaling solution in equations (22.29) and (22.30) is a reasonable approximation to the mass distribution outside the core of a rich cluster of galaxies, even reproducing the slow decrease of the velocity dispersion with increasing radius. The steep shoulder at larger radius is quite unrealistic: it is not seen in the galaxy distribution around the Coma cluster in figure 3.5, or in the mean galaxy density as a function of distance from a cluster, as measured by $\xi_{cg}(r)$ (eq. [19.35]). There are several ways out. First, we could be observing clusters at a special early epoch, before they have collapsed and reached the scaling solution in figure 22.2. That is the case in the adiabatic cold dark model to be discussed in section 25. There may be the problem that the excess of early-type galaxies in the central parts of clusters suggests these galaxies have been in a cluster environment for some time, sufficient to allow them to express their differences from field galaxies. This would seem to be difficult to arrange if the central parts of clusters were being assembled now. Consistent with this are the observations of systems that look like classical rich clusters at redshifts $z \sim 1$. A second possibility is that clusters are growing out of material with a positive local initial density contrast, which causes the density run to grow in advance of accretion. A third possibility is that the growth of aspherical motion serves to slow the radial collapse that is responsible for the low density of infalling material as it approaches the region where the mass distribution is in approximate dynamical equilibrium.

The last effect is the previrialization conjecture (Davis and Peebles 1977) introduced as a way around the analogous problem in the scaling solution for the growth of the mass autocorrelation function $\xi(x,t)$: if newly forming mass clumps collapse before reaching statistical equilibrium, the slope of the mass autocorrelation function increases at separations where $\xi(x,t) \sim 1$, an effect not observed in the galaxy two-point correlation function. The proposed remedy is that clumps form with appreciable nonradial motions, because a protoclump is distinctly nonspherical, and the nonradial kinetic energy opposes the collapse that raises the density and produces a shoulder in $\xi(r)$. The effect clearly operates at some level, for clusters of galaxies are far from spherically symmetric, and N-body model computations show vivid examples of nonradial motion in nascent mass clumps (Villumsen and Davis 1986). The possible size of the effect is still under discussion. A simplified model that helps illustrate what is involved goes as follows.

Consider an ensemble of gravitationally growing mass concentrations. To simplify the calculation, we will imagine that the material in each system is collecting within the same given central gravitational potential well, with inward-pointing acceleration $g(r,t)$.

The probability that in a given one of the realizations there is a particle at proper radius r in the range dr, with radial momentum per unit mass in the range $v_r = \dot{r}$ to $v_r + dv_r$, and angular momentum

$$l = rv_\theta, \qquad (22.32)$$

in the range dl, is

$$dP = f(l,r,v_r,t)\, dl\, dr\, dv_r. \qquad (22.33)$$

The radial acceleration of this particle is

$$\frac{dv_r}{dt} = \frac{l^2}{r^3} - g(r). \qquad (22.34)$$

This equation of motion can be derived from a potential, so we know from Liouville's theorem that the distribution function f at fixed l is constant along orbits in the two-dimensional phase space r and v_r. That is, the distribution satisfies the collisionless Boltzmann equation

$$\frac{\partial f}{\partial t} + v_r \frac{\partial f}{\partial r} + \frac{\partial f}{\partial v_r} \left[\frac{l^2}{r^3} - g(r) \right] = 0. \qquad (22.35)$$

Since f gives the mean number of particles per unit increment of radius r, the ensemble average three-dimensional number density $n(r,t)$ satisfies

$$4\pi n(r,t)r^2 = \int f \, dv_r dl \,. \tag{22.36}$$

With this definition, the result of integrating the Boltzmann equation (22.35) over v_r and l is

$$\frac{\partial n}{\partial t} + \frac{1}{r^2}\frac{\partial}{\partial r}r^2 n\langle v_r \rangle = 0 \,, \tag{22.37}$$

where the mean radial streaming velocity is

$$\langle v_r \rangle = \frac{\int f \, v_r \, dv_r dl}{\int f \, dv_r dl} \,. \tag{22.38}$$

This is the usual particle conservation law for the spherical symmetry appropriate to the ensemble average. The result of multiplying the Boltzmann equation by v_r, integrating over v_r and l, and rearranging, is

$$\frac{\partial}{\partial t}n\langle v_r \rangle + \frac{\partial}{\partial r}n\langle v_r{}^2 \rangle + \left(2\langle v_r{}^2 \rangle - \langle v_\theta{}^2 \rangle \right)\frac{n}{r} + n(r,t)g(r,t) = 0 \,. \tag{22.39}$$

The transverse velocity is $v_\theta = l/r$. If the system is static and the velocity distribution isotropic (so $\langle v_\theta{}^2 \rangle = 2\langle v_r{}^2 \rangle$ because there are two transverse directions), the third term vanishes, and this reduces to the pressure support equation (3.31). If the system is static and the velocity distribution anisotropic, the equation for gravitational equilibrium depends on two velocity dispersions, radial and transverse, as discussed in analyses of star clusters and clusters of galaxies (e.g., The and White 1986).

The result of eliminating $\partial n/\partial t$ from equations (22.37) and (22.39) is

$$\frac{\partial\langle v_r \rangle}{\partial t} + \langle v_r \rangle\frac{\partial\langle v_r \rangle}{\partial r} + \frac{1}{r^2 n}\frac{\partial}{\partial r}r^2 n\langle \delta v_r{}^2 \rangle - \frac{\langle v_\theta{}^2 \rangle}{r} + g = 0 \,, \tag{22.40}$$

where the velocity dispersion relative to the mean radial flow is

$$\langle \delta v_r{}^2 \rangle = \langle (v_r - \langle v_r \rangle)^2 \rangle \,. \tag{22.41}$$

The first two terms in equation (22.40) are the usual convective time derivative of the streaming velocity. The next two are the effective pressure gradient force terms.

For the present purpose the significance of this model result is that it illustrates the order-of-magnitude condition for previrialization. Suppose at time t the material at radius r is breaking away from the general expansion. That means the radial velocity has changed by the amount $\sim r/t$. In a protocluster that is breaking away from the general expansion, one might expect the transverse velocity to be on the order of the change in the radial velocity, $v_\theta \sim r/t$. Then the outward pressure acceleration term in equation (22.40) is $\langle v_\theta^2 \rangle / r \sim r/t^2 \sim G\rho_b r$. In a protocluster with density contrast on the order of unity, this is on the order of the inward gravitational acceleration, $g \sim G\bar{\rho} r$. This is the condition that nonradial motions stop the collapse and prevent the development of an unwanted shoulder in $\xi(r)$ or $\xi_{cg}(r)$.

Origin of the Rotation of Galaxies

The similarity of a spiral galaxy to a turbulent eddy naturally leads one to ask whether galaxies might be the remnants of turbulence in the early universe (von Weizsacker 1947; Gamow 1952). We will see that the turbulence could not be primeval, something issuing out of the unknown physics of very high redshifts, because the turbulent velocities would be supersonic at any reasonable matter temperature after decoupling, meaning primeval turbulent eddies would dissipate into shocks in a crossing time, producing unacceptably dense objects (Jones 1973; Jones and Peebles 1972). The circulation in galaxies could be produced in oblique shocks during pancake collapse (Chernin 1970; Doroshkevich 1973), or by the stress of a primeval magnetic field (Wasserman 1978). In scenarios where structure grows in a hierarchy by gravitational assembly of clumps out of subclumps, the commonly adopted picture is that protogalaxies acquire their angular momentum as they formed by the tidal torques of neighboring protogalaxies (Strömberg 1934; Hoyle 1949; Peebles 1969a, 1971b; Doroshkevich 1970). The possible implications of this picture for the epoch of formation of the disks of spiral galaxies are explored in section 25.

To see why the primeval turbulence picture is believed to fail, suppose the circulation typical of a large spiral galaxy is present at decoupling. This amounts to velocity $v \sim 200$ km s^{-1} at radius $r \sim 10h^{-1}$ kpc within a system of density $n_g \sim 1h^2$ protons cm^{-3}. At redshift $z_{\rm dec} = 1400$, the mean baryon density is

$$n_{\rm dec} = 3 \times 10^4 \Omega_B h^2 \, {\rm cm}^{-3} , \qquad (22.42)$$

where Ω_B is the density parameter in baryons. At this density the radius is decreased by the factor $(n_{\rm dec}/n_g)^{1/3}$. Since circulation is conserved in the absence of nongravitational forces, this increases the velocity of circulation at decoupling by the same factor, to

$$v_{\rm dec} = 6 \times 10^3 \Omega_B^{1/3} \, {\rm km \ s}^{-1} . \qquad (22.43)$$

If the material at decoupling were neutral, the velocity of sound would be much less than this circulation velocity. Thus, in an eddy turnaround time the motion would lead to shocks that rapidly cool by radiation drag (eq. [6.136]), leaving gravitationally bound systems with some remnant of the initial circulation, but with densities comparable to the mean in equation (22.42), which is much greater than that typical of galaxies. The familiar cascade of turbulent energy through the hierarchy of eddies does not happen here because the motions are supersonic, so the only way to deflect an eddy is in a shock that dissipates it. If the matter were hot enough to make the flow subsonic, the matter would be ionized and radiation drag would dissipate the circulation (eq. [6.141]). This is one of the reasons for believing the universe prior to decoupling had to have been a tranquil place.

In hierarchical scenarios, where protogalaxies are assembled by gravity out of lesser systems, the usual assumption is that the rotational angular momenta of the galaxies came from tidal torques. As we noted in connection with the pre-virialization hypothesis, a protogalaxy in the process of breaking away from the general expansion is unlikely to be spherically symmetric; a better picture would be a messy blob moving away from an irregular boundary separating it from other developing protogalaxies. The unequal pull of neighboring mass concentrations on the material within a blob produces a velocity shear that in general leaves the protogalaxy with angular momentum of rotation.

The final angular momentum L about the center of mass of a protogalaxy of mass $\mathcal{M}$ and net energy (gravitational plus kinetic) $-E$ is characterized by the dimensionless number

$$\lambda = \frac{LE^{1/2}}{G\mathcal{M}^{5/2}} = 0.05 \,. \tag{22.44}$$

The numerical value in the tidal torque picture is best found from large-scale numerical N-body computations; the number here is the median from the results of Barnes and Efstathiou (1987).

The parameter λ is a measure of the degree of rotational support of the system, for the ratio of the centrifugal acceleration $g_\phi \sim v_\phi^2/r$ from the streaming motion v_ϕ of rotation to the gravitational acceleration $g \sim G\mathcal{M}/r^2$ is

$$\frac{g_\phi}{g} \sim \frac{v_\phi^2}{r}\frac{r^2}{G\mathcal{M}} \sim \lambda^2 \,, \tag{22.45}$$

where the angular momentum is $L \sim \mathcal{M}rv_\phi$ and the binding energy is $E \sim G\mathcal{M}^2/r$. The small value of λ in equation (22.44) is consistent with the slow rotation of the spheroid components of large galaxies. If the disk angular momentum originated this way, the disk material would have to have contracted by dissipation of energy (but not angular momentum) to reach rotational support.

The contraction factor for the disk material depends on the model. In the simplest picture the disk is gravitationally bound by its own mass. This would mean the energy E of the protodisk varies inversely as its radius R. During dissipative contraction that conserves L, the parameter λ would scale as $R^{-1/2}$, and to reach rotational support at $\lambda \sim 1$ from the initial value $\lambda = 0.05$ would require a contraction factor ~ 100 in radius. The collapse time scales as $R^{3/2}$, and is about 10^8 y in the disk of the Milky Way at its present radius. If the disk had contracted by two orders of magnitude in radius, the initial collapse time would be greater than the present age of the universe, which is not acceptable.

The way out was shown by Fall and Efstathiou (1980), who built on the remark by White and Rees (1978) that since spiral galaxies are dominated by dark halos the disk might be expected to have formed as a secondary component that dissipatively contracted within the gravitational potential well of the dark matter. Suppose the density run in the dark mass is $\rho(r) \propto 1/r^2$, so the circular velocity v_c for rotational support is independent of radius. We see from equation (22.45) that the initial streaming velocity of rotation of disk material that originates at radius r_i is

$$v_\phi \sim (G\mathcal{M}_i \lambda^2/r_i)^{1/2} \sim \lambda v_c. \tag{22.46}$$

If the disk material dissipatively contracts without losing much angular momentum, it contracts in radius to rotational support by the factor $\sim \lambda^{-1}$, rather than the contraction factor $\sim \lambda^{-2}$ needed in the case of a self-gravitating disk. A typical scale height in equation (3.26) for the exponential disk of an L_* spiral galaxy is $a \sim 4h^{-1}$ kpc, so this would say the disk material collected from initial radius

$$r_i \sim a/\lambda \sim 100 h^{-1}\,\text{kpc}. \tag{22.47}$$

It may be a significant coincidence that the wanted initial radius r_i is comparable to the halo radii inferred from the dynamical measures of the mass of the material clustered on scales of a few megaparsecs (eq. [20.31]). This is one of the reasons for thinking that the disk is made of material that fell from the halo (Ostriker and Thuan 1975; Gunn 1982).

The Epoch of Galaxy Formation

There need not have been a single characteristic epoch at which galaxies were assembled: giants and dwarfs may have formed at different times, and we have seen that the tidal torque picture for the rotation of galaxies seems to demand that disks appear after the formation of the dark halos. We get upper limits on the redshifts of assembly of the components that dominate the mass of a galaxy from two almost equivalent conditions: there has to have been room for the components, and their densities have to have been larger than the mean in the background from which the material was drawn.

A characteristic number density for the L_* galaxies that contribute most of the luminosity density (adding a little to n_* in eq. [5.144] to account for all the dwarfs) is $n_g \sim 0.03 h^3$ Mpc^{-3}. If galaxies are conserved, the mean distance between them at redshift z is

$$d_g \sim n_g^{-1/3} \sim 3 h^{-1}(1+z)^{-1}\,\text{Mpc}. \tag{22.48}$$

If we allow a radius of $30h^{-1}$ kpc for each galaxy, not counting the massive dark halo, we see that there was room for the bright parts of the galaxies at redshift

$$z_g \sim 50 . \tag{22.49}$$

A related bound follows from the condition that whatever produced the mass concentrations in galaxies surely did so by collecting material no less dense than the cosmological mean. That is, protogalaxies formed at densities larger than the background. Dynamical relaxation after formation increases the velocity within a protogalaxy (for recall that at equilibrium the kinetic energy $\mathcal{M}v^2/2$ is half the magnitude of the negative binding energy, E, so energy loss increases E and v). This means that a reasonable upper bound on the velocity v_c for a circular orbit in a newly formed protogalaxy is the value in a present-day L_* galaxy. Then the ratio of the mass density $\bar{\rho}$ averaged over radius r within the protogalaxy to the cosmological mean density ρ_b at redshift z_f in equation (5.67) is

$$\begin{aligned}
\frac{\bar{\rho}}{\rho_b} &= \frac{3}{4\pi r^3} \frac{v_c^2 r}{G} \frac{8\pi G}{3H_o^2 \Omega (1+z_f)^3} \\
&= \frac{2 v_c^2}{\Omega (H_o r)^2 (1+z_f)^3} > 1 .
\end{aligned} \tag{22.50}$$

With $v_c \sim 200$ km s^{-1}, this gives

$$z_f \lesssim 40\, \Omega^{-1/3} \quad \text{at} \quad r = 10 h^{-1}\,\text{kpc} . \tag{22.51}$$

Since the dynamical estimates in section 20 indicate the density parameter is in the range $0.05 \lesssim \Omega \lesssim 1$, this is close to the redshift in equation (22.49) at which protogalaxies start to fill space. The difference depends on how much mass there is between the galaxies.

A more detailed and perhaps more realistic version of the constraint in equation (22.50) uses a spherical model for the collection of the mass. Under spherical symmetry, we can assign a proper radius $r(t)$ to the shell that contains mass $\mathcal{M}$. In the solution in equation (22.21), the shell that contains mass $\mathcal{M}$ reaches maximum expansion when the parameter is $\eta = \pi$, the time is $t = \pi B$, the radius is $r = 2A$, and the mean density within the shell is $\bar{\rho} = 3\mathcal{M}/4\pi r^3$. At redshifts of interest here, the mass density ρ_b dominates the expansion rate, so the cosmological mean mass density ρ_b at world time t is given by the Einstein–de Sitter relation (5.21),[12]

---

[12] The time variables in the solutions in equations (22.21) and (22.52) can differ by a constant value, Δt. We are assuming the mass concentration has grown by gravity out of small density fluctuations released at the decoupling of matter and radiation, so Δt is negligibly small compared to the world time t when the concentration breaks away from the general expansion. In an explosion picture, Δt could be appreciable, but the spherical model for the gravitational collection of the mass would be even less believable.

$$\rho_b = \frac{1}{6\pi G t^2} .$$
(22.52)

It follows that at maximum expansion in the spherical model, the mean density within the shell relative to the background is

$$\frac{\bar{\rho}}{\rho_b} = \frac{9\pi^2}{16} = 5.6 .$$
(22.53)

In the original application of this approach (Partridge and Peebles 1967a) it was argued that the spherical model might be a useful if rough approximation when applied to about the nominal epoch of maximum expansion. At this point the radial peculiar velocity would have canceled the Hubble flow, it would be reasonable to expect the transverse motion is similar, and if so the system would be distinctly aspherical and might be expected to have enough random kinetic energy to support itself. With $\bar{\rho}/\rho_b$ in equation (22.53) replacing the bound in equation (22.50), and $v_c \sim 200$ km s^{-1}, we get the redshift bounds for the assembly of parts of a protogalaxy,

$$z_f \lesssim \begin{cases} 25\Omega^{-1/3} & \text{at } r = 10h^{-1} \text{ kpc}, \\ 10\Omega^{-1/3} & \text{at } r = 30h^{-1} \text{ kpc}, \\ 5\Omega^{-1/3} & \text{at } r = 100h^{-1} \text{ kpc}. \end{cases}$$
(22.54)

The assembly epochs would be at lower redshifts if the material contracted by an appreciable factor after maximum expansion, as for the dissipative contraction that is thought to have spun up the disk material. The numbers in equation (22.54) suggest the bright spheroids of galaxies could be in place at $z \sim 10$, and massive halos could be in place at redshift $z \sim 5$, allowing protodisks to be form at $z \sim 1$, which would agree with the age of the disk of the Milky Way. This is discussed in more detail in section 25, in connection with table 25.1.

Scaling

Consider the Einstein-de Sitter limit, where the universe expands as the power of time $a \propto t^{2/3}$. Suppose the primeval spectrum of mass fluctuations is a power law,

$$\mathcal{P} \propto k^n ,$$
(22.55)

with the power law index n a constant greater than -3, and suppose nongravitational forces may be neglected. Then there are no fixed characteristic quantities, so the structure of the mass distribution evolves according to a scaling law. (The alternative is a singular evolution in which the memory of initial conditions such as the starting coherence length is never erased.) The scaling solution could be relevant to the large-scale structure of the galaxy distribution, which resembles a power law clustering hierarchy (sections 19 and 21). The analysis of the scaling solution is a rich problem (for example, Fry 1984; Schaeffer 1984); only a few basic elements are presented here.

The rms value of the mass density contrast smoothed through a window of comoving radius x at time t, $\bar{\delta}(t,x)$, is an integral over the power spectrum at $k \lesssim 1/x$ (eq. [21.53]). In the power law model with $-3 < n < 4$ we have $\bar{\delta}^2 \propto x^{-(3+n)}$. This relation applies at scales x large enough or redshifts high enough so that nonlinear effects have not appreciably perturbed the fluctuation spectrum from the primeval form in equation (22.55). In this linear limit, the spectrum evolves according to the linear perturbation relation in equation (22.1), $\mathcal{P} \propto t^{4/3}$ at fixed comoving wave number k, and the rms contrast scales with time and separation as

$$\bar{\delta} \propto t^{2/3} x^{-(3+n)/2} \quad \text{at } x \gtrsim x_o(t).\tag{22.56}$$

The characteristic clustering length $x_o(t)$ marks the transition from small fluctuations on large scales to nonlinear small-scale structures. Since the rms density contrast on this transition scale is on the order of unity, $\bar{\delta}(t, x_o(t)) \sim 1$, the clustering length scales as

$$x_o(t) \propto t^{4/(9+3n)},$$
$$r_o(t) = a(t)x_o(t) \propto t^{(10+2n)/(9+3n)}.\tag{22.57}$$

The characteristic mass of the nonlinear structure contained within the clustering length grows as

$$\mathcal{M}_{\mathrm{nl}} \propto \rho_b r_o{}^3 \propto x_o{}^3 \propto t^{4/(3+n)}.\tag{22.58}$$

This assumes that the initial spectrum (22.55) is less steep than $\mathcal{P} \propto k^4$. Nonlinear clustering would fill in a steeper initial spectrum to $n \to 4$, giving the minimum rate of growth $\mathcal{M}_{\mathrm{nl}} \propto t^{4/7}$ in the Einstein-de Sitter model (eq. [22.3]).

At separation $x \sim x_o(t)$, gravity is gathering into bound systems material that already has been collected into subsystems, producing a clustering hierarchy of the kind discussed in sections 7 and 19. If the mean value of the mass density within a typical cluster in the hierarchy does not evolve much once formed, we can find the fractal dimension by the following argument.

The mass concentrations forming at time t have typical size $r \sim r_o(t)$ (eq. [22.57]), and mean density on the order of the background mean value at the time the concentrations break away from the general expansion, $\rho(r) \propto t^{-2}$. The result of eliminating t from these two expressions is

$$\rho(r) \propto r^{-\gamma}, \qquad \gamma = \frac{9+3n}{5+n}\tag{22.59}$$

(Peebles 1965, 1974b). The velocity dispersion scales with cluster size as

$$\sigma \sim (G\mathcal{M}/r)^{1/2} \propto (\rho r^2)^{1/2} \propto r^{(1-n)/(10+2n)}, \qquad (22.60)$$

and the cluster mass scales with the velocity dispersion as

$$\mathcal{M} \propto \rho r^3 \propto \sigma^{12/(1-n)} \qquad (22.61)$$

(Faber 1982). The mass autocorrelation function in this clustering hierarchy is $\xi \propto r^{-\gamma}$ (eqs. [19.24] and [19.25]); the fractal dimension is $D = 3 - \gamma$.

In numerical N-body model computations of the scaling solution, the mass two-point correlation function is approximately consistent with the power law $\xi(r) \propto r^{-\gamma}$ (eq. [22.59]) at $\xi \gtrsim 10$ (Peebles 1985; Efstathiou et al. 1988). Merging might be expected to dissipate the clustering hierarchy on small scales, but numerical experiments indicate the hierarchical forms of the three- and four-point mass correlation functions are preserved for a crossing time of the largest scale of the hierarchy (Peebles 1978). In the numerical solutions the correlation function has a shoulder at $\xi \sim 1$, where the slope is steeper than the power law. As in figure 22.2, this is because newly forming levels in the hierarchy have to contract to gain enough internal kinetic energy to support themselves.

We saw in section 19 that the galaxy space distribution resembles a bounded power law clustering hierarchy, on scales 10 kpc $\lesssim hr \lesssim$ 10 Mpc. With $n = 0$ the scaling solution is $\xi \propto r^{-1.8}$, close to the observations in figures 7.2 and 7.3, and the relative velocity dispersion (22.60) scales with separation as $\sigma \propto r^{0.1}$, close to the results in figure 20.1. There are some significant differences from a simple clustering hierarchy, however. One is the great clusters of galaxies, some of which look like monolithic structures rather than clustering hierarchies. These might be regions where relaxation that eliminates subclustering is unusually advanced because the mass concentration is unusually great. The monolithic structure of the mass distribution within a galaxy is not part of the clustering hierarchy either, but one might imagine that, as in clusters, relaxation has partially erased initial substructures. This would agree with figure 20.1, which shows that the mass density run within an L_* galaxy is continuous with the mean density run with distance from a galaxy in the clustering hierarchy. However, as Faber (1982) notes, the relation between velocity dispersion and luminosity in galaxies, $\sigma \sim \mathcal{M}^{0.2}$ (eqs. [3.35] and [3.39]), indicates $n \sim -1.5$ in equation (22.61), while the fit to γ and the galaxy relative velocity dispersion, $n = 0$, would predict $\sigma \sim \mathcal{M}^{0.08}$. This may be a result of the scatter around the simple single-parameter description in equations (22.59) to (22.61). Finally, the shoulder in the numerical scaling solutions for the mass autocorrelation function at $\xi \sim 1$ is not seen in the galaxy two-point correlation function. This could be because the power law form of the galaxy two-point correlation function is an accidental transient effect, not a reflection of scaling behavior. Or it may be that the numerical scaling solution is missing the previrialization effect, perhaps because it assumes Gaussian initial density fluctuations.

23. Young Galaxies and the Intergalactic Medium

The theme for this section is that the intergalactic medium, and objects that look like young galaxies or their near ancestors, were already in place at expansion factor $1 + z \sim 6$. This tells us that the structure formation processes in operation at smaller expansion factors had to have worked with material that already was strongly disturbed by what happened earlier.

An example of the main source of evidence is shown in figure 23.1. The broad peak is the $2p \rightarrow 1s$ $L\alpha$ emission line from atomic hydrogen in the spectrum of the quasar Q1215+333, shifted from the emission wavelength $\lambda = 1216$ Å to the observed wavelength $\lambda = 4380$ Å by the redshift factor $1 + z_{em} = 3.60$ (Wolfe et al. 1993). The broad absorption trough at 3650 Å wavelength is caused by resonant $L\alpha$ scattering in a gas cloud along the line of sight between us and the quasar, at redshift $z_{abs} = 2.00$. The column density in the line through this cloud is 1×10^{21} atoms cm^{-2}, comparable to the column density in the disk in a present-day spiral galaxy. The identification as $L\alpha$ absorption is unambiguous, because there are absorption lines at the same redshift from common elements heavier than helium. The heavy element abundances in such high column density, high redshift clouds generally are estimated to be about 10% of the present cosmic abundance in the interstellar matter in the Milky Way. As will be discussed in this section, the total amount of atomic hydrogen in these clouds at redshift $z \sim 3$ is comparable to what is found now in the gas and stars in the bright central parts of galaxies. The column densities, the heavy element abundances, and the total mass of hydrogen in these clouds are consistent with the picture that these are young galaxies or their ancestors, gas-rich massive star clusters.

The figure shows a prominent forest of narrow absorption lines. Lynds (1971) and Arons (1972) suggested most are caused by $L\alpha$ scattering in lower surface density clouds along the line of sight between us and the quasar, because the forest of lines appears to the short wavelength side of the $L\alpha$ emission line of the quasar. For the stronger lines in the forest, this is confirmed by the observation of the shorter wavelength $L\beta$ and higher ($1s \rightarrow np$) lines in the Lyman series. (A few lines in the forest are from heavy elements in clouds with greater surface densities; other heavy element lines are seen on the long wavelength side of the quasar $L\alpha$ emission line.) The column densities in these clouds range down at least to the limiting observational sensitivity at about 10^{13} hydrogen atoms cm^{-2}, eight orders of magnitude below the surface density in the cloud that produced the broad absorption trough in this example. The comoving space number density of the clouds in this Lyman-α forest varies roughly as $(1 + z)^2$ at $z \sim 3$, meaning

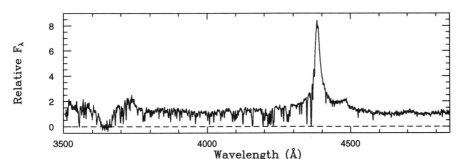

Figure 23.1. Spectrum of a quasar at redshift $z_{em} = 2.6$ (Wolfe et al. 1993). The broad peak is the atomic hydrogen Lα emission line from the quasar. Most of the absorption features at shorter wavelengths are produced by resonant Lα scattering in gas clouds along the line of sight between us and the quasar.

the clouds are dissipating. We will see that the mass in atomic hydrogen in the Lα forest is quite small, and there is very little atomic hydrogen in diffuse gas between the clouds. The clouds do contain a large ionized mass fraction, however, in an amount comparable to what is observed in the neutral high surface density clouds and to what is observed at the present epoch in baryons in gas and stars.

Not appearing in the figure is a Lyman-limit cloud, with column density $\Sigma \gtrsim 3 \times 10^{17}$ atoms cm^{-2}, high enough to be opaque to ionizing radiation at the threshold. This produces an absorption feature that is relatively prominent even for Lyman-limit clouds at low redshifts (observed above the atmosphere).

The abundance of these clouds, from the Lα forest to the Lyman-limit clouds to the densest observed, decreases with increasing surface density; a useful fitting formula at $z \sim 3$ is given in equation (23.47).

These observations show that structure formation was well advanced when the universe was just one-sixth its present size, at the highest presently observed quasar redshifts. At this epoch there was an intergalactic medium in a cloudy form that exhibits a continuous evolution to its present state, with cloud surface densities ranging up to those characteristic of galaxies, and, in the highest surface density clouds, significant heavy element abundances from earlier generations of stars. What caused all this activity is a matter of debate; at the time of this writing most of the clues are enigmatic. For example, the line-of-sight distribution of redshifts of the clouds in the Lyman-α forest is close to uniform, in striking contrast to the very clumpy present-day galaxy distribution illustrated in the figures in section 3. Could this mean the galaxy distribution at $z \sim 3$ was smooth? Perhaps more likely is the idea that these low surface density clouds are fragile, and at $z \gtrsim 3$ are crowded (eq. [23.69]), so they could only have existed where there were no galaxies, in the voids that fill most of space at the present epoch. The absorption lines from the heavier elements observed in higher surface density clouds do cluster along the line of sight, about as strongly as expected if the young galaxies or protogalaxies were distributed in the clumpy fashion of present-day galaxies within the smoother froth of the $L\alpha$ forest (Sargent, Boksenberg, and Steidel 1988).

Equally enigmatic is the high degree of isotropy of the thermal cosmic background radiation. Whatever happened in the early universe to bring structure formation to the advanced state observed at $z \sim 5$ cannot have been so violent as to have greatly perturbed the cosmic background radiation, or, if it were, the optical depth for absorption and scattering of the radiation had to have been high enough to have relaxed the CBR back to an exceedingly close approximation to a homogeneous thermal sea of radiation.

The purpose of this section is to explore the use of the absorption line spectra to obtain order-of-magnitude measures of the state and evolution of the intergalactic medium and the clouds of matter in it. The next section deals with constraints from the spectrum and isotropy of the thermal 3 K and X-ray cosmic radiation backgrounds.

We continue the practice of writing the mass density parameter as Ω (eq. [5.55]), with Ω_B the part in baryons (that are not locked up in black holes or otherwise inactive). Thus the mean baryon number density at redshift z is (eq. [5.68])

$$n = 1.12 \times 10^{-5} \Omega_B h^2 (1+z)^3 \text{ protons cm}^{-3} . \tag{23.1}$$

The world time at high redshift is usefully approximated by the Einstein-de Sitter limit in equation (5.61),

$$t = \frac{2}{3} \frac{1}{H_o \Omega^{1/2}(1+z)^{3/2}} = 2.06 \times 10^{17} \Omega^{-1/2} h^{-1} (1+z)^{-3/2} \text{ s} . \tag{23.2}$$

If space curvature is negligible and $\Omega \gtrsim 0.1$ this is reasonably close even at $z \sim 1$. And for almost all purposes the error introduced by using this simple approximation is small compared to the main observational and theoretical uncertainties in establishing the interesting orders of magnitude.

The discussion in this section centers on the most abundant (and simplest) component of the matter, atomic and ionized hydrogen. For further details of this and of the heavy elements the first place to look, as usual, is in the proceedings of recent conferences.

Gunn-Peterson Test

The scattering cross section integrated through the resonance in the $1s$ to $2p$ atomic hydrogen $L\alpha$ transition is large enough for observable absorption to be produced by relatively small column densities of neutral hydrogen. This test for smoothly distributed intergalactic hydrogen was first applied to the spectra of quasars at $z \gtrsim 2$, where the Lα resonance at the quasar is redshifted into the optical band observable from the ground (at $\lambda \gtrsim 3200$ Å), as illustrated in figure 23.1. Observations above the atmosphere extend the test to the density of intergalactic hydrogen at low redshifts, to the $\lambda 584$ Å $1s^2 - 1s2p$ resonance line in neutral helium, and even the $\lambda 304$ Å $1s - 2p$ resonance in singly ionized helium.

When the time is right for the application of a measure such as this, it tends to occur to more than one person. The remarkably sensitive probe of neutral hydrogen was discovered at least three times, by Shklovsky (1964), Scheuer (1965), and Gunn and Peterson (1965). The last hit on it at the right time, with the data on hand for a clean application.

The easy way to compute the effect of scattering by a smoothly distributed sea of intergalactic hydrogen is to write the scattering cross section as a function of frequency through the Lα resonance as

$$\sigma(\omega) = C\Delta(\omega - \omega_\alpha), \tag{23.3}$$

where Δ approximates a Dirac delta function with unit area,

$$\int_{-\infty}^{\infty} \Delta(x)\, dx = 1. \tag{23.4}$$

The shape of the resonance, in equation (23.43), will be used in the analysis of scattering in gas clouds, but here we need only the normalizing constant, C. The easy way to get it is from the rate of spontaneous radiative decay from the $2p$ to $1s$ energy level,

$$\Lambda = 6.25 \times 10^8 \, \text{s}^{-1}. \tag{23.5}$$

Suppose a gas of hydrogen atoms is in thermal equilibrium with radiation at temperature T. Then the ratio of numbers of atoms in the $2p$ and $1s$ levels is

$$\frac{n_2}{n_1} = 3e^{-\hbar\omega_\alpha/kT}. \tag{23.6}$$

The energy level difference is $\hbar\omega_\alpha$, and the factor of three counts the $m = 0, \pm 1$ quantum numbers for the $2p$ wave functions. The rates of radiative decay and excitation, which have to agree at thermal equilibrium, are

$$3e^{-\hbar\omega_\alpha/kT} \cdot \Lambda \cdot (1 + \mathcal{N}_\alpha) = \int \sigma(\omega)\frac{8\pi\omega^2 d\omega}{(2\pi c)^3}\mathcal{N}c. \tag{23.7}$$

In the decay rate on the left side, the first factor is the ratio of numbers of atoms in the excited and ground levels. The last factor is the correction for stimulated emission. We are assuming thermal equilibrium, so the photon occupation number is

$$\mathcal{N}_\alpha = \frac{1}{e^{\hbar\omega_\alpha/kT} - 1}, \qquad 1 + \mathcal{N}_\alpha = \frac{e^{\hbar\omega_\alpha/kT}}{e^{\hbar\omega_\alpha/kT} - 1}. \qquad (23.8)$$

In the expression for the rate of excitation on the right-hand side of equation (23.7), the photon flux density is the product of the occupation number per mode and the number of modes per unit volume and frequency interval (eq. [6.8]). The resonance scattering cross section is narrow, so in the integral we can replace $\mathcal{N}$ with the value $\mathcal{N}_\alpha$ on the resonance, and we see from equation (23.8) that this eliminates the temperature-dependent parts of equation (23.7). This was Einstein's original reason for introducing the stimulated emission term. The remaining part gives

$$\sigma = C\Delta(\omega - \omega_\alpha), \qquad C = \frac{3}{4}\Lambda\lambda_\alpha^2, \qquad (23.9)$$

where the wavelength on resonance is

$$\lambda_\alpha = 2\pi c/\omega_\alpha = 1216\,\text{Å}. \qquad (23.10)$$

Now let us get the optical depth for absorption of light that passes through the local Lα resonance, as the radiation moves toward us through a smoothly distributed sea of neutral atomic hydrogen. The element of path length along the line of sight is $c\,dt = c\,da/\dot{a}$. The optical depth for scattering in the resonance at observed frequency ω_o is

$$\tau = \int \sigma(\omega_o a_o/a(t))\, n_I(t)\frac{cH_o^{-1}da}{a[\Omega(a_o/a)^3 + \Omega_R(a_o/a)^2 + \Omega_\Lambda]^{1/2}}. \qquad (23.11)$$

The proper space number density of hydrogen atoms at the redshift where the radiation passes through the resonance is n_I. (This follows the convention that HI refers to neutral atomic hydrogen and HII to ionized hydrogen.) The denominator is the general expression for the expansion rate $\dot{a}$ (eq. [5.53]). We will use the Einstein–de Sitter limit in equation (23.2), which assumes the redshift z and the density parameter Ω are large enough so that the expansion rate is dominated by the first term, representing the mass density. With the expression in equation (23.9) for the resonance scattering cross section, the optical depth is

$$\tau = \frac{3\Lambda\lambda_\alpha^3 n_I}{8\pi H_o\Omega^{1/2}}(1+z)^{-3/2}, \qquad (23.12)$$

$$n_I = 2.4 \times 10^{-11}\Omega^{1/2}h(1+z)^{3/2}\tau\,\text{cm}^{-3}.$$

The second line gives the proper number density n_I of hydrogen atoms, in an assumed homogeneously distributed component, in terms of the optical depth τ for extinction by scattering. This number density is evaluated at the redshift $z = \omega_\alpha/\omega_o - 1$ at which the radiation at observed frequency ω_o passes through the resonance.

Between the discrete absorption lines of the $L\alpha$ forest clouds, quasar spectra show no very pronounced decrease from what would be expected from an extrapolation of the long wavelength side of the $L\alpha$ quasar emission line, where the spectrum is not affected by resonant scattering by atomic hydrogen. To establish orders of magnitude, a conservative nominal bound on τ is sufficient. Following Steidel and Sargent (1987), we will use

$$\tau \lesssim 0.1 , \tag{23.13}$$

at $z \lesssim 4$. Then equation (23.12) indicates that the atomic hydrogen number density in a smoothly distributed component has to satisfy the remarkably tight bound,

$$n_I \lesssim 2 \times 10^{-12} \Omega^{1/2} h (1+z)^{3/2} \, \text{cm}^{-3} . \tag{23.14}$$

The ratio of this expression to the critical Einstein-de Sitter mass density (23.1) scaled to the epoch z is the Gunn-Peterson limit on the density parameter in homogeneously distributed atomic hydrogen,

$$\Omega(n_I) \lesssim 2 \times 10^{-7} \Omega^{1/2} h^{-1} (1+z)^{-3/2} . \tag{23.15}$$

A variant of the Gunn-Peterson bound uses observations at shorter wavelengths of sources at higher redshifts, where the frequency at the source is above the threshold for ionization. The photoionization cross section as a function of frequency near the threshold is usefully approximated by the expression

$$\sigma_{\text{pi}} = \sigma_1 (\omega_1/\omega)^3 , \qquad \sigma_1 = 7.9 \times 10^{-18} \, \text{cm}^2 , \tag{23.16}$$

at ω greater than the limiting frequency ω_1 for the Lyman series $1s \rightarrow np$. The threshold wavelength is $\lambda_1 = 2\pi c/\omega_1 = 912$ Å. Following the calculation in equation (23.11), one sees that the optical depth for absorption by ionization of atomic hydrogen in a homogeneous intergalactic medium is

$$\tau = \frac{2}{3} \sigma_1 n_I c H_o^{-1} \Omega^{-1/2} (1+z_1)^{-3/2}$$

$$= 5 \times 10^{10} n_I (\text{cm}^{-3}) h^{-1} \Omega^{-1/2} (1+z_1)^{-3/2} . \tag{23.17}$$

Here z_1 is the redshift at which the observed frequency passes the threshold for ionization, and as before n_I is the proper number density of atomic hydrogen, in units of cm^{-3}, at z_1. Along some lines of sight the spectrum has an absorption trough, produced by a Lyman-limit cloud with column density $\Sigma \gtrsim 3 \times 10^{17} \, cm^{-2}$ that is optically thick to ionizing radiation near the threshold. Along other lines of sight the source is visible through the limit for ionization. We see that this implies a bound on n_I comparable to what follows from the absence of appreciable resonant scattering at lower frequencies.

The strikingly low limit in equation (23.15) on the density of neutral smoothly distributed hydrogen applies at redshifts near zero to $z \sim 5$. At the largest observed redshifts, crowding of the $L\alpha$ lines weakens the limit on τ, but still we can conclude that at $z \sim 5$ the intercloud neutral hydrogen density is at least six orders of magnitude below the estimate for the baryon density from light element production (eq. [6.27]). There is a similar bound on the density of molecular hydrogen, from the $1s - 2p$ electronic transition, at wavelength ~ 1000 Å, and a comparable bound also on neutral helium from the $1s^2 - 1s2p$ resonance line at $\lambda584$ Å, which is observable in space at redshift $z \gtrsim 1$.

This beautiful result tells us that at expansion factor $1 + z \sim 6$ some combination of depletion by the formation of stars and protogalaxies and ionization of whatever is left had produced an exceedingly low density of neutral matter between the $L\alpha$ clouds. We consider next some orders of magnitude for the ionization.

Ionization

The transitions to be considered are the radiative ones,

$$H + \gamma \leftrightarrow p + e, \tag{23.18}$$

and electron collisions,

$$H + e \rightarrow p + e + e. \tag{23.19}$$

The first step in estimating the neutral fraction in the intergalactic medium is to get the rate coefficient for electron capture, $p + e \rightarrow H + \gamma$. The cross section, σ_r, follows from the ionization cross section (23.16) by a detailed balance argument, in which one imagines that a proton and an electron are placed in a dissipationless box and one considers the probability of finding that the box contains a hydrogen atom and a photon. The proton is relatively heavy, so we can imagine it is fixed in space and the electron has kinetic energy in the range E to δE. The ratio of the probabilities P_h and P_p that the box is found to contain a hydrogen atom and photon, or proton and electron, is the ratio of the number of quantum states available to each case, as given by equation (6.8):

$$\frac{P_h}{P_p} = \frac{2 \cdot 4 \cdot p_\gamma^2 dp_\gamma}{4 \cdot p_e^2 dp_e} = \frac{\sigma_r v_e}{\sigma_{pi} c}. \tag{23.20}$$

The numerical factors count spin states for the particles. The momentum intervals are $dp_\gamma = dE/c$ and $p_e dp_e = mdE$. The last expression is the ratio of rates of recombination and ionization. Collecting, we have

$$\frac{\sigma_r}{\sigma_{pi}} = \frac{2p_\gamma^2}{p_e^2} = \frac{2\hbar^2\omega^2}{c^2 p_e^2} \, . \tag{23.21}$$

In astrophysics this expression (for a general number of angular momentum states for the ions) is known as the Milne relation.

The rate of the recombination reaction $p + e \rightarrow H + \gamma$ per unit volume is

$$\frac{dn_I}{dt} = \alpha n_e n_p \, , \tag{23.22}$$

where n_e and n_p are the number densities of electrons and protons. The recombination coefficient α is the product of the electron capture cross section σ_r and the electron velocity, averaged over a thermal distribution,

$$\alpha = \langle \sigma_r v_e \rangle = \int \frac{4\pi p^2 dp \, e^{-p^2/2mkT}}{(2\pi mkT)^{3/2}} \frac{p}{m} \frac{2\hbar^2\omega^2}{c^2 p^2} \frac{\sigma_1 \omega_1^3}{\omega^3} \, . \tag{23.23}$$

The first factor is the normalized probability distribution in the electron momentum. (The exponential is the occupation number in eq. [6.5], with eq. [6.8] for the probability distribution in phase space.) The photon energy is the sum of the electron kinetic energy and the hydrogen atom binding energy χ_1,

$$\hbar\omega = \frac{p^2}{2m} + \chi_1 \, , \qquad \chi_1 = 13.6 \, \text{eV}. \tag{23.24}$$

This brings equation (23.23) to

$$\alpha = \frac{8\pi\sigma_1\chi_1^3 e^{\chi_1/kT}}{(2\pi mkT)^{3/2} c^2} \int_{\chi_1/kT}^{\infty} \frac{dx}{x} e^{-x} \, . \tag{23.25}$$

If χ_1 is large compared to kT, the integral is reasonably well approximated as e^{-x}/x, with $x = \chi_1/kT$, giving

$$\alpha = \frac{8\pi\sigma_1\chi_1^3}{(2\pi mkT)^{3/2}} \frac{kT}{c^2\chi_1} = \frac{2.07 \times 10^{-13}}{T_4^{1/2}} \, \text{cm}^3 \, \text{s}^{-1} \, , \tag{23.26}$$

with

$$T = 10^4 T_4 \, \text{K} \, . \tag{23.27}$$

This calculation uses approximations to the integral and the photoionization cross section σ_{pi}, and it includes only recombinations to the ground state. A better interpolation formula for the recombination coefficient at $T \sim 10^4$ K, summed over capture to all states of the atom, is

$$\alpha = 4 \times 10^{-13} T_4^{-0.7} \, \text{cm}^3 \, \text{s}^{-1} \tag{23.28}$$

(Osterbrock 1989, table 2.1). As Osterbrock explains, the temperature $T_4 \sim 1$, which is $kT \sim 1$ eV, is commonly encountered. If the matter is ionized by radiation the photoelectron energy tends to be a modest fraction of the ionization energy, 13.6 eV for hydrogen. That means the energy deposited by photoionization cannot make the kinetic temperature much greater than $T \sim 10^4$ K. The thermostat tending to prevent the temperature from falling much below 10^4 K is recombination and the suppression of collisional excitation and radiative decay.

Now let us consider how the Gunn-Peterson limit on the neutral mass density in smoothly distributed material between the $L\alpha$ clouds translates to a bound on the plasma density. The ionized fraction is determined by the balance between the recombination rate in equation (23.28) and the rate of ionization by radiation or collisions.

When the rate of collisional ionization, $H + e \rightarrow p + e + e$, is in equilibrium with the recombination rate, $p + e \rightarrow H + \gamma$, the number densities of neutral and ionized hydrogen satisfy

$$\alpha n_p n_e = \langle \sigma_{ci} v \rangle n_e n_I . \tag{23.29}$$

The cross section for collisional ionization reaches its maximum value, $\sigma_{ci} \cong 2a_o^2$, where a_o is the Bohr radius, at electron energy $\epsilon \sim 100$ eV (Bely and van Regemorter 1970). At higher energies the cross section decreases as $\sigma_{ci} \sim \ln \epsilon / \epsilon$. The equilibrium neutral fraction n_I / n_p thus is at its minimum value at $\epsilon \sim 100$ eV, corresponding to temperature $T \sim \epsilon / k \sim 10^6$ K, where

$$\sigma_{ci} v = 3 \times 10^{-8} \, \text{cm}^3 \, \text{s}^{-1} . \tag{23.30}$$

With equations (23.28) and (23.29), the equilibrium ratio of neutral to ionized number density at this temperature is

$$\frac{n_I}{n_p} = 5 \times 10^{-7} . \tag{23.31}$$

At this neutral fraction the Gunn-Peterson bound in equation (23.15) translates to plasma mass density

$$\Omega_{IGM} = \frac{n_p}{n_{crit}} \lesssim \frac{0.4}{(1+z)^{3/2}} \frac{\Omega^{1/2}}{h} , \tag{23.32}$$

where the bound is referred to the critical Einstein-de Sitter density (23.1). At the assumed temperature $T \sim 10^6$ K this is about at the upper bound of the pressure allowed by the Lα forest clouds (eq. [23.78]).

If the matter temperature were well below 100 eV, the dominant source of ionization could be electromagnetic radiation. We will frame the discussion in terms of the ionizing radiation from quasars because this radiation is observed, redshifted into the visible. Since quasars and damped Lα systems at $z \sim 3$ contain heavy elements, we know massive stars are another likely source. And there has been considerable interest in the idea that dark matter particles can decay with the emission of ionizing radiation (Rephaeli and Szalay 1981; Melott 1984; Sciama 1990b).

The observed optical flux density from a quasar at $z \sim 3$ is ionizing radiation at the source, so the mean sky brightness from the counts and flux densities from the quasars, extrapolated back in time using the brightness theorem (eq. [9.53]), is the mean intensity of ionizing radiation from these sources, independent of the parameters of the cosmological model. The contribution from sources at epochs much earlier than $z \sim 3$ is limited by absorption in clouds that are optically thick to ionizing radiation near the threshold. The redshift distribution for the Lyman-limit clouds (the probability that a line of sight intersects one of these clouds at redshift z to $z + dz$) is (eq. [23.46] below)

$$dP \sim 1\,(1+z)\,dz\,. \tag{23.33}$$

This means that at $z = 3$ the mean free path for an ionizing photon near threshold is only $\delta z \sim 0.3$, so ionizing radiation produced at much higher redshifts is efficiently absorbed.

There is a check on the intensity of ionizing radiation at $z \sim 3$ from the inverse or proximity effect in the redshift distribution of the absorption lines in the Lα forest. At redshifts not close to that of the quasars, the distributions along different lines of sight are consistent with the assumption that the cloud positions are drawn from a universal redshift-dependent population. This agrees with the idea that the Lα forest lines are produced in intergalactic clouds, rather than being intrinsic to the quasars. The density of these lines increases with increasing redshift. Weymann, Carswell, and Smith (1981), Carswell et al. (1982), and Murdoch et al. (1986) noted that the density decreases at redshifts near that of the quasar. The standard interpretation of this proximity or inverse effect is that the ionizing radiation from the quasar reduces the neutral fraction in nearby clouds. It would mean that at the distance from the quasar where the density of lines is half the universal value, the observed flux of ionizing radiation from the quasar is comparable to the mean ionizing background. The idea was noted by Weymann, Carswell, and Smith (1981), and worked out in detail by Bajtlik, Duncan, and Ostriker (1988).

A standard normalization for the mean radiation energy density near the ionizing threshold ν_1 is

$$i_\nu = i_{21} \frac{\nu_1}{\nu} \times 10^{-21} \text{ erg cm}^{-2} \text{s}^{-1} \text{Hz}^{-1} \text{ster}^{-1}. \tag{23.34}$$

The spectrum approximates that of a quasar. The photon number density per logarithmic bandwidth is

$$n_\gamma = \nu n_\nu = \frac{4\pi i_\nu}{hc} = 6 \times 10^{-5} i_{21} \frac{\nu_1}{\nu} \text{ photons cm}^{-3}. \tag{23.35}$$

For the purpose of establishing orders of magnitude, a reasonable number for the dimensionless factor is

$$i_{21} \sim 1 \quad \text{at } z \sim 3. \tag{23.36}$$

This is about what is indicated by the analysis of the proximity effect (Bajtlik, Duncan, and Ostriker 1988; Lu, Wolfe, and Turnshek 1991), and somewhat larger than is found from counts of quasars (Bechtold et al. 1987), but people are continuing to find more quasars at $z \gtrsim 3$, and as we have noted it is easy to imagine there is a significant contribution from stars in young galaxies.

If radiation ionizes the bulk of the baryonic matter between the Lα clouds and within the low surface density ones, something has to have provided enough photons to dissociate all the atoms. The ratio of the numbers of ionizing photons to baryons is (eqs. [23.1] and [23.35]) $n_\gamma/n_B \sim 0.1 \, i_{21} \Omega_B^{-1} h^{-2}$ at $z = 3$. As discussed by Shapiro and Giroux (1987), if the density parameter Ω_B in intergalactic atomic matter were near unity, the observed number of ionizing photons from the quasars would be insufficient to ionize most of the atoms. That could mean the matter between the clouds has not been ionized, but instead quite thoroughly cleared away and placed in clouds and protogalaxies, or else that something else provided the photons. We will be considering relatively low values of $\Omega_B h^2$, where there are more photons than atoms, and, for larger values of $\Omega_B h^2$, we can suppose the initial ionization was accomplished by earlier generations of stars or quasars, perhaps now hidden by Lyman-limit clouds.

With equation (23.34) for the ionizing radiation background and equation (23.16) for the cross section, the ionization rate per hydrogen atom is

$$\lambda_{\text{pi}} = \int_{\nu_1}^\infty \frac{4\pi i_\nu}{h\nu} d\nu \, \sigma_{\text{pi}} = \frac{\sigma_1 i(\nu_1)}{2\hbar} = 4 \times 10^{-12} i_{21} \text{ s}^{-1}. \tag{23.37}$$

The mean life of an atom is $\lambda_{\text{pi}}^{-1} \sim 10^4$ y. Since this is short compared to the expansion time, we can compute the equilibrium ionization (except near a quasar, where the ionizing radiation may fluctuate on shorter timescales).

Neglecting collisional ionization, the equilibrium equation expressed in terms of the mean number density $\langle n_I \rangle$ of hydrogen atoms is

$$\lambda_{\text{pi}} \langle n_I \rangle = \alpha \langle n_e n_p \rangle = \alpha C \langle n_p \rangle^2 . \tag{23.38}$$

The local proton and electron number densities are n_p and n_e, the brackets are the space average of the product, and the mean values of the densities are nearly equal, $\langle n_p \rangle = \langle n_e \rangle$. The factor $C > 1$ takes account of the clumping of the space distribution. If the plasma uniformly filled a fraction $1/C$ of space, with the rest of space empty, the mean square density would reproduce the second part of equation (23.38).

With equation (23.28) for the recombination coefficient α, and equation (23.37) for the ionization rate, we see that the ratio of the mass in the neutral and ionized components is

$$\frac{\langle n_I \rangle}{\langle n_p \rangle} = \frac{\alpha C \langle n_p \rangle}{\lambda_{\text{pi}}} = 1 \times 10^{-6} \frac{C \Omega_{\text{IGM}} h^2}{i_{21} T_4^{0.7}} (1+z)^3 . \tag{23.39}$$

This says that smoothly distributed matter at $z \sim 3$ would be highly ionized. With the HI Gunn-Peterson bound in equation (23.14), the limit on the density in a smoothly distributed intergalactic plasma is

$$\Omega_{\text{IGM}} \lesssim 0.4 \frac{\Omega^{1/4} T_4^{0.35} i_{21}^{1/2}}{h^{3/2} C^{1/2} (1+z)^{9/4}} . \tag{23.40}$$

This has been expressed as a cosmological density parameter, Ω_{IGM}, as in equation (23.1).

The conclusion is that at $z \sim 3$ the density parameter in a smoothly distributed intergalactic medium is not likely to be more than $\Omega_{\text{IGM}} \sim 0.03$. The bound for collisionally ionized plasma uses the temperature at maximum ionization (eq. [23.32]). The upper bound on the density of photoionized baryonic matter between the clouds (eq. [23.40]) is reduced if the plasma is appreciably clumped. It would be increased if i_{21} were well above unity, but that would require some new explanation for the proximity effect. The bound at $z \sim 3$ is less than the critical Einstein-de Sitter density, but comparable to the net baryon density in the homogeneous model for light element production (eq. [6.27]).

We will see in the next subsections that the mass densities in the neutral and ionized matter associated with the discrete clouds that cause the absorption lines in quasar spectra are comparable to these upper bounds on a homogeneous component.

Column Densities in Clouds

The gas clouds described at the beginning of this section are assigned three names, based on the appearance of the absorption features they produce in the spectra of background objects. In damped Lyman-α systems[13] the HI column or surface densities are $\Sigma \gtrsim 10^{20}$ atoms cm^{-2}, comparable to the surface density in interstellar gas in a spiral galaxy at the present epoch. The high column density produces a very distinctive absorption trough at the Lα resonance, as one sees in figure 23.1. The more abundant Lyman-limit clouds have column densities $\Sigma \sim 3 \times 10^{17}$ cm^{-2}, high enough to block radiation near the photoionization edge at the limit of the Lyman series of resonance absorption lines (eq. [23.16]), again a prominent effect. Gas clouds at HI column densities $\Sigma \sim 10^{14}$ cm^{-2} are abundant enough at $z \sim 3$ to produce the distinctive forest of absorption lines to the short wavelength side of Lα at the source. These classifications have some physical basis, for at the Lyman limit the surface density becomes high enough to shield the material in the center of the cloud from extragalactic ionizing radiation, which may be related to the fact that the evolution of the density with redshift is considerably more rapid for the lower column density clouds. However, there is not a marked discontinuity in the abundance of clouds as a function of HI column density at $z \sim 3$, where the distribution is well described by the simple power law in equation (23.47) below.

The standard measure of the strength of an absorption line is the equivalent width W defined by the equation

$$Wf_o = \int (f_o - f) d\lambda, \tag{23.41}$$

where f is the observed spectrum as a function of wavelength across the line, and f_o is the interpolation to what the spectrum would have been in the absence of the line. If the line were square and black, W would be the line width.

When the HI column density is low enough or the Doppler broadening of the line large enough to make the optical depth for scattering small at the line center, the equivalent width at HI column density Σ follows from the integral of the cross section in equation (23.9):

$$W = \int \tau d\lambda = \Sigma \int \sigma d\lambda = \frac{3\Lambda}{8\pi} \frac{\lambda_\alpha{}^4}{c} \Sigma. \tag{23.42}$$

This is a useful approximation at the low surface density end of the range of detectable Lα forest clouds. Thus, at $\Sigma = 10^{13}$ cm^{-2} the equivalent width is $W = 0.05$ Å in the rest frame of the absorber. The observed equivalent width for a line at redshift z is stretched to $W_o = W(1 + z)$, or $W_o = 0.2$ Å at $\Sigma = 10^{13}$ cm^{-2} and $z = 3$, about at the current practical limit for detection.

---

[13] The name comes from the fact that the absorption line extends well into the Lorentzian wings of the scattering cross section in equation (23.43). These are said to be radiation-damped wings, in analogy to the spread in line width caused by collisional deexcitation or damping of an excited state.

When the optical depth in the line center is large, the line shape is a convolution of the scattering cross section with the distribution of Doppler shifts from the dispersion of velocities along the line of sight and the instrumental resolution. Deducing Σ from W depends on the subtle art of modeling all the effects.

The column densities of the Lyman-limit systems are distinctive, and at the very high column densities of the damped $L\alpha$ systems the situation again is simple because the shape of the line is dominated by the wings of the scattering cross section. As shown at the end of this section (eq. [23.97]), the $L\alpha$ scattering cross section is well approximated as

$$\sigma = \frac{3\lambda_\alpha^2}{8\pi} \frac{\Lambda^2}{(\omega - \omega_\alpha)^2 + \Lambda^2/4} . \tag{23.43}$$

The integral over frequency agrees with equation (23.9). The line width is determined by the decay rate Λ (eq. [23.5]). At high surface density the equivalent width is

$$W = \int d\lambda \left(1 - e^{-\tau}\right) = \int d\omega \frac{\lambda^2}{2\pi c} \left[1 - \exp\left(-\frac{3\lambda_\alpha^2 \Lambda^2 \Sigma}{8\pi(\omega - \omega_\alpha)^2}\right)\right] . \tag{23.44}$$

The prefactor converts the variable of integration, which traditionally is the wavelength, to frequency. We can ignore the second term in the denominator in equation (23.43) for the cross section, because the line center is black, and we can replace λ with λ_α, because the factor in brackets is peaked at the resonance. Then the result of changing the variable of integration is

$$W = \left(\frac{3}{8\pi}\right)^{1/2} \frac{\lambda_\alpha^3 \Lambda \Sigma^{1/2}}{2\pi c} \int_{-\infty}^{\infty} dx \left(1 - e^{-1/x^2}\right)$$

$$= 7.3 \left(\frac{\Sigma}{10^{20} \, \text{cm}^{-2}}\right)^{1/2} \text{Å} . \tag{23.45}$$

The damped $L\alpha$ system in figure 23.1 has $\Sigma = 1 \times 10^{21}$ cm^{-2} at $z = 2$, meaning the observed equivalent width is $W_o = (1 + z)W = 70$ Å. The grownup analysis fits the line shape in the integrand of equation (23.44) to the measured line, as discussed by Turnshek et al. (1989).

The distribution of HI column densities and redshifts is characterized by the probability dP that a line of sight intersects a cloud with density Σ in the range $d\Sigma$ at redshift z in the range dz,

$$dP = g(\Sigma, z) \, d\Sigma \, dz . \tag{23.46}$$

At redshift $z \sim 3$ a useful fitting function for the distribution function g is the power law form

$$g = A\Sigma^{-\beta}, \qquad A = 10^{8.4}, \qquad \beta = 1.46 \qquad (23.47)$$

(Carswell et al. 1984; Tytler 1987), for $10^{13} \lesssim \Sigma \lesssim 10^{22}$ cm^{-2}. In this expression the column density Σ is measured in units of hydrogen atoms per square centimeter, so $A\Sigma^{-\beta}$ has units of cm^2. The parameters are from Sargent, Steidel, and Boksenberg (1989), Lanzetta (1991), and Lanzetta et al. (1991). There has to be an upper cutoff in this power law model because the integral for the mean HI mass density diverges. The logarithmic derivative of g steepens to $\beta = 1.7 \pm 0.2$ at $\Sigma \sim 2 \times 10^{20}$ cm^{-2} (Lanzetta et al. 1991), and the indications are that β is considerably steeper at $\Sigma \sim 10^{22}$ cm^{-2}. At redshift $z \sim 3$ this simple power law model fits the observations, to a factor of about three, over the remarkably wide range of column densities from the Lα forest to the damped Lα systems.

If the clouds at a given surface density are conserved, with fixed comoving space number density and fixed cross section, the redshift distribution is given by equation (13.42). If the universe is cosmologically flat, the expansion rate at $z = 3$ is close to the Einstein-de Sitter limit, and the redshift distribution for conserved clouds is predicted to be

$$g(\Sigma, z) \propto (1+z)^3 \frac{dt}{dz} \propto (1+z)^{1/2}. \qquad (23.48)$$

The redshift distributions of the Lyman-limit and damped Lα clouds are not greatly different from this relation, indicating evolution is not rapid (Lanzetta 1991; Lanzetta et al. 1991). The rate of increase of g with z in the Lα forest is considerably faster than this (eq. [23.55]), that is, the low surface density clouds are dissipating. That means the shape of the distribution $g(\Sigma, z)$ as a function of the surface density Σ has to be evolving, and the fitting function in equation (23.47) can only be a useful approximation in a limited range of redshifts around $z \sim 3$.

Young Galaxies

The damped Lα systems have surface densities comparable to those characteristic of present-day spiral galaxies. The following numbers indicate that the rate of interception of the clouds along a line of sight is comparable to what would be expected if the clouds had the size and comoving number density characteristic of present-day galaxies (Wolfe 1986).

To translate the distribution $g(\Sigma, z)$ of column densities to the mean mass density in atomic hydrogen, write the mean value of the number $dN_{\rm I}$ of hydrogen atoms found in a column with unit proper face area and depth dz as

$$dN_{\rm I} = dz \int \Sigma g(\Sigma, z) \, d\Sigma = \frac{\langle n_{\rm I} \rangle c}{H_o \Omega^{1/2}} \frac{dz}{(1+z)^{5/2}}. \qquad (23.49)$$

The last expression is the product of the cosmic mean number density $\langle n_I \rangle_z$ of hydrogen atoms belonging to the clouds and the proper distance $c\,dt$ along the line of sight from z to $z + dz$, in the Einstein-de Sitter limit. With equation (23.47) for $g(\Sigma, z)$, the mean density of atomic hydrogen scaled to zero redshift is

$$\frac{\langle n_I \rangle}{(1+z)^3} = \frac{A\Sigma_{\text{max}}^{2-\beta}}{2-\beta} \frac{H_o \Omega^{1/2}}{c(1+z)^{1/2}} . \tag{23.50}$$

This assumes the power law has a sharp cutoff at Σ_{max}. The density referred to the Einstein-de Sitter value is

$$\Omega_I \sim 0.002 (\Sigma_{\text{max},22})^{0.54} \Omega^{1/2} h^{-1} , \tag{23.51}$$

at $z \sim 3$, with the cutoff for the power law expressed in units of 10^{22} cm^{-2}. This uses the Einstein-de Sitter limit for the expansion rate. More detailed estimates of Ω_I are given by Lanzetta et al. (1991).

In the range of surface densities well described by the power law model in equation (23.47), most of the neutral hydrogen is in the clouds with the largest surface densities. The effective cutoff for the power law seems to be at about $\Sigma = 10^{22}$ cm^{-2}, meaning the density parameter in the atomic hydrogen in the clouds at $z \sim 3$ is about $\Omega_I \sim 0.002 \, \Omega^{1/2}/h$. This is comparable to the mean density in the baryonic material in the luminous parts of galaxies at the present epoch (eq. [5.150]).

In the power law model (23.47) the number of clouds with surface densities greater than Σ per unit increment of redshift is

$$\frac{dN}{dz} = \frac{A\Sigma^{1-\beta}}{\beta - 1} = 0.3 \, \Sigma_{20}^{-0.46} . \tag{23.52}$$

This can be compared to the expected rate of intersection of young L_* galaxies under the assumption that galaxies are conserved back to $z \sim 3$ at number density $n_g(1+z)^3$, with $n_g = 0.01 h^3$ Mpc^{-3}. The probability that a line of sight intercepts one of these galaxies at impact parameter $r_g = 10h^{-1}$ kpc and redshift z is

$$\frac{dN}{dz} = \frac{\pi r_g^2 n_g c}{H_o \Omega^{1/2}} (1+z)^{1/2} \tag{23.53}$$

$$\sim 0.02 \, \Omega^{-1/2} .$$

The observed rate at $\Sigma > 10^{20}$ cm^{-2} in equation (23.52) is a factor three to ten times larger, depending somewhat on the cosmological model. The larger observed rate could be taken to mean that at $z \sim 3$ galaxies were immature, tending to be surrounded by more extended HI disks or gas clouds that since have been blown away or have merged.

Finally, let us note that the mass per unit area in the disk of the Milky Way at our position, converted to the equivalent in protons, is (eq. [18.27])

$$\Sigma_g = 55 \pm 10 \mathcal{M}_\odot \, pc^{-2} = 10^{21.8} \, \text{protons cm}^{-2} \, . \tag{23.54}$$

This is comparable to the largest commonly observed HI column densities at $z \sim 3$ (Lanzetta et al. 1991), consistent with the idea that the damped Lα systems are young galaxies or their near relatives.

Lyman-Alpha Forest Clouds

We consider here three notable properties of the clouds in the Lα forest, at HI surface densities $\Sigma \sim 10^{14}$ cm^{-2}. First, the net mass in plasma in the clouds is comparable to the amount of neutral hydrogen in the high surface density clouds and what is seen now in galaxies. Second, the comoving number density of these low surface density systems shows a marked evolution with redshift, indicating the clouds are dissipating (Peterson 1978). Third, at a given epoch the space distribution in these clouds is quite close to uniform, in striking contrast to the clumpy distribution of galaxies at the present epoch and, one expects, at $z \sim 3$ (Sargent et al. 1980).

The redshift distribution in the Lα forest clouds is approximated by the expression (Lu, Wolfe, and Turnshek 1991)

$$\frac{dN}{dz} = B(1+z)^\gamma \, , \tag{23.55}$$

with

$$B = 3.5 \, , \qquad \gamma = 2.75 \, , \tag{23.56}$$

at

$$\Sigma \geq 10^{14} \, cm^{-2} \quad \text{and} \quad 2 \lesssim z \lesssim 4 \, . \tag{23.57}$$

If the clouds were conserved, the power law index would be $\gamma = 0.5$ in the Einstein-de Sitter limit, $\gamma = 1$ in the limit $\Omega = 0$ with no cosmological constant (eq. [13.42]). The considerably larger value of γ at the low surface densities in the Lα forest shows that these clouds are dissipating.

The mean proper distance between clouds along the line of sight is

$$L = c \, \frac{dt}{dz} \frac{dz}{dN} \sim \frac{c}{H_0 \Omega^{1/2} B (1+z)^{\gamma+5/2}} \tag{23.58}$$
$$\sim 1000 \, h^{-1} \Omega^{-1/2} (1+z)^{-5.25} \, \text{Mpc} \, .$$

This amounts to $L \sim 0.6 h^{-1} \Omega^{-1/2}$ Mpc at $z = 3$.

The kinetic temperature of the hydrogen atoms in a cloud is estimated from the line shape, which is a convolution of the natural shape in equation (23.43) with the Doppler distribution and instrumental resolution. The standard model for the one-dimensional velocity distribution is

$$\frac{dP}{dv} \propto e^{-v^2/2\sigma^2} = e^{-v^2/b^2} = e^{-m_p v^2/2kT}, \tag{23.59}$$

where σ is the rms velocity dispersion, b is the conventional measure, and the last expression is the thermal Boltzmann distribution. A typical value for the low density clouds is

$$b = 30\,\text{km s}^{-1} \tag{23.60}$$

(Carswell et al. 1987). The kinetic temperature belonging to this velocity dispersion is

$$T = m_p b^2/2k \sim 5 \times 10^4\,\text{K}. \tag{23.61}$$

This is about what would be expected if the plasma were heated by photoionization by radiation at wavelengths close to the threshold, yielding photoelectrons with energies that are a modest fraction of the ionization energy for hydrogen, 13.6 eV (where $13.6\,\text{eV}/k = 1.5 \times 10^5$ K). That is, it is plausible to assume equation (23.61) is a useful approximation to the kinetic temperature of the matter, rather than a measure of Doppler broadening by streaming motions.

We see from equation (23.51) that the density parameter belonging to the atomic hydrogen in clouds with surface densities $\Sigma \sim 10^{14}$ cm^{-2} is

$$\Omega(n_\text{I}) \sim 1 \times 10^{-7}\Omega^{1/2}h^{-1}. \tag{23.62}$$

This is comparable to the very low bound for smoothly distributed hydrogen (eq. [23.15]). But since the material is exposed to the ionizing radiation from the quasars, we must consider the ionized fraction in the clouds, as in equation (23.39).

The estimate of the ionized fraction depends on the cloud size, which converts the observed surface density of neutral hydrogen to the space density. We will write the characteristic physical size as

$$l = 10\,l_{10}h^{-1}\,\text{kpc}, \qquad 1 < l_{10} \lesssim 10, \tag{23.63}$$

so we can track the sensitivity of the following numbers to the parameter l_{10}.[14]

---

[14] We ought to take account of the fact that the deduced value of l depends on the cosmological model as well as the Hubble parameter, but that would further complicate results that already are burdened with a large number of parameters.

The lower bound is a very conservative interpretation of the rate of coincidences of cloud redshifts along neighboring lines of sight (Foltz et al. 1984; Smette et al. 1992). The upper bound is indicated by the redshift measurements and by the condition that the clouds not overfill space (eq. [23.67] below). It will be noted that the region occupied by a cloud can be considerably broader than the path length through it if the cloud has an irregular structure. Also, we will see that the distance between clouds is comparable to the cloud size, so it may be better to think of l as a characteristic coherence length in the clumpy HI distribution, rather than a measure of the sizes of physically isolated and regularly shaped clouds.

For the purpose of establishing vital statistics let us consider clouds with HI column densities $\Sigma \sim 10^{14}$ atoms cm^{-2}, at redshift $z \sim 3$, where the abundance along the line of sight is given by equation (23.55).

The characteristic density of hydrogen atoms within a single cloud is

$$n_I \sim \Sigma/l \sim 3 \times 10^{-9} h/l_{10} \text{ atoms cm}^{-3} . \tag{23.64}$$

The characteristic mass of neutral hydrogen in a cloud is

$$\mathcal{M}_I \sim \Sigma l^2 m_p \sim 100 \, l_{10}^2 h^{-2} \mathcal{M}_\odot . \tag{23.65}$$

It is startling that objects a few hundred times the mass of the Sun are detected at redshifts well above unity.

Next, let us estimate the mean fraction of space occupied by clouds. Let $\langle n_I \rangle$ be the mean space number density of hydrogen atoms from clouds with mean surface density $\langle \Sigma \rangle$ that appear along the line of sight at the rate dN/dz. Then in the Einstein-de Sitter limit for dt/dz,

$$\frac{c\langle n_I \rangle}{H_o \Omega^{1/2}(1+z)^{5/2}} = \langle \Sigma \rangle \frac{dN}{dz} . \tag{23.66}$$

The mean density of atoms within a cloud is $n_I = \langle \Sigma \rangle /l$, so the fraction of space filled by the clouds is

$$\begin{aligned}
\frac{1}{C} = \frac{\langle n_I \rangle}{n_I} &= \Omega^{1/2}(1+z)^{5/2} \frac{dN}{dz} \frac{H_o l}{c} \\
&= 1 \times 10^{-5}(1+z)^{5.25} l_{10} \Omega^{1/2} \\
&= 0.02 \, l_{10} \Omega^{1/2} \quad \text{at } z = 3 .
\end{aligned} \tag{23.67}$$

Here C is the clumping factor used in equation (23.38). The second line uses equation (23.55) for dN/dz. The characteristic number density of clouds is

$$n_{\text{clouds}} = \frac{\langle n_I \rangle}{n_I l^3} = \frac{1}{C l^3} = 2 \times 10^4 \frac{h^3 \Omega^{1/2}}{l_{10}^2} \text{ Mpc}^{-3} . \tag{23.68}$$

The width of the cube that on average contains one cloud is

$$w_{cl} = (n_{clouds})^{-1/3} = 40 h^{-1} \Omega^{-1/6} l_{10}^{2/3} \text{ kpc} , \qquad (23.69)$$

at $z = 3$.

The last line of equation (23.67) says that with the nominal parameters the clouds fill between 2% and 10% of space, which is to say the mean distance between clouds is quite similar to the cloud size. This is about as tightly packed as could be expected if the cloud shapes are irregular. That is, $l_{10} \sim 1$ to 10 gives a not unreasonable picture of a cloudy intergalactic medium.

At the temperature in equation (23.61) collisional ionization is negligible, so the neutral fraction in the cloud material is fixed by equations (23.28) for the recombination rate, equation (23.37) for the ionization rate, and equation (23.64) for the density of neutral atoms in the cloud:

$$f \equiv \frac{n_I}{n_p} = \frac{\alpha n_p}{\lambda_{pi}} = \frac{\alpha}{\lambda_{pi}} \frac{\Sigma}{f l} . \qquad (23.70)$$

Thus the neutral ratio is

$$f \sim \left(\frac{\alpha \Sigma}{\lambda_{pi} l} \right)^{1/2} \sim \frac{2 \times 10^{-5}}{T_4^{0.35}} \left(\frac{h}{i_{21} l_{10}} \right)^{1/2} , \qquad (23.71)$$

the plasma density within a cloud is

$$n_p \sim \frac{\Sigma}{lf} \sim 2 \times 10^{-4} T_4^{0.35} \left(\frac{i_{21} h}{l_{10}} \right)^{1/2} \text{cm}^{-3} , \qquad (23.72)$$

and the characteristic cloud mass is

$$\mathcal{M}_{II} \sim \frac{\Sigma l^2 m_p}{f} \sim 10^7 T_4^{0.35} i_{21}^{1/2} \left(\frac{l_{10}}{h} \right)^{5/2} \mathcal{M}_{\odot} . \qquad (23.73)$$

We note in section 25 that it could be significant that $\mathcal{M}_{II}$ is comparable to the mass of a globular star cluster and to the primeval baryon Jeans mass (eq. [6.146]).

The contribution to the mean cosmological mass density by neutral hydrogen in the Lα forest clouds is given by equation (23.62). This density divided by the neutral ratio f in equation (23.71) gives

$$\Omega_{II} \sim 0.005 \, T_4^{0.35} (i_{21} l_{10} \Omega / h^3)^{1/2} , \qquad (23.74)$$

at $\Sigma \sim 10^{14} \text{ cm}^{-2}$. This plasma mass is comparable to the neutral mass observed in the high surface density clouds (eq. [23.51]), and comparable to the prediction of the homogeneous model for light element production (eq. [6.27]).

Carswell and Rees (1987) point out that it would be of considerable interest to know whether the clouds in the forest fill the voids that may be present in the distribution of galaxies or protogalaxies at $z \sim 3$. This is discussed further by Duncan, Ostriker, and Bajtlik (1989). The orders of magnitude go as follows.

The common voids seen in figure 3.6 have diameters $d_o \sim 30h^{-1}$ Mpc. If the galaxy distribution expanded with the homogeneous Hubble flow, this would scale back to a characteristic diameter $d = d_o/(1+z) = 8h^{-1}$ Mpc at $z = 3$. If the clustering pattern were growing, the diameter at $z \sim 3$ would be smaller. As a simple model, equation (22.57) with $n = 0$ says the comoving clustering length scales as $(1+z)^{2/3}$. That would bring the characteristic void size to $d \sim 3h^{-1}$ Mpc. The ratio of this number to the mean distance between $L\alpha$ forest clouds along a line of sight (eq. [23.58]) is

$$\frac{d}{L} \sim 5\Omega^{1/2} , \tag{23.75}$$

at $z \sim 3$. If $\Omega = 1$ the mean distance between clouds is appreciably smaller than this model for the voids, so if the clouds avoided the voids one would have expected to have seen the effect in the redshift distribution. If $\Omega = 0.1$ the test is marginal.

The autocorrelation function of the forest lines along a line of sight, and the cross correlation of line redshifts in neighboring lines of sight, show that the position autocorrelation function of the $L\alpha$ forest clouds is considerably smaller than that of the galaxies extrapolated to $z \sim 3$ using the mass scaling law in equation (22.57) (Sargent et al. 1980; Sargent, Young, and Schneider 1982). That is, we are invited to picture the low surface density clouds as nearly uniformly spread through the space between the young galaxies, or their ancestors, the damped $L\alpha$ systems. A partial explanation for this effect is that the clouds come close to filling space (eq. [23.69]), so their local space density is not allowed to rise much above the mean, and their space autocorrelation function therefore cannot be much larger than unity at small separations.

The mean trend of the distribution of forest clouds with redshift (eq. [23.55]) indicates that the low surface density clouds are dissipating on a timescale comparable to that of the expansion of the universe. This could be because the clouds are in pressure equilibrium with a hotter, less dense intercloud plasma, for the expansion of the plasma naturally would cause the clouds to dissipate at about the same rate (Sargent et al. 1980; Ikeuchi and Ostriker 1986). Other possibilities are that the clouds are gravitationally confined (Rees 1986, 1988; Bond, Szalay, and Silk 1988), or inertially confined, that is, expanding by pushing away surrounding material, or some combination of the three.

A cloud freely expanding at speed $\sim b$ (eq. [23.60]) for an expansion time t at epoch z changes size by the fractional amount

$$\frac{\delta l}{l} \sim \frac{bt}{l} \sim \frac{3}{l_{10}\Omega^{1/2}} , \tag{23.76}$$

at $z = 3$. This number suggests that the pressure gradient force would tend to re-arrange the cloudy HI distribution on a timescale comparable to the cosmological expansion time at $z \sim 3$, roughly in line with what is observed. This is not true of the clouds at much lower redshifts, which are thought to be gravitationally confined. Gravitational binding requires cloud mass

$$\mathcal{M} \sim \frac{kTl}{Gm_p} = 2 \times 10^8 T_4 l_{10} h^{-1} \mathcal{M}_{\odot}.$$ (23.77)

It is suggestive that this mass is about an order of magnitude larger than the plasma mass in equation (23.73), in line with the idea discussed in section 18 that there may be nonbaryonic dark matter with mass an order of magnitude larger than that of the baryons (Rees 1986). Pressure confinement may also be important. At pressure balance, the confining plasma would satisfy (eq. [23.72])

$$P/k = Tn_p \sim 2T_4^{1.35}(i_{21}h/l_{10})^{1/2} \, \mathrm{K \, cm^{-3}}.$$ (23.78)

In the next section we compare this to the pressure of a hot intergalactic plasma that would make an interesting contribution to the X-ray background.

Summary

We are offered a picture of the way things were at $z \sim 3$ that is schematic at best in many respects, but remarkably detailed in others. Structure formation at $z \sim 3$ had advanced to the point of opening deep holes in the neutral hydrogen distribution (eq. [23.15]), leaving clouds with neutral hydrogen surface densities that range over at least eight orders of magnitude (eq. [23.47]). Most of the neutral mass at this epoch is in the clouds at the highest surface densities, the damped Lα systems (eq. [23.51]). The neutral hydrogen mass is comparable to that of the baryonic matter seen now in the bright parts of galaxies. The neutral mass in a Lα forest cloud is remarkably small, but these clouds are highly ionized, and the plasma mass in the Lα forest is comparable to the neutral mass in the Lα clouds (eq. [23.74]). The Gunn-Peterson test allows a similar amount of mass in a smoothly distributed plasma (eq. [23.40]). Within the considerable uncertainties, one can imagine that these components account for the baryon mass predicted in the standard model for the origin of the light elements (eq. [6.27]).

Are the damped Lα clouds at $z \sim 3$ the long-sought primeval galaxies? We have two pieces of information from their absorption lines: these clouds have about the right HI column densities Σ_g (eq. [23.54]), and they contain about the right amount of baryonic mass. The rate dN/dz for intersection of the clouds along a line of sight also is about right (eqs. [23.52] and [23.53]). This is not a separate piece of information, for if galaxies were broken into many parts, each with column density Σ_g, it would not affect dN/dz. However, one might wonder whether it is reasonable to imagine that the pieces out of which galaxies are assembled have the surface densities characteristic of galaxies. The heavy elements in these systems show that they contain stars. The observations of quasars and radio sources (as in the redshift-magnitude diagram in figure 5.3) tell us that some galaxies at $z \sim 3$ hosted central engines that produced what look like the radio lobes of mature galaxies. The radio galaxies observed at $z \gtrsim 1$ tend to look immature, with irregular and extended optical shapes. This is consistent with their youth, and with the idea that structures are rapidly evolving at this epoch, as indicated by the behavior of the $L\alpha$ forest clouds. One must bear in mind, however, that these galaxies are observed because they are active. The discussion of what all this might mean is continued in section 26.

Resonance Line Shape

The resonant scattering line shape in equation (23.43) is widely used, but its origin perhaps is not as familiar as it might be. A derivation, following Weisskopf and Wigner (1930), goes as follows.

We will consider the scattering cross section through the $L\alpha$ resonance, and we will simplify the computation by ignoring the other bound states of the atom. We can ignore the electron and proton spins, and the complication of many-photon states. That leaves us with the three excited states $|m\rangle$, with $m = -1$, 0 and 1 representing the angular momentum components in the $2p$ level, and the states $|k'\rangle$ representing the atom in the ground state with a photon with definite spin and momentum. Thus, the state vector is

$$|\psi\rangle = \sum_m c_m |m\rangle e^{-i\omega_a t} + \sum_{k'} c' |k'\rangle e^{-i\omega' t} . \qquad (23.79)$$

We are imagining the system is in a very large box with periodic boundary conditions, so the photon momenta are discrete.

The initial state contains a photon with momentum k, energy $\omega = k$. We are seeking the probability that at time t later the photon has a different momentum, k'. The probability amplitudes c' evolve by the matrix element $V_{k'm}$ that connects a state with a photon to an excited state of the atom. Schrödinger's equation for the amplitudes is

$$i\dot{c}_m = V_{mk} e^{i(\omega_a - \omega)t} + \sum_{k' \neq k} c' V_{mk'} e^{i(\omega_a - \omega')t} , \qquad (23.80)$$

and

$$i\dot{c}' = \sum_m c_m V_{k'm} e^{i(\omega' - \omega_a)t} . \qquad (23.81)$$

In the first equation the coefficient $c(k) = 1$ represents the initial photon. We are seeking the probability $|c'|^2$ that at time t later the photon is found scattered into another state.

The system is in a very big box, so the transition rate is low, and the transition probability therefore is appreciable only when ω' is very close to ω. That means the interesting part of the transition amplitude looks like

$$c' = A' \left[e^{i(\omega'-\omega)t} - 1 \right] , \tag{23.82}$$

with A' independent of time. This form in equation (23.81) gives

$$c' = -\sum_m \frac{c_m V_{k'm}}{\omega' - \omega} \left[e^{i(\omega'-\omega_a)t} - e^{i(\omega-\omega_a)t} \right] . \tag{23.83}$$

This expression in the right-hand side of equation (23.80) gives a differential equation for the c_m that involves the sum

$$\sum_{k'} V_{mk'} V_{k'm'} \frac{1 - e^{i(\omega-\omega')t}}{\omega - \omega'} \propto \delta_{m'm} . \tag{23.84}$$

At large t the second factor acts as the delta function of energy conservation. Isotropy tells us the sum vanishes unless $m = m'$. This brings equation (23.80) to

$$i\dot{c}_m = V_{mk} e^{i(\omega_a-\omega)t} - c_m \sum_{k'} \frac{|V_{mk'}|^2}{\omega' - \omega} \left[1 - e^{i(\omega-\omega')t} \right] . \tag{23.85}$$

The odd denominator $\omega' - \omega$ eliminates the real part of the expression in brackets, and the imaginary part gives

$$\frac{\sin(\omega-\omega')t}{\omega - \omega'} = \pi\delta(\omega - \omega') , \tag{23.86}$$

at large t, so equation (23.85) is

$$i\dot{c}_m = V_{mk} e^{i(\omega_a-\omega)t} - i\pi c_m \sum_{k'} |V_{mk'}|^2 \delta(\omega - \omega') . \tag{23.87}$$

We can simplify this expression by considering the rate of spontaneous decay from the excited state of the atom in linear perturbation theory. Here one of the c_m is unity, and the probability amplitude for finding a photon after time t is, from equation (23.81),

$$c(k) = -\frac{V_{km}}{\omega - \omega_\alpha} \left[e^{i(\omega-\omega_a)t} - 1 \right] . \tag{23.88}$$

The result of squaring the absolute value, summing over the photon states, and using

$$\frac{\sin^2(\omega - \omega_\alpha)t/2}{(\omega - \omega_\alpha)^2} = \frac{\pi t}{2}\delta(\omega - \omega_\alpha), \tag{23.89}$$

is the decay rate,

$$\Lambda = 2\pi \sum_{k'} |V_{k'm}|^2 \delta(\omega' - \omega_\alpha). \tag{23.90}$$

This sum appears in the last term in equation (23.87), at frequency ω different from ω_α. We are considering an electric dipole transition, so $|V_{km}|^2 \propto \omega$. (The simple way to see this is to note that the potential energy is proportional to the electric field strength, E, and the energy in the single-photon state is $\hbar\omega \propto E^2$.) The volume element in the conversion of the sum over photon momenta to an integral is proportional to $\omega^2 d\omega$, and the matrix element for an electric dipole transition gives another power of ω, so equation (23.90) generalizes to

$$\Lambda\frac{\omega^3}{\omega_\alpha^3} = 2\pi \sum_{k'} |V_{k'm}|^2 \delta(\omega' - \omega). \tag{23.91}$$

Equation (23.91) in equation (23.87) gives

$$i\dot{c}_m = V_{mk}e^{i(\omega_\alpha - \omega)t} - \frac{i}{2}\frac{\omega^3}{\omega_\alpha^3}c_m\Lambda. \tag{23.92}$$

The integral of this expression in equation (23.83) gives the wanted probability amplitude,

$$c' = \frac{\sum_m V_{k'm}V_{mk}}{\omega_\alpha - \omega - i\Lambda(\omega/\omega_\alpha)^3/2}\left[\frac{e^{i(\omega'-\omega)t} - 1}{\omega' - \omega}\right]. \tag{23.93}$$

The square $|c|^2$ summed over photon states is the transition probability, with

$$\frac{P}{t} = \frac{\Lambda\omega^3}{\omega_\alpha^3}\frac{\sum_m |V_{mk}|^2}{(\omega_\alpha - \omega)^2 + \Lambda^2(\omega/\omega_\alpha)^6/4}. \tag{23.94}$$

The last step uses equation (23.91).

The final step is to note that the result of averaging the matrix element in this expression over photon direction and spin appears in equation (23.91):

$$\Lambda\frac{\omega^3}{\omega_\alpha^3} = 2\pi\frac{2 \cdot 4\pi V_u}{(2\pi)^3}\omega^2\langle|V_{k'm}|^2\rangle. \tag{23.95}$$

The second factor on the right-hand side is the sum over photon polarization and momentum, with V_u the volume of the box within which we put the system. The photon density in the initial state is $1/V_u$, so the transition probability is

$$P = \sigma(\omega)t/V_u, \tag{23.96}$$

where σ is the scattering cross section at frequency ω. Collecting, we get

$$\sigma = \frac{3\lambda_\alpha{}^2\Lambda^2}{8\pi} \frac{(\omega/\omega_\alpha)^4}{(\omega - \omega_\alpha)^2 + \Lambda^2(\omega/\omega_\alpha)^6/4}, \tag{23.97}$$

the factor of three coming from the sum over m in equation (23.94).

At ω close to ω_α this is the resonance form in equation (23.43). At $\omega \ll \omega_\alpha$ the cross section is proportional to ω^4. This is Rayleigh scattering from the induced polarization of the atom (but the coefficient is a little low, because we have ignored scattering through all the other excited states of the atom). It is interesting that the line produced by a damped Lα cloud is broad enough to make the asymmetry of the line shape marginally appreciable.

24. Diffuse Matter and the Cosmic Radiation Backgrounds

The gas and dust in and between evolving protogalaxies and clusters of galaxies interact with the X-ray and thermal (CBR) cosmic radiation backgrounds, leaving signatures in the spectra and angular distributions. Here we consider some constraints on the picture for cosmic evolution from the background radiation produced by thermal bremsstrahlung by optically thin plasma, the scattering of the CBR by electrons, and the absorption and emission by dust at CBR wavelengths. We continue to use equations (23.1) and (23.2) for the plasma density and expansion timescale.

Thermal Bremsstrahlung

Free electrons accelerated in the electric fields of ions in a plasma produce and absorb radiation by the reactions

$$p + e \leftrightarrow p' + e' + \gamma. \tag{24.1}$$

For a nonrelativistic plasma in kinetic equilibrium (with a Maxwell-Boltzmann velocity distribution) the process is termed thermal bremsstrahlung, or free-free emission and absorption (because the electron is free in the initial and final states). Free-free emission is detected at radio wavelengths from clouds of ionized gas (HII regions) in the interstellar medium and as a flux of X-rays from the much hotter plasma within rich clusters of galaxies (Kellogg et al. 1971). The bounds on an isotropic free-free radiation background constrain the picture discussed in the last section for the physical state of the intergalactic medium. We begin in this subsection with the physics of the free-free opacity and luminosity of an optically thin plasma.

The general form of the rate equation for the production and absorption of electromagnetic radiation by a plasma at temperature T, expressed in terms of the occupation number per mode, $\mathcal{N}$, as a function of frequency ω, is

$$
\begin{aligned}
\frac{d\mathcal{N}}{dt} &= A\left[(1+\mathcal{N})e^{-\hbar\omega/kT} - \mathcal{N}\right] \\
&= A\left[e^{-\hbar\omega/kT} - \left(1 - e^{-\hbar\omega/kT}\right)\mathcal{N}\right].
\end{aligned}
\tag{24.2}
$$

In the first line the term $A\mathcal{N}$ is the rate of absorption of photons from the mode, in the reverse reaction in equation (24.1). Detailed balance tells us the rate coefficients for the reactions (24.1) going either way are the same (because the matrix elements are the same), and thermodynamics says the population of states in the first term is down from the second by the Boltzmann factor for energy $\hbar\omega$. Finally, the factor $(1+\mathcal{N})$ in the first line gives the sum of spontaneous and stimulated rates of emission of a photon, as in equation (23.7). The rearrangement in the second line shows that the net rate of absorption, corrected for stimulated emission, is

$$
A' = A\left(1 - e^{-\hbar\omega/kT}\right).
\tag{24.3}
$$

The radiative transfer equation expressed in terms of the surface brightness i_ω (the energy flux per unit area, solid angle and frequency interval) is

$$
\frac{di_\omega}{dr} = j_\omega - \kappa_\omega' i_\omega.
\tag{24.4}
$$

The left side is the rate of change of surface brightness with respect to position along the path of a packet of radiation (and expressed in locally Minkowski coordinates). In the right-hand side, j_ω is the energy radiated per unit volume, time, frequency interval, and solid angle, and κ' is the opacity (the fractional decrease in surface brightness due to absorption per unit proper displacement along the path). Following equation (24.3), we see that the opacity has to be of the form

$$
\kappa' = \kappa\left(1 - e^{-\hbar\omega/kT}\right) \sim \kappa\frac{\hbar\omega}{kT},
\tag{24.5}
$$

where κ is the opacity one would compute ignoring stimulated emission, and the last expression applies in the classical limit, $\hbar\omega \ll kT$.

The easiest way to arrive at the order of magnitude for the thermal bremsstrahlung luminosity density j_ω starts from the rate of absorption of radiation by the plasma, computed in the classical limit where we can suppose each mode of oscillation of the applied electromagnetic field has a definite electric field strength,

$$\mathbf{E} = \mathbf{E}_o \cos \omega t = -m\dot{\mathbf{v}}/e , \tag{24.6}$$

where $\mathbf{E}_o$ is constant. The last expression is the contribution of this mode to the acceleration of an electron with charge $-e$ and mass m. The electron velocity as a function of time t following a collision with an ion at time $t = 0$ is

$$\mathbf{v} = \mathbf{v}_i - \frac{e\mathbf{E}_o}{m\omega} \sin \omega t , \tag{24.7}$$

and the mean value of the kinetic energy of the electron after the collision is

$$\epsilon = \frac{1}{2}m\langle \mathbf{v}^2 \rangle = \epsilon_i + \frac{e^2 E_o{}^2}{4m\omega^2} . \tag{24.8}$$

This assumes the electron displacement during the collision with the ion is small compared to the wavelength in the mode. The initial energy is $\epsilon_i = mv_i^2/2$. The addition to the energy is a sum of terms like the second on the right-hand side, one for each mode. We can ignore the cross terms in the square of the velocity, because the phases in the modes are not correlated with each other or with v_i. Thus, the mean value of the energy absorbed from the mode when an electron collides with an ion is

$$\delta\epsilon = \frac{e^2 E_o{}^2}{4m\omega^2} . \tag{24.9}$$

The impact parameter, r_s, for a collision with an ion of charge ze that substantially changes the electron velocity satisfies

$$\frac{1}{2}mv_{\text{th}}{}^2 \sim \frac{ze^2}{r_s} , \tag{24.10}$$

where a characteristic electron velocity is the rms thermal value,

$$v_{\text{th}} \sim (3kT/m)^{1/2} , \tag{24.11}$$

for a plasma at temperature T. The collision cross section is $\sigma_s \sim \pi r_s^2$, and the rate of absorption of energy from the mode per unit volume is

$$\sigma_s n_i n_e v_{\text{th}} \delta\epsilon \sim c\kappa' E_o^2/(8\pi) . \tag{24.12}$$

The left side is the product of the energy loss $\delta\epsilon$ per collision (eq. [24.9]) with the rate of collisions per unit volume in a plasma with electron density n_e and ion density n_i. The right-hand side is the opacity defined in equation (24.4) multiplied by the energy density in the mode and by the velocity of light (because κ' has units of reciprocal length; the fractional absorption of energy per unit time is $c\kappa'$). The expression from equation (24.12) for the opacity is derived in the classical limit, $\hbar\omega \ll kT$, so we can rewrite it as

$$\kappa' \sim \frac{z^2 e^6 n_i n_e}{c\hbar\omega^3 m^{3/2}(kT)^{1/2}} \left(1 - e^{-\hbar\omega/kT}\right). \tag{24.13}$$

The last factor cancels $kT/\hbar\omega$ from the first part of the right-hand side, bringing κ' to what follows from equation (24.12). It is rewritten this way to bring it to the standard form of equation (24.5).

When the plasma and radiation are in thermal equilibrium, the surface brightness has to be constant, so the transfer equation (24.4) says the luminosity per unit volume, solid angle, and frequency interval is

$$\begin{aligned} j_\omega = \kappa' P_\omega &= \kappa' \frac{2\omega^2}{8\pi^3 c^2} \frac{\hbar\omega}{e^{\hbar\omega/kT} - 1} \\ &\sim \frac{z^2 e^6 n_e n_i}{m^{3/2} c^3 (kT)^{1/2}} e^{-\hbar\omega/kT}. \end{aligned} \tag{24.14}$$

Here P_ω is the surface brightness of blackbody radiation at temperature T (eq. [6.12]).

There is an exponential cutoff in the luminosity density (24.14) because the typical electron energy is $\sim kT$, so the electrons cannot produce photons with energy $\hbar\omega$ much greater than kT. This quantum effect does not follow from the classical calculation; we arrived at it by writing the classical opacity (24.13) in a way that matches the known form in equation (24.5). At $\hbar\omega \ll kT$ the luminosity density j_ω in equation (24.14) is independent of frequency, because we have modeled the electron acceleration in the collision with the ion as a pulse much narrower than ω^{-1}, so the Fourier transform of the radiation rate is flat.

A more careful analysis multiplies equation (24.14) by the numerical factor $(2^7\pi/27)^{1/2}$ (Kramers 1923). A still more careful computation follows the electron orbit through the collision. This produces a correction factor g which is a function of the plasma temperature and the radiated frequency. The correction is close to unity, and for the purpose of establishing orders of magnitude we can set $g = 1$. In the same spirit, we will ignore the small corrections for helium and heavier elements, setting the ion charge to $z = 1$ and the ion number density to the free proton and electron number densities, $n_i = n_e = n_p$. Then numerical results for the opacity and luminosity density are

$$\kappa' = 2.0 \times 10^{-23} \lambda^2 n_e^2 T^{-3/2} \text{ cm}^{-1},$$

$$j_\nu = 5.4 \times 10^{-39} n_e^2 T^{-1/2} e^{-h\nu/kT} \text{ erg cm}^{-3} \text{ s}^{-1} \text{ ster}^{-1} \text{ Hz}^{-1}. \tag{24.15}$$

The angular frequency is $\nu = \omega/2\pi$. The temperature is expressed in units of degrees K, the particle density in cm^{-3}, and the wavelength λ in centimeters.

The net free-free luminosity in this approximation is

$$J = 4\pi \int j_\nu d\nu = 1.42 \times 10^{-27} T^{1/2} n_e^2 \text{ erg cm}^{-3} \text{ s}^{-1}. \tag{24.16}$$

The thermal energy per unit volume in a fully ionized plasma is $u = 3n_e kT$, so the characteristic cooling time[15] due to thermal bremsstrahlung is

$$t_{\text{ff}} \equiv u/J = 2.9 \times 10^{11} T^{1/2} n_e^{-1} \text{ s}. \tag{24.17}$$

In the interstellar medium, an HII region ionized and heated by radiation from massive stars has a temperature of about 10^4 K. As we noted in the last section, this is the typical energy of a photoelectron produced by ionizing starlight, so the photons cannot make the plasma much hotter, and at much lower matter temperatures the rate of loss of energy is considerably reduced. At $T \sim 10^4$ K the rate of loss of energy by collisional excitation and radiative decay of fine-structure states of ions of heavy elements is an order of magnitude larger than the free-free emission rate (Osterbrock 1989). At higher temperatures, or at higher redshifts where the heavy element abundance is lower, the free-free time t_{ff} gives a useful measure of the timescale for energy loss.

The free-free cooling time sets the character of the evolution of structure in some interesting situations. Suppose that in a system with mass $\mathcal{M}$ and radius r a fraction ϵ of the mass is in a plasma with about the same spatial distribution as the net mass, and with the rest of the mass in dark matter. Then the characteristic plasma density and temperature for pressure support are

$$n_e \sim \frac{\epsilon \mathcal{M}}{r^3 m_p}, \qquad T \sim \frac{G\mathcal{M}m_p}{kr}, \tag{24.18}$$

and the characteristic orbit period or collapse time is

$$t_c \sim \frac{r}{v_c}, \qquad v_c \sim \left(\frac{G\mathcal{M}}{r}\right)^{1/2}. \tag{24.19}$$

---

[15] In a self-gravitating system the net energy is $E = K + U \sim -K$, where K and U are the kinetic and gravitational energies, and the last expression follows from the virial theorem. When the system loses energy, the net energy E becomes more negative, so K increases and the material gets hotter. The reference to a cooling time thus can be somewhat misleading.

The ratio of the free-free time (24.17) to the orbit time, with the radius expressed in kiloparsecs, is

$$\frac{t_{ff}}{t_c} \sim \frac{0.01\, r_{kpc}}{\epsilon}, \tag{24.20}$$

This number is independent of the mass, in the limit where T is hot enough that the matter is ionized and free-free emission dominates. More detailed analyses of cooling times are given by Rees and Ostriker (1977) and Silk (1977).

Our galaxy has a high baryon fraction, $\epsilon \sim 1$, within our radius $r \sim 10$ kpc. It is unlikely that this matter ever was a diffuse gas cloud supported by pressure at about the present radius, because the gas would be collisionally ionized, and we see from equation (24.20) that the plasma would lose pressure support in a small fraction of an orbit time.

The plasma in a cluster of galaxies has core radius $r \sim 100h^{-1}$ kpc, and the plasma mass fraction is $\epsilon \sim 0.1$. This gives $t_{ff}/t_c \sim 10$, meaning the plasma can stay in pressure support. The cooling times in cores of clusters are comparable to the Hubble time, with the result that in some cases the intracluster plasma in the central parts of the cluster either must have an energy source to balance the radiation loss, or else the plasma must be contracting and converting to stars or some other dissipationless state (Fabian 1988; Fabian, Nulsen, and Canizares 1991).

Equation (24.20) defines the characteristic radius

$$r_{ff} = 100\epsilon \text{ kpc} \tag{24.21}$$

(Ostriker 1974). If protogalaxies originated as pressure-supported gas spheres, those with radii much smaller than r_{ff} would promptly lose pressure and collapse at nearly the free-fall rate. Those at larger initial radii would contract in a quasistatic way until at radius $r \sim r_{ff}$ the collapse became free and the non-axisymmetric substructure could start to grow, perhaps leading to the formation of the first generation of stars. It is an interesting coincidence that r_{ff} is comparable to the characteristic halo radius of L_* galaxies (eq. [20.31]), as would be expected if protogalaxies originated as pressure-supported gas spheres with initial radii greater than r_{ff}.

For the Lyman-α forest clouds, with the characteristic plasma density in equation (23.72), the cooling timescale (24.17) is

$$\frac{t_{ff}}{t(z)} \sim 0.7\,(1+z)^{3/2}T_4^{0.15}(\Omega l_{10}h/i_{21})^{1/2}, \tag{24.22}$$

where the expansion timescale t uses the Einstein-de Sitter limit (23.2). We see that at $2 \lesssim z \lesssim 4$, where the forest is prominent, energy loss by free-free radiation is not significant.

Consider next the integrated background of free-free emission from an intergalactic plasma. The surface brightness of the background produced by luminosity density $j(t,\nu)$ per steradian is

$$i(\nu) = \int c\,dt\, j(t, \nu a_o/a)(a/a_o)^3 . \qquad (24.23)$$

The expansion parameter is $a(t)$, and a_o is the present value, so the redshift is $z = a_o/a - 1$ (and this equation differs from the expression in eq. [5.159] because here j is the luminosity density per steradian). A model for the evolution of structure may predict that there is an epoch, z_e, at expansion time t_e, when the bulk of the background at a chosen frequency originates. That would allow us to write

$$i(\nu) \sim \frac{c t_e j(\nu(1 + z_e))}{(1 + z_e)^3}$$

$$\qquad (24.24)$$

$$= 4 \times 10^{-23} \frac{C}{T_4^{1/2}} \frac{h^3 \Omega_{\text{IGM}}^2}{\Omega^{1/2}} (1 + z_e)^{3/2} \, \text{erg cm}^{-2} \, \text{s}^{-1} \, \text{ster}^{-1} \, \text{Hz}^{-1} .$$

The second line uses equation (24.15) for the thermal bremsstrahlung luminosity density at plasma temperature T_4 (in units of 10^4 K), and it neglects the quantum cutoff at $h\nu \sim kT$. The density parameter in the intergalactic plasma is Ω_{IGM}, and the clumping factor for the plasma distribution is $C = \langle n_e^2 \rangle / \langle n_e \rangle^2$, as in equation (23.38).

The radiation background $i(\nu)$ at long wavelengths is minimum at about 30-cm wavelength, where there are about equal contributions from the thermal 3 K cosmic background and from synchrotron emission by cosmic ray electrons in our galaxy. The thermal bremsstrahlung background from the intergalactic medium could not exceed the surface brightness of these components,

$$i_\nu \lesssim 3 \times 10^{-19} \, \text{erg cm}^{-2} \, \text{s}^{-1} \, \text{ster}^{-1} \, \text{Hz}^{-1}, \quad \text{at } \lambda = 30 \, \text{cm} . \qquad (24.25)$$

We know that at redshift $z \lesssim 5$ there is optically thin plasma in the Lα forest clouds at temperature $T \sim 10^4$ K (eq. [23.61]). Let us check that the free-free background from this plasma is well below the measured bound. With the clumping factor C in equation (23.67) we have from equation (24.24)

$$i_\nu \sim 1 \times 10^{-20} \frac{h^3 \Omega_\alpha^2}{l_{10} \Omega} \, \text{erg cm}^{-2} \, \text{s}^{-1} \, \text{ster}^{-1} \, \text{Hz}^{-1}, \qquad (24.26)$$

at $z_e = 3$. If the density parameter for the baryons in the Lα forest is $\Omega_\alpha \lesssim 0.03$, as is indicated by the model for light element production (eq. [6.27]) and by the density of absorption lines in the forest (eq. [23.74]), this is well below the measured bound in equation (24.25). The possible contribution to the radio background from thermal bremsstrahlung at higher redshifts is discussed below (eq. [24.74]).

Let us consider finally the diffuse X-ray background pictured in figure 3.11. Its surface brightness is

$$i_\nu = 3 \times 10^{-26} \, \text{erg cm}^{-2} \, \text{s}^{-1} \, \text{ster}^{-1} \, \text{Hz}^{-1} \tag{24.27}$$

(Boldt 1987), at energy

$$\epsilon = 3 \, \text{keV} . \tag{24.28}$$

To see the orders of magnitude, suppose a fraction ι comes from the intergalactic plasma,

$$i_{\text{IGM}} = \iota i_\nu , \tag{24.29}$$

and write the plasma temperature at redshift z as

$$T = 10^8 T_8 (1 + z) \, \text{K} . \tag{24.30}$$

Then the quantum cutoff for the observed thermal bremsstrahlung spectrum is $\epsilon \sim kT \sim 10 T_8$ keV, and equations (24.24) and (24.29) say the density parameter in the plasma that would produce the background is

$$\Omega_{\text{IGM}} = \frac{0.3\iota^{1/2}}{(1+z_e)^{1/2}} \frac{T_8^{1/4} \Omega^{1/4}}{C^{1/2} h^{3/2}} . \tag{24.31}$$

We see that a sensible value for the density parameter Ω_{IGM} in a reasonably smooth plasma at a not unreasonable temperature produces an X-ray background comparable to what is observed. The picture fails, however, because the pressure in this plasma would be much greater than the pressure within the clouds in the Lα forest. The ratio of equation (23.78) for the pressure in the Lα forest to the pressure given by equations (24.30) and (24.31) is

$$\frac{(nkT)_{\text{L}\alpha}}{(nkT)_{\text{IGM}}} \sim \frac{0.001}{\iota^{1/2}} \frac{i_{21}^{1/2} C^{1/2}}{T_8^{5/4} \Omega^{1/4} l_{10}^{1/2}} , \tag{24.32}$$

at $z \sim 3$. This says that a smoothly distributed plasma that produces an appreciable part of the X-ray background, so $\iota \sim 1$, would have unacceptably high pressure if it were to intermingle with the Lα forest clouds. If the X-ray emitting plasma were confined to clouds, the two components could be separated; but since the Lα clouds fill most of space at $z \sim 3$ the X-ray sources would have to have to be in compact clumps, which would violate the bound on angular fluctuations in the background unless the X-ray emitting clouds were numerous, in which case they would tend to collapse to systems such as protogalaxies. Balancing these conditions with the known contributions to the X-ray background from quasars and other active galaxies is a subtle art that is not yet in a presentable state. Recent examples are in Barcons, Fabian, and Rees (1991), Daly (1991a), Loeb (1991), and Rogers and Field (1991).

Compton-Thomson Scattering

Scattering of the 3 K cosmic background radiation by free electrons,

$$e + \gamma \leftrightarrow e' + \gamma', \tag{24.33}$$

can have important observable effects. If the electron velocities are negligibly small, the scattering smooths the CBR without affecting the mean spectrum. The streaming motion of a plasma produces a Doppler shift in the scattered radiation, tending to make the CBR anisotropic. The thermal velocities of the electrons in a plasma with no streaming motion cause a random walk in the photon frequencies, again perturbing the CBR angular distribution and spectrum. The use of this last effect as a constraint on the history of the intergalactic medium was explored by Weymann (1966), Zel'dovich and Sunyaev (1969), and Chan and Jones (1975). As will be disucussed in the following subsections, important new developments are the precision measurements of the CBR spectrum, which strongly constrain scenarios for the temperature history of the intergalactic medium, and the detection of the random walk effect in the cosmic background radiation that passes through the plasma in a rich cluster of galaxies. We consider here some elements of the physics of electron scattering, and in the next subsections we obtain some order-of-magnitude constraints on the history of the diffuse matter. The equation describing the effect on the CBR of scattering by isotropic nonrelativistic electron motions was derived by Kompaneets (1957). Its application in cosmology is called the Sunyaev-Zel'dovich effect, after their pioneering analyses.

Before dealing with the Kompaneets equation, let us note that if the electron velocities and photon energies are so small that the scattering produces negligible changes in the photon frequencies, then the correction to the scattering rate for stimulated emission disappears. To see this, write the net rate for the reaction (24.33) from left to right as

$$\frac{\partial \mathcal{N}'}{\partial t} = A \left[\mathcal{N}(1 + \mathcal{N}') - \mathcal{N}'(1 + \mathcal{N}) \right]$$
$$= A \left[\mathcal{N} - \mathcal{N}' \right] . \tag{24.34}$$

The first term on the right-hand side of the first line is proportional to the number $\mathcal{N}$ of photons available for scattering and to the sum of the rates for spontaneous plus stimulated emission of the scattered photon, which is proportional to $(1 + \mathcal{N})$, as in equation (23.7). The second term is the rate going the other way. If the photon energy is much less than the electron mass we can ignore the electron recoil, so the same population of electron states appears in both terms. (That is, unlike eq. [24.2] for thermal bremsstrahlung, there is no Boltzmann factor for the electrons.) Also, we are assuming we can ignore the motion of the electron, so the transition from photon momentum $\mathbf{k}$ to $\mathbf{k'}$ is just the time reversal of $\mathbf{k'} \to \mathbf{k}$, with the same rate coefficient, A. As indicated in the second line, in this limit the stimulated emission terms just cancel.

Next, let us suppose the electrons have no streaming motion and are moving with an isotropic distribution of velocities that are nonrelativistic but large enough to cause the CBR spectrum to evolve by the random walk of photons scattered and Doppler-shifted by the electrons. This is often called Compton-Thomson scattering, though the name is somewhat misleading because in cases of interest, where the kinetic temperature of the electrons is large compared to the radiation temperature, the frequency shifts of the photons are caused by the motions of the electrons, not by electron recoil. (That is, for given electron velocity dispersion, $\langle v^2 \rangle$, the walk of photon energies, $\delta\nu/\nu$, is independent of the photon energy.) The derivation of the equation describing the evolution of the spectrum is a strengthening but lengthy exercise in Lorentz transformations, so it is placed in the end of the section. The result is

$$\frac{1}{\sigma_t n_e c} \frac{\partial \mathcal{N}}{\partial t} = \frac{\langle v^2 \rangle}{3c^2} \left[\nu^2 \frac{\partial^2 \mathcal{N}}{\partial \nu^2} + 4\nu \frac{\partial \mathcal{N}}{\partial \nu} \right]$$

$$+ \frac{h\nu}{m_e c^2} \left[4\mathcal{N}(1+\mathcal{N}) + (1+2\mathcal{N})\nu \frac{\partial \mathcal{N}}{\partial \nu} \right]. \tag{24.35}$$

This assumes that the radiation, with photon occupation number $\mathcal{N}(\nu, t)$ that is a function of time and frequency ν, is scattered at cross section σ_t, conserving photons, by a homogeneous isotropic gas of nonrelativistic electrons with number density n_e and mean square velocity $\langle v^2 \rangle$. If the electrons have a Maxwell-Boltzmann energy distribution at temperature T_e, the dispersion is

$$\langle v^2 \rangle = 3kT_e/m_e. \tag{24.36}$$

The Thomson cross section is σ_t (eq. [6.120]).

The Kompaneets equation is made more compact by following Zel'dovich and Sunyaev (1969) in using the scaled variables

$$x = \frac{h\nu}{kT_e}, \qquad dy = \frac{kT_e}{m_e c^2} \sigma_t n_e c \, dt = \frac{\langle v^2 \rangle}{3c^2} \sigma_t n_e c \, dt. \tag{24.37}$$

This brings equation (24.35) to

$$\frac{\partial \mathcal{N}}{\partial y} = x^2 \frac{\partial^2 \mathcal{N}}{\partial x^2} + \left(x^2 + 4x + 2x^2 \mathcal{N} \right) \frac{\partial \mathcal{N}}{\partial x} + 4x \mathcal{N} (1 + \mathcal{N})$$

$$= \frac{1}{x^2} \frac{\partial}{\partial x} \left[x^4 \left(\frac{\partial \mathcal{N}}{\partial x} + \mathcal{N} + \mathcal{N}^2 \right) \right] . \tag{24.38}$$

The parameter y can be interpreted in several ways. It is proportional to the pressure of the electron gas, $n_e k T_e$. It is the product of the scattering optical depth and a weighted average of the electron temperature in units of the electron mass:

$$y = \int^t \frac{\langle v^2 \rangle}{3c^2} \sigma_t n_e \, c \, dt = \frac{k \langle T_e \rangle}{m_e c^2} \int^t \sigma_t n_e c \, dt . \tag{24.39}$$

The Doppler shifts caused by scattering by electrons with an isotropic velocity distribution add incoherently, so y is the integrated mean square frequency shift, $y = (\delta v / v)^2$. We discuss below the effect of scattering by plasma with a mean streaming velocity v, where $\delta v / v$ is the product of the optical depth and the first power of v (eq. [24.59]).

Let us consider some properties of the Kompaneets equation. The number of photons per unit volume, n_γ, is the integral of the occupation number $\mathcal{N}$ over phase space (eq. [6.8]),

$$n_\gamma \propto \int_0^\infty \mathcal{N} v^2 dv . \tag{24.40}$$

On multiplying the second line of equation (24.38) by x^2 and integrating over x, one sees that n_γ is independent of time. That checks, because the equation describes scattering that conserves photons.

The energy density in the radiation is

$$u_\gamma \propto \int \mathcal{N} v^3 dv . \tag{24.41}$$

The result of multiplying equation (24.35) by v^3, integrating over frequencies, and eliminating the derivatives of $\mathcal{N}$ by integrating by parts, is the energy equation

$$\frac{d}{dt} \int \mathcal{N} v^3 dv = \sigma_t n_e c \left[\frac{4}{3} \frac{\langle v^2 \rangle}{c^2} \int \mathcal{N} v^3 dv - \frac{h}{m_e c^2} \int v^4 dv \mathcal{N} (1 + \mathcal{N}) \right] . \tag{24.42}$$

If the radiation spectrum is close to blackbody at temperature T, the occupation number is

$$\mathcal{N} = \frac{1}{e^{hv/kT} - 1} , \tag{24.43}$$

and the last factor in the last integral in equation (24.42) is

$$\mathcal{N}(1+\mathcal{N}) = \frac{e^{h\nu/kT}}{\left(e^{h\nu/kT} - 1\right)^2} = -\frac{kT}{h} \frac{\partial \mathcal{N}}{\partial \nu} . \tag{24.44}$$

This expression in equation (24.42) brings the rate of change of the radiation energy density to

$$\frac{1}{u_\gamma} \frac{du_\gamma}{dt} = \frac{4}{3} \frac{\sigma_t n_e}{c} \left(\langle v^2 \rangle - \frac{3kT}{m_e} \right)$$

$$= \frac{4\sigma_t n_e k}{m_e c} (T_e - T) . \tag{24.45}$$

The first term was derived in equation (6.133), from the radiation drag force on a moving electron. The second term from the fluctuating radiation pressure gives the correct limit at thermal equilibrium.

A static solution to equation (24.38) is

$$\mathcal{N}(x) = \frac{1}{e^{x+x_o} - 1} , \tag{24.46}$$

where x_o is a constant. Again, we already knew this, for when the photon number is conserved the spectrum relaxes to a Bose-Einstein distribution at chemical potential proportional to $-x_o$.

As an example of relaxation to a Bose-Einstein distribution, imagine an idealized box with perfectly reflecting walls in which we place some electrons (and protons to keep the gas neutral), a definite amount of energy, and a definite conserved number of photons. Then the photon energy distribution relaxes to the equilibrium Bose-Einstein form in equation (24.46). If there are too few photons for the given energy to make a Planck blackbody spectrum, for which $x_o = 0$, the equilibrium distribution is equation (24.46) with the parameter $x_o > 0$ fixed by the given number of photons (relative to the number required to make a Planck distribution at the given energy). If there are too many photons for the given energy for a blackbody spectrum, the distribution relaxes to a Bose-Einstein condensation, where the chemical potential is less than but very close to the lowest mode energy in the box, so as to put the excess photons into this lowest mode.

In the real world, long wavelength modes are readily damped by electron collisions, so one would not expect to observe Bose-Einstein condensation. One could imagine a situation at high redshift in which the CBR received energy, perhaps from annihilation of pockets of antimatter, and then relaxed to statistical equilibrium without a significant increase in the photon number. The result would be the "bosonized" spectrum in equation (24.46), with negative chemical potential (positive x_o). This is a special case, however, because the condition for bosonization,

$y \sim 1$, is up to a factor of order unity the same as the condition under which the soft thermal bremsstrahlung photons are promoted in energy to produce a black-body spectrum. To see this, suppose there is an epoch at which the CBR has a bosonized spectrum over some range of wavelengths. Since the free-free opacity of the plasma varies as λ^2 at long wavelength, where $h\nu \ll kT$ (eq. [24.15]), we know that at long wavelengths the radiation is thermally coupled to the plasma and the spectrum has to be blackbody. That leaves an intermediate range of wavelengths where the spectrum dips below blackbody. We see from the derivative terms in the Kompaneets equation (24.35) that if the spectrum drops below an equilibrium form in a dip that spreads from frequency x_1 to $2x_1$, the timescale for the dip to diffuse away is $y \sim 1$, independent of the value of x_1. Therefore, the timescale for removal of a dip in a spectrum that is close to blackbody at long and short wavelengths is $y \sim 1$, up to a logarithmic term, the same as the timescale to produce the bosonized spectrum in the first place. This means that only a relatively narrow range of the parameter y is large enough to produce a bosonized CBR spectrum but not large enough to erase it. Danese et al. (1990) and Burigana, De Zotti, and Danese (1991) show useful numerical examples of the relaxation toward a blackbody spectrum.

The dissipation of pressure waves by diffusion of the CBR has an effect on the radiation spectrum that is approximately equivalent to that of hot electrons. If the energy density ϵ is dissipated in the CBR with energy density u_γ, the equivalent y parameter is $y = \epsilon/(4u_\gamma)$ (eq. [24.45]). The effect is discussed by Daly (1991b).

Sunyaev-Zel'dovich Effect in Clusters

The plasma in a rich cluster of galaxies is much hotter than the CBR. In this limit, equation (24.35) is

$$\frac{\partial \mathcal{N}}{\partial y} = \nu^2 \frac{\partial^2 \mathcal{N}}{\partial \nu^2} + 4\nu \frac{\partial \mathcal{N}}{\partial \nu} . \tag{24.47}$$

Since the measured CBR spectrum is very close to blackbody, it is an excellent approximation to use the thermal form (24.43) in the right-hand side of this equation, to get

$$\frac{\delta \mathcal{N}}{\mathcal{N}} = y \left[\frac{x^2 e^x (e^x + 1)}{(e^x - 1)^2} - \frac{4xe^x}{e^x - 1} \right]$$

$$\rightarrow \begin{cases} -2y & \text{at } x \ll 1, \\ x^2 y & \text{at } x \gg 1. \end{cases} \tag{24.48}$$

A more complete solution is given by Bernstein and Dodelson (1990). The conclusion is that the perturbed spectrum at long wavelength, $x \ll 1$, where $\mathcal{N} \propto T$, has the thermal Rayleigh-Jeans form with effective temperature lowered by the fractional amount

$$\frac{\delta T}{T} = -2y. \tag{24.49}$$

The hot electrons tend to upscatter the photons, increasing the surface brightness at short wavelengths, and lowering the effective temperature at the long wavelength end of the spectrum of the radiation that passes through a cluster.

Sunyaev and Zel'dovich (1972) pointed out that the plasma in a rich cluster of galaxies is hot enough and at a high enough surface density to cause a significant perturbation to the CBR surface brightness across the face of the cluster. The effect has been observed (Birkinshaw 1990; Uson and Wilkinson 1988) and provides an important measure of the structures of clusters, adding to those discussed in section 20. Rephaeli (1981) noted that if clusters and intracluster plasma are present back to redshift $z \sim 1$, the integrated effect of the electron scattering makes an important contribution to the CBR anisotropy on the scale of one arc minute.

We can understand the orders of magnitude for these effects by considering a simple model for the intracluster plasma, with a uniform temperature, T_x, and electron number density as a function of radius

$$n(r) = \frac{n_c}{1 + r^2/r_c^2}. \tag{24.50}$$

In this model the optical depth for scattering at projected distance x from the cluster center is

$$\tau = \sigma_t \int n(r)\, dl = \frac{\tau_o}{(1 + x^2/r_c^2)^{1/2}}, \tag{24.51}$$

$$\tau_o = 0.0064\, n_c(\text{cm}^{-3})\, r_c(\text{kpc}),$$

where the central electron number density n_c is measured in units of cm^{-3} and the core radius r_c in kiloparsecs. The thermal bremsstrahlung luminosity of the cluster, in the approximation of equation (24.16), is

$$L_x = \int J\, d^3 r \tag{24.52}$$

$$= 1.4 \times 10^{42} n_c(\text{cm}^{-3})^2\, r_c(\text{kpc})^3\, T_x(\text{keV})^{1/2}\ \text{erg s}^{-1}.$$

A typical value for the cluster X-ray luminosity,

$$L_x \sim 1 \times 10^{44} h^{-2}\ \text{erg s}^{-1}, \tag{24.53}$$

and for the core radius and temperature of the plasma,

$$r_c \sim 200 h^{-1}\ \text{kpc}, \qquad T_x \sim 4\ \text{keV}, \tag{24.54}$$

in equation (24.52) give the central electron number density,

$$n_c \sim 0.003 h^{1/2} \text{ electrons cm}^{-3},$$ (24.55)

and the scattering optical depth through the cluster center,

$$\tau_o \sim 0.003 h^{-1/2}.$$ (24.56)

At the long wavelength microwave side of the CBR spectrum, the plasma lowers the effective temperature seen through the cluster center by the fractional amount (eqs. [24.39] and [24.48])

$$\frac{\delta T}{T} = -2\tau_o \frac{kT_x}{m_e c^2} \sim -5 \times 10^{-5} h^{-1/2}.$$ (24.57)

This temperature decrement is roughly constant over the core radius, which subtends angular diameter

$$\theta \sim \frac{2H_o r_c}{cz} \sim \frac{0.5}{z} \text{ arc min}$$ (24.58)

in a cluster at redshift z. Detecting this signal through the noise from radio sources in and behind the cluster is a severe challenge that has been met (Birkinshaw 1990; Uson and Wilkinson 1988). As discussed in section 20, this adds to the evidence that we do have a believable picture for the mass concentrations in the great clusters of galaxies. At wavelengths shorter than the peak of the CBR spectrum, the scattering is predicted to increase the CBR surface brightness. Detecting this aspect of the Sunyaev-Zel'dovich effect in clusters would be of considerable interest.

Since the measurement of the plasma temperature T_x from the cluster X-ray spectrum is independent of the cluster distance, the CBR temperature perturbation $\delta T/T$ (or the y parameter) in principle measures the cluster distance (through the Hubble parameter h in eq. [24.57]; Silk and White 1978; Cavaliere, Danese, and De Zotti 1979), though a precision measurement requires a heroic mapping of the plasma density and temperature through the cluster.

If the cluster is moving toward us at peculiar velocity v relative to the preferred rest frame in which the CBR is isotropic, the Doppler shift of the scattered radiation perturbs the CBR temperature across the face of the cluster by the amount

$$\frac{\delta T}{T} = 2\tau \frac{v}{c},$$ (24.59)

independent of frequency (Sunyaey and Zel'dovich 1980). A cool plasma with streaming velocity $v = 1000$ km s^{-1} produces the same perturbation to the CBR temperature as the perturbation at long wavelengths by a stationary plasma at temperature $T = 1$ keV.

Let us consider finally the integrated effect of the perturbations by intracluster plasma on the angular distribution of the CBR. With the cluster number density n_{cl} in equation (20.33), the probability a line of sight to redshift $z \sim 1$ intercepts the core of a cluster at the radius in equation (24.54) is

$$P \sim \pi r_c^2 n_{cl} c / H_o \sim 0.004 \,. \tag{24.60}$$

With equation (24.57), the rms perturbation to the CBR at long wavelengths is

$$\frac{\delta T}{T} = 2y P^{1/2} \sim 3 \times 10^{-6} h^{-1/2} \,. \tag{24.61}$$

Clusters are observed at redshift $z \sim 1$. If the intracluster plasma at $z \sim 1$ is comparable to what is present in clusters at lower redshifts, equation (24.61) is a reasonably secure lower bound on the CBR rms anisotropy on the angular scale ~ 1 arc min subtended by the clusters (Rephaeli 1981).

Constraints on the Intergalactic Medium

Scenarios for the history of diffuse matter in the intergalactic medium and in optically thin clouds are constrained by the effects on the radiation backgrounds. Some consequences of thermal bremsstrahlung emission have already been considered. Electron scattering in a clumpy medium perturbs the angular distribution of the CBR, as we have discussed for the intracluster plasma. Here we consider the effect of electron scattering on the mean CBR spectrum.

In interesting situations stars or quasars or gravitational collapse or some other energy source at redshifts $z \lesssim z_{dec} = 1400$ have ionized diffuse matter, making it significantly hotter than the CBR. The resulting perturbation to the CBR spectrum by the thermal motions of the electrons is given by equation (24.48). Recent spectrum measurements by the COBE satellite indicate a reasonable bound on the parameter y is

$$y \lesssim 1 \times 10^{-4} \,. \tag{24.62}$$

This is an integral over the mean electron pressure through the epochs when the plasma was appreciably hotter than the radiation.

Let us consider first the energy needed to produce this value of y. We are assuming the plasma is hotter than the CBR, so the energy equation (24.45), with equation (24.37) for y, is

$$\dot{u}_\gamma = 4 u_\gamma \dot{y} = \rho_m c^2 \dot{\epsilon} \,. \tag{24.63}$$

In the last expression, $\dot{\epsilon}\delta t$ is the fraction of the mass density ρ_m of the matter that is being annihilated to increase the value of y by the amount $\dot{y}\delta t$. This equation is

$$\frac{d\epsilon}{dy} = 4\frac{\rho_\gamma}{\rho_m} = \frac{1 \times 10^{-4}(1+z)}{\Omega_m h^2} , \qquad (24.64)$$

so the bound on the mass fraction that has been transferred to the CBR is

$$\epsilon = 4y\frac{\rho_\gamma}{\rho_m} \lesssim \frac{1 \times 10^{-8}(1+z)}{\Omega_m h^2} . \qquad (24.65)$$

The gravitational binding energy per unit mass in structures with velocity dispersion v is $\epsilon \sim (v/c)^2$, so this translates to

$$v \lesssim 30(1+z)^{1/2}\Omega_m^{-1/2}h^{-1} \text{ km s}^{-1} . \qquad (24.66)$$

At the baryon density $\Omega_m h^2 \sim 0.01$ indicated by light element production, and for structure formation at $z \sim 10$, this is $v \sim 1000$ km s^{-1}, in line with observed peculiar velocities in the large-scale structure.

If the energy came from nuclear burning that produced heavy element abundance Z_m, the energy release would be $\epsilon \sim 0.007\, Z_m$, giving

$$Z_m \lesssim 10^{-6}(\Omega_m h^2)^{-1}(1+z). \qquad (24.67)$$

With $\Omega_m h^2 \sim 0.01$, nuclear burning at this limit at redshift $z \sim 10$ would make a heavy element mass fraction $Z_m \sim 0.001$, about at the allowed bound for early element production. The conclusion is that it is easy to see how to find the energy to perturb the CBR spectrum at the bound in equation (24.62). Let us consider now some details of how electron scattering might transfer the energy.

The contribution to y by plasma with mean pressure $n_e k T_e$ at epoch z, where the expansion timescale is t, is

$$\begin{aligned} y &\sim \sigma_t n_e ct \frac{kT_e}{m_e c^2} \\ &\sim \frac{7 \times 10^{-7}}{(1+z)^{3/2}} \frac{n_e T_e}{h\Omega^{1/2}} , \end{aligned} \qquad (24.68)$$

with $n_e T_e$ expressed in units of cm^{-3} K. The evidence discussed in the last section is that at $z = 3$ the Lα forest clouds contain plasma at the pressure in equation (23.78). At this pressure, the contribution to the y parameter is

$$y \sim 10^{-7} T_4^{1.35} \left(\frac{i_{21}}{hl_{10}\Omega}\right)^{1/2} \qquad (24.69)$$

at $z = 3$. That is, an intracloud plasma at the pressure in the Lα forest clouds could be present back to redshifts considerably in excess of what is now observed without having an appreciable effect on the CBR spectrum.

In some of the scenarios discussed in the next section, galaxies form at high redshifts, and the young galaxies contribute to y through the peculiar motions of the plasma within the galaxies. If the rms line-of-sight peculiar velocity dispersion of the plasma in the young galaxies is σ at redshift z, the contribution to y is

$$y = \sigma_t n_e ct (\sigma/c)^2/3$$
$$\sim 10^{-8} \sigma_{100}{}^2 h \Omega_p \Omega^{-1/2} (1+z)^{3/2}.$$
(24.70)

Here σ_{100} is the velocity dispersion in units of 100 km s^{-1}, and Ω_p is the density parameter belonging to the plasma in young galaxies. For internal motions, as in young protodisks, a reasonable number is $\sigma_{100} = 1.5$. The present characteristic galaxy peculiar velocity (in one dimension) is $\sigma_{100} \sim 3$, but if the motions were produced by gravity they would have been smaller in the past. Thus a reasonable number is $\sigma_{100} = 1.5$, and with $\Omega_p h^2 \sim \Omega h^2 \sim 0.1$ and a galaxy formation redshift $z \sim 30$ we get $y \sim 10^{-6}$, well below feasible spectrum measurements. The plasma in rich clusters of galaxies has line-of sight velocity dispersion $\sigma_{100} \sim 10$, but the density parameter in plasma in clusters is $\Omega_p \sim 0.003$ (30% of the entry in table 20.1), so even at $z = 10$, the earliest one might imagine clusters forming, the contribution to y is small.

An intergalactic plasma hot enough to make a substantial contribution to the X-ray background would make an appreciable contribution to y. Following equation (24.30), suppose intergalactic plasma at redshift z_e and temperature $T = 10^8 T_8 (1 + z)$ K contributes a fraction ι of the 3 keV X-ray background. Then the density parameter Ω_{IGM} in the plasma is given by equation (24.31), and equation (24.68) amounts to

$$y \sim 2 \times 10^{-4} (1+z_e)^2 \frac{\iota^{1/2} T_8{}^{5/4}}{C^{1/2} h^{1/2} \Omega^{1/4}}.$$
(24.71)

Unless the clumping factor C is large and the redshift z_e is not much greater than unity, a plasma dense and hot enough to produce an appreciable part of the X-ray background violates the bound on y. This constraint from the effect on the CBR spectrum is less restrictive but more direct than the condition that the X-ray emitting plasma not crush the clouds in the Lα forest (eq. [24.32]).

At redshifts beyond the most distant observed quasars there are no direct observations of the intergalactic medium, but the bound on y from the CBR spectrum constrains scenarios for how processes such as gravitational collapse and the ionizing radiation and blast waves from young galaxies might have affected the thermal history of the diffuse matter. Suppose that for an expansion time at redshift z_p, with $5 \lesssim z_p \lesssim 1000$, matter with density parameter Ω_{IGM} was ionized at temperature τ times the background CBR value. Then equation (24.68) for the contribution to y is

$$y \sim 2 \times 10^{-11}(\tau - 1)(1 + z_p)^{5/2} h \Omega_{\text{IGM}} \Omega^{-1/2}. \tag{24.72}$$

This replaces the plasma temperature $T_e = \tau T$ with $T_e = (\tau - 1)T$, to take account of the condition that there is no effect on the spectrum when T_e is equal to the CBR temperature. At $z_p = 1000$ the bound in equation (24.62) requires

$$\Omega_{\text{IGM}}(\tau - 1) \lesssim 0.1 h^{-1} \Omega^{1/2}. \tag{24.73}$$

When the plasma is significantly hotter than the CBR the ratio of the contribution to the sky brightness at $h\nu \lesssim kT/(1 + z)$ by thermal bremsstrahlung (eq. [24.24]) to the bound i_o in equation (24.25) is

$$\frac{i_{\text{ff}}}{i_o} \sim 0.01(1 + z_p)\frac{(\tau - 1)}{\tau^{1/2}}\frac{Ch^3\Omega_{\text{IGM}}^2}{\Omega^{1/2}}. \tag{24.74}$$

For a specific example, if $\Omega h^2 \sim \Omega_{\text{IGM}} h^2 \sim 0.1$ the bounds at $z \sim 1000$ are $\tau \lesssim 1.3$ and $i_{\text{ff}}/i_o \sim 0.1C\tau^{-1/2}$. Increasing Ω and decreasing Ω_{IGM} relaxes both bounds. The conclusion is that if the clumping factor C is not large and the plasma is not considerably warmer than the CBR, then sources of ionization could have kept the bulk of the baryons ionized and in optically thin clouds at $z \lesssim 1000$ without violating the constraints from the CBR spectrum and the radio background (Bartlett and Stebbins 1991).

Finally, let us find the characteristic redshift at which a perturbed CBR spectrum can relax back to the observed very close approximation to a thermal Planck function. The dominant relaxation is through photon production and annihilation at long wavelengths by free-free transitions, followed by the redistribution of the photon energies by electron scattering. We have seen that the former process is rapid, and the characteristic number for the latter is $y \sim 1$. With standard estimates of the cosmological parameters, y reaches unity in the radiation-dominated epoch, where the age of the universe is

$$t = \left(\frac{3c^2}{32\pi GaT^4}\right)^{1/2} = \frac{3.1 \times 10^{19}}{(1 + z)^2} \text{ s}. \tag{24.75}$$

Then equation (24.68) with T_e equal to the CBR temperature says that the redshift at a given value of the parameter y is

$$z \sim 2 \times 10^4 (\Omega_B h^2)^{-1/2} y^{1/2}. \tag{24.76}$$

With the density parameter in baryons from the standard model for the production of the light elements, $\Omega_B h^2 \sim 0.015$ (eq. [6.27]), this is

$$z \sim 1 \times 10^5 y^{1/2}. \tag{24.77}$$

At $y \sim 10$ relaxation to a blackbody spectrum is well advanced, the redshift in equation (24.77) is $z \sim 5 \times 10^5$, and the CBR temperature at this redshift is $kT \sim 100$ eV. At this epoch and earlier, the plasma can heal mild disruptions from a thermal CBR spectrum, as might be caused by annihilation of pockets of anti-matter, or by flux reconnection in small-scale primeval magnetic fields, or by the dissipation of primeval pressure waves in the matter-radiation fluid (Daly 1991b).

Scattering the Cosmic Background Radiation at Redshifts $z \lesssim 100$

We know the universe at the present epoch is transparent at CBR wave-lengths, because radio galaxies are visible at redshift $z \sim 3$ with no pronounced suppression of the radio flux density. In the standard cosmological model, we can be sure that at $z > z_{\text{dec}} \sim 1400$ the mean free path for scattering by free electrons is much less than the Hubble length. Between these epochs, it may be that the CBR has been perturbed only by the gravitational field of the growing irregularities in the mass distribution, or it may be that the CBR has been scattered or absorbed and reemitted by material spread along the line of sight. In the latter case irregulari-ties in the space distribution of the CBR that might have been produced in earlier phases in the evolution of structure would have been erased, and the scattering or absorption and reemission would insert new perturbations. This subsection deals with some elements of the possible effects of free electrons and dust.

Suppose at the epoch with redshift z early generations of galaxies had filled space with diffuse plasma with density parameter Ω_{IGM}, so the mean number density of free electrons is

$$\langle n_e \rangle = 1.1 \times 10^{-5} \Omega_{\text{IGM}} h^2 \, \text{cm}^{-3}. \tag{24.78}$$

(This excludes opaque things such as stars that may contain an appreciable part of the matter but intercept a negligible fraction of the line of sight back to de-coupling.) In the Einstein-de Sitter limit in equation (23.2), the optical depth for scattering by the free electrons is

$$\tau_s = \int \sigma_t n_e c \, dt$$
$$= 0.046 \, h \Omega_{\text{IGM}} \Omega^{-1/2} (1+z)^{3/2}, \tag{24.79}$$

and the optical depth is unity at redshift

$$1 + z_s = 7.8 \, \Omega^{1/3} (h \Omega_{\text{IGM}})^{-2/3}. \tag{24.80}$$

Ignoring the motions of the electrons, which will be discussed further below, scattering makes the observed CBR surface brightness across the sky a convolution of the initial space distribution of the CBR through a window roughly as wide as the Hubble length at the time of last scattering. That is, in the Rayleigh-Jeans limit, where the surface brightness is proportional to the temperature, an initial distribution of the CBR,

$$\tau(\mathbf{x}) = \frac{\delta T}{T}, \tag{24.81}$$

is observed as a brightness fluctuation

$$\tau_0 = \int W(\mathbf{x})\tau(\mathbf{x})\, d^3x, \tag{24.82}$$

where $W(\mathbf{x})\, d^3x$ is the probability that a photon received along the line of sight traces back through the random walk of scatterings to initial position $\mathbf{x}$ in the range d^3x. Figure 24.1 shows an example of the shape of this window function (Peebles 1987a). This assumes baryons dominate the mass density. The detected photons are moving up in the figure (toward decreasing x) from the direction of the small circle. The coordinates x and y parallel and perpendicular to the line of sight are converted to physical coordinates, expanded to the present epoch, by multiplying by

$$\Delta = 2cH_o^{-1}\Omega^{-1/2}. \tag{24.83}$$

The contours of constant W are normalized to unity at the peak, at the small circle. The contours spread more broadly at the near side, for if the last scattering occurred at unusually low redshift, when the optical depth was small, there likely is an unusually large coordinate displacement between the last scattering and the next to the last one.

The characteristic width of this smoothing window is the mean free path at the mean epoch z_s of last scattering, which is comparable to the Hubble length ct_s at z_s. Thus, the observed surface brightness of the CBR along a given line of sight is an average over the initial values within a region of present size $L_s \sim ct_s(1+z_s)$. With equation (24.80) for z_s, this is

$$hL_s \sim 700 \left(\frac{h\Omega_{IGM}}{\Omega^2}\right)^{1/3} \text{Mpc}. \tag{24.84}$$

The angle θ_s subtended by this smoothing length at the present epoch depends on the angular size distance, as plotted in figure 13.3. To get some numbers, let us assume the cosmological constant is negligible. Then the angular size distance

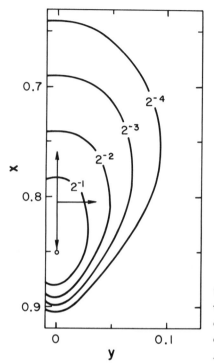

Figure 24.1. Distribution of initial positions of the CBR photons last scattered at $y = 0$ into the direction of decreasing x. The coordinates are converted to proper lengths by multiplying by Δ in equation (24.83).

at high redshift is $r_\infty = 2c/(H_o a_o \Omega)$ (eq. [13.39]), and the angular smoothing scale is

$$\theta_s \sim 0.1(\Omega_{\text{IGM}}\Omega h)^{1/3} \text{ radians}. \tag{24.85}$$

For example, the parameters

$$h = 1, \qquad \Omega_{\text{IGM}} = \Omega = 0.1, \tag{24.86}$$

give

$$z_s \sim 20, \qquad L_s \sim 1500 \text{ Mpc}, \qquad \theta_s \sim 2° \text{ to } 5°, \tag{24.87}$$

with the angular size distances from figure 13.3. In this picture, the smoothing length L_s is much larger than any known structure in the galaxy distribution. This means that whatever disturbances to the space distribution of the CBR may have accompanied the formation of the known objects would have been erased and replaced by whatever was happening at the epoch of last scattering, such as Doppler shifts from motions in the plasma and emission from the young galaxies.

Before considering the anisotropy inserted by scattering by moving plasma, let us estimate the redshift at which the universe might be optically thick to the absorption and scattering of the CBR by dust. We will consider two models for the possible properties of dust at high redshifts: the conducting whisker model in section 7, and an extrapolation from the known opacity of the dust in the interstellar medium in our galaxy.

In the whisker model, the material in the grains is assigned conductivity σ. If the aspect ratio of a typical grain satisfies equation (7.19), with ω the radiation frequency at redshift z, we can ignore the electrical polarization of the grains and write the current density as $j = \sigma E$, where E is the component of the electric field along the grain. Then the rate of dissipation of energy per unit volume in the material is σE^2, and the opacity is (eq. [7.21])

$$\kappa_g = \frac{4\pi}{3} \frac{\sigma}{c} v_g n_g . \tag{24.88}$$

Here v_g is the volume per grain and n_g is the mean number of grains per unit volume. If the fraction of the baryonic mass density ρ_b in these grains at redshift z is Z_g, and the mean mass density of the grain material is ρ_g, the fraction of space occupied by the grains is

$$v_g n_g = Z_g \rho_b / \rho_g . \tag{24.89}$$

Following Wright (1982), let us suppose the grains are made of graphite, for carbon has a high present cosmic abundance and it readily condenses to needle-like grains with the wanted large aspect ratio, and with a useful electrical conductivity. The density of graphite is $\rho_g = 2$ g cm^{-3}, a reasonable value for the conductivity $\sigma = 10^{14}$ s^{-1}, and the above equations for the optical depth for absorption at redshift z are

$$\kappa_g c t \sim 1000 \, h\Omega_B \Omega^{-1/2} Z_g (1+z)^{3/2} . \tag{24.90}$$

This is unity at redshift

$$1 + z_{\text{abs}} \sim \frac{0.01 \, \Omega^{1/3}}{(h\Omega_B Z_g)^{2/3}} . \tag{24.91}$$

The present cosmic abundance of carbon is $Z_g \sim 0.003$ by mass. Let us suppose one percent of that is present in graphite grains at high redshift, so the mass fraction when galaxies were young is $Z_g = 3 \times 10^{-5}$. Then with the choice of parameters in equation (24.86) the optical depth for absorption is unity at redshift

$$z_{\text{abs}} \sim 20 . \tag{24.92}$$

For every photon absorbed from the CBR another is scattered (by diffraction in the shadow). The absorbed photon has to be reradiated at the same temperature as the background radiation if the coupling to the CBR dominates the grain temperature. The result is that in this model the CBR is smoothed and thermalized at a redshift comparable to that for electron scattering (eq. [24.87]).

Another approach extapolates from the opacity of the interstellar dust in our galaxy. In the semiempirical model of Draine and Lee (1984), the opacity at wavelength $\lambda > 100\mu$ is

$$\frac{\kappa}{n_{\rm I}} = 10^{-24} \left(\frac{100\mu}{\lambda}\right)^2 \, {\rm cm}^2 . \tag{24.93}$$

This is the local ratio of the opacity (optical depth per unit length) to the number density $n_{\rm I}$ in atomic hydrogen. Suppose that in young galaxies the heavy element abundance is down from the present cosmic value by the factor Z/Z_o, and that the opacity relative to the nucleon number density is down by the same factor. Then the optical depth for absorption at redshift z and wavelength $\lambda(1+z)^{-1}$, where λ is the observed wavelength, is

$$\tau \sim \frac{10^{-5}(1+z)^{7/2}}{\lambda_{\rm cm}^2} \frac{h\Omega_B}{\Omega^{1/2}} \frac{Z}{Z_o} , \tag{24.94}$$

where Ω_B is the density parameter in the baryonic material in which the heavy element abundance is Z. The optical depth in this expression reaches unity at redshift

$$1 + z_{\rm abs} \sim 30 \frac{\lambda_{\rm cm}^{4/7}\Omega^{1/7}}{(h\Omega_B)^{2/7}} \left(\frac{Z_o}{Z}\right)^{2/7} . \tag{24.95}$$

With $Z = 0.1 Z_o$ and the numbers in equation (24.86), this is $z_{\rm abs} \sim 80$ at observed wavelength $\lambda \sim 1$ cm, three times the redshift at unit optical depth for the conducting grain model in equation (24.92) and the redshift for electron scattering in a low-density cosmological model (eq. [24.87]).

The conclusion is that it is not difficult to see how the CBR could have been scattered by free electrons, or processed through dust, at redshifts in the range $z \sim$ 20 to 100. The numbers are sensitive to the models for the cosmology, the plasma, the dust, and the evolution of the structures that might have produced the plasma and dust. The observations discussed in the last section show that at expansion factor $1 + z = 6$, structure formation was well advanced: the intergalactic medium was in place, with a highly ionized Lα forest, and the quasars and damped Lα clouds contained appreciable abundances of heavy elements. Whether it is judged reasonable to assume that at another factor of six in expansion factor back in time structure formation had advanced to the state postulated in these estimates of the opacity at CBR wavelengths depends in part on whether one feels there is a credible case for one of the models to be discussed in the next section. One can note that the epoch $z \sim 30$ in equations (24.87) and (24.92) is within the range of redshifts at which spheroids of galaxies can have been assembled, according to the discussion in section 22 (eq. [22.54]). The present-day disks of spiral galaxies are not thought to have existed at such high redshifts (for the reasons discussed in the next section), but it is easy to imagine the disk material was being prepared in the spheroids, and since disks are made of material already polluted with heavy elements it is not unreasonable to imagine that dust and plasma were present in amounts large enough to have made the universe opaque to the CBR.

If there were a large optical depth for absorption and reradiation of the CBR by dust, the dust would tend to relax to thermal equilibrium with the radiation, and the processing of the radiation through the dust would serve to erase any blemishes in the CBR spectrum and space distribution. If the dust received enough energy from starlight to make it warmer than the CBR when the optical depth is appreciable but not very large, the dust would tend to distort the CBR spectrum, contrary to what is observed. Constraints from this effect are surveyed by Bond, Carr, and Hogan (1986), and Wright (1992). The Doppler shifts from peculiar motions of the dust would perturb the angular distribution of the CBR, leaving a coherence length comparable to the angular size of a galaxy, if the dust is in galaxies. The small coherence length makes the effect relatively hard to detect, but this has not yet been discussed in much detail.

Electron Scattering and Anisotropies in the CBR

If dust is unimportant, the CBR would have been last scattered by free electrons, either at decoupling or in plasma clouds along the line of sight at lower redshifts. Either process would be expected to leave a signature of some sort in the angular distribution of the radiation. We have already discussed the effect of scattering by hot plasma in clusters of galaxies (eq. [24.61]), which inserts fluctuations on the angular scale of arc minutes, and the Sachs-Wolfe gravitational effect, which is an important constraint on mass fluctuations on the scale of the present Hubble length (eq. [21.107]). In the cold dark matter model discussed in the next section, matter is not likely to be ionized at the redshift $z_s \sim 50$ in equation (24.80) for electron scattering at the parameters in this model. Thus one expects that in this model the CBR anisotropy on angular scales ~ 10 arc minutes to $1°$ is dominated by the fluctuations $\delta T/T \sim 3 \times 10^{-6}$ that are put in at decoupling by the motion of the recombining plasma and the amplitude of the oscillating fluctuations in the density of matter and radiation. In the baryon isocurvature model the anisotropy on large scales is the assumed primeval irregularity in the space distribution of the entropy per baryon. The mass density is lower in this model, which lowers z_s, and the large density fluctuations on the scale of galaxies cause protogalaxies to form early. The likely result is that in this model the CBR is scattered at $z \gtrsim z_s \sim 20$, and smoothed on scales $\theta \lesssim \theta_s \sim 2°$ to $5°$ (eq. [24.87]).

The central topic for this subsection is the constraint from the isotropy of the CBR on models in which plasma scatters the CBR after decoupling. The scattering plasma has to have peculiar streaming motions, from the gravitational fields that are forming structures, and likely also from blast waves from supernovae in the active young galaxies. The Doppler shifts of the peculiar streaming motion of the plasma perturb the CBR temperature. The following discussion of this effect has many elements in common with what is done in the analysis of the perturbations introduced to the CBR at decoupling.

The character of the effect on the angular distribution of the CBR depends on the space distribution of the plasma. If the plasma were in clouds, and if the mean free path between clouds were larger than the correlation length of the peculiar velocity field, the coherence length in the CBR angular fluctuations would be the typical angular size subtended by a gas cloud. That could be small and difficult to detect. If the scattering plasma were smoothly distributed, the Doppler shifts from the large-scale peculiar velocity field would make the CBR anisotropic on observationally interesting angular scales. This was first discussed in detail by Sunyaev (1978); a recent survey of the state of understanding of the effect is given by Efstathiou (1988). The analysis in linear perturbation theory goes as follows.

We will confine the discussion to the CBR anisotropy on scales small compared to the scattering length L_s in equations (24.84) and (24.87), and we further simplify the calculation by assuming a flat spectrum of mass density fluctuations.

In the Rayleigh-Jeans limit the CBR surface brightness is proportional to the temperature, and the radiation transfer equation expressed as a perturbation τ to the temperature,

$$T = (1 + \tau)T_b ,\qquad\qquad (24.96)$$

is

$$\frac{d}{dt}\tau(t, \mathbf{x}(t)) = \frac{\partial \tau}{\partial t} + \frac{\gamma_\alpha}{a} \frac{\partial \tau}{\partial x^\alpha} \qquad\qquad (24.97)$$
$$= \sigma_t n_e(\tau_r + \gamma_\alpha v^\alpha - \tau) .$$

The γ_α are the direction cosines for the propagation of the radiation along the line of sight. The optical depth per unit length for scattering is $\sigma_t n_e$, where n_e is the free electron number density, and we can take the Thomson scattering cross section to be isotropic. The last term in the parentheses is the rate of scattering out of the beam element per unit path length, and the other two terms represent scattering into the beam. The angular average of the temperature perturbation is

$$\tau_r(t, \mathbf{x}) = \int \frac{d\Omega}{4\pi}\tau . \qquad\qquad (24.98)$$

This is the mean temperature (in the Rayleigh-Jeans limit) seen by an electron at **x**, and the mean value of the radiation temperature the electron scatters back into the line of sight. The term $\gamma_\alpha v^\alpha$ is the Doppler shift due to the streaming velocity v^α of the plasma. We are considering scales small compared to the Hubble length, so we can ignore the perturbation to τ by the gravitational field of the growing mass density fluctuations (in the term that appears in eq. [21.8]). Further details of this transfer equation, which was derived by Peebles and Yu (1970), and properties of its solutions, are given in LSS, §92.

We are computing the anisotropy at small angular scales that subtend distances small compared to the Hubble length at the epoch of last scattering. Kaiser (1984b) pointed out that in this limit the velocity term in the transfer equation (24.97) tends to average out, because the matter velocities on either side of a mass concentration are equal and opposite. This means we get a useful approximation by dropping the velocity term and considering only the contribution from the isotropic part τ_r. Also, Kaiser showed that in this small-scale limit there is a simple relation between τ_r and the mass distribution, as follows.

The angular average of the transfer equation (24.97) is

$$\frac{\partial \tau_r}{\partial t} + \frac{1}{a}\frac{\partial f_\alpha}{\partial x^\alpha} = 0, \tag{24.99}$$

where

$$f_\alpha = \int \frac{d\Omega}{4\pi} \gamma_\alpha \tau. \tag{24.100}$$

Since $4\tau_r = 4\delta T/T$ is the fractional perturbation to the radiation energy density ρ_γ, this is the equation of local energy conservation, with energy flux density $4f_\alpha \rho_\gamma$.

The result of multiplying the transfer equation by the direction cosines γ_α and averaging over angles is

$$\frac{\partial f_\alpha}{\partial t} + \frac{1}{a}\frac{\partial}{\partial x^\beta}\int \frac{d\Omega}{4\pi}\gamma_\alpha \gamma_\beta \tau = \sigma_t n_e \left(\frac{v^\alpha}{3} - f_\alpha\right). \tag{24.101}$$

This is the momentum equation. For density fluctuations on scales small compared to the Hubble length t, the dominant parts are the pressure gradient force in the integral in the second term, because of the space derivative, and the velocity term, which drives the radiation perturbation. We can approximate the integral as

$$\int \frac{d\Omega}{4\pi}\gamma_\alpha \gamma_\beta \tau = \frac{1}{3}\delta_{\alpha\beta}\tau_r. \tag{24.102}$$

Then the momentum equation at small scales is

$$\frac{1}{a}\frac{\partial \tau_r}{\partial x^\alpha} = \sigma_t n_e v^\alpha. \tag{24.103}$$

Since the radiation drag force is negligible at the redshifts of interest here, the equation of motion for the peculiar velocity of the plasma is (eq. [5.110])

$$\frac{\partial \mathbf{v}}{\partial t} + \frac{\dot{a}}{a}\mathbf{v} = -\frac{1}{a}\nabla\phi,$$

(24.104)

where ϕ is the gravitational potential of the matter distribution. In the Einstein-de Sitter limit, $\mathbf{v} \propto t^{1/3}$ and $a \propto t^{2/3}$, so this equation is $\mathbf{v} = -t\nabla\phi/a$, and equation (24.103) is

$$\tau_r = -\sigma_t n_e t\phi.$$

(24.105)

This says that where the gravitational potential is negative, the converging matter flow drags in radiation.

With this equation for τ_r, and neglecting the velocity term, the transfer equation (24.97) is

$$\tau = -\int_0^{t_o} dt\, \sigma_t n_e e^{-\int_t^{t_o} \sigma_s n_e\, dt'}(\sigma_t n_e t\phi).$$

(24.106)

This is the perturbation to the radiation temperature observed at epoch t_o, expressed in terms of the gravitational potential along the line of sight.

The source for ϕ is the matter density contrast $\delta(\mathbf{r}, t) = \delta_o(\mathbf{r})D(t)/D_o$, where δ_o is the contrast extrapolated to the present epoch in linear perturbation theory, $D(t)$ is the time evolution of the contrast in linear perturbation theory and normalized to $D = a/a_o$ in the Einstein-de Sitter limit, and D_o is the present value. With these conventions we can write the potential as

$$\phi = -\frac{G\rho_o a_o^2}{D_o}\int \frac{\delta_o}{r}d^3r.$$

(24.107)

We will compute the second moment of τ under the assumption that the spectrum of the mass density fluctuations δ_o is flat. To suppress the large-scale fluctuations in ϕ we will follow Uson and Wilkinson (1984) and Efstathiou (1988) in considering the second difference of the temperatures observed along neighboring lines of sight,

$$\tau_1 - 2\tau_2 + \tau_3 = \int A\, dt(\phi_1 - 2\phi_2 + \phi_3),$$

(24.108)

where the points are in a line on the sky at equal angular separations θ. The factor A represents all the rest of the integrand in equation (24.106).

The mean square value of this second difference is of the form

$$\langle(\tau_1 - 2\tau_2 + \tau_3)^2\rangle = \int AA'\, dt dt' B,$$

(24.109)

where

$$B = \langle 6\phi_1\phi_{1'} - 8\phi_1\phi_{2'} + 2\phi_1\phi_{3'} \rangle$$
$$= 8\langle \phi_1(\phi_{1'} - \phi_{2'}) \rangle - 2\langle \phi_1(\phi_{1'} - \phi_{3'}) \rangle . \tag{24.110}$$

The points 1 and $1'$ are along the same line of sight at separation $x = \int dt/a$ from t to t'. The points 1 and $2'$ have the same separation as 1 and $1'$ along the line of sight and are separated by angle θ, and 1 and $3'$ are at angular separation 2θ. Thus the coordinate distances x_2 between 1 and $2'$ and x_3 between 1 and $3'$ are

$$x_2 = (x^2 + x_\sigma^2)^{1/2}, \qquad x_3 = (x^2 + 4x_\sigma^2)^{1/2}, \tag{24.111}$$

where the projected coordinate separation of neighboring lines of sight at the epoch of last scattering is

$$r_\sigma = \theta r_s, \tag{24.112}$$

with r_s the angular size distance at large redshift.

For equation (24.110), we need

$$\langle \phi_1(\phi_{1'} - \phi_{2'}) \rangle = \left(\frac{G\rho_o a_o^2}{D_o} \right)^2 \int d^3s \, d^3t \langle \delta_o(\mathbf{s})\delta_o(\mathbf{t}) \rangle \frac{1}{s} \left(\frac{1}{|\mathbf{x} - \mathbf{t}|} - \frac{1}{|\mathbf{x}_2 - \mathbf{t}|} \right) \tag{24.113}$$
$$= \left(\frac{G\rho_o a_o^2}{D_o} \right)^2 \frac{4\pi J_3}{a_o^3} K .$$

The second line introduces the assumption that on the length scale subtended by the observations the spectrum of density fluctuations is flat. That means the mass autocorrelation function acts as a delta function, with J_3 the integral over the small-scale part of the present value of the function (eq. [21.57]). The remaining integral is

$$K = \int d^3s \frac{1}{s} \left(\frac{1}{|\mathbf{x} - \mathbf{s}|} - \frac{1}{|\mathbf{x}_2 - \mathbf{s}|} \right) \tag{24.114}$$
$$= 2\pi(x_2 - x) .$$

The second line follows on observing that this is the potential energy difference between $\mathbf{x}$ and $\mathbf{x}_2$ for the spherical charge distribution $\rho = 1/s$. The charge within radius s is $2\pi s^2$, so the force is 2π, and the potential difference is 2π times the difference in radii.

Equation (24.114) says that the integrand in equation (24.110) is

$$B \propto 8(x^2 + x_\sigma^2)^{1/2} - 6x - 2(x^2 + 4x_\sigma^2)^{1/2} . \tag{24.115}$$

This expression is flat at $x \ll x_\sigma$ and decreases as $B \propto x^{-3}$ at $x \gg x_\sigma$. This means we can rewrite the variables of integration in equation (24.109) as $dt\,dt' = a\,dx\,dt$, set $t' = t$ everywhere except in the part of B in equation (24.115), and then compute separately the integrals over t and x. With the changes of variable $x/x_\sigma = \sinh\theta$ and $2\sinh\theta$, the latter is

$$\int_{-\infty}^{\infty} dx\,[8(x^2 + x_\sigma^2)^{1/2} - 6x - 2(x^2 + 4x_\sigma^2)^{1/2}] = 8x_\sigma^2 \ln 2. \quad \textbf{(24.116)}$$

Now it is an easy if tedious exercise to work the integral over t in the Einstein-de Sitter limit. The result is

$$\left(\frac{\delta T}{T}\right)^2 = \langle(\tau_1 - 2\tau_2 + \tau_3)^2\rangle$$

$$= 2\ln 2 \left(\frac{10}{3}\right)! \left(\frac{3}{4}\right)^{10/3} \frac{J_3 H_o^3}{D_o^2} \left(\frac{H_o}{\sigma_t n_o}\right)^{1/3} \Omega^{8/3} (H_o a_o r_s \theta)^2. \quad \textbf{(24.117)}$$

With equation (21.100) for J_3, equation (23.1) for the electron number density in terms of the density parameter Ω_B in the plasma, and the angle θ between successive lines of sight in the triple beam expressed in degrees, the numerical result is

$$\frac{\delta T}{T} = \frac{1 \times 10^{-5}}{D_o} \frac{\Omega^{8/6} y_s}{(h\Omega_B)^{1/6}} \theta^\circ. \quad \textbf{(24.118)}$$

The dimensionless distance $y_s = H_o a_o r_s$ at large redshift is shown in figure 13.3, and the growth factor D_o is shown in figure 13.13. This result applies at 2θ less than the smoothing length θ_s in equation (24.85). With the parameters in equation (24.87), θ_s is about one degree and $\delta T/T$ in equation (24.118) is about at the present measurements.

Vishniac (1987) showed that nonlinear corrections to this calculation can produce an appreciable increase in the predicted anisotropy. Efstathiou's (1988) estimates indicate that the relation between the plasma velocity and the plasma density in second-order perturbation theory produces an anisotropy that for some otherwise reasonable-looking choices of parameters violates the measurements of the CBR isotropy. In the case we have been considering, where the spectrum of fluctuations is flat and $\Omega = \Omega_B \sim 0.1$, Efstathiou's estimate of the nonlinear correction still is an order of magnitude below the measured limit on $\delta T/T$. The calculation does depend on the details of the model for the diffuse plasma, however. For example, one might imagine that where the baryon density is high, the electron density contrast is even higher, because the high density encourages formation of the stars that ionize the matter, or that it is low, because star formation locally depletes the diffuse matter. Also, if the plasma distribution is patchy it reduces the coherence length of the background temperature fluctuations, as we noted for the case of absorption and scattering by dust. The interpretation of the velocity effect in the anisotropy of the CBR thus is likely to depend on the analysis of details of specific models.

Kompaneets Equation

The purpose of this subsection is to derive equation (24.35), which describes the evolution of the spectrum of a homogeneous isotropic sea of radiation that is elastically scattered by a homogeneous isotropic nonrelativistic gas of electrons. The integrated effect of many scatterings by the moving electrons produces a random walk in the energy of each photon (while preserving the photon number), which can produce an observable effect on the CBR spectrum when its temperature is different from the kinetic temperature of the scattering electrons.

Quantities in the laboratory frame, where the electron and photon distributions are isotropic, are written without subscripts, quantities in the initial rest frame of an electron (before it scatters a photon) have the subscript 1, and a prime indicates the new value of a quantity after a scattering. In the rest frame of an electron moving at speed v along the x axis in the laboratory frame, the Lorentz transformation for the position coordinates is

$$t = \gamma(t_1 + v x_1), \qquad x = \gamma(x_1 + v t_1).   \tag{24.119}$$

The same transformation applies to the energy and x component of momentum of a photon moving at angle θ to the x axis,

$$\nu = \gamma \nu_1 (1 + v \cos \theta_1), \qquad \nu \cos \theta = \gamma \nu_1 (\cos \theta_1 + v).   \tag{24.120}$$

The energy transformation the other way is

$$\nu_1 = \gamma \nu (1 - v \cos \theta),   \tag{24.121}$$

so we have

$$\gamma(1 - v \cos \theta) = \frac{1}{\gamma(1 + v \cos \theta_1)}.   \tag{24.122}$$

The ratio of the relations for energy and the x component of momentum in equation (24.120) gives the transformation law for the angle of motion of the photon relative to the x axis,

$$\cos \theta = \frac{\cos \theta_1 + v}{1 + v \cos \theta_1}.   \tag{24.123}$$

The derivative of this expression gives the transformation equation for the solid angle $d\Omega = d\cos\theta d\phi$ of a beam of photons,

$$d\Omega = \frac{d\Omega_1}{\gamma^2(1 + v\cos\theta_1)^2} \, . \tag{24.124}$$

The last of the transformation relations we will need applies to the photon occupation number. Since $\mathcal{N}$ is the distribution in single-particle phase space, the brightness theorem (eq. [9.50]) says that in the absence of collisions $\mathcal{N}$ is constant along a photon orbit. Since that is true even in a curvilinear coordinate system, $\mathcal{N}$ has to be a scalar:

$$\mathcal{N}_1(\theta_1, \nu_1, t_1, x_1) = \mathcal{N}(\nu = \gamma\nu_1(1 + v\cos\theta_1), t = \gamma(t_1 + vx_1)) \, . \tag{24.125}$$

The occupation number in the laboratory frame on the right-hand side is a function of frequency and time alone. The Lorentz transformations make the occupation number $\mathcal{N}_1$ in the electron rest frame a function of photon direction and position x_1 along the x axis.

The next step is to write down the Boltzmann collision equation for $\mathcal{N}_1$,

$$\frac{\partial\mathcal{N}_1}{\partial t_1} + \cos\theta_1 \frac{\partial\mathcal{N}_1}{\partial x_1} = R_1 \, . \tag{24.126}$$

The right-hand side is the rate R_1 of scattering of photons into the beam, less the rate of scattering out of it, as measured by an observer at rest at the electron. The left side is the derivative moving with the photon beam. We can use equation (24.125) to write the two derivatives in terms of the occupation number in the laboratory frame,

$$\frac{\partial\mathcal{N}_1}{\partial t_1} = \gamma\frac{\partial\mathcal{N}}{\partial t} \, , \qquad \frac{\partial\mathcal{N}_1}{\partial x_1} = \gamma v\frac{\partial\mathcal{N}}{\partial t} \, . \tag{24.127}$$

This brings the collision equation to

$$\gamma\frac{\partial\mathcal{N}}{\partial t}(1 + v\cos\theta_1) = R_1 \, . \tag{24.128}$$

With equation (24.122), this is

$$\frac{\partial\mathcal{N}}{\partial t} = \langle \gamma(1 - v\cos\theta)R_1 \rangle_\theta \, . \tag{24.129}$$

The brackets indicate that we have to average over photon directions to get the final expression for the time evolution of the occupation number in the laboratory frame.

In writing down the expression for the scattering rate, R_1, we have to take account of the electron recoil (because we are computing to terms of order v^2, and at statistical equilibrium this is of the same order as the Compton shift in the photon frequency). In the initial electron rest frame the electron momentum vanishes before the scattering, and after the scattering it is

$$m\mathbf{v}_1' = \vec{\nu}_1 - \vec{\nu}_1'. \tag{24.130}$$

The square of this expression gives the final kinetic energy of the electron, which is the energy lost by the photon. To lowest nontrivial order, this is the Compton shift

$$\nu_1 - \nu_1' = \delta\nu_1 = \frac{\nu_1^2}{m}(1 - \cos\Theta). \tag{24.131}$$

The time rate of change of the number of photons in a beam element of solid angle $d\Omega_1$ and bandwidth ν_1 to $\nu_1 + d\nu_1$, measured in the initial rest frame of the electron and averaged over an ensemble of trials, is proportional to [16]

$$R_1\nu_1^2 d\nu_1 d\Omega_1 = \frac{d}{dt_1}\mathcal{N}_1(\theta_1,\nu_1)\nu_1^2 d\nu_1 d\Omega_1 = \int \frac{d\sigma}{d\Omega_s} I, \tag{24.132}$$

where

$$I = [1 + \mathcal{N}_1(\theta_1,\nu_1)] \cdot \mathcal{N}_1(\theta_1',\nu_1^+) \cdot (\nu_1^+)^2 d\nu_1^+ d\Omega_1' \cdot d\Omega_1$$
$$- [1 + \mathcal{N}_1(\theta_1',\nu_1^-)] \cdot \mathcal{N}_1(\theta_1,\nu_1) \nu_1^2 d\nu_1 d\Omega_1 \cdot d\Omega_1', \tag{24.133}$$

with

$$\nu_1^{\pm} = \nu_1 \pm \delta\nu$$
$$= \nu_1 \left[1 \pm \frac{\nu_1}{m}(1 - \cos\Theta)\right]. \tag{24.134}$$

The second line in equation (24.133) gives the rate of scattering out of the beam. The second factor in this line is the number of photons in the beam element. In the stimulated emission factor in front of it, the occupation number is evaluated at the direction and frequency of the scattered photon. The frequency of the scattered photon is lowered by the Compton effect, to ν_1^-, as defined in equation (24.134). The photons are scattered into the solid angle $d\Omega_1'$, with cross section $d\sigma/d\Omega_s$, at

---

[16] The derivation given by Rybicki and Lightman (1979) starts with an equation that looks like this one, but without some of the factors. That is because the term $d\sigma/d\Omega$ in the Rybicki-Lightman expression is a transition probability, rather than a cross section as conventionally defined and used here. Another derivation is given by Bernstein (1988). The derivation presented here by Lorentz transformations is from PC.

scattering angle Θ. The first line of equation (24.133) is the rate of scattering into the beam element. Here the Compton effect requires that the incident photon has energy $\nu_1{}^+ = \nu_1 + \delta\nu$. This means the photons are scattered into the beam element from a larger volume of phase space, as is indicated in the third factor in the first line. Here the incident beam, with solid angle $d\Omega_1'$, is scattered into the solid angle $d\Omega_1$. We are ignoring the energy dependence of the cross section, because we are computing in a nonrelativistic limit, to order ν_1/m.

The next step is to write the occupation number that appears in equation (24.133) in terms of the laboratory function, $\mathcal{N}(\nu)$. We need the Compton-shifted photon energy in the laboratory frame for a photon moving at angle θ_1' to the x axis in the initial electron rest frame. With equation (24.120) this is

$$\nu^{\pm} = \nu_1{}^{\pm}\gamma(1 + v\cos\theta_1')$$
$$= \nu\frac{\nu_1{}^{\pm}}{\nu_1}\frac{(1 + v\cos\theta_1')}{(1 + v\cos\theta_1)}\,. \tag{24.135}$$

The result of expanding this expression to order v^2 and ν/m, with equation (24.134) for $\nu_1{}^{\pm}$, is

$$\nu^{\pm} = \nu[1 \pm (\nu/m)(1 - \cos\Theta) + v(\cos\theta_1' - \cos\theta_1)$$
$$+ v^2(\cos^2\theta_1 - \cos\theta_1'\cos\theta_1)] \tag{24.136}$$
$$\equiv \nu + \Delta^{\pm}\,.$$

We have then

$$\mathcal{N}_1(\theta_1, \nu_1) = \mathcal{N}(\nu)\,, \qquad \mathcal{N}_1(\theta_1', \nu_1{}^{\pm}) = \mathcal{N}(\nu + \Delta^{\pm})\,. \tag{24.137}$$

The antipenultimate step is to write out the expression for the rate R_1 in equations (24.128) and (24.132) for $d\mathcal{N}/dt$. With equation (24.134), the momentum volume element that appears in the first line of equation (24.133) is

$$(\nu_1{}^+)^2 d\nu_1{}^+ = \left[1 + 4\frac{\nu_1}{m}(1 - \cos\Theta)\right]\nu_1{}^2 d\nu_1\,. \tag{24.138}$$

Then we have from equations (24.129) and (24.133) that the time derivative of the occupation number in the laboratory frame is

$$\frac{\partial\mathcal{N}}{\partial t} = \int \gamma(1 - v\cos\theta)\frac{d\Omega}{4\pi}\frac{d\sigma}{d\Omega_s}d\Omega_1'(A + B)\,, \tag{24.139}$$

where

$$A = \frac{4\nu}{m}(1 - \cos\Theta)\mathcal{N}(\nu)[1 + \mathcal{N}(\nu)]\,, \tag{24.140}$$

and

$$B = [1 + \mathcal{N}(\nu)]\mathcal{N}(\nu + \Delta^+) - [1 + \mathcal{N}(\nu + \Delta^-)]\mathcal{N}(\nu). \qquad \textbf{(24.141)}$$

In the term A, from the Compton correction to the momentum volume element in equations (24.133) and (24.138), we can write the occupation number as a function of the laboratory frequency ν, because we are computing to order ν/m. The first integral in equation (24.139) is the average over the photon direction θ in equation (24.129). The second integral is over the photon direction θ_1' that increases or decreases the beam element by scattering into or out of the element. Since we are computing to second order in the electron speed v, we can expand B to second order in $\Delta^{\pm}$:

$$\begin{aligned}
B = (1 + \mathcal{N}) &\left[\mathcal{N} + \frac{d\mathcal{N}}{d\nu}\Delta^+ + \frac{1}{2}\frac{d^2\mathcal{N}}{d\nu^2}(\Delta^+)^2 \right] \\
&- \mathcal{N}\left[1 + \mathcal{N} + \frac{d\mathcal{N}}{d\nu}\Delta^- + \frac{1}{2}\frac{d^2\mathcal{N}}{d\nu^2}(\Delta^-)^2 \right],
\end{aligned} \qquad \textbf{(24.142)}$$

which rearranges to

$$B = \frac{d\mathcal{N}}{d\nu}[(1+\mathcal{N})\Delta^+ - \mathcal{N}\Delta^-] + \frac{1}{2}\frac{d^2\mathcal{N}}{d\nu^2}[(1+\mathcal{N})(\Delta^+)^2 - \mathcal{N}(\Delta^-)^2]. \qquad \textbf{(24.143)}$$

Here and below $\mathcal{N} = \mathcal{N}(\nu, t)$ is the laboratory occupation number at frequency ν. With equation (24.136) for $\Delta^{\pm}$, we get from equations (24.140) and (24.143)

$$\begin{aligned}
A + B = {} & (1 - \cos\Theta)\frac{\nu}{m}\left[4\mathcal{N}(1+\mathcal{N}) + \nu\frac{d\mathcal{N}}{d\nu}(1 + 2\mathcal{N}) \right] \\
&+ \nu\frac{d\mathcal{N}}{d\nu}[v(\cos\theta_1' - \cos\theta_1) + v^2(\cos^2\theta_1 - \cos\theta_1\cos\theta_1')] \qquad \textbf{(24.144)} \\
&+ \frac{\nu^2}{2}\frac{d^2\mathcal{N}}{d\nu^2}v^2(\cos\theta_1' - \cos\theta_1)^2.
\end{aligned}$$

This expression for $A + B$ is first order in v, so we only need to keep terms of order v in the factor multiplying it in equation (24.139). Using equation (24.124) for the solid angle, we have to this order

$$(1 - v\cos\theta)d\Omega = (1 - 3v\cos\theta_1)d\Omega_1. \qquad \textbf{(24.145)}$$

We can use this to write equation (24.139) as an integral over the photon directions θ_1 and θ_1' in the initial electron rest frame.

Since $A + B$ is of order v we can use the classical Thomson scattering cross section. It is symmetric under $\theta_1 \to \pi + \theta_1$ and $\theta_1' \to \pi + \theta_1'$, so all the terms odd in $\cos \theta_1$ or $\cos \theta_1{'}$ vanish in the integral. Also, we have

$$\int d\Omega_1 d\Omega_1{'} \cos^2 \theta_1 (d\sigma/d\Omega_s) = 4\pi\sigma_t/3 \, . \tag{24.146}$$

All this brings equation (24.139) to

$$\frac{\partial \mathcal{N}}{\partial t} = \int \frac{d\Omega_1 d\Omega_1{'}}{4\pi} \frac{d\sigma}{d\Omega_s}(1 - 3v\cos\theta_1)(A + B) = \sigma_t C \, , \tag{24.147}$$

where

$$C = \frac{v}{m} \left[4\mathcal{N}(1 + \mathcal{N}) + (1 + 2\mathcal{N})v\frac{d\mathcal{N}}{dv} \right]$$
$$+ \frac{4}{3}\frac{d\mathcal{N}}{dv}vv^2 + \frac{1}{3}\frac{d^2\mathcal{N}}{dv^2}v^2 v^2 \, . \tag{24.148}$$

This is the expectation value computed for scattering by a single electron. For a gas of electrons with number density n_e and mean square velocity $\langle v^2 \rangle$, the equation becomes

$$\frac{1}{\sigma_t n_e}\frac{\partial \mathcal{N}}{\partial t} = \frac{v}{m}\left[4\mathcal{N}(1 + \mathcal{N}) + (1 + 2\mathcal{N})v\frac{\partial \mathcal{N}}{\partial v} \right]$$
$$+ \frac{4}{3}\langle v^2 \rangle v\frac{\partial \mathcal{N}}{\partial v} + \frac{1}{3}\langle v^2 \rangle v^2\frac{\partial^2 \mathcal{N}}{\partial v^2} \, . \tag{24.149}$$

This is equation (24.35).

25. Galaxy Formation

Since the beginning of modern cosmology, people have recognized that Einstein's cosmological principle is at best a useful approximation to the real world of galaxies and clusters of galaxies, and that a more complete cosmology would explain why matter is organized in this clumpy fashion. This section deals with clues to the puzzle and some of the proposed resolutions. The first subsection offers an interpretation of what happened, in terms of a timetable for structure formation. Next we consider the main physical elements of the commonly discussed models for structure formation. This is followed by a more detailed survey of the models and the lessons we learn from their problems, real or apparent. We ought to include an assessment of the evidence on the value of the mean mass density, because low and high density cosmological models suggest rather different pictures for structure formation. The state of the evidence is highly uncertain, however, so the discussion of the cosmological parameters is reserved for the really controversial conclusions to be entered in the next section.

Before starting on this long road, let us consider how one approaches the puzzle of the galaxies. Prior to about 1980 the usual way was to ask whether one can find initial conditions at high redshift that would evolve to a present state that looks at least roughly like what is observed. If this could be done in a way that is physically sensible and agrees with all the observational constraints, we would at least have a demonstration of consistency of the expanding world picture, and perhaps also hints from the character of the initial conditions to the nature of the seeds of structure. During the 1980s a remarkable flow of ideas from other parts of physics, including those summarized in sections 15 to 18, led to the introduction of scenarios motivated by physical principles rather than phenomenology. The approach has an honorable history, for it is how Einstein hit on the very successful cosmological principle discussed in sections 2 and 3, and it certainly is worth trying again. At the time of this writing, however, there is no credible empirical evidence that any of the major *ab initio* scenarios really is on the right track, or even that any correlate particularly well with the observations beyond what went into the construction.

One reason progress in substantiating the new ideas has been so slow, even by the relaxed standards of cosmology, is the difference in character of the structure problem and what was involved in the discovery of the expansion of the universe. In the large-scale average the observable universe is constructed on a wonderfully simple plan, as is seen in figures 3.8 to 3.11. This is what Einstein anticipated in the cosmological principle. Perhaps after the puzzle of the galaxies is resolved we will see that they too are constructed on a plan that could have been (or perhaps has been) revealed through pure thought and general physical principles. Galaxies and their space distribution do look a good deal more complicated than the large-scale structure of the universe, however, so it is reasonable to anticipate that a convincing solution will require that we sort through the clues Nature has chosen to offer us, as well as the hints from fundamental theory.

Unless we are lucky enough to hit on an unambiguous signature of what drives structure formation, such as the distinctive linear features in the CBR that would be produced by a network of long cosmic strings, we are not likely to resolve the puzzle by pure deduction from the observations, no matter how rich the detail they reach, because too many key pieces are likely to be at high redshifts where the observational probes have to be schematic at best. And since the *ab initio* approach from fundamental theory seems likely to be confused by the abundance of possibilities, we may have to see what grows out of the cross-fertilization of empirical studies with trial and error fits to the many scenarios available within the standard model. The process is messy, but it has worked in other areas of physical science, and it is the guiding philosophy behind the following discussion.

Several cautions and explanations are in order. First, the discussion emphasizes problems as a guide to the way to improve our ideas. Successes certainly are a relevant measure of the models for galaxy formation, but they can be a seductive trap, because the only models that are widely publicized are those that can be adjusted to fit some subset of the data. Second, there is no generally accepted list of the most serious challenges for each of the more popular models. The selections presented here are a personal choice of which are most interesting in the sense that they seem unlikely to be resolved by relatively easy adjustments of model parameters and therefore may be teaching us something of value. Third, the astronomical study of galaxies for clues to how they formed is a rich subject that is only sparsely sampled here. An example of the richness of detail available to us is to be found in *The Milky Way as a Galaxy* (Gilmore, King, and van der Kruit 1990). Larson (1990), Schweizer (1990), and Efstathiou (1990) give useful readings of the issues in the astronomy and physics. And finally, there is no attempt here to be systematic in the documentation of the major recent observational developments, because in this rapidly evolving subject the best review almost certainly will have appeared after this is written. Instead, the reader is invited to consult and consider the proceedings of recent conferences and adjust the estimates accordingly.

Timetable for Structure Formation

Table 25.1 lists some estimates of epochs in the sequence of formation of structures.[17] This is meant to be an outline; it assigns characteristic epochs to what very likely are processes that may operate over considerable ranges of redshifts. And as already noted, it has to be a personal assessment of very uncertain issues that will evolve as the evidence accumulates.

Most entries in the table are chosen because there is thought to be something useful and positive to be said; a few are in the table because they seem to be fundamental but particularly enigmatic. Notable among the latter is the origin of cosmic magnetic fields. The magnetic field in the interstellar medium in our Milky Way galaxy is observed in the Zeeman effect in the 21-cm line in atomic hydrogen, in the rotation measures of polarized radiation from pulsars and extragalactic radio sources, from polarization of starlight and the polarized thermal emission by dust grains aligned in the magnetic field, and from the polarized synchrotron emission by relativistic electrons spiraling in the magnetic field. Most of these effects are observed also in varying degrees of detail in other galaxies. Of particular interest are the indications that the damped Lyman-α systems at redshifts $z \sim 3$ insert appreciable rotation measures in the radio flux from the background quasars. Can a dynamo operate on a timescale short enough to account for the existence of magnetic fields with coherence length comparable to the size of a galaxy in these young systems? If this proves impossible, indicating magnetic fields existed before galaxies, it will be a fascinating hint to what happened in the early universe and a considerable challenge to conventional ideas.

---

[17] This table was developed in collaboration with Craig Hogan at the June 1991 Astrophysics Workshop at the Aspen Center for Physics.

Table 25.1
Timetable for Formation

Gravitational potential fluctuations	$z \gtrsim 10^3$
Spheroids of galaxies	$z \sim 20$
The first engines for active galactic nuclei	$z \gtrsim 10$
The intergalactic medium	$z \sim 10$
Dark matter	$z \gtrsim 5$
Dark halos of galaxies	$z \sim 5$
Angular momentum of rotation of galaxies	$z \sim 5$
The first 10% of the heavy elements	$z \gtrsim 3$
Cosmic magnetic fields	$z \gtrsim 3$
Rich clusters of galaxies	$z \sim 2$
Thin disks of spiral galaxies	$z \sim 1$
Superclusters, walls and voids	$z \sim 1$

The central engines that power quasars and radio galaxies are known to be less active now than at redshift $z \sim 2$, for the comoving number density of quasars is considerably lower at lower redshifts, and it is presumed that many galaxies at the present epoch harbor quasar engines that are dormant because they have exhausted the fuel supply (which usually is thought to be interstellar gas captured in a tight hot disk around a massive black hole). The evolution of the comoving number density of quasars earlier than redshift $z = 3$ is still under discussion, but it is known that the most luminous exhibit a roughly constant comoving abundance from $z = 5$ to $z = 3$. As discussed in section 23, the radiation from quasars is a significant contributor to the ionization of the Lα forest clouds and the intergalactic medium, which already were in place at $z = 5$. On the principle that cosmic evolution happens on a cosmic expansion timescale, it is reasonable to allow a factor of two expansion from the time the first quasars turned on to the completed formation of an intergalactic medium. This suggests the epoch of appearance of the first quasar engines is at redshift $z \gtrsim 10$.

The redshift assigned to the formation of an intergalactic medium that contains little neutral hydrogen is based on the increase in the density of Lα forest clouds and Lyman-limit clouds from $z = 2$ to $z = 5$. The straightforward extrapolation of this trend suggests that at redshift $z = 10$ the Lα forest clouds are merged or close to it, the Lyman-limit clouds absorb a large part of the ionizing radiation, and the neutral fraction therefore is considerably greater than at $z \sim 5$. Contributing to this is the dissipation of the plasma energy by radiation drag, which becomes important at $z \gtrsim 10(\Omega h^2)^{1/5}$ (eq. [6.136]). Thus the reasonable guess is that the intergalactic medium at $z = 10$ is quite different from what is observed in the absorption spectra of quasars at $z < 5$.

The assignment of an epoch for the origin of the heavy elements is an exceedingly schematic summary of a complex process. Quasars at the largest observed redshifts (now $z \sim 5$) have prominent emission lines from the more common heavy elements. As far as is known, only stars make elements heavier than boron in significant amounts, so the presence of these elements in quasars is taken to mean that at $z \sim 5$ galaxies of stars massive enough and evolved enough to have accumulated heavy elements hosted the quasar engines. The damped Lα clouds at $z \sim 3$ show absorption lines from heavy elements, at roughly 10% of the present "cosmic" abundance in the interstellar medium and in relatively young stars such as the Sun. As discussed in section 23, these high surface density clouds have properties one might look for in young galaxies or their near ancestors. This is the basis for the entry in the table for the epoch of production of the first 10% of the heavy elements. The lower surface density Lyman-α forest clouds have lower heavy element abundances. This is in line with the positive correlation of heavy element abundance with luminosity in present-day galaxies; the standard interpretation is that more massive systems are better able to promote star formation and more capable of retaining the heavy elements ejected when stars explode as supernovae.

In assigning an epoch to the formation of rich clusters of galaxies we have to bear in mind that the great clusters could form by the merging of littler ones, in a process still in operation. The characteristic epoch $z \sim 2$ adopted in the table is based in part on the observations of clusters at redshifts $z \sim 0.5$ with velocity dispersions characteristic of the most massive present-day clusters (White, Silk, and Henry 1981; Gunn 1989). This tells us that the formation of these large systems is not exclusively a recent process. An upper bound on the epoch of assembly follows from the dynamical arguments in section 22. The mean mass density within the Abell radius r_a of a rich cluster, relative to the present mean background density, is (eq. [20.39])

$$\frac{\bar{\rho}}{\rho_o} = \frac{4}{\Omega} \left(\frac{\sigma}{H_o r_a} \right)^2 \sim \frac{100}{\Omega}. \tag{25.1}$$

In the spherical model in equation (22.53), a growing mass fluctuation stops expanding at density contrast $\bar{\rho}/\rho_b = 5.6$. We noted in section 22 why one might suspect that by the time the mean density contrast in a real protocluster has reached this value, nonradial motions are becoming important. If so, this is a reasonable contrast for the development of dynamical equilibrium at the Abell radius and at the value of the velocity dispersion characteristic of present-day clusters. After the cluster has reached dynamical equilibrium and the mean density $\bar{\rho}$ is no longer rapidly evolving, the contrast grows roughly as the cube of the expansion factor. This would put the epoch of formation of a cluster at the present contrast in equation (25.1) at redshift

$$1 + z_f \sim 2.5\Omega^{-1/3}. \tag{25.2}$$

Rich clusters cannot have been present much before this because the cosmic mean density approaches that of the cluster. If the density contrast when the mass distribution in the cluster had become stable were larger than the value $\bar{\rho}/\rho_b \sim 6$ assumed here, it would lower z_f. In this case, however, the collapse would tend to increase the slope of $\xi_{cg}(r)$, as indicated in figure 22.2, which is contrary to the observations (eq. [19.35]). This problem may be avoided by assuming the great clusters are only now collapsing, as in the cold dark matter model to be discussed below, and we will see that the interesting test is the predicted rapid evolution in the abundance of clusters at redshifts less than unity.

The voids and sheets in the galaxy distribution, as illustrated in figures 3.3 to 3.7, have been assigned an assembly redshift somewhat less than that of the rich clusters, because they look like less dense and less well advanced parts of the same hierarchical clustering phenomenon. An alternative picture is that the walls are concentrations of galaxies but not of mass. For example, one might imagine that long cosmic strings triggered galaxy formation in their sheetlike wakes, leaving the voids undisturbed. That would place a strong galaxy concentration on a much weaker concentration of mass. A similar effect is produced by the cold dark matter biasing scheme to be discussed in the next subsection. In these pictures, the sheets of galaxies are present at high redshifts as the pattern in which those galaxy seeds that can germinate are sown. This is an attractive way to account for the discrepancy between the dynamical estimates of the mean mass density and the critical Einstein-de Sitter value. However, we will see that such pictures are challenged by the observations that dwarf galaxies cluster with giants, as illustrated in figure 3.3. Is it reasonable to assume that the many cosmic strings that would have passed through the voids marked out by L_* galaxies left so few dwarf and irregular galaxies in the voids?

In yet another picture for structure formation, galaxies formed by a "top-down" process in which protosheets fragmented to produce galaxies. The numbers in the table assume structure formation operated in the opposite direction, in a "bottom-up" hierarchical growth of structure by gravity, because it will be argued below that we seem to have clear examples of the latter process in operation, in the recent formation of the Local Group out of old galaxies, and in the present assembly of the Local Supercluster out of such groups.

The hierarchical calculation in equations (25.1) and (25.2) gives the estimate in the table for the epoch of assembly of the stellar spheroid components of galaxies. This follows the classical analyses by Hoyle (1953), Eggen, Lynden-Bell, and Sandage (1962), Mestel (1963), and Oort (1965), who argue that the spheroid Population II stars are remnants of the generations of stars that formed as the spheroid broke away from the general expansion of the universe. For a numerical example, consider the system of globular star clusters around our galaxy. Most are within distance $R = 10$ kpc from the center of the galaxy. Let us suppose the cosmic time t_f at which these clusters first were in place is the circular orbit time, $2\pi R/v_c$,

$$t_f = \frac{2\pi R}{v_c} = \frac{2}{3H_o\Omega^{1/2}}\frac{1}{(1+z_f)^{3/2}} \cdot \tag{25.3}$$

The last expression is the Einstein-de Sitter limit (5.61). With $R = 10$ kpc and $v_c = 220$ km s^{-1}, this says the epoch of assembly of the globular cluster system is

$$1 + z_f = \frac{8}{h^{2/3}\Omega^{1/3}} \cdot \tag{25.4}$$

An Einstein-de Sitter universe with $h \sim 0.5$, for a reasonable timescale, and a low-density model with $\Omega \sim 0.1$ and $h \sim 1$, give $z_f \sim 15$ in this model. The slightly larger redshift entered in the table follows from the argument in section 22 that one might not expect a spherical collapse model to be a useful approximation much past the epoch of maximum expansion (Partridge and Peebles 1967a). The entry also agrees with the thought that the engines for active galactic nuclei might be expected to have formed after the spheroid components of giant galaxies had been assembled.

As for clusters of galaxies, one could easily imagine that a spheroid forms by accretion or merging of gas clouds or gas-rich star clusters in a process that operates over a considerable range of time. Evidence that this is what actually happened comes from the estimates that there is a spread of about 4 Gy in the stellar evolution ages of globular star clusters (Demarque, Deliyannis, and Sarajedini 1991). Larson (1990) notes that the spread in the globular cluster ages increases with increasing distance from the center of the galaxy, the younger clusters tending to be at larger radii, consistent with the idea that after the bulk of the spheroid formed, perhaps at about the timescale of equation (25.4), there was a rain of material from larger radii that mainly was deposited at greater distances from the center of the galaxy. There is ample evidence from the disturbed appearances of elliptical galaxies that merging of galaxies and settling of the gaseous components is happening, at a low rate, at the present epoch (Schweizer 1990).

Opinion still is divided on whether elliptical galaxies could have formed by the merging of spirals. Toomre (1977) gives the classical case for this picture. Perhaps the clearest argument against it is van den Bergh's (1990b) point that it would be difficult to understand the high abundance of metal-poor globular star clusters around some giant elliptical galaxies if these galaxies had formed by merging of metal-rich spiral galaxies that tend to contain relatively fewer globular clusters. The striking similarity of the spheroids of spirals and ellipticals suggests both formed by the same process, which certainly could have included the merging of immature gas-rich spiral objects. But the following discussion assumes most of this happened before spirals became adorned with their present mature disks.

If the gaseous protospheroid material dissipatively settled by an appreciable factor before fragmenting into stars that can remember where they were formed, it would decrease the implied initial density of the protospheroid and so lower the redshift at which the material was assembled. There are a few arguments that constrain the dissipation picture (along with all the other models for galaxy formation). The first is the low dimensionality of the families of galaxies, as is illustrated by the remarkably small scatter from the fundamental plane for ellipticals in figure 3.14, and the striking success of the Tully-Fisher relation for spirals. This leaves little room for the accidental variations in properties of the final product one might have expected would have accompanied a large contraction factor. Second, the angular momentum parameter λ (eq. [22.44]) is comparable to what is expected from tidal torques. Even with angular momentum transfer to the outer regions, it would seem surprising that giant elliptical galaxies rotate as slowly as they do if dissipation had caused them to contract by a large factor. Third, the galaxy relative velocity dispersion as a function of separation (figure 20.1) is continuous with the run of velocity dispersion within a typical large spiral galaxy. If the velocity dispersion in the spheroid had been substantially increased by dissipative settling relative to that of systems of galaxies, it would add another element to the conspiracy illustrated in figure 3.12 that is responsible for making the rotation curves within galaxies so remarkably flat. The straightforward reading of the flat run of velocity dispersion, from spheroids to dark halos to the relative motions of galaxies, is that all experienced about the same amount of settling. A fourth even more enigmatic clue is the large core mass densities ρ_c in giant elliptical galaxies (eq. [18.39]). The ratio of the core density to the present mean cosmological density belonging to baryon density parameter Ω_B is

$$\frac{\rho_c}{\rho_o(B)} = \frac{10^9}{\Omega_B} .$$ (25.5)

If the core material were assembled at low redshifts it would require a considerable contraction that produced a surprisingly uniform set of cores. An alternative is that the cores were assembled as soon as it was allowed, at $z_{\text{dec}} \sim 1000$. That would bring the ratio at formation to

$$\frac{\rho_c}{\rho_{\text{dec}}(B)} \sim \frac{1}{\Omega_B} ,$$ (25.6)

meaning the cores could form with a modest collapse factor. It could be argued that this would not be unexpected in the baryon isocurvature model outlined below. Perhaps more reasonable is an intermediate picture, where spheroids were assembled at about the redshift in the table.

There are several lines of argument that suggest the disks of spiral galaxies formed toward the end of the merging or accretion phase that created the spheroids out of material that dissipatively settled by a large contraction factor to rotational support (Ostriker and Thuan 1975; Gunn 1982; White and Frenk 1991).

First, the star distribution in a disk is remarkably thin, an arrangement that would be easily disrupted by the addition of massive lumps of material. Thus it seems likely that the thin disks formed by the settling of gas after the large fluctuations in the gravitational field that are likely to have accompanied the main assembly of the mass (Quinn and Goodman 1986; Gunn 1987; Carlberg and Hartwick 1989; Ostriker 1990). Second, if the angular momentum in the disk came from the tidal torques (section 22), the disk material had to have contracted by a considerable factor, from an initial radius comparable to that of the dark halo (eq. [22.47]), and likely within the $\rho(r) \propto 1/r^2$ density run in a previously deposited dark halo and spheroid. The large initial radius translates to a long collapse time and a low redshift of formation. Third, the evidence from white dwarf cooling times (eq. [18.20]) and stellar evolution ages is that the disk stars in the Milky Way are younger than the spheroid. A commonly discussed age for the local part of the disk is about 9 ± 2 Gy, while standard estimates of the ages of the oldest globular star clusters in the spheroid are about 15 Gy (Winget et al. 1987; Freeman 1989; Demarque, Deliyannis, and Sarajedini 1991). These numbers would put the red-shift of formation of the disk at $z \sim 1$

We have an important guide to the picture for the assembly of galaxies and their disks from the counts of galaxies as a function of absolute magnitude and redshift at $z \lesssim 1$. The interpretation depends on the still highly controversial sub-ject of the parameters in the cosmological model, however, so we reserve a dis-cussion of the main elements for the next section. It is sufficient to note here that the absolute magnitudes of the galaxies at $L \gtrsim L_*$ have not changed by more than about 50% back to $z \sim 1$. The effect is illustrated in figure 5.7. It is discussed by Broadhurst, Ellis, and Shanks (1988), Colless et al. (1990), Lilly, Cowie, and Gardner (1991), Cowie, Songaila, and Hu (1991). The straightforward in-terpretation is that this is because bright galaxies were not greatly different at $z \sim 1$. The problem is that the counts of galaxies as a function of redshift are a factor of about three larger than predicted in an Einstein-de Sitter model in which galaxies are not evolving. The resolution may be that the galaxies are in pieces that coincidentally are bright enough to masquerade as ordinary whole galaxies, consistent with the observation that many of the excess galaxies are blue and have relatively low luminosities. Or perhaps we do not live in an Einstein-de Sitter universe.

The redshift assigned to the assembly of the massive extended dark halos of galaxies again is based on a hierarchical picture, in which these components are added at a lower density and lower redshift, after the assembly of the spheroid. As we have noted, the redshift of formation of the spheroid is reduced if it is assembled along with the dark halo at considerably lower density and redshift than what is assigned in the table. And we have noted the problems this picture may encounter.

The evidence for the existence of massive halos of galaxies is unambiguous if conventional gravity physics is a useful approximation (fig. 13.12). The best test of the physics for this purpose is the luminous gravitational arcs (eq. [20.50]), which appear to be an impressive success for general relativity theory. While the debate on the issue of the reality of the dark mass certainly is well worth following, the best place to concentrate our effort is in the search for the nature of this stuff, under the assumption it really is there. The options are that it is baryonic: low mass stars or star remnants; or nonbaryonic: massive neutrinos or something even more exotic. If baryonic, when did the matter become "dark"? The amount of diffuse baryonic matter observed at $z \sim 3$ in damped $L\alpha$ systems, and inferred to be present as the ionized fraction in the $L\alpha$ forest, is comparable to what is seen now in stars (eqs. [23.51] and [23.74]), and it does not seem to be as large as the dynamical estimates of what is present now as dark matter, $\Omega_d \gtrsim 0.1$ (table 20.1). As indicated in table 25.1, this has been taken to be a hint that the matter became "dark" during or possibly a long time before the assembly of the massive halos. The related issue of how much dark mass there might be is discussed below, in connection with the biased cold dark matter model, and in section 26.

The first line in table 25.1 refers to the effect of the inhomogeneous mass distribution on irregularities in the curvature of spacetime. The issue here is whether the seeds of structure formation are present in the very early universe as fixed wrinkles in spacetime, as in adiabatic scenarios; as inhomogeneities in the distributions of different mass components or stresses, as in isocurvature scenarios; or in seeds for explosions. In adiabatic scenarios the initial local value of the entropy per conserved quantum number is a universal value, meaning the local number densities of photons and baryons fluctuate together, along with all the other significant contributions to the mass density. The growing mass fluctuations perturb spacetime (eq. [10.127]), and if dissipation is unimportant these wrinkles end up as the dimples in spacetime at the potential wells of the bound systems. The primeval wrinkles would have to have originated outside the standard cosmological model, in the physics of the very early universe, perhaps in an inflation epoch. In primeval isocurvature models the mass distribution in the early universe is homogeneous, meaning the curvature of spacetime is not perturbed, but there is an initially irregular distribution in a component such as the baryons, or in the stresses left over from a cosmic phase transition that produced one of the fields discussed in section 16.[18] The formation of this component out of the locally available energy would not disturb an assumed initially homogeneous mass distribution, and the expansion of the universe tends to preserve homogeneity, so when nongravitational forces can be neglected the mass remains homogeneous, the

---

[18] One can assume a mixture of primeval adiabatic and isocurvature fluctuations, but it would require a very special arrangement to make both play a significant role in the formation of the structure observed at the present epoch, because the amplitude of an adiabatic perturbation grows by a large factor from the very early universe to the present, while the amplitude of an isocurvature fluctuation on an interesting length scale is frozen until relatively recent times.

local fluctuations in the distributions of the components canceling each other. In these scenarios the potential fluctuations we see in the large-scale structure originate at relatively low redshifts, from nongravitational stresses.

An example of the isocurvature picture is the baryonic dark matter scenario, in which one assumes that the baryon distribution at high redshifts is a fluctuating function of position — the energy to make the baryons being taken from the local radiation — so the net energy density is homogeneous. The Jeans length for matter and radiation acting as a single fluid (eq. [25.81]) sets the scale within which radiation pressure can rearrange the mass distribution. As the universe becomes dominated by matter, the matter and radiation tend to move so as to put the radiation in a clumpy distribution with the now dominant baryons nearly smoothly distributed. This means that on scales larger than the matter-radiation Jeans length, which subtends an angle of a few degrees, the mass distribution remains nearly homogeneous. Thus there is no large-scale Sachs-Wolfe effect on the CBR, and one instead sees the assumed primeval irregularities in the space distribution of the entropy per baryon. On scales less than the Jeans length, radiation pressure forces the radiation to end up nearly homogeneous, and the radiation drags the matter with it, leaving the matter in the initial clumpy distribution on small scales. This would be the origin of the present curvature fluctuations.

Yet another possibility is that the curvature fluctuations originated in explosions well after decoupling. Nongravitational forces certainly play an essential role in determining the arrangement of matter within galaxies in stars and thin disks, and it is reasonable to consider the idea that they operated on larger scales in explosions or pressure waves from the first generations of stars that rearranged the large-scale mass distribution. The seeds would have to have come out of the new physics of still earlier epochs, but the important message from such models is that there need not be a close relation between the seeds and what grew out of them. The point was made by Doroshkevich, Zel'dovich, and Novikov (1967), Ostriker and Cowie (1981), Ikeuchi (1982), and Hogan and Kaiser (1983). We will argue that there are observational problems with straightforward applications of this idea for galaxy formation, but this is true of all the others that have been explored in any detail.

Ingredients for a Model

To build a theory for galaxy formation one must decide on the parameters for the Friedmann-Lemaître model: Are space sections to be curved or cosmologically flat? Is the mass density to be high or low? A cosmologically flat model agrees with inflation, the best picture we have for the very early universe, though there is little empirical evidence that the picture is right. The straightforward reading of the evidence is that the density parameter is less than unity, but one might not accept that because it would mean we have appeared at a special time, as the universe is making the transition away from being dominated by its mass density. The Einstein-de Sitter seems more reasonable, because it eliminates this special epoch.

The cosmological parameters fix physical conditions, such as the maximum age of a galaxy at redshift $z = 1$, and whether a loose concentration of mass at $z = 1$ is self-gravitating and growing into a compact system. The parameters also influence the answer to the next question: What are the dominant contributions to the mass of the universe? We know there are baryons in abundance. Are there also nonbaryonic particles? Is there something that acts like a cosmological constant? If the mass density is low it is tempting to try working with baryons alone, for what is allowed by nucleosynthesis is not far from what is demanded by dynamics. It is difficult to see how the Einstein-de Sitter mass density could be hidden in baryons. In this high-density case the offer from particle physics — of neutrinos or something more exotic — is hard to resist.

What are the stresses that caused material to gather into the observed structures? Most models assume gravity is the dominant force. We see it happen in the present formation of the Local Group and the Local Supercluster, and there is no known acceptable alternative for the formation of the other large-scale structures. Nongravitational processes are essential to the rearrangement of mass within galaxies, in the the formation of stars, thin disks, and galactic winds. The dividing line at which nongravitational stresses are a secondary effect usually is put at the scale of an individual galaxy. This agrees with the continuity of the velocity dispersion within an L_* galaxy and the galaxy relative velocity dispersion in the clustering pattern on larger scales that very likely was assembled by gravity (fig. 20.1).

What are the seeds for the gravitational growth of structure? The exponential instability of the motion of oil in a pipeline means we can be sure that turbulence develops whatever the initial conditions; there is no need to specify the seed. The gravitational growth of mass fluctuations in an expanding universe is a power law instability, meaning there are specific constraints on the initial conditions. The characteristic density contrast δ_h appearing at the Hubble length must not be too large, to avoid overproduction of black holes, nor too small, so something interesting can happen before the mean mass density becomes unreasonably low. Some models minimize this special arrangement by taking δ_h to be scale-invariant, a constant value that might be computable from fundamental physics. In other models the fit to the observations requires that the value of δ_h is a function of redshift. Under this arrangement the fundamental theory has to produce several numbers, including a normalizing amplitude at a chosen mass and a characteristic number where the scaling of the perturbation with mass is broken.

Are the primeval departures from homogeneity adiabatic, or isocurvature, or a mixture? This might be determined by the picture for their origin. If the departures from homogeneity are produced during inflation, it is reasonable but not required to assume that the fluctuations are adiabatic, Gaussian, and scale-invariant. In this case all that remains for the fundamental theory is to tell us why the constant amplitude δ_h is what fits the observations. Cosmic fields produce fluctuations that have a specific non-Gaussian character, determined by the number of field components. This could be an advantage, for it means the fluctuations in the tail of the distribution can be causing interesting things to happen on scales where the rms fluctuations still are small. It can also imprint distinctive features in the anisotropy of the CBR.

At present there is no natural-looking picture for the origin of an acceptable pattern of baryon isocurvature fluctuations; it must be assumed ad hoc. The possible advantage of this case is that it avoids the constraint on adiabatic fluctuations, where the primeval density contrast must be small enough everywhere to avoid overproducing black holes. By construction this does not apply to primeval isocurvature fluctuations, so one can arrange the spectrum to assemble galaxies much earlier than in an adiabatic scenario. Of course, this is a real advantage only if one accepts the early galaxy assembly entered in table 25.1.

Now let us turn to some specific models. The simplest and most elegant combination of these ingredients arguably is the adiabatic scale-invariant Gaussian hot dark matter (HDM) model in an Einstein-de Sitter universe. The dark matter would be a family of neutrinos with a rest mass of a few tens of electron volts. The primeval density fluctuations could be produced by a reasonable-looking model for inflation. The scale invariance of the cosmological model and of the fluctuation spectrum mean the fundamental theory only has to get the fluctuation amplitude right. We will argue that this picture convincingly fails, however, because it makes galaxies out of protoclusters, which almost certainly is not what happened.

The next choice might be the cold dark matter (CDM) model, which replaces the neutrinos with nonbaryonic matter with negligible pressure (prior to orbit crossings). There is more convincing evidence for the existence of neutrinos, but particle theory offers reasonable candidates for cold dark matter. We will argue that the major challenge for this model is to find a convincing explanation for how the large mass density in this model is so cunningly hidden from the small-scale dynamical tests.

The transient mass concentrations in cosmic strings or global monopoles or textures can produce a scale-invariant spectrum of density fluctuations. The reason is different from CDM, but the advantage is the same: the fundamental physics has to produce only one number. The specific non-Gaussian character of the perturbations can be an advantage, for the CDM model may be challenged to produce clusters of galaxies at redshifts where they are seen to exist. The major problem again is the assumption of an Einstein-de Sitter mass density. It is characteristic of these models that relatively small field structures are responsible for the mass perturbations that develop into galaxies, while larger field structures produce clusters of galaxies. Since the field structures know about the Hubble length, larger field structures tend to operate later. Since the field structures that produce clusters are not likely to know where the earlier generations put the galaxies, it would be reasonable to expect that the large cosmic field structures on average gather fair samples of galaxies and mass into clusters. But if that were so, we know from the well-determined masses of clusters that the mean density would have to be well below the Einstein-de Sitter value.

Perhaps the mass problem in one of these models will be shown to be only apparent, or maybe there is another more convincing approach within the Einstein-de Sitter model, or possibly the mass density really is low. This last option lowers the gravitational growth of density fluctuations, and the challenge is to find a way structure can form without overperturbing the CBR. The opportunity is that the larger angular size distance at high redshift in a low density model means there is a greater separation between the scales of the density fluctuations that produce superclusters and the scales that can affect the CBR. An easy way to meet the challenge assumes isocurvature fluctuations, because the amplitude can be taken to be quite large on small scales, so galaxies form early, while the amplitude is small on the large scales that affect the measured anisotropy of the CBR. This is ad hoc, but it does allow us to follow the naive interpretation of what is observed.

Now we will evaluate at some length these and other models for structure formation. The most useful lessons may come from the problems, for they are surely telling us how our ideas might be improved. We begin with the cold dark matter model, because it has been examined in the greatest detail.

The Cold Dark Matter Model

This model assumes an Einstein-de Sitter universe, in which structure grows out of Gaussian adiabatic (curvature) fluctuations with the scale-invariant spectrum $P_k \propto k$. That leaves one parameter, the normalization of the spectrum, which can be fixed by the measured large-scale anisotropy of the CBR. Since the density parameter in baryons is thought to be well below unity (from the fit in eq. [6.27] for the observationally successful model for the origin of the light elements, and the measures of the diffuse baryon density at $z \sim 3$ in eqs. [23.51] and [23.74]), the model assumes the mass is dominated by nonbaryonic matter, and it adopts the simple case where the matter initially is cold. As will be discussed, a major lesson from this model is that the matter in the very nearby systems of galaxies does not seem to be behaving this way. But no matter how the issue is resolved, the model will be remembered for its central role in the development of ideas on how to find and test models for the origin of structure.

To begin, let us note that the cold dark matter (CDM) model is motivated by the inflation scenario, but not uniquely predicted by it. If inflation is to account for the large-scale homogeneity of the universe on the scale of the present Hubble length, the mean space curvature has to make a negligibly small contribution to the present cosmological expansion rate (eq. [17.23]). That leaves either a term in the expansion rate equation that acts like a cosmological constant, with density parameter $\Omega \sim 0.1$ in movable matter, which is the straightforward interpretation of most of the observations, or matter with density parameter $\Omega = 1$ that can move but has a coherence length broad enough to have escaped detection in the dynamical tests described in section 20. The latter is the assumption of the CDM model, and it certainly has the better pedigree from particle physics. Simple versions of the inflation model discussed in section 17 do produce the Gaussian, adiabatic, scale-invariant spectrum of density fluctuations assumed in the CDM model, but other inflation models can yield quite different forms for the spectrum, or isocurvature fluctuations, where primeval irregularities in the dark matter distribution are locally compensated by opposing fluctuations in the baryons and radiation (Efstathiou and Bond 1986; Kofman and Linde 1987; Bond, Salopek, and Bardeen 1988). Finally, it is worth emphasizing that the CDM model is not the same thing as the idea that the dark matter revealed by Newtonian dynamics is nonbaryonic, perhaps with initially negligible pressure. The CDM model adds to this cold dark matter a very definite set of assumptions about how much dark matter there is and how it is distributed in the early universe.

The CDM model was introduced (Peebles 1982a) as a combination of particularly simple cases: the scale-invariant Einstein-de Sitter model requires no coincidences of cosmic timescales with the world time at which we have come on the scene (section 15); the density fluctuation spectrum $\mathcal{P} \propto k$ can be extrapolated to large and small scales without catastrophic perturbations to spacetime; and cold dark matter has the minimum number of free parameters. Blumenthal et al. (1984) noted that this picture offers interesting possibilities for the interpretation of observations on the scale of galaxies, some aspects of which are considered below. Davis, Efstathiou, Frenk, and White (1985) pioneered the detailed numerical study of the evolution of the cold dark matter distribution and motion. Their first computations assumed galaxies trace the distribution of dark matter. When the mass clustering amplitude in the model is adjusted to agree with the fluctuations in galaxy counts, it makes the galaxy velocities much too large. This is expected, for we saw in section 20 that the dynamical measures consistently show that the mass fraction that clusters with galaxies has density parameter $\Omega \sim 0.1$. The remedy is the biasing concept, that galaxies might naturally be expected to be more strongly clustered than the fluctuating mass distribution from which they emerge. This idea evolved out of a workshop in 1984 at the Institute for Theoretical Physics in Santa Barbara. It showed that the CDM model is worth studying in detail, and it greatly influenced the way we have come to think about the puzzle of galaxy formation.

The biasing concept was inspired by Kaiser's (1984a) discussion of the observation that clusters of galaxies cluster a good deal more strongly than do galaxies, in the sense that the dimensionless cluster two-point correlation function $\xi_{cc}(r)$ is much larger than the galaxy two-point function. Kaiser noted the possible connection to the following property of a Gaussian random process.

Let $\delta(\mathbf{x}) = \delta\rho/\rho$ be the mass density contrast evaluated at an epoch before galaxies have formed, when the fluctuations are still small. We are assuming $\delta(\mathbf{x})$ is a Gaussian random process, fully characterized by its autocorrelation function

$\xi(r)$ (eq. [21.40]). Let δ_1 and δ_2 be the values of the contrast at positions 1 and 2 at separation r. The second moments are

$$\langle \delta_1{}^2 \rangle = \langle \delta_2{}^2 \rangle = \xi(0), \qquad \langle \delta_1 \delta_2 \rangle = \xi(r).$$ (25.7)

The easy way to find the joint distribution in δ_1 and δ_2 is to introduce the combinations

$$\delta_+ = \delta_1 + \delta_2, \qquad \delta_- = \delta_1 - \delta_2.$$ (25.8)

These are Gaussian variables, because they are linear combinations of Gaussian variables. Their product is

$$\langle \delta_+ \delta_- \rangle = \langle \delta_1{}^2 - \delta_2{}^2 \rangle = 0,$$ (25.9)

so they are statistically independent. The variances are

$$\langle \delta_\pm{}^2 \rangle = 2[\xi(0) \pm \xi(r)].$$ (25.10)

Since δ_+ and δ_- are independent, their Gaussian distributions multiply:

$$
\begin{aligned}
G(\delta_+, \delta_-) \propto F(\delta_1, \delta_2) &\propto \exp -\frac{1}{2} \left[\frac{(\delta_1 + \delta_2)^2}{2(\xi(0) + \xi(r))} + \frac{(\delta_1 - \delta_2)^2}{2(\xi(0) - \xi(r))} \right] \\
&= \exp -\frac{1}{2} \left[\frac{\delta_1{}^2 \xi(0) - 2\delta_1 \delta_2 \xi(r) + \delta_2{}^2 \xi(0)}{\xi(0)^2 - \xi(r)^2} \right].
\end{aligned}
$$ (25.11)

This gives the joint distribution $F(\delta_1, \delta_2)$ in the density contrast at points separated by distance r (because the Jacobian for the change of variables from δ_+, δ_- to δ_1, δ_2 is constant). On rearranging, we see that the distribution in δ_2 for given δ_1 is

$$F(\delta_2) \propto \exp -\frac{1}{2} \frac{\xi(0)}{\xi(0)^2 - \xi(r)^2} \left[\delta_2 - \delta_1 \frac{\xi(r)}{\xi(0)} \right]^2.$$ (25.12)

This says that if the density contrast δ_1 at a chosen point is nonzero, the density contrast at distance r away is biased away from zero by the amount

$$\langle \delta_2 \rangle = \delta_1 \xi(r)/\xi(0).$$ (25.13)

Kaiser (1984a) noted that this relation might explain the strong clustering of clusters of galaxies (eq. [19.39]). If the primeval spectrum of mass fluctuations had a positive autocorrelation function $\xi(r)$ extending to large scales, then clusters of galaxies, which might be expected to form where the primeval density contrast is high, on average would be surrounded by regions where the contrast is biased high and so are likely to produce a higher than average density of other clusters. The result would be that clusters have a larger position autocorrelation function than do galaxies, as observed.

Kaiser (1986) and Bardeen (1986) saw that the same argument shows that galaxies when formed are more strongly clustered than the underlying mass distribution, which is what we need if the mean mass density is to be large. This segregation of galaxies from mass is not a permanent condition, for gravity acts alike on dark matter and galaxies to produce a growing hierarchy of clustering that tends to draw together galaxies and dark mass. Thus, if biasing is to explain the low apparent values of Ω derived from the small-scale dynamical tests in section 20, there cannot have been much rearrangement of the relative positions of galaxies since formation. In the CDM mass fluctuation spectrum to be discussed next, this condition may be satisfied because the spectrum decreases at short wavelengths.

The point was tested by Davis et al. (1985) in their numerical computations of the evolution of the dark-matter distribution. They assigned initial model galaxy positions to the places where there is an unusually large upward fluctuation in the initial mass distribution smoothed over the scale of a galaxy mass. Because the initial CDM mass distribution has a positive autocorrelation function $\xi(r)$ on this scale, the initial positions of the model galaxies are clustered. When this initial distribution of galaxies and mass is evolved forward numerically to the epoch when the rms fluctuations in the model galaxy positions about agree with the observations, the mass fluctuations remain lower than that of the galaxies, so the model galaxy velocities are smaller than what would follow if galaxies traced mass, and closer to what is observed. This encouraging result naturally led to considerable further explorations of the model. Impressive examples of the state of the art are given in Frenk et al. (1989) and Park (1990). White and Frenk (1991) give a useful history of the ideas in the development of the CDM model.

Let us consider now some of the main features of the CDM model. We will focus on the conventional and simplest case in which the density parameter is unity, the mass is dominated by the dark matter, and the spectrum of Gaussian mass fluctuations at very high redshifts has the scale-invariant form $P \propto k$ obtained in equation (17.85).

To understand how the shape of the mass density fluctuation spectrum evolves through the epoch z_{eq} of equality of the mass densities in nonrelativistic and relativistic matter (eq. [6.81]), consider a Fourier component $\delta_{\mathbf{k}}$ of the mass distribution, with wavenumber k and proper wavelength $2\pi a(t)/k$. At the epoch t_k when this wavelength is comparable to the Hubble length, the wavenumber satisfies $a(t_k)/k \sim t_k$. If this happens at time $t_k < t_{eq}$, when the expansion parameter is scaling with time as $a \propto t^{1/2}$, then we see that the time at which the wavelength is equal to the Hubble length scales with the wavenumber as

$$t_k \propto k^{-2}. \tag{25.14}$$

The amplitude of the Fourier component grows with time as $\delta_{\mathbf{k}} \propto t$ (eqs. [5.90] and [10.123]) until $t \sim t_k$, and then the radiation pressure causes the fluid of baryons and radiation to oscillate as a sound wave. At $t < t_{eq}$ the dark matter acts as a trace component, so this stops the gravitational growth of fluctuations in the dark matter distribution until redshift $z \sim z_{eq}$, when the dark matter becomes dominant and self-gravitating. Thereafter, the amplitude $\delta_{\mathbf{k}}$ for the dark matter grows as $\delta_{\mathbf{k}} \propto t^{2/3}$. Thus the net growth factor for this Fourier component of the dark matter distribution is, in linear perturbation theory,

$$\delta_{\mathbf{k}} \propto t_k \propto k^{-2}. \tag{25.15}$$

This applies at $t_k \lesssim t_{eq}$, for the Fourier components that become smaller than the Hubble length at $z > z_{eq}$. At much longer wavelengths the final amplitude is independent of the wavenumber, because the gravitational growth of density fluctuations never was impeded by the pressure gradient force. At the wave number k_x separating these limiting cases, $t_k \sim t_{eq}$, or

$$\frac{1}{H_o \Omega^{1/2} z_{eq}^{3/2}} \sim t_{eq} \sim \frac{a_{eq}}{k_x} = \frac{a_o}{z_{eq} k_x}. \tag{25.16}$$

The first expression is the Einstein-de Sitter limit in equation (5.61). Since $z_{eq} \propto \Omega h^2$ (eq. [6.81]), the comoving wave number at the break in the spectrum scales with the cosmological parameters as

$$k_x \propto \Omega h^2. \tag{25.17}$$

The conclusion is that the primeval scale-invariant spectrum $\mathcal{P}_i = |\delta_{\mathbf{k}}|^2 \propto k$ evolves into the mass fluctuation spectrum

$$\mathcal{P} \sim \begin{cases} A(k_x/k)^3 & \text{for } k \gg k_x \propto \Omega h^2, \\ Ak/k_x & \text{for } k \ll k_x. \end{cases} \tag{25.18}$$

This applies at $z \ll z_{eq}$ and before the fluctuations in the mass distribution become nonlinear.

A more detailed calculation treats the coupled baryons and radiation as a fluid with mean mass density ρ_r, pressure $\rho_r/3$ if we ignore the baryon mass density, and density contrast $\delta_r(t)$ in the Fourier component with comoving wavenumber k. The dark matter is represented as a pressureless fluid with mean density $\rho_x(t)$ and density contrast $\delta_x(t)$ in the mode. Then equation (10.116) for the perturbation to the line element is

$$\frac{d^2 h}{dt^2} + 2 \frac{\dot{a}}{a} \frac{dh}{dt} = 8\pi G (2\rho_r \delta_r + \rho_x \delta_x), \tag{25.19}$$

the energy conservation equation (10.110) for the dark matter is

$$\frac{d\delta_x}{dt} = \frac{1}{2}\frac{dh}{dt} , \tag{25.20}$$

and for the matter-radiation fluid the energy-momentum conservation equation (10.107) works out to

$$\frac{d\delta_r}{dt} = \frac{4}{3}\left(\frac{kv}{a} + \frac{1}{2}\frac{dh}{dt}\right) , \qquad \frac{dv}{dt} = -\frac{k\delta_r}{4a} , \tag{25.21}$$

where the real parts of $\delta_r e^{i\mathbf{k}\cdot\mathbf{x}}$ and $iv e^{i\mathbf{k}\cdot\mathbf{x}}$ are the matter-radiation density contrast and the fluid velocity relative to the dark matter. The results of a numerical integration of these equations are adequately represented by the expression[19]

$$\mathcal{P}(k) = \frac{Ak}{(1 + \alpha k + \beta k^2)^2} , \tag{25.22}$$

where the constants are

$$\alpha = 8/(\Omega h^2)\,\mathrm{Mpc} , \qquad \beta = 4.7/(\Omega h^2)^2\,\mathrm{Mpc}^2 . \tag{25.23}$$

This is the form of the power spectrum at $z \ll z_{eq}$ and before the development of nonlinear structures. The numerator, Ak, is the primeval form of the spectrum; the denominator is the transfer function. An even more careful computation takes account of the mass density in baryons, and the details of the diffusion of the radiation through the baryons when the plasma recombines at $z \sim z_{dec} = 1400$, but these corrections do not affect the points to be discussed here.

The mass autocorrelation function $\xi(r)$ is the Fourier transform of the power spectrum (eq. [21.40]). Since the CDM spectrum approaches zero at long wavelengths ($k \to 0$), the autocorrelation function has to have a zero (eq. [21.56]). With the transfer function in equation (25.22) the mass autocorrelation function is negative at separation $r > r_\xi$, with

$$r_\xi = 16/(\Omega h^2)\,\mathrm{Mpc} . \tag{25.24}$$

In the linear approximation to the evolution of the mass distribution, r_ξ is the maximum scale on which things are correlated. This correlation length scales as h^{-2}, while the sizes of structures observed in redshift maps scale as h^{-1}. Thus, if

---

[19] This form is from Peebles (1982a), corrected for the mass density in the neutrinos that were neglected in the original computation, and to the new measurements of the CBR temperature. Still more detailed fitting functions are given by Blumenthal and Primack (1983), Davis et al. (1985), and Bardeen et al. (1986), but the simpler form used here gives quite similar numbers for the observational tests.

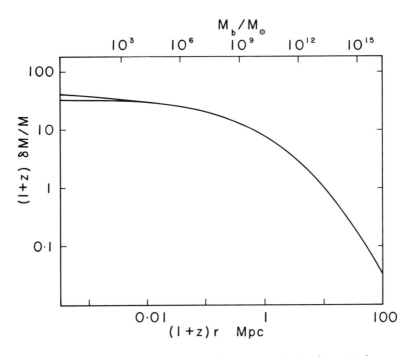

Figure 25.1. Mass fluctuations at $z \ll z_{eq}$ in linear perturbation theory in the CDM model. The rms fluctuation in the mass in a randomly placed sphere of radius r is $\delta M / M$. The normalization is to the observed large-scale fluctuations in the galaxy distribution. The mass scale is the mean value of the baryon mass in a sphere of radius r. The upper curve refers to the fluctuations in the dark matter, the lower to the baryons.

the Hubble parameter were as small as $h = 0.5$, r_χ would not be out of line with the observed scales of correlation in the galaxy space distribution discussed in section 19.

The rms fluctuation $\delta M / M$ in the mass found within a sphere of given radius is an integral over the power spectrum (eq. [21.53]); figure 25.1 shows an example of the results for the CDM spectrum (Peebles 1984b). The slope at the large-scale end is $\delta M / M \propto r^{-2}$, steeper than for the uncorrelated fluctuations of white (shot) noise because of the anticorrelation at $r > r_\xi$ (eq. [17.89]). The slope is shallow at the small-scale end, where the spectrum varies as $\mathcal{P} \sim k^{-3}$ and the variance per octave is proportional to the logarithm of the smoothing length.

At the mass scales characteristic of galaxies the slope of $\delta M / M$ as a function of the smoothing scale is shallower than for white noise, meaning the initial mass autocorrelation function is positive. There is some division of opinion on whether this is a good thing. The scaling argument in section 22 for the development of the mass clustering hierarchy indicates that, if the initial mass fluctuation spectrum is approximated as the power law $\mathcal{P} \propto k^n$, clumps of mass $\mathcal{M}$ form with velocity dispersion $\sigma \propto \mathcal{M}^{(1-n)/12}$, and the velocity dispersion scales with clump size as $\sigma \propto r^{(1-n)/(10+2n)}$ (eqs. [22.60] and [22.61]). If $n \sim -2$, the first of these relations is $\sigma \sim \mathcal{M}^{0.25}$. With equation (3.39) this is the Faber-Jackson/Tully-Fisher relation (eqs. [3.34] and [3.35]), and the index $n = -2$ is close to the logarithmic derivative of the spectrum in the CDM model on the scale of galaxies. This elegant result was one of the factors that led to the great interest in the model (Faber 1982; Gunn 1982; Blumenthal et al. 1984). However, with $n \sim -2$ the second relation is $\sigma \sim r^{0.5}$, which is much steeper than the estimates in figure 20.1. A better fit here is $n \sim 0$. The discrepancy is in part a result of the fact that the scaling relation is the mean trend for a distribution with an appreciable width that is being cut in different ways by the two observational measures. A fuller interpretation awaits further analysis of how the clustering hierarchy evolves through the deeply nonlinear sector to galaxies.

The CDM mass fluctuation spectrum might be normalized at large r, where linear perturbation theory applies through to the present epoch, to agree with the measured large-scale anisotropy in the CBR (eq. [25.31] below). Then we can estimate the present abundance of mass concentrations such as galaxies and clusters of galaxies, and the redshift at which the first structures are expected to form. As will be discussed, the last issue is central to one of the leading crises for the model.

To see how the normalization goes, consider the integral of the mass autocorrelation function (eq. [21.62]),

$$\int_0^r r^2 dr \xi = J_3(r) = 4\pi \int_0^\infty \mathcal{P}(k)(\sin kr - kr \cos kr) dk / k . \tag{25.25}$$

At large r, where the spectrum approaches $\mathcal{P} = Ak$, we need the integral

$$\int_0^\infty dy(\sin y - y \cos y) e^{-\lambda y} = 2 , \tag{25.26}$$

in the limit $\lambda \to 0$. The exponential is a model for the cutoff of the real spectrum; the fact that the result is independent of λ as long as it is small tells us that the form of the cutoff is not important. Collecting, we have

$$J_3(r) = \frac{8\pi A}{r} . \tag{25.27}$$

(Another way to arrive at this expression uses eq. [21.110] with $J_3(\infty) = 0$.) The estimates discussed in section 19 suggest we write

$$J_3(20h^{-1} \, \text{Mpc}) = \frac{750}{h^3 b^2} \, \text{Mpc}^3 . \tag{25.28}$$

The numerical value is from the galaxy two-point correlation function. If mass is more smoothly distributed than the galaxies on scales $\sim 20h^{-1}$ Mpc, this overestimates J_3 for the mass; the CDM parameter b takes this into account. Estimates for b currently under discussion are in the range $b = 1$ to 3. We have from these two equations the amplitude of the CDM spectrum at the present epoch in linear perturbation theory,

$$A = \frac{600}{h^4 b^2} \text{ Mpc}^4 . \tag{25.29}$$

In the approximation $\mathcal{P} = Ak$, the multipole moments of the CBR anisotropy are given by equation (21.111),

$$a_l^2 = \frac{2\pi^2 H_o^4 A}{l(l+1)} , \tag{25.30}$$

and with the above normalization the quadrupole moment is

$$a_2 = \frac{5 \times 10^{-6}}{b} . \tag{25.31}$$

The anticorrelated large-scale mass fluctuations in the CDM model make the quadrupole moment considerably smaller than for a flat spectrum (eq. [21.107]). This was one of the reasons for introducing the model (Peebles 1982a). And it is noteworthy that if b is not far from unity, the predicted large-scale anisotropy is quite similar to the present measurements. This is discussed in section 21, under the assumption that the large-scale fluctuations in the galaxies trace the mass (eqs. [21.108] and [21.115]). If $\Omega = 1$ the assumption $b = 1$, that is, that galaxies trace mass, cannot be quite right. For example, it would predict that the virgocentric/tidal flows around the Virgo cluster are considerably larger than observed. An easy remedy assumes that the primeval fluctuation spectrum varies with wavenumber slightly more slowly than $\mathcal{P} \sim k$.

The CDM model prediction for CBR angular fluctuations on smaller scales was explored by Vittorio and Silk (1984) and Bond and Efstathiou (1984), and is reviewed by Bond (1990); here we only take note of a few of the order-of-magnitude estimates that go before a detailed computation. The expansion time at decoupling, at redshift $z_{\text{dec}} \sim 1400$, is well approximated by the Einstein-de Sitter limit. This gives the comoving distance radiation travels in an expansion time,

$$l_{\text{dec}} = \frac{2c}{3H_o(1 + z_{\text{dec}})^{1/2}} \sim 50h^{-1} \text{ Mpc} . \tag{25.32}$$

Fluctuations in the distribution of the CBR on much smaller scales are strongly erased by scattering in the recombining plasma. A matter density fluctuation

$\delta \mathcal{M}/\mathcal{M}$ on the scale l_{dec} is accompanied by an adiabatic temperature fluctuation $\delta T/T \sim (\delta \mathcal{M}/\mathcal{M})/3$, which we observe with little smoothing. The angular size distance is $y/c = H_o a_o r/c = 2$ (eq. [13.39]), so the comoving length l_{dec} subtends the angle

$$\theta_l = H_o l_{dec}/y \sim 30 \text{ arc min}. \tag{25.33}$$

We get a reasonable approximation to the rms fluctuation in the mass within a sphere of radius l_{dec} at decoupling by using equation (21.57) with equations (25.27) and (25.29) for $J_3(l)$:

$$\frac{\delta \mathcal{M}}{\mathcal{M}} \sim \frac{(24\pi A)^{1/2}}{(l_{dec})^2(1 + z_{dec})} \sim \frac{5 \times 10^{-5}}{b}. \tag{25.34}$$

The temperature fluctuation is about one third of this, or, with $b \sim 2$, about one part in 10^5 on the scale of θ_l, which is in line with the observations. At smaller scales the mass fluctuation amplitude at decoupling is larger, but the temperature fluctuation is strongly suppressed by diffusion through the recombining plasma. Finding even the right order of magnitude for the residual $\delta T/T$ on small scales requires a more detailed computation than is appropriate to work through here; the result at the time this is written is that the predictions for $\delta T/T$ as a function of angular scale are about at the level of the measurements.

We will consider next cluster formation and the segregation of galaxies from mass in the CDM picture. Related to the latter effect is the character of the galaxy peculiar velocity field. It has been known for some time that the dispersion in the galaxy relative peculiar velocities in our neighborhood is less than about 100 km s^{-1}; the effect is illustrated in figure 5.5. It is reasonably well established that the CBR dipole anisotropy is telling us that the mean peculiar velocity in our neighborhood is 600 km s^{-1}. That is, the peculiar velocity field appears to have a coherence length no smaller than about $10h^{-1}$ Mpc. Ostriker and Suto (1990) note that this is flow with a large Mach number: the velocity dispersion is small compared to the streaming motion. They show that accounting for this effect in the CDM model is a serious challenge.

Formation of Clusters of Galaxies

When the nongravitational forces on the baryons can be modeled or neglected, the CDM model with a fixed normalization gives a definite picture for the origin and evolution of structures such as massive galaxy halos and the space distribution of the halos. The results of numerical studies of the model predictions, which include some impressive successes, are surveyed by Efstathiou (1990) and Frenk (1991). We will consider two topics that illustrate some interesting analytic methods and observational issues. This subsection deals with the formation of mass concentrations that might be compared to the great clusters of galaxies. In the next subsection we consider the formation of the first generation of bound systems, as a prelude to the discussion of biasing.

In the CDM model the density fluctuations at decoupling are Gaussian, with a given spectrum. We can write the normalization in the form of equation (25.29). Then, with a definite choice for the parameter b, the model predicts the number density of concentrations in which the mass within the Abell radius $r_a = 1.5h^{-1}$ Mpc exceeds the observed Abell cluster mass (eq. [20.39]),

$$\mathcal{M}(<r_a) = 3 \times 10^{14} h^{-1} \mathcal{M}_\odot . \tag{25.35}$$

This predicted number density can be computed by numerical N-body methods. We will use an analytic approach that has proved to give a useful approximation to the numerical results. The analytic method is based on some rough estimates; this is a case where sensible intuition has given us a tool that is more accurate than one might have anticipated. And the analytic method allows us to understand how the orders of magnitude for the number density of protoclusters come about.

The material now within the Abell radius of an Abell cluster would have been in a region with a quite irregular boundary at decoupling. The first step is to replace the true boundary with a sphere that contains the cluster mass. The comoving sphere radius normalized to the present epoch is

$$r_s = \left[\frac{3\mathcal{M}(<r_a)}{4\pi \rho_o} \right]^{1/3} = 6.4h^{-1} \text{ Mpc} . \tag{25.36}$$

Let the mass density contrast at decoupling, averaged through a spherical window with the radius r_s, be

$$\Delta(\mathbf{r}) = \delta\rho/\rho . \tag{25.37}$$

Below we discuss the minimum value Δ_c of the contrast at a peak of $\Delta(\mathbf{r})$ that causes the material within r_s to collapse to a compact system at the present epoch. Given that, the predicted number density of clusters is the number density of peaks of $\Delta(\mathbf{r})$ at contrast greater than the critical value Δ_c in the original Gaussian random process. This is a problem one can analyze; the approach in cosmology was pioneered by Doroshkevich (1970), and a detailed survey of methods is given by Bardeen et al. (1986). We will use a simpler approximation found by Press and Schechter (1974).

Let the standard deviation of the density contrast smoothed through the comoving window r_s at decoupling be

$$\sigma = \langle \Delta^2 \rangle^{1/2} . \tag{25.38}$$

Then, since the distribution is Gaussian, the probability that the contrast in a randomly placed window exceeds Δ_c is

$$P(>\Delta_c) = \frac{1}{(2\pi)^{1/2}\sigma} \int_{\Delta_c}^{\infty} e^{-\Delta^2/2\sigma^2} d\Delta . \tag{25.39}$$

The standard deviation is a function of the mass, $\sigma = \sigma(\mathcal{M})$, for $\mathcal{M}$ fixes the window radius. Equation (25.39) approximates the probability that the window has been placed on a protocluster with mass $\mathcal{M}$ or smaller. The difference between this expression and the probability computed at $\sigma(\mathcal{M} - \delta\mathcal{M})$ thus approximates the probability that the window has been placed on a protocluster with mass between $\mathcal{M}$ and $\mathcal{M} - \delta\mathcal{M}$. This probability is the product of the comoving number density, $\delta n = (dn/d\mathcal{M})\delta\mathcal{M}$, and the comoving window volume, $V_s = \mathcal{M}/\rho_o$. Press and Schechter recommend multiplying the result by a factor of two to take account of the mass from the underdense regions that falls onto the protoclusters. The result of carrying through these operations is the Press-Schechter approximation,

$$\mathcal{M}\frac{dn}{d\mathcal{M}} = \left(\frac{2}{\pi}\right)^{1/2} \frac{d\ln\sigma^{-1}}{d\ln\mathcal{M}} \frac{\rho_o}{\mathcal{M}} \nu_c e^{-\nu_c^2/2} . \tag{25.40}$$

We are computing in comoving coordinates, so ρ_o is the present mean mass density, and the left side is the number of clusters per unit logarithmic interval of mass and per unit volume extrapolated to the present epoch. The derivative of the standard deviation with respect to the mass contained by the window comes from differentiating equation (25.39) with respect to $\mathcal{M}$, after which the integral can be simplified by integration by parts. Finally,

$$\nu_c = \frac{\Delta_c}{\sigma} \tag{25.41}$$

is the minimum number of standard deviations at a fluctuation that has collapsed.

The critical contrast Δ_c in equation (25.41) is estimated in the spherical model in equation (20.18), where the physical radius $r(t)$ of the shell that contains mass $\mathcal{M}$ is

$$r(t) = A(1 - \cos\eta), \qquad t = B(\eta - \sin\eta), \tag{25.42}$$

with

$$A^3 = G\mathcal{M}B^2 . \tag{25.43}$$

This shell collapses to zero radius at $\eta = 2\pi$, when the time is

$$t_c = 2\pi B . \tag{25.44}$$

The mean physical density within the shell is $\bar{\rho} = 3\mathcal{M}/4\pi r^3$, the mean density in the background Einstein-de Sitter model is $\rho_b = (6\pi G t^2)^{-1}$, and the ratio gives the density contrast. At $\eta \ll 1$, the contrast is small, and a series expansion of equation (25.42) gives

$$\Delta = \frac{\bar{\rho}}{\rho_b} - 1 = \frac{3\eta^2}{20} = \frac{3}{20}\left(\frac{12\pi t}{t_c}\right)^{2/3}. \tag{25.45}$$

This says the density contrast extrapolated to $t = t_c$ in linear perturbation theory is $\Delta_c \to 1.69$. In linear perturbation theory in the Einstein-de Sitter model the contrast scales as $\Delta \propto (1 + z)^{-1}$. The analytic approximation assumes that if the density contrast at a peak of $\Delta(\mathbf{x})$ at decoupling is

$$\Delta_c = 1.69/(1 + z_{\text{dec}}), \tag{25.46}$$

then the peak has just collapsed to a dense concentration.

In these approximations, the present number density of compact concentrations in the mass range $\mathcal{M}$ to $\mathcal{M} + d\mathcal{M}$ that developed out of Gaussian fluctuations with the rms value $\sigma(\mathcal{M})$ is given by equations (25.41) and (25.46) for ν_c, in equation (25.40) for $dn/d\mathcal{M}$. The provenance of this result should not be examined too critically; one uses it because it is in reasonable agreement with N-body computations (Efstathiou et al. 1988).

Now let us estimate the predicted space number density of the mass concentrations observed to be present in rich clusters of galaxies. We can set the scale for σ from the observation that the rms fluctuation $\delta N/N$ in the galaxy count in a randomly placed sphere is unity when the sphere radius is (eq. [7.73])

$$r_n = 8h^{-1}\,\text{Mpc}. \tag{25.47}$$

Following the normalization in equation (25.29), we will let the present rms mass contrast be $\sigma_o = 1/b$ at this radius. As discussed in section 22, the value of σ at a fixed comoving radius scales back in time as $\sigma \propto 1/(1 + z)$ in an Einstein-de Sitter model. We need σ at the slightly smaller window radius in equation (25.36). We can write the scaling of σ with the window radius as

$$\sigma \propto r^{-(3+\beta)/2}, \qquad \beta = -0.85, \tag{25.48}$$

where β is the logarithmic derivative of the primeval fluctuation spectrum with respect to the wavenumber, and the value applies to the CDM spectrum (25.22) at the normalization scale. Collecting, we have that an upward density fluctuation at decoupling in the comoving window r_s that has collapsed at the present epoch is larger than the rms fluctuation by the ratio

$$\nu_c = 1.69 b (r_s/r_n)^{(3+\beta)/2}. \tag{25.49}$$

Finally, we need to choose the normalization, b. There is no consensus on the best value for this parameter, but we do have one reasonably well fixed result. We saw in section 20 that if the mass followed the concentration of IRAS galaxies around the Virgo cluster, and if Ω were unity, as we are assuming in the CDM model, it would make the peculiar velocities around the cluster much larger than seems to be indicated by the quiet local relative Hubble flow of the galaxies. Since this measures the mass fluctuations on length scales comparable to the normalization scale in equation (25.47), it is reasonable to take $b \sim 2$, which would reduce the gravitational accelerations around the Virgo cluster by a factor of two to an observationally more acceptable level.

With $b = 2$, equations (25.36), (25.47), (25.48), and (25.49) give $\nu_c = 2.7$, and the Press-Schechter approximation (25.40) is

$$\mathcal{M} \frac{dn}{d\mathcal{M}} = 2 \times 10^{-5} h^3 \, \text{Mpc}^{-3}, \qquad (25.50)$$

which is quite close to the observed value (eq. [20.33]).

This result is not sensitive to the shape of the power spectrum, because the spectrum is normalized at a radius r_n close to the radius r_s that contains the cluster mass. The important assumptions in the calculation are that the density parameter is close to unity, the matter does not resist gravitational collapse on the scale of a cluster of galaxies, and the spectrum of Gaussian mass fluctuations on this scale is close to flat at about half the rms value seen in the galaxy distribution.

With $\Omega = 1$, the rms mass fluctuation at a fixed comoving scale varies with redshift as $\sigma \propto (1 + z)^{-1}$, so the parameter ν_c varies as $(1 + z)$. In the approximation of equation (25.40), this determines the evolution of the comoving number density of clusters. With the normalization in equation (25.50), the model indicates that half the clusters now present would have collapsed at $z \lesssim 0.1$, and 90% would have collapsed at $z \lesssim 0.3$. Frenk et al. (1990) find similar rapid evolution in cluster-size mass concentrations in numerical N-body computations in the CDM model. They note that the rapid evolution means optically selected catalogs may be expected to contain many systems that have not yet collapsed, that appear to be compact in projection and to have a large velocity dispersion because of the large velocity difference across a relatively low-density collapsing system. As we have noted in section 20, the correlation of galaxy velocity dispersion with the X-ray temperature of the intracluster plasma suggests that most Abell clusters are physically compact entities, and the gravitational lensing effect shows there are rich compact clusters at $z \sim 0.5$. Thus it should be possible to find unambiguous tests of the predicted very rapid evolution of physical clusters and of the related projection effect.

If the two predictions are unambiguously ruled out there are several options. We have noted in section 22 the reasons to doubt that the strong collapse in the spherical model happens in the formation of the galaxy clustering pattern (fig. 22.2). Perhaps the growth of σ is reflected in the development of less dense superclusters rather than in the very recent formation of new generations of compact clusters. This hierarchical development might be aided by the non-Gaussian primeval density fluctuations produced by one of the cosmic fields considered in section 16.

If the density parameter is reduced to $\Omega \sim 0.1$, the radius of the window that contains the cluster mass (eq. [25.36]) is increased, and it is reasonable to lower b to about unity because the relative motions within the Local Supercluster are about what would be expected if galaxies traced mass and $\Omega \sim 0.1$. This would leave ν_c in equation (25.49) nearly unchanged. The time evolution of σ at low redshift is slower in a low-density cosmological model, which reduces the predicted rate of formation of clusters at low redshift, which seems to be the more promising reading of the observations. However, the lower mass density in equation (25.40) lowers the predicted number density to $dn/d\mathcal{M} \sim 10^{-6}h^3$ Mpc^{-3}, a factor of about ten below the observations. If numerical solutions confirm that the low-density model does predict too few rich clusters, an option to consider will be non-Gaussian initial conditions.

The First Generation in the CDM Model

The next subsection deals with the postulated segregation of mass from galaxies on scales less than about 10 Mpc. For this discussion it is useful to have the predicted epoch of formation of the first gravitationally bound clouds of baryons.

After the primeval plasma combines to almost completely neutral atoms, and before the density fluctuations have gone nonlinear, the Fourier transform of the density contrast $\delta_m(\mathbf{x}, t) = \delta\rho_m/\rho_m$ in the baryon distribution evolves according to the equation

$$\frac{d^2\delta_m}{dt^2} + \frac{4}{3t}\frac{d\delta_m}{dt} = \frac{2}{3}\frac{\delta_x}{t^2} - \frac{kT}{m_p}\frac{k^2}{a^2}\delta_m. \tag{25.51}$$

This follows from equation (5.124) by replacing the gravitational source term with the density contrast δ_x in the dark matter, which we are assuming is the dominant component. We are using the Einstein-de Sitter model, where $a \propto t^{2/3}$, and the density contrast in the dark matter is growing as

$$\delta_x \propto t^{2/3}. \tag{25.52}$$

The coupling of the CBR to the residual ionization of the baryonic matter keeps the baryon temperature T close to that of the CBR, $T \propto 1/a(t)$ (eq. [6.138]). Thus the baryon pressure satisfies $dp/d\rho = kT/m_p$ (and we are relying on context to distinguish Boltzmann's constant from the wavenumber). With $T \propto 1/a$, equation (25.51) is homogeneous in time, with the solution

$$\delta_m = \delta_x (1 + \gamma k^2)^{-1} , \qquad (25.53)$$

where

$$\gamma = \frac{3}{2} \frac{kT}{m_p} \frac{t^2}{a^2} = \frac{1.5 \times 10^{-6}}{a_o^2 \Omega h^2} \, \text{Mpc}^2 . \qquad (25.54)$$

The comoving length $\gamma^{1/2}$ is the baryon Jeans length; it defines the Jeans mass enclosed by the coherence length of the baryon mass distribution (eq. [6.146]).

When the rms mass density contrast is small, the power spectrum at fixed comoving wave number varies with time as $(1 + z)^{-2}$ (eq. [22.1]), so the spectra of the dark matter and baryons after decoupling and before structure forms are

$$|\delta_x(k)|^2 = \frac{\mathcal{P}(k)}{(1+z)^2} , \qquad |\delta_m(k)|^2 = \frac{|\delta_x|^2}{(1 + \gamma k^2)^2} . \qquad (25.55)$$

Because of the short wavelength cutoff in the denominator in the second equation, the baryon distribution has the coherence length $\sim \gamma^{1/2}$. This makes $\delta \mathcal{M}/\mathcal{M}$ independent of the window size on small scales, as shown in the lower curve in figure 25.1. The variance $\xi(0) = \langle \delta^2 \rangle$ of the baryon distribution is the integral of the spectrum. The numerical result, with equation (25.22) for $\mathcal{P}(k)$, equation (25.29) for the normalization, and equation (25.54) for the Jeans cutoff, and taking the Hubble parameter to be $h = 0.75$, is

$$\delta_m = \frac{\delta \rho_m}{\rho_m} = \frac{1}{b} \frac{25}{1+z} . \qquad (25.56)$$

This expression for the standard deviation of the mass distribution applies at redshifts large enough for the density fluctuations still to be in the linear regime, $\delta_m \lesssim 1$.

On scales larger than but comparable to the Jeans length, the fluctuation spectrum in the CDM model is $\mathcal{P} \sim k^{-3}$, so the variance of the mass distribution has equal contribution per octave of wavelength (eq. [21.53]), meaning there are mass fluctuations of comparable amplitude on all length scales. The density contrast evaluated along a line in a three-dimensional process with this k^{-3} spectrum has the one-dimensional spectrum $\mathcal{P} \propto k^{-1}$. (To see this, imagine computing the Fourier transform of $k^{-3/2}$, with random phases, along the line $y = z = 0$. The sum of $[k_x^2 + k_y^2 + k_y^2]^{-3/4}$ over k_y and k_z, with random phases, adds as a random walk, leaving an integrand proportional to $k_x^{-1/2}$.) This is the commonly encountered $1/f$ noise spectrum, which again has the property that something interesting

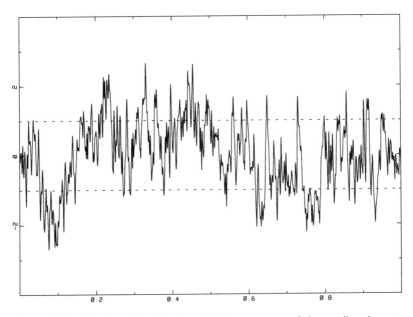

Figure 25.2. Gaussian $1/f$ noise. This is the character of the small-scale mass distribution in the CDM model before the formation of the first nonlinear structures.

is happening on all length scales along the line. Figure 25.2 shows a realization of $1/f$ Gaussian noise.

The most prominent feature of this CDM model for the initial small-scale distribution of the hydrogen is the oscillations at the coherence length defined by the short-wavelength cutoff of the $1/f$ spectrum. (The wiggles are prominent because the power spectrum of the derivative of the density has two extra powers of f, so most of the variance of the derivative is at the cutoff at the Jeans length.) The dashed lines are at one standard deviation. Since 2σ fluctuations are common, equation (25.56) says that at the epoch

$$z_i = \frac{50}{b}, \qquad t_i = 2 \times 10^7 \frac{b^{3/2}}{h\Omega^{1/2}} \, \text{y}, \qquad (25.57)$$

a substantial fraction of the hydrogen has been placed in gravitationally bound clouds with masses comparable to the Jeans mass, $\mathcal{M}_J \sim 10^5 \mathcal{M}_\odot$. These clouds contain enough molecular hydrogen to radiate their gravitational binding energy in a Hubble time. (The formation of molecular hydrogen is discussed in section 6; the energy loss rate through the molecular hydrogen is computed in Peebles and Dicke 1968). That is, the CDM model predicts that at redshift $z_i \sim 50/b$ much of the hydrogen is in gas clouds that are contracting to form regular stars, or possibly supermassive ones, that can reheat the remainder of the gas.

The second prominent feature of figure 25.2 is the tendency of the clouds to appear in clumps on all scales (consistent with the fact that there is equal contribution to the variance in each octave of wavelength). This is an extreme example of the Kaiser effect in equation (25.13). We see that, in the CDM model, as Jeans mass clouds are forming they are collecting in gravitationally bound systems of clouds in the clustering hierarchy discussed in section 19.

The important point for our purpose is this. It would be reasonable to expect that at $z \sim 10$ structure formation is well advanced everywhere, with much of the hydrogen gravitationally bound in these $10^5 \mathcal{M}_\odot$ clouds or in bound collections of clouds (assuming they are not all destroyed by supernovae). This is not unlike what is seen at redshift $z \sim 5$, for the cloud masses in the Lα forest are not much larger than the Jeans mass (eq. [23.73]), and it is not difficult to imagine that at $z \sim 5$ the more massive of the systems of clouds have coalesced to gas-rich star systems similar to the Lyman-limit and damped Lα clouds discussed in section 23. And it would be quite reasonable to expect that star formation in these systems at $z \sim 10$ is far enough advanced to have been capable of ionizing the intergalactic medium, even in the absence of quasars.

Now let us consider the main complication, the condition that the mean mass density is equal to the critical Einstein-de Sitter value.

Biasing and the Mean Mass Density

The observations discussed in section 20 show that if $\Omega = 1$, as assumed in the CDM model and many others, most of the mass is not clustered with the galaxies: it has to be in the voids. It was the improving observational evidence on this point in the early 1980s that led to the development of the biasing concept that made CDM a viable model for structure formation. The CDM biasing picture is illustrated in figure 25.2, where we see that the high peaks tend to appear in clumps. Thus, if L_* galaxies form only at unusually high peaks, they form already clustered, with most of the mass outside the regions where the protogalaxies are concentrated. If the amplitude of the mass fluctuation spectrum is low enough so that gravity has not strongly rearranged the mass distribution on a scale of about 10 Mpc, emptying the voids between the original galaxy concentrations, then only a small fraction of the total mass is counted in the dynamical measures of the masses of the galaxy concentrations. This is the biasing picture we now examine.

It will be noted that the mass distribution pictured in figure 25.2 is far from smooth in the regions between the concentrations of unusually high peaks: the small-scale part of the fluctuation spectrum extends down to the Jeans cutoff at $10^5 \mathcal{M}_\odot$. Thus the CDM model predicts that the voids defined by the L_* galaxies are filled with lower mass objects, on scales ranging from globular clusters to giant molecular clouds to dwarf galaxies. The expected physical properties of these void dwarfs are explored by Dekel and Silk (1986).

The problem with this picture is that it violates the observed character of the galaxy distribution. Highly luminous elliptical and cD galaxies prefer the centers of rich clusters of galaxies, and with the clusters these galaxies have a larger two-point correlation function than for L_* galaxies (Hamilton 1988b). The IRAS galaxies mapped in figures 3.7 and 3.8 are gas-rich and avoid the dense central parts of rich clusters where galaxies tend not to contain much gas, a presumed result of stripping by collisions and ram pressure from the intracluster plasma. Such special cases aside, L_* galaxies, dwarf galaxies, and IRAS galaxies have strikingly similar space distributions, all avoiding the voids. Observations demonstrating this are discussed by Oemler (1987), Valls-Gabaud, Alimi, and Blanchard (1989), Binggeli (1989), Binggeli, Tarenghi, and Sandage (1990), Weinberg et al. (1991), Thuan and Alimi (1991), and earlier references therein. The effect is illustrated in figures 3.3 and 3.4. Figure 3.3, showing the space distribution of the very nearby galaxies, gives a particularly useful picture of how the galaxies are arranged, because at these small distances the sample is complete to quite low luminosities.

The volume of the cube in figure 3.3 is $V = 500h^{-3}$ Mpc^3. In the mean a cube this size contains $\phi_* V \sim 10$ bright galaxies at $L \gtrsim L_*$. As shown in the third panel, the box contains about this number of large galaxies, meaning we are in a region of roughly average galaxy density. There are some two hundred fainter galaxies in this region. As indicated by the luminosity function in equation (5.129), this is in the expected direction, that there are many more faint galaxies than bright ones (though it is not known whether the very large local number is typical because such faint galaxies are difficult to detect farther away). Most of the galaxies in this local sample are concentrated in a sheet, the nearby part of the de Vaucouleurs Local Supercluster. Just off the sheet (and away from the zone of avoidance at $|Y| \lesssim 1h^{-1}$ Mpc) there are regions in the box with distinctly low numbers of galaxies. The possibilities are that these local voids contain few galaxies because they contain relatively little mass, or because something has suppressed the formation of readily observable galaxies within the voids. If the density parameter were $\Omega = 1$, and the voids were empty because the mass had been swept into the sheets, it would spoil the model for the motions of the galaxies in the Local Group (fig. 20.3), and it would be surprising that this mass swept up with the galaxies is not detected in nearby groups and in the galaxy relative velocity dispersion as a function of separation (fig. 20.1). Also, if the mass clustered with the galaxies, it would seem surprising that the large gravitational effect of this mass is not seen in the quiet local Hubble flow (fig. 5.5), or in relative velocity field around the Virgo cluster, or in the great clusters of galaxies. That is, it appears that in the CDM model the local voids have to contain most of the local mass, and, we will now argue, large numbers of failed seeds of galaxies.

In the CDM model the seeds of galaxies are the upward fluctuations in a random Gaussian process. Since figure 3.3 contains some two hundred visible galaxies, it follows that the voids in the figure would have to have contained literally hundreds of failed seeds. The failure rate in the voids is not complete, for they do contain some galaxies. One surely would expect that if in the voids the survival odds are on the order of one percent, then the galaxies that did form there are irregular or otherwise show the signs of an exceedingly traumatic early environment. Contrary to this, one of the more isolated of the very nearby galaxies, NGC 6946, looks like a normal if gas-rich spiral (Sandage and Bedke 1988). The local sample does contain unusual objects; particularly interesting is the "dark galaxy" of Carignan and Freeman (1988), DDO 154. Its luminosity is low: Carignan and Freeman estimate $L \sim 5 \times 10^7 \, L_{\odot}$, about three times that of the most luminous of the dwarf spheroidal satellites of the Milky Way. Since DDO 154 has a relatively high mass-to-light ratio, it is an excellent model for dark matter and for what one might expect is produced by a nearly failed seed. But this galaxy is in the plane of the Local Supercluster, with most of the others, and there are not many candidates for similar objects in the very nearby voids. Again, this is quite contrary to the idea that something was suppressing galaxy formation in the voids.

In rich clusters of galaxies, the mass-to-light ratio is 10% to 20% of the cosmic mean for an Einstein-de Sitter universe (eq. [20.37]). The computations of West and Richstone (1988), Carlberg, Couchman, and Thomas (1990), and Carlberg (1991) show that the low cluster mass-to-light ratios (for $\Omega = 1$) could be a result of dynamical relaxation that has driven dark matter out of the concentrations of galaxies. There are two important tests for biasing in rich clusters. First, if the dark matter were driven out of the cluster center, one would expect to find it on the outskirts. The same would result from gravitational accretion: if clusters formed out of material rich in galaxies, it would follow that the material the cluster accretes after formation is rich in mass. By either argument, one would expect that the mass-to-light ratio is an increasing function of radius in a cluster. That certainly is not suggested by the tendency for the galaxy line-of-sight velocity dispersion to decrease with increasing radius (Regös and Geller 1989), but more definite dynamical analyses will be welcome. A second test for biasing by dynamical relaxation is that the same process would be predicted to segregate the giant galaxies from the dwarfs. Giant cD galaxies do prefer the centers of clusters; one can interpret this as dynamical settling, or the result of accretion of debris by the best-situated elliptical (Hausman and Ostriker 1978), or the effect of the condensation of intracluster plasma to star formation at the bottom of the potential well (Fabian, Nulsen, and Canizares 1991). But for our purpose the important effect is that there does not seem to be an appreciable shortage of dwarfs in the central parts of clusters, contrary to what one would expect if dynamical relaxation were important. And it is difficult to see how there could be dynamical segregation of galaxies from the dark mass in the Local Group, where the derived mass per galaxy is quite similar to what is found in the great clusters, because the two dominant members of the Local Group are thought to be approaching each other for the first time.

The segregation of mass from galaxies tends to be a transient effect, because the instability of the expanding universe gravitationally draws together galaxies and mass alike. The idea of biasing at the present epoch has thus naturally led to considerable interest in the epoch at which the mass concentrations characteristic of galaxies first were in place. An excellent picture of the state of thinking on this issue is given in *The Epoch of Galaxy Formation* (Frenk et al. 1989). As discussed in section 7, in connection with the steady-state cosmology, galaxies do show prominent evidence of evolution at $z < 1$. If that means galaxies have been assembled fairly recently, it is reasonable to expect they have not moved much since formation, and that the biasing effect in equation (25.13) therefore is present now. As we have argued, however, this picture would predict that dwarf galaxies still are segregated from giants, contrary to what is observed. An alternative interpretation is that we happen to have come on the scene as the common spiral nebulae are exhausting their gas supplies (Kennicutt 1983), just as an observer at redshift $z \sim 2$ would be at the special epoch where quasars are exhausting the ready supply of fuel for their engines. It is useful to note Oke's (1984) remark: "When one looks at the spectra of first-ranked cluster galaxies over the whole range of z covered [to redshift $z \sim 0.8$] one is impressed by the fact that the vast majority are very similar to each other and to nearby ellipticals." This suggests that it is not absurd to imagine these galaxy mass concentrations had been put in place still earlier, as proposed in table 25.1. If so, there would be ample time to erase the initial segregation of giants and dwarfs, consistent with the observations, and maybe also to erase mass segregation.

Other problems for the CDM model, and their remedies, are surveyed in Efstathiou (1990) and Rubin and Coyne (1988). Some depend on relatively fine points in the shape and Gaussian character of the fluctuation spectrum that may be subject to negotiation as explorations of the possibilities offered by the inflation scenario continue. A few substantial rearrangements of the picture are worth mentioning.

The conventional CDM picture assumes an Einstein-de Sitter universe dominated by nonbaryonic matter whose primeval pressure may be ignored, and one with a roughly flat spectrum of Gaussian adiabatic density fluctuations. Blumenthal, Dekel, and Primack (1988) and Efstathiou, Sutherland, and Maddox (1990) note the advantages of a CDM model with $\Omega \sim 0.1$ in cold dark matter, perhaps 10% of that in baryons, consistent with the number from light element production, and the rest in space curvature or a term that acts like a cosmological constant. This would say the voids contain little mass, which certainly is the straightforward interpretation of the galaxy space distribution and the bulk of the dynamical estimates in section 20. An added advantage is that it increases the correlation length (25.24) to a reasonable value for a reasonable value of the Hubble parameter. But this must be balanced against the unappealing prospect of introducing hypothetical matter that does not even serve to close the universe. Another possibility is that the seeds of galaxies are not mass fluctuations that can be approximated as a random Gaussian process, or otherwise are not uniformly sown throughout the mass; we will consider an example from cosmic field defects. It may be that a component of the dark matter has (or has developed through decay) pressure high to prevent it from clustering on scales less than about 10 Mpc. It may be that a density parameter $\Omega \sim 0.1$, all in baryons, is all there is. Or possibly galaxy formation is a good deal more subtle than this analysis has given it credit for.

The Adiabatic Hot Dark Matter Model

In section 18 we considered the idea that the dark matter that might close the universe is a family of neutrinos with a rest mass of a few tens of electron volts (eq. [18.4]). In its simplest form, the adiabatic hot dark matter (HDM) model replaces the cold dark matter of the CDM model with this family of neutrinos. This candidate for the dark matter is termed "hot" because at the redshift z_{eq} of equality of mass densities in relativistic and nonrelativistic components the neutrino peculiar velocities are close to relativistic. The present characteristic peculiar velocity of the neutrinos in the absence of perturbations from homogeneity is small (eq. [18.63]) and would be overwhelmed by whatever kinetic energy the neutrinos acquire from the gravitational acceleration of the present irregular mass distribution.

As for the CDM model, the adiabatic HDM model is an elegant implementation of the simplest ideas from inflation, with the added advantage that neutrinos really are known to exist. (The mass of the family that dominates nuclear beta decay is too low, but there are two more chances from the other families.) Its very distinctive feature is the effect of the thermal motions of the freely moving neutrinos in smoothing the primeval irregularities in the neutrino distribution on the scale of the Hubble length at redshift $z \sim z_{eq}$ (eq. [25.62]). Since radiation pressure smooths the baryon distribution on similar scales, as will be discussed in connection with baryonic dark matter, in this model the mass distribution after decoupling has a broad coherence length. The first nonlinear structures thus would form by pressureless collapse to the pancakes illustrated in figure 22.1. These pancakes certainly look like the sheets of galaxies described in section 3. Other attractive features of the adiabatic HDM model are discussed by Centrella et al. (1988). The central point, however, is that since the characteristic pancake mass is comparable to that of a rich cluster, galaxies would have to be fragments of protoclusters. We will argue that this does not look like what is happening in our neighborhood. Before discussing this in the next subsection, let us establish the order of magnitude for the initial mass coherence length fixed by smoothing by free streaming of the neutrinos.

For a first approximation we can ignore the gravitational effect of the growing mass fluctuations on the motions of the neutrinos. Then the momentum of a neutrino scales with time as (eq. [5.43])

$$p = m_\nu v_o a_o / a(t) \, , \tag{25.58}$$

where v_o is the present peculiar velocity (eq. [18.63]). The relation between momentum and peculiar velocity is

$$v(t) = a(t) \frac{dx}{dt} = \frac{pc}{(p^2 + m_\nu^2 c^2)^{1/2}} \, . \tag{25.59}$$

Most of the comoving displacement is at redshift $z \sim z_{\rm eq}$, where space curvature and a cosmological constant have negligible effect on the expansion rate. And since the median neutrino velocity at $z = z_{\rm eq}$ is $v/c = 0.12$, we can take the neutrino mass density to vary as $\rho_\nu \propto a(t)^{-3}$. Then the expansion rate equation (5.18) at $z \sim z_{\rm eq}$ is

$$\left(\frac{\dot{a}}{a} \right)^2 = \Omega H_o^2 \left[\frac{a_o^3}{a^3} + \frac{a_{\rm eq} a_o^3}{a^4} \right] , \tag{25.60}$$

where the first term in brackets represents the mass densities in the baryons and massive neutrinos, and the second term the relativistic mass in radiation and the other (assumed low mass) neutrino families. Equations (25.58) to (25.60) give the net comoving displacement

$$\begin{aligned}
a_o x_s &= \int_0^{t_o} \frac{a_o}{a} v dt \\
&= \frac{v_o z_{\rm eq}^{1/2}}{\Omega^{1/2} H_o} \int_0^{z_{\rm eq}} \frac{dy}{[(y^2 + \beta^2)(1+y)]^{1/2}} ,
\end{aligned} \tag{25.61}$$

where $z_{\rm eq}$ is given in equation (6.81) and $\beta = v_o z_{\rm eq}/c = 0.12$ is the median neutrino velocity at $z_{\rm eq}$ (eq. [18.63]). The numerical result is

$$r_s = a_o x_s = 10/(\Omega h^2) \, {\rm Mpc} \, . \tag{25.62}$$

This is the characteristic smoothing length, evaluated at the present epoch. The value of β is independent of the mass density $\propto \Omega h^2$ in the massive neutrinos, and $z_{\rm eq}$ is proportional to Ωh^2, so r_s varies inversely as Ωh^2.

A more detailed calculation treats the neutrinos as a free-streaming gas and takes account of the gravitational effect of the mass fluctuations on the motions of the mass components. The numerical results are usefully approximated by an exponential transfer function

$$\mathcal{P}(k) \propto \mathcal{P}_i(k) e^{-\lambda k} , \qquad \lambda = 12/(\Omega h^2) \, {\rm Mpc} , \tag{25.63}$$

(Peebles 1982b; Bond and Szalay 1983), where k is the comoving wave number measured in reciprocal megaparsecs at the present epoch, $\mathcal{P}_i(k)$ is the primeval spectrum of adiabatic fluctuations coming out of the very early universe, and $\mathcal{P}(k)$ is the fluctuation spectrum well after the universe has become dominated by the neutrinos and before the formation of nonlinear structures.

For the scale-invariant primeval spectrum $\mathcal{P}_i \propto k$, the spectrum at $z \ll z_{eq}$ is

$$\mathcal{P}(k) = Ake^{-\lambda k}, \qquad (25.64)$$

and the Fourier transform of this expression is the mass autocorrelation function,

$$\xi(r) = 8\pi A \frac{3\lambda^2 - r^2}{(\lambda^2 + r^2)^3}. \qquad (25.65)$$

This refers to the mass distribution before it has been perturbed by strongly non-linear structures. The zero is at separation

$$r_\xi = 3^{1/2}\lambda = 20/(\Omega h^2)\,\text{Mpc}. \qquad (25.66)$$

For comparison, a flat primeval spectrum gives

$$\mathcal{P}(k) = Be^{-\lambda k} \qquad (25.67)$$

at $z \ll z_{eq}$, with mass autocorrelation function

$$\xi(r) = \frac{8\pi\lambda B}{(\lambda^2 + r^2)^2}. \qquad (25.68)$$

The scaling of the transfer function with Ω and h and the characteristic length scale at which the spectrum bends away from its primeval shape are similar here and in the cold dark matter model (eq. [25.24]) because these characteristic lengths are set by the displacement at redshift $z \sim z_{eq}$ of the nearly relativistic neutrinos or the nearly relativistic coupled matter-radiation fluid.

The transfer function for the density fluctuation spectrum on scales less than r_ξ is much smaller here than in the cold dark matter model (eq. [25.22]). In the latter case, small-scale fluctuations are suppressed by a factor that varies about as k^{-4}. This means that if the primeval spectrum is $\mathcal{P}_i \propto k$, the relevant coherence length is the small-scale baryon Jeans length, as shown in figure 25.1. In the HDM model, the nearly exponential suppression of small-scale fluctuations means the coherence length of the mass distribution at decoupling is on the order of λ (eq. [25.63]). This is much larger than the baryon Jeans length, so the first nonlinear structures are the Zel'dovich pancakes illustrated in figure 22.1. With the initially scale-invariant spectrum in equation (25.64), the distance between pancakes is $\sim 2r_\xi \sim 40/(\Omega h^2)$ Mpc. That is, the model naturally predicts the formation of sheetlike concentrations of mass at separations comparable to the sheetlike concentrations of galaxies illustrated in figures 3.3 and 3.6. This is impressive, for the prediction was made when the tendency for galaxies to lie on sheets was not as manifest in the data as it is now (Bond, Efstathiou, and Silk 1980; Zel'dovich, Einasto, and Shandarin 1982). Furthermore, Uson, Bagri, and Cornwall (1991) have discovered a massive system of atomic hydrogen observed in emission in the 21-cm line at redshift $z = 3.4$, and they point out that the cloud has the properties one would look for in a Zel'dovich pancake.

In this model the characteristic distance between Zel'dovich pancakes is fixed by the Hubble length at z_{eq}, which is considerably larger than the distance between galaxies. Thus we are led to consider a "top-down" picture in which galaxies form by the fragmentation of pancakes. We will argue in the next subsection that this is not a promising way to interpret the observations. A related problem is that it is difficult to arrange for galaxies to form at a reasonable time before the present without making the galaxies cluster a good deal more strongly than is observed (Peebles 1982b; White, Frenk, and Davis 1983). One can reduce the separation between pancakes by adopting a primeval density fluctuation spectrum $\mathcal{P}_i(k)$ that decreases sufficiently rapidly with increasing wavelength, reducing r_ξ, but that makes the initial mass fluctuations anticorrelated on scales where the galaxy positions are observed to be correlated, which is not reasonable.

A more interesting way out assumes that structures formed in pancake collapse when the mass of the universe is dominated by neutrinos with masses several times that of the conventional model. The neutrinos are assumed to be unstable and to have decayed after structure formation, the excess mass density being redshifted away in the relativistic decay products. The greater mass of the original neutrinos would reduce the distance between pancakes. After decay, when the mass of the universe is dominated by the high-velocity dark matter, the growth of mass clustering would be suppressed. Both effects would help relieve the problem with late galaxy formation. And some fraction of the decay products might be ionizing radiation that could serve to keep the intergalactic medium ionized (section 23). The basic elements of this picture seem to have first been considered by Dicus, Kolb, and Teplitz (1978); interest developed with the recognition of the problems with the simpler HDM model based on stable massive neutrinos (Fukugita and Yanagida 1984; Hut and White 1984; Turner, Steigman, and Krauss 1984). The state of the idea is reviewed by Klypin and Doroshkevich (1989). A notable feature is that the rearrangement of the mass when the original dark matter decays ensures a clean separation from the galaxies without segregating giant galaxies from dwarfs. In the commonly discussed versions of this picture the decay products are relativistic, so they can be redshifted away. That would mean the dynamical measures in the last line of table 20.1 ought to yield an apparent value for the density parameter consistent with the number from smaller scales.

In the stable particle version of the adiabatic HDM model, galaxies would form by the fragmentation of Zel'dovich pancakes. This leads to an issue of general interest: Can galaxies have formed in a "top-down" sequence, by the fragmentation of protoclusters?

"Top-down" and "Bottom-up" Galaxy Formation

The debate on the relative merits of the "top-down" picture, and a "bottom-up" hierarchical picture for the growth of structure in which smaller objects gather together to form larger ones, traces back to the 1930s. Lemaître (1933, 1934) suggested that in the gravitational instability picture it would be natural to suppose the dense mass concentrations in galaxies were assembled at high redshift, when the mean density was high, and before the formation of the less dense concentrations of mass in clusters. Hubble (1936, p. 81) pointed out that the tendency of early-type (gas-poor) galaxies to be concentrated in clusters could be taken to mean that these are preferential sites for galaxy formation, and perhaps that the field was populated by the dissolution of clusters. This could be related to Hubble's sequence of galaxy types. Though Hubble was careful to stress that he devised his sequence as a way to classify galaxies, he noted that one could imagine that galaxies evolve along the sequence, from early-type ellipticals to late-type spirals, as Jeans (1928) had suggested. That would mean spirals escaped from clusters and then evolved to their present forms in a "top-down" sequence of structure formation.

The evidence as now understood is that this is not what happened in our neighborhood. The Virgo cluster is the nearest massive source for galaxies, but if the galaxies around us were shot out of the Virgo cluster, one would expect that they still have considerable peculiar motions. This is contrary to the evidence in figure 5.5, which shows that the nearby relative velocity field just outside the Local Group is quite close to the universal Hubble flow. It is contrary also to the evidence that the galaxies in our neighborhood have a peculiar streaming motion toward the Virgo cluster, not away from it (eq. [20.57]). Within the Local Group, the crossing time (the ratio of galaxy separation to relative velocity) is comparable to the Hubble time, suggesting this group is forming now. The same picture follows from the more detailed model in figures 20.2 and 20.3. Our galaxy had to have formed at a moderately high redshift, because it contains old stars, and we are therefore in a galaxy that is older than the group now forming around it, a direct example of the "bottom-up" process. The local sheet of galaxies in figure 3.3 does look very much like a pancake, but since the crossing times of peculiar motions across the sheet are comparable to the Hubble time, as in the Local Group, this appears to be a sheet that is forming now, out of old galaxies, in another example of the "bottom-up" sequence.

One might imagine that galaxies in dense regions formed by pancake collapse, while the galaxies in our neighborhood and the objects that filled space with Lα forest clouds at redshift $z \sim 5$ formed in some other way. However, this does not agree with the observation that galaxies define a few low-dimension sequences. If bright galaxies formed by more than one process, why is luminosity so well correlated with the velocity dispersion in the dark halo, as in the Tully-Fisher relation for spirals, and how could there be such a tight relation between the spheroid luminosity, the core radius, and the central velocity dispersion, as shown in figure 3.14? The reasonable interpretation is that, apart from special cases such as the cD galaxies at the centers of clusters, the masses in galaxies were assembled by a universal process. And in our neighborhood we have clear examples of galaxies that were assembled before the formation of the systems in which they now are found. Thus it appears that the classical "top-down" picture is ruled out, and with it the adiabatic hot dark matter model (Peebles 1984a).

Isocurvature Hot Dark Matter Scenarios

As in the case of cold dark matter, it is well to note that the problem with the specific HDM model we have been considering is not generic to the idea that the dark mass is a family of neutrinos. Another route to structure formation out of HDM is an isocurvature picture.

The starting assumptions are that in the beginning the mass distribution was homogeneous, that whatever made the baryons deposited many of them in clumps, and that we live in an Einstein-de Sitter universe dominated by a nondegenerate family of neutrinos. The first assumption is in line with the strong constraint from the gravitational instability of the standard model, which means the mass distribution had to have evolved out of a remarkably homogeneous initial state. The second assumption is quite ad hoc, but the physics of the very early universe is not so well known that we should exclude scenarios that seem to be observationally acceptable.

In this picture, where baryons are produced the energy had to be taken out of what is locally available, a process that cannot have perturbed the mass distribution or spacetime curvature on scales of interest at the present epoch. The free streaming of the neutrinos at redshift $z \sim z_{eq}$ would keep their distribution smooth on the scale of λ in equation (25.63). After that, as the neutrinos cool they would tend to gravitationally collect around the clumps of baryons. If the baryons are concentrated in well-separated islands, the accretion is close to spherically symmetric, and we can find the scaling law for the density run in the neutrino halo around a primeval baryon concentration as follows.

The typical neutrino velocity at expansion parameter $a(t)$ is $v = v_o a_o/a$, where the present value is $v_o = 1.5h^{-2}$ km s^{-1} (eq. [18.63]). The comoving displacement in an expansion time t is $\sim vt/a \propto t^{-1/3}$, which decreases from the scale of the Hubble length at z_{eq}. A neutrino becomes bound to a mass concentration m at coordinate distance x at expansion parameter a_x when the potential energy is comparable to the kinetic energy,

$$v^2 \sim v_o^2 \frac{a_o^2}{a_x^2} \sim \frac{Gm}{a_x x}, \tag{25.69}$$

or

$$\frac{a_x}{a_o} \sim \frac{v_o^2 a_o x}{Gm} . \qquad (25.70)$$

The neutrino is then bound in the halo at physical radius $r \sim a_x x$. A neutrino at larger impact parameter sees that these inner neutrinos have had their orbits rearranged without affecting the net mass, so the same scaling applies to its capture at a later time. The neutrino mass within radius $r = x a_x$ is

$$\mathcal{M}(r) \sim \rho_o(a_o x)^3 \sim \rho_o(Gmr)^{3/2}/v_o^3 , \qquad (25.71)$$

the density run is

$$\rho(r) \sim \frac{\mathcal{M}(r)}{r^3} \sim \frac{\rho_o(Gm)^{3/2}}{v_o^3 r^{3/2}} , \qquad (25.72)$$

and the velocity of a particle bound in a circular orbit is

$$v_c \sim \rho_o^{1/2} G^{5/4} m^{3/4} v_o^{-3/2} r^{1/4} . \qquad (25.73)$$

This scaling argument is from Peebles (1983); a more detailed computation of the gravitational accretion of HDM by a moving mass concentration is given by Dobyns (1988).

The circular velocity (25.73) scales as $v_c \propto r^{1/4}$; the slow increase with increasing radius is not out of line with what is seen in isolated spiral galaxies. A dark halo that supports rotation velocity $v_c = 200$ km s^{-1} at radius $r = 10h^{-1}$ kpc is accreted by mass

$$m \sim 10^{10} h^{-5} \mathcal{M}_\odot . \qquad (25.74)$$

It is interesting and maybe significant that this is comparable to the baryonic mass associated with the starlight in an L_* galaxy. Since the inner part of the neutrino halo would be attached at high redshift, without strong collapse, one would expect the phase space bound in equation (18.74) to be saturated, as observed in large spirals (though we noted that a special arrangement is needed for some dwarfs). And in this picture, stars can form in the central parts of galaxies and ionize the intergalactic medium at a reasonably large redshift, well before the outer parts of the dark halo are attached.

Since the dominant contributions to the mass are in radiation and neutrinos, and they are smoothly distributed to begin with, the perturbation to the CBR by structure formation is limited to the relatively small growing gravitational fields and velocity fields around the mass concentrations around clumps of baryons.

The large-scale velocity field would have to have grown out of large-scale fluctuations in the baryon distribution; the phenomenological models in section 21 suggest it may not be difficult to find an observationally acceptable prescription.

All of this is suggestive, but it is not a model for structure formation because we have no guide to how to place the accreting masses. It does show that if laboratory experiments demonstrated the existence of a family of neutrinos with a mass of several tens of electron volts, there would not seem to be much difficulty in finding a place to put them.

Explosions and the Large-Scale Velocity Field

In most scenarios the seeds of structure formation are present at high redshifts as wrinkles in spacetime or as structures in the material or stresses in an isocurvature model. The exception is the explosion picture, where nongravitational forces produced by early generations of stars move matter enough to trigger the gravitational formation of large-scale structure.

Ostriker has given a useful way to understand the scale on which explosions within young galaxies could rearrange matter. Consider an isolated galaxy surrounded by material at the cosmological mean mass density ρ_b. Suppose supernovae cause a flow of material out of the galaxy that carries net kinetic energy E integrated over a time short compared to the timescale t for the expansion of the universe. This pushes on the material outside the galaxy, piling it into a shell of radius $R(t)$. We see from dimensional analysis that the radius of the mass shell after an expansion time t is

$$R \sim \left(\frac{Et^2}{\rho_b} \right)^{1/5} \sim (GEt^4)^{1/5}. \tag{25.75}$$

A much more detailed analysis of this self-similar solution is given by Ostriker and McKee (1988).

To get some numbers, suppose a newly assembled L_* galaxy cycles $\sim 10^{11}$ solar masses of baryons through stars to produce $\sim 10^{10}$ supernovae, each of which emits a blast wave with energy $\sim 10^{50}$ erg, producing a total energy release $E \sim 10^{60}$ erg. For an explosion at a fairly recent epoch, we might take the expansion time to be $t \sim 10^{17}$ s, and the mean mass density outside the galaxy to be $\rho_b \sim 10^{-30}$ g cm^{-3}. Then the shell radius is

$$R \sim 3 \, \text{Mpc}. \tag{25.76}$$

Ostriker and Cowie (1981) point out that since this is comparable to the mean distance between galaxies, even at low redshifts, the birth of a galaxy certainly can influence its neighbors. As a limiting case, one could imagine that galaxy formation is a chain reaction, in which a seed protogalaxy explodes and piles up material in a ridge, which fragments to form a new generation of galaxies, which explode and pile up more material for the next generation. One inspiration for this model is the evidence that this happens in star formation in the interstellar medium, and it certainly is sensible to consider the possibility that the analogous process operates on the scale of galaxies.

The most pressing challenge to the picture is the local peculiar velocity field, which does not look like the expected effect of explosions (Peebles 1988a). Figure 5.5 shows the redshift-distance relation referred to the Local Group for very nearby galaxies with infrared Tully-Fisher distances. The line is the extrapolation from the redshift-distance relation for clusters at redshifts five to ten times the maximum distance shown in this figure. The small scatter from the universal mean Hubble flow is striking, particularly when one recalls that in the standard interpretation of the CBR dipole anisotropy the region represented in this graph has a mean streaming velocity of 600 km s^{-1} (table 6.1), and that the galaxy distribution in our neighborhood is distinctly inhomogeneous (fig. 3.3). Ostriker and Cowie (1981) point out that the galaxies that form on a shell of material pushed out by an explosion would define a sheetlike structure, and that the galaxies on the shell would share a streaming velocity that is large compared to the relative velocities of neighboring galaxies on the shell. All of this agrees with what is observed in our neighborhood, but there are two problems. First, our streaming velocity relative to the CBR is not normal to the local sheet of galaxies, as the bubble model would predict; the larger component is tangent to the sheet. Second, many galaxies in our neighborhood are not on the dominant local sheet at $H_o|Z| \lesssim 100$ km s^{-1}. One would presume that the galaxies not on this sheet were produced in other explosions, consistent with the idea that there would have to have been many explosions to have made all the galaxies. If our peculiar velocity is typical for the exploding shells, there ought to be galaxies from other shells at distances $H_o R \lesssim 1000$ km s^{-1} that have peculiar velocities relative to us ~ 1000 km s^{-1}. Outside the core of the Virgo cluster, this is not observed.

As discussed in section 21, the conventional picture is that the local streaming velocity is driven by the gravitational field of mass fluctuations with low amplitude and large extent, and that the small-scale relative velocities are small despite the large fluctuations in the galaxy distribution because the galaxies carry relatively little mass. It is difficult to see how the large-scale mass fluctuations could be produced in a pure explosion picture without destroying the galaxies over which the large-scale wave would have to pass to rearrange mass on large scales. Ostriker and Suto (1990) have emphasized that accounting for the broad coherence length of the galaxy peculiar velocity field is a challenge none of the adiabatic models proposed so far has met either. This is a problem for these theories, but, it is worth emphasizing, not for the standard cosmological model, for we saw in section 21 that the velocity field is at least roughly comparable to what one would expect from the measured fluctuations in the distribution of galaxies, if the mass fluctuations are traced by the galaxies (eqs. [21.105] and [21.116]).

That is, we seem to have a reasonably well founded picture for gravitationally driven peculiar velocities, and we are lacking the *ab initio* theory for why the mass fluctuations are what they are. The baryon isocurvature picture to be considered below is an example of a scenario in which one can tune the primeval mass density fluctuation spectrum, by hand, to get what seems to be an observationally acceptable scheme.

One might assume that explosions produced the galaxies while something else produced the large-scale mass fluctuations that drive the peculiar velocities. This has to be at least part of the story. For example, it is thought that dwarf galaxies have low heavy element abundances because star formation was disrupted by supernovae. The main reason to think the masses of the galaxies were assembled by gravity is the continuity of the velocity dispersion statistic in figure 20.1, which shows that the density run $\rho \sim r^{-2}$ typically found within galaxies is continuous with the mean density run in the galaxy clustering that we have argued very likely was assembled by gravity.

Cosmic Fields and the Segregation of Galaxy Seeds

In the processes discussed in section 16, a cosmic field that carries an interesting amount of energy acquires a spatially inhomogeneous value in the early universe. If the universe is homogeneous to begin with, the large-scale mass distribution remains homogeneous after the field acquires its value, with the field energy taken from what is available within the Hubble length at this early epoch, as in the isocurvature HDM model above. The tension in the field, or the pressure in the CBR, moves material and could trigger structure formation. Cosmic fields offer fascinating possibilities for structure formation, but despite considerable work in the past five years there is not yet an accepted and presentable model, so we shall confine the discussion to some general issues.

Cosmic fields may unambiguously reveal themselves through a unique signature in the angular distribution of the cosmic background radiation. Long and reasonably straight cosmic strings produce distinctive linear steps in the CBR surface brightness from the difference in the Doppler shifts on either side of a moving string (Kaiser and Stebbins 1984). The frequency shift caused by a topological texture event results in distinctive isolated circular patches at lower or higher surface brightness (Turok and Spergel 1990). Other less direct tests for the role of cosmic fields in structure formation come from issues in the phenomenology of the structure.

The arguments in section 16 indicate that cosmic fields tend to produce a density fluctuation spectrum that is similar to the CDM model. The character of the fluctuations is quite different, however, for the field energy can appear in a variety of structures, depending on the number of field components: $N = 2$ gives cosmic strings, $N = 3$ global monopoles, $N = 4$ textures, and larger N produces fluctuations that are close to Gaussian. This could be significant, for it is striking that the galaxy maps in section 3 show very pronounced structures on scales where the galaxy two-point correlation function is quite small. That is, Nature has arranged the large-scale pattern of the galaxy distribution so that the second moment is small on scales where the fluctuations are not statistically independent. It is easy to imagine how this could have happened in a cosmic string scenario, for long strings could have produced patterns that run across the present Hubble length, even though the Kibble scaling argument (eq. [16.46]) shows that the mass fluctuation spectrum they produce is scale-invariant, that is, anticorrelated on large scales.

The highly non-Gaussian and sporadic character of the mass distribution in cosmic field defects could offer a way to account for the strong clustering of galaxies relative to mass within an Einstein-de Sitter universe: perhaps voids contain most of the mass but few of the galaxies because there were few seeds in the protovoids. The way is not easy, however, for the seeds of dwarf galaxies must know to avoid the voids defined by the seeds of L_* galaxies. It would be natural to suppose that dwarf galaxies formed out of smaller defects, or systems of defects, than giants, and it might be natural to suppose smaller systems of defects formed at higher redshifts, when the comoving Hubble length was smaller. That would not do, however, unless there were some arrangement to assure that these different generations of seeds know how to end up in the same concentrations at the present epoch.

The same consideration applies to rich clusters of galaxies. As we have seen in the discussion of the "top-down" picture, the evidence is that these systems formed out of preexisting galaxies. In a cosmic field picture one might imagine that early generations of defects produced galaxies and later generations gravitationally gathered material to make clusters of galaxies. The two generations of defects would have been produced at different epochs, so it is reasonable to expect that the positions of the different generations of defects are not much correlated. However, that would mean the cluster seeds typically gather a fair sample of galaxies and dark mass. If $\Omega = 1$ that is contrary to the observations summarized in section 20, which show that the mass-to-light ratios in clusters are well below the Einstein-de Sitter value.

These problems are of course eliminated if one is willing to live in a universe with density parameter $\Omega \sim 0.1$ in movable matter, in which clusters are fair samples and voids are empty because gravity has pulled the giants and dwarfs into groups and clusters. But then the biasing function of the cosmic fields is not needed.

We have another useful constraint from the tight correlation between circular velocity and luminosity in spiral galaxies (eq. [3.34]), and the exceedingly small scatter around the fundamental plane for ellipticals (figure 3.14). In a long cosmic string picture, it would be reasonable to expect that galaxies are labeled by a parameter that represents the speed with which the string happens to be moving, because the speed determines the strength of the perturbation to the matter distribution. Another parameter might be required for the epoch at which the string passed by, because that determines whether the string is working with dilute or dense material. Thus, the galaxies produced by a string that was operating at an earlier than average epoch might be expected to be exceptionally dense for their luminosities, with higher than average circular velocities. The redshift-distance relations in figures 5.4 and 5.5 are based on the assumption that there is a universal relation between circular velocity and luminosity. The feature in figure 5.5 in the neighborhood of Centaurus perhaps is an example of a local excursion from this relation (Silk 1989), but the effect certainly is not large.

The message is that the physical properties of galaxies do not exhibit much evidence of a second location-dependent parameter. The straightforward picture is that a galaxy seed carries a principal number that assigns the galaxy a luminosity and a central velocity dispersion or a circular velocity in its dark halo. Secondary numbers may be needed to determine whether the spheroid is to be adorned with a disk — though environment may do that — or where the galaxy will sit on the fundamental plane. If the seed is a density fluctuation, the principal number could be the amplitude of the primeval density fluctuation, and a secondary number may be the shear. We have argued that in a long cosmic strings picture it would be reasonable to suspect that there are several principal numbers, with values that are correlated with location. The same is true of "top-down" pictures, where the second parameter records whether the galaxy is born in a rich cluster, in a group, or all alone in the field. But this does not agree with the straightforward reading of the observations.

Primeval Magnetic Fields

It is worth pausing to take note of yet another cosmic field. There is no consensus on the origin of the magnetic fields in galaxies; they might be products of dynamos working on seed fields from exploding stars within galaxies, or they could be primeval, present before galaxies. The state of ideas is reviewed in Beck, Kronberg, and Wielebinski (1990), Ratra (1992), and Kulsrud and Anderson (1992); the evidence from Faraday rotation for the presence of magnetic fields in the high surface density damped Lα clouds at redshifts $z \gtrsim 2$ is developed in Kronberg and Perry (1982) and Wolfe, Lanzetta, and Oren (1992). Let us estimate the effect a primeval magnetic field would have on the large-scale mass distribution.

If a primeval magnetic field homogeneously expanded to the present epoch has present characteristic value B_o for the component with present characteristic length l_o, the mass density perturbation produced by the field stress at decoupling is

$$\delta_i \sim \frac{B_o^2(1+z)^4}{4\pi} \cdot \frac{l_o^2}{(1+z)^2} \cdot \frac{1}{\rho_o l_o^3} \cdot \frac{4}{9\Omega H_o^2(1+z)^3} \cdot \frac{1+z}{l_o} \cdot \frac{\Omega_B}{\Omega} \quad (25.77)$$

(Wasserman 1978). The first factor is the magnetic force per unit area. The force multiplied by the proper area of the coherence length l_o and divided by the mass is the fluid acceleration. The product with the square of the world time is the characteristic displacement in an expansion time. That divided by the characteristic proper length of the field fluctuation is the density contrast δ_i produced in the baryonic matter at decoupling. The last factor indicates that the net mass contrast is reduced by the ratio Ω_B/Ω of the baryon density to the net density in movable matter.

Let us assume the dark mass is baryonic, $\Omega = \Omega_B$, as in the scenarios to be discussed next. Then the gravitational growth factor of δ to the present is about $(1 + z)$ if the universe is cosmologically flat and $\Omega \gtrsim 0.1$ (fig. 13.13), and the magnetic field needed to produce the present density contrast δ_o on the scale l_o measured in megaparsecs works out to

$$B_o \sim \frac{l_o H_o^2 \Omega \delta_o^{1/2}}{G^{1/2}(1+z_{\text{dec}})^{1/2}}$$

$$\sim 10^{-9} \delta_o^{1/2} l_{\text{Mpc}} h^2 \Omega \text{ gauss} . \quad (25.78)$$

Typical magnetic fields in galaxies are $B \sim 10^{-6}$ gauss at matter densities $\sim$ 1 proton cm^{-3}. If this were isotropically expanded by six orders of magnitude, to the background density, it would bring the flux density to

$$B_o \sim 10^{-10} \text{ gauss} . \quad (25.79)$$

Here again is a conceivably significant coincidence: that an interesting magnetic flux density, if primeval, would do interesting things to the mass distribution.

The energy density in the cosmic background radiation is equivalent to $B_\gamma \sim$ 10^{-6} gauss (eq. [6.60]), and the CBR energy density scales with redshift in the same way as a homogeneously expanding primeval magnetic field, so the energy in a primeval magnetic field would be a fixed fractional perturbation to the mass density at high redshift,

$$\delta\rho/\rho \sim (B_o/B_\gamma)^2 \sim 10^{-8} , \quad (25.80)$$

for the present flux density in equation (25.79). This is an unacceptably large perturbation to the net mass density at very high redshift, where fluctuations in the total density grow as $\delta\rho/\rho \propto t \propto a(t)^2$, because after an expansion factor $\sim 10^4$ the fluctuation (25.80) on scales larger than the Hubble length would grow to $\delta\rho/\rho \sim 1$ and then collapse to a black hole. This is avoided if the field were created from the locally available energy in the isocurvature initial condition standard for other cosmic fields.

Baryonic Dark Matter

In this picture the important actors are baryons and their electrons, the cosmic background radiation, and the known three families of neutrinos, all assumed to be nondegenerate and to have negligible rest mass. Since the dark mass in the halos of galaxies and in clusters of galaxies has to be baryonic, perhaps very low mass stars, this may be termed the baryonic dark matter (BDM) picture. We will consider two versions, adiabatic and isocurvature. The former very likely is only of historical interest, for without exceedingly special arrangements it violates the isotropy of the cosmic background radiation and the condition that galaxies formed before clusters of galaxies. The isocurvature scenario is of practical interest, because it can be adjusted to fit what seem to be the main outlines of the phenomena as they are understood at the time this is written.

The dynamical estimates in section 20 suggest the mean mass density is about 10% of the Einstein-de Sitter value and very likely is larger than the number in equation (6.27) for the baryon density parameter $\Omega_B h^2$ derived from the standard model for light-element nucleosynthesis. Thus we must begin with a special assumption that the estimate of $\Omega_B h^2$ from nucleosynthesis is biased low. One possibility is that the baryon distribution at $z \sim 10^{10}$ is clumpy, in a small-scale version of the isocurvature model to be discussed below (where we will want $\Omega \sim 0.1$ and $h \sim 0.75$). The presently understood possibilities for reconciling the numbers for the dynamical estimates of Ω and what nucleosynthesis allows are surveyed by Kurki-Suonio et al. (1990), and Fuller, Boyd, and Kalen (1991).

In the adiabatic version of the BDM scenario the local entropy per baryon (in effect, the local ratio of the number densities of photons and baryons) is a fixed value at high redshift. Small amplitude fluctuations in the initial mass distribution grow by gravitational instability on scales larger than the Hubble length and oscillate as pressure waves on scales smaller than the Hubble length at z_{eq}. The result is that the mass distribution after decoupling has a broad coherence length, comparable to that of the adiabatic HDM model. This led to the first pancake picture (Doroshkevich, Sunyaev, and Zel'dovich 1974) and to the issue of the sequence of structure formation.

The transfer function for the power spectrum of the baryon distribution after decoupling in the adiabatic BDM model is shown in figure 25.3. The transfer function has been multiplied by k^4 to make the envelope of the fluctuations closer to flat; in an application one would divide the function plotted here by k^4 and multiply by the assumed primeval spectrum to get the fluctuation spectrum after decoupling and before the development of bound systems. This transfer function can be compared to the CDM function in equation (25.22); the main features can be understood as follows.

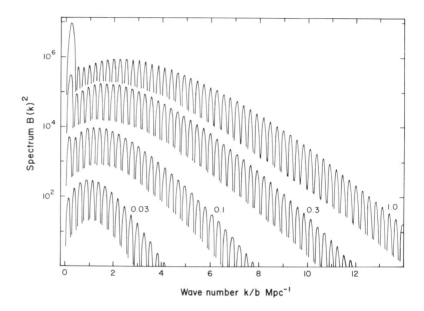

Figure 25.3. Transfer function for the adiabatic baryonic dark matter model (Peebles 1981b). The parameter is $\Omega_B h^2$, where the nonrelativistic mass density is entirely in baryons. The wave number is normalized to proper units at the present epoch, and the scale factor b (not to be confused with the CDM parameter) has been adjusted to line up the zeros. Values are $b = 0.39$ at $\Omega_B h^2 = 1.0$, $b = 0.18$ at $\Omega_B h^2 = 0.3$, $b = 0.10$ at $\Omega_B h^2 = 0.1$, and $b = 0.065$ at $\Omega_B h^2 = 0.03$. The curves have been shifted vertically to separate them.

At very high redshift the short mean free path for Thomson scattering causes matter and radiation to act as a single fluid. The Fourier amplitude $\delta_k(t)$ of the mass distribution at fixed comoving wavenumber k grows until the wavelength becomes comparable to the Hubble length, and then $\delta_k(t)$ oscillates as a pressure wave until decoupling, when the baryon part is released and its amplitude can grow again. The final amplitude after decoupling thus is an oscillating function of the wavenumber k, with zeros where the phase of $\delta_k(t)$ is just such as to produce a pure decaying mode in the pressureless baryon distribution after decoupling. Thus, the number of zeros in the transfer function is twice the total number of oscillations of $\delta_k(t)$ at given k. If the pressure waves suffered no dissipation up to decoupling, the transfer function for the power spectrum for these waves would be proportional to k^{-4} by the argument in equation (25.18). That is why the envelope of the transfer function in the figure, which has been multiplied by k^4, has a nearly flat part. The matter-radiation fluid has viscosity from photon diffusion, which causes the decrease in amplitude at large k, an effect first analyzed by Silk (1968, 1974).

At small comoving wavenumber the proper wavelength becomes equal to the Hubble length after the epoch z_{eq} of equality of mass densities in matter and radiation. The wavelengths of these modes are longer than the Jeans length for the matter-radiation fluid, so the gravitational growth of $\delta_k(t)$ is not much impeded by the radiation pressure gradient force. In a high-density cosmological model, z_{eq} is earlier than the epoch of decoupling, and these long wavelength modes grow in the interval $z_{eq} > z > z_{dec}$, while the modes with shorter wavelengths are stored as oscillating waves at nearly fixed amplitude. This produces the first large peak in the top curve in figure 25.3, where the wavelength is larger than the matter-radiation Jeans length at $z > z_{dec}$.

The scaling of the comoving matter-radiation Jeans length with the cosmological density parameter follows by the same arguments used for the adiabatic CDM and HDM models. The Jeans length is on the order of the wavelength at the first zero of the transfer function, which can be approximated as

$$\lambda_0 = \frac{30}{\Omega h^2} \text{ Mpc} . \tag{25.81}$$

The transfer functions for the CDM and adiabatic BDM models are roughly similar: at wavelengths large compared to the matter-radiation Jeans length, both functions are close to unity, and at shorter wavelengths they vary roughly at k^{-4}. However, for $\Omega h^2 \sim 1$ the transfer function at shorter wavelengths levels off at a considerably larger value in CDM than BDM, because the shorter wavelength fluctuations grow in the interval $z_{eq} > z > z_{dec}$ in CDM, and oscillate as pressure waves at fixed amplitude in BDM. The smaller transfer function means the BDM model with $\Omega = 1$ requires a higher primeval amplitude than the CDM model with the same shape for the primeval fluctuation spectrum. If the initial fluctuation spectrum is scale-invariant or close to it, this makes the residual large-scale fluctuations in the cosmic background radiation unacceptably large (Wilson and Silk 1981). The dominant part of the CBR anisotropy is in the first lobe of the transfer function in figure 25.3, because in the lobes at shorter wavelengths the radiation diffusion during decoupling strongly smooths the radiation relative to the baryons. Thus, one could contrive an adiabatic BDM model that produces small-scale structure without large perturbations to the CBR by tuning the primeval spectrum to eliminate mass fluctuations in the first peak; however, it would likely prevent the formation of interesting large-scale structures.

We avoid the problem of overperturbing the CBR, or at least we greatly change the situation, by going to an isocurvature BDM model, where at high redshift the mass distribution is homogeneous and the local value of the entropy per baryon is a random function of position,

$$s(\mathbf{x}) \propto T^3/n = C[1 - \epsilon(\mathbf{x})] , \tag{25.82}$$

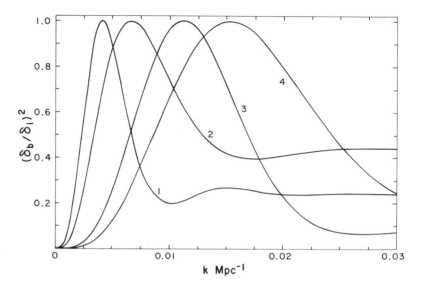

Figure 25.4. Transfer function for the linear isocurvature BDM model. The wavenumber is measured in radians Mpc^{-1} at the present epoch. The transfer function is unity at short wavelengths; the normalization is chosen to put all the peaks at the same height. For curves 1 and 2, the density parameter is $\Omega h^2 = 0.025$; for 3 and 4, $\Omega h^2 = 0.1$. Curves 1 and 3 assume that early star formation keeps the matter fully ionized; curves 2 and 4 assume that the ionization is the larger of thermal equilibrium and 10% ionization.

where C is a constant and $\epsilon(\mathbf{x})$ represents the initial entropy fluctuations. In what follows we will simplify the analysis by assuming the fractional perturbation ϵ is small, so we can use linear perturbation theory. Then at high redshift, where the mass density is dominated by the radiation, the condition that the net density is homogeneous means the baryon distribution is

$$\frac{\delta n}{n} = \epsilon(\mathbf{x}) . \qquad (25.83)$$

Examples of the transfer function for the baryon distribution well after decoupling are shown in figure 25.4. This was discussed by Doroshkevich, Zel'dovich, and Sunyaev (1978); the transfer function in linear perturbation theory is analyzed in detail in Peebles (1987b) and Efstathiou and Bond (1987). The main features are the following.

At wavelengths much larger than the matter-radiation Jeans length in equation (25.81), the pressure gradient force can be neglected, and the universe evolves so as to keep space curvature uniform and the mass density homogeneous. This means that at low redshift, where the baryon density dominates, its large-scale distribution is nearly homogeneous, as is indicated by the zero of the transfer function at $k \to 0$. In this limit, the radiation temperature in equation (25.82) satisfies

$$\frac{\delta T}{T} = -\frac{\epsilon(\mathbf{x})}{3}.$$

(25.84)

In the absence of scattering, this is the observed CBR temperature fluctuation produced by the large-scale entropy fluctuation.

On scales small compared to the Jeans length for the coupled matter and radiation, the motion of the material tends to bring the baryons to a homogeneous distribution, and electron scattering brings the radiation with the baryons. This produces a combination of an entropy wave and a pressure wave. Photon diffusion damps the pressure wave, leaving a smooth sea of radiation with the original irregular baryon distribution, $\delta n/n = \epsilon(\mathbf{x})$, which can gravitationally grow after radiation drag becomes negligible. This is the flat part of the transfer function at large k, where at decoupling $\mathcal{P}(k) = \mathcal{P}_\epsilon(k)$. The peak of the transfer function occurs where the matter-radiation fluid has completed about half an oscillation at decoupling, releasing the baryons with motion that adds to the growing amplitude. The length scale at the peak is fixed by the matter-radiation Jeans length in equation (25.81). At shorter wavelengths, photon diffusion heavily damps the oscillations, so the velocity kick is appreciable only at the first oscillation.

To see what this scenario can offer, let us consider a specific case. We will adopt an open cosmological model, with

$$\Omega = 0.1, \qquad h = 0.6, \qquad \Lambda = 0,$$

(25.85)

and with Ω dominated by baryons. This makes the age of the universe reckoned from high redshift $t_o \sim 14$ Gy (fig. 13.1), toward the lower end of the range of estimates from stellar evolution ages (eq. [5.77]). Since the Hubble parameter also is toward the lower end of the range of current estimates, we may be led to a cosmologically flat model with a cosmological constant. This would not greatly affect the numbers presented here, and it would agree with the prediction of inflation. A universe with $\Omega = 1$ in baryons would be even more elegant, but we have seen that it is difficult to reconcile the high density with the dynamical mass estimates, and it appears to be quite difficult to get a consistent picture for the origin of the light elements.

We will assume the primeval entropy per baryon has a flat fluctuation spectrum, because that gives reasonable-looking large-scale peculiar velocity fields and large-scale anisotropy in the CBR. Again, this is not the preferred choice, which would be the scale-invariant form $\mathcal{P}_\epsilon \propto k^{-3}$. (This is the analog of the scale-invariant spectrum $\mathcal{P} \propto k$ for adiabatic fluctuations: the integrated perturbation to space curvature, for adiabatic fluctuations, and the primeval baryon distribution, for isocurvature fluctuations, diverge only as the logarithm of the length scale, so the power law cutoffs can be placed at very small and large wavelengths.)

Finally, we will see that with this spectrum it is reasonable to assume that star-forming mass concentrations are abundant at redshift $z \sim 100$. We must assume newly formed stars suppress further star formation to save material for the considerable activity that is observed to continue to $z < 1$, and we will assume that much of the matter is kept in dilute clouds ionized by the stars. Then equation (6.141) for the characteristic redshift at which radiation drag releases the plasma is

$$z_i \sim 100 . \tag{25.86}$$

With the flat spectrum of initial entropy fluctuations, the spectrum at decoupling of matter and radiation is flat at short wavelengths, has a bump at the matter-radiation Jeans length, and approaches zero at very long wavelength. The bump can be approximated as

$$\mathcal{P}_\epsilon \sim A[1 + K\delta(k - k_o)] , \tag{25.87}$$

where

$$K = 0.01 \, \text{Mpc}^{-1} , \qquad k_o = 0.01 \, \text{Mpc}^{-1} . \tag{25.88}$$

The delta function approximates the area of the bump, and k_o sets its length scale. This does not correctly represent the behavior at very long wavelengths, where the spectrum approaches zero.

At wavelengths shorter than the matter-radiation Jeans length $\sim k_o^{-1}$, the original baryon distribution is released from radiation drag at rest to grow as

$$\delta = \alpha D(t) + \beta/t , \tag{25.89}$$

where $D(t)$ is the growing solution to the linear density perturbation equation. Since $D \propto t^{2/3}$ at high redshift, the initial condition $\dot{\delta} = 0$ gives

$$\delta_k = \frac{3}{5} \frac{D(t)}{D_i} \epsilon_k . \tag{25.90}$$

Now we can set the normalization. Observations of the galaxy distribution on scales smaller than k_o^{-1} would miss the effect of the bump in the spectrum and detect the flat part, with initial autocorrelation function

$$\xi_i = \int d^3k P_\epsilon e^{i\mathbf{k}\cdot\mathbf{x}} = (2\pi)^3 A\delta(\mathbf{x}), \tag{25.91}$$

the delta function representing the assumed initially uncorrelated small-scale entropy fluctuations. With the growth factor in equation (25.90), the present value of the integral of the correlation function is

$$4\pi J_3 = (2\pi)^3 a_o{}^3 A \cdot \frac{9}{25}\left(\frac{D_o}{D_i}\right)^2. \tag{25.92}$$

This fixes the normalization, A, in equation (25.87) in terms of the integral J_3 of the galaxy two-point correlation function (eq. [21.100]).

The present autocorrelation function contributed by the delta function in equation (25.87) that represents the bump is

$$\xi_b(r) = \frac{2}{\pi} K k_o{}^2 J_3 \frac{\sin k_o r}{k_o r} \sim 0.001 \frac{\sin k_o r}{k_o r}. \tag{25.93}$$

The first zero of this function is at present wavelength

$$r_\xi = \pi/k_o \sim 300 \text{ Mpc}. \tag{25.94}$$

This says the mass density is fluctuating by the rms amount $\delta_b = \xi_b(0)^{1/2} \sim 0.03$ on scales $r_\xi \sim 300$ Mpc. This would produce large-scale streaming motions on the order of $H_o r_\xi \delta_b \Omega^{0.6} \sim 300$ km s^{-1} (eq. [5.119]). More careful estimates are in Peebles (1987b). The magnitude of the velocity is reasonable. Its coherence length seems large, but the observational constraints are still somewhat schematic.

In this model there are not appreciable mass fluctuations on scales large compared to the matter-radiation Jeans length (where $\mathcal{P} \to 0$ at $k \to 0$ in figure 25.4), so the CBR is not perturbed by a large-scale Sachs-Wolfe effect. That leaves as the dominant large-scale perturbation to the CBR the entropy fluctuations in equation (25.84). The mean square fluctuation in the temperature averaged through a square window with comoving diameter W_s larger than r_ξ is

$$\left(\frac{\delta T}{T}\right)^2 = \frac{25}{9}\left(\frac{D_i}{D_o}\right)^2 \frac{4\pi J_3}{9} \frac{6}{\pi W_s^3}. \tag{25.95}$$

An estimate of the integral J_3 is given in equation (21.100). The angular size distance is $H_o a_o r/c = 20$ (eq. [13.39]), so the angular scale $\theta = 3°$ subtends comoving length

$$W_s = a_o r \theta \sim 5000 \text{ Mpc}. \tag{25.96}$$

This is larger than the cutoff of the CBR temperature fluctuations at the matter-radiation Jeans length, and about equal to the smoothing length from electron scattering (eq. [24.87]). These numbers in equation (25.95) give an estimate of the predicted large-scale anisotropy in the CBR from the primeval entropy fluctuations. The somewhat more careful estimate in Peebles (1987a) is

$$\frac{\delta T}{T} = 1 \times 10^{-5} \quad \text{at } \theta_c = 3°. \tag{25.97}$$

As for the CDM model, this is comparable to the measured value in equation (21.115).

The model prediction for the quadrupole moment from the primeval entropy fluctuations depends on a long extrapolation of the fluctuation spectrum from its normalization, so it does not appear to be a useful test in the unfortunately loose present state of this picture. The CBR is perturbed by fluctuations of the gravitational potential on the scale of the peak of the power spectrum at the Jeans length; the effect is computed by Gouda, Sugiyama, and Sasaki (1991). It may be reduced by electron scattering, for the smoothing length can be larger than the Jeans length (eq. [24.87]). Motions in the plasma would reinsert smaller scale anisotropy in the CBR. As discussed in the last section, the effect seems to be small for the parameters we have adopted here, but it does depend on the details of the model for the plasma.

On scales smaller than the matter-radiation Jeans length, the rms fluctuation in the mass found within a sphere that contains mean baryon mass $\mathcal{M}$ at redshift z is

$$\delta_s{}^2 = \frac{4\pi J_3}{V} \frac{D^2}{D_o{}^2} \sim \frac{10^{15} \mathcal{M}_\odot}{(1+z)^2 \mathcal{M}}. \tag{25.98}$$

This extrapolates to $\delta_s = 1$ at redshift $z = 1000$ at

$$\mathcal{M}_{\text{nl}} \sim 10^9 \mathcal{M}_\odot. \tag{25.99}$$

At the mass $\mathcal{M} \sim 10^{11} \mathcal{M}_\odot$ characteristic of the bright part of an L_* galaxy, the rms contrast is $\delta_s = 0.5$ at redshift

$$z_* \sim 100. \tag{25.100}$$

This is the epoch at which two standard deviation upward fluctuations in the mass distribution are producing concentrations comparable to spheroid masses. The density in these concentrations would be comparable to the background at z_*, $\rho_* \sim 1$ proton cm^{-3}, about right for the brighter parts of galaxies. A second possibly significant coincidence is that z_* is comparable to the redshift at which radiation drag on the plasma becomes ineffective on the scale of the Hubble time (eq. [25.86]). All this could be taken as an invitation to consider the following scenario (Peebles 1988b).

At redshift $z \sim 1000$ neutral clouds of hydrogen at the nonlinear mass in equation (25.99) break away from the general expansion and the first stars form. We must assume star formation is inefficient, because the density is too high for galaxies (outside the cores of giant ellipticals; eq. [25.6]), but we might expect there are enough stars to keep the bulk of the matter ionized. That would recouple the plasma to the radiation drag, forcing the matter to keep expanding until the massive stars have died and allowed the plasma to recombine and collapse again. This feedback might keep the gas fraction high, so that as new generations of gas clouds form they erase their substructures. The situation would change at redshift $z \sim z_* \sim 100$, when radiation drag becomes unimportant. This could trigger an epoch of heavy star formation, maybe producing protospheroids at about the wanted mean density.

In this model we have been forced to introduce some inelegant and arguably unnatural elements, including a universe that is not Einstein-de Sitter, and a flat spectrum of entropy fluctuations that has a characteristic mass $\mathcal{M}_{nl} \sim 10^9 \mathcal{M}_\odot$. We must assume this physical scale was built in by the physics of the early universe. There is no fundamental theory for the origin of the assumed entropy fluctuations, or for why their spectrum might be roughly flat, or for whether the large-scale fluctuations in the entropy might be Gaussian. An extrapolation of the postulated flat spectrum to shorter wavelengths produces nonlinear fluctuations on scales less than $\mathcal{M}_{nl}$. The fluctuations thus could be Gaussian only if the baryon number were allowed to change sign, to pockets of antimatter. We must assume this does not happen, because it almost certainly would violate the remarkable purity of the material in rich clusters of galaxies (eq. [18.114]). All of this is unattractive. On the other hand, the model does seem to be capable of being adjusted to fit the main outlines of the observations as they are understood at the time this is written. That is, it serves as a demonstration that the existence of galaxies is not inconsistent with the standard cosmological model described in section 1.

26. Lessons and Issues

This final section presents a set of estimates of what we have learned about the open issues in cosmology, and where the evidence seems to be leading. As a part of this exercise, we will consider which parts of the subject are thought to be well enough understood that it may be possible to anticipate the kinds of surprises that could be in store. This necessarily is a personal assessment on uncertain ground, because we are considering work in progress. But it is a useful way to build an outline within which we can try assembling the impressions of what has happened in cosmology in the past decade. The fact that it is easy to list problems in fitting all the pieces together is a declaration of opportunity, not despair. It would be pleasant if resolutions to the central conundrums of cosmology were to be found in the scenarios that have been most closely studied, but considering the number of options yet to be examined in detail it would have to be counted as surprising.

The Large-Scale Structure of the Universe

The distribution of matter in the observable universe, averaged through a spherical window with diameter comparable to the Hubble length, has to be isotropic to better than one percent to agree with the isotropy of the X-ray background and the deep radio source counts. In the standard relativistic cosmological model, the mass distribution has to be isotropic to better than one part in 10^4 on the scale of the Hubble length to make the gravitational potential perturbations to the thermal cosmic background radiation consistent with the measured large-scale fluctuations, $\delta T/T \sim 1 \times 10^{-5}$ (eq. [21.115]). The rms mass fluctuation averaged through a spherical window with diameter $l = 60h^{-1}$ Mpc is found to be $\delta \mathcal{M}/\mathcal{M} \sim 0.3$, from the fluctuations in galaxy counts and from the large-scale peculiar velocity field of the galaxies. We saw in sections 21 and 25 that in the standard model, consistency with the constraint from the large-scale isotropy of the CBR means the fluctuations in the mass distribution on scales $\gtrsim 60h^{-1}$ Mpc have to be anticorrelated.

This is a remarkably sanguine view of our state of understanding of the large-scale structure of the universe; could things really be so simple and our understanding so complete? The answer, at least in part, is that the conclusion is tightly circumscribed, leaving ample room for surprises. Let us list a few possibilities.

The universe in the condition we observe may have edges, as predicted in the inflation scenario, and perhaps as Einstein had in mind in writing down equation (2.5); we can only say that the edges have to be outside the redshifts of useful observations (z on the order of five for galaxies, $z \gtrsim 20$ for the CBR, depending on when it was last scattered; eqs. [24.87] and [24.92]). By construction, however, this is not a surprise we may expect to encounter. The observations do not allow topological connections of the kind provided by a wormhole or jug handle in spacetime, unless the transit time through a wormhole is negligibly small, for otherwise there is no way to avoid differences in redshifts that would make the CBR anisotropic. One can imagine that the universe is periodic on a scale small compared to the Hubble length (though the topological connection would have to preserve parity to prevent the possibility that neutrinos return with the wrong helicity). Dyer (1987) notes that a small periodic universe is one way to eliminate the particle horizon and make the X-ray surface brightness and radio source counts close to isotropic (because a line of sight to high redshift wraps around a small periodic universe many times, averaging out the irregularities). The tests, which are not straightforward, are to establish whether the same objects are seen in different orientations in different directions in the sky, or whether a spatial periodicity reveals itself in deep redshift maps, as discussed by Broadhurst et al. (1990). A universe that is periodic within the Hubble length seems contrived, but then so do the other proposed remedies for the particle horizon.

Might we be at the center of an inhomogeneous but spherically symmetric universe? The success of the nucleosynthesis computation described in section 6 is an important check on the assumption that the cosmic background radiation we receive from the Hubble distance is homogeneous with the radiation that was the active ingredient in our neighborhood when the universe was young. The direct test from galaxy counts as a function of apparent magnitude or redshift indicates that the number density of galaxies at $z \sim 0.5$ is within a factor of about three of the local density. We are not likely to have a much tighter constraint from counts until it is learned how to distinguish a radial gradient in density from the effects of evolution and spacetime curvature. Meanwhile, the best argument against a spherically symmetric inhomogeneous universe is that the Milky Way does not appear to be a special galaxy, nor does it seem to be in a special place.

If there were large fluctuations in density, with a coherence length large compared to our Hubble length, it could explain why the universe behaves as if the mass density were lower than the Einstein-de Sitter value: maybe we are near a low spot. If such fluctuations were adiabatic, the local density gradient would be equivalent to an unobservable velocity transformation, so the lowest order observable effect would be a quadrupole anisotropy in the CBR. This is not observed: the quadrupole anisotropy seems to be continuous with the other low-order moments beyond the dipole. We noted in section 6 the idea that a uniform isocurvature gradient in the entropy per baryon could produce the observed dipole anisotropy, and that the anisotropy would be in the thermodynamic temperature, as observed. There are two tests for the conventional interpretation of the dipole as the result of the large-scale galaxy velocity field: our motion relative to the preferred frame in which the CBR is isotropic ought to agree with our motion relative to the frame in which the galaxy redshifts are isotropic on average, and we ought to be able to identify the galaxy fluctuations that trace the mass that gravitationally drives our motion. The preliminary evidence is that both agree with the velocity field interpretation, but the development of the tests will be followed with interest.

A homogeneous model universe can expand with shear and rotation, where the Cartesian components of the relative velocity field are of the form

$$v^\alpha = \sum_\beta H^{\alpha\beta} r^\beta . \tag{26.1}$$

The Hubble tensor $H^{\alpha\beta}$ can have a symmetric anisotropic part representing expansion and shear, and an antisymmetric part representing rotation. In this model the redshift-distance relation at small distances r^α is linear, but the shear and rotation introduce a quadrupole anisotropy in the Hubble parameter. This describes the local effect of the large-scale velocity field that is thought to be the cause of the dipole anisotropy in the CBR. The homogeneous anisotropic solutions allowed by general relativity are a very useful tool for the study of departures from the Robertson-Walker line element (Ryan and Shepley 1975; Zel'dovich and Novikov 1983). As a realistic model for our universe, however, these solutions seem to be of limited interest, for they require very special initial conditions: if the physics of the early universe allowed appreciable shear, why would it not also allow appreciable inhomogeneities?

The standard cosmological model allows surfaces of fixed world time to be curved. We have no reasonably definite alternative to the inflation concept which requires that space curvature is negligibly small, but since this is a wishful argument it may be well to bear in mind the possibility of open negative space curvature. We will return to this point in connection with the state of the observational tests of the cosmological models.

The tendency of galaxies to lie on sheets has been under discussion since the studies by de Vaucouleurs (1953), and became manifest even to theorists in the 1980s, from the first of the large-scale redshift surveys. The Great Wall in figure 3.6 is at least $100h^{-1}$ Mpc wide. Figure 3.3 shows that the galaxies in our immediate neighborhood, at distances less than about $5h^{-1}$ Mpc, define a plane, the nearby part of the de Vaucouleurs Local Supercluster. The sheet of galaxies in panel (a) of figure 3.7 running down to the right from the Virgo cluster at $15h^{-1}$ Mpc distance is part of this plane. The last panel in this figure shows that the nearby clusters of galaxies, at distances $\lesssim 60h^{-1}$ Mpc, prefer the same plane. Shaver (1991) finds a similar tendency in the radio galaxies at this distance, and Tully (1987b) finds indications that the same is true of still more distant Abell clusters.

For all we know, such wall-like structures or other patterns could run across the Hubble length. This is not to say that a substantial fraction of the matter within redshift $z \sim 1$ could lie on a few sheets, for that would violate the striking large-scale isotropy of the visible universe. Thus the large-scale continuation of the local sheet of galaxies in the distributions of nearby clusters and radio sources is seen to be a pattern running through the much smoother general distribution of galaxies in panel (b) of figure 3.7 and in figure 3.8. It is easy to find an observationally acceptable prescription for this effect. Place galaxies in bounded irregular cluster balls, as in the construction in figures 19.1 and 19.2, with the cluster balls placed independently and at random. Next, rotate each of the cluster balls that lie near a given plane so the long axes of each of these clumpy structures are parallel to the plane. This produces a feature that could be quite prominent in galaxy maps, and could run across the Hubble length without appreciably perturbing the cosmic background radiation, because there has been negligible large-scale rearrangement of mass. The wake of a long, roughly straight cosmic string could produce a similar effect by promoting the formation of clusters and the radio sources that prefer dense regions. The galaxies would have to have formed prior to the wake, for as we have discussed in the last section the walls do not seem to contain a distinct population of galaxies that might have been seeded by the strings capable of making the walls. The Zel'dovich pancake collapse illustrated in figure 22.1 produces wall-like structures, though again there are special conditions. In a fully developed pancake collapse much of the mass is in the network of pancakes, which is not what we want. That is, the walls would have to be pancakes that have collapsed only in the densest spots, in a flow that is coherent over a scale of tens of megaparsecs. If the flow is produced by Gaussian primeval density fluctuations, it would require that the fluctuation spectrum has a large-scale peak, as in the model in equation (25.87).

A major task for the new generations of deeper redshift surveys is to sort out the nature of the large-scale galaxy distribution. Will very deep redshift maps show patterns different from what would be expected from a mildly anticorrelated distribution of cluster balls? The challenge may be to distinguish patterns from the statistical accidents the eye is so remarkably adept at finding even in noise. If real, we may be lucky enough to see that the patterns suggest a dynamical process. Thus, in models such as CDM, the departures from Gaussian fluctuations on large scales are those due to the onset of nonlinear growth of initially Gaussian fluctuations, to be distinguished from a non-Gaussian scenario such as cosmic string wakes. The gravitational growth of departures from an initially Gaussian distribution is discussed in LSS (§18), Melott (1991), Coles and Frenk (1991), and Bouchet et al. (1992).

Another result from deeper redshift surveys may be a test of the prediction from the measured large-scale anisotropy of the CBR that the galaxy distribution is anticorrelated on large scales (assuming galaxies trace mass). The test may come first from the large-scale structure of the peculiar velocity field, rather than directly from the counts, because the velocity is much more sensitive to large-scale low amplitude fluctuations in the mass distribution.

Could the universe be evolving toward order out of chaos? General relativity theory allows us to assign the distribution and the peculiar motion of the mass on an initial hypersurface. In linear perturbation theory the choice of these two functions is equivalent to the assignment of the growing and decaying modes of the departure from homogeneity. We are allowed to choose initial conditions in which only the decaying mode is present and the universe is expanding from chaos to homogeneity. But since the universe is observed to be quite close to homogeneous in the large-scale average, and the growing mode of the density perturbation is allowed, the physically reasonable condition within the standard model surely is that the growing mode dominates, that is, that peculiar velocities and the departures from a homogeneous mass distribution are growing larger. This is assumed in the discussions of the relations between the fluctuations in the mass distribution, the large-scale peculiar velocity field, and the anisotropy of the CBR (section 21). The tests for a consistent picture for these effects also test the prediction that the universe is growing more chaotic, rather than the other way around. This prediction is what allows us to analyze the physical state of the early universe.

Unless the gravitational instability of the mass distribution has been temporar-

ily countered by recent explosive disruptions of galaxies or the loss of gravitational binding through the decay of dark matter, deep redshift surveys ought to
be consistent with a growing clustering of galaxies, the integral of the two-point
correlation function varying as (eq. [22.1])

$$J_3(r) \propto D(t)^2 , \tag{26.2}$$

where the function $D(t)$ is given by equation (13.78). It will be interesting to
see whether the power law form of the galaxy two-point correlation function
$\xi(r)$ is a transient accident, as one finds in numerical simulations in the $\Omega = 1$
CDM model, or whether the low-order correlation functions evolve according to
a scaling law, as might be suggested by the observation that the present form of
$\xi(r)$ is remarkably close to a power law. Other aspects of cosmic evolution are
discussed in the following subsections.

The Contents

The broad variety of the known classes of galaxies surely is bounded by
what can be observed with available angular resolution and sensitivity to surface
brightness. The point was emphasized by Arp (1965), who cited as examples the
dwarf spheroidal satellites of the Milky Way, with mean surface brightnesses so
low that these objects are known only because the individual stars are detected;
the quasars or quasi-stellar radio sources, then recently discovered, and so named
because they are hard to distinguish from stars; and Arp's own observations of
compact dwarf galaxies that again are distinguished from stars only with difficulty. This line of thought is developed further by Disney and Phillipps (1985).
There are several reasons why one might have expected that irregular or low surface brightness galaxies prefer the voids marked out by the regular L_* galaxies.
First, the voids would be a less hostile environment for the galaxies with low
binding energy or low density that might be most easily disrupted by the tidal
fields or blast waves from massive young galaxies. Second, the CDM biasing picture has taught us that the voids would be an excellent place to put most of the
mass, along with most of the failed seeds of galaxies. Third, the voids would seem
to be a good place to put the remnants of the Lα forest. But this is theory. Figures 3.3 and 3.4 show that the evidence from our immediate neighborhood, where
inconspicuous galaxies are best sampled, is that there is little difference between
the space distributions of L_* galaxies and the much more numerous dwarfs. Section 25 reviews the extensive literature showing that this seems to be a general
result.

The upper bound on the surface brightness of the extragalactic sky allows the
mean luminosity density to be an order of magnitude larger than the known contribution from cataloged galaxies (eq. [5.168]). That is, this constraint allows
ample room for the discovery of new classes of objects that are ultracompact

or have low surface density or low luminosity, including individual intergalactic stars. If uncataloged classes of objects have the mass-to-light ratio characteristic of normal galaxies, the upper limit on the mean luminosity density implies the mean mass density from these objects is an order of magnitude below the critical Einstein-de Sitter value. However, since there is no need to suppose an unknown class of galaxies has the mass-to-light ratio characteristic of known ones, the more important constraint comes from the dynamical mass measures to be discussed in the next subsection. It is easy to imagine that an unknown class of extragalactic objects is distributed in space in a quite different way from that of the known galaxies. That certainly is not the straightforward extrapolation of the observations, however, which show the dwarf and irregular galaxies cluster with the bright ones on scales greater than about $1h^{-1}$ Mpc. Thus, the sanguine and reasonably well founded conclusion is that we have a fair sample of the character of the space distribution of luminous matter.

Could there be a significant class of dark galaxies, with masses comparable to L_* galaxies but none of the light? If the luminous baryons were striped from an L_* galaxy, it likely would make the surface density lower than the critical value for gravitational lensing, so such dark galaxies might be detectable only through their contributions to the mean mass density and to the dynamics of systems of galaxies. It seems unlikely that there are any massive dark galaxies in the Local Group, because that would spoil the successful model in figures 20.2 and 20.3. If there were appreciable amounts of mass in dark galaxies in the voids, the puzzle would be why they do not cluster with all the known luminous classes of galaxies.

Suppose a sphere with diameter $D = 60h^{-1}$ Mpc is placed at random within the Hubble distance. The estimates developed in section 21 say that if the trial is repeated many times, the rms fluctuation in the mass within the sphere is about 30%. Voids this large are not known, but the statistic says they may exist at three standard deviations. Assuming the sphere is not placed on a large void, are the galaxies it contains a statistically fair sample of the population within the Hubble length? It could not be entirely so, for the sphere may contain several rich clusters, or none. Would the L_* galaxies be a fair sample? The lesson from the success of the Tully-Fisher relation and related methods in predicting galaxy distances is that galaxies are close to uniform in the global physical properties reflected in their mass concentrations and luminosities, and that in this respect the samples would be fair. Would there be the same fractional departures from the mean in the mass and the number of L_* galaxies in the sphere? As we have noted, the test may come from studies of the large-scale velocity field.

Antimatter exists in the laboratory, and it would be elegant if there were regions in the observable universe where our sphere lands on galaxies made of antimatter. The isotropy of the CBR constrains the mass per unit area in domain walls that might separate regions of matter and antimatter (section 16). The domain walls would not promote linear patterns of galaxy concentrations, because in conventional physics domain walls gravitationally repel both matter and antimatter. That would be useful, however, for it would help suppress the annihilation of matter and antimatter. We noted in section 18 that the regions of matter and antimatter would have to be large enough so that they rarely mix in the plasma in clusters of galaxies, for the purity of the matter or antimatter in intracluster plasma is striking (eq. [18.114]).

Magnetic fields exist in galaxies and in damped Lα clouds at redshifts $z \sim 3$, and we saw in the last section that a primeval magnetic field could play an interesting dynamical role in galaxy formation. If a seed magnetic field predates the galaxies, where did the field come from? Turbulence could act as a dynamo (Harrison 1970b). We argued in section 22 that if there were turbulent motions at very high redshift, the turbulence would have to have dissipated before decoupling, for otherwise the dissipation at decoupling would have deposited the mass in clumps as dense as the cores of galaxies. Other options are that the field is the remnant of galaxies from the last cycle of a phoenix universe, or it is the product of physics yet to be discovered that operates in the very early stages of expansion of this cycle.

Gravitational waves exist (Taylor and Weisberg 1989), and it would be easy to imagine they were important actors in the very early universe. The remarkable constraint from the theory of light element production is that the energy density in gravitational waves at redshift $z \sim 10^{10}$ is less than that in the thermal energy density in a family of neutrinos. The gravitational waves that would be produced during an epoch of inflation are bounded by the quadrupole anisotropy of the CBR (eq. [17.31]). Thus in a nearly scale-invariant inflation that produces interesting adiabatic density fluctuations, the relict strain on the scale λ is $h \sim 10^{-4} H_o \lambda / c \sim 10^{-27}$ at one kilometer wavelength. This is impressively small, but maybe not out of experimental reach. And there is the exciting chance of detection of the gravitational waves produced during one of the phase transitions of particle physics or in the dissipation of cosmic strings.

A lesson from condensed matter and particle physics is that it is easy to imagine, and perhaps likely, that there are other relict fields frozen as the universe expanded and cooled through phase transitions. This could have left an interesting amount of mass in fields in the forms of strings, global monopoles, textures, low-pressure dark matter, or a homogeneous component that masquerades as a cosmological constant.

Black holes exist in general relativity theory and very likely exist in Nature in massive star remnants and the nuclei of galaxies. Could there be supermassive black holes in places where something went very wrong with the expansion of the universe? The mass of a cluster of galaxies could not be dominated by a single black hole, because the velocity dispersion as a function of radius is that characteristic of a distributed mass rather than a pointlike concentration. If a black hole with the mass characteristic of a cluster of galaxies, $\mathcal{M} \sim 10^{15} \mathcal{M}_\odot$, were placed in a nearby void, it would be detectable by the hole it makes in the CBR, though the size of the hole would only be one arc second at 10 Mpc distance, and it would be a powerful gravitational lens. However, a highly special arrangement would be needed to explain why such a massive object would not make itself highly visible by gravitationally attracting and destroying galaxies. People have on occasion debated the merits of white holes or other naked singularities as possibly useful models for what is happening in quasars and other active nuclei. The picture is ad hoc, but maybe it is worth bearing in mind at least until we have better foundations for a theory of galaxy formation along more conventional lines. And black holes with masses less than about $10^6 \mathcal{M}_\odot$, to avoid overheating the thin disk of our galaxy by gravitational perturbations, would be an acceptable solution to the dark matter puzzle (Lacey and Ostriker 1985).

Nonbaryonic dark matter exists in neutrinos, and there is a good case for the existence of other kinds of particles that could make an appreciable contribution to the mass of the universe. Nonbaryonic matter could resolve the marginally significant discrepancy between the dynamical mass estimates and the baryon density predicted by the standard homogeneous model for the origin of the light elements, and it could provide the mass predicted by the Einstein-de Sitter model. Structure formation does not require nonbaryonic matter, as shown by the isocurvature baryonic scheme, but galaxy formation in an Einstein-de Sitter model dominated by nonbaryonic matter does seem to be the more elegant possibility. And it certainly is an encouraging coincidence that, quite independent of the problems in cosmology, particle physics has produced at least three credible types of candidates for dark matter: neutrinos, axions, and supersymmetric particles.

It is easy to list the classes of solutions to the dark matter puzzle (sections 18 and 20): (1) the mass is in a nonbaryonic form with stress small enough to allow it to be confined by the gravitational field of a galaxy; (2) the mass is in baryons locked up in an inert form such as black holes, low mass stars, or stellar remnants; (3) the problem is in the misapplication of Newtonian mechanics. We will argue in the next subsection that the successes of the standard model suggest the third option is the least likely but is certainly not ruled out. The laboratory searches for tests of the first option and the astronomical searches for the second are central topics for cosmology in the coming decade. The odds on whether the solution will be found in the first or second class depend on the amount of dark mass. If the cosmological tests indicated that the mean mass density agrees with the Einstein-de Sitter model, the mismatch with the baryon density predicted by the model for the origin of the light elements would argue for nonbaryonic matter. And although small black holes would do, the offer from particle physics would be difficult to resist. If the density in movable matter were shown to be about 10% of the Einstein-de Sitter value, it would be tempting to look for a resolution that uses baryons. The state of the measurements of the density parameter is considered below.

We have a considerable list of things, from primeval magnetic fields and gravitational waves to cosmic fields and nonbaryonic matter, whose discovery would not be entirely surprising, in the sense that there is good reason to think it could happen. A convincing detection of any of these relicts would be a welcome and wonderfully important result for cosmology.

The Physics

In the standard cosmological model, general relativity theory fixes the relation among the parameters of the Robertson-Walker line element: the rate of expansion of the universe, the radius of curvature of space sections at fixed world time, and the mean mass density, which might consist of ordinary movable material and something that behaves like a cosmological constant. One also uses the gravity theory of the standard model to relate the irregularities in the mass distribution to the galaxy peculiar velocities and the large-scale anisotropy of the CBR. The heavy interdependence of theory and observation is the essential glue in a mature and believable physical science. But in a developing field such as cosmology it adds a layer of uncertainty, for we must be prepared to encounter new physics on the poorly explored ground of extragalactic astronomy and cosmology. The conventional wisdom, which is adopted here, is that new physics is encountered so rarely that we do well to devote most of our collective attention to the conventional and accepted variety until we are driven away from it. Examples of the absence of surprises in physics are the remarkable stability of the line ratios in the spectra of extragalactic objects, indicating that the atomic and nuclear physics that determine them is little affected by time or place; the concordance of stellar evolution ages and radioactive decay ages, still a crude test to be sure, but evidence that stars have been evolving and nuclei decaying by the weak or strong interactions at a roughly constant relative rate for a very long time; and the success of the theory of the origin of the light elements, which indicates that nuclear physics and the low-energy limit of the weak interactions have held steady since the universe was a few seconds old. This last example is a straightforward computation but an enormous extrapolation from the universe we see. One pays close attention to the results because the present evidence is that they are observationally successful. Helium is the simplest case; the predictions and tests for the less abundant isotopes are more subtle and for that reason seem more likely to be debated. If it could be shown that there is a low helium abundance in a damped $L\alpha$ system, or some other object that looks to be a fair sample of cosmic matter, it would be a serious problem for the standard cosmological model. But that has not happened yet.

If the model for the origin of the light elements continues to be as successful as now appears, it is evidence that the relativistic relation between the mass density and the expansion rate is a good approximation back to redshift $z \sim 10^{10}$. Since the universe at this redshift is compact and likely to be very close to homogeneous within the part we can observe now, this need not be equivalent to a test of the prediction that the inverse square law of gravity applies at large separations. An exciting development here is the observation of gravitational lensing, which tests gravity physics on a scale of about 100 kpc in clusters of galaxies. The present indications are that general relativity theory gives a better fit than a quasi-Newtonian theory without the factor of two in the index of refraction (eq. [10.89]). It will be of considerable interest to see the refinements of this comparison of the lensing effect with the prediction from detailed analyses of the cluster mass distribution based on the motions of the galaxies and the density and temperature runs in the intracluster plasma. On still larger scales, the inverse square law does give a reasonable-looking account of the large-scale velocity field in terms of the gravitational acceleration of the mass fluctuations traced by the galaxies, though again this is a developing field. Even at the present relatively crude level of understanding of these two tests, it is striking to see how well conventional gravity physics has done in an extrapolation of more than ten orders of magnitude in length scale from the precision tests in the Solar System and binary pulsar systems. This is another example of the remarkable rarity of surprises in physics.

The Cosmological Tests

Sections 15 and 17 detail the reasons for a strong theoretical preference for the Einstein-de Sitter model. If the observations show that this model fits the cosmological tests, it will be a triumph for the standard physics and for the procedure of looking to the simplest possibility first. The procedure does not always work: a cautionary example is the contrary behavior of the space distribution of dwarf galaxies. If the Einstein-de Sitter model fails and the observations are fit by another set of parameters in the standard model, it will still be a triumph for the physics, but we will have to conclude that Nature has found another path that we are likely to persuade ourselves is even more elegant. There is no clear case for which it will be, though at the present time the Einstein-de Sitter model is not the straightforward interpretation of the bulk of the evidence.

The dynamical measures reviewed in section 20 show that the contribution to the mean mass density by the material concentrated in and near galaxies is within a factor of two of $\Omega_d = 0.1$. The situation is illustrated in figures 3.3 and 5.5, which show the situation in our neighborhood. The galaxy distribution is very clumpy, while the Hubble flow is quite smooth. This is not what one would expect if the mass density of the Einstein-de Sitter model clustered with the galaxies on the scale of this map.

The indications from the large-scale velocity field are that the density parameter of the mass component clustered with galaxies on scales greater than about $20h^{-1}$ Mpc is close to unity. If this is so, galaxies are segregated from mass, as in the CDM biasing picture. But since the large-scale dynamical measurements are a difficult art — and as we will discuss there are some problems with this high density — we will bear in mind the alternative, that what is measured on small scales, $\Omega_d \sim 0.1$, is all there is.

If the density parameter is unity, and this is revealed in the large-scale veloc-ity field, then there has to be a characteristic length scale, $\Xi_{g\rho}$, beyond which galaxies cluster with mass. There are several arguments indicating $\Xi_{g\rho}$ could not be as small as $3h^{-1}$ Mpc. First, the map in figure 3.3 shows a region $8h^{-1}$ Mpc across, in which there is strong clustering in galaxies but not in mass if $\Omega = 1$ (for the relative peculiar velocities in fig. 5.5 are small). Second, if $\Xi_{g\rho}$ were as small as $3h^{-1}$ Mpc one would expect that mass clusters with IRAS galaxies around the Virgo cluster. But if the large virgocentric flow that would imply (eq. [20.59]) is canceled by a tidal field, why does the shear not show up in peculiar veloc-ities in other directions, as probed in figure 5.5? Third, the mass-to-light ratio within the Abell radius of a rich cluster is a factor of five to ten below the global value in an Einstein-de Sitter universe. It is difficult to see how the falling ve-locity dispersions in the outer parts of the clusters could be reconciled with the assumption that the extra order of magnitude in mass is in a halo with a core ra-dius of $3h^{-1}$ Mpc. If $\Omega = 1$ and this mass density is detected in the large-scale velocity field then $\Xi_{g\rho}$ cannot be much larger than scale of the measurements, $\sim 30h^{-1}$ Mpc. The conclusion is that if Ω is close to unity the characteristic length $\Xi_{g\rho}$ on which mass clusters with galaxies is

$$\Xi_{g\rho} \sim 10h^{-1} Mpc . \qquad (26.3)$$

The broad characteristic length for the dominant component of the mass might result from a pressure or other stress too large to allow the mass to cluster with the galaxies on smaller scales, or it could be that galaxies were formed in islands in a less strongly disturbed background mass distribution. The latter is the prediction of the CDM biasing picture. Challenges in this model are to explain why $\Xi_{g\rho}$ is so large, why the clouds in the baryon distribution observed at $z \sim 3$ did not populate the voids with observable remnants, and why the dwarf galaxies show so little evidence of segregation from the giants. The last point is detailed in section 25.

We noted in section 25 that cosmic field scenarios for the origin of structure also have a problem with the biasing required in an Einstein-de Sitter universe. In these pictures the gravitational perturbation caused by an inhomogeneous distri-bution of cosmic field energy causes the assembly of mass concentrations such as galaxies and clusters of galaxies. Unless the seeds for the clusters know where the galaxies are, the cluster seeds gather a fair sample of mass and galaxies. The well-established measurements of the mass per galaxy in clusters show us that in this case the density parameter is well below unity. An option is that galaxies form already clustered, from seeds placed in a clustered pattern. The difficulty is with the perturbations from field structures present after galaxy formation that tend to draw together fair samples of the global mixture of mass and galaxies. Under the postulate that the large-scale velocity field reveals that the density parameter is unity, we want this to happen on scales greater than $\Xi_{g\rho}$. The conundrum is why it did not happen on the scales intermediate between galaxies and the galaxy-mass clustering length $\Xi_{g\rho}$.

Another way to biasing by nongravitational stresses assumes that some fraction of the dark matter has decayed. We noted (in the discussion in section 25 of the adiabatic hot dark matter model) that this can cleanly remove mass from the galaxies without separating the giants from the dwarfs. If the decay products were relativistic and the redshift at the decay large enough to have allowed the growing density perturbation mode to become dominant again, the density parameter derived from large-scale peculiar motions would agree what is observed on smaller scales, $\Omega \sim 0.1$. A more baroque but observationally interesting scenario assumes the decay products are nonrelativistic, so they are deposited with coherence length $\Xi_{g\rho}$, in line with the present auguries.

The age t_o computed from high redshift offers a powerful constraint on the cosmological parameters within the standard model, for t_o can be no less than the ages of the stars and elements. It is thought that the oldest stellar evolution ages are not likely to be less than about 14 Gy years (eq. [5.77]). If the density parameter Ω were close to unity, and the material nonrelativistic, the reasonable presumption would be that the universe is well approximated by the Einstein-de Sitter model, and the age would be predicted to be $t_o = 2/3H_o$. That means Hubble's constant could be no more than about 50 km s^{-1} Mpc^{-1}. If the density parameter and the magnitude of the cosmological constant were small, the product $H_o t_o$ could approach unity, and the age constraint would allow $H_o = 70$ km s^{-1} Mpc^{-1}. If $H_o t_o$ were larger than unity it would require the presence of a term in the expansion rate equation that acts like a cosmological constant. This is one of the reasons for the intense interest in the calibration of the distance scale.

Other constraints on the parameters of the Robertson-Walker line element follow from the classical cosmological tests discussed in sections 13 and 14. The exploration of what one learns from deep redshift surveys promises to be one of the major advances in cosmology in the coming decade, and the major complication almost certainly will continue to be the evolution of the objects one would like to use as markers for the geometry and the time history of the expansion rate. As an example, let us compare the predicted count per steradian as a function of redshift for a class of conserved objects (constant comoving number density) in two cosmologically flat models:

$$\frac{d\mathcal{N}}{dz} = \begin{cases} 4C\left[1 - (1+z)^{-1/2}\right]^2 (1+z)^{-3/2}, & \text{for } a \propto t^{2/3}, \\ Cz^2, & \text{for } a \propto e^{H_o t}. \end{cases} \quad (26.4)$$

The first is the Einstein-de Sitter model and the second is the limiting low-density case, the de Sitter solution. These expressions have the same normalization at low redshift, $d\mathcal{N}/dz = Cz^2$. At redshift $z = 0.25$, the predicted value of $d\mathcal{N}/dz$ in the second model is twice the prediction of the Einstein-de Sitter model. This is an easily measured difference at a redshift where there already are useful estimates of $d\mathcal{N}/dz$. And the counts are well above the Einstein-de Sitter prediction (Broadhurst, Ellis, and Shanks 1988; Colless et al. 1990; Lilly, Cowie, and Gardner 1991; Cowie, Songaila, and Hu 1991). The problem is that even at this relatively modest redshift the fraction of blue galaxies is larger than is typical at low redshift, as was first observed in deep imaging (Kron 1980; Tyson 1988). Since this tells us there has been appreciable evolution since $z = 0.25$, how does one distinguish the effects of the cosmology from evolution of the galaxies?

The way out of the conundrum may be to seek a consistent picture from over-constrained measures of galaxy evolution and the cosmological parameters. Perhaps many of the now inconspicuous dwarfs were lit by short-lived generations of massive stars at $z \sim 0.25$. If so, it ought to be possible to show where the remnants are. If they merged with L_* galaxies, as is suggested by the CDM model (White and Frenk 1991), it would be surprising that the pieces are not strongly clustered at $z \sim 0.25$ (Efstathiou et al. 1991), and that the mergers did not strongly disturb the spirals, which are the most abundant bright galaxies. (That is why the redshift for disk formation in table 25.1 is taken to be $z \sim 1$.) If there was a formerly bright population of galaxies which now are dwarfs, it could be possible to find evidence of it in the star populations in nearby dwarfs (Babul and Rees 1992). The debate on these points has developed into a rich subject; examples of the art are in Broadhurst, Ellis, and Glazebrook (1992), Guiderdoni and Rocca-Volmerange (1990a,b), and Yoshii and Fukugita (1991). The conclusions are not yet fit for review, but a few observational results are of evident general interest. At redshift $z = 1$ the brightest galaxies or their ancestors were not much more luminous than the brightest galaxies at low redshift. At $z \sim 0.5$ the distribution of redshifts of galaxies selected by apparent magnitude is about what would be expected if the luminosity function were roughly similar to what is observed at low redshift, after allowance for the larger numbers of blue galaxies at luminosities close to but less than L_*. This is seen in figure 5.6. And as we noted in connection with figure 3.2, the count $d\mathcal{N}/dz$ is roughly three times what would be expected for an unevolved population in an Einstein-de Sitter universe.

The omens from the flow of preprints at the time this is written may be summarized as follows. Current estimates of the distance scale favor a value of Hubble's constant of about $H_o = 80$ km s^{-1} Mpc^{-1} (Jacoby et al. 1992; eq. [5.78]). If this were close to the correct value the likely interpretation would be that the density parameter Ω is well below unity. (The alternative within the standard model is that Ω and Ω_Λ in eq. [5.53] both are large and accidentally close to balancing each other, an even uglier solution.) The low value of the density parameter would help account for the large counts $d\mathcal{N}/dz$ (fig. 13.8). It would agree with the small-scale dynamical estimates of the mass density, and in this case the dynamical tests on larger scales would be exceedingly interesting as a probe of consistency of the physics. The absolute magnitudes of the most luminous objects selected by apparent magnitude show little evolution from $z = 1$ to the present. The straightforward interpretation is that this is because these are nearly ordinary galaxies. A low density parameter in a cosmologically flat universe would be beneficial to this interpretation, for it would allow the galaxies at $z \sim 1$ ample time to pass through the expected early phase of rapid luminosity evolution (figure 13.1). However, the low-density, cosmologically flat case is not consistent with the preliminary reading of the Fukugita-Turner test in figure 13.12. The conflict is considerably eased if $\Lambda = 0$, though this conflicts with inflation.

It would be difficult to overstate the effect a laboratory detection of nonbaryonic dark matter would have on cosmology in general and the approach to the cosmological tests in particular. Absent this development, and if there continue to be indications that the density parameter is low, attention surely will turn to the material we know is present in abundance: baryons (Shanks et al. 1991; Daly and McLaughlin 1992). At the current large estimates of Hubble's constant, the assumption that the dark matter is baryonic likely would require an adjustment of the model for light-element production (table 20.1), but the idea has the appeal of minimum hypotheses.

All of this discussion is based on omens, not the interlocking network of firmly established measurements that we may hope to see emerge in the next decade. It has been sixty years since people recognized that within the standard cosmological model one can predict how the universe ends. The exciting news is that a rich suite of constraints is developing that might be capable of being refined into a believable case for which it is — whether collapse to a big crunch, expansion into the indefinite future with many little crunches as bound systems collapse, the option in inflation that the universe ultimately reveals its primeval chaos, or something completely different. Since people have been working on the problem for more than sixty years, perhaps the most surprising result would be that in the next decade a consistent and believable picture for the values of the cosmological parameters is at last established.

Cosmic Evolution

The properties of extragalactic objects show a clear correlation with redshift. Examples are the Butcher-Oemler effect in the colors of galaxies, the alignment of radio lobes with optical images of radio galaxies at higher redshifts, the increase in the fraction of galaxies with high star formation rate with increasing redshift, and the marked difference in the abundance of $L\alpha$ forest clouds at high and low redshifts (sections 7 and 23). The general trend of these effects is in the direction one would expect if objects at higher redshifts were younger. We have emphasized the bad news — that evolution greatly complicates the cosmological tests. The good news is that these observations provide manifest evidence that the contents of the universe really are evolving, as has to be the case in the standard cosmological model. The rate of evolution is higher than was anticipated in many of the discussions prior to the discovery of the Butcher-Oemler effect, but in retrospect it seems reasonable enough that the timescale for the evolution of the contents of the universe should prove to be comparable to the Hubble timescale for the expansion. And we surely are being taught something of value about the way galaxies formed from the way they evolve. Let us consider some of the elementary lessons.

As we remarked in connection with the cosmological tests, the rapid evolution observed at redshifts less than unity could be taken to mean that galaxies as we know them were assembled fairly recently. There are cluster members at redshifts in the range $0.5 \lesssim z \lesssim 1$ that have the spectra and luminosities one would expect if they were normal giant elliptical galaxies, but one might assume that these are the earliest products of galaxy assembly that for the most part is taking place at $z \lesssim 1$. Galaxy formation by the collapse of a nearly smooth gas cloud out of material that is close to the primeval state is not an option if galaxies are assembled at $z \lesssim 1$, because we know substructure is well developed at $z \sim 5$. Thus the simple pictures for galaxy assembly at low redshift are the merging of gas-rich dwarf systems or the fragmentation of larger ones. Fragmentation is theoretically attractive but observationally problematic. Thus we remarked in the last section that the Local Group is forming now by the assembly of galaxy-size pieces, and if other spirals were formed by fragmentation rather than hierarchical assembly, one surely would not have expected to see the striking continuity of the Faber-Jackson/Tully-Fisher relations. The merging of galaxy fragments at low redshifts is suggested by the numerical computations of the evolution of the mass distribution in the $\Omega = 1$ CDM model, and by the large galaxy counts $d\mathcal{N}/dz$ at $z \sim 0.5$. However, the easy interpretation of the near normal luminosities of the galaxies at this redshift, and of the other evidence summarized in connection with table 25.1, is that present-day disks of spiral galaxies were in place at $z \sim 1$, and that the mass within the disks had been assembled before that. That is, the straightforward reading of the evidence is that the mass concentrations in L_* galaxies at radii less than about $10h^{-1}$ Mpc were in place at $z \sim 1$.

Observations of radio galaxies and quasar absorption line spectra have given a surprisingly detailed picture of what the universe was like at redshifts $z \sim 3$ to 5 (section 23). There are three very notable results for our purpose. First, at $z \sim 5$ there is no undisturbed ground: much of the baryonic matter has been gathered into clouds, and whatever remains between the clouds has been heavily ionized. Second, a froth of atomic hydrogen with close to primeval composition nearly fills space and is accompanied by a much greater amount of ionized matter in a clumpy distribution with a mean comoving mass density on the order of what is present now in the bright parts of galaxies. Third, we have a reasonable case for interpreting the damped $L\alpha$ clouds, those with the highest column densities, as the long-sought primeval young galaxies: they have about the right column density, a reasonable interception rate along the line of sight, and a reasonable heavy element abundance. Some contain magnetic fields, and some host active nuclei that have produced mature-looking radio lobes. There is evidence that some have thin disks, but these need not be the same as the ones present now, and there seems to be no reason to believe that an astronomer transported in time back to redshift $z \sim 3$ would be willing to classify the damped Lα systems as protoellipticals and protospirals. The images of the very luminous high redshift radio galaxies show that some of what may be a very heterogeneous class of primeval galaxies are irregular and puffy. Some might resemble present-day gas-rich dwarf galaxies that are merging to form L_* galaxies. And perhaps some would be recognizable as ordinary younger L_* galaxies.

Primeval galaxies or their ancestors are detected at $z \sim 3$ in emission, as radio galaxies, Lα emitters (Spinrad 1989), and 21-cm emitters (Uson, Bagri, and Cornwell 1991). Measures of the spatial clustering of these young galaxies are of great interest and may be within reach. The absorption lines from heavy elements tend to appear in clusters along the lines of sight in quasar spectra, suggesting the higher surface density clouds in which they appear are clustered in space. The lesson from the CDM model is that if galaxies are being assembled at $z \sim 3$ to 5 out of density fluctuations that are close to Gaussian, one would expect that the giants are forming already clustered in the denser regions, while the protodwarfs favor lower density spots. If this were so, the gravitational growth of clustering since then would have to have erased the segregation of giants and dwarfs to produce the present situation illustrated in figures 3.3 and 3.4, in which giants and dwarfs are well mixed. The observations at $z \sim 3$ may not readily distinguish between this picture and galaxy assembly at a considerably higher redshift, however, for the example from low redshifts suggests the active galaxies most likely to be detected as emitters at $z \sim 3$ are more strongly clustered than the average galaxies. Zel'dovich pancakes may be present at $z \sim 3$ to 5. They would have to consist of material that already has been disturbed by something else to produce the structure in the Lα forest, and we have argued that typical L_* galaxies are not likely to be fragments of these pancakes. A test would be welcome from a measurement of the space distribution and abundance of the HI emitters. Perhaps it will be seen that they outline the structures now traced by the great clusters of galaxies.

The conventional picture is that the rotational angular momenta of the galaxies come from tidal interactions with neighboring protogalaxies. The arguments in sections 22 and 25 suggest that the circulation now present in the disk of a spiral galaxy came out of material that at $z \sim 3$ is settling from the region of the massive halo of the protogalaxy. This picture is not known to be wrong, but some positive evidence would be welcome. One test is that the tidal torque tends to cause the spin axis of a spiral galaxy to be perpendicular to the direction to the neighbor mainly responsible for transferring the angular momentum (Gott and Thuan 1978). This means the angle between the spin axes of the two galaxies would be predicted to be biased away from $\theta \sim 90°$, relative to an isotropic distribution. The brightest members of clusters of galaxies do tend to line up with the long axis of the galaxy distribution, in what may be the shear version of the tidal effect (Rhee and Katgert 1987; Lambas, Groth, and Peebles 1988). The general distribution of galaxy position angles in the sky is remarkably close to isotropic (Hawley and Peebles 1975; Djorgovski 1987), but there is considerable noise from pairs seen accidentally close in projection. Measures of the distribution of relative position angles of neighboring spirals should be considerably improved in the large-scale redshift surveys.

One can ask whether the expansion of the universe traces back much beyond the largest observed for quasars, for the relativistic model does allow a bounce if the cosmological constant is large enough. Given the density parameter Ω, and assuming the material pressure may be neglected, the maximum value of the redshift at the bounce is given by equation (13.22). There are quasars at $z = 5$, so a bounce requires $\Omega < 0.01$, which is unacceptable because there is this much mass in the great clusters alone (table 20.1). The Lemaître model (eq. [13.23]) can agree with the standard model for helium production, and the dwell phase could happen at a lower redshift and a larger value of Ω. But if there were an appreciable quasistatic phase at $z < 5$, when the acceleration $\ddot{a}$ is small, one would wonder why it is not apparent in the redshift distribution in the Lα forest. That is, these models do not look promising. We noted in section 15 that a bounce at high redshift would be very interesting, but discussions of how it might happen still are exploratory.

At expansion factor $1 + z = 6$ the quasars illuminate a highly disturbed distribution of baryonic matter. What were things like another factor of three back in expansion, at $z \sim 20$? One option is a Great Desert of neutral atomic hydrogen little disturbed from a smooth primeval condition at decoupling. If this were so, there would have to have been a considerable burst of activity to have cleared the intergalactic medium of neutral hydrogen by $1 + z \sim 6$. The $\Omega = 1$ CDM model with bias parameter $b \sim 2$ predicts that at $z \sim 20$ star formation is just commencing in collapsing gas clouds at the baryon Jeans mass (eq. [25.56]). A still more advanced state of structure formation is suggested by some of the isocurvature models discussed in section 25. These include textures with cold dark matter and the isocurvature hot dark matter and baryonic dark matter scenarios.

We argued in the last section, in connection with table 25.1, that the characteristic mass densities within galaxies offer some hints as to when they might have been assembled. For example, the core density in a large elliptical galaxy is comparable to the mean baryon density at decoupling (eq. [25.5]), which one could take as a hint that that is when structure formation commenced. The mean mass density within a sphere of radius $r_g = 10h^{-1}$ kpc centered on an L_* galaxy, relative to the mean baryon density at redshift z, is

$$\frac{\bar{\rho}(< r_g = 10h^{-1}\,\mathrm{kpc})}{\rho_B} = \frac{2}{\Omega_B(1+z)^3}\left(\frac{v_c}{H_o r_g}\right)^2 \sim \frac{10^5}{\Omega_B(1+z)^3} . \tag{26.5}$$

This says that the mass in the bright parts of L_* galaxies cannot have been assembled much before $z \sim 20$. If the mass were assembled much later, it would require a considerable collapse factor, which would appear to make it difficult to understand the striking rarity of galaxies with v_c greater than twice the value $v_c = 200$ km s^{-1} typical of L_* galaxies.

A more direct observational basis for a choice between the Great Desert at $z \sim 20$, or the active young universe of isocurvature scenarios, or something intermediate, may come from the character of the anisotropy of the CBR, as measured by the angular autocorrelation function of the sky temperature and the tests for departures from Gaussian fluctuations. One way to get early galaxy formation out of a scale-invariant fluctuation spectrum assumes non-Gaussian density fluctuations that produce isolated mass concentrations when the rms density fluctuations still are small. This may produce non-Gaussian perturbations to the CBR, as has been studied in most detail in the topological textures model (Turok and Spergel 1990).

A second diagnostic is the effect of the scattering by plasma or dust in early generations of protogalaxies on the autocorrelation function of the angular distribution of the CBR. If the density parameter is unity and the baryon density has the value required by the standard model for light element production, the universe is opaque to scattering by free electrons only if the matter is ionized at $z \sim 100$ (eq. [24.80]). In scenarios similar to the $\Omega = 1$ CDM model this does not happen, and early accumulation of dust may be unlikely because star formation is just commencing at $z \sim 20$. In this case the anisotropy of the CBR probes the details of what happened at decoupling. The scale is set by the Hubble length at decoupling, which subtends an angle of about 30 arc min in an Einstein-de Sitter universe (eq. [25.33]). On smaller scales the anisotropy produced at decoupling depends on the details of the competition between the smoothing of the CBR by diffusion through the recombining plasma and the perturbation by the peculiar motion of the plasma. The limiting noise in exploring what is happening at decoupling may be the Sunyaev-Zel'dovich effect in clusters of galaxies (eq. [24.61]).

In models in which the density parameter is well below unity the universe might be opaque to the CBR at $z \sim 20$ because of scattering by free electrons or absorption and scattering by dust (eqs. [24.87] and [24.92]). That would erase whatever small-scale structure was present in the radiation distribution at higher redshifts and insert new perturbations from the Doppler shifts of the motion of the material in the young galaxies and from the thermal emission of starlight absorbed by the dust. Elements of these effects are estimated in sections 23 and 24. The signature of scattering at $z \sim 20$ may be a feature in the autocorrelation function of the sky temperature at an angular scale in the range of 3° to 10°, the angle subtended by the Hubble length at last scattering, for it would be difficult to see what other process could rearrange the radiation on such large angular scales.

The Very Early Universe

In the standard model the universe at redshifts $z \gtrsim 10^4$ has to expand in a nearly homogeneous way, with few of the complications that characterize recent epochs. There are departures from near thermal equilibrium, at the epoch of production of light elements, at the earlier phase transitions of the particle physics that produced ordinary baryonic matter out of quarks, separated the electric and weak interactions, and maybe adjusted the local value of the entropy per baryon, and perhaps still earlier at the phase transitions that produced cosmic strings or textures or nonbaryonic dark matter. We have relicts in the light elements; it would be exceedingly interesting to find others.

The predicted tranquility of the early universe follows because the pressure is on the order of the mass density, meaning the Jeans length is comparable to the Hubble length. Thus a density fluctuation large enough to be gravitationally bound is large enough to collapse to a black hole with the mass contained in the Hubble length,

$$\mathcal{M} \sim 10^{24}(1+z)^{-2}\mathcal{M}_\odot. \tag{26.6}$$

Black holes that collapsed near the end of the radiation-dominated epoch, at redshift $z \sim 10^4$ and mass comparable to that of a rich cluster, are not common. Black holes that formed much earlier than this cannot be common because the radiation energy they store is not redshifted away. The conclusion is that the mass density within the comoving region now observed at the Hubble length had to have been very close to uniform in the early universe and all the way back to the new physics of an inflation epoch or a quantum cosmology or something completely different.

The matter can have been irregularly distributed relative to the radiation in the very early universe, provided the density fluctuations in the matter are compensated by the radiation, so the net mass density is close to homogeneous. Radiation drag prevents the baryons from doing anything very interesting at $z \gtrsim 1000$. Matter that does not strongly couple to radiation does not do much either at $z \gtrsim z_{eq}$, because the mass density in the matter is subdominant and self-gravitating only in very dense patches.

There can be primeval mass currents with a peculiar velocity field $\mathbf{v}(\mathbf{x})$ that is a function of comoving position alone when $p = \rho/3$ and the crossing time is larger than the expansion time.[20] The currents can develop into primeval incompressible turbulence, but this would have to have dissipated before $z \sim 1000$, for at decoupling turbulent motion becomes supersonic and would dissipate by shocks into clumps too dense to be galaxies. Turbulence that dissipates at $z \gtrsim 10^6$ would have little effect on the CBR, because the radiation can relax back to thermal equilibrium (eq. [24.77]).

The simplicity of the early universe in the standard model is what makes possible the theory of light element production, and the success of the theory is the

---

[20] The easy way to see this is to note that the angular momentum in a sphere with proper radius $l \propto a(t)$ and volume $V \propto a^3$ is $L \propto \rho V l v$, so conservation of angular momentum requires that $\mathbf{v}$ is constant when $\rho \propto a^{-4}$.

main check that the standard model can be believed all the way back to $z \sim 10^{10}$. There are a few free parameters in the standard model for light element production. In addition to the possibility that the baryons are not homogeneously distributed, one can imagine that entropy is being dumped by primeval magnetic or velocity fields; that some neutrino families are degenerate, with Fermi energies that could be functions of position; that a primeval magnetic field is strong enough to cause local perturbations to the neutron-proton abundance ratio; or that gamma rays from decaying dark matter are capable of photodissociating helium. Careful examinations of what is on this list and what might be added to it will continue, and certainly will become of intense interest if there should develop a substantial observational problem with the simplest case in which these complications are ignored.

The phenomenological basis of physical science can be indirect, and has to be particularly so in the physics of the very early universe. The dread limiting case is a picture of what happened in the remote past which seamlessly agrees with physical theory that is well motivated by the evidence from the laboratory and extragalactic astronomy at low redshifts, but is tested by no empirical evidence except what went into its construction. Would this picture be a part of physical cosmology? Perhaps it is fortunate that there is no need to decide, for the pictures now under discussion are far from seamless. Cosmology is in an exciting state because we have a rich and growing list of problems and a growing observational base that may allow us to find a few solutions. The historical record here and in other physical sciences suggests that as the puzzles and conundrums we know about are laid to rest, they will be replaced by still more interesting ones.

REFERENCES

Aaronson, M. 1983. *Ap. J.* **266**, L11 (§18).

Aaronson, M., et al. 1982. *Ap. J. Suppl.* **50**, 241 (§5).

Aaronson, M., et al. 1986. *Ap. J.* **302**, 536 (§§3, 5, 6).

Aaronson, M., Huchra, J., and Mould, J. 1979. *Ap. J.* **229**, 1 (§5).

Aaronson, M., and Olszewski, E. W. 1987. *A. J.* **94**, 657 (§18).

Abbott, L. F., and Wise, M. B. 1984. *Nucl. Phys.* **B244**, 541 (§17).

Abell, G. O. 1958. *Ap. J. Suppl.* **3**, 211 (§§3, 7, 20).

Abell, G. O. 1962. In *Problems of Extragalactic Research*, ed. G. C. McVittie, 232. New York: Macmillan (§5).

Abell, G. O. 1965. *Ann. Rev. Astron. Ap.* **3**, 1 (§3).

Adams, W. S. 1941. *Ap. J.* **93**, 11 (§6).

Albrecht, A., and Steinhardt, P. J. 1982. *Phys. Rev. Lett.* **48**, 1220 (§17).

Alcock, C., and Paczyński, B. 1979. *Nature* **281**, 358 (§13).

Alfvén, H. 1966. *Worlds-Antiworlds*. San Francisco: Freeman (§7).

Allen, C. W. 1973. *Astrophysical Quantities*. 3d ed. London: Athlone Press; reprinted with corrections, 1976 (§3).

Alpher, R. A. 1948. *Phys. Rev.* **74**, 1577 (§6).

Alpher, R. A., Bethe, H. A., and Gamow, G. 1948. *Phys. Rev.* **73**, 803 (§6).

Alpher, R. A., and Herman, R. 1948. *Nature* **162**, 774 (§6).

Alpher, R. A., and Herman, R. 1988. *Phys. Today*, August, 24 (§6).

Applegate, J. H., and Hogan, C. 1985. *Phys. Rev.* **D31**, 3037 (§6).

Arons, J. 1972. *Ap. J.* **172**, 553 (§23).

Arp, H. C. 1965. *Ap. J.* **142**, 402 (§26).

Arp, H. C. 1987. *Quasars, Redshifts, and Controversies*. Berkeley: Interstellar Media (§7).

Arp, H. C., Burbidge, G., Hoyle, F., Narlikar, J. V., and Wickramasinghe, N. C. 1990. *Nature* **346**, 807 (§7).

Baade, W. 1956. *Publ. Astron. Soc. Pacific* **68**, 5 (§5).

Baade, W., and Spitzer, L. 1951. *Ap. J.* **113**, 413 (§3).

Babul, A., Paczyński, B., and Spergel, D. N. 1987. *Ap. J.* **316**, L49 (§6).

Babul, A., and Rees, M. J. 1992. *M.N.R.A.S.* **255**, 346 (§26).

Bahcall, J. N., Flynn, C., and Gould, A. 1992. *Ap. J.* **389**, 234 (§18).

Bahcall, J. N., and Schmidt, M. 1967. *Phys. Rev. Lett.* **19**, 1294 (§5).

Bahcall, N. A. 1975. *Ap. J.* **198**, 249 (§20).

Bahcall, N. A. 1988. *Ann. Rev. Astron. Ap.* **26**, 631 (§§3, 19, 20).

Bahcall, N. A., and Soneira, R. M. 1983. *Ap. J.* **270**, 20 (§19).

Bahcall, N. A., and West, M. J. 1992. *Ap. J.* **392**, 419 (§19).

Bajtlik, S., Duncan, R. C., and Ostriker, J. P. 1988. *Ap. J.* **327**, 570 (§23).

Barcons, X., Fabian, A. C., and Rees, M. J. 1991. *Nature* **350**, 685 (§24).

Bardeen, J. M. 1986. In *Inner Space/Outer Space*, ed. E. W. Kolb et al., 212. Chicago: University of Chicago Press (§25).

Bardeen, J. M., Bond, J. R., Kaiser, N., and Szalay, A. S. 1986. *Ap. J.* **304**, 15 (§25).

Bardeen, J. M., Steinhardt, P. J., and Turner, M. S. 1983. *Phys. Rev.* D**28**, 679 (§17).

Barnes, J., and Efstathiou, G. 1987. *Ap. J.* **319**, 575 (§22).

Barriola, M., and Vilenkin, A. 1989. *Phys. Rev. Lett.* **63**, 341 (§16).

Barrow, J. D., Bhavsar, S. P., and Sonoda, D. H. 1985. *M.N.R.A.S.* **216**, 17 (§19).

Bartlett, J. G., and Stebbins, A. 1991. *Ap. J.* **371**, 8 (§24).

Bean, A. J., et al. 1983. *M.N.R.A.S.* **205**, 605 (§20).

Bechtold, J., Weymann, R. J., Lin, Z., and Malkan, M. A. 1987. *Ap. J.* **315**, 180 (§23).

Beck, R., Kronberg, P. P., and Wielebinski, R. 1990. *Galactic and Intergalactic Magnetic Fields*. Dordrecht: Kluwer (§25).

Bekenstein, J. D. 1975. *Phys. Rev.* D**11**, 2072 (§15).

Bely, O., and van Regemorter, H. 1970. *Ann. Rev. Astron. Ap.* **8**, 329 (§23).

Bennett, D. P., and Bouchet, F. R. 1989. *Phys. Rev. Lett.* **63**, 2776 (§16).

Bennett, D. P., and Rhie, S.-H. 1990. *Phys. Rev. Lett.* **65**, 1709 (§16).

Bernstein, J. 1988. *Kinetic Theory in the Expanding Universe*. Cambridge, U.K.: Cambridge University Press (§24).

Bernstein, J., Brown, L. S., and Feinberg, G. 1989. *Rev. Mod. Phys.* **61**, 25 (§6).

Bernstein, J., and Dodelson, S. 1990. *Phys. Rev.* D**41**, 354 (§24).

Bernstein, J., and Feinberg, G. 1986. *Cosmological Constants*. New York: Columbia University Press (§12).

Bertotti, B. 1966. *Proc. R. Soc.* A**269**, 195 (§14).

Bertschinger, E. 1985. *Ap. J. Suppl.* **58**, 39 (§22).

Bertschinger, E., and Dekel, A. 1989. *Ap. J.* **336**, L5 (§20).

Binggeli, B. 1989. In *Large Scale Structure and Motions in the Universe*, ed. M. Mezzetti, G. Giuricin, F. Mardirossian, and M. Ramella, 47. Dordrecht: Kluwer (§25).

Binggeli, B., Sandage, A., and Tammann, G. A. 1988. *Ann. Rev. Astron. Ap.* **26**, 509 (§5).

Binggeli, B., Tammann, G. A., and Sandage, A. 1987. *A. J.* **94**, 251 (§3).

Binggeli, B., Tarenghi, M., and Sandage, A. 1990. *Astron. Ap.* **228**, 42 (§25).

Birkhoff, G. D. 1923. *Relativity and Modern Physics*. Cambridge, Mass.: Harvard University Press (§§4, 5, 11).

Birkinshaw, M. 1990. In *The Cosmic Microwave Background: 25 Years Later*, ed. N. Mandolesi and N. Vittorio, 77. Dordrecht: Kluwer (§24).

Blackman, R. B., and Tukey, J. W. 1958. *The Measurement of Power Spectra*. New York: Dover (§19).

Blanford, R. D., and Kochanek, C. S. 1987. In *Dark Matter in the Universe*, ed. J. N. Bahcall, T. Piran, and S. Weinberg, 133. Singapore: World Scientific (§14).

Bludman, S. A., and Ruderman, M. 1977. *Phys. Rev. Lett.* **38**, 255 (§17).

Blumenthal, G. R., Dekel, A., and Primack, J. R. 1988. *Ap. J.* **326**, 539 (§25).

Blumenthal, G. R., Faber, S. M., Primack, J. R., and Rees, M. J. 1984. *Nature* **311**, 517 (§25).

Blumenthal, G. R., and Primack, J. R. 1983. In *Fourth Workshop on Grand Unification*, ed. H. A. Weldon, P. Langacker, and P. J. Steinhardt, 256. Boston: Birkhauser (§25).

Boldt, E. 1987. *Phys. Reports* **146**, 215 (§§7, 24).

Bond, J. R. 1990. In *The Cosmic Microwave Background: 25 Years Later*, ed. N. Mandolesi and N. Vittorio, 45. Dordrecht: Kluwer (§24).

Bond, J. R., Carr, B. J., and Hogan, C. J. 1986. *Ap. J.* **306**, 428 (§24).

Bond, J. R., and Efstathiou, G. 1984. *Ap. J.* **285**, L45 (§25).

Bond, J. R., and Efstathiou, G. 1987. *M.N.R.A.S.* **226**, 655 (§21).

Bond, J. R., Efstathiou, G., and Silk, J. 1980. *Phys. Rev. Lett.* **45**, 1980 (§25).

Bond, J. R., Salopek, D., and Bardeen, J. M. 1988. In *Large-Scale Motions in the Universe*, ed. V. C. Rubin and G. V. Coyne, 115. Princeton: Princeton University Press (§§17, 25).

Bond, J. R., and Szalay, A. S. 1983. *Ap. J.* **274**, 443 (§25).

Bond, J. R., Szalay, A. S., and Silk, J. 1988. *Ap. J.* **324**, 627 (§23).

Bondi, H. 1960. *Cosmology.* 2d ed., 166. Cambridge, U.K.: Cambridge University Press (§§1, 5, 7, 15).

Bondi, H., and Gold. T. 1948. *M.N.R.A.S.* **108**, 252 (§7).

Bonnor, W. B. 1957. *M.N.R.A.S.* **117**, 104 (§5).

Bouchet, F. R., Juszkiewicz, R., Colombi, S., and Pellat, R. 1992. *Ap. J.* **394**, L5 (§26).

Bowyer, S., and Leinert, C. 1990. *The Galactic and Extragalactic Background Radiation.* Dordrecht: Kluwer (§5).

Brans, C., and Dicke, R. H. 1961. *Phys. Rev.* **124**, 925 (§18).

Broadhurst, T. J., Ellis, R. S., and Glazebrook, K. 1992. *Nature* **355**, 55 (§26).

Broadhurst, T. J., Ellis, R. S., Koo, D. C., and Szalay, A. S. 1990. *Nature* **343**, 726 (§§19, 26).

Broadhurst, T. J., Ellis, R. S., and Shanks, T. 1988. *M.N.R.A.S.* **235** 827 (§§25, 26).

Brown, R. W., and Stecker, F. W. 1979. *Phys. Rev. Lett.* **43**, 315 (§17).

Burbidge, E. M., Burbidge, G. R., and Hoyle, F. 1963, *Ap. J.* **138**, 873 (§7).

Burigana, C., De Zotti, G., and Danese, L. 1991. *Ap. J.* **379**, 1 (§24).

Butcher, H., and Oemler, A. 1978. *Ap. J.* **219**, 18, and **226**, 559 (§7).

Carignan, C., and Freeman, K. C. 1988. *Ap. J.* **332**, L33 (§25).

Carlberg, R. G. 1991. *Ap. J.* **367**, 385 (§§20, 25).

Carlberg, R. G., Couchman, H.M.P., and Thomas, P. A. 1990. *Ap. J.* **352**, L29 (§25).

Carlberg, R. G., and Hartwick, F.D.A. 1989. *Ap. J.* **345**, 196 (§25).

Carroll, S. M., Press, W. H., and Turner, E. L. 1992. *Ann. Rev. Astron. Ap.* **30** (§§13, 18).

Carswell, R. F., and Rees, M. J. 1987. *M.N.R.A.S.* **224**, 13P (§23).

Carswell, R. F., et al. 1982. *M.N.R.A.S.* **198**, 91 (§23).

Carswell, R. F., et al. 1984. *Ap. J.* **278**, 486 (§23).

Carswell, R. F., et al. 1987. *Ap. J.* **319**, 709 (§23).

Cavaliere, A., Danese, L., and De Zotti, G. 1979. *Astron. Ap.* **75**, 322 (§24).

Cavaliere, A., and Fusco-Femiano, R. 1976. *Astron. Ap.* **49**, 137 (§20).

Cen, R. Y., Ostriker, J. P., Spergel, D. N., and Turok, N. 1991. *Ap. J.* **383**, 1 (§16).

Centrella, J. M., Gallagher, J. S., Melott, A. L., and Bushouse, H. A. 1988. *Ap. J.* **333**, 24 (§25).

Centrella, J. M., and Melott, A. L. 1983. *Nature* **305**, 196 (§22).

Chambers, K. C., and McCarthy, P. J. 1990. *Ap. J.* **354**, L9 (§§7, 13).

Chan, K. L., and Jones, B.J.T. 1975. *Ap. J.* **195**, 1 (§24).

Chandrasekhar, S., and Henrich, L. R. 1942. *Ap. J.* **95**, 288 (§6).

Chapline, G. F. 1975. *Nature* **253**, 251 (§18).

Charlier, C.V.L. 1908. *Arkiv. för Mat. Astron. Fys.* **4**, 1 (§§2, 7).

Charlier, C.V.L. 1922. *Arkiv. för Mat. Astron. Fys.* **16**, 1 (§§2, 3, 7).

Chernin, A. D. 1970. *Zh. E.T.F. Lett.* **11**, 317 (§22).

Chincarini, G., and Rood, H. J. 1975. *Nature* **257**, 294 (§3).

Chuang, I., Durrer, R., Turok, N., and Yurke, B. 1991. *Science* **251**, 1336 (§16).

Clutton-Brock, M., and Peebles, P.J.E. 1981. *A. J.* **86**, 1115 (§§3, 6, 21).

Code, A. D., and Welch, G. A. 1982. *Ap. J.* **256**, 1 (§5).

Cohen, M. H., et al. 1988. *Ap. J.* **329**, 1 (§13).

Coleman, P. H., Pietronero, L. 1992. *Physics Reports* **213**, 311 (§7).

Coles, P., and Frenk, C. S. 1991. *M.N.R.A.S.* **253**, 727 (§26).

Colless, M., Ellis, R. S., Taylor, K., and Hook, R. N. 1990. *M.N.R.A.S.* **244**, 408 (§§25, 26).

Compton, A., and Getting, I. 1935. *Phys. Rev.* **47**, 817 (§6).

Condon, J. J. 1991. Private communication (§3).

Conklin, E. K. 1969. *Nature* **222**, 971 (§6).

Cowan, J. J., Thielemann, F., and Truran, J. W. 1991. *Ann. Rev. Astron. Ap.* **29**, 447 (§5).

Cowie, L. L. 1991. In *Observational Tests of Cosmological Inflation*, ed. T. Shanks et al., 257. Dordrecht: Kluwer (§§7, 13).

Cowie, L. L., Songaila, A., and Hu, E. M. 1991. *Nature* **354**, 460 (§§25, 26).

Cowsik, R., and McClelland, J. 1973. *Ap. J.* **180**, 7 (§18).

Curtis, H. D. 1918. *Publ. Lick Obs.* **13**, 11 (§3).

da Costa, L., et al. 1991. *Ap. J. Suppl.* **75**, 935 (§20).

Daly, R. A. 1991a. *Ap. J.* **379**, 37 (§24).

Daly, R. A. 1991b. *Ap. J.* **371**, 14 (§24).

Daly, R. A., and McLaughlin, G. C. 1992. *Ap. J.* **390**, 423 (§§18, 26).

Danese, L., Burigana, C., Toffolatti, L., De Zotti, G., and Franceschini, A. 1990. In *The Cosmic Microwave Background: 25 Years Later*, ed. N. Mandolesi and N. Vittorio, 153. Dordrecht: Kluwer (§24).

Dar, A. 1991. *Ap. J.* **382**, L1 (§7).

Dar, A. 1992. *Nucl. Phys.* B (Suppl.) **28A**, 321 (§20).

Davis, M. 1988. In *Cosmology and Particle Physics*, ed. L.-Z. Fang and A. Zee, 65. New York: Gordon and Breach (§20).

Davis, M., Efstathiou, G., Frenk, C. S., and White, S.D.M. 1985. *Ap. J.* **292**, 371 (§25).

Davis, M., and Geller, M. J. 1976. *Ap. J.* **208**, 13 (§19).

Davis, M., Huchra, J., Latham, D. W., and Tonry, J. 1982. *Ap. J.* **253**, 423 (§3).

Davis, M., and Peebles, P.J.E. 1977. *Ap. J. Suppl.* **34**, 425 (§§21, 22).

Davis, M., and Peebles, P.J.E. 1983. *Ap. J.* **267**, 465 (§§7, 20, 21).

Davis, R. L. 1987. *Phys. Rev.* D**35**, 3705 (§16).

Dekel, A., and Silk, J. 1986. *Ap. J.* **303**, 39 (§25).

de Lapparent, V., Geller, M. J., and Huchra, J. P. 1986. *Ap. J.* **302**, L1 (§3).

Demarque, P., Deliyannis, C. P., and Sarajedini, A. 1991, in *Observational Tests of Cosmological Inflation*, ed. T. Shanks et al., 111. Dordrecht: Kluwer (§§5, 25).

de Sitter, W. 1916. *M.N.R.A.S.* **77**, 181 (§2).

de Sitter, W. 1917. *M.N.R.A.S.* **78**, 3 (§§4, 5).

de Sitter, W. 1930. *Bull. Astron. Inst. Netherlands* **185**, 157 (§5).

de Sitter, W. 1931. *Nature* **128** 706 (§2).

de Vaucouleurs, G. 1953. *A. J.* **58**, 29 (§§7, 26).

de Vaucouleurs, G. 1970. *Science* **167**, 1203 (§7).

de Vaucouleurs, G., and de Vaucouleurs, A. 1973. *Astron. Ap.* **28**, 109 (§3).

de Vaucouleurs, G., de Vaucouleurs, A., and Corwin, H. G. 1976. *Second Reference Catalogue of Bright Galaxies*. Austin: University of Texas Press (§3).

de Vaucouleurs, G., and Peters, W. L. 1968. *Nature* **220**, 868 (§6).

Dicke, R. H. 1946. *Rev. Sci. Instr.* **17**, 268 (§6).

Dicke, R. H. 1961. *Nature* **192**, 440 (§§6, 15).

Dicke, R. H. 1962. *Nature* **194**, 329 (§5).

Dicke, R. H. 1968. *Ap. J.* **152**, 1 (§6).

Dicke, R. H. 1970. *Gravitation and the Universe*. Philadelphia: American Philosophical Society (§§1, 5, 7, 15).

Dicke, R. H., Beringer, R., Kyhl, R. L., and Vane, A. B. 1946. *Phys. Rev.* **70**, 340 (§6).

Dicke, R. H., and Peebles, P.J.E. 1979. In *General Relativity: an Einstein Centenary Survey*, ed. S. W. Hawking and W. Israel. London: Cambridge University Press (§15).

Dicke, R. H., Peebles, P.J.E., Roll, P. G., and Wilkinson, D. T. 1965. *Ap. J.* **142**, 414 (§6).

Dicus, D. A., Kolb, E. W., and Teplitz, V. L. 1978. *Ap. J.* **221**, 327 (§25).

Dicus, D. A., Kolb, E. W., Teplitz, V. L., and Wagoner, R. V. 1978. *Phys. Rev.* **D17**, 1529 (§6).

Dingle, H. 1933a. *Proc. N.A.S.* **19**, 559 (§11).

Dingle, H. 1933b. *M.N.R.A.S.* **94**, 134 (§11).

Disney, M., and Phillipps, S. 1985. *M.N.R.A.S.* **216**, 53 (§26).

Djorgovski, S. 1987. In *Nearly Normal Galaxies*, ed. S. M. Faber, 227. Berlin: Springer (§26).

Djorgovski, S., and Davis, M. 1987. *Ap. J.* **313**, 59 (§3).

Dmitriev, N., and Zel'dovich, Ya. B. 1963. *JETP* **45**, 1150 (§21).

Dobyns, Y. H. 1988. *Ap. J.* **329**, L5 (§25).

Doroshkevich, A. G. 1970. *Astrofizika* **6**, 581 (§§22, 25).

Doroshkevich, A. G. 1973. *Ap. Lett.* **14**, 11 (§22).

Doroshkevich, A. G., and Novikov, I. 1964. *Dokl. Akad. Nauk. S.S.S.R.* **154**, 809 (§6).

Doroshkevich, A. G., Sunyaev, R. A., and Zel'dovich, Ya. B. 1974. In *Confrontation of Cosmological Theories with Observational Data*, ed. M. S. Longair, 213. Dordrecht: Reidel (§25).

Doroshkevich, A. G., Zel'dovich, Ya. B., and Novikov, I. D. 1967. *Astron. Zh.* **44**, 295 (§25).

Doroshkevich, A. G., Zel'dovich, Ya. B., and Sunyaev, R. A. 1978. *Soviet Astron.* **22**, 523 (§25).

Doroshkevich, A. G., et al. 1980. *M.N.R.A.S.* **192**, 321 (§22).

Draine, B. T., and Lee, H. M. 1984. *Ap. J.* **285**, 89 (§24).

Dressler, A. 1978. *Ap. J.* **226**, 55 (§20).

Dressler, A. 1987. In *Nearly Normal Galaxies*, ed. S. M. Faber, 276. Berlin: Springer (§7).

Dressler, A., and Gunn, J. E. 1983. *Ap. J.* **270**, 7 (§7).

Dressler, A., et al. 1987. *Ap. J.* **313**, 42 (§3).

Duncan, R. C., Ostriker, J. P., and Bajtlik, S. 1989. *Ap. J.* **345**, 39 (§23).

Dyer, C. C. 1987. In *Theory and Observational Limits in Cosmology*, ed. W. R. Stoeger. Vatican: Pontifical Academy of Sciences (§26).

Dyer, C. C., and Roeder, R. C. 1973. *Ap. J.* **180**, L31 (§14).

Eddington, A. S. 1914. *Stellar Motions and the Structure of the Universe.* London: Macmillan (§2).

Eddington, A. S. 1924. *The Mathematical Theory of Relativity.* 2d ed. Cambridge, U.K.: Cambridge University Press (§4).

Eddington, A. S. 1930. *M.N.R.A.S.* **90**, 668 (§§4, 5).

Eddington, A. S. 1931a. *M.N.R.A.S.* **91**, 413 (§5).

Eddington, A. S. 1931b. *Nature* **127**, 447 (§5).

Edge, A. C. 1989. Ph.D. diss., University of Leicester (§20).

Efstathiou, G. 1988. In *Large-Scale Motions in the Universe*, ed. V. C. Rubin and G. V. Coyne, 115. Princeton: Princeton University Press (§24).

Efstathiou, G. 1990. In *Physics of the Early Universe*, ed. J. A. Peacock, A. F.

Heavens, and A. T. Davies, 361. Edinburgh, U.K.: SUSSP, Publishers (§§10, 25).

Efstathiou, G. 1991. *Physica Scripta* T36, 88 (§§3, 21).

Efstathiou, G., Bernstein, G., Katz, N., Tyson, J. A., and Guhathakurta, P. 1991. *Ap. J.* **380**, L47 (§26).

Efstathiou, G., and Bond, J. R. 1986. *M.N.R.A.S.* **218**, 103 (§25).

Efstathiou, G., and Bond, J. R. 1987. *M.N.R.A.S.* **227**, 33P (§25).

Efstathiou, G., Ellis, R. S., and Peterson, B. A. 1988. *M.N.R.A.S.* **232**, 431 (§5).

Efstathiou, G., Frenk, C. S., White, S.D.M., and Davis, M. 1988. *M.N.R.A.S.* **235**, 715 (§§22, 25).

Efstathiou, G., Sutherland, W. J., and Maddox, S. J. 1990. *Nature* **348**, 705 (§25).

Eggen, O. J., Lynden-Bell, D., and Sandage, A. R. 1962. *Ap. J.* **136**, 748 (§25).

Ehlers, J., Perry, J. J., and Walker, M. 1980. *Ann. N. Y. Acad. Sci.* **336** (§18).

Einasto, J., Kaasik, A., and Saar, E. 1974. *Nature* **250**, 309 (§18).

Einasto, J., et al. 1982. *M.N.R.A.S.* **206**, 529 (§19).

Einstein, A. 1917. *S.-B. Preuss. Akad. Wiss.* **142** (§4).

Einstein, A. 1922. *Ann. Phys.* **69**, 436 (§2).

Einstein, A. 1923. *Z. Phys.* **16**, 228 (§5).

Einstein, A. 1933. *Structure Cosmologique de l'Espace.* Paris: Hermann et Cie (§2).

Einstein, A. 1945. *The Meaning of Relativity.* 2d ed., 127. Princeton: Princeton University Press (§5).

Einstein, A., and de Sitter, W. 1932. *Proc. N.A.S.* **18**, 213 (§§5, 18).

Ellis, G.F.R. 1980. *Ann. N.Y. Acad. Sci.* **336**, 130 (§1).

Ellis, G.F.R. 1985. In *Theory and Observational Limits in Cosmology*, ed. W. R. Stoeger, 47. Vatican: Pontifical Academy of Sciences (§§1, 6).

Ellis, G.F.R., Maartens, R., and Nel, S. D. 1978. *M.N.R.A.S.* **184**, 439 (§7).

Ellis, R. S. 1991. In *Observational Tests of Cosmological Inflation*, ed. T. Shanks et al., 243. Dordrecht: Kluwer (§13).

Emden, R. 1907. *Gaskugeln.* Leipzig and Berlin: Teubner (§§3, 18).

Evans, D. S. 1984. *Ann. N. Y. Acad. Sci.* **422** (§18).

Faber, S. M. 1982. In *Astrophysical Cosmology*, ed. H. A. Brück, G. V. Coyne, and M. S. Longair, 191. Vatican: Pontifical Academy of Sciences (§§22, 25).

Faber, S. M., and Burstein, D. 1988. In *Large-Scale Motions in the Universe*, ed. V. C. Rubin and G. V. Coyne, 115. Princeton: Princeton University Press (§5).

Faber, S. M., and Gallagher, J. S. 1979. *Ann. Rev. Astron. Ap.* **17**, 135 (§5).

Faber, S. M., and Jackson, R. E. 1976. *Ap. J.* **204**, 668 (§§3, 5).

Faber, S. M., et al. 1989. *Ap. J. Suppl.* **69**, 763 (§§3, 13).

Fabian, A. C. 1988. *Cooling Flows in Clusters and Galaxies.* Dordrecht: Kluwer (§24).

Fabian, A. C., Nulsen, P.E.J., and Canizares, C. R. 1991. *Astron. Ap. Rev.* **2**, 191 (§§3, 20, 24, 25).

Fabricant, D., and Gorenstein, P. 1983. *Ap. J.* **267**, 535 (§3).

Fall, S. M. 1975. *M.N.R.A.S.* **172**, 23P (§21).

Fall, S. M., and Efstathiou, G. 1980. *M.N.R.A.S.* **193**, 189 (§22).

Fall, S. M., and Rees, M. J. 1987. In *Globular Cluster Systems in Galaxies*, ed. J. E. Grindlay and A.G.D. Philip. Dordrecht: Reidel (§6).

Felten, J. E. 1965. *Phys. Rev. Lett.* **15**, 1003 (§6).

Felten, J. E. 1966. *Ap. J.* **144**, 241 (§5).

Felten, J. E., and Isaacman, R. 1986. *Rev. Mod. Phys.* **58**, 689 (§§5, 13).

Field, G. B., Arp, H. C., and Bahcall, J. N. 1973. *The Redshift Controversy*. Reading, Mass.: Benjamin (§7).

Fischler, W., Ratra, B., and Susskind, L. 1985. *Nucl. Phys.* **B259**, 730 (§17).

Fitchett, M. J. 1990. In *Clusters of Galaxies*, ed. W. R. Oegerle, M. J. Fitchett, and L. Danly, 111. Cambridge, U.K.: Cambridge University Press (§20).

Fitchett, M. J., and Webster, R. 1987. *Ap. J.* **317**, 653 (§19).

Foltz, C. B., Weymann, R. J., Röser, H.-J., and Chaffee, F. H. 1984. *Ap. J.* **281**, L1 (§23).

Fowler, W. A. 1970. *Comments Ap. Space Phys.* **2**, 134 (§6).

Fowler, W. A. 1989. *Ann. N.Y. Acad. Sci.* **571**, 68 (§5).

Freeman, K. C. 1970. *Ap. J.* **160**, 811 (§§3, 18).

Freeman, K. C. 1989. In *The Epoch of Galaxy Formation*, 331. Dordrecht: Kluwer (§25).

Freese, K., et al. 1987. *Nucl. Phys.* **B287**, 797 (§18).

Frenk, C. S. 1991. *Physica Scripta* **T36**, 70 (§§21, 25).

Frenk, C. S., White, S.D.M., Efstathiou, G., and Davis, M. 1990. *Ap. J.* **351**, 10 (§25).

Frenk, C. S., et al. 1989. *The Epoch of Galaxy Formation*. Dordrecht: Kluwer (§25).

Friedmann, A. 1922. *Z. Phys.* **10**, 377 (§§5, 12).

Friedmann, A. 1924. *Z. Phys.* **21**, 326 (§§5, 12).

Fry, J. N. 1983. *Ap. J.* **267**, 483 (§19).

Fry, J. N. 1984. *Ap. J.* **277**, L5 (§§19, 22).

Fry, J. N., and Peebles, P.J.E. 1978. *Ap. J.* **221**, 19 (§19).

Fry, J. N., and Peebles, P.J.E. 1980. *Ap. J.* **238**, 785 (§19).

Fukugita, M., Futamase, T., and Kasai, M. 1990. *M.N.R.A.S.* **246**, 24P (§13).

Fukugita, M., and Turner, E. L. 1991. *M.N.R.A.S.* **253**, 99 (§13).

Fukugita, M., and Yanagida, T. 1984. *Phys. Lett.* **144B**, 386 (§25).

Fuller, G. M., Boyd, R. N., and Kalen, J. D. 1991. *Ap. J.* **371**, L11 (§25).

Futamase, T., and Sasaki, M. 1989. *Phys. Rev.* **D40**, 2502 (§14).

Gamow, G. 1942. *J. Wash. Acad. Sci.* **32**, 353 (§6).

Gamow, G. 1946. *Phys. Rev.* **70**, 572 (§6).

Gamow, G. 1948a. *Phys. Rev.* **74**, 505 (§6).

Gamow, G. 1948b. *Nature* **162**, 680 (§6).

Gamow, G. 1952. *Phys. Rev.* **86**, 251 (§22).

Gamow, G. 1956. In *Vistas in Astronomy*, ed. A. Beers, vol. 2, 1726. New York: Pergamon (§6).

Geller, M. J. 1990. In *Clusters of Galaxies*, ed. W. R. Oegerle et al., 25. Cambridge, U.K.: Cambridge University Press (§19).

Geller, M. J., and Huchra, J. P. 1988. In *Large-Scale Motions in the Universe*, 3. Princeton: Princeton University Press (§3).

Geller, M. J., and Huchra, J. P. 1989. *Science* **246**, 897 (§3).

Geller, M. J., and Peebles, P.J.E. 1972. *Ap. J.* **174**, 1 (§§5, 7).

Geller, M. J., and Peebles, P.J.E. 1973. *Ap. J.* **184**, 329 (§20).

Gerhard, O. E., and Spergel, D. N. 1992. *Ap. J.* **389**, L9 (§18).

Gershtein, S. S., and Zel'dovich, Ya. B. 1966. *JETP Lett.* **4**, 174 (§18).

Gilmore, G., King, I. R., and van der Kruit, P. C. 1990. *The Milky Way as a Galaxy*. Mill Valley, Calif.: University Science Books (§25).

Giovanelli, R., Haynes, M. P., Rubin, V. C., and Ford, W. K. 1986. *Ap. J.* **301**, L7 (§3).

Gödel, K. 1949. *Rev. Mod. Phys.* **21**, 447 (§2).

Gooding, A. K., Spergel, D. N., and Turok, N. 1991. *Ap. J.* **372**, L5 (§16).

Górski, K. 1988. *Ap. J.* **332**, L7 (§21).

Gott, J. R. 1975. *Ap. J.* **201**, 296 (§22).

Gott, J. R. 1981. *Ap. J.* **243**, 140 (§18).

Gott, J. R. 1985. *Ap. J.* **288**, 422 (§11).

Gott, J. R., Gunn, J. E., Schramm, D. N., and Tinsley, B. M. 1974. *Ap. J.* **194**, 543 (§18).

Gott, J. R., Melott, A. L., and Dickinson, M. 1986. *Ap. J.* **306**, 341 (§19).

Gott, J. R., Park, M.-G., and Lee, H. M. 1989. *Ap. J.* **338**, 1 (§13).

Gott, J. R., and Thuan, T. X. 1978. *Ap. J.* **223**, 426 (§26).

Gott, J. R., and Turner, E. L. 1979. *Ap. J.* **232**, L79 (§19).

Gouda, N., Sugiyama, N., and Sasaki, M. 1991. *Progr. Theor. Phys.* **85**, 1023 (§§21, 25).

Gould, R. J., and Schréder, G. 1966. *Phys. Rev. Lett.* **16**, 252 (§6).

Gradshteyn, I. S., and Ryzhik, I. M. 1965. *Table of Integrals, Series, and Products*, trans. A. Jeffrey, 692. New York: Academic Press (§21).

Greenstein, G. 1969. *Nature* **223**, 938 (§6).

Gregory, P. C., and Condon, J. J. 1991. *Ap. J. Suppl.* **75**, 1011 (§3).

Greisen, K. 1966. *Phys. Rev. Lett.* **16**, 748 (§6).

Grishchuk, L. P. 1988. *Soviet Phys.-Uspekhi* **31**, 940 (§17).

Grishchuk, L. P., and Zel'dovich, Ya. B. 1978. *Soviet Astron.-A. J.* **22**, 125 (§6).

Groth, E. J. 1992. Private communication (§3).

Groth, E. J., and Peebles, P.J.E. 1977. *Ap. J.* **217**, 385 (§§7, 19).

Groth, E. J., and Peebles, P.J.E. 1986. *Ap. J.* **310**, 507 (§§7, 19).

Groth, E. J., Juszkiewicz, R., and Ostriker, J. P. 1989. *Ap. J.* **346**, 558 (§21).

Guhathakurta, P., Tyson, J. A., and Majewski, S. R. 1990. *Ap. J.* **357**, L9 (§5).

Guiderdoni, B., and Rocca-Volmerange, B. 1990a. *Astron. Ap.* **227**, 362 (§26).

Guiderdoni, B., and Rocca-Volmerange, B. 1990b. *M.N.R.A.S.* **247**, 166 (§26).

Gunn, J. E. 1967. *Ap. J.* **150**, 737 (§14).

Gunn, J. E. 1982. In *Astrophysical Cosmology*, ed. H. A. Brück, G. V. Coyne, and M. S. Longair, 233. Vatican: Pontifical Academy of Sciences (§§22, 25).

Gunn, J. E. 1987. In *Nearly Normal Galaxies*, ed. S. M. Faber, 455. Berlin: Springer (§25).

Gunn, J. E. 1989. In *The Epoch of Galaxy Formation*, ed. C. S. Frenk et al., 167. Dordrecht: Kluwer (§25).

Gunn, J. E., and Dressler, A. 1988. In *Towards Understanding Galaxies at Large Redshifts*, ed. R. G. Kron and A. Renzini, 227. Dordrecht: Kluwer (§7).

Gunn, J. E., and Gott, J. R. 1972. *Ap. J.* **176**, 1 (§§3, 22).

Gunn, J. E., and Oke, J. B. 1975. *Ap. J.* **195**, 255 (§5).

Gunn, J. E., and Peterson, B. A. 1965. *Ap. J.* **142**, 1633 (§23).

Gunn, J. E., and Tinsley, B. M. 1975. *Nature* **257**, 454 (§18).

Gursky, H., et al. 1971. *Ap. J.* **167**, L81 (§20).

Gush, H. P., Halpern, M., and Wishnow, E. 1990. *Phys. Rev. Lett.* **65**, 537 (§6).

Guth, A. 1981. *Phys. Rev.* D23, 347 (§17).

Guth, A., and Pi, S.-Y. 1982. *Phys. Rev. Lett.* **49**, 1110 (§17).

Hale-Sutton, D., Fong, R., Metcalfe, N., and Shanks, T. 1989. *M.N.R.A.S.* **237**, 569 (§20).

Hamilton, A.J.S. 1988a. *Ap. J.* **332**, 67 (§19).

Hamilton, A.J.S. 1988b. *Ap. J.* **331**, L59 (§25).

Harrison, E. R. 1970a. *Phys. Rev.* D1, 2726 (§16).

Harrison, E. R. 1970b. *M.N.R.A.S.* **147**, 279 (§26).

Harrison, E. R. 1987. *Darkness at Night: A Riddle of the Universe*. Cambridge, Mass.: Harvard University Press (§§4, 5).

Harrison, E. R. 1990. In *The Galactic and Extragalactic Background Radiation*, ed. S. Bowyer and C. Leinert, 3. Dordrecht: Kluwer (§§4, 5).

Hauser, M. G., et al. 1991. In *After the First Three Minutes*, ed. S. S. Holt et al., 161. New York: American Institute of Physics (§6).

Hauser, M. G., and Peebles, P.J.E. 1973. *Ap. J.* **185**, 757 (§§19, 21).

Hausman, M. A., and Ostriker, J. P. 1978. *Ap. J.* **224**, 320 (§§3, 25).

Hawking, S. W. 1982. *Phys. Lett.* **115B**, 295 (§17).

Hawkins, I., and Wright, E. L. 1988. *Ap. J.* **324**, 46 (§6).

Hawley, D. L., and Peebles, P.J.E. 1975. *A. J.* **80**, 477 (§26).

Hayashi, C. 1950. *Progr. Theor. Phys.* **5**, 224 (§6).

Haynes, M. P., and Giovanelli, R. 1986. *Ap. J.* **306**, L55 (§3).

Hegyi, D. J., and Olive, K. A. 1986. *Ap. J.* **303**, 56 (§18).

Henry, P. S. 1971. *Nature* **231**, 516 (§6).

Hill, C. T., Schramm, D. N., and Fry, J. N. 1989. *Comments on Nuclear and Particle Physics* **19**, 25 (§16).

Hill, C. T., Schramm, D. N., and Walker, T. P. 1986. *Phys. Rev.* D34, 1622 (§6).

Hogan, C. J. 1978. *M.N.R.A.S.* **185**, 889 (§6).

Hogan, C. J., and Kaiser, N. 1983. *Ap. J.* **274**, 7 (§25).

Houston, B. P., Wolfendale, A. W., and Young, E.C.M. 1984. *J. Phys.* G**10**, L147 (§18).

Hoyle, F. 1948. *M.N.R.A.S.* **108**, 372 (§7).

Hoyle, F. 1949. In *Problems of Cosmical Aerodynamics*, ed. J. M. Burgers and H. C. van de Hulst, 195. Dayton, Ohio: Central Air Documents Office (§22).

Hoyle, F. 1953. *Ap. J.* **118**, 513 (§§18, 25).

Hoyle, F. 1965. *Phys. Rev. Lett.* **15**, 131 (§6).

Hoyle, F., and Narlikar, J. V. 1966. *Proc. R. Soc. London* A**290**, 162 (§7).

Hoyle, F., and Tayler, R. J. 1964. *Nature* **203**, 1108 (§6).

Hubble, E. 1925. *Observatory* **48**, 139 (§3).

Hubble, E. 1926a. *Ap. J.* **63** 236 (§3).

Hubble, E. 1926b. *Ap. J.* **64** 321 (§§3, 4, 18).rho

Hubble, E. 1929. *Proc. N.A.S.* **15**, 168 (§5).

Hubble, E. 1934. *Ap. J.* **79**, 8 (§3).

Hubble, E. 1936. *The Realm of the Nebulae.* New Haven: Yale University Press (§§3, 5, 7, 25).

Hubble, E., and Humason, M. 1931. *Ap. J.* **74**, 43 (§§3, 5).

Hubble, E., and Tolman, R. C. 1935. *Ap. J.* **82**, 302 (§§5, 7).

Huchra, J. P. 1985. In *The Virgo Cluster*, ed. O. Richter and B. Binggeli, 181. Garching, Germany: ESO (§3).

Huchra, J. P. 1991. Private communication (§3).

Huchra, J. P. 1992. *Science* **256**, 321 (§5).

Huchra, J. P., and Brodie, J. 1987. *A. J.* **93**, 779 (§3).

Huchra, J. P., et al. 1985. *A. J.* **90**, 691 (§14).

Hughes, J. P. 1989. *Ap. J.* **337**, 21 (§3).

Hut, P., and White, S.D.M. 1984. *Nature* **310**, 637 (§25).

Ikeuchi, S. 1982. *Publ. Astron. Soc. Japan* **33**, 211 (§25).

Ikeuchi, S., and Ostriker, J. P. 1986. *Ap. J.* **310**, 522 (§23).

Irvine, W. M. 1961, Ph.D. diss., Harvard University (§21).

Iso, K., Kodama, H., and Sato, K. 1986. *Phys. Lett.* **169**B, 337 (§6).

Jacoby, G. H., et al. 1992. *Publ. Astron. Soc. Pacific* **104**, 599 (§§5, 26).

Jahnke, E., and Emde, E. 1945. *Tables of Functions.* 4th ed., 116. New York: Dover (§21).

Jahoda, K., et al. 1992. Private communication (§3).

Jaki, S. L. 1967. *The Paradox of Olbers' Paradox.* New York: Herder and Herder (§§4, 5).

Jeans, J. H. 1928. *Astronomy and Cosmogony.* Cambridge, U.K.: Cambridge University Press (§§5, 25).

Jôeveer, M., Einasto, J., and Tago, E. 1978. *M.N.R.A.S.* **185**, 357 (§3).

Jones, B.J.T. 1973. *Ap. J.* **181**, 269 (§22).

Jones, B.J.T., and Peebles, P.J.E. 1972. *Comments Ap. Space Phys.* **4**, 121 (§22).

Jones, C., and Forman, W. 1984. *Ap. J.* **276**, 38 (§20).

Juszkiewicz, R., Górski, K., and Silk, J. 1987. *Ap. J.* **323**, L1 (§21).

Juszkiewicz, R., Vittorio, N., and Wyse, R.F.G. 1990. *Ap. J.* **349**, 408 (§20).

Kahn, C., and Kahn, F. 1975. *Nature* **257**, 451 (§2).

Kahn, F. D., and Woltjer, L. 1959. *Ap. J.* **130**, 705 (§20).

Kaiser, N. 1984a. *Ap. J.* **284**, L9 (§§19, 25).

Kaiser, N. 1984b. *Ap. J.* **282**, 374 (§24).

Kaiser, N. 1986. In *Inner Space/Outer Space*, ed. E. W. Kolb et al., 258. Chicago: University of Chicago Press (§25).

Kaiser, N. 1987. *M.N.R.A.S.* **227**, 1 (§§19, 20).

Kaiser, N. 1991. In *After the First Three Minutes*, ed. S. Holt et al., 248. New York: American Institute of Physics (§20).

Kaiser, N., and Stebbins, A. 1984. *Nature* **310**, 391 (§25).

Kantowski, R. 1969. *Ap. J.* **155**, 89 (§14).

Kapahi, V. K. 1989. *A. J.* **97**, 1 (§13).

Kazanas, D. 1980. *Ap. J.* **241**, L59 (§17).

Kellogg, E., et al. 1971. *Ap. J.* **165**, L49 (§24).

Kennicutt, R. C. 1983. *Ap. J.* **272**, 54 (§25).

Kent, S. M., and Gunn, J. E. 1982. *A. J.* **87**, 945 (§3).

Kerszberg, P. 1990. In *Einstein and the History of General Relativity*, ed. D. Howard and J. Stachel, 362. Berlin: Birkhäuser (§2).

Kiang, T. 1961. *M.N.R.A.S.* **122**, 263 (§3).

Kibble, T.W.B. 1976. *J. Phys.* A**9**, 1387 (§16).

Kibble, T.W.B. 1980. *Phys. Reports* **67**, 183 (§16).

King, C. R., and Ellis, R. S. 1985. *Ap. J.* **288**, 456 (§13).

Kirshner, R. P., Oemler, A., Schechter, P. L., and Shectman, S. A. 1981. *Ap. J.* **248**, L57 (§3).

Klein, O. 1971. *Science* **171**, 339 (§7).

Klypin, A. A., and Doroshkevich, A. A. 1989. In *The Epoch of Galaxy Formation*, ed. C. S. Frenk et al., 327. Dordrecht: Kluwer (§25).

Klypin, A. A., and Koplyov, A. I. 1983. *Sov. Astron. Lett.* **9**, 41 (§19).

Klypin, A. A., and Shandarin, S. F. 1983. *M.N.R.A.S.* **204**, 891 (§22).

Kofman, L. A., and Linde, A. D. 1987. *Nucl. Phys.* B**282**, 555 (§25).

Kofman, L. A., Pogosyan, D., and Shandarin, S. 1990. *M.N.R.A.S.* **242**, 200 (§22).

Kofman, L. A., and Starobinsky, A. A. 1985. *Soviet Astron. Lett.* **11**, 271 (§§13, 21).

Kolb, E. W., and Turner, M. S. 1990. *The Early Universe*. Redwood City, Calif.: Addison-Wesley (§§6, 16, 17, 18).

Kompaneets, A. S. 1957. *Soviet Phys.–JETP* **4**, 730 (§24).

Koo, D. C. 1989. In *The Epoch of Galaxy Formation*, ed. C. S. Frenk et al., 71. Dordrecht: Kluwer (§13).

Kormendy, J. 1982. In *Morphology and Dynamics of Galaxies*, ed. L. Martinet and M. Mayor, 113. Sauverny: Geneva Obs. (§3).

Kormendy, J. 1987. In *Nearly Normal Galaxies*, ed. S. Faber, 163. New York: Springer (§3).

Kormendy, J., and Djorgovski, S. 1989. *Ann. Rev. Astron. Ap.* **27**, 235 (§3).

Kormendy, J., and Knapp, G. R. 1987. *Dark Matter in the Universe*. Dordrecht: Reidel (§18).

Kramers, H. A. 1923. *Phil. Mag.* **46**, 836 (§24).

Krolik, J. H. 1990. *Ap. J.* **353**, 21 (§6).

Kron, R. 1980. *Ap. J. Suppl.* **43**, 305 (§26).

Kronberg, P. P., and Perry, J. J. 1982. *Ap. J.* **263**, 518 (§25).

Kuhn, J. R., and Uson, J. M. 1982. *Ap. J.* **263**, L47 (§19).

Kuhn, T. S. 1962. *The Structure of Scientific Revolutions*. Chicago: University of Chicago Press (§5).

Kuijken, K., and Gilmore, G. 1991. *Ap. J.* **367**, L9 (§18).

Kulsrud, R. W., and Anderson, S. W. 1992. *Ap. J.* **396**, 606 (§25).

Kulsrud, R. W., and Loeb, A. 1992. *Phys. Rev.* D**45**, 525 (§10).

Kumar, S. S. 1963. *Ap. J.* **137**, 1126 (§16).

Kurki-Suonio, H., Matzner, R. A., Olive, K. A., and Schramm, D. N. 1990. *Ap. J.* **353**, 406 (§§20, 25).

Lacey, C. G., and Ostriker, J. P. 1985. *Ap. J.* **299**, 633 (§26).

Lahav, O. 1987. *M.N.R.A.S.* **225**, 213 (§3).

Lahav, O. 1991. In *After the First Three Minutes*, ed. S. Holt et al., 421. New York: American Institute of Physics (§20).

Lambas, D. G., Groth, E. J., and Peebles, P.J.E. 1988. *A. J.* **95**, 996 (§26).

Lanzetta, K. M. 1991. *Ap. J.* **375**, 1 (§23).

Lanzetta, K. M., et al. 1991. *Ap. J. Suppl.* **77**, 1 (§23).

Larson, R. B. 1990. *Publ. Astron. Soc. Pacific* **102**, 709 (§25).

Lauer, T. R. 1985. *Ap. J.* **292**, 104 (§18).

Layzer, D. 1963. *Ap. J.* **138** (§21).

Layzer, D., and Hively, R. 1973. *Ap. J.* **179**, 361 (§6).

Leavitt, H. S. 1912. *Harvard College Obs. Circ.* **173** (§3).

Lee, B. W., and Weinberg, S. 1977. *Phys. Rev. Lett.* **39**, 165 (§18).

Lemaître, G. 1925. *J. Math. Phys.* **4**, 188 (§§5, 12).

Lemaître, G. 1927. *Ann. Soc. Sci. Bruxelles* **47A**, 49 (§§4, 5, 7, 13).

Lemaître, G. 1931a. *M.N.R.A.S.* **91**, 483 (§5).

Lemaître, G. 1931b. *M.N.R.A.S.* **91**, 490 (§5).

Lemaître, G. 1931c. *Nature* **127**, 706 (§5).

Lemaître, G. 1931d. *Nature* **128**, 704 (§5).

Lemaître, G. 1931e. *La Revue des Questions Scientifiques*, 4e série, **20**, 391 (§§6, 7).

Lemaître, G. 1933. *Ann. Soc. Sci. Bruxelles* A**53**, 51 (§§5, 6, 11, 15, 18, 25).

Lemaître, G. 1934. *Proc. N.A.S.* **20**, 12 (§25).

Lepp, S., and Shull, J. M. 1984. *Ap. J.* **280**, 465 (§6).

Liebert, J. 1980. *Ann. Rev. Astron. Ap.* **18**, 363 (§18).

Liebes, S. 1964. *Phys. Rev.* B**133**, 835 (§18).

Lifshitz, E. M. 1946. *J. Phys. USSR* **10**, 116; *Zh. E.T.F.* **16**, 587 (§5).

Lilje, P. B., and Efstathiou, G. 1988. *M.N.R.A.S.* **231**, 635 (§§3, 19).

Lilly, S. J. 1991. In *Observational Tests of Cosmological Inflation*, ed. T. Shanks et al., 233. Dordrecht: Kluwer (§13).

Lilly, S. J., Cowie, L. L., and Gardner, J. P. 1991. *Ap. J.* **369**, 79 (§§25, 26).

Lilly, S. J., and Longair, M. S. 1982. *M.N.R.A.S.* **199**, 1053 (§5).

Limber, D. N. 1953. *Ap. J.* **117**, 134 (§7).

Linde, A. D. 1974. *JETP Lett.* **19** 183 (§17).

Linde, A. D. 1982. *Phys. Lett.* **108**B, 389 (§17).

Linde, A. D. 1983. *Phys. Lett.* **129**B, 177 (§17).

Linde, A. D. 1990. *Particle Physics and Inflationary Cosmology*. New York: Harwood (§§16, 17).

Livio, M., and Shaviv, G. 1986. *Ann. N.Y. Acad. Sci.* **470** (§18).

Loeb, A. 1991. In *After the First Three Minutes*, ed. S. Holt et al., 329. New York: American Institute of Physics (§24).

Loh, E. D., and Spillar, E. J. 1986. *Ap. J.* **307**, L1 (§13).

Loveday, J., Peterson, B. A., Efstathiou, G., and Maddox, S. J. 1992. *Ap. J.* **390**, 338 (§5).

Lu, L., Wolfe, A. M., and Turnshek, D. A. 1991. *Ap. J.* **367**, 19 (§23).

Lubimov, V. A., et al. 1980. *Phys. Lett.* **94**B, 266 (§16).

Lucchin, F., and Matarrese, S. 1985. *Phys. Rev.* D**32**, 1316 (§17).

Lundmark, K. 1925. *M.N.R.A.S.* **85**, 865 (§5).

Luo, X., and Schramm, D. N. 1992. *Science* **256**, 513 (§19).

Lynden-Bell, D., et al. 1988. *Ap. J.* **326**, 19 (§§5, 6).

Lynds, R. 1971. *Ap. J.* **164**, L73 (§23).

Lynds, R., and Petrosian, V. 1986. *Bull. A.A.S.* **18**, 1014 (§§14, 20).

McCarthy, P. 1992. Private communication (§5).

McCrea, W. H., and Milne, E. A. 1934. *Q. J. Math.* Oxford **5**, 73 (§§4, 7).

Mach, E. 1893. *The Science of Mechanics*. La Salle, Ill.: Open Court Publishers, 6th ed. of English trans., 1960 (§2).

McKellar, A. 1941. *Publ. Dominion Astrophys. Obs.* **7**, 251 (§6).

Maddox, S. J., Efstathiou, G., Sutherland, W. J., and Loveday, J. 1990. *M.N.R.A.S.* **242**, 43P (§§3, 7, 19).

Maffei, P. 1968. *Publ. Astron. Soc. Pacific* **80**, 618 (§3). 6).

Mandelbrot, B. B. 1975a. *Les Objets Fractals*. Paris: Flammarion (§7).

Mandelbrot, B. B. 1975b. *C. R. Acad. Sci.* (Paris) A**280**, 1551 (§7).

Marx, G., and Szalay, A. S. 1972. In *Neutrino '72* **1**, 191. Budapest: Technoinform (§18).

Matese, J. J., and O'Connell, R. F. 1970. *Ap. J.* **160**, 451 (§6).

Mather, J. C., et al. 1990. *Ap. J.* **354**, L37 (§§6, 7).

Matsumoto, T. 1990. In *The Galactic and Extragalactic Background Radiation*, ed. S. Bowyer and C. Leinert, 317. Dordrecht: Kluwer (§5).

Matthews, T. A., Morgan, W. W., and Schmidt, M. 1964. *Ap. J.* **140**, 35 (§3).

Mattig, W. 1958. *Astron. Nachr.* **284**, 109 (§13).

Mattila, K. 1990. In *The Galactic and Extragalactic Background Radiation*, ed. S. Bowyer and C. Leinert, 257. Dordrecht: Kluwer (§5).

Meiksin, A., Szapudi, I., and Szalay, A. S. 1992. *Ap. J.* **384**, 87 (§19).

Melott, A. L. 1984. *Soviet Astron.* **28**, 478; *Astron. Zh.* **61**, 817 (§23).

Melott, A. L. 1991. In *Observational Tests of Cosmological Inflation*, ed. T. Shanks et al., 389. Dordrecht: Kluwer (§26).

Melott, A. L. 1992. Private communication (§22).

Melott, A. L., and Shandarin, S. F. 1990. *Nature* **346**, 633 (§22).

Mestel, L. 1952. *M.N.R.A.S.* **112**, 583 (§18).

Mestel, L. 1963. *M.N.R.A.S.* **126**, 553 (§25).

Meyer, B. S., and Schramm, D. N. 1986. *Ap. J.* **311**, 406 (§5).

Milne, E. A. 1934. *Q. J. Math.* Oxford **5**, 64 (§7).

Milne, E. A. 1935. *Relativity, Gravitation and World Structure*. Oxford: Clarendon Press (§§1, 2, 5, 7).

Misner, C. W. 1968. *Ap. J.* **151**, 431 (§15).

Mitchell, R. J., et al. 1976. *M.N.R.A.S.* **176**, 29P (§20).

Miyama, S., and Sato, K. 1978. *Prog. Theor. Phys.* **60**, 1703 (§6).

Morgan, W. W. 1958. *Publ. Astron. Soc. Pacific* **70**, 364 (§3).

Morgan, W. W., and Lesh, J. R. 1965. *Ap. J.* **142**, 1364 (§3).

Mould, J. R., et al. 1991. *Ap. J.* **383**, 467 (§5).

Murdoch, H. S., Hunstead, R. W., Pettini, M., and Blades, J. C. 1986. *Ap. J.* **309**, 19 (§23).

Mushotzky, R. F. 1991. In *After the First Three Minutes*, ed. S. Holt et al., 394. New York: American Institute of Physics (§20).

Narlikar, J. V., and Wickramasinghe, N. C. 1968. *Nature* **217**, 1236 (§6).

Noh, H.-R., and Scalo, J. 1990. *Ap. J.* **352**, 605 (§18).

Novikov, I. D. 1964. *Zh. E.T.F.* **46**, 686 (§5).

O'Dell, C. R., Peimbert, M., and Kinman, T. D. 1964. *Ap. J.* **140**, 119 (§6).

Oegerle, W. R., and Hoessel, J. G. 1991. *Ap. J.* **375**, 15 (§3).

Oemler, A. 1987. In *Nearly Normal Galaxies*, ed. S. M. Faber, 213. Berlin: Springer (§25).

Ohm, E. A. 1961. *Bell Syst. Tech. J.* **40**, 1065 (§6).

Oke, J. B. 1984. In *Clusters and Groups of Galaxies*, ed. F. Mardirossian et al., 99. Dordrecht: Reidel (§25).

Oke, J. B., and Sandage, A. 1968. *Ap. J.* **154**, 21 (§7).

Olive, K. A. 1990. *Phys. Reports* **190**, 307 (§17).

Olive, K. A., Schramm, D. N., Steigman, G., and Walker, T. P. 1990. *Phys. Lett.* **236**B, 454 (§6).

Olive, K. A., Steigman, G., and Walker, T. P. 1991. *Ap. J.* **380**, L1 (§6).

Oort, J. H. 1932. *Bull. Astron. Inst. Netherlands* **6**, 249 (§18).

Oort, J. H. 1965. In *Galactic Structure*, ed. A. Blaauw and M. Schmidt, 455. Chicago: University of Chicago Press (§25).

Oort, J. H. 1983. *Ann. Rev. Astron. Ap.* **21**, 373 (§3).

Öpik, E. 1922. *Ap. J.* **55**, 406 (§3).

Osterbrock, D. E. 1989. *Astrophysics of Gaseous Nebulae and Active Galactic Nuclei*. Mill Valley, Calif.: University Science Books (§§23, 24).

Osterbrock, D. E., and Rogerson, J. B. 1961. *P.A.S.P.* **73**, 129 (§6).

Ostriker, J. P. 1974. Unpublished (§24).

Ostriker, J. P. 1990. In *Evolution of the World of Galaxies*, ed. R. G. Kron, 25. San Francisco: Astron. Soc. Pacific (§25).

Ostriker, J. P., and Cowie, L. L. 1981. *Ap. J.* **243**, L127 (§25).

Ostriker, J. P., and McKee, C. F. 1988. *Rev. Mod. Phys.* **60**, 1 (§25).

Ostriker, J. P., and Peebles, P.J.E. 1973. *Ap. J.* **186**, 467 (§18).

Ostriker, J. P., Peebles, P.J.E., and Yahil, A. 1974. *Ap. J.* **193**, L1 (§18).

Ostriker, J. P., and Suto, Y. 1990. *Ap. J.* **348**, 378 (§25).

Ostriker, J. P., and Thuan, T. X. 1975. *Ap. J.* **202**, 353 (§§22, 25).

Özer, M., and Taha, M. O. 1986. *Phys. Lett.* **171**, 363 (§18).

Paczyński, B. 1986. *Ap. J.* **304**, 1 (§18).

Paczyński, B. 1987. *Nature* **325**, 572 (§20).

Paczyński, B., and Piran, T. 1990. *Ap. J.* **364**, 341 (§6).

Pagel, B. 1991. *Physica Scripta* **T36**, 7 (§6).

Pais, A. 1982. *'Subtle Is the Lord.'* Oxford: Clarendon Press (§10).

Park, C. 1990. *M.N.R.A.S.* **242**, 59P (§25).

Partridge, R. B., and Peebles, P.J.E. 1967a. *Ap. J.* **147**, 868 (§§22, 25).

Partridge, R. B., and Peebles, P.J.E. 1967b. *Ap. J.* **148**, 377 (§5).

Patterson, C. 1956. *Geochim. Cosmochim. Acta* **10**, 230 (§5).

Pearson, T. J., and Zensus, J. A. 1987. In *Superluminal Radio Sources*, ed. J. A. Zensus and T. J. Pearson, 1. Cambridge, U.K.: Cambridge University Press (§13).

Peebles, P.J.E. 1965. *Ap. J.* **142**, 1317 (§§6, 22).

Peebles, P.J.E. 1966. *Ap. J.* **146**, 542 (§6).

Peebles, P.J.E. 1967. *Ap. J.* **147**, 859 (§5).

Peebles, P.J.E. 1968. *Ap. J.* **153**, 1 (§6).

Peebles, P.J.E. 1969a. *Ap. J.* **155**, 393 (§22).

Peebles, P.J.E. 1969b. *Ap. J.* **157**, 1075 (§6).

Peebles, P.J.E. 1971a. *Physical Cosmology*. Princeton: Princeton University Press (PC).

Peebles, P.J.E. 1971b. *Astron. Ap.* **11**, 377 (§22).

Peebles, P.J.E. 1974a. *Astron. Ap.* **32**, 391 (§22).

Peebles, P.J.E. 1974b. *Ap. J.* **189**, L51 (§22).

Peebles, P.J.E. 1976. *Astrophys. Space Sci.* **45**, 3 (§20).

Peebles, P.J.E. 1978. *Astron. Ap.* **68**, 345 (§§19, 22).

Peebles, P.J.E. 1979. *A. J.* **84**, 730 (§20).

Peebles, P.J.E. 1980. *The Large-Scale Structure of the Universe*. Princeton: Princeton University Press (LSS).

Peebles, P.J.E. 1981a. *Ap. J.* **243**, L119 (§21).

Peebles, P.J.E. 1981b. *Ap. J.* **248**, 885 (§25).

Peebles, P.J.E. 1982a. *Ap. J.* **263**, L1 (§§21, 25).

Peebles, P.J.E. 1982b. *Ap. J.* **258**, 415 (§25).

Peebles, P.J.E. 1983. In *The Origin and Evolution of Galaxies*, ed. B.J.T. Jones and J. E. Jones, 143. Dordrecht: Reidel (§25).

Peebles, P.J.E. 1984a. *Science* **224**, 1385 (§§20, 25).

Peebles, P.J.E. 1984b. *Ap. J.* **277**, 470 (§25).

Peebles, P.J.E. 1984c. *Ap. J.* **284**, 439 (§13).

Peebles, P.J.E. 1985. *Ap. J.* **297**, 350 (§22).

Peebles, P.J.E. 1987a. *Ap. J.* **315**, L73 (§§24, 25).

Peebles, P.J.E. 1987b. *Nature* **327**, 210 (§25).

Peebles, P.J.E. 1988a. *Ap. J.* **332**, 17 (§§5, 25).

Peebles, P.J.E. 1988b. In *The Early Universe*, ed. W. G. Unruh and G. W. Semenoff, 203. Dordrecht: Reidel (§25).

Peebles, P.J.E. 1990. *Ap. J.* **362**, 1 (§20).

Peebles, P.J.E., and Dicke, R. H. 1965. *Space Sci. Rev.* **4**, 419 (§6).

Peebles, P.J.E., and Dicke, R. H. 1968. *Ap. J.* **154**, 891 (§§6, 25).

Peebles, P.J.E., and Groth, E. J. 1975. *Ap. J.* **196**, 1 (§19).

Peebles, P.J.E., and Groth, E. J. 1976. *Astron. Ap.* **53**, 131 (§22).

Peebles, P.J.E., Melott, A. L., Holmes, M. R., and Jiang, L. R. 1989. *Ap. J.* **345**, 108 (§20).

Peebles, P.J.E., Schramm, D. N., Turner, E. L., and Kron, R. G. 1991. *Nature* **352**, 769 (§7).

Peebles, P.J.E., and Wilkinson, D. T. 1968. *Phys. Rev.* **174**, 2168 (§6).

Peebles, P.J.E., and Yu, J. T. 1970. *Ap. J.* **162**, 815 (§§16, 24).

Penzias, A. A., and Wilson, R. W. 1965. *Ap. J.* **142**, 419 (§6).

Peterson, B. A. 1978. In *The Large Scale Structure of the Universe*, ed. M. S. Longair and J. Einasto, 389. Dordrecht: Reidel (§23).

Petrosian, V. 1976. *Ap. J.* **209**, L1 (§5).

Petrosian, V., Salpeter, E. E., and Szekeres, P. 1967. *Ap. J.* **147**, 1222 (§13).

Polyakov, A. M. 1974. *JETP Lett.* **20**, 194 (§16).

Preskill, J. P. 1979. *Phys. Rev. Lett.* **43**, 1365 (§16).

Press, W. H., and Gunn, J. E. 1973. *Ap. J.* **185**, 397 (§§14, 18).

Press, W. H., Ryden, B. S., and Spergel, D. N. 1989. *Ap. J.* **347**, 590 (§16).

Press, W. H., and Schechter, P. 1974. *Ap. J.* **187**, 425 (§25).

Primack, J. R., Seckel, D., and Sadoulet, B. 1988. *Ann. Rev. Nucl. Part. Sci.* **38**, 751 (§18).

Pryor, C. 1992. In *Morphological and Physical Classification of Galaxies*, ed. G. Busarello et al., 163. Dordrecht: Kluwer (§18).

Quinn, P. J., and Goodman, J. 1986. *Ap. J.* **309**, 472 (§25).

Ramaty, R., and Jones, F. C. 1981. *Ann. N.Y. Acad. Sci.* **375** (§18).

Rana, N. C. 1991. *Ann. Rev. Astron. Ap.* **29**, 129 (§18).

Ratra, B. 1985. *Phys. Rev.* D**31**, 1931 (§17).

Ratra, B. 1988. *Phys. Rev.* D**38**, 2399 (§10).

Ratra, B. 1991. *Phys. Rev.* D**43**, 3802 (§17).

Ratra, B. 1992. *Ap. J.* **391**, L1 (§§6, 25).

Ratra, B., and Peebles, P.J.E. 1988. *Phys. Rev.* D**37**, 3406 (§18).

Rees, M. J. 1966. *Nature* **211**, 468 (§13).

Rees, M. J. 1978. *Nature* **275**, 35 (§6).

Rees, M. J. 1986. *M.N.R.A.S.* **218**, 25P (§23).

Rees, M. J. 1988. In *QSO Absorption Lines—Probing the Universe*, ed. J. C. Blades, D. A. Turnshek, and C. A. Norman, 319. Cambridge, U.K.: Cambridge University Press (§23).

Rees, M. J., and Ostriker, J. P. 1977. *M.N.R.A.S.* **179**, 541 (§24).

Rees, M. J., and Sciama, D. W. 1968. *Nature* **217**, 511 (§21).

Refsdal, S. 1964. *M.N.R.A.S.* **128**, 295 (§18).

Refsdal, S. 1970. *Ap. J.* **159**, 357 (§14).

Regös, E., and Geller, M. J. 1989. *A. J.* **98**, 755 (§§20, 25).

Rephaeli, Y. 1981. *Ap. J.* **245**, 351 (§24).

Rephaeli, Y., and Szalay, A. S. 1981. *Phys. Lett.* **106**B, 73 (§23).

Rhee, G.F.R.N., and Katgert, P. 1987. *Astron. Ap.* **183**, 217 (§26).

Richstone, D. O., and Tremaine, S. 1986. *A. J.* **92**, 72 (§18).

Roberts, M. S. 1972. In *External Galaxies and Quasi-Stellar Objects*, ed. D. S. Evans, 12. Dordrecht: Reidel (§5).

Roberts, M. S. 1976. *Comments Ap. Space Phys.* **6**, 105 (§18).

Roberts, M. S., and Rots, A. H. 1973. *Astron. Ap.* **26**, 483 (§18).

Robertson, H. P. 1928. *Phil. Mag.* **5**, 835 (§§5, 12).

Robertson, H. P. 1933. *Rev. Mod. Phys.* **5**, 62 (§7).

Robertson, H. P. 1935. *Proc. N.A.S.* **15**, 822 (§5).

Robertson, H. P. 1955. *Publ. A.S.P.* **67**, 82 (§15).

Rogers, R. D., and Field, G. B. 1991. *Ap. J.* **366**, 22 (§24).

Rogstad, D. H. 1971. *Astron. Ap.* **13**, 108 (§18).

Rogstad, D. H., and Shostak, G. S. 1972. *Ap. J.* **176**, 315 (§18).

Roll, P. G., and Wilkinson, D. T. 1966. *Phys. Rev. Lett.* **16**, 405 (§6).

Rood, H. J. 1976. *Ap. J.* **207**, 16 (§19).

Rowan-Robinson, M. 1985. *The Cosmological Distance Ladder*. New York: Freeman (§5).

Rubakov, V. A., Sazhin, M. V., and Veryaskin, A. V. 1982. *Phys. Lett.* **115**B, 189 (§17).

Rubin, V. C. 1954. *Proc. N.A.S.* **40**, 541 (§7).

Rubin, V. C., and Coyne, G. V. 1988. *Large-Scale Motions in the Universe*. Princeton: Princeton University Press (§§6, 20, 21, 25).

Rubin, V. C., and Ford, W. K. 1970. *Ap. J.* **159**, 379 (§18).

Rubin, V. C., Thonnard, N., Ford, W. K., and Roberts, M. S. 1976. *A. J.* **81**, 719 (§6).

Rudnicki, K., et al. 1973. *Acta Cosmologica* **1**, 7 (§7).

Rutherford, E. 1929. *Nature* **123**, 313 (§5).

Ryan, M. P., and Shepley, L. C. 1975. *Homogeneous Relativistic Cosmologies.* Princeton: Princeton University Press (§26).

Rybicki, G. B., and Lightman, A. P. 1979. *Radiative Processes in Astrophysics*, 213. New York: Wiley (§24).

Sachs, R. K. 1961. *Proc. R. Soc. London* A**264**, 309 (§14).

Sachs, R. K., and Wolfe, A. M. 1967. *Ap. J.* **147**, 73 (§21).

Sahni, V., Feldman, H., and Stebbins, A. 1992. *Ap. J.* **385**, 1 (§13).

Salpeter, E. E. 1955. *Ap. J.* **121**, 161 (§13).

Sandage, A. 1958. *Ap. J.* **127**, 513 (§5).

Sandage, A. 1961. *The Hubble Atlas of Galaxies.* Washington, D.C.: Carnegie Institution (§3).

Sandage, A. 1972a. *Q.J.R.A.S.* **13**, 282 (§§3, 5).

Sandage, A. 1972b. *Ap. J.* **173**, 485 (§§3, 5).

Sandage, A. 1987. In *Observational Cosmology*, ed. A. Hewitt et al., 1. Dordrecht: Kluwer (§§5, 13).

Sandage, A. 1988. *Ann. Rev. Astron. Ap.* **26**, 561 (§13).

Sandage, A., and Bedke, J. 1988. *Atlas of Galaxies.* Washington, D.C.: U.S. Government Printing Office (§25).

Sandage, A., and Perelmuter, J.-M. 1991. *Ap. J.* **370**, 455 (§§5, 7).

Sandage, A., and Tammann, G. A. 1990. *Ap. J.* **365**, 1 (§§3, 5, 20).

Sandage, A., Tammann, G. A., and Hardy, E. 1972. *Ap. J.* **172**, 253 (§7).

Sanders, R. H. 1991. *Astron. Ap. Rev.* **2**, 1 (§18).

Sanford, R. F. 1917. *Bull. Lick Obs.* **9**, 80 (§3).

Sargent, W.L.W., et al. 1978. *Ap. J.* **221**, 731 (§3).

Sargent, W.L.W., Boksenberg, A., and Steidel, C. C. 1988. *Ap. J. Suppl.* **68**, 539 (§23).

Sargent, W.L.W., Steidel, C. C., and Boksenberg, A. 1989. *Ap. J. Suppl.* **69**, 703 (§23).

Sargent, W.L.W., Young, P. J., Boksenberg, A., and Tytler, D. 1980. *Ap. J. Suppl.* **42**, 41 (§23).

Sargent, W.L.W., Young, P. J., and Schneider, D. P. 1982. *Ap. J.* **256**, 374 (§23).

Saslaw, W. C. 1985. *Gravitational Physics of Stellar and Galactic Systems.* Cambridge, U.K.: Cambridge University Press (§19).

Saslaw, W. C., and Zipoy, D. 1967. *Nature* **216**, 967 (§6).

Sato, K. 1981a. *Phys. Lett.* **99**B, 66 (§17).

Sato, K. 1981b. *M.N.R.A.S.* **195**, 467 (§17).

Saunders, W., et al. 1991. *Nature* **349**, 32 (§3).

Savedoff, M. P. 1956. *Nature* **178**, 688 (§5).

Scalo, J. M. 1986. *The Stellar Initial Mass Function.* New York: Gordon and Breach (§18).

Schaeffer, R. 1984. *Astron. Ap.* **134**, L15 (§§19, 22).

Schechter, P. 1976. *Ap. J.* **203**, 297 (§5).

Scheuer, P.A.G. 1965. *Nature* **207**, 963 (§23).

Schneider, D. P., et al. 1988. *A. J.* **95**, 1619 (§14).

Schneider, P., and Weiss, A. 1988. *Ap. J.* **327**, 526 (§14).

Schramm, D. N. 1991. *Physica Scripta* T**36**, 22 (§6).

Schwarzschild, K. 1900. *Vierteljahrschr. der Astron. Ges.* **35**, 337 (§4).

Schweizer, F. 1990. In *Dynamics and Interactions of Galaxies*, ed. R. Wielen, 60. Berlin: Springer (§25).

Sciama, D. W. 1967. *Phys. Rev. Lett.* **18**, 1065 (§6).

Sciama, D. W. 1990a. In *The Cosmic Microwave Background: 25 Years Later*, ed. N. Mandolesi and N. Vittorio, 1. Dordrecht: Kluwer (§6).

Sciama, D. W. 1990b. *Phys. Rev. Lett.* **65**, 2839 (§23).

Scott, E. L. 1957. *A. J.* **62**, 248 (§5).

Scott, E. L., Shane, C. D., and Swanson, M. D. 1954. *Ap. J.* **119**, 91 (§19).

Seeliger, H. 1895. *Astron. Nachr.* **137**, 129 (§4).

Segal I. E. 1976. *Mathematical Cosmology and Extragalactic Astronomy.* New York: Academic Press (§5).

Seldner, M., and Peebles, P.J.E. 1977. *Ap. J.* **215**, 703 (§§3, 19).

Seldner, M., Siebers, B., Groth, E. J., and Peebles, P.J.E. 1977. *A. J.* **82**, 249 (§3).

Shafer, R. A., and Fabian, A. C. 1983. In *Early Evolution of the Universe and Its Present Structure*, ed. G. O. Abell and G. Chincarini, 333. Dordrecht: Reidel (§6).

Shandarin, S. F. 1988. In *Large Scale Structures of the Universe*, ed. J. Audouze, M. Ch. Peleton, and A. Szalay, 273. Dordrecht: Kluwer (§22).

Shane, C. D., and Wirtanen, C. A. 1967. *Publ. Lick Obs.* **22**, part 1 (§§3, 7).

Shanks, T. 1990. In *The Galactic and Extragalactic Background Radiation*, ed. S. Bowyer and C. Leinert, 269. Dordrecht: Kluwer (§§3, 13).

Shanks, T. 1991. Private communication (§§3, 5).

Shanks, T., et al. 1991. In *Observational Tests of Cosmological Inflation*, ed. T. Shanks et al., 205. Dordrecht: Kluwer (§26).

Shanks, T., Stevenson, P.R.F., Fong, R., and MacGillivray, H. T. 1984. *M.N.R.A.S.* **206**, 767 (§3).

Shapiro, P. R., and Giroux, M. L. 1987. *Ap. J.* **321**, L107 (§23).

Shapley, H. 1925. *Harvard College Obs. Circ.* **280** (§3).

Shapley, H. 1934. *M.N.R.A.S.* **94**, 791 (§7).

Shapley, H. 1938. *Proc. N.A.S.* **24**, 282 (§7).

Sharp, N. A., Bonometto, S. A., and Lucchin, F. 1984. *Astron. Ap.* **130**, 79 (§19).

Shaver, P. A. 1991. *Australian J. Phys.* **44**, 759 (§§3, 26).

Shaya, E. J., Tully, R. B., and Pierce, M. J. 1992. *Ap. J.* **391**, 16 (§20).

Shklovsky, I. S. 1964. *Astron. Zh.* **41**, 408 (§23).

Shklovsky, I. S. 1967. *Ap. J.* **150**, L1 (§13).

Shvartsman, V. F. 1969. *Pis' ma Zh. E.T.F.* **9**, 315 (§6).

Silberstein, L. 1924. *M.N.R.A.S.* **85**, 285 (§5).

Silk, J. 1968. *Ap. J.* **151**, 459 (§25).

Silk, J. 1974. In *Confrontation of Cosmological Theories with Observational Data*, ed. M. S. Longair, 175. Dordrecht: Reidel 175 (§25).

Silk, J. 1977. *Ap. J.* **211**, 638 (§24).

Silk, J. 1989. *Ap. J.* **345**, L1 (§25).

Silk, J. 1991. *Science* **251**, 537 (§18).

Silk, J., and Vilenkin, A. 1984. *Phys. Rev. Lett.* **53**, 1700 (§16).

Silk, J., and White, S.D.M. 1978. *Ap. J.* **226**, L103 (§24).

Smette, A., et al. 1992. *Ap. J.* **389**, 39 (§23).

Smirnov, Yu. 1964. *Soviet Astron.-A. J.* **41**, 1084 (§6).

Smith, S. 1936. *Ap. J.* **83**, 23 (§18).

Smoot, G. F., et al. 1991. *Ap. J.* **371**, L1 (§§6, 21).

Smoot, G. F., et al. 1992. *Ap. J.* **396**, L1 (§21).

Soifer, B. T., Houck, J. R., and Neugebauer, G. 1987. *Ann. Rev. Astron. Ap.* **25**, 187 (§3).

Soneira, R. M. 1978, Ph.D. diss., Princeton University (§19).

Soneira, R. M. 1979. *Ap. J.* **230**, L63 (§5).

Soneira, R. M., and Peebles, P.J.E. 1978. *A. J.* **83**, 845 (§19).

Songaila, A., Cowie, L. L., and Lilly, S. J. 1990. *Ap. J.* **348**, 371 (§5).

Soucail, G., and Fort, B. 1991. *Astron. Ap.* **243**, 23 (§20).

Soucail, G., et al. 1988. *Astron. Ap.* **191**, L19 (§§14, 20).

Spinrad, H. 1989. In *The Epoch of Galaxy Formation*, ed. C. S. Frenk et al., 39. Dordrecht: Kluwer (§26).

Spinrad, H., and Djorgovski, S. 1987. In *Observational Cosmology*, ed. A. Hewitt et al., 129. Dordrecht: Kluwer (§5).

Starobinsky, A. A. 1979. *JETP Lett.* **30**, 682 (§17).

Starobinsky, A. A. 1980. *Phys. Lett.* **91B**, 99 (§17).

Starobinsky, A. A. 1982. *Phys. Lett.* **117B**, 175 (§17).

Stebbins, J., and Whitford, A. E. 1948. *Ap. J.* **108**, 413 (§7).

Stecker, F. W. 1989. In *Proceedings of the Gamma Ray Observatory Workshop*, ed. W. N. Johnson, 4–73. Greenbelt, Md.: Goddard Space Flight Center (§18).

Steidel, C. C., and Sargent, W.L.W. 1987. *Ap. J.* **318**, L11 (§23).

Steigman, G. 1976. *Ann. Rev. Astron. Ap.* **14**, 339 (§§7, 18).

Steinhardt, P. J. 1990. *Nature* **354**, 47 (§17).

Stevenson, D. J. 1991. *Ann. Rev. Astron. Ap.* **29**, 163 (§18).

Stinebring, D. R., Ryba, M. F., Taylor, J. H., and Romani, R. W. 1990. *Phys. Rev. Lett.* **65**, 285 (§17).

Stoeger, W. R. 1987. *Theory and Observational Limits in Cosmology*. Vatican: Pontifical Academy of Sciences (§7).

Strauss, M. A. 1992. Private communication (§3).

Strauss, M. A., and Davis, M. 1988. In *Large-Scale Motions in the Universe*, ed. V. C. Rubin and G. V. Coyne, 255. Princeton: Princeton University Press (§6).

Strauss, M. A., Davis, M., Yahil, A., and Huchra, J. P. 1990. *Ap. J.* **361**, 49 (§3).

Strauss, M. A., Davis, M., Yahil, A., and Huchra, J. P. 1992a. *Ap. J.* **385**, 421 (§§3, 20).

Strauss, M. A., et al. 1992b. *Ap. J.* (§§3, 20).

Strömberg, G. 1934. *Ap. J.* **79**, 460 (§22).

Struble, M. F., and Rood, H. J. 1991. *Ap. J. Suppl.* **77**, 363 (§20).

Strukov, I. A. 1990. In *The Cosmic Microwave Background: 25 Years Later*, ed. N. Mandolesi and N. Vittorio, 95. Dordrecht: Kluwer (§21).

Sunyaev, R. A. 1978. In *Large Scale Structure of the Universe*, ed. M. S. Longair and J. Einasto, 393. Dordrecht: Reidel (§24).

Sunyaev, R. A., and Zel'dovich, Ya. B. 1972. *Comments Ap. Space Sci.* **4**, 173 (§24).

Sunyaev, R. A., and Zel'dovich, Ya. B. 1980. *M.N.R.A.S.* **190**, 413 (§24).

Suto, Y., Górski, K., Juszkiewicz, R., and Silk, J. 1988. *Nature* **332**, 328 (§21).

Szalay, A. S., and Marx, G. 1976. *Astron. Ap.* **49**, 437 (§18).

Szalay, A. S., and Schramm, D. N. 1985. *Nature* **314**, 718 (§19).

Szapudi, I., Szalay, A. S., and Boschán, P. 1992. *Ap. J.* **390**, 350 (§19).

Tarter, J. C. 1975, Ph.D. diss., University of California, Berkeley (§18).

Taylor, J. H., and Weisberg, J. M. 1989. *Ap. J.* **345**, 434 (§26).

The, L. S., and White, S.D.M. 1986. *A. J.* **92**, 1248 (§22).

't Hooft, G. 1974.. *Nucl. Phys.* B**79**, 276 (§16).

Thuan, T. X., and Alimi, J.-M. 1991. In *Physical Cosmology*, ed. A. Blanchard et al., 261. Gif-sur-Yvette: Editions Frontières (§25).

Tinsley, B. M. 1972. *Ap. J.* **178**, 319 (§13).

Tolman, R. C. 1930. *Proc. N.A.S.* **16**, 511 (§§5, 7).

Tolman, R. C. 1934. *Relativity Thermodynamics and Cosmology*. Oxford: Clarendon Press (§§1, 5, 6, 11, 15).

Toomre, A. 1977. In *The Evolution of Galaxies and Stellar Populations*, ed. B. M. Tinsley and R. B. Larson, 401. New Haven: Yale University Observatory (§25).

Tremaine, S., and Gunn, J. E. 1979. *Phys. Rev. Lett.* **42**, 407 (§18).

Tubbs, A. D., and Wolfe, A. M. 1980. *Ap. J.* **236**, L105 (§5).

Tully, R. B. 1987a. *Ap. J.* **321**, 280 (§20).

Tully, R. B. 1987b. *Ap. J.* **323**, 1 (§26).

Tully, R. B., and Fisher, J. R. 1977. *Astron. Ap.* **54**, 661 (§§3, 5).

Tully, R. B., and Fisher, J. R. 1987. *Nearby Galaxies Atlas*. Cambridge, U.K.: Cambridge University Press (§3).

Turner, E. L. 1980. *Ap. J.* **242**, L135 (§14).

Turner, E. L. 1990. *Ap. J.* **365**, L43 (§13).

Turner, E. L., and Gott, J. R. 1976. *Ap. J. Suppl.* **32**, 409 (§19).

Turner, E. L., Ostriker, J. P., and Gott, J. R. 1984. *Ap. J.* **284**, 1 (§13).

Turner, M. S. 1991. *Phys. Rev.* D**44**, 3737 (§6).

Turner, M. S., Steigman, G., and Krauss, L. M. 1984. *Phys. Rev. Lett.* **52**, 2090 (§25).

Turnshek, D. A., et al. 1989. *Ap. J.* **344**, 567 (§23).

Turok, N. 1985. *Phys. Rev. Lett.* **55**, 1801 (§16).

Turok, N. 1989. *Phys. Rev. Lett.* **24**, 2625 (§16).

Turok, N. 1991. *Physica Scripta* T**36**, 135 (§16).

Turok, N., and Spergel, D. N. 1990. *Phys. Rev. Lett.* **64**, 2736 (§§25, 26).

Turok, N., and Spergel, D. N. 1991. *Phys. Rev. Lett.* **66**, 3093 (§16).

Tyson, J. A. 1988. *A. J.* **96**, 1 (§26).

Tyson, J. A. 1990. In *The Galactic and Extragalactic Background Radiation*, ed. S. Bowyer and C. Leinert, 245. Dordrecht: Kluwer (§§5, 13).

Tyson, J. A., Valdes, F., and Wenk, R. A. 1990. *Ap. J.* **349**, L1 (§20).

Tytler, D. 1987. *Ap. J.* **321**, 49 (§23).

Uson, J. M., Bagri, D. S., and Cornwell, T. J. 1991. *Phys. Rev. Lett.* **67**, 3328 (§§25, 26).

Uson, J. M., Boughn, S. P., and Kuhn, J. R. 1990, *Science* **250**, 539 (§3).

Uson, J. M., and Wilkinson, D. T. 1984. *Ap. J.* **277**, L1 (§24).

Uson, J. M., and Wilkinson, D. T. 1988. In *Galactic and Extragalactic Astronomy*. ed. G. L. Verschuur and K. I. Kellermann, 2d ed., 603. Berlin: Springer (§§6, 24).

Valls-Gabaud, D., Alimi, J.-M., and Blanchard, A. 1989. *Nature* **341**, 215 (§25).

van Albada, T. S., and Sancisi, R. 1986. *Phil. Trans. R. Soc. London* A**320**, 447 (§3).

van den Bergh, S. 1968. *J. Roy. Astron. Soc. Canada* **62**, 1 (§3).

van den Bergh, S. 1990a. *Ap. J.* **348**, 57 (§3).

van den Bergh, S. 1990b. *Q.J.R.A.S.* **31**, 153 (§25).

Veryaskin, A. V., Rubakov, V. A., and Sazhin, M. V. 1983. *Soviet Astron.* **27**, 16 (§17).

Vietri, M. 1985. *Ap. J.* **293**, 343 (§14).

Vilenkin, A. 1981a. *Phys. Rev.* D**23**, 852 (§11).

Vilenkin, A. 1981b. *Phys. Rev. Lett.* **46**, 1169 (§16).

Vilenkin, A. 1983. *Phys. Lett.* **133**B, 177 (§11).

Vilenkin, A. 1985. *Phys. Reports* **121**, 265 (§16).

Vilenkin, A. 1991. Private communication (§11).

Vilenkin, A., and Ford, L. H. 1982. *Phys. Rev.* D**26**, 1231 (§17).

Vilenkin, A., and Shellard, E.P.S 1993. *Cosmic Strings and Other Topological Defects*. Cambridge, U.K.: Cambridge University Press (§16).

Villumsen, J. V., and Davis, M. 1986. *Ap. J.* **308**, 499 (§22).

Vishniac, E. T. 1987. *Ap. J.* **322**, 597 (§24).

Vittorio, N. 1988. In *Large-Scale Motions in the Universe*, ed. V. C. Rubin and G. V. Coyne, 397. Princeton: Princeton University Press (§20).

Vittorio, N. and Silk, J. 1984. *Ap. J.* **285**, L39 (§25).

von Weizsacker, C. F. 1947. *Z. Ap.* **24**, 181 (§22).

Wagoner, R. V., Fowler, W. A., and Hoyle, F. 1967. *Ap. J.* **148**, 3 (§6).

Walker, A. G. 1936. *Proc. London Math. Soc.* **42**, 90 (§5).

Walker, T. P., et al. 1991. *Ap. J.* **376**, 51 (§6).

Wasserburg, G. J., Papanastassiou, D. A., Tera, F., and Huneke, J. C. 1977. *Phil. Trans.* A**285**, 7 (§5).

Wasserman, I. 1978. *Ap. J.* **224**, 337 (§§22, 25).

Watanabe, K., and Tomita, K. 1990. *Ap. J.* **355**, 1 (§14).

Weinberg, D. H, Szomoru, A., Guthathakurta, P., and van Gorkom, J. H. 1991. *Ap. J.* **372**, L13 (§25).

Weinberg, S. 1972. *Gravitation and Cosmology*. New York: Wiley (§1).

Weinberg, S. 1975. *Ann. N.Y. Acad. Sci.* **262**, 409 (§6).

Weisskopf, V., and Wigner, E. 1930. *Z. Phys.* **63**, 54 (§23).

West, M. J., and Richstone, D. O. 1988. *Ap. J.* **335**, 532 (§25).

Weyl, H. 1922. *Space Time Matter*, 4th ed., trans. H. L. Brose. New York: Dover (§4).

Weyl, H. 1923. *Phys. Z.* **24**, 230 (§§5, 12).

Weymann, R. 1966. *Ap. J.* **145**, 560 (§24).

Weymann, R. J., Carswell, R. F., and Smith, M. G. 1981. *Ann. Rev. Astron. Ap.* **19**, 41 (§23).

White, S.D.M., and Frenk, C. S. 1991. *Ap. J.* **379**, 52 (§§25, 26).

White, S.D.M., Frenk, C. S., and Davis, M. 1983. *Ap. J.* **274**, L1 (§25).

White, S.D.M., and Rees, M. J. 1978. *M.N.R.A.S.* **183**, 341 (§22).

White, S.D.M., Silk, J., and Henry, J. P. 1981. *Ap. J.* **251**, L65 (§25).

Whitrow, G. J., and Yallop, B. D. 1964. *M.N.R.A.S.* **127**, 301 (§5).

Wilkinson, D. T. 1992. Private communication (§6).

Wilson, M. L., and Silk, J. 1981. *Ap. J.* **243**, 14 (§25).

Winget, D. E., et al. 1987. *Ap. J.* **315**, L77 (§§18, 25).

Wirtz, C. 1924. *Astron. Nachr.* **222**, 21 (§5).

Wolfe, A. M. 1986. *Phil. Trans. R. Soc.* A**320**, 503 (§23).

Wolfe, A. M., et al. 1985. *Ap. J.* **294**, L67 (§5).

Wolfe, A. M., Brown, R. L., and Roberts, M. S. 1976. *Phys. Rev. Lett.* **37**, 179 (§5).

Wolfe, A. M., Lanzetta, K. M., and Oren, A. L. 1992. *Ap. J.* **388**, 17 (§25).

Wolfe, A. M., Turnshek, D. A., Lanzetta, K. M., and Lu, L. 1993. *Ap. J.* **404** (§23).

Wright, E. L. 1982. *Ap. J.* **255**, 401 (§§7, 24).

Wright, E. L. 1992. In *The Infrared and Submillimeter Sky After COBE*, ed. M. Signore and C. Dupraz, 231. Dordrecht: Kluwer 231 (§§7, 24).

Yahil, A., and Beaudet, G. 1976. *Ap. J.* **206**, 26 (§6).

Yahil, A., Tammann, G. A., and Sandage, A. 1977. *Ap. J.* **217**, 903 (§6).

Yoshii, Y., and Fukugita, M. 1991. In *Observational Tests of Cosmological Inflation*, ed. T. Shanks et al., 267. Dordrecht: Kluwer (§26).

Young, P. 1981. *Ap. J.* **244**, 756 (§18).

Yu, J. T., and Peebles, P.J.E. 1969. *Ap. J.* **158**, 103 (§21).

Zatsepin, G. T., and Kuz'min, V. A. 1966. *Pis' ma Zh. E.T.F.*, **4**, 114 (§6).

Zel'dovich, Ya. B. 1964. *Soviet Astron.-A. J.* **8**, 13 (§14).

Zel'dovich, Ya. B. 1965. *Adv. Astron. Ap.* **3**, 241 (§§6, 22).

Zel'dovich, Ya. B. 1968. *Soviet Phys.-Uspekhi* **95**, 209 (§§4, 17).

Zel'dovich, Ya. B. 1970. *Astrofizika* **6**, 319; *Astron. Ap.* **5**, 84 (§22).

Zel'dovich, Ya. B. 1972. *M.N.R.A.S.* **160**, 1P (§16).

Zel'dovich, Ya. B. 1980. *M.N.R.A.S.* **192**, 663 (§16).

Zel'dovich, Ya. B., Einasto, J., and Shandarin, S. F. 1982. *Nature* **300**, 407 (§25).

Zel'dovich, Ya. B., and Khlopov, M. Yu. 1978. *Phys. Lett.* **79B**, 239 (§16).

Zel'dovich, Ya. B., Kobzarev, I. Yu., and Okun, L. B. 1974. *Zh. E.T.F.* **67**, 3; *Soviet Phys.-JEPT* **40**, 1 (§16).

Zel'dovich, Ya. B., Kurt, V. G., and Sunyaev, R. A. 1969. *Soviet Phys.-JEPT* **28**, 146 (§6).

Zel'dovich, Ya. B., and Novikov, I. D. 1983. *The Structure and Evolution of the Universe*, ed. G. Steigman, 99. Chicago: University of Chicago Press (§§4, 26).

Zel'dovich, Ya. B., and Sunyaev, R. A. 1969. *Ap. Space Sci.* **4**, 301 (§24).

Zwicky, F. 1929. *Proc. N.A.S.* **15**, 773 (§7).

Zwicky, F. 1933. *Helv. Phys. Acta* **6**, 110 (§18).

Zwicky, F. 1942. *Phys. Rev.* **61**, 489 (§5).

Zwicky, F. 1957. *Morphological Astronomy*. Berlin: Springer (§3).

Zwicky, F., Herzog, E., Wild, P., Karpowicz, M., and Kowal, C. T. 1961–68. *Catalogue of Galaxies and Clusters of Galaxies*, in 6 vols. Pasadena: California Institute of Technology (§§3, 7).

INDEX

Abell clusters, 32, 490; number density of, 491; velocity dispersions in, 491. *See also* clusters of galaxies
Abell radius, 491
Absolute magnitude, 21, 57
acceleration parameter, 314
action, 244, 246; for cosmic field, 375; for cosmic string, 381; for dust, 263; for gravity, 265; for Local Group, 485; for scalar field, 257
active gravitational mass, 63, 269, 283, 453
adiabatic scenarios, 617, 655; from inflation, 619
aether drift, 151. *See also* CBR dipole anisotropy; Compton-Getting effect
affine parameter, 253, 351
Andromeda nebula, 18
angular correlation functions, 216, 458; and power spectrum, 517; scaling law for, 219
angular power spectrum, 517
angular size distance, 319
angular size-distance relation, 61
angular size-redshift relation, 325
anisotropic homogeneous cosmologies, 15, 665
antenna temperature, 147
anthropic condition, 186, 363, 365
antimatter, 207, 393, 455, 669
APM catalog, 42, 220
apparent magnitude, 21, 57
atomic hydrogen collisional ionization, 556
atomic hydrogen photoionization, 553
atomic hydrogen recombination coefficient, 172, 556
autocorrelation function, 509; for mass, 520; and power spectrum, 509, 519; for velocity, 515. *See also* galaxy correlation function, two-point correlation function

balloon analogy, 81
baryon adiabatic scenario, 655
baryon density, 123, 424, 556, 559; in dark matter, 476; and light element production, 151, 195, 422, 569; in Lyman-alpha forest, 567; in Wolfe clouds, 563
baryon isocurvature scenario, 187, 618, 657
baryon Jeans length and mass, 179, 567, 636
baryonic matter, 424
biasing, 423n, 613, 638; in clusters, 640; in the cold dark matter model, 613, 622, 668; in cosmic field scenarios, 652; and the mass clustering length, 674
big bang, 6, 40
big crunch, 3, 363, 418n
Birkhoff's theorem, 63, 109, 289
black dwarfs. *See* brown dwarfs
black holes, 392, 425, 670, 682
blackbody radiation, 134, 158; spectrum of, 137. *See also entries for* CBR
Boltzmann equation, 539
Bose-Einstein condensation, 584
bosonized spectrum, 584
bottom-up picture, 613, 646. *See also* clustering hierarchy
Bran-Dicke theory, 453
brightness theorem, 157, 253
brown dwarfs, 432, 476; mass limit of, 427; microlensing by, 436
Butcher-Oemler effect, 202, 677

caustic, 353, 531
CBR, 131; energy density of, 138, 160; magnetic field of, 160; and steady-state cosmology, 203; temperature of, 131; transfer equation for, 599
CBR angular fluctuations, 132, 525, 681; and absorption and scattering by dust, 203, 595; in baryon isocurvature model, 659, 662; in cold dark matter model, 629; and gravitational radiation, 404; in plasma universe, 208; power spectrum of, 519; and scattering by free electrons, 174, 592, 597. *See also* Rees-Sciama effect; Sachs-Wolfe effect; Sunyaev-Zel'dovich effect